Ground Movements and Structures

ORGANISING COMMITTEE

Department of Civil Engineering and Building Technology, UWIST

J D Geddes, BSc, PhD, CEng, FICE, FASCE, MIHE, FGS (Chairman)
D O M Davies, BSc, CEng, MIStructE, MICE (Secretary)
P Sutcliffe, BScTech, CEng, MICE, FIPHE, FRSA
M V Symons, BSc, PhD, FGS

The Institution of Structural Engineers

C J Evans, MA, CEng, FIStructE, FICE, FIHE, FASCE, MConsE
(Chairman Wales Branch)
J Blackshaw, BEng, CEng, FIStructE, MICE, MConsE
L Slowikowski, BA, CEng, FIStructE, MConsE

GROUND MOVEMENTS AND STRUCTURES

Proceedings of the 2nd International Conference
held at
The University of Wales Institute of Science and Technology
Cardiff, April 1980

Edited by
James D Geddes
The University of Wales Institute of Science and Technology

Sponsored by
The Department of Civil Engineering and Building Technology,
The University of Wales Institute of Science and Technology
and
The Institution of Structural Engineers

PENTECH PRESS
London : Plymouth

First published 1981
by Pentech Press Limited
Estover Road, Plymouth, Devon

ISBN 0–7273–0701–0

British Library Cataloguing in Publication Data

Ground movements and structures.
1. Earth movements and building – Congresses
I. Geddes, J. D.
624.1′76 TH1094

ISBN 0–7273–0701–0

Reproduced from copy supplied
printed and bound in Great Britain
by Billing and Sons Limited and Kemp Hall Bindery
Guildford, London, Oxford, Worcester

FOREWORD

It is with great pleasure that I write this Foreword to the Proceedings of the Second Conference on Ground Movements, that UWIST organised jointly with the Institution of Structural Engineers as part of the Golden Jubilee Celebrations of the Wales Branch of the Institution in 1980. It follows an earlier conference held in 1977 also at UWIST titled "Large Ground Movements and Structures" which was very successful and the Proceedings of which have become established as a standard reference on this subject.

The interaction of ground movements and structures is one of the most important aspects of structural engineering and this second conference has been widened to include not only large ground movements but also dynamic movements, a topic which has grown in importance in the research field in recent years.

The range and quality of papers presented were of the highest standard, and as with its predecessor, the Proceedings will become a standard reference for some time to come.

C.J. Evans
Chairman Wales Branch Institution of
Structural Engineers 1979-80
Vice-President, Institution of Structural Engineers

PREFACE

This book contains the fifty papers presented at the Conference held in Cardiff during the period 14th - 17th April 1980, together with some of the discussion which took place.

In 1980 the Wales Branch of the Institution of Structural Engineers celebrated its Golden Jubilee and the Department of Civil Engineering and Building Technology at UWIST was invited to collaborate in mounting this Conference, to form part of the special programme of events marking the occasion. We were encouraged by the earlier success of the 1977 Conference on Large Ground Movements and Structures, and a joint committee was soon set up to make the necessary arrangements. It was decided that, as for the first conference, the approach should be a multi-disciplinary one involving structural, civil, municipal, highway and mining engineers, engineering geologists, mineral, mining and land surveyors and others with professional interests in ground movements and/or their effects on structures. The scope was widened to include dynamic movements and those which could not be described as large. The emphasis throughout was on movements initiated by factors other than purely structural loadings.

Many papers were submitted for inclusion. Those finally accepted dealt with such topics as the estimation and measurement of surface and near-surface ground movements due to the extraction of coal, tunnelling, the presence of old underground workings, excavations, reclaimed and backfilled areas, soil shrinkage and heave, and seismic and dynamic forces and the effects of such movements on structures, including pipelines. The selection was based on the need to cater for the professional interests and expertise of a wide range of participants. Representatives of sixteen countries were present as authors or delegates.

The papers were presented in seven half-day sessions. Sessions were prefaced by an introduction, given by an invited speaker, on the general theme of the session and the papers within it and this was followed by authors' introductions. Space limitations have not allowed these to be reproduced here, but a record of some of the discussion is presented.

Finally I record my thanks to members of the Organising Committee; to colleagues of the geotechnical engineering group in my Department, Dr. S.P. Bentley, Dr. D.N. Cathie, Miss L. Cooper, Mr. M. Carter, Mr. C.A.M. Franks and Dr. M.V. Symons; to my secretary Mrs. G. Lewis; to my wife Ann for her contribution to the production of preprints and this Proceedings volume; and to the many others, too numerous to mention individually, who contributed in many varied ways to the mounting of the Conference.

James D. Geddes
Chairman, Organising Committee

CONTENTS

SESSION II (Contd.)

SESSION III

SESSION IV

SESSION IV (Contd.)

SESSION V

SESSION VI

SESSION VII

SESSION VII (Contd.)

DISCUSSION

SESSION I

Chairman: Professor J D Geddes, BSc, PhD, CEng, FICE, FASCE, FGS
Department of Civil Engineering and Building Technology
University of Wales Institute of Science and Technology

Introduction to Session: Dr D N Cathie, BSc, PhD
Department of Civil Engineering and Building Technology
UWIST

SESSIONAL PAPERS

REMOTE MEASUREMENT OF GROUND MOVEMENTS BY SURVEYING TECHNIQUES

Professor V. Ashkenazi
Dr. A. H. Dodson
Dr. R. M. Sykes
Mr. S. A. Crane

Department of Civil Engineering, University of Nottingham

1 INTRODUCTION

A large variety of civil engineering structures have to be monitored for movement both during and after construction. In some cases, the monitoring extends to the ground on which the structure is built, and its surroundings. Reservoir embankments, bridges, tall structures and special (high precision) mechanical machinery fall in this category. Of these, reservoirs have become particularly important with the proposed implementation of the 1975 Reservoirs Act (Hoyle, 1979).

There is a wide variety of instrumentation for monitoring structural deformations (Penman and Charles, 1973). Inclinometers, extensometers and settlement gauges may be used in earth embankments, dams and foundations. Strain gauges and pendulums would be more appropriate for use in concrete structures. However, all of these devices, which of necessity are connected in some way to the structure under observation, only provide data on relative displacements. A surveying network, based on a number of horizontal and vertical control stations, is the only method which allows remote monitoring of absolute movements to very high accuracies.

Modern surveying equipment, including high precision electronic distance measuring (EDM) instruments and geodetic theodolites, coupled with modern data processing and analysis techniques, can lead to positioning accuracies of the order of 1 to 2 mm. The method, which is very flexible, has the following advantages.

(a) The system is 3-dimensional in the full (geodetic) sense of the word, involving the periodical monitoring of small changes in the x, y and z coordinates of a series of points. Several independent monitoring systems may be combined together into one geodetic coordinate system.

(b) Detail points on the structure can be observed remotely, by positioning the control (observation) pillars either inside or outside the area of influence of the structure. In addition to detail points, all other visible monitoring devices may be linked to this absolute surface reference system.

(c) The number and type of observations can be varied, stations in the control network can be abandoned and new stations added, without affecting the continuity of the system.

(d) The system is flexible in terms of positioning the reference pillars to suit the topography and avoid naturally occurring obstructions and barriers. This is advantageous for construction purposes as pillars can be erected close to vehicular access routes and can be founded on stable ground.

A brief theoretical background and details of the Nottingham data processing and analysis programs are given in § 2. The instrumentation and observational techniques are described in § 3. This is followed, in § 4, by an illustration of the method using one of the projects recently processed at Nottingham.

2 THEORY AND NOTTINGHAM PROGRAMS

A surveying network is made up of a set of stations which are interconnected by a series of observations. The aim of an adjustment procedure is to produce a set of coordinates for the stations which best fits the observational data. Clearly, the strength of the adjusted network depends on the accuracy and quantity of the various types of observation which make up the network. In surveying networks, the observations include distances and theodolite angles. Occasionally, (gyro-theodolite) orientation observations may also be incorporated into a network to provide additional strength. In general, the position of one station (the origin) and the direction of one line are arbitrarily held fixed to position and orientate the network.

For most applications the adjustment procedure adopted should not only yield a consistent set of station positions but also an estimate of the accuracy of the adjusted coordinates. This is particularly important when two sets of coordinates are to be compared and the significance of any differences determined. The adjustment procedure must also be flexible to allow different weights to be assigned to the various observations. This ensures that each observation contributes its appropriate share to the adjustment. The most suitable method for adjusting engineering surveying control networks is "variation-of--coordinates". This procedure, which leads to the most probable values of the station coordinates, can be extended to incorporate a statistical error analysis based on the covariance matrix of the adjusted quantities.

The network analysis technique outlined above can also be used to optimize the field-work required, while meeting the required standards of accuracy. The computational process involved is relatively inexpensive and it may save a great deal of field-work.

The suite of Nottingham Deformation software comprises the following programs:-

(a) Horizontal network adjustment and error analysis program, based on the "variation-of-coordinates" method and the covariance matrix respectively. The input may consist of observed theodolite angles (or directions), EDM distances and (measured or arbitrarily held fixed) position and orientation observations. The initial estimates of the a priori (observational) standard errors may be revised by an iterative procedure. In addition to the observations, the input also includes a set of approximate station positions (usually scaled from a plan).

The output is made up of the final coordinates of the adjusted stations, the corresponding a priori (adjusted) positional standard errors, the residual error in each observation, the grid bearing and length of any line in the adjusted network and the corresponding standard errors and the absolute (point) and relative (line) error ellipses. The program may be used either with real data, following a set of field observations, or in the simulation mode during the initial network design and optimization stages.

Because of the curvature of the earth and the resulting non-parallelism of the verticals at the various station pillars the measured distances should be reduced to an arbitrary height datum above mean sea level. However, in practice, this reduction need only be applied when the height differences between the network stations are large.

(b) Vertical network adjustment and error analysis program, based on similar principles to those adopted in the horizontal network program, but dealing with the vertical rather than the plan positions of the control stations. The observational data may consist of any combination of vertical theodolite angles and spirit levelled height differences. The program contains an option to include in the adjustment a refraction parameter which leads to the average optimum value of the coefficient of refraction for the period covered by the observations.

(c) A detail point computation program which uses either one or two of the control stations (the positions of which have already been determined, see (a) and (b) above) and any feasible combination of observed distances, horizontal and vertical angles and spirit levelled height differences to fix the position of individual detail points.

(d) Programs producing lineprinter and graph-plotter outputs giving changes in the adjusted horizontal and vertical positions, resulting from two different sets of observations, of both the control stations and detail points. In the case of control stations, the lineprinter output also includes the standard error of the corresponding differences of coordinates and a statistical indication of the significance of these differences in terms of movement.

The control point comparison program allows the introduction and computation of unknown systematic scale and orientation parameters between the two sets of coordinates. This enables the removal of any remaining systematic influences, such as EDM frequency errors and atmospheric refraction, before proceeding with the computation of coordinate differences.

3 EQUIPMENT AND OBSERVATIONS

3.1 Instrumentation

Horizontal and vertical angles for both the control network and the detail points are observed with a geodetic precision theodolite whereas the lengths are measured using an EDM instrument. The resolution of these instruments may be of the order of 0.2 seconds-of-arc and 0.1 mm respectively. In order to maintain this *potential* accuracy, it is essential to use a high precision system of instrument and target centring. Instruments are mounted on solid (concrete) pillars through a "forced centring" arrangement. A suitable design of pillar (as used in the Nottingham baseline, Ashkenazi and Dodson, 1975) is shown in Figure 1.

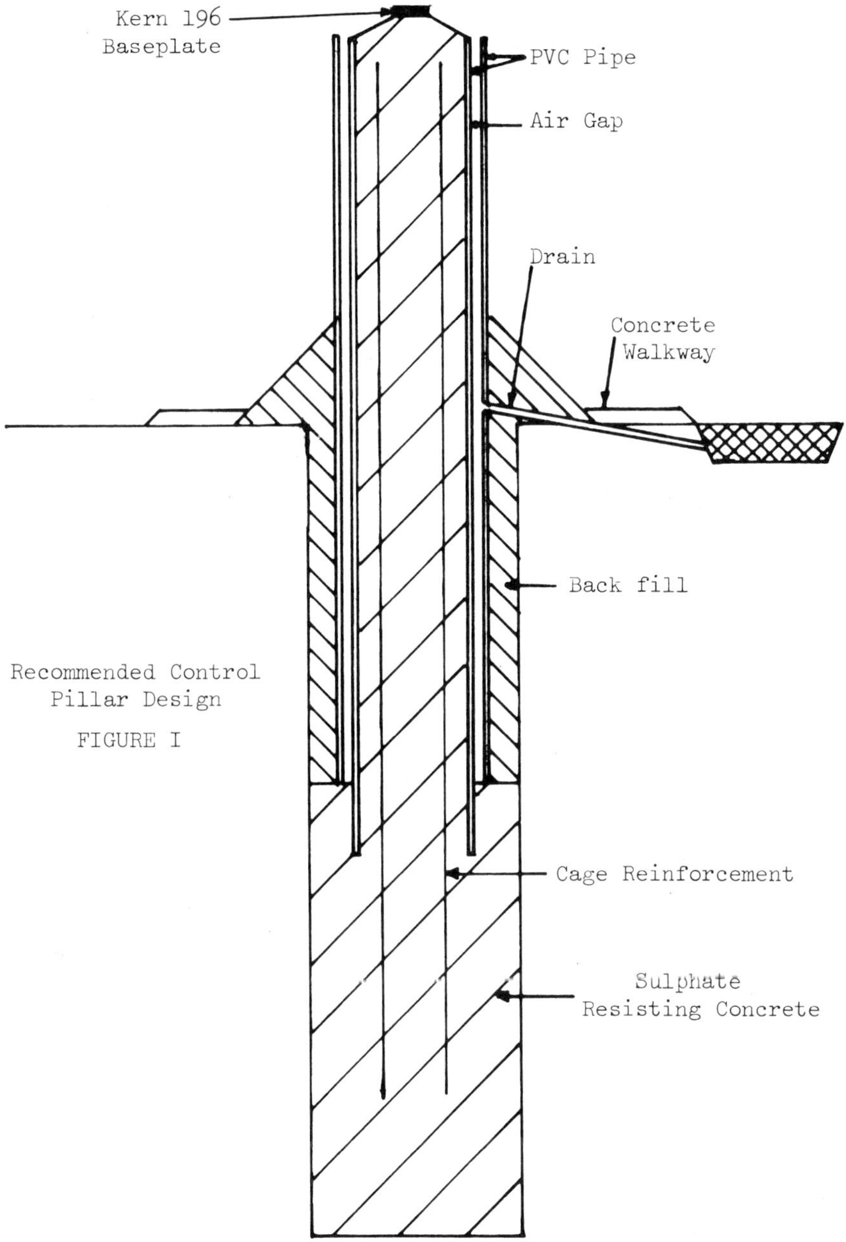

Recommended Control
Pillar Design

FIGURE I

The most important feature of this design is the insulating air gap between the inner concrete pillar and the protective outer sleeve, which ensures that the core is kept free from contact with the surrounding soil (where diurnal and seasonal movements are most likely to occur) and also provides thermal insulation. The outer sleeve also offers some protection against accidental collisions.

The heights of control and detail points may be obtained either by trigonometric or by spirit levelling, as part of a network layout (see § 2). The choice of instrumentation for heighting will be influenced by the terrain, the type of structure which is being monitored, and by the order of accuracy required. The inherent accuracy of vertical theodolite angle measurements may be reduced significantly by atmospheric refraction. In such cases, the alternative is spirit levelling which is not only time consuming and expensive, but may not even be practicable.

3.2 Calibration

There are three types of systematic error in an EDM instrument (Ashkenazi and Dodson, 1977), namely zero error, cyclic error and proportional (scale) error. Each one of these may exceed the nominal accuracy of the instrument which cannot, therefore, achieve its full potential without proper calibration. Both the cyclic error and the frequency (scale) error can be determined by a laboratory calibration whereas the zero error (additive constant) is best evaluated in the field. However, these should be supplemented by a full field calibration, on a multi-pillar baseline (Ashkenazi and Dodson, 1975). This will lead not only to the zero error and the scale error (including contributions other than pure frequency effects), but also to the observational accuracy of the instrument, under field calibration conditions. This value may then be used as an estimate for weighting the measured distances (see § 2). Of the three systematic errors associated with observed distances, the scale error is most important since, however large, it would remain undetected in a network adjustment. Although it is now possible to include a scale bias into a network coordinate-difference analysis (see § 2) this should not be used as an alternative to calibration, but rather as a means of detecting residual effects. A proper calibration procedure may up-grade the nominal accuracy of a standard EDM instrument, thus making it acceptable for use in deformation measurements.

The calibration of the EDM instrument is accompanied by that of the (wet and dry) thermometers and barometer used for determining the refractive index of the air. Small errors in the measurement of the temperature and pressure can lead to significant scale errors in distance observations.

An absolute temperature error of 1°C or a pressure error of 3 mbar will lead to a significant 1 ppm scale error.

Theodolites are also tested for any significant maladjustment. Small instrumental errors are eliminated by using an appropriate observational procedure. Moreover, the elevation of the geometric centre of the theodolite (as well as of the electro-optical centre of the EDM instrument) above the height reference point of the pillar is carefully determined to ensure proper reduction of the measurements. The same also applies to visual and light-reflecting (corner-cube prism) targets.

3.3 Field Observational Procedure

When using a geodetic theodolite on pillars, with forced centring and good targeting, it is not necessary to observe more than 3 rounds of horizontal angles (Ashkenazi, 1975 and Ashkenazi *et al.*, 1979). Accuracies of the order of 0.5 seconds-of-arc may consistently be achieved and this is unlikely to be improved upon by making further rounds of observations.

Unlike horizontal angles, vertical angle measurements are very susceptible to refractional influences. At present (1979) this limits their widespread use for height determination. Consequently, whenever possible, spirit levelling should be used to complement vertical angle observations. The accuracy of trigonometric levelling may be improved by adopting one or more (spirit levelled) height reference points and using these for empirically correcting for the effects of changes in the refractive index of the atmosphere. Otherwise a refraction model may be employed in the vertical network adjustment to produce an overall refraction bias parameter. One could also use more complex atmospheric refraction models, but this could involve more field-work than straightforward spirit levelling.

Distance measurements are affected by the atmosphere to a lesser extent than vertical angles. The standard procedure for removing the atmospheric effects is to measure the variable parameters (temperature, barometric pressure and humidity) at the two ends of the line, and apply a mean computed correction to the measured distance.

4 DATA PROCESSING AND ANALYSIS

4.1 Castleshaw Reservoirs Project

The method discussed in the previous two sections will be illustrated by describing the monitoring of the two Castleshaw reservoirs belonging to the North West Water Authority.

A full description of the project, including the complete analysis of the results, is given in Ashkenazi *et al.* (1979).

The horizontal network adopted after computer simulations (aimed at achieving the highest possible accuracies), consisted of 10 control stations connected by 19 distances and 45 horizontal angles (see Figure 2). Six of the control stations were located in the valley in order to monitor the detail points on the two embankments whereas the remaining four stations were sited on the surrounding hills and constituted a stable reference network.

The height control was provided by a combination of precise spirit and trigonometric levelling. The 6 valley stations were precision levelled, and vertical angles taken from all the stations at which horizontal angles were observed. These observations were used both for reducing the observed distances to a horizontal datum and for monitoring the vertical movements of the valley stations. The 10 control station pillars were constructed to the design and specification outlined in § 3.1. The reference point structures for the 90 detail points were founded approximately 1 metre below the surface of the embankments, with the actual reference points at ground level.

All the angle observations were taken with a Wild T3 theodolite. In order to attain the high standard of centring accuracy required a special adaptor, with a ball centring device, was used to mount the theodolite onto the Kern 196 baseplates on the pillar tops. The distances were measured with a Kern ME 3000 Mekometer. A great deal of care was also taken in the design of the prism and target mountings in order to ensure repeatability of centring on both the control stations and the detail points.

A full calibration of the equipment was carried out prior to each set of observations. This included calibration of the Mekometer in the laboratory and on the Nottingham multi-pillar baseline (Ashkenazi and Dodson, 1975), standard collimation checks on the theodolite, and accurate measurement of the heights of the various instruments and targets used, above the reference marks on the pillars.

4.2 First Set of Observations

The angle and distance observations of the horizontal network were found to have a priori (observational) standard errors of 0.54 seconds-of-arc and 0.2 mm + 2 ppm respectively. These values, which emerged as a product of the adjustment procedure, are an indication of the internal consistency of the network.

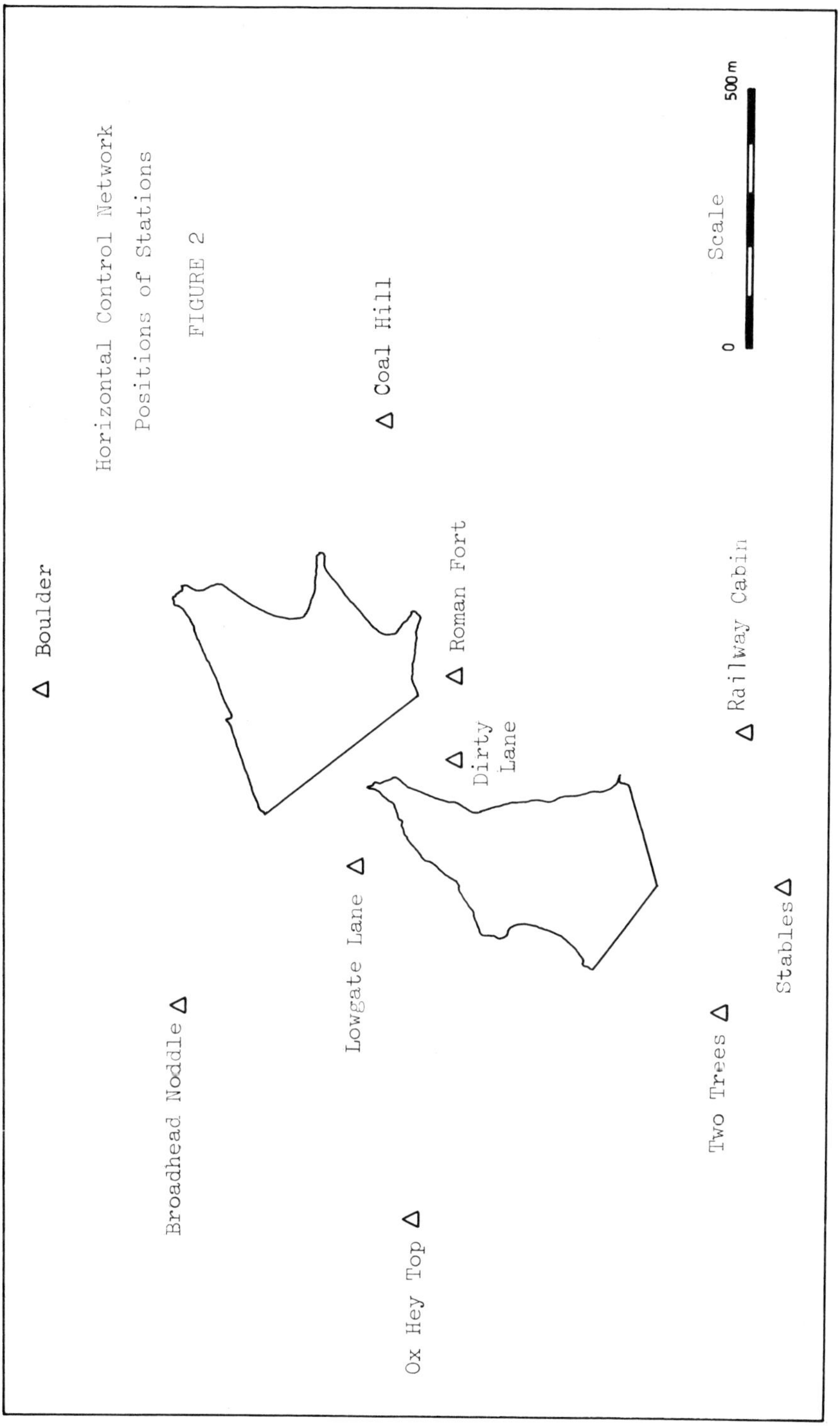

Horizontal Control Network
Positions of Stations
FIGURE 2

The standard error for the observed distances was the same as that used during simulation whereas that for the angles was much better than the corresponding 0.8 sec. As a result, the network was more accurate than planned, with absolute positional standard errors averaging 0.9 mm and 1.2 mm in Northings and Eastings respectively. All but two of the angle residuals were less than 1 second, with an average absolute value of 0.35 sec. Only two of the 14 distance residuals just exceeded 2 mm with an average absolute value of 0.7 mm.

The vertical network was also well-determined although the trigonometrically heighted stations were not as accurate as those which were established using precise levelling. Over a nominal 500 metre long line, the a priori standard error of trigonometric levelling was found to be 4.6 mm for a reciprocally observed line (6.5 mm for a single-direction observation). The corresponding value for spirit levelling was 0.6 mm per kilometre of level run. The a posteriori standard error of the adjusted heights of the pillars averaged 1.6 mm, with only one value over 4 mm, whereas the adjusted height differences between any two pillars had a mean standard error of 1.9 mm, with a maximum value of 4.8 mm.

The accuracy of the horizontal and vertical positioning of the detail points is dependent upon the corresponding accuracies of the control pillars used for observation, the sighting accuracies and, most important, errors due to refraction. It is estimated that detail points were positioned with an accuracy of 2 to 3 mm horizontally and 7 to 8 mm (in the case of trigonometric levelling) or 1 to 2 mm (when precision levelled) vertically.

4.3 Second Set of Observations

The processing of the second set of observations presented some difficulties. The first computation of the vertical network led to an initial a priori standard error of 10.6 mm for reciprocal trigonometric heighting of a nominal 500 m long line. A second computation including an unknown parameter for the coefficient of refraction produced a considerable improvement. After rejecting two observations with large residual errors, the a priori standard error settled down to an acceptable 6.5 mm. For the sake of consistency the vertical network corresponding to the first set of observations was also recomputed with a refraction parameter, although this hardly affected the corresponding vertical and horizontal networks.

The horizontal angles of the second set also appeared to have a lower accuracy than the first. The average a priori standard error of the angles was 0.8 seconds as opposed to 0.54 seconds for the first set, whereas the standard error of the distances remained unchanged at 0.2 mm + 2 ppm. These values led to a posteriori positional standard errors averaging 1.1 mm and 1.3 mm in Northings and Eastings respectively. The corresponding vertical standard errors averaged 3.0 mm with a maximum value of 6.6 mm. Although the accuracies obtained in set 2 are slightly lower than the corresponding values in set 1, they are compatible with the accuracy criteria used during simulation and are therefore deemed to meet the accuracy requirements.

4.4 Interpretation of Results

The first comparison of the two sets of results was made between the control station co-ordinates. The analysis indicated the existence of a significant scale bias of 2.73 ± 0.90 ppm but not an orientation difference. This was supported by the evidence of the laboratory calibrations of the Mekometer before and after each set of observations (Ashkenazi et al., 1979, page 15, table 2) which indicated that there was a scale uncertainty in the second set of measured distances. Thus a scale correction was applied to the distances and the second set of horizontal control and detail point coordinates were recomputed. Further evidence of a scale change was produced by the apparent movement of the station at Roman Fort. This pillar, whose apparent movement is almost entirely eliminated by the application of the scale correction, is unlikely to have moved because it probably has one of the best (rock) foundations of all the control stations.

Tables I and II show the horizontal and vertical movements of the control stations with their corresponding standard errors. The maximum standard error in an overall horizontal movement determination is 2.3 mm. At the 95% significance level there appear to have been movements of stations Coal Hill, Boulder and Lowgate Lane, with deformations of the order of 5 mm. There have been no corresponding significant vertical movements of the control stations.

There appear to have been very significant horizontal movements of many of the detail points on both the upper and lower dams (see Figure 3). These coordinate differences are not due to the random observational error in the measurements from the adjacent control stations, which will be of the order of 2 mm for the ranges involved. In general it seems that the largest movements occurred for the points in the centre of the embankments, becoming smaller towards the ends. On the Upper Dam, the points on the road along the crest have much smaller and statistically insignificant movements.

No.	STATION NAME	ΔN (mm)	ΔE (mm)	Δ (mm)	σ (mm)	Δ/σ	Prob.
1	BOULDER	5.5	-1.0	5.6	2.0	2.82	1.00
2	LOWGATE LANE	-3.6	1.4	3.9	1.5	2.65	0.99
3	DIRTY LANE	-2.7	-0.4	2.7	1.5	1.85	0.94
4	ROMAN FORT	-1.0	0.5	1.1	1.5	0.75	0.55
5	OX HEY TOP	0.9	3.2	3.3	2.0	1.64	0.90
6	TWO TREES	0.2	4.2	4.2	2.3	1.85	0.94
7	STABLES	-0.9	2.9	3.0	2.3	1.29	0.80
8	RAILWAY CABIN	0.7	2.9	3.0	2.2	1.40	0.84
9	COAL HILL	1.4	-4.0	4.2	1.3	3.15	1.00
10	BROADHEAD NODDLE	ORIGIN (HELD FIXED)					

TABLE I

Control Station Movements: Horizontal

No.	STATION NAME	ΔH (mm)	σ (mm)	Δ/σ	Prob.
1	BOULDER	-6.1	7.1	0.85	0.61
2	LOWGATE LANE	-0.9	1.7	0.54	0.41
3	DIRTY LANE	0.6	0.8	0.74	0.54
4	ROMAN FORT	ORIGIN (HELD FIXED)			
5	OX HEY TOP	2.3	6.5	0.35	0.27
6	TWO TREES	2.1	2.0	1.07	0.71
7	STABLES	-0.2	1.9	0.10	0.08
8	RAILWAY CABIN	0.6	1.8	0.32	0.25
9	COAL HILL	-12.1	7.9	1.52	0.87
10	BROADHEAD NODDLE	-3.9	4.0	0.97	0.67

TABLE II

Control Station Movements: Vertical

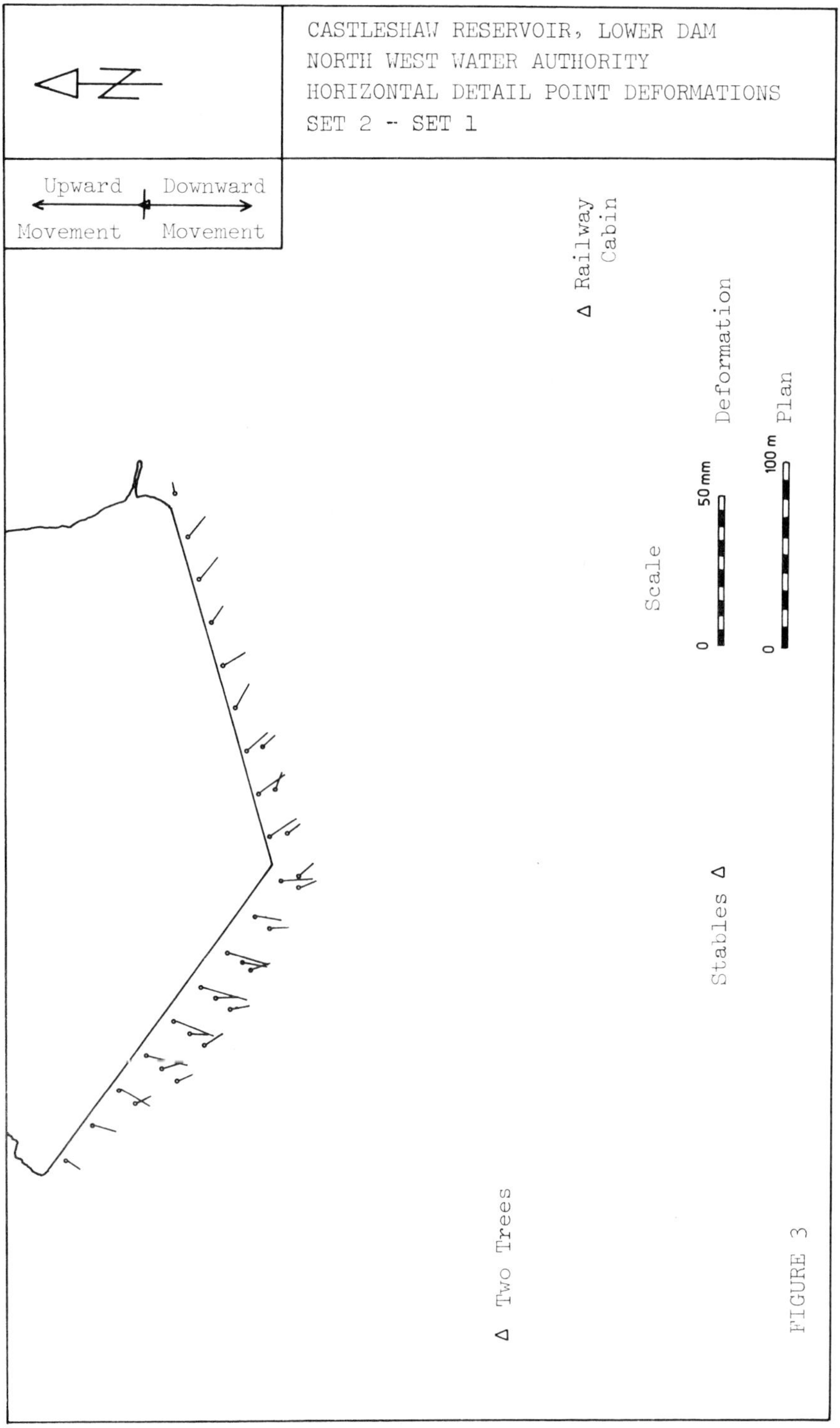
CASTLESHAW RESERVOIR, LOWER DAM
NORTH WEST WATER AUTHORITY
HORIZONTAL DETAIL POINT DEFORMATIONS
SET 2 - SET 1
Upward Movement
Downward Movement
Railway Cabin
Scale
Deformation
Plan
0
50 mm
0
100 m
Stables
Two Trees

FIGURE 3

The vertical movements of the detail points (see Figure 4) are less easily interpreted. Those points which have been spirit levelled (along the top of both embankments) show small height differences between the two sets whereas those which have been determined using only vertical angles show much larger (but not necessarily significant) differences. Clearly, because of the nature of the topography in the area, the uncertainty due to refraction is significant and can only be fully overcome by spirit levelling. A lesser alternative would be to use a refraction correction based on vertical angle measurements taken to targets, which have been spirit levelled, from the observation station and to use this as a standard criterion for correcting all vertical angles taken at that station.

5 CONCLUSIONS

5.1 A geodetic reference system is capable of monitoring horizontal and vertical surface movements to a very high degree of accuracy.

5.2 The a posteriori horizontal positioning accuracy of the control stations at Castleshaw averages 1 mm in any direction, with a maximum value of 2 mm.

5.3 The corresponding a posteriori vertical positioning accuracy of the control stations averages 2.5 mm, with 1.5 mm for spirit levelled stations and up to 6.5 mm for purely trigonometrically heighted pillars.

5.4 The positional accuracies of the detail point intersections at Castleshaw are a little worse than the corresponding values for the control stations, i.e. 2 mm for horizontal positions, 1.5 mm for levelled heights and 7 mm for trigonometrically determined elevations.

5.5 Clearly, the only possible improvement to the system would be to increase the amount of spirit levelling or to devise appropriate physical models to correct for the atmospheric refraction effects on vertical angles.

5.6 The results of the two sets of observations at Castleshaw show likely horizontal movements of the order of 5 mm for three control stations, namely Boulder, Lowgate Lane and Coal Hill. There do not appear to be any corresponding significant vertical movements.

5.7 Several of the detail points on both embankments appear to have moved horizontally by significant amounts, but not vertically.

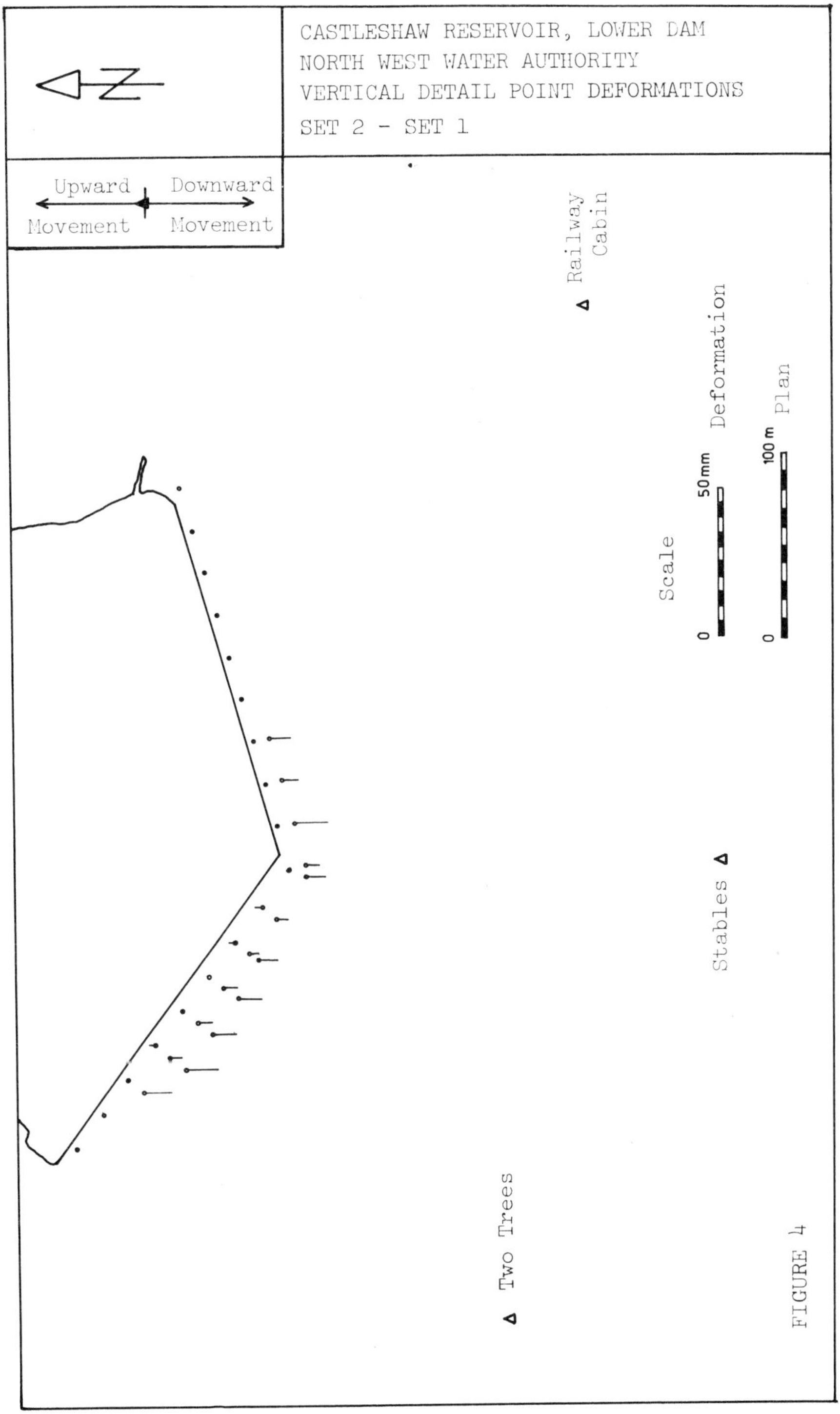
CASTLESHAW RESERVOIR, LOWER DAM
NORTH WEST WATER AUTHORITY
VERTICAL DETAIL POINT DEFORMATIONS
SET 2 - SET 1
Upward Movement
Downward Movement
Railway Cabin
Scale
Deformation
0
50 mm
Plan
0
100 m
Stables
Two Trees

FIGURE 4

5.8 The a priori horizontal angular accuracies achieved with the Wild T3 theodolite were 0.5 and 0.8 seconds-of-arc for Set 1 and Set 2 respectively.

5.9 The a priori standard error of the distances measured with the Kern ME 3000 Mekometer emerged as 0.2 mm + 2 ppm for both sets of observations. However, the two sets were found to differ by a systematic scale factor of 2.7 ppm.

ACKNOWLEDGEMENTS

The authors are grateful to the North West Water Authority for permission to quote the results of deformation measurements at the Castleshaw reservoirs. In particular, the authors wish to express their thanks to Mr. C. W. P. Heaton-Armstrong with whom they collaborated during the planning, design and execution of the project.

BIBLIOGRAPHY

V. Ashkenazi - Adjustment of control networks for precise engineering surveys, Chartered Surveyor, Jan. 1970.

V. Ashkenazi - The weighting of geometric control networks, Proceedings of the International Conference on Methods and Instruments for Densification of Geodetic Networks, Budapest, (1970).

V. Ashkenazi - The measurement of spatial deformations by geodetic methods, Proceedings of the British Geotechnical Society Symposium on Field Instrumentation, Part I, London, (1973) 1-12.

V. Ashkenazi - Discussion on 'principles of measurement' and 'application', Proceedings of the British Geotechnical Society Symposium on Field Instrumentation, Part II, London, (1973), 551-2 and 649-50.

V. Ashkenazi - Optimum methods for the measurement of structural deformations by surveying techniques, Proceedings of the XIV F.I.G. Congress of Surveyors, Washington, D.C., (1975).

V. Ashkenazi - Deformation measurements at the Empingham dam site, special publication by Messrs. Wild U.K. (1977).

V. Ashkenazi and A. H. Dodson - The Nottingham multi-pillar baseline, Journal of the Association of Surveyors in Civil Engineering, June 1977.

V. Ashkenazi and A. H. Dodson - The calibration and evaluation of EDM instruments, Proceedings of the XV F.I.G. Congress of Surveyors, Stockholm, (1977).

V. Ashkenazi, A. H. Dodson and C. W. P. Heaton-Armstrong, - Deformation measurements at Castleshaw Reservoirs - Presented at a meeting of the Institute of Water Engineers and Scientists, Northern Section, Blackpool, (1979).

British National Commission on Large Dams, Proceedings of a Symposium on the Inspection, Operation and Improvement of Existing Dams, University of Newcastle-upon-Tyne, (1975).

N. Hoyle - Safety of reservoirs, Presented at a meeting of the Institute of Water Engineers and Scientists, Northern Section, Blackpool, (1979).

International Commission on Large Dams - Note on the application of geodetic methods to the determination of movements of dams, Bulletin No. 23, (1972).

A. D. M. Penman and J. A. Charles - Measuring movements of embankment dams, Proceedings of the British Geotechnical Society Symposium on Field Instrumentation, Part I, London, (1972) 341-358.

MEASUREMENTS BY TERRESTRIAL PHOTOGRAMMETRY - AN INTERMEDIATE TECHNOLOGY APPROACH

David A Cook, MSc CEng MICE MIStructE

Lecturer in Building Engineering, University of Bath

THEME

Whilst a significant increase in usage of air and satellite photography has occurred in the last two decades, application of the same basic techniques to terrestrial work has generally remained in the hands of a few specially equipped firms. Perhaps one reason for this is that the high cost of acquiring good quality air photos justifies a comparable investment in equipment for their analysis and interpretation. Unfortunately much air photo plotting equipment will not adjust to the range of focal length and depth of field common to terrestrial photographs. Also, terrestrial photogrammetry requires to show a cost benefit against traditional survey methods and/or the current generation of distance measuring instruments. Except for its specific application to situations where accessibility or speed of recording override other factors (eg tall buildings, snow depths, traffic accidents, matching steel fabrications) (1)(2), references to its use as a direct survey substitute are comparatively few (3)(4). However good quality photographic records act as an insurance against the loss of incidental information which is omitted from a normal survey through lack of time, the weather, or plain misjudgement of its usefulness. How often do such reasons complement each other! In this context it is often forgotten that, except for the control survey and photography itself, the main body of work is independent of weather conditions.

When comprehensive measured coverage of the object field is required then obviously the use of costly plotting equipment is justified. But where the area of interest can be quantified or compared at a few selected points, or the final amount of required detail not known, the work can be carried out quite accurately using a pocket magnifier, a fine glass scale and programmable calculator. It is hoped therefore that the resume of the theory and simple examples of fieldwork, presented here, will encourage engineers to experiment with terrestrial photo techniques themselves even though the work might extend to a requirement for more powerful equipment.

Comparison with air photography

Normal air photography attempts to provide a sequence of overlapping photographs from a constant flying height on perfectly vertical axes. Departures from this ideal are back-calculated from known ground reference stations which enable the photographs to be 'rectified'. This may be done analytically, or, more usually by (analogue) reduction on a stereo plotting instrument. At the risk of over-simplification the calculation is shown within Figure 1.

By geometry $$(H-h) = \frac{B}{p}(f) \qquad (1)$$

where p = displacement of image relative to line of flight (taken as the x-axis) and called the parallax

$= (x-x')$

Plotting instruments create an optical model of Figure 1 and contour the terrain by observing parallax differences between a base reference level and points such as, a, e. For air photography the plotting instrument seldom requires to accommodate a depth of field of more than five per cent of flying height.

By contrast in terrestrial work the camera stations themselves provide a reference grid as these are located with comparable accuracy to the ground stations of a flight programme. Departures from parallel, horizontal axes are deliberate and the depth of field may well be some multiple of the distance to the nearest object in view. Indeed inclination of the camera axes may improve accuracy if it permits a closer approach to the object and it enables the whole camera plate to be utilised (Figure 2).

With building structures the sighting is often governed by the surrounding space restrictions which give no choice but to angle the camera.

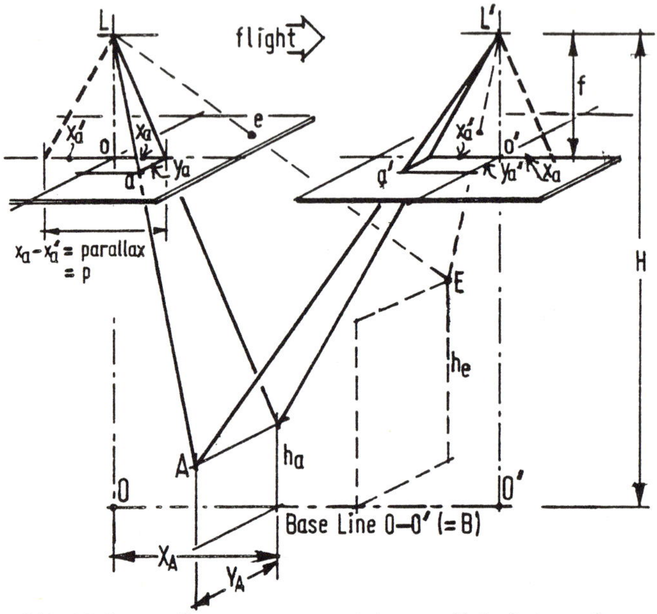

FIGURE 1: Parallax measurement for parallel photo pairs

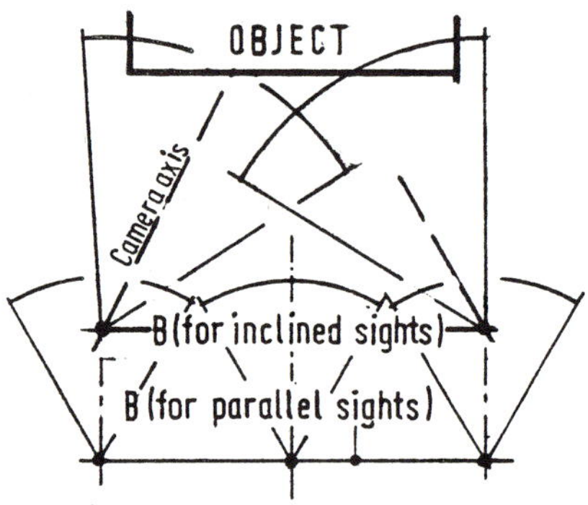

FIGURE 2: Relative coverage from parallel and inclined photos

Photographs on angled sights

The geometry for computation of coordinates from angled photos is given in Figures 3 and 4. Figure 3 shows a pair of inclined photographs set on horizontal axes.

Referencing to the Left Hand camera station the coordinates of an object are given by

$$X_A = \frac{B}{1-\cot(\delta+\alpha)\tan(\delta'+\alpha')} \qquad (2(a))$$

$$Y_A = \frac{B}{\tan(\delta+\alpha)-\tan(\delta'+\alpha')} \qquad (2(b))$$

V_A = Height above/below camera axis

$$= \frac{B\cos(\delta'+\alpha')\tan\beta}{\sin\{(\delta+\alpha)-(\delta'+\alpha')\}} \qquad (2(c))$$

where

$$\tan\beta = \frac{y}{\sqrt{f^2+x^2}} \text{ and } \tan\alpha = \frac{x}{f}$$

for the corresponding values x, y in each photo.

If the camera axis cannot be set horizontal then its inclination can be accurately measured during the survey. In Figure 4 the camera axis is shown depressed by an angle θ at its exposure station L. If the axes of the fiducial marks remain vertical and horizontal then inclination of the camera involves no change in the (photo) coordinate measurement of x_a, relative to the image position. However the true horizontal and vertical angles, α and β between L and the object field, must be referenced to the 'horizon' line and not the x axis of the photograph plate. Referenced to Figure 4, these angles then derive from the following expressions

$$y_a^* = y_a + f\tan\theta \qquad (3(a))$$

$$\tan\alpha_a = \frac{x_a}{(f\sec\theta - y_a^*\sin\theta)} \qquad (3(b))$$

$$\tan\beta_a = \frac{y_a^*}{x_a}\ \frac{\cos\theta}{\operatorname{cosec}\alpha} \qquad (3(c))$$

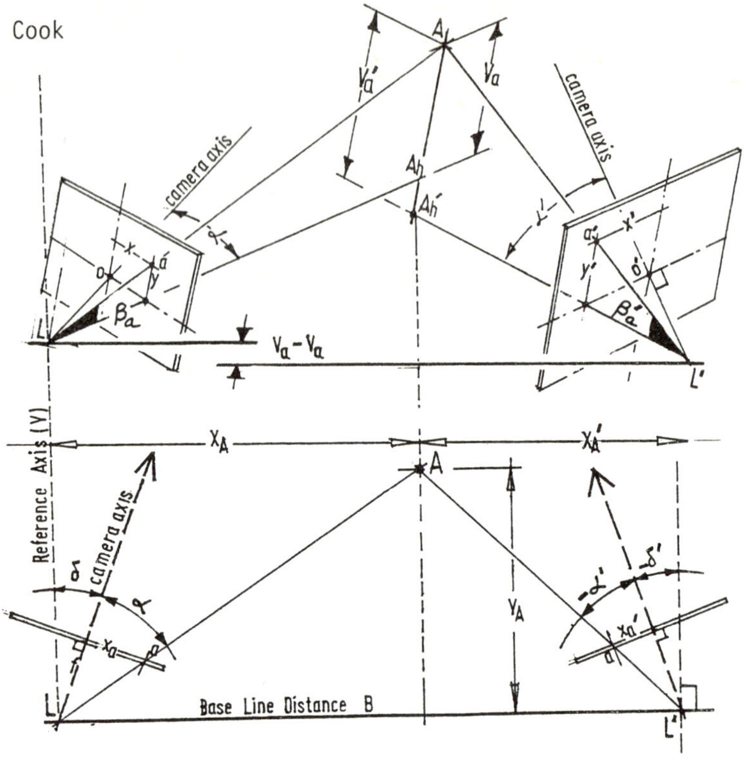

FIGURE 3 : HORIZONTAL PHOTO PAIRS NOT PARALLEL

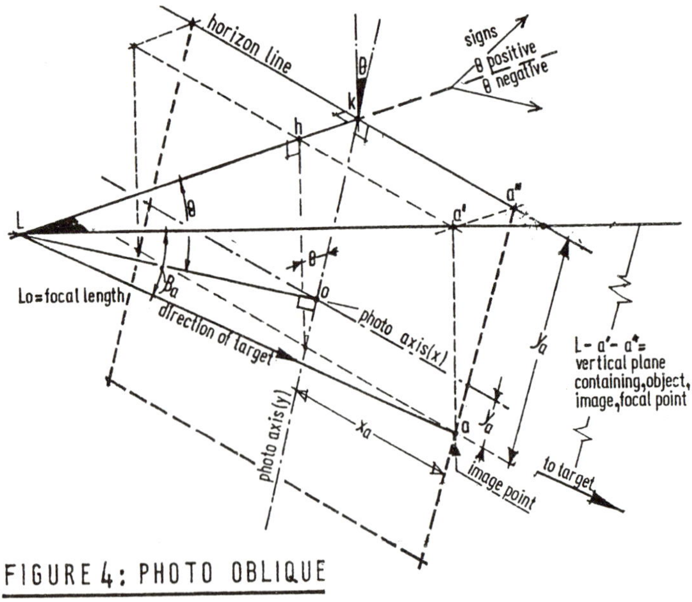

FIGURE 4: PHOTO OBLIQUE

Coordinates in the object field may then be calculated from the same intersection procedure as given by Figure 3, but using the values of α_a and β_b from equations (3), in equations (2(a)), (2 (b)), (2(c)). Sign convention is most important - the algebraic sign of α and β are determined from the sign of x_a and y_a* which in turn follow the normal rule for any x, y, coordinate system. The inclination of the camera (θ) is positive/elevated, negative/depressed, and for clarity in Figure 3 it is shown depressed. A fuller derivation of equations (2) and (3) is given in Appendix 'A'.

Determination of photo coordinates x and y

Where the survey plotting is done analytically then the final accuracy is governed by the ability to resolve the complementary coordinates x, y, at the image points. The determination of x, y is therefore a crucial office computation. As already mentioned there are a range of expensive comparators on the market to do this. They can however be measured to a very acceptable accuracy using a simple 0.1mm glass scale and magnifier, by measurement from the four or five fixed fiducial marks on the photograph (Sibson and Bowyer, 1980).

FIGURE 5

Measurement to point from photo fiducial marks

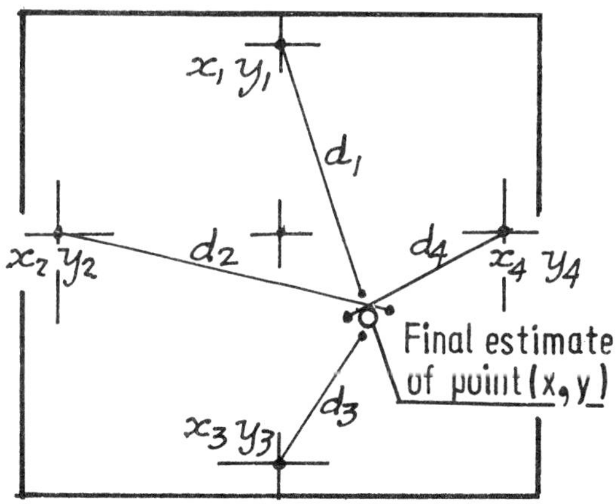

In Figure 5, if the image point is referenced to three points which are not collinear then a good first approximation is obtained from the equations of circles of radius d_i.

$$d_i^2 = (x-x_i)^2 + (y-y_i)^2 \qquad i = 1, 2, 3 \tag{4(a)}$$

These equations over-determine x, y since there are three equations and two unknowns. In each of them, x, y occur linearly and as the sum x^2+y^2. If x^2+y^2 is eliminated as though it was a third variable then two linear equations remain which can be solved for x and y by normal algebra. On writing

$$\ell_i = \tfrac{1}{2}(x_i^2 + y_i^2 - d_i^2)$$

we have

$$x = \frac{(\ell_2-\ell_1)(y_3-y_1) - (\ell_3-\ell_1)(y_2-y_1)}{(x_2-x_1)(y_3-y_1) - (x_3-x_1)(y_2-y_1)} \tag{4(b)}$$

$$y = \frac{(\ell_2-\ell_1)(x_3-x_1) - (\ell_3-\ell_1)(x_2-x_1)}{(y_2-y_1)(x_3-x_1) - (y_3-y_1)(x_2-x_1)} \tag{4(c)}$$

Whilst a permutation of equations (4 (b)), (4(c)), through every measurement of d, would give a reasonable 'average' value of x, y, the best fit is only obtainable by a Least Squares reduction of errors. Writing x_n, y_n, for the coordinates of the fiducial points (n = 1, ... N) and x, y for the final estimate of the survey point then the achieved distance from each pair of fiducial marks

$$e_n = \sqrt{(x-x_n)^2 + (y-y_n)^2} \tag{5}$$

The sum of the squares of the errors between e_n and the measured distance d_n will be

$$S = \sum_{n=1}^{N} (d_n-e_n)^2 \tag{6}$$

x and y must be chosen to minimise S

$$\frac{\partial S}{\partial x} = \frac{\partial S}{\partial y} = 0 \tag{7}$$

It is clear that no explicit solution to equation (7) can be expected and the problem has to be tackled by numerical methods. With two variables, the Newton-Raphson method can be applied.

Sibson's numerical method for optimising x and y is set out in Appendix 'B'. The speed of convergence depends only on a good first estimate of the point, made from three non-collinear measurements of d_n, substituted into eqns (4(b)), (4(c)).

The method is of practical interest because the related mathematics for this and that in Appendix 'A' is well within the capacity of the current generation of programmable calculators. Sibson's point reduction on the pocket size Texas Instruments Tl-59 occupies about 400 program steps and 17 + 3N memory registers for N-fiducial marks. The older Hewlett Packard HP9810 performs similarly. If the initial measurements of d_n are made to an accuracy of ±0.05mm (scale graduations 0.10mm) then the errors in x and y will only be of slightly greater magnitude than those due to lens distortion and film shrinkage.

Fieldwork

As a preliminary checking exercise a photogrammetry programme was compared against a simple conventional survey. This was done by setting out orthogonal base lines on a local hillside slope of 10°. About a dozen ranging poles were set out in the enclosed space between 30m to 90m from the base lines. Usual care was taken in fixing the reference pegs for the camera stations on each base line, after which the ranging poles were located by sightings on one face only from a pair of theodolites at the 50m mark on each base line. Photographs were taken with a theodolite mounted 'Wild P32' camera from six points, with photos and markers mutually positioned so that each of the latter were visible from four photo positions (Figure 6). A typical combination showing also the topography, are reproduced in Plate I. As a genuine practical comparison was wanted which embodied the usual survey difficulties or hazards, no attempt was made to idealise the working conditions on the site. In this respect it was necessary to complete the photography facing west into a late autumn sun; but in spite of the glare the photographs were usable.

From the results it was possible to establish a measure of consistency within the survey and photogrammetry. This was done by using all combinations of photos within the bundle of four available in each case. Figure 7 shows these results in terms of

- maximum variation in the calculated values of the same x or y coordinate
- amount of departure from the actual survey measurement

plotted against the corresponding x or y coordinate value. Note, a distinction is drawn between coordinates calculated from photo pairs on parallel sightings and those on inclined sights by combining photos on opposing base lines.

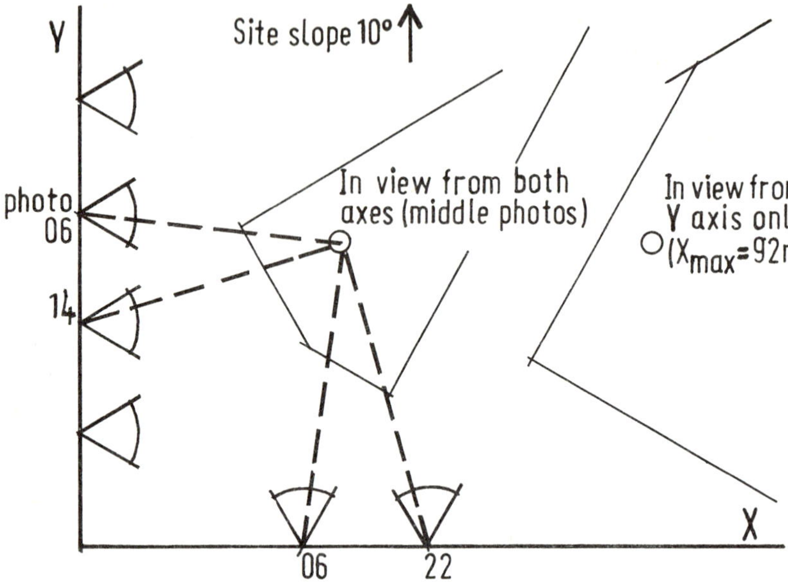

FIGURE 6: Disposition of camera points (see also Photo Plate I).

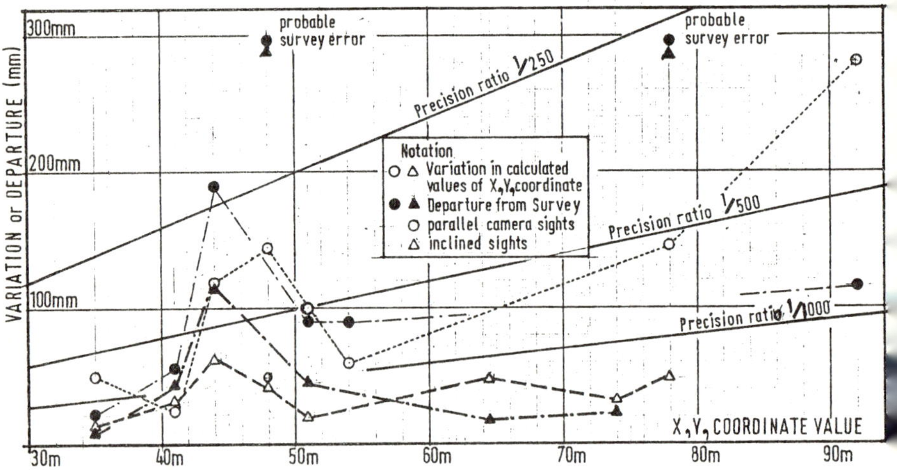

FIGURE 7: Variations in calculated coordinates and departures from survey versus X, Y, val

PLATE I: Experimental survey at, Lime Kiln Lane.

In conclusion it was considered

- the repetition of coordinate computations was amply sufficient to confirm freedom from mistakes, but not enough to provide a reliable field statistic of probable error.
- in the majority of cases the variations between point measurements, and departures from the survey, maintained a Precision Ratio of $^1/_{500}$. The divergency from the survey at x = 48m, y = 74m, was thought to be indicative of a large survey error.
- there was more consistency between coordinate values using photos on opposing base lines (inclined sights) than parallel sights on the same base line. This was mainly due to the increased base line length provided in the former case (Figure 3).

The system of measurement has been used on a number of occasions to survey building faces for (undergraduate) architectural drawing work, to monitor movement in a landslip area and establish the deflected profile of a flexible airhouse structure. The latter task is currently proceeding and will be briefly described in the conference discussion. It has also been applied to occasional instances of severe structural movement, two of which are worthy of mention

- a retaining wall at Lockshill, Frome (Figure 8 and photo plate II)
- a cottage at Minsterworth, Gloucester (Figure 9 and photo plate III).

The photographs and figures serve to describe the situations better than words and a brief commentary only is therefore necessary. For plotting, the photo coordinates are transferred to a convenient reference axis within each structure and movements relative to this are shown to an exaggerated scale. These are figured in millimetres in Figures 8 and 9.

At Lockshill ownership of the wall was in dispute (nobody wanted it!) and it became necessary to establish a complete record of its attitude before being strengthened. Two photographs were necessary at each camera station in order to cover the whole length of wall. A fifth exposure was also made at an acute angle, 16m beyond the right-hand station. Altogether there was sufficient overlap to provide a choice of at least three photographs from which to fix position.

At Minsterworth an earlier site investigation had revealed a deep bed of peat under half the cottage. If it had been built only 10m forward, 60 years ago, there should have been little or no settlement. Nevertheless, in spite of its alarming inclination, it had been lived in continuously. The best (albeit most drastic) solution to the problem was nullified by a local preservation order which left little option but to underpin the existing fabric.

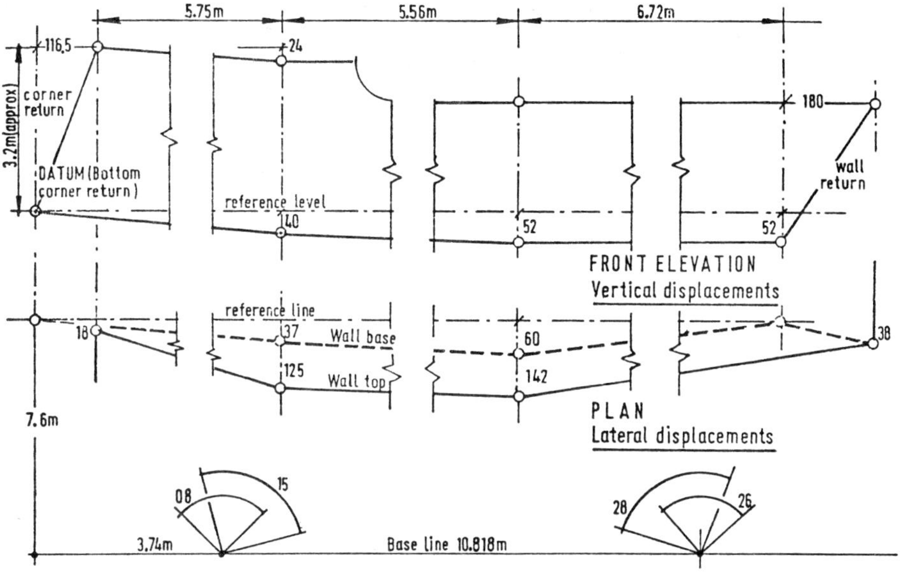

FIGURE 8 : Displacement Diagram for wall at Lockshill, Frome (see also photo Plate II)

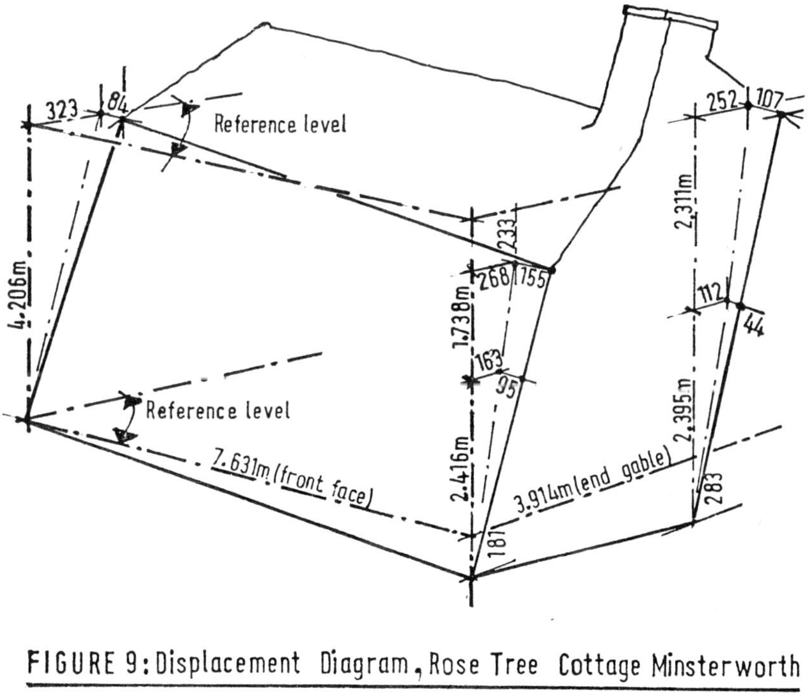

FIGURE 9: Displacement Diagram, Rose Tree Cottage Minsterworth (see also photo Plate III)

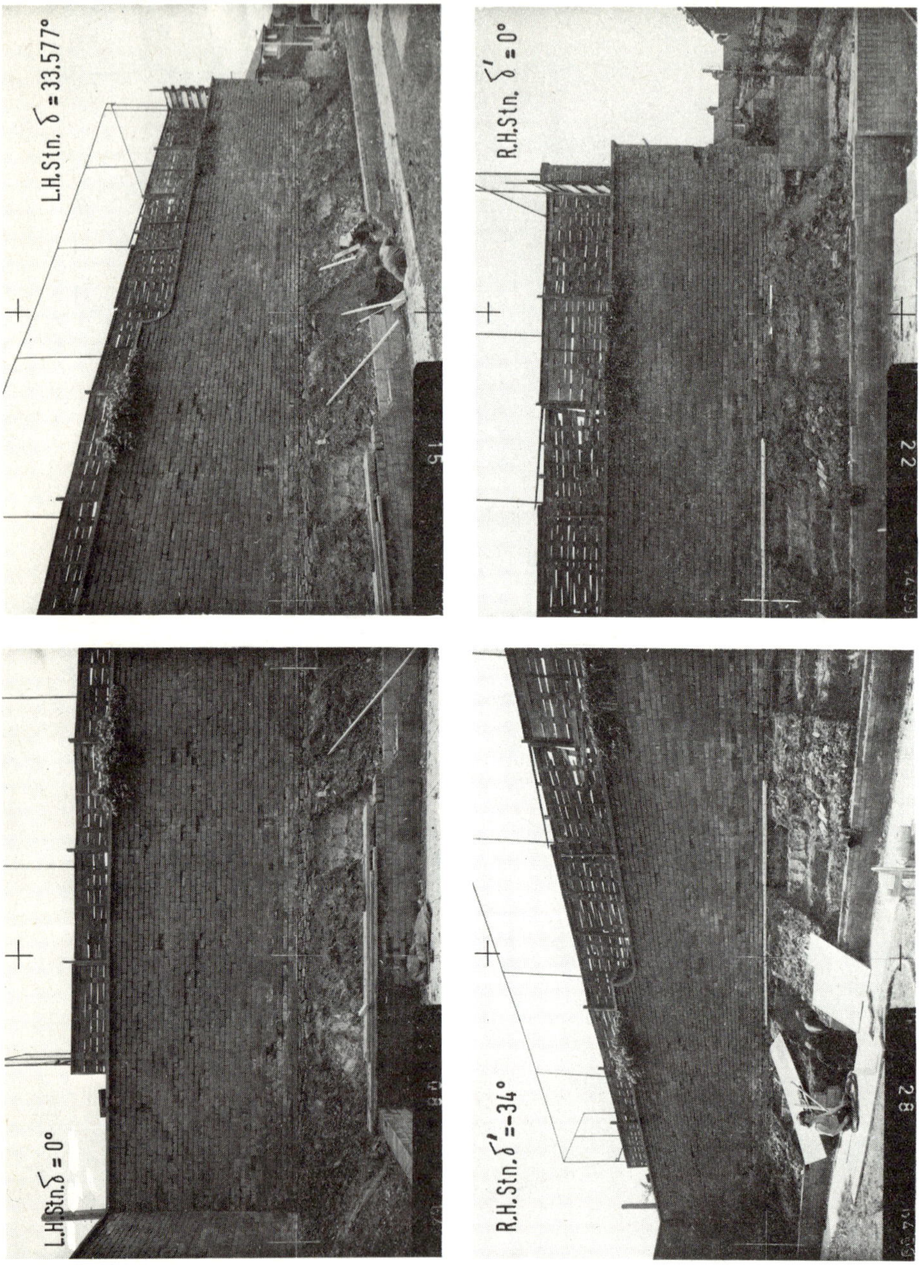

PLATE II; Lockshill retaining wall, Frome.

PLATE III: Rose Cottage, Walrow Moor, Minsterworth

SUMMARY and CONCLUSIONS

(1) Air and satellite photography has received a more widespread adoption during the last two decades than the same techniques for terrestrial photogrammetry. In the latter case the skills have largely remained in the hands of specialist companies who are engaged for tasks for which there is seldom any substitute.

(2) Often, the high cost of photogrammetric plotting equipment has proved a deterrent to the latter's usage. However where good qualitative information is of advantage, photography cannot be improved on. Where also the field of interest can be adequately defined by a few points, then the benefits of photogrammetry may be achieved very cheaply, through a simple glass scale and magnifier.

(3) Once the image coordinates of a point have been accurately established from the photographs, their coordinate positions in the object field can be determined from the direct solution of the intersection equations. However the ultimate accuracy which is achieved depends largely on this initial resolution of the photo image points. If the distances from the point to the four or five fiducial marks on the photo plate can be measured to 0.05mm, on a simple scale, then Sibson's method of solution (1980) achieves values of x and y to an acceptable accuracy for a quite wide range of work.

(4) The mathematics at all stages lies within the capacity of the current generation of pocket and small desk programmable calculators. These have been used to evaluate the fieldwork examples described in the text.

ACKNOWLEDGEMENTS

The author wishes to thank Professor Robin Sibson for developing the mathematics set out in the text and for his patient tuition. The Knightstone Housing Association and Mr Charles Foote are thanked for permission to reproduce the photographs. Thanks are also due to Janet Gingell for the handling of a difficult typescript.

REFERENCES

(1) K Blyth, M Cooper, N Lindsey, R Painter - Snow depth measurement with terrestrial photos, Photogrammetric Engineering XL No 8 (1974) 937-942.

(2) L Hardegen - Photogrammetric evidence recording, Schweizerische Technische Zeitschrift No 4 (1974).

(3A) J Burland, T Longworth, J Moore - A study of ground movement and progressive failure caused by a deep excavation in Oxford Clay, Geotechnique 27 (1977) 557-591.

(3B) W Heath, L Paisley, J Dowling - Terrestrial photogrammetric surveys of unstable terrain in Colombia, Transport and Road Research Laboratory Report No 816.

(4) Paul R Wolf - Elements of Photogrammetry, McGraw Hill (1974).

(5) R Sibson and A Bowyer - Trilateration and Scaling Methods in Surveying, to be published (1980).

APPENDIX 'A'

EXPRESSIONS FOR X, Y, Z FOR PHOTOS INCLINED TO BASE LINE

Refering to the diagram in Figure 3

$$\phi = 90 - (\delta+\alpha) \quad \text{and} \quad \phi' = 90 - (\delta'+\alpha')$$

$$\phi'' = \phi'-\phi = (\delta+\alpha) - (\delta'+\alpha')$$

Measuring coordinates with respect to origin at LEFT HAND camera station

$$LA = \frac{B}{\sin\phi''} \sin(180-\phi')$$

$$= \frac{B\cos(\delta'+\alpha')}{\sin\{(\delta+\alpha) - (\delta'+\alpha')\}} \qquad \text{(i)}$$

$$X_A = LA\cos\phi = LA\sin(\delta+\alpha)$$

$$\text{and } Y_A = LA\sin\phi = LA\cos(\delta+\alpha)$$

$$X_A = \frac{B}{1-\cot(\delta+\alpha)\tan(\delta'+\alpha')} \qquad \text{(ii)}$$

$$\text{and } Y_A = \frac{B}{\tan(\delta+\alpha) - \tan(\delta'+\alpha')} \qquad \text{(iii)}$$

If the origin is taken from the RIGHT HAND camera station

$$\text{then } X'_A = \frac{B}{\tan(\delta+\alpha)\cot(\delta'+\alpha') - 1} \qquad \text{(iv)}$$

and the expression for Y_A remains unchanged.

$$V_A = LA\tan\beta \text{ (or } L'A\tan\beta') \qquad \text{(v)}$$

The value of LA is derived from eqn (i).

Where the camera axis is depressed, but the x and y photo axes remain horizontal and vertical, then from Figure 4:-
x_a and y_a are measurable from the photo and θ from the inclination of the camera.
LK is horizontal on the lens axis, intersecting the photo at K.
The line on the photo through K and parallel to the photo axis will be horizontal. This is termed the horizon line, x^*.
a-a'-a'' is constructed as a vertical plane, with a' and a'' in the horizontal plane of the exposure station.

α_a = the true horizontal angle of the line of site will be KLa

β_a = the true vertical angle of the line of site in the vertical plane a-L-a

y_a = coordinate of image point with respect to horizon line

$$y_a^* = y_a + f\tan\theta$$

$$\tan\alpha_a = \frac{ha'}{(LK-hk)}$$

$$= \frac{x_a}{(f\sec\theta - y_a^*\sin\theta)} \qquad \text{putting, } hk = a'a'' \qquad \text{(vi)}$$

$$\tan\beta_a = \frac{aa'}{La'}$$

$$= \frac{y_a^*\cos\theta}{(Lk-hk)\sec\alpha_a}$$

$$= \frac{y_a^*}{x_a}\cdot\frac{\cos\theta}{\operatorname{cosec}\alpha_a} \qquad \text{(vii)}$$

Having obtained the true directions of the image point in terms of α_a and β_a, coordinates are obtainable from equations (ii)-(v) above.

APPENDIX 'B'

POINT LOCATION BY DISTANCE MEASUREMENTS (after Sibson and Bowyer 1980)

The location of points by distance measurements has a number of applications in normal surveying and ecology as well as photogrammetry. In the last case the problem is usually referred to as tri-lateration. The common feature is that sets of measurements to a low level of accuracy require to be coordinated to a limited number of accurately established fixed points.

Equations 4(a)-4(c) in the preceeding text show that a good estimate of a point is obtainable from the intersection of circles of radii equal to the distance measurements (d_n) from three fixed points. From equations 6, 7, if all measurements of d_n are of equal weight, then the sum of the squares of the errors between the measurements d_n and the 'achieved' value of the distance, e_n, is shown to be

$$S = \sum (d_n - e_n)^2$$

which will be a minimum at

$$\frac{\partial S}{\partial x} = \frac{\partial S}{\partial y} = 0$$

(Note, the function of x and y is therefore the first derivative of S in this case.)

If one variable only were present, then the Newton Raphson iterative step would be

$$x_1 \text{ to } x_0 - \left\{ f''(x_0)^{-1} \cdot f'(x_0) \right\} \qquad \text{(i)}$$

For two variables

$$\begin{bmatrix} x_1 \\ y_1 \end{bmatrix} = \begin{bmatrix} x_0 \\ y_0 \end{bmatrix} - \begin{bmatrix} \partial^2 S/\partial x^2 & \partial^2 S/\partial x \partial y \\ \partial^2 S/\partial x \partial y & \partial^2 S/\partial y^2 \end{bmatrix}^{-1} \begin{bmatrix} \partial S/\partial x \\ \partial S/\partial y \end{bmatrix} \qquad \text{(ii)}$$

As d_n is not dependent on x and y the first and second derivatives of S can be evaluated explicitly.

$$\frac{\partial S}{\partial x} = 2\sum_{n=1}^{N} (1 - d_n/e_n)(x - x_n) = \theta \qquad \text{(iii)}$$

$$\frac{\partial S}{\partial y} = 2\sum (1 - d_n/e_n)(y - y_n) = \phi \qquad \text{(iv)}$$

$$\frac{\partial^2 S}{\partial x^2} = 2\sum \left(1 - d_n/e_n + (d_n/e_n^3)(x - x_n)^2\right) = a \qquad \text{(v)}$$

$$\frac{\partial^2 S}{\partial x \partial y} = 2\sum (x - x_n)(y - y_n)(d_n/e_n^3) = b \qquad \text{(vi)}$$

$$\frac{\partial^2 S}{\partial y^2} = 2\sum \left(1 - d_n/e_n + (d_n/e_n^3)(y - y_n)^2\right) = c \qquad \text{(vii)}$$

Simultaneous values of x and y are obtained from a normal solution to the matrix as follows

$$x_1 = x_0 - \frac{1}{ac - b^2}\{c\theta - b\phi\} \qquad \text{(vii)}$$

$$y_1 = y_0 - \frac{1}{ac - b^2}\{-b\theta + a\phi\} \qquad \text{(viii)}$$

For the method to operate successfully it is essential to have a good first estimate of the final point at x, y, in order to avoid convergence toward some local optima. As mentioned this is provided by a preliminary calculation to solve eqns 4(a)-4(c), measuring from three reference points which are non-collinear. Once obtained the solution to the derivates of S can follow and thus a second estimate of the point using equations (vii)-(viii). In practice it has been found the method is rapidly convergent, to an acceptable order of accuracy after two or three iterations.

PHOTOGRAMMETRIC MEASUREMENT OF THE DEFORMATION OF HIGH EARTH EMBANKMENTS

Lech Kolondra, MSc
Antoni Przewdzing, MSc

District Geodetic and Cartographic Enterprise
Katowice

Mirosław Zak, DSc

Deputy Director for Research and Development
District Geodetic and Cartographic Enterprise
Gdańsk

INTRODUCTION

Large individual hydraulic engineering structures are often present on industrial sites, especially in the vicinity of steam power plants, mines and various processing plants. The structures consist of settling tanks in which after sedimentation, sludges, slag and ash are collected. As they are situated in zones of earth movement they are subject to considerable additional loadings. Therefore an obvious need arises to investigate the condition of such tanks on a regular basis. Failure of such a dam type structure may result in very dangerous consequences because of the typically high density of population and industry in the nearby region.

Surveying of the structure is carried out periodically to monitor its safety. Such surveys are normally made by levelling, to determine elevational changes of targets marked on the structure. When horizontal movements must also be determined, a control network composed of targets on the structure and observational and control points is set up. Common surveying methods utilizing theodolites only or theodolites and electronic distance measuring equipment have proved to be labour-and time-consuming. Due to these shortcomings, the photogrammetric method was considered since it reduces the time taken for a survey and also simplifies the work. Although the technique is more economical than other survey methods, to be successful it requires additional information to fully orientate the stereogram. The necessary input may be obtained by providing the stereogram with a supporting geodetic network which is measured in the field.

From a geometric view-point the procedure for orientating a stereogram requires a solution of the problem of three-dimensional resection based on the known coordinates of some photographed points. The points must be suitably distributed on the photograph to make the resection geometrically correct. This requirement gives rise to a substantial amount of field work. This is an argument against an application of the standard photogrammetric technique, especially for elongated objects such as dams, as in this paper. Thus an attempt to develop a more economical method of surveying was made and the method developed is presented below.

THE STRUCTURE MEASURED

An earth embankment was investigated. It was situated between a main reservoir and a surge reservoir, which together formed a wet slag and ash storage yard. Water is drained from the main reservoir to the surge reservoir through two outlet boxes. The reservoirs are separated by the dam 300 m long and 8 m high, composed of granular materials, namely gravel and sand. The dam has side slopes of 1:3 (Figures 1 and 2). Because of an approximately 2 m head it was reasonable to ssumme that water seeped through the dam causing displacement. Moreover a railway line runs along the crest of the dam, providing extra loading on the structure

DAM DISPLACEMENT SURVEYS BASED ON CONVENTIONAL METHODS

Due to the loadings produced by mine workings, seepage and the rail track it was necessary to provide periodic observations of the dam displacements i.e. the vertical and horizontal components. For this purpose observation stations were established on the dam, at 18 sections spaced at 15 m intervals (Figure 4). The stations were plastic pegs 5 cm in diameter and 1.3 m long set into the ground. The projecting part of each peg (30 cm long) was painted alternately red and yellow (Figure 3).

Vertical movements (settlements) were determined by precise levelling, periodical observations of the network (Figure 5) being made.

Horizontal displacements were surveyed by means of a traverse network diagrammatically shown on Figure 4. On this drawing, observation stations are marked and directions to the control points distributed around the structure are indicated. For the measurements a Wild T-3 theodolite was used, its mean angular error being

Fig. 1 View of slope

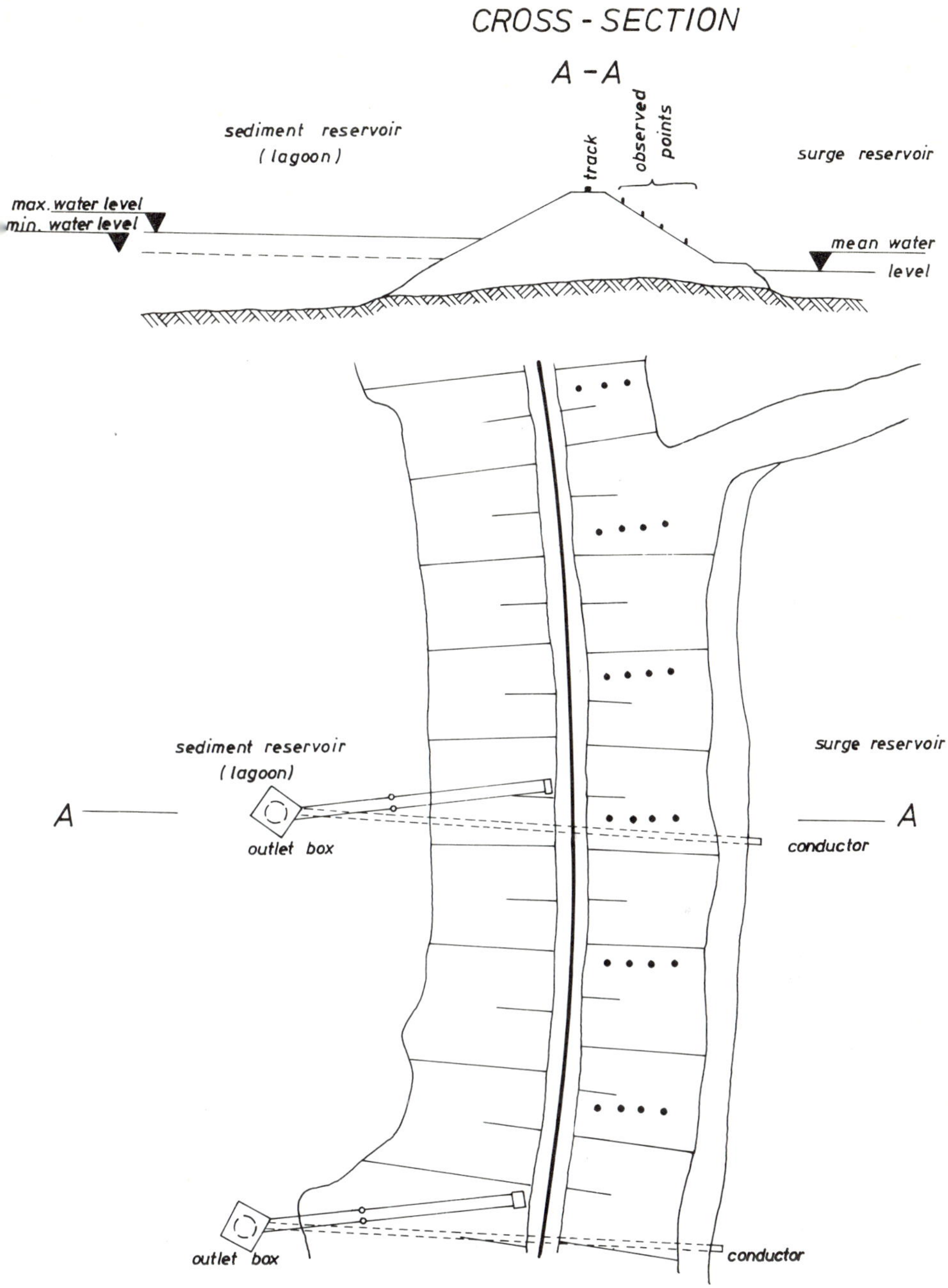
CROSS - SECTION
A - A
sediment reservoir
(lagoon)
track
observed points
surge reservoir
max. water level
min. water level
mean water level
sediment reservoir
(lagoon)
surge reservoir
A
A
outlet box
conductor
outlet box
conductor

Fig. 2

Fig. 3 Example of line measuring points

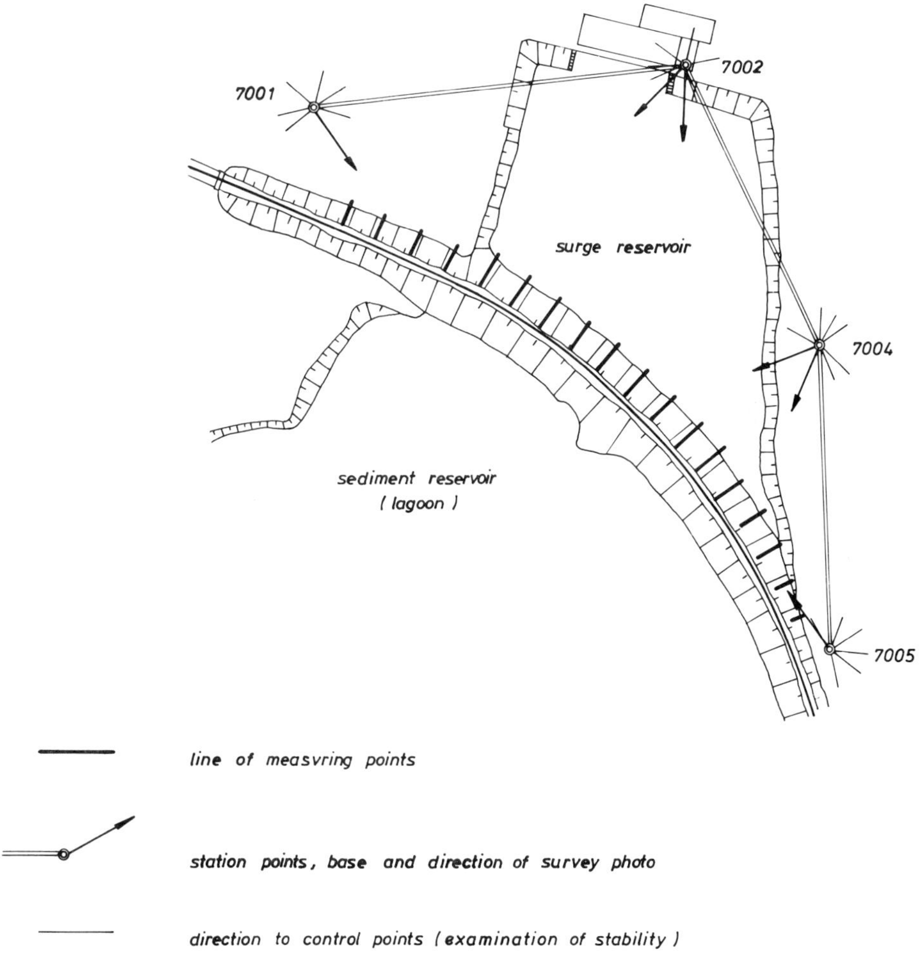
7001
7002
7004
7005
surge reservoir
sediment reservoir
(lagoon)
line of measvring points
station points, base and direction of survey photo
direction to control points (examination of stability)

Fig. 4

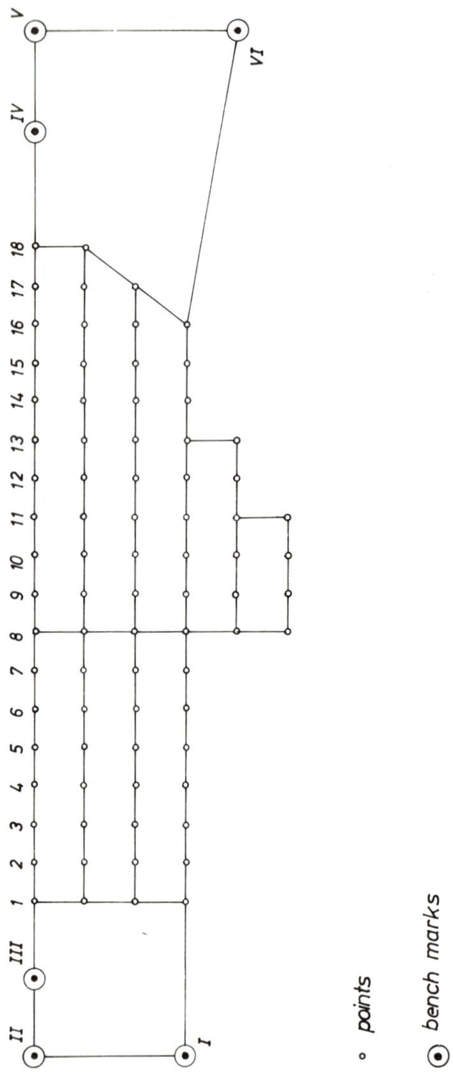
V
VI
IV
18
17
16
15
14
13
12
11
10
9
8
7
6
5
4
3
2
1
III
II
I
points
bench marks

Fig. 5

$$m_k = \pm 1^{cc}$$

together with a ME 3000 stadia, characterized by a mean error in distance of

$$m_l = \pm 0.2 \text{ mm.}$$

It will be appreciated that these estimates of accuracy were made to enable a comparison, as far as possible, between results obtained by the conventional method with those obtained by the photogrammetric method to be presented.

THE MODIFIED PHOTOGRAMMETRIC METHOD

The correct plotting of every photograph depends on a correct location of the beam of rays used to take the photograph, in the coordinate system used. This was made possible by the fact that the field coordinates of some points shown on the photograph were also obtained by survey. Thus, a relation between these field coordinates and the photograph coordinates (picture coordinates) could be found. On this basis it was then possible to determine the coordinates of the other photographed points.

In this method there is no need to determine the field coordinates of many points on every stereogram. Only the angles are measured from the observation station to some chosen points photographed on the stereogram. Such an easy procedure enables one to obtain a complete space orientation of every stereogram (Figure 6) from which the coordinates of any photographed point can be determined. With this method over 50% of the field-work can be eliminated. Thus it is noteworthy from the view-point of economy. Technical arguments in support of the new method may also be provided by an accuracy analysis.

COMPARISON BETWEEN THE PROPOSED PHOTOGRAMMETRIC METHOD AND THE TRADITIONAL PROCEDURE

Conventional investigations were carried out very exactly and carefully and valuable data was obtained for comparison with that of the photogrammetric method.

The mean error of a test point position (in a horizontal plane) amounted to

$$M_p = \pm 2.8 \text{ mm .}$$

The mean error in a settlement determination was equal to

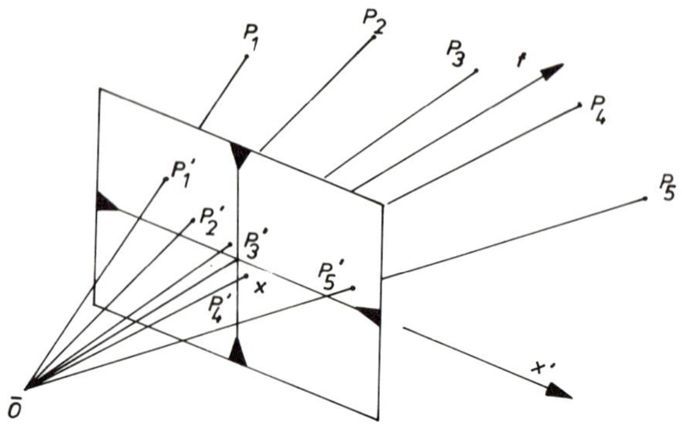
P_1
P_2
P_3
f
P_4
P_5
P_1'
P_2'
P_3'
x
P_4'
P_5'
x'
$\bar{O}$

Fig. 6

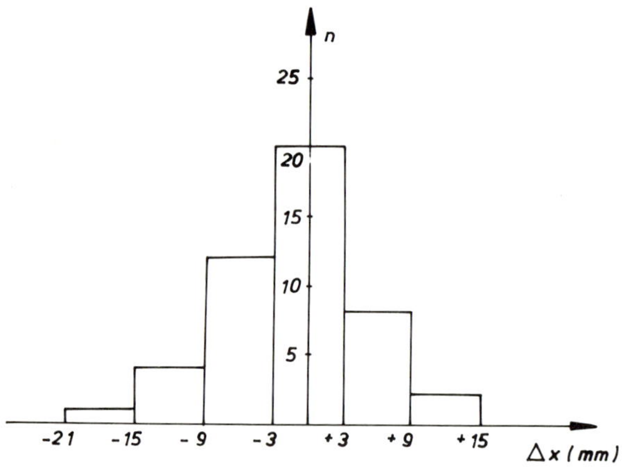
n
25
20
15
10
5
-21
-15
-9
-3
+3
+9
+15
Δx (mm)

Fig. 7a

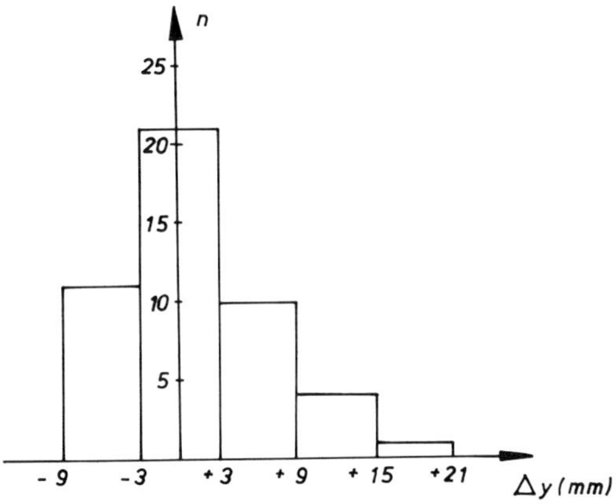

Fig. 7b

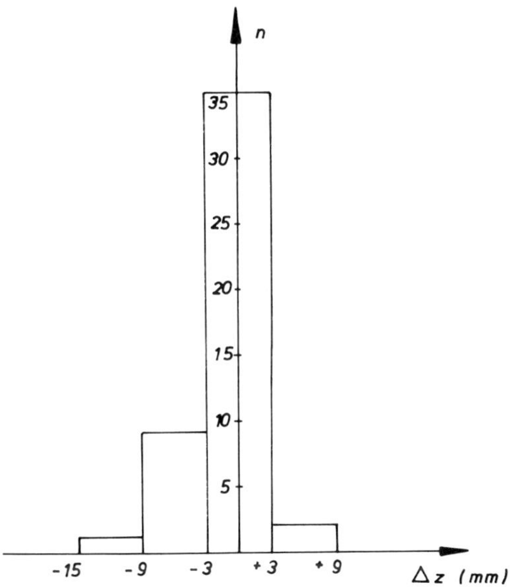

Fig. 7c

$$M_H = \pm 0.55 \text{ mm} .$$

It is reasonable to accept that these results from the standard method are practically faultless. These results were compared with the photogrammetric ones. To obtain the latter, photographs were obtained in the field with a Zeiss Photheo 19/1318 photo-theodolite of focal length 195.17 m. An ortochromatic TO 1 film was used (the photographic plates were 13 x 18 cm in size). To take the photographs the same observation stations were used as for the traverse investigations (Figure 2). The groundwork coordinates were measured with a Zeiss Stecometr stereocomparator and the results calculated with an Odra 1204 computer. The comparison between the two methods (conventional and photogrammetric) was based on a difference analysis of the components: x, y and z using the formula defining the error of a so-called observation pair, as follows:

$$m_R = \pm \sqrt{\frac{(\Delta \Delta)}{n}}$$

where

Δ is a difference of coordinate
n is a number of the differences.

The values of the errors were:

$$m_x = \pm 6.7 \text{ mm}$$

$$m_y = \pm 6.0 \text{ mm}$$

$$m_z = \pm 3.5 \text{ mm} .$$

A statistical specification of the differences is presented in Figures 7a, b, c.

CONCLUSIONS

The proposed photogrammetric method has fully come up to expectation as being particularly useful for displacement determination with long objects. The method has the virtues of being accurate and cost and time saving.

The maximum discrepancies between the standard and photogrammetric results in the case discussed were

$$\Delta x = 17\text{mm} , \quad \Delta y = 17\text{mm} , \quad \Delta z = 9\text{mm} .$$

This accuracy is entirely acceptable. In Poland, the method presented is called "The independent beams method".

REFERENCE

W. Bychawski and W. Mizerski - Analytical plotting of a single stereogram of terrestrial photographs optionally averted, Works of Institute of Land Surveying and Cartography, (1971) 1.

THE INTRODUCTION OF MATHEMATICAL MODELS FOR THE PURPOSE OF PREDICTING SURFACE MOVEMENTS DUE TO MINING

Donald Burton

Area Mining Engineer, British Railways Board, Derby

INTRODUCTION

Given the history of subsidence investigation in this country it is possible to look upon the construction of a mathematical model with undoubted scepticism. The same remarks could apply to the idea of a machine making subsidence predictions. In the light of experience it may be assumed that the investigator who constructs such a model, intends to use it at the expense of reality and somewhat regardless of the information collected by observers.

Mathematical systems which allow an engineer to predict surface movements in the vicinity of important surface structures and installations have been in use for some time. For example, the mining engineering staff of the British Railways Board are able to use calculating machines to predict ground movements above longwall panels as a matter of routine. This raises a number of obvious questions concerning the investigator's use of a mathematical model, and his regard for detail in each particular case.

The modern calculating machine is such an efficient tool for generating the numbers used in prediction that when it is employed successfully, previous descriptions of mining subsidence account for very little. The facts are the same, but the way in which they are collected, described, and used in the control of future predictions is different. For this reason, the manner in which a mathematical model is introduced is a subject of importance to the professional investigator.

It is a problem which has interested the author for many years, outside of the day to day activities in protecting surface interests against the effects of mining, and the following text outlines a particular solution for engineers who wish to use a mathematical model in the prediction of surface movement.

A difficult task facing the subsidence engineer is to predict a change in surface length as the face of a panel advances, and emphasis is placed upon this problem in the paper. This is not to suggest that vertical displacements are less important, and in many cases where mine workings affect railway property it is the surface lowering which receives most attention.

The word "machine" in the context of this paper refers to a calculator with a memory of 8K, or above, combined with an X-Y plotter.

COLLECTIONS OF FACTUAL INFORMATION

Engineers working in mining areas will be familiar with the description of subsidence provided by observers who have fitted graphs to the results of observation. The description deals with isolated features of subsidence, such as subsidence and strain profile shapes for panels of varying width/depth ratio, and it provides a set of average quantities which can be applied to particular surface directions.

Such a description can be replaced by a few equations and simple rules which are contained within a small part of the machine's memory, but before this is accepted, the engineer will probably want to see that the machine can generate a set of average quantities. To demonstrate this, without drawing a large number of separate graphs, the factual information can be arranged into a single pattern which the machine will simulate. For example, at the first Cardiff Conference on Large Ground Movements and Structures, the author illustrated the average vertical and lateral displacement for lines across panels of varying width/depth ratio, in the form of a static set (Burton, 1977).

Figure 1 shows the pattern of average vertical and lateral displacement for points in lines across panels of varying width/depth ratio.

Figure 2 illustrates a corresponding pattern of development for the line of advance of a panel, the ratio of length/depth increasing.

These drawings were produced by a HP 9825A calculator and plotter, using B.R. subsidence model No. 2. The same pattern is to be obtained from the graphs which are familiar to engineers by plotting vertical displacement against lateral movement. The lateral displacement is obtained from the strain profile by adding together small changes in length from

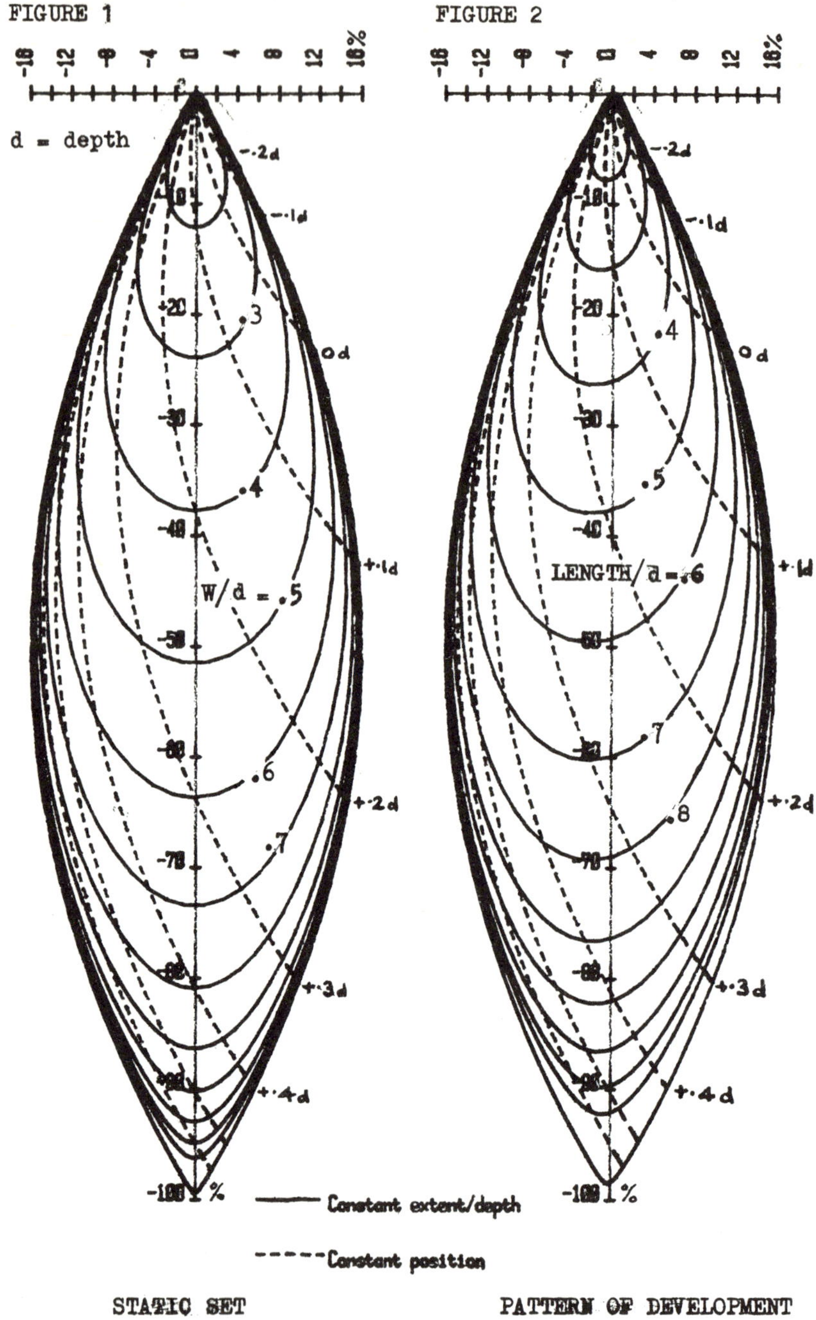

Model 2

Drawings which illustrate the average vertical and lateral displacement for various width/depth and length/depth ratios.

one end of the line to the other. Each loop relates to a particular extent/depth ratio, as indicated by a number, whilst the broken lines in Figure 1 illustrate the increase in the displacement of points which are a given distance, expressed as a fraction of the depth, from the left-hand edge of each panel.

100% vertical displacement in Figure 1 represents the full amount of subsidence to be observed above an extensive extraction, when the edges are too far away to have any effect. It is noted that in this set of average quantities the maximum lateral displacement is 16% of the full amount of subsidence.

When constructing a mathematical model it is possible to use a static set, formed out of a collection of factual information, as a test-piece. If the model will not generate such a pattern, with quantities which are reasonable in terms of the average values, then it will be discarded.

GRAPHICAL OPTIONS

It is not the purpose of the machine to generate average quantities but before looking at its actual role, the engineer will need to see that it is capable of producing information in a convenient form. As an example the subsidence programs used for routine prediction by B.R. staff allow the machine operator to select any of the following options.

(1) A subsidence profile for any series of surface lines, for any face position, the face static or moving.
(2) A strain profile for the same lines.
(3) A graph showing the change in length of any surface line, plotted against face position. The machine normally advances the face at intervals of 0.02 x depth.
(4) The relative movement of any two points, as seen in a plan view, keeping either point stationary.
(5) The lateral displacement of any point, as seen in plan.

The operator may allow the machine to select its own maximum and minimum X and Y values for each graph it draws, in which case it will calculate the maximum displacement, or the maximum differential displacement, depending upon which option is chosen.

At this stage the machine is working beyond the normal scope of the subsidence engineer who solves subsidence problems by reference to a series of graphs which represent a collection of average values. Not only does it work in three dimensions but the results are obtained at high speed. For example it would draw the graph of the change in length of a surface line placed at say 45° to the ribside of a panel in less than sixty seconds. It might be necessary for the engineer to produce a plan showing the horizontal strain at regular intervals throughout the floor space of a large factory, these strains applying not to lines which are at right angles to the edges of the mine working, but to lines which are parallel to the sides of the building, or some other direction associated with the internal structure. The engineer sees that the machine has no difficulty in generating the information required to produce the strain contour map.

Figure 3 illustrates the strains above an advancing face, assuming that the observer is looking along lines which are in the direction of advance. Figure 4 shows the same situation, except that the observer is now looking along lines which are at 30° to the ribside, as indicated by the arrow. These drawings would be produced in order to illustrate to the engineer the importance of surface direction when considering structural damage.

The obvious weakness in a collection of factual information and the familiar description of subsidence is that the graphs generally relate to lines running across the mine workings. Therefore students are given the impression that a structure placed in the path of a panel will be subject to tension before compression, or that structural damage may be related to a strain profile shape in the line across the panel. If this was applied to buildings, railway bridges and railway lines, it would be necessary to invent all manner of unknown factors to account for the actual differential displacement of the structure. If the machine does nothing else, at least it prevents the observer from failing to take into account the direction of the structure in relation to the edges of the panel.

The early railway engineers had to overcome this particular problem when dealing with longwall workings by relating structural damage to the depth and thickness of extraction, and by visualising the position of the structure in relation to the subsidence basin. If the bridge ran across the subsidence contours, one abutment would be drawn in more quickly than the other and the structure would be pulled apart; broadside on to the contours and both abutments would be drawn in together, the bridge being compressed. As the source of the disturbance moved forwards with the advance of the face, the situation changed. The reader may visualise the same situation by illustrating the shape of the subsidence basin which forms during a short period of time, and moving this forwards across the line of a structure as the face advances.

CONTROL OF SUBSIDENCE PREDICTION

As a result of observations in different parts of the country, the observer is aware that variations in geology, and more important, differences in the mining history of each site, produce marked variations in the surface movement above panels

FIGURE 3

Strain values in mm/m

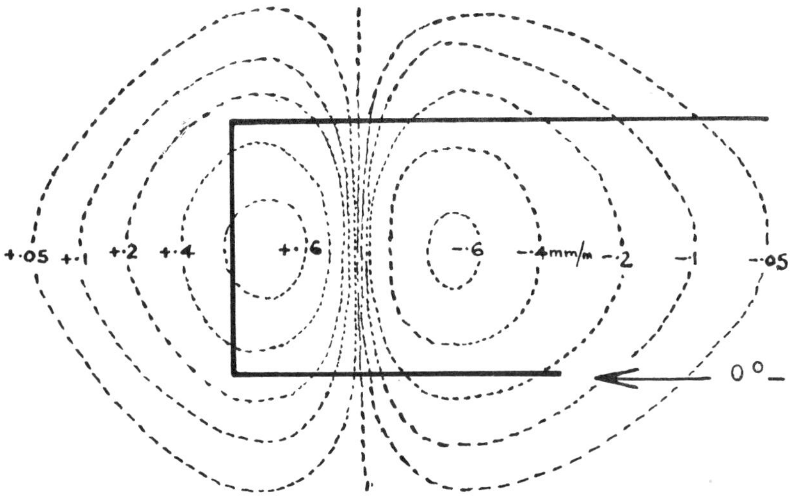

Strain contour maps obtained from B.R. model No 3.
The difference in the two patterns illustrates the importance of surface direction.
Width of panel = 200m. Depth = 400m. Max. Sub. = 0.5m.
Lateral subsidence factor = 0.2

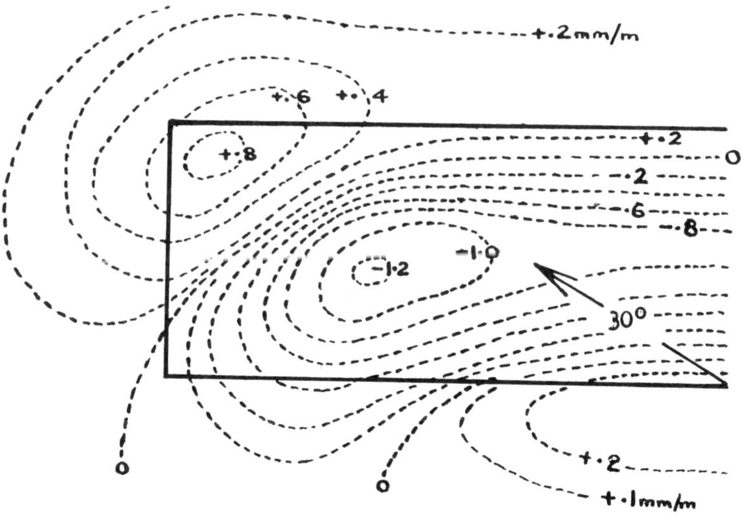

FIGURE 4

of given extent, thickness and depth. From one locality to another these differences may well represent an error in prediction of 100%.

The problem of introducing a mathematical model is to prevent the engineer from assuming that a set of average quantities will suffice. The machine, or the engineer using the machine, needs to be in a position where it is possible to learn from experience.

It may be supposed that if the model was constructed more realistically, then the predictions would be more accurate, and the control of the model would be that much more effective. However, to add realism from this point onwards, is to add complexity, and the returns for added realism rapidly diminish (Bellman, 1964). A solution is to design the system such that it can be controlled by a feedback of information. The information may come from a variety of sources, and may relate to conditions observed underground or studies of the behaviour of rock. The information which is most useful and readily available is derived from the previous measurement of surface movement.

Figure 5 illustrates a simple control system devised by the author in relation to a three dimensional model of surface movement above longwall panels. The shape of the model is controlled by four numbers, Sf, L, v1 and v2. The letter T represents the thickness of extraction.

Sf is the full subsidence, or the vertical subsidence factor. In a previously mined area it has an average value of 0.9 but may vary considerably in the vicinity of the edges of earlier workings. L is the maximum lateral displacement, or the lateral subsidence factor. It has an average value of 0.16 x Sf, but again this varies considerably. For example in the South Midlands Area near Coalville the author estimates that an observer would obtain an average value for L of 0.28.

The numbers v1 and v2 represent the vertical displacement at two convenient points on a subsidence profile across a single edge. In this case they are placed 0.2 x depth outside, and 0.2 x depth inside the edge, respectively. They are expressed as fractions of Sf, and allow the machine to distinguish between one edge of a panel and another. For example when v1 = 0.05 and v2 = 0.64, the machine treats that edge as a typical rib of a panel. When v1 = 0.04 and v2 = 0.44, the machine treats the edge as a typical moving face.

The four numbers Sf, L, v1 and v2, are the minimum required to control the shape of a three dimensional model. In practice however, the subsidence model becomes "five-sided" by programming the machine to accept different values for

FIGURE 5

Curve representing the vertical and lateral displacement across a single edge.

Sf = full subsidence (expressed as a fraction of thickness extracted)

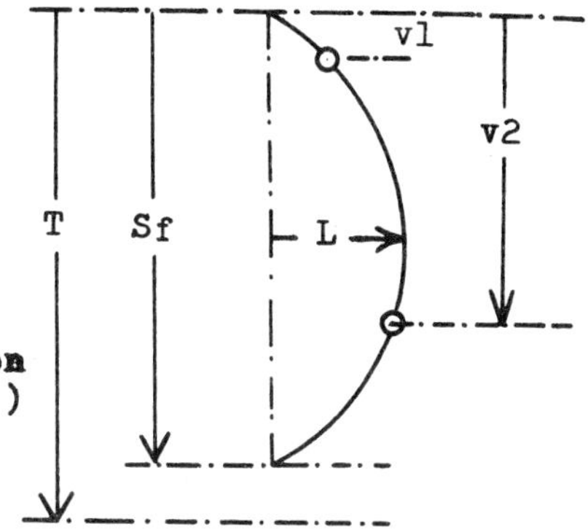

FIGURE 6

L = maximum lateral displacement, expressed as a fraction of full subsidence.

d = depth

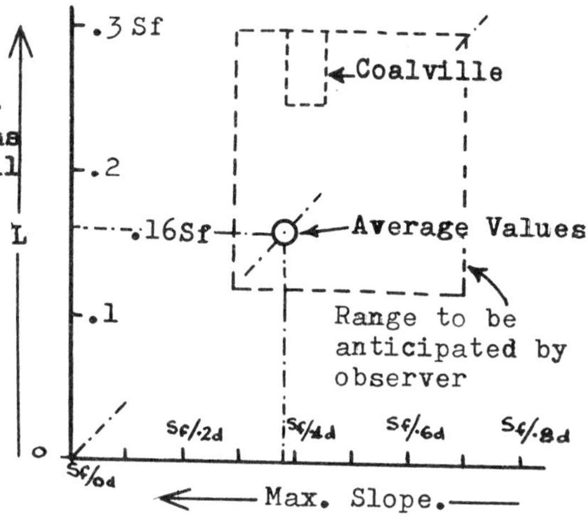

FIGURE 7

Graph showing change in length of line AB.

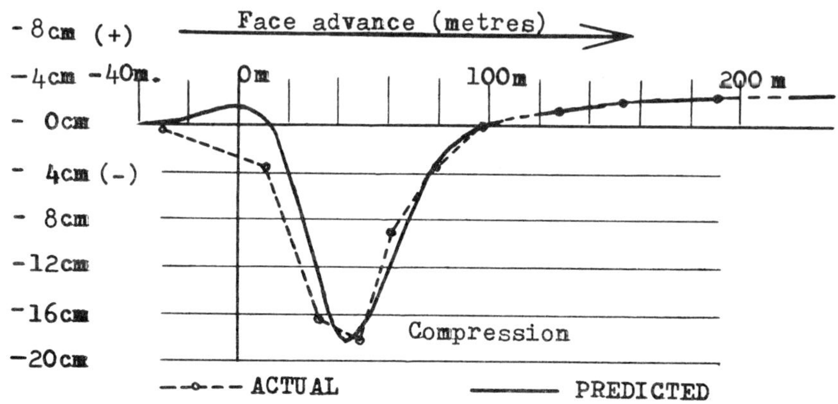

L, v1 and v2, for each side of a panel. The fifth edge is a moving face as distinct from a static face.

Observed subsidence profile shapes, and measured changes in surface length, can now be matched by quantities generated by the machine. Using observations from different localities, with previous panels in different positions with respect to the current extraction, information is extracted from past cases, recorded, and used to control future predictions.

This simple control system is probably the most effective for students and engineers, the speed of the machine allowing a large amount of useful information to be extracted from previous observations within a short period of time. Outside of this practice however, in which the mathematical model is used in a purely descriptive role, and predictions are controlled simply in terms of past events, the machine and the subsidence model has a wider application.

In calculating the displacement of a surface point above a longwall panel the model automatically accounts for the basic features of subsidence. In illustrating the difference in vertical and lateral displacement between one width/depth horizon and another, it automatically describes the changes which are required within the rock in order to account for the observed or predicted surface movement. The machine can be programmed to illustrate the pattern of strain between the extraction and the surface which would account for the known vertical and lateral surface movement. This is not to say that the description provided by the machine and the behaviour of the rock below the surface are identical, but it is the information required by the constructive investigator to determine the relative importance of geology and previous mining in the control of the three dimensional model.

As a result of these studies the investigator will allow previous mining to take precedence over geology in the majority of control problems. Surface features and conditions underground take priority.

Returning to the control of predictions in terms of past events, the constructive investigator may also develop the subsidence program such that the machine itself controls the process of prediction. The engineer who wishes to take advantage of the experience of the investigator may find this attractive but there are difficulties. To take a simple example, no sooner has the machine learned to predict the effect of a panel worked on retreat alongside a previous extraction, than the investigator finds that the roadway serving both extractions has not only been protected by packs, but for several metres either side the thickness of extraction has been reduced by say 40%. On the whole, the mining staff of B.R. find it most practical to use the "five-sided" model, controlling the predictions in terms of the information which is available at the time, and using the speed of the machine to examine the effect of possible error in prediction upon vertical and lateral displacements in the vicinity of important structures. Control with this model is extremely good, and not only is it possible to anticipate differences between one ribside and another, but also anticipate changes in the shape of the subsidence basin as a face moves away from, and approaches the edges of previous workings.

Figure 6 is drawn to illustrate the need for the control of the shape of the three dimensional model. It shows the range in values for maximum lateral displacement, L, and maximum tilt of a subsidence profile associated with a single static edge, which the observer can expect to meet in the major coalfields. Values outside of this range are known but would not normally be anticipated for a flat surface.

A particular problem facing railway engineers provides an interesting, if somewhat unusual example, of high speed calculation and feedback control. Normally mine workings are not allowed to proceed under track where continuous welded rail has been installed but it is possible under certain circumstances. There are a number of factors to be taken into account prior to the event, and to obtain a clear view of what is entailed, a machine is used to predict the pattern of movement associated with the advancing face.

As the working approaches the railway, measurements of vertical displacement, change in surface length, and change in rail force, or stress from temperature, as determined by rail force transducers inserted in the track, provide a continual feedback of information. The actual and predicted movements are compared and when the predictions move outside a prescribed tolerance, the shape of the model is altered. Throughout the period of active subsidence the model may be continually reshaped, the tolerance being small, but at any given moment in time the engineer has the best possible prediction of future events in terms of the available information. This is feedback control after the event, and it illustrates that not all problems of accurate prediction are necessarily solved in advance.

CHANGES IN SURFACE LENGTH

Figures 7 and 8 illustrate the actual changes in length of two lines AB and CD, together with the predicted differential lateral displacement extracted from graphs drawn by a machine. In each case the pattern of movement is determined by the direction of the surface line in relation to the edges of the mine working, and whereas each prediction only takes a few minutes with the aid of the machine, a subsidence engineer could take several hours to arrive at a

FIGURE 8 Graph showing change in length of line CD.

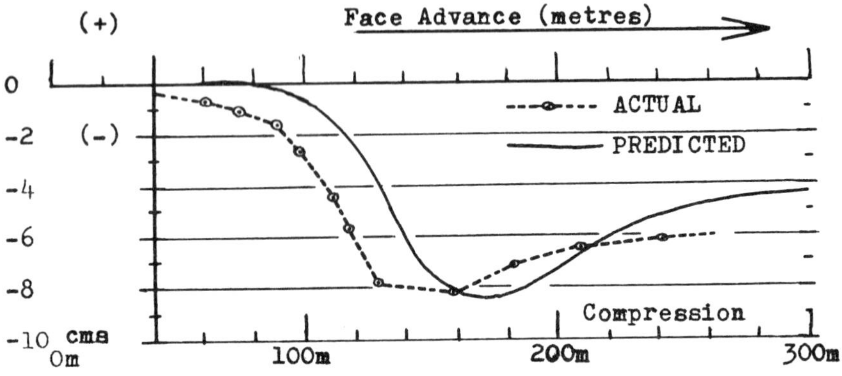

FIGURE 9 Smoile Seam; Whitwick Colliery. Subsidence profiles observed along line of railway.

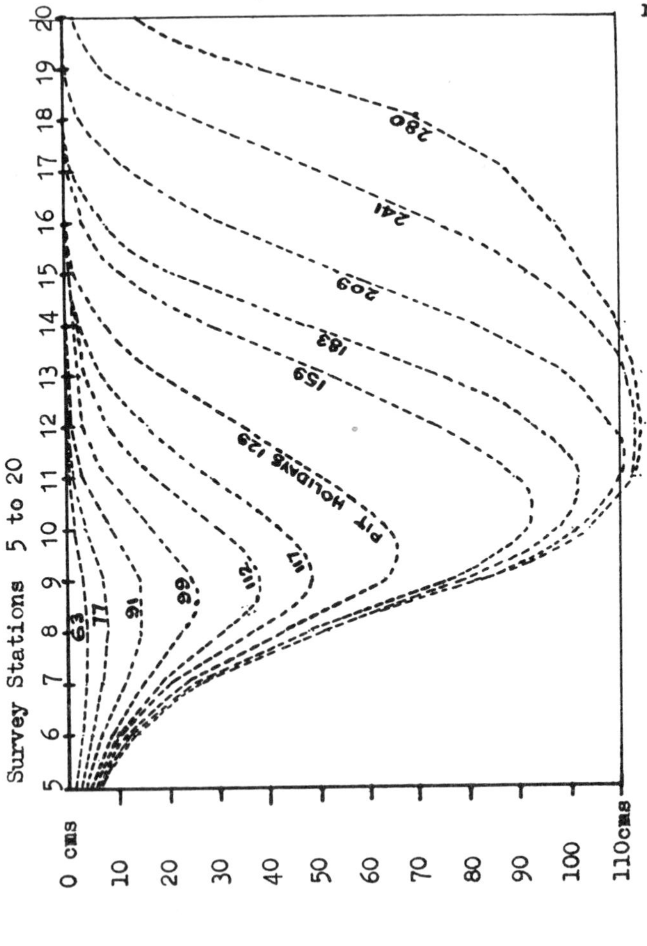

graphical solution.

The machine is simple to operate, but an attempt to describe why it is successful in predicting these changes in length can be confusing, particularly if the engineer has become familiar with the description of subsidence which uses subsidence and strain profile shapes for lines across longwall panels. The best way to understand the system is to use it, and the obvious advantages of the machine can be illustrated in a simple experiment involving the prediction of the change in length of two surface lines, as seen in Figures 7 and 8.

As mentioned earlier, the familiar description of subsidence can give the impression that if a structure is placed in the path of a panel it will be subject to tension followed by compression, or that structural damage can be related to strain in a line across the panel. That this is not necessarily the case may be gathered from a recent article which described the effect of mining upon a road bridge (Hayward, 1979). To understand the behaviour of such a structure, it is necessary to work in three dimensions, and look at a particular situation involving a change in surface length.

In the two examples which follow, the lines AB and CD are placed above panels worked in the South Midlands near Coalville. The surface points are located with reference to a convenient point of origin on the left-hand rib of each panel. The X co-ordinate is measured in the direction of advance, the Y co-ordinate in a line across the panel. The direction of full dip is measured in a clockwise direction from the left-hand rib, 0^0 being in the direction of advance. In both cases it may be assumed that the starting and finishing positions of the face do not affect the predictions, although the machine asks for these positions.

<u>EXAMPLE 1</u> (AB) MINGE SEAM: SOUTH LEICESTER COLLIERY

Point A X = 0m Y = 175.3m Point B X = 18.2m Y = 164.9m
Depth at origin = 102m Dip = 0.048 in 1 Direction of dip = 216^0 Thickness = 1.829m Width of panel = 181m .

<u>EXAMPLE 2</u> (CD) SMOILE SEAM: WHITWICK COLLIERY

Point C X = 86m Y = 39m Point D X = 98m Y = 51m
Depth = 193m Dip = 0.058 in 1 Direction of dip = 160^0
Thickness = 1.07m Width of panel = 186m .

Before asking for the above information, the machine needs to know how the prediction will be controlled and requests values of Sf, L, v1 and v2. With reference to Figure 6, it may be assumed that L = 0.28, as against the national average value of 0.16. This means that the predicted changes in length will be 75% above those obtained from a set of average quantities. Given no additional information relating to the particular sites, apart from the fact that both have been affected by previous mining, it may be assumed that v1 and v2 for each rib have the average values of 0.05 and 0.64, whilst the moving faces have the values of 0.04 and 0.44, respectively. Since both sites have been underworked it could be assumed that Sf = 0.9 in both cases. However the line CD is positioned close to the edges of workings in other seams, which would warrant an increase in this value (Actual observations provided a value of 1.08 in example 2; 0.9 in example 1).

The immediate impression to be obtained from the results illustrated in Figures 7 and 8 is that the machine accurately predicts the pattern and quantities of differential lateral displacement. It is evident that a line, or a structure placed in these surface lines, would not be subject to tension followed by compression. It is also evident that structural damage in these lines could not be sensibly related to a strain value obtained from a profile across the subsidence basin. If it was sensible, then reference to a strain profile as presented in the National Coal Board Subsidence Engineers' Handbook for example, (NCB, 1975), would place the centre of the line AB in a zone of tension, and on this basis the length would increase by approximately 2½ cms. To be realistic the engineer must predict a compression of 18 cm. Added to this would be the reaction of the structure to the ground movement.

It is seen that even in a comparatively simple situation, the ability to work in three dimensions, and the ability to control the prediction with reference to information collected in the past, means the difference between surface movement being predictable, and unpredictable. The machine eliminates the need for calculation but still requires experience. The most important step in the prediction is the fact that the machine asks for values for Sf, L, v1 and v2. If the operator can only supply average values, he is immediately made aware of the possible errors involved, and he will need to take account of these in any decisions he may take.

In illustrating the problems of subsidence prediction to the student or engineer who is less familiar with the subject, the author generally begins with a simple exercise as illustrated in Figure 7 in which the surface line is not at right angles to the edges of a panel. This removes misconceptions which may have been derived from the familiar description of subsidence, before looking at the more important subject of control. In practice the experienced investigator has many cases to refer to, these cases illustrating the full range in profile shape for given extent/depth ratios, and the full range in differential lateral displacement for a given surface distance. It is a short step from this point to the

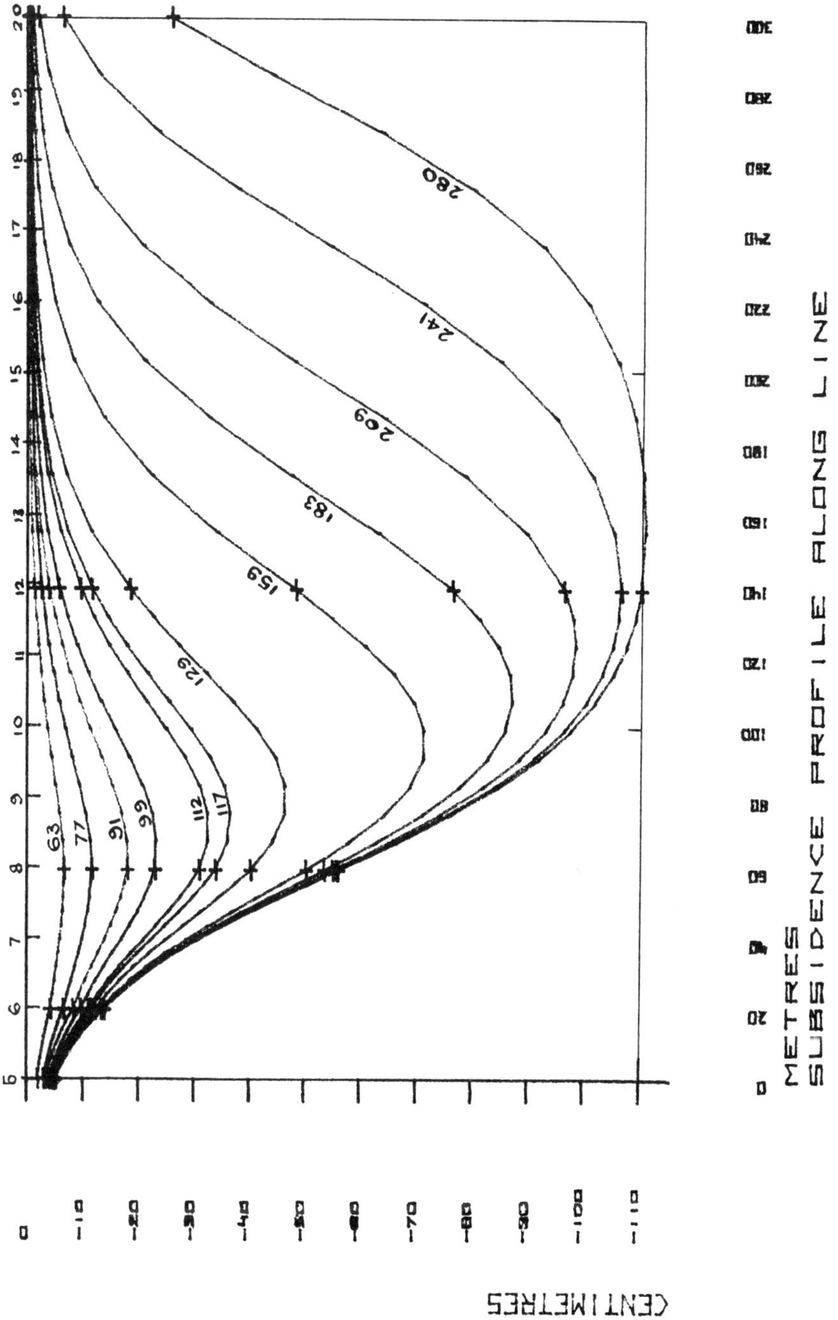

FIGURE 10 Predicted pattern of movement along line of railway. B.R. Model No 2. HP 9821A calculator.

behaviour of structures and the study of surface features such as ditches and embankments.

Figure 9 illustrates the vertical displacement at a series of survey stations placed 20 metres apart in the line of a railway affected by the Smoile panel from Whitwick Colliery. Each profile relates to a particular face position as denoted by a number which gives the distance in metres of the face past the chosen point of origin. Figure 10 illustrates the predicted pattern of movement as might have been drawn by the machine prior to the event, assuming the times of observation were known.

CONCLUSION

Although an observer could begin with the assumption that a mathematical model will prevent anyone from keeping a sensible balance between the general description of mining subsidence, and the peculiarities of each case, the actual effect of introducing such a model is known to be quite different. By taking advantage of the memory and the speed of the calculating machine, the investigator gains control over predictions in particular situations, rather than losing it.

For practical purposes the process of subsidence prediction is divided into two parts, the construction of a suitable model which is required to generate the numbers for prediction, and the subsequent control of predictions with a feedback of information. Given a simple and effective system it should not be difficult for the engineer to adopt a satisfactory policy for dealing with the uncertainties of the environment.

REFERENCES

R. Bellman - Control Theory, Scientific American, Sept. (1964) 186-200.

D. Burton - A three dimensional system for the prediction of surface movements due to mining, Proceedings of the Conference on Large Ground Movements and Structures, UWIST, Cardiff, (1977) 209-228.

D. Hayward - M1 bridge movements, a mystifying problem, New Civil Engineer Feb. (1979) 16-17.

National Coal Board - Subsidence Engineers' Handbook, 2nd Edition (1975) NCB, Hobart House, London.

ACKNOWLEDGEMENT

This paper is presented with the permission of Mr. A. Wilson, Mining Engineer, Chief Civil Engineering Department, British Railways Board.

A BOUNDARY ELEMENT METHOD FOR THE ANALYSIS OF SUBSIDENCE ASSOCIATED WITH LONGWALL MINING

Yehoshua Tsur-Lavie, D.Sc and Saul Denekamp, D.Sc

Faculty of Civil Engineering, Technion, Haifa, Israel

ABSTRACT

A boundary element method for the evaluation of ground subsidence associated with longwall coal mining is presented. The method is based on a fundamental solution of stresses and deformation around a single rectangular indentation at the boundary of an infinite elastic half plane. An important attribute of the fundamental solution, not to be found in other comparable solutions given in the literature, is that it holds both for compressible $(\nu<\frac{1}{2})$ and incompressible material $(\nu=\frac{1}{2})$. The model represents a state of uniform vertical displacement of the roof of the mined area, equal to the width of the mined coal seam (causing total closure of the opening). A state of zero stress at the ground surface is obtained in the model by application of a known technique of superimposing a complementary imaginary mirror model which counteracts any normal stress generated at ground surface.

The model was used for an analysis of ground subsidence as a function of the span and height of the longwall opening, assuming various Poisson's ratios. Comparison with published measurements of ground subsidence as compiled by the N C B shows that for deep mines use of small Poisson's ratios renders close agreement between calculated and measured subsidence, whereas for shallow mines use of large Poisson's ratios gives the best agreement. These results lead to the conclusion that subsidence in shallow mines is associated with a state of failure extending to the ground surface. In deep mines failure is confined to a relatively limited zone above the subsiding roof.

INTRODUCTION

Comprehensive studies of ground surface distortions resulting from the collapse of underground mine workings were performed by the National Coal Board. These studies were confined to the English coal fields where longwall mining is extensively used. Ground subsidence was found to occur in a dish-shaped pattern and an empirical relationship was established between maximum subsidence and the thickness, depth and width of the collapsed coal seam.

Whittaker and Breeds (1977) analysed surface subsidence associated with coal mining in three major rock formations and found no significant differences between observations of subsidence for the different rock types.

Mabry (1972) compared results of finite element analyses with the N C B data, assuming total closure of the seam in the finite element models and a Poisson's ratio of 0.25. Settlements calculated in finite element models were found to be much less than those given by the N C B and do not exhibit the dish-shaped profile.

Dahl (1972) analysed mine subsidence in two-and three-dimensional finite element models assuming both elastic and elastoplastic rock behaviour. The N C B data and the results of additional field surveys performed in the U.S. served as a basis for comparison. Isotropic elastic properties, a Young's modulus of 765 MPa and a Poisson's ratio of 0.25 were assumed for the models in the pre-failure state and the Coulomb criterion was used for the extension of the analysis into the post-failure state. It was found by Dahl that models of relatively low strength approximate more closely the field subsidence results.

The present paper presents a mathematical model for the analysis of subsidence above a longwall mine using a boundary element method.

METHOD

Boundary element methods were applied in the past to problems related to underground openings involving conditions of boundary stress (Moris, 1975) i.e. problems associated with underground openings where a state of zero stress exists along the periphery of the excavation. The method applied in the present study makes use of steplike boundary displacements for the presentation of the collapsing roof of a longwall stope. Propagation of displacements throughout the surrounding rock and the resulting deformation at the groundsurface is then followed by means of the elastic model. The method is based on an elementary solution for a step-like displacement at the boundary of an infinite elastic half plane, given by Golecki (1974).

Elementary Solution

The solution for a single step-like displacement consisting of a discontinuous and uniform displacement at the boundary of an infinite half plane (see Fig. 1) was given by Golecki as follows :

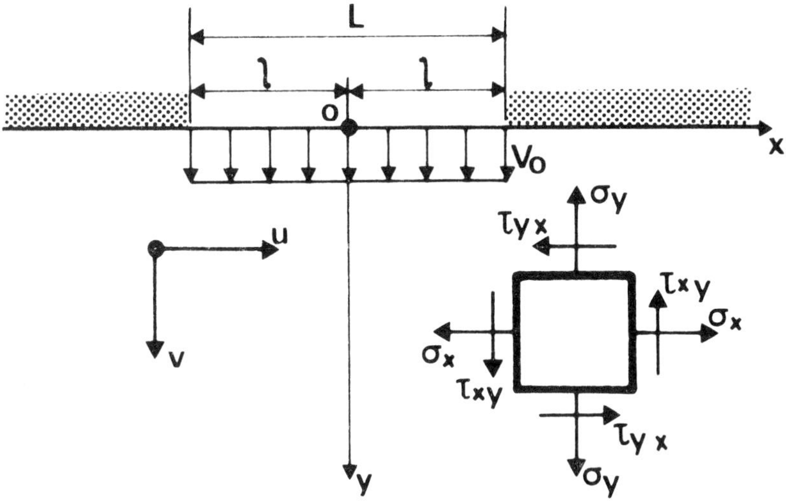

Figure 1 Elastic half plane; discontinuous vertical displacement as boundary condition

$$u = \frac{K\nu}{\pi} v_0 \, y^2 \left[\frac{1}{(l+x)^2 + y^2} - \frac{1}{(l-x)^2 + y^2} \right] \qquad (1)$$

$$v = -\frac{V_0}{\pi} \left\{ arctg \, \frac{l+x}{y} + arctg \, \frac{l-x}{y} + K_\nu y \left[\frac{l+x}{(l+x)^2+y^2} + \frac{l-x}{(l-x)^2+y^2} \right] \right\} \qquad (2)$$

$$\sigma_x = \frac{4K\nu G}{\pi} v_0 \left\{ \frac{(l+x)\left[\nu(l+x)^2 - (1-\nu)\,y^2\right]}{\left[(l+x)^2 + y^2\right]^2} + \frac{(l-x)\left[\nu\,(l-x)^2 - (1-\nu)\,y^2\right]}{\left[(l-x)^2 + y^2\right]^2} \right\} \qquad (3)$$

$$\sigma_y = \frac{4K\nu G}{\pi} v_0 \left\{ \frac{(l+x)\left[(1-\nu)\,(l+x)^2 + (2-\nu)\,y^2\right]}{\left[(l+x)^2 + y^2\right]^2} + \frac{(l-x)\left[(1-\nu)\,(l-x)^2 + (2-\nu)\,y^2\right]}{\left[(l-x)^2 + y^2\right]^2} \right\} \qquad (4)$$

$$\tau_{xy} = \frac{4K\nu G}{\pi} v_0\, y \left\{ \frac{\nu\,(l+x)^2-(1-\nu)y^2}{\left[(l+x)^2+y^2\right]^2} - \frac{\nu\,(l-x)^2-(1-\nu)y^2}{\left[(l-x)^2+y^2\right]^2} \right\} \qquad (5)$$

$K\nu$ and G are defined as:

$$K\nu = \frac{1}{3-4\nu} \qquad G = \frac{E}{2(1+\nu)} \qquad (6)$$

Where ν is the Poisson's ratio and E is the modulus of elasticity.

The solution shows singularity of stress at the edges of the step element at the boundary of the model ($x=-\ell$, $x=\ell$). The singularity however disappears rapidly with distance from the boundary. An important attribute of this solution, not to be found in other comparable solutions given in the literature, is that it holds both for compressible ($\nu<\frac{1}{2}$) and for incompressible material ($\nu=\frac{1}{2}$).

Evaluation of Displacements Associated with the Caving of the Roof of a Longwall Opening

The subsiding roof of the longwall stope is simulated in the model by a uniform vertical displacement along a portion of the boundary of an infinite half plane. Only in the case where caving of the roof takes place at great depth below the surface, stresses and displacements computed in an infinite elastic medium may represent those in the rock mass surrounding the longwall opening. In the case involving the collapse of a roof at a relatively small depth below ground surface this model is no longer applicable because of the proximity of the ground surface, which is a stress free boundary above the longwall stope.

In order to adjust the solution, to cases involving a stress free surface, use is made of a complementary "imaginary" model which is superimposed on the infinite model (Figure 2).

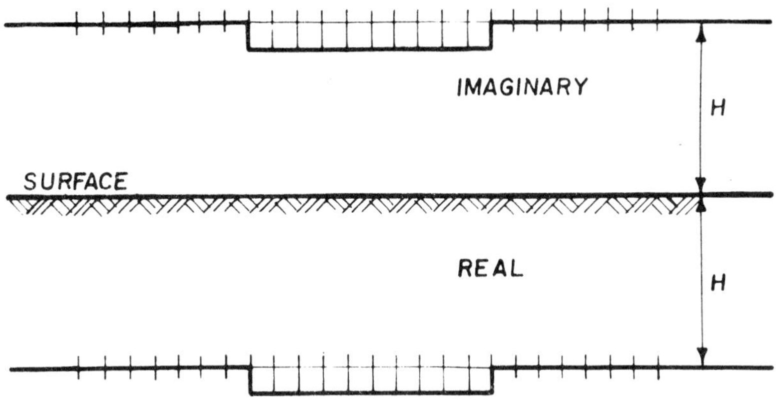

Figure 2 Real and imaginary model

The role of the "imaginary" model is to counteract stresses existing in the infinite model at the horizon representing the ground surface.

The "imaginary" model is the exact opposite of the initial "real" model; the displacements at the boundary of the "real" model are of vertical compression whereas the displacements at the boundary of the "imaginary" model are of vertical extension. The two opposing and quantitatively equal models produce a "net" effect of zero normal stress along an horizon between, and at equal distance, from the two boundaries. The "composite" model resulting from superposition of the "real" and the "imaginary" models, both of which are infinite, is therefore representative of a situation involving a finite elastic medium limited on one side by a stress free boundary.

One problem associated with the above described method of producing a stress free surface is that the superposition of the "imaginary" model affects stresses and displacements throughout the entire elastic medium, not only at the horizon representing the ground surface. As a consequence, stresses and displacements in the "composite" model, including those at the boundaries, deviate from those in the "real" model. In order to correct these deviations, boundary displacements in the "real" model and the "imaginary" model should be applied so that their superposition will create a composite model with boundary displacements of the desired form and magnitude.

The method employed for the calculation of the appropriate boundary displacements in the "real" and the "imaginary" models is demonstrated in Fig. 2. The boundaries of these models are discretized into an equal number of N boundary elements. The algorithm used for the evaluation of the boundary displacements in the "composite" model is based on the principle that the displacement of each boundary element equals its displacement in the "real" model and the sum of displacements induced by all boundary elements in the "imaginary" model.

The above algorithm can be formulated for each of the N elements:

$$\overset{C}{v}_i = \overset{R}{v}_i + \sum_{j=1}^{N} C_{ij} \overset{I}{v}_j \tag{7}$$

Where:

$\overset{C}{v}_i$ - vertical displacement of element i^{th} in the Composite model.

$\overset{R}{v}_i$ - vertical displacement of element i^{th} in the Real model.

$\overset{I}{v}_j$ - vertical displacement of element j^{th} in the Imaginary model.

C_{ij} - "influence coefficient" of displacement $\overset{I}{v}_j$ on the displacement at the center element of ith in the real model. C_{ij} is derived from Equation 2.

Values of $\overset{C}{v}_i$ are known and equal:

$$\overset{C}{v}_i = \text{thickness of the seam} \quad \text{- mined section}$$

$$\overset{C}{v}_i = 0 \quad \text{- unmined section} \tag{8}$$

It follows from the equality of the "imaginary" and the "real" models that:

$$\overset{R}{v}_i = \overset{I}{v}_i \tag{9}$$

Solution of Equations (7), using substitutions from Equations (8) and (9) gives the N unknown values of v_i.

From the calculated values of boundary displacements (v_i) in the real and imaginary models, stresses and displacements can be calculated using Equations 1-5.

Adjustments

One undesired side effect of the superposition of the real and the imaginary models is that horizontal displacements are created throughout the model. The displacements at the boundary are therefore no longer exclusively vertical but acquire also horizontal components. In addition shear stresses are generated at the surface of the model. These phenomena disagree with the phenomenon of roof caving in reality.

A more complete solution permitting the adjustment of horizontal displacements at the boundary can be achieved by introducing a set of N horizontal displacements at the boundary of the real and imaginary models. The algorithm for the N elements is a follows:

$$\overset{C}{v}_i = \overset{R}{v}_i + \sum_{j=1}^{N} C_{ij} \overset{I}{v}_j + \sum_{j=1}^{N} D_{ij} u_j \qquad (10)$$

$$\overset{C}{u}_i = \overset{R}{u}_i + \sum_{j=1}^{N} F_{ij} \overset{I}{v}_j + \sum_{j=1}^{N} K_{ij} u_j \qquad (11)$$

where:

u_i - horizontal displacement of element i.

D_{ij} - influence coefficient of horizontal displacement $\overset{I}{u}_j$ on the vertical displacement at the center of the ith element in the real model.

F_{ij} - influence coefficient of the vertical displacement $\overset{I}{v}_j$ on the horizontal displacement at the center of the ith element in the real model.

K_{ij} - influence of the horizontal displacement $\overset{I}{u}_j$ on the horizontal displacement at the center of the ith element in the real model.

The set of $2N$ linear equations with $2N$ unknowns can be solved numerically. Then the calculated values for v_i and u_i are used as input for the calculation of stresses and displacements.

RESULTS

The above technique was applied to the analysis of ground subsidence as a function of span and height of an horizontal opening assuming various Poisson's ratios. The results of the analysis are summarized in Fig. 3 giving *(S/M)*, the ratio of maximum displacement at the surface to the thickness of the seam, as a function of *(W/H)*, the ratio between the span of the longwall to its depth below ground surface. Results are given for different Poisson's ratios. Fig. 3 also includes comparable results obtained by field measurements conducted by the National Coal Board.

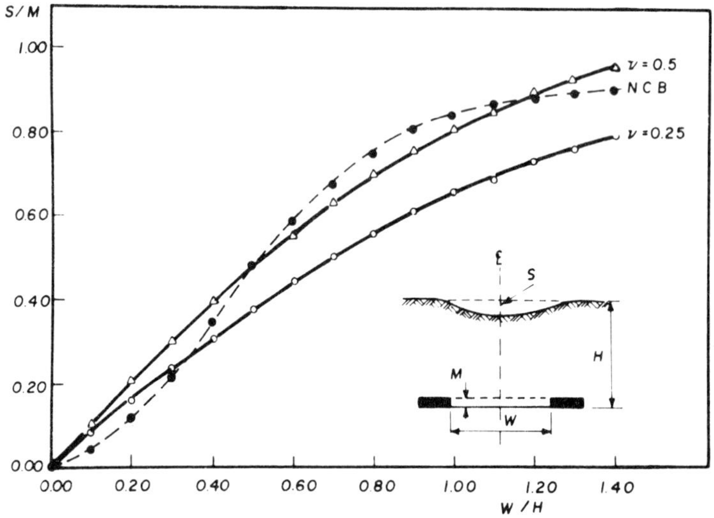

Figure 3 Calculated subsidence and N C B measurements

DISCUSSION

Application of the elastic model to the study of ground subsidence is based on the hypothesis that even though the behaviour of the rock mass is not elastic itself, the displacement pattern can be closely represented by an elastic model (Golecki, 1979; Diest, 1965).

Results of field measurements as given by the N C B imply that subsidence is dependent on the depth and the dimensions of the longwall openings, not on the Young's modulus of the rock. A similar conclusion was reached by Wittaker and Breeds (1977) on the basis of a study of ground subsidence associated with different rock types. The elastic model is in agreement with the above observations in that deformations are dependent both on the dimensions of the longwall stope and the Poisson's ratio of the material, but not on the modulus of elasticity, as is apparent from the elementary solution.

Results show that relatively large calculated values of subsidence are associated with large Poisson ratios. It is furthermore apparent from the elementary solution for a single step-like displacement that subsidence in a model with a large Poisson's ratio must be associated with large horizontal displacements.

Comparison of calculated values of maximum subsidence with field measurements as given by the N.C.B shows that for a Poisson ratio of 0.25 and less, calculated results come close to measured results for longwall openings at great depth $(W/H<0.3)$, but deviate from measured results for openings at shallow depth. For a Poisson's ratio of 0.5 the calculated results come close to the measured results for openings at shallow depth $(W/H=0.4-1.2)$. An explanation for these results can be given as follows: Roof caving in a longwall opening at great depth is associated with a limited zone of collapse above the roof, whereas most of the overlying rock remains intact. Overall rock properties are therefore best represented by a model with a low Poisson's ratio and hence the correspondence between measured values of subsidence and those calculated in the model. Roof caving at shallow depth involves a much larger zone of failure in the overlying rock, extending to the surface. It is therefore best represented by a model with a high Poisson's ratio.

ACKNOWLEDGEMENTS

We wish to express our thanks to Prof. J.J. Golecki of the Faculty of Civil Engineering of the Technion - Israel Institute of Technology for his guidance and helpful discussions of the subject, and Mr. J.A. Sebel for his valuable assistance.

REFERENCES

H.D. Dahl — Two and three dimensional elastic - elastoplastic analyses of mine subsidence, Fifth Strata Control Conference, London (1972) 28,1 - 5.

F.H. Deist — A nonlinear continuum approach to the problem of fracture zones an rockbursts, Symp. on Rock Mechanics and Strata Control in Mines, Johannesburg (1965) 568-588.

J.J. Golecki — Integration of differential equations in elastostatic through determination of the mean stress, Ingenieur-Archiv 43 (1974) 331 - 344.

J.J. Golecki Stresses in rock outside buckled layer, Int. Jour. of Rock Mechanics and Mining Sciences 16 (1979) 93-105.

R.E. Mabry An evaluation of mine subsidence potential, Fourteenth Symp. on Rock Mechanics, Pennsylvania State Univ. (1972) 263-297.

J.P.E Moris SHAMIP a computer program for elastic analysis of large size, tabular, monoplanar, shallow, hard orebodies, Chamber of Mines of South Africa, Res. Rept. 51/75 (1975).

N.C.B Subsidence Engineers Handbook, N.C.B Mining Department, London (1966).

B.N. Wittaker & Breeds The influence of surface geology on the character of mining subsidence, The geotechnics of structurally complex formations, Capri (1977) 1, 459-468.

THE MECHANICS OF GROUND DEFORMATION ABOVE A CAVING LONGWALL FACE

I.W. Farmer, BSc, PhD, CEng, MICE, MIMinE

Reader in Mining Engineering
University of Newcastle upon Tyne

P.F.R. Altounyan, MSc

Research Associate, Department of Mining Engineering
University of Newcastle upon Tyne

SYNOPSIS

An examination of contours of vertical strain computed from three investigations of ground deformation behind caving longwall faces, shows that an extensive zone of fractured and dilated rock is created above the caved area. The boundaries of the fractured zone and the magnitude of residual dilation throughout the zone are strongly influenced by the lithology of the overlying strata.

INTRODUCTION

Ground deformation caused by caving of worked areas behind advancing and retreating longwall faces in coal mines is most commonly observed as surface subsidence. Although various attempts have been made to explain the mechanics of such deformation (Wiggill, 1965; Denkhaus, 1965; Farmer and Attewell, 1975; Shadbolt, 1977), these have been limited by a lack of detailed deformation measurements during caving and of detailed descriptions of proximate geology and geotechnics. In most cases, the information available is confined to the solid geometry of the mine workings and of the resultant subsidence trough. Thus the conventional approach to estimation of ground deformation and subsidence above a longwall working is usually wholly empirical (National Coal Board, 1975), and models of ground deformation tend to be phenomenological rather than mechanical. Typical phenomenological models are reviewed by Voight and Pariseau (1970) and Berry (1977). Berry concludes that an elastic model will simulate most ground deformations with reasonable accuracy, the most obvious difficulty being the measurement or estimation of the elastic constants for massive, layered rocks. Voight and Pariseau suggest a combination of an elastic model with a plastic yield zone simulating the fractured and highly deformed rocks immediately above the excavation. In both cases the key to improved models of ground deformation lies in an accurate description of the mechanics of ground deformation above the caved area. This can only be obtained through interpretation of actual deformation measurements.

DEFORMATION MEASUREMENTS ABOVE LONGWALL FACES

Numerous attempts have been made to measure ground deformation immediately above caving longwall faces. These have involved direct measurements such as levelling and banding at the surface and at intermediate underground levels where access was feasible, and remote measurements through vertical boreholes drilled from the same locations. Remote measurements have usually been confined to vertical movements relative to the borehole mouth using anchor extensometer systems (Hedley, 1969), although more sophisticated vertical and horizontal magnetic extensometer and inclinometer systems of the type used in soft ground tunnel instrumentation (Attewell and Farmer, 1974) have recently been used by the authors with limited success.

Because of the large scale deformations associated with fracture zones above caving longwall faces, and because of the difficulty of access, the length of anchor wires involved and the cost and commitment of resources required, only limited success has been obtained in most measurement programmes. A detailed review of work carried out by and commissioned by the National Coal Board in Britain in the past has revealed only three case histories with sufficient detail to warrant consideration. These comprise an investigation at Wearmouth Colliery described by Johnson (1963), Leigh (1963) and Potts (1964), an investigation at Markham Colliery described by Batchelor (1972) and by King et al. (1972) and a continuing investigation at Lynemouth Colliery, partly described by Hodkin et al. (1979).

The face and seam geometry and instrumentation layouts of each case are described in detail in the appropriate references, and are summarised briefly in Table 1. At Wearmouth Colliery vertical boreholes for anchor extensometers were drilled into the floor and roof from a roadway constructed 30 years earlier in the Hutton seam goaf area, 56 m above the proposed workings which were at a depth from the surface of 457 m in the Harvey seam. The roadway ran at an angle of 17° to the face centreline and boreholes containing anchors were located over the central 100 m of the face. Additional vertical movement data was obtained from boreholes drilled into the ribside and goaf from the face access roads in the Harvey seam.

At Markham vertical boreholes for anchor extensometers were drilled into floor and roof from a roadway in the 1st Piper

Table I Case History Details

Location	Colliery Seam Face	Wearmouth Co.Durham Harvey NW6	Markham Derbyshire Threequarters L58	Lynemouth Northumberland Brass Thill K4(K6)
Face Geometry	Depth from surface (m)	457	587	165 (205m below MHWL)
	Azimuth of face centre line	44°	114°	75° (345)
	Dip and dip direction	4°/225°	3°/90°	1°/95°
	Seam extraction thickness (m)	1.1	0.9	1.6
	Face width (m) (rib - rib)	140 advance	180 advance	180 retreat (advance)
	Width/depth ratio	0.31	0.31	1.09
	Type of instrument	anchor wire	anchor wire	anchor wire (magnet reed switch)
Instrumentation	Height of instruments above seam roof (m)	(max) 69 (min) 23	71 5	77 (78) 8 (7)
	No. of instrument levels	6	8	7 (8)
	No. of boreholes	12	4	2 (3)

Principal references: Johnson (1963); Leigh (1963)
Batchelor (1972); Hodkin (1978);
Farmer (1979)

seam goaf area worked 5 years previously and 45m above the proposed workings, which were at a depth of 587m from the surface in the Threequarters Seam. The roadway was at an angle of 65° to the face centre line and four boreholes containing anchors were drilled at positions over the centre line, 45m from the centre line, directly over the ribside and 15m over the solid on the northern side of the face. An additional single borehole was drilled into the roof over the face centre line.

At Lynemouth Colliery access was obtained to pillar workings in the Main Seam, 77m above the proposed undersea Brass Thill Seam workings. These were 165 below sea bed and 205m below MHWL. A total of 5 boreholes were bored into the intervening strata, which also contained some Yard Seam pillar workings, 12m below the Main Seam. The instrumentation was installed over two faces, K4 and K6, at right angles to each other. Three boreholes were drilled over the face centre lines, one 45m from the centre line and one 10m over the solid ribside.

In the presentation of data from these cases, emphasis tended to be placed on settlement profiles above the faces, both parallel to the face centre line and normal to the centre line, and the effect of geology and the mechanics of deformation were not seriously considered. The present paper is concerned with presenting the available data in the form of contours of vertical strain in such a way that the mechanics of deformation and their dependence on lithology are emphasised.

STRAIN CONTOURS

Data from the three case histories are presented as contours of percentage vertical strain plotted from computed values of vertical extension (positive) or compression (negative) between anchors. Figures 1, 3 and 5 plot strain over the face centre-lines respectively for the Wearmouth, Markham and Lynemouth cases. Figures 2, 4 and 6 plot strain normal to the face centre-line at 200m behind the face for the same cases. The former are based largely on continuous measurement as the face passed a centre-line borehole, the latter on a degree of extrapolation between boreholes across the face half-line.

Figure 7, based on the Lynemouth horizontal and vertical geometry but assuming an idealised continuum not subject to fracture or yield, gives an indication of the vertical strains which would be expected across the edge of a non-yielding ribside. The strains postulated by the diagram could not exist in a rock material. The compression strains would be relieved by yielding of the solid edge and shear fracture of the rocks above it and the tensile strains by fracturing in tension above the caved area. In practice, as is illustrated in Figures 1-6, this occurs, but the strain patterns also illustrate numerous

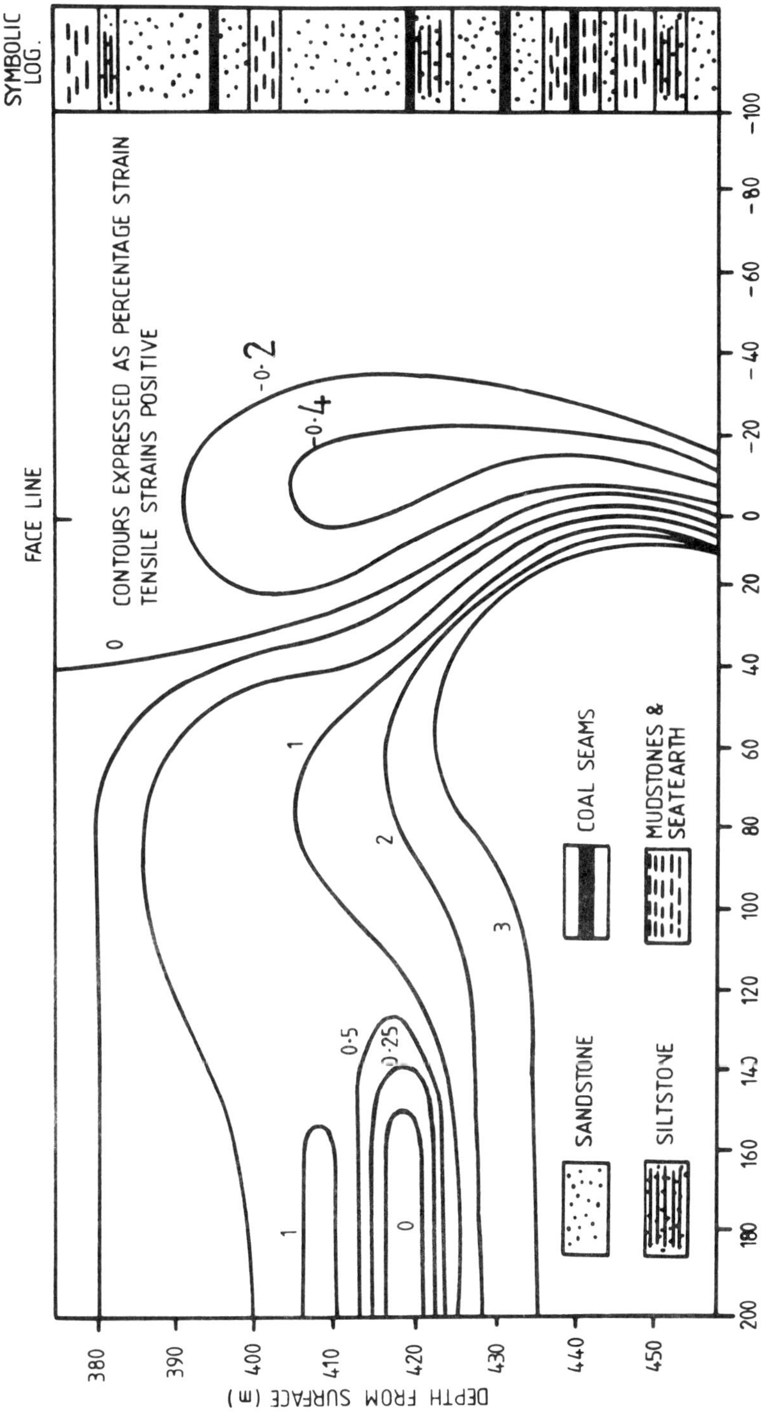

Figure 1 Wearmouth Colliery - Contours of vertical strain along the face centre line

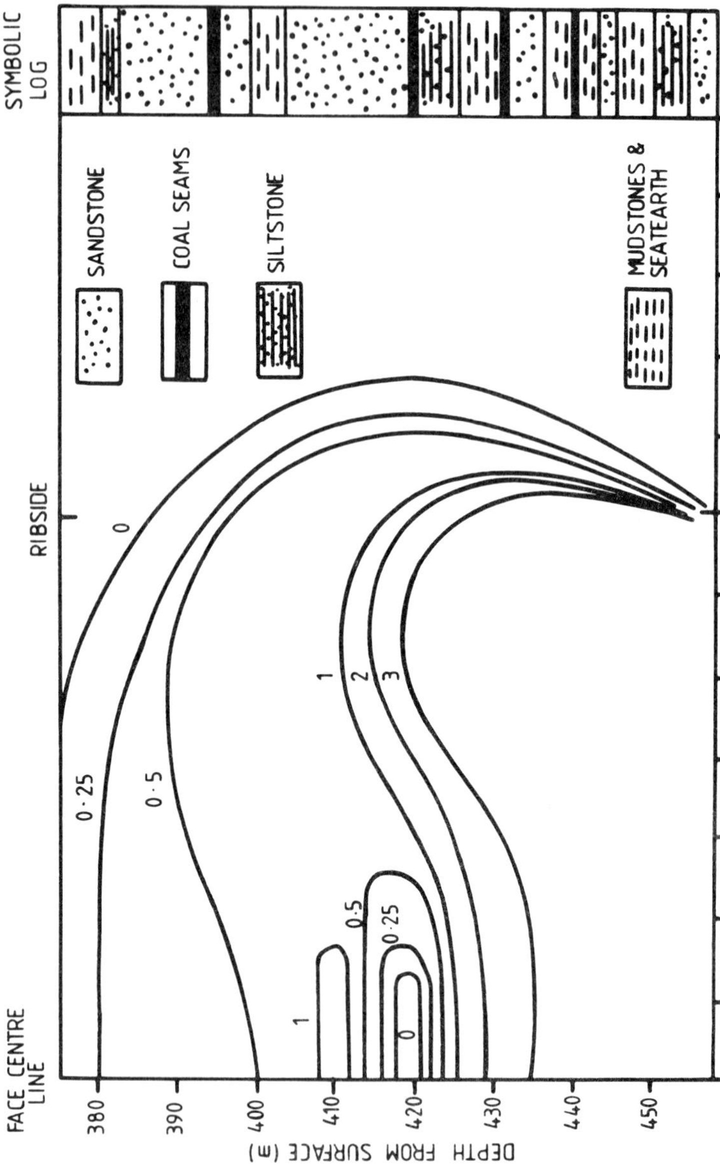

Figure 2 Wearmouth Colliery - Contours of vertical strain normal to the face line

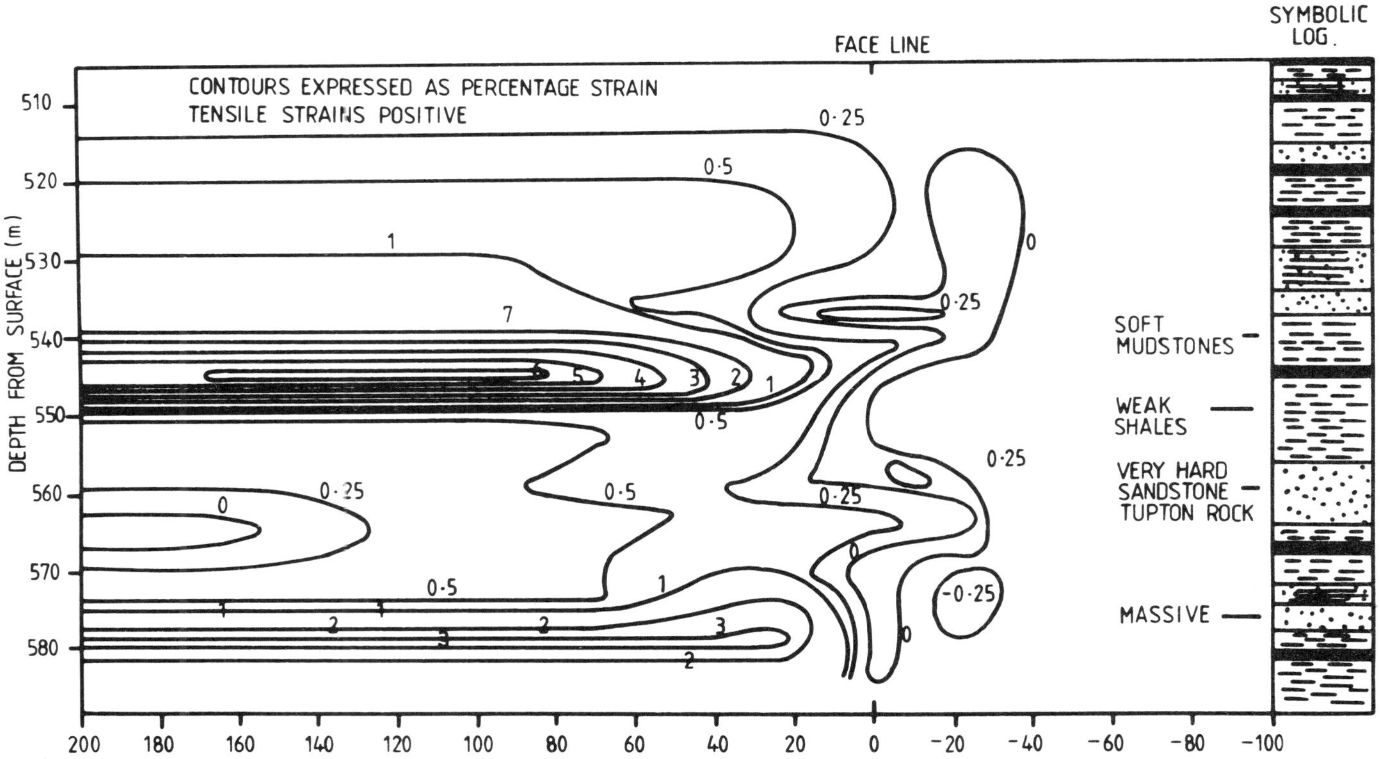

Figure 3 Markham Colliery - Contours of vertical strain along the face centre line

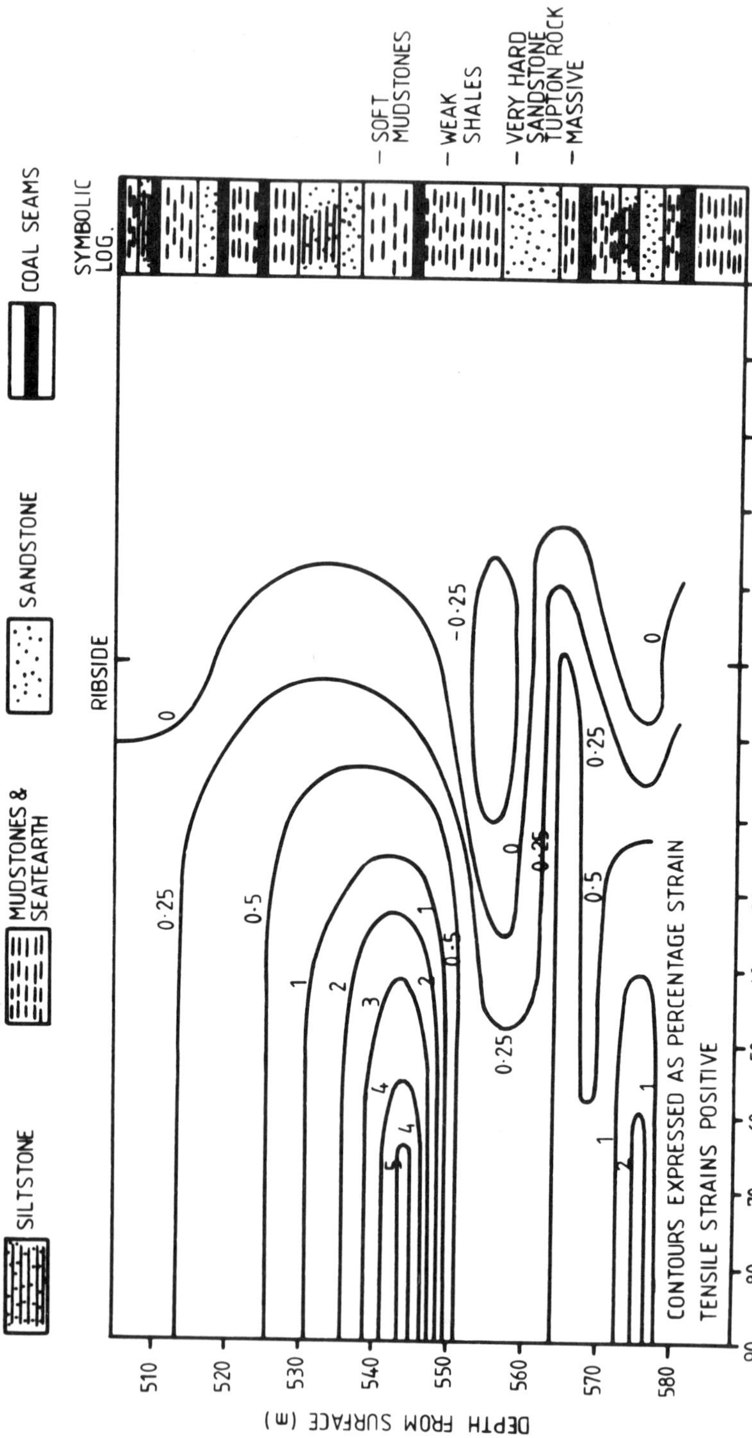

Figure 4 Markham Colliery - Contours of vertical strain normal to the face centre line

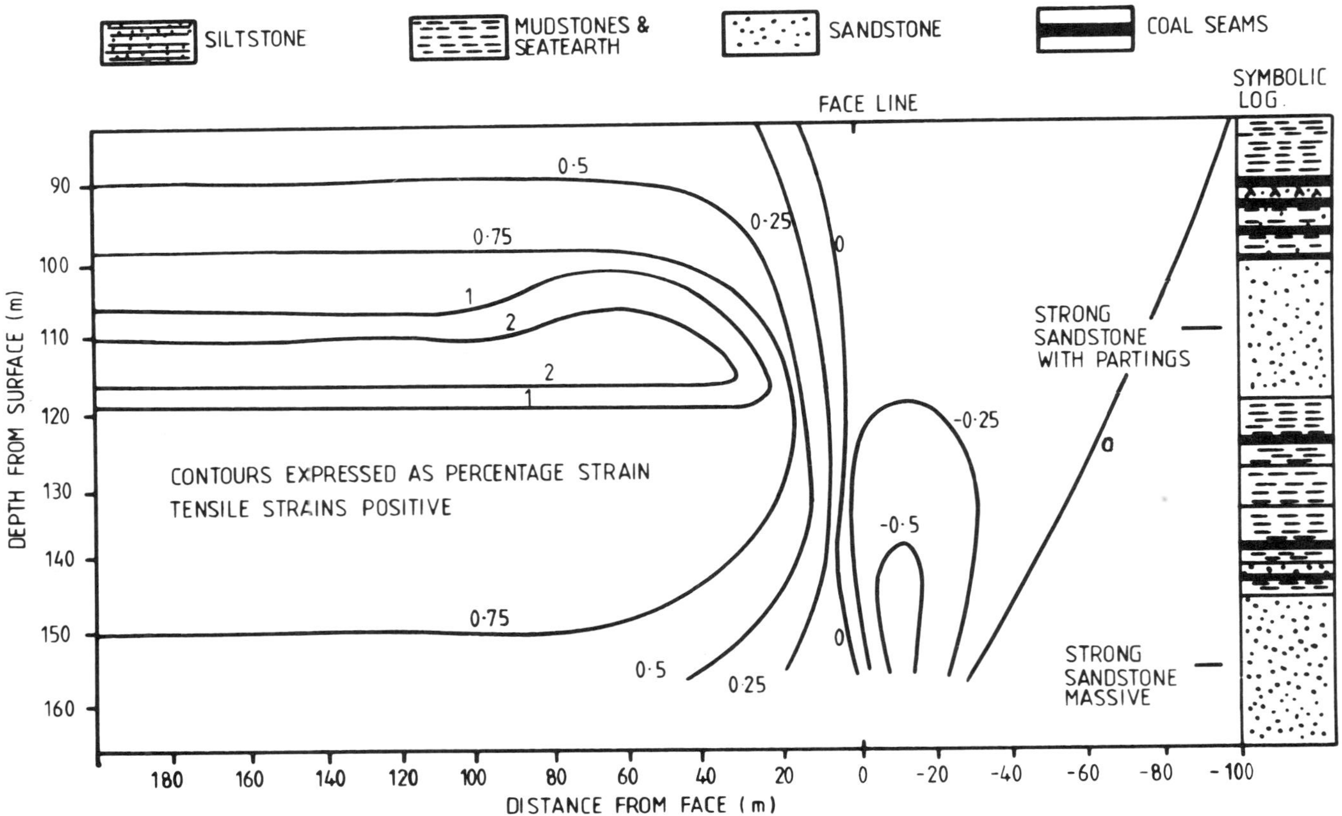

Figure 5 Lynemouth Colliery - Contours of vertical strain along the face centre line

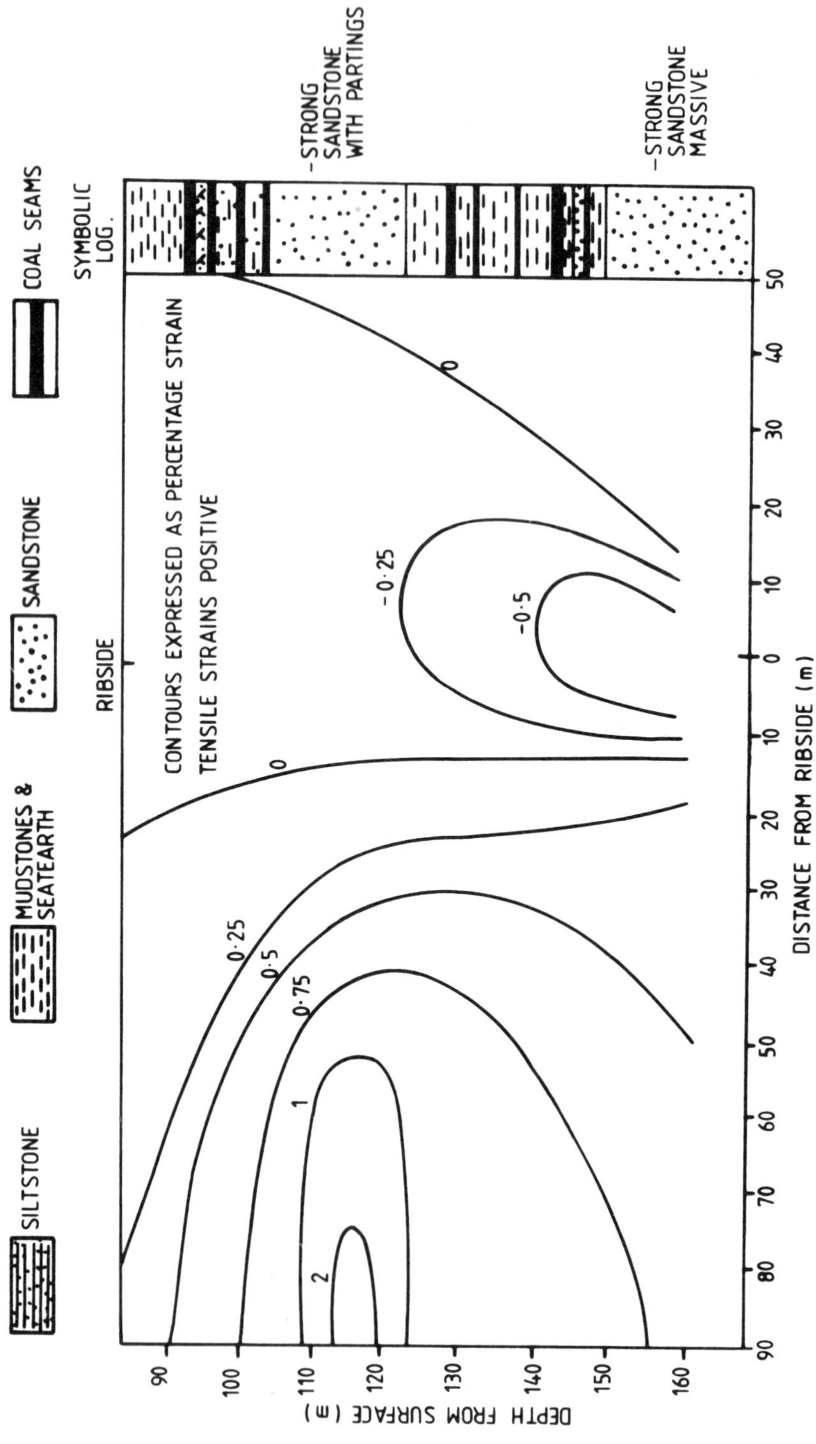

Figure 6 Lynemouth Colliery - Contours of vertical strain normal to the face line

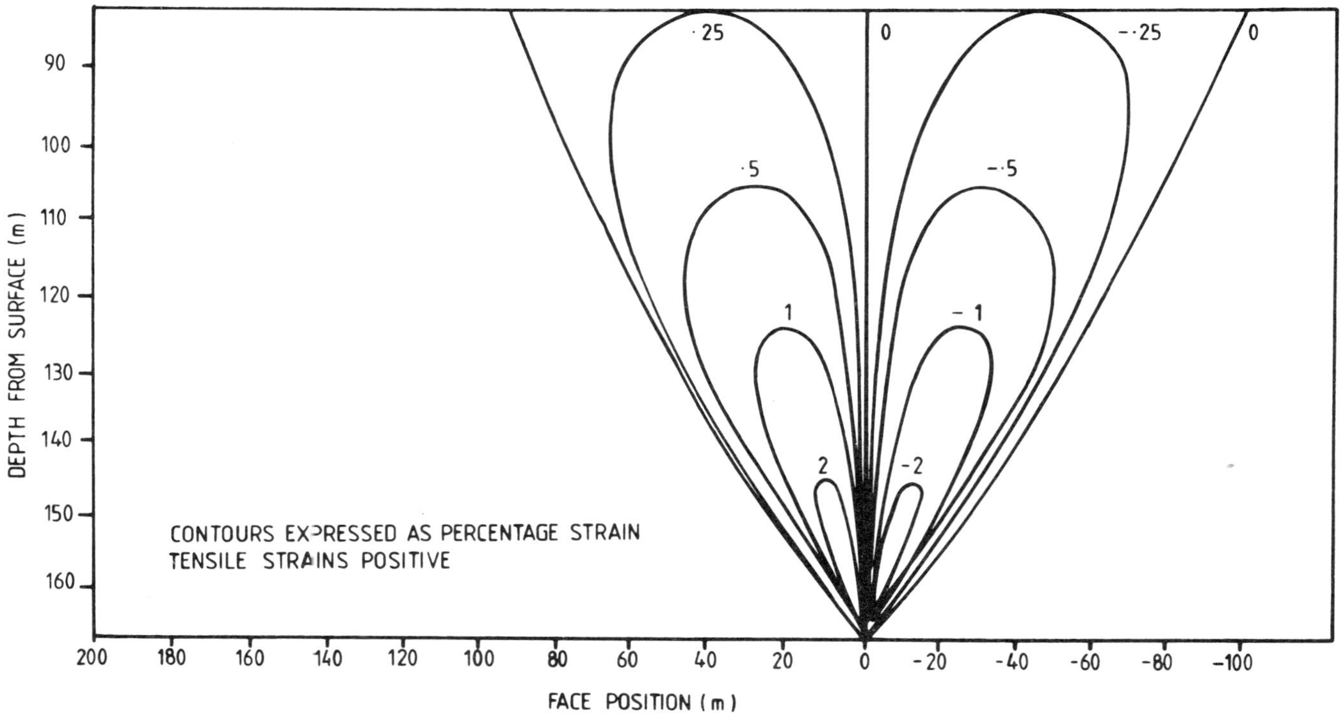

Figure 7 Contours of vertical strain along a face centre line for an idealised material using stochastic analysis based on the Lynemouth Colliery geometry

anomalies, principally due to the stratigraphy of the Coal Measures cyclothem and the mechanical behaviour of the rocks in the individual units of the cylothem. The particular points worth noting are:

(a) There is evidence of yielding of the coal at the face line and ribsides, indicated by extended and relatively low compression strain zones ahead and to the sides of the face.

(b) In the Markham (Figures 3 and 4) and Lynemouth (Figures 5 and 6) cases, tensile strains are confined to an approximate rectangular prism above the caving area. Compression strains exist in a vertical direction over the ribsides and slightly over the face. At Wearmouth (Figures 1 and 2) tensile strains extend some distance over the face and ribside abutment.

(c) Tensile strains tend to be much more closely related to stratigraphy than to proximity to the face. At Lynemouth (Figure 5) peak tensile strains (2%) are associated with a layer of sandstone containing strong bedding plane shears. At Markham (Figure 3) and Wearmouth (Figure 1) very low strains are associated with sandstone beds, and much higher strains with weaker rocks.

(d) With the exception of Markham (Figure 3) where compression strain peaks of 0.25% are associated with sandstone layers ahead of the face, compression strains appear less dependent on stratigraphy.

(e) Tensile strains - effectively dilation in a vertical direction - do not show recovery with passage of the face. Each of the contoured sections (Figures 2, 4 and 6) normal to the face centre-line is taken at 200m from the face. At Lynemouth and Wearmouth, where observations are available respectively 2 and 3 years later, there is no evidence of significant recompression. This corroborates the evidence of Wilson and Ashwin (1972) among others who have observed long term pillar stresses in the ribside pillars of longwall faces. It is also supported by the evidence of the Subsidence Engineers Handbook (National Coal Board, 1975) that significant dilation (see Figure 8) occurs during settlement. The estimated 10% volume loss between extraction level and surface for a 'critical width' subsidence trough would, however, represent a relatively conservative ground loss estimate based on the cases considered in the present paper.

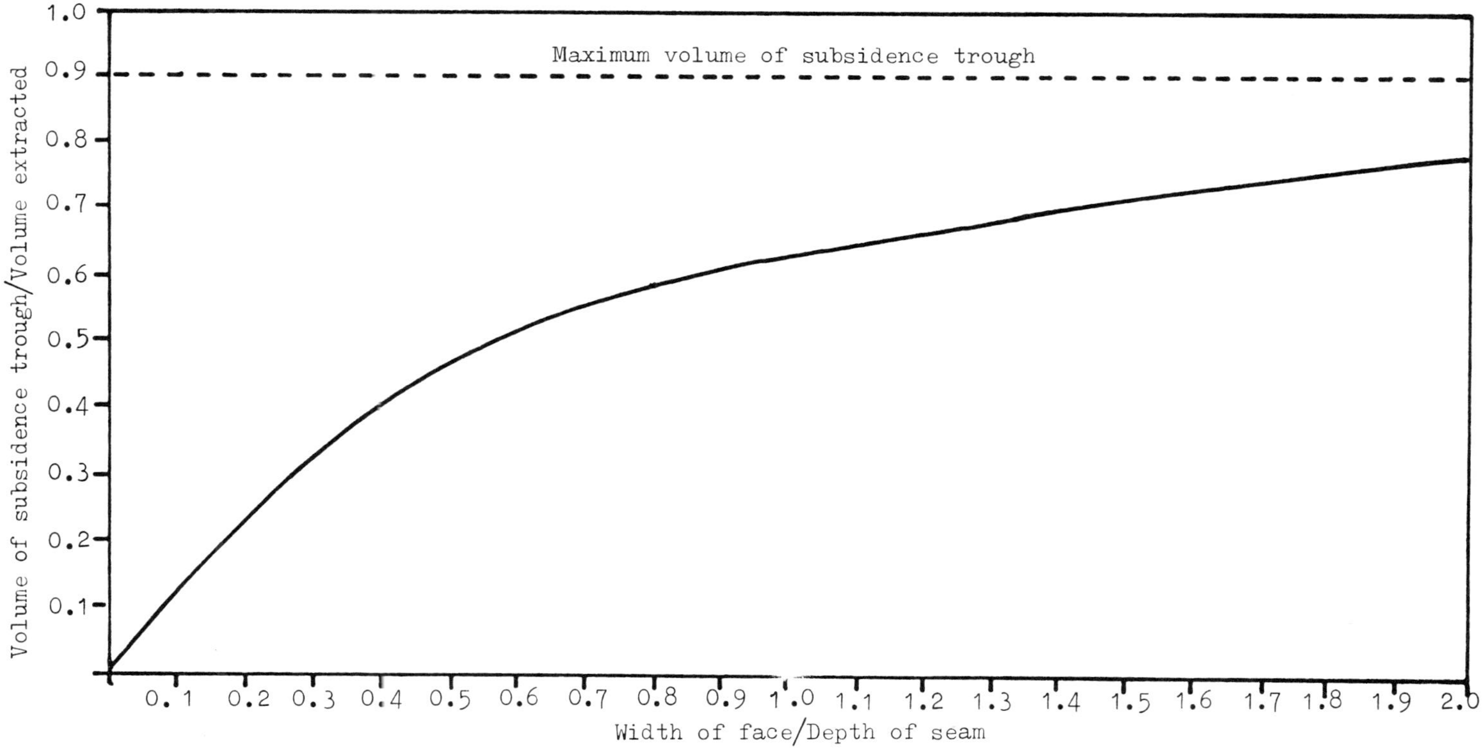

Figure 8 Graph showing relation between extracted volume and volume of subsidence trough for different depths as predicted by the Subsidence Engineers Handbook

(f) A particular feature in the Markham case (Figures 3 and 4) is the low tensile strain in the sandstone beds 30-35 m above the seam and large dilations of 7% in the mudstones immediately above. The large dilations are associated with further post yield failure of the mudstones around a previously worked out face in the 1st Piper Seam. The low strains in the sandstone indicate that the sandstone beds have collapsed in single units exacerbating the dilation in the mudstones above. The particular sandstone known as the Tupton Rock is of a similar nature to the Parkgate sandstone where Phillips (1944) describes how spectacular releases of strain energy associated with collapse of this sandstone were responsible for the Barnborough rockburst in 1943.

MECHANICS OF DEFORMATION

Under the geostatic pressures existing at the depths considered in the case histories (200-600 m), confining pressures in the range 4000 - 20,000 kN/m^2 might be expected. These would induce the compressive strains of the order of magnitude measured but would not be sufficient to induce shear fracture or significant horizontal dilation in typical Coal Measures rocks (see Price, 1979; Price and Farmer, 1979).

A linear compression strain of 0.8 - 1% would normally be needed at low confining pressures to cause shear fracture and significant dilation, and this does not occur at Markham or Lynemouth, although horizontal extension would be expected of a relatively low magnitude above the abutments. A much lower tensile strain would be needed to cause fracture and dilation. Although very little experimental data exists, the ratio between tensile and compressive strengths of 1 : 10 would indicate that fracture in tension would occur at a linear strain of about 0.1%. The indications are, therefore, that a fractured zone of considerable extent exists over the caving longwall workings in each of the cases.

The mechanics of deformation immediately above a caving excavation are well known. Removal of coal and face support is followed by collapse of the immediate roof layers subjected to a combination of bending and tensile stresses. These broken rocks (the height of which they break is known as the caving height and is estimated by Kenny, 1969 and Wilson, 1975, to be between 2 and 4 times the extraction height) partially support the superincumbent roof layers. However, the degree of support is insufficient to restore the initial geostatic stresses, with the result that deviatoric stresses in the rock mass are significantly increased - particularly by increase in the minor (tensile) principal stress above the caved zone. The resultant fracture and associated dilation may result in relatively small strains, but the rock strength is effectively reduced from a peak to a residual with loss in load bearing capacity and resultant stress redistribution. In addition, strongly laminated rocks - of which there are many in the Coal Measures cyclothem - may be affected by bed separation, also increasing dilation in the massive rock.

The tensile strain contours in Figures 1 - 6, particularly those above 0.25%, may be said to represent fractured rock which has deformed 'non-elastically'. The compression strain contours represent unfractured rock which, for modelling purposes, can be assumed to have deformed 'elastically'.

In all cases it is interesting to note that the zone of fracture defined by the extent of dilation extends at least half the face width above the seam level, and probably to a rather higher level than this. The strength of rocks in the fracture zone will be reduced to the residual level with equivalent changes in deformability, and stresses will tend to be permanently redistributed to the abutments. In modelling ground deformation above a caving longwall face, therefore, a rectangular prism of fractured ground should be introduced above the caved area in order to simulate more closely the mechanics of ground deformation. Where stronger sandstone layers exist above the face, the potential for stored strain energy is illustrated by the low strain contours in these layers.

ACKNOWLEDGEMENTS

The authors acknowledge support from the National Coal Board and the Research Committee of the European Coal and Steel Community. Valuable discussions with Mr. A.H. Wilson, Dr. G. Johnson, Dr. N.H. Glossop and Professor E.L.J. Potts are acknowledged.

REFERENCES

Attewell P.B. and Farmer I.W. - Ground deformation resulting from shield tunnelling in London Clay, Can. Geotech.JL., 11 (1974) 380-395.

Batchelor A.S. - The correlation of roadway displacement with the stress redistribution and strata movements caused by longwall mining. Ph.D. Thesis, University of Nottingham (1972).

Berry D.S. - Progress in the analysis of ground movements due to mining. Large ground movements and structures. Ed. J.D. Geddes, Pentech, London (1977) 781-811.

Denkhaus H.G. - Critical review of strata movement theories and their application to practical problems, Symp.Rock Mechanics and Strata Control in Mines, S. African Inst. Min.Met. (1965) 204-226.

Farmer I.W. - Ground deformation observations - Lynemouth Colliery. Interim report on Research Project 7220-AC/806 to E.C.S.C. Research Committee (1979).

Farmer I.W. and Attewell P.B. - A note on the similarities between ground movement around soft ground tunnels and longwall mining excavations, Mining Engineer 134 (1975) 397-403.

Hedley D.G.F. - Design Criteria for multi-wire boreholes extensometer systems, Mining Research Centre, Department of Energy, Mines and Resources, Ottawa. Int.Report (1969) MR69/68ID.

Hodkin D.L. - Interaction between longwall and pillared workings at Lynemouth Colliery, Ph.D. Thesis, University of Newcastle upon Tyne (1978).

Hodkin D.L., Dunham R.K. and Farmer I.W. - Deformation of Coal Measures Strata above a retreating longwall face. 20th U.S. Symposium on Rock Mechs, Austin (1979) 517-524.

Johnson G. - Study of rock pressures round mining excavations Ph.D. Thesis, University of Durham (1963).

Kenny P. - The caving of the waste on longwall faces. Int.J. Rock Mech. and Min.Sci. 6 (1969) 541-555.

King H.J., Whittaker B.N. and Batchelor A.S. - The effects of interaction in Mine layouts. 5th Int.Strata Control Conf. London (1972) Paper 17.

Leigh R.D. - Strata pressures and rock mass movements induced by longwall mining, Ph.D. Thesis, University of Durham (1963).

National Coal Board - Subsidence Engineers Handbook 2nd Ed. N.C.B. Mining Department, London (1975).

Phillips D.W. - Rock bursts or "bumps" in coal mines, Transactions of the Institution of Mining Engineers 104 (1944) 55-84.

Potts E.L.J. - Current investigations in rock mechanics and strata control, 4th International Conference on Strata Control and Rock Mechanics, Columbia University, New York (1964) 29-45.

Price A.M. - The Effect of Confining Pressure on the Post-Yield Deformation Characteristics of Rocks, PhD Thesis, University of Newcastle upon Tyne (1979).

Price A.M. and Farmer I.W. - Application of yield models to rock, International Journal of Rock Mechanics and Mining Science and Geomechanics Abstracts 16 (1979) 157-159.

Shadbolt C.H. - Mining subsidence, Proceedings of the Conference on Large Ground Movements and Structures, Cardiff, Pentech Press (1977) 705-748.

Voight B. and Pariseau W. - State of predictive art in subsidence engineering, Journal of the Soil Mechanics and Foundation Engineering Division, ASCE 96 (1970) 721-750.

Wiggill R.B. - The effects of different support methods on strata behaviour around stoping excavations, Symposium on Rock Mechanics and Strata Control in Mines, South African Institute of Mining and Metallurgy (1965) 1-35.

Wilson A.H. - Support load requirements on longwall faces, Mining Engineer 134 (1975) 479-491.

Wilson A.H. and Ashwin D.P. - Research into the determination of pillar size, Mining Engineer 131 (1972) 409-430.

THE INFLUENCE OF GROUND SURFACE DEFORMATIONS CAUSED BY UNDERGROUND MINING ON THE STATE OF STRESS IN EMBANKMENTS

Leszek Litwinowicz, Dr Eng

Assist Prof, Technical University of Lublin, Poland

INTRODUCTION

An examination of the damage to earth embankments (especially those of great length) in mining regions, where there are large displacements of the ground surface has shown that this is caused by a reduction in the density of the soil forming the embankments. This reduction is produced by the horizontal strain, ε_R, on the convex part of the mining subsidence profile (Litwinowicz, 1976).

Bronsztajn and Chadzinow (1963) found that a considerable reduction in the degree of compaction of the ground could be caused by these strains. For example, a strain of $\varepsilon_R = 3$ to 10 mm/m caused 20 to 30 percent loss of density for an alluvial clay containing more than 50 percent of fines. They also discovered that the cohesive strength diminished (by about 10 percent). Many measurements and investigations have been made in monitoring the condition of embankments in mining areas, mainly those of industrial water supply reservoirs. These also proved the influence of ε_R on embankments. It has been generally concluded that a mining strain of $\varepsilon_R \geqslant 6$ to 9 mm/m can cause damage (Litwinowicz, Malcharek and Rosikon, 1975; Szumierz, 1977; Wytyczne projektowania, 1976).

A new method for calculating the local factor of safety (in terms of strength) in embankments as a function of ε_R is proposed in this paper.

METHOD OF CALCULATION

In embankments with no arching effect, described by Trollope (1957), decreasing density of the soil structure at a given point is associated with decreases in the horizontal stress component σ_x, under constant vertical stress σ_z.

The intergranular stress in an embankment depends on the variable coefficient of lateral earth pressure $\xi_R = f(\varepsilon_R)$. Because of discontinuities in the soil mass produced by weakening of the structural bonds between the soil grains, continuum medium models should not be used. It is better to use a granular soil model (Litwinowicz, 1977).

The method is based on relating the coefficient ξ_R (known from the ground conditions) with the Kandaurow - Muller (1965) parameter which characterises the expandability of a granular medium. From this we obtain:

$$\frac{\partial \sigma_z}{\partial z} = \xi_R z \frac{\partial^2 \sigma_z}{\partial x^2} - \gamma \quad (1)$$

$$\sigma_x = \xi_R \sigma_z + \xi_R^2 z^2 \frac{\partial^2 \sigma_z}{\partial x^2} \quad (2)$$

$$\tau_{xz} = \xi_R z \frac{\partial \sigma_z}{\partial x} \quad (3)$$

where γ is the unit weight of the soil. To evaluate σ_z, σ_x, τ_{xz}, in the absence of external loading, it is advisable to make some simplifying assumptions (Dobrow and Choruzyj, 1971).

Taking into account the load increment $\gamma dz'$ at any level z' above a given point A(x,z) we get:

$$\sigma_z = \frac{\gamma}{2} \left\{ z + \int_0^z \Phi\left(\sqrt{\xi_R^{-1}} \frac{x-d}{z-z'} \right) dz' \right\} \quad (4)$$

$$\sigma_x = \xi_R \sigma_z - \gamma \sqrt{\frac{\xi_R}{2\pi}} \int_0^z \frac{x-d}{z-z'} \exp\left(- \frac{(x-d)^2}{2\xi_R (z-z')^2} \right) dz' \quad (5)$$

$$\tau_{xz} = \gamma \sqrt{\frac{\xi_R}{2\pi}} \int_0^z \exp\left(- \frac{(x-d)^2}{2\xi_R (z-z')^2} \right) dz' \quad (6)$$

where

$$d = \frac{H-z'}{\tan\alpha}$$

H is the height of the embankment

α is the angle of the sloping face of the embankment.

The subsequent equations may be simplified if a new parameter β is introduced, such that,

$$\tan \beta = \frac{1}{2.5\sqrt{\xi_R}} \quad (7)$$

Components σ_z, σ_x and τ_{xz} are calculated by superposition of the trapezoidal load occurring in the upper part of the embankment having thickness z_z and the load occurring in the lower part having thickness z_L.

$$\sigma_z = \frac{\gamma z_z}{2}\left\{\left(1 - \frac{x-x_2}{x_1}\right)\Phi\left(\frac{x-z\cot\alpha}{z_L\sqrt{\xi_R}}\right) + \frac{x-x_2}{x_1}\Phi\left(\frac{x-x_2}{z_L\sqrt{\xi_R}}\right) + \right.$$
$$+ 1 - \frac{z_L\sqrt{2\xi_R}}{x_1\sqrt{\pi}}\left[\exp\left(-\frac{(x-z\cot\alpha)^2}{2z_L^2\xi_R}\right) - \right.$$
$$\left.\left.- \exp\left(-\frac{(x-x_2)^2}{2z_L^2\xi_R}\right)\right]\right\} + \gamma z_L \qquad (8)$$

$$\sigma_x = \xi_R\left\{\frac{\gamma z_z}{2}\left[\left(1 - \frac{x-x_2}{x_1}\right)\Phi\left(\frac{x-z\cot\alpha}{z_L\sqrt{\xi_R}}\right) + \frac{x-x_2}{x_1}\Phi\left(\frac{x-x_2}{z_L\sqrt{\xi_R}}\right) + \right.\right.$$
$$+ 1 - 2\frac{z_L\sqrt{2\xi_R}}{x_1\sqrt{\pi}}\left[\exp\left(-\frac{(x-z\cot\alpha)^2}{2z_L^2\xi_R}\right) - \right.$$
$$\left.\left.\left.- \exp\left(-\frac{(x-x_2)^2}{2z_L^2\xi_R}\right)\right]\right] + \gamma z_L\right\} \qquad (9)$$

$$\tau_{xz} = \xi_R\frac{\gamma z_z z_L}{2x_1}\left[\Phi\left(\frac{x-x_2}{z_L\sqrt{\xi_R}}\right) - \Phi\left(\frac{x-z\cot\alpha}{z_L\sqrt{\xi_R}}\right)\right] \qquad (10)$$

where:

$$\Phi(m) = \frac{2}{\sqrt{2\pi}}\int_0^m e^{-\frac{t^2}{2}}\,dt\ ,$$

$$z_z = z - z_L\ ,$$

$$z_L = \frac{\sin\alpha\sin\beta}{\sin(\alpha+\beta)}\cdot x\ ,$$

$$x_1 = z_z\cot\alpha\ ,$$

$$x_2 = (z - z_z)\cot\alpha\ .$$

For a given deformation ε_R, ξ_R can be computed approximately (Litwinowicz, 1977):

$$\xi_R = \xi_o - (\xi_o - \xi_e)\left[1 - \exp(-A_R\varepsilon_R)\right] \qquad (11)$$

where:

ξ_o = lateral earth pressure coefficient for $\varepsilon_R = 0$,

ξ_e = lateral earth pressure coefficient for the lower limiting value of density,

A_R = an empirical parameter.

The method of testing the soil described by Glinko and Kwiatek (1975) is most convenient for use with equation (11).

It is then possible to compute a stability number which characterises the relationship between the strength of the ground and the actual shear stress on the potential surface of failure (Trollope, 1957) from the components of stress at the point A (x,z):

$$f_A = \frac{2c + \left[(\sigma_z + \sigma_x) - \sqrt{(\sigma_z - \sigma_x)^2 + 4\tau_{xz}^2} \cdot \sin\phi\right] \tan\phi}{\sqrt{(\sigma_z - \sigma_x)^2 + 4\tau_{xz}^2} \cdot \cos\phi} \qquad (12)$$

where:

ϕ = angle of internal friction,

c = cohesion of the ground.

The proposed method has been used in examining the embankments of a sedimentation tank.

Figure 1 shows values of

$f_A^{\xi_o} = 0.51$ and $f_A^{\xi_R} = 0.29$

within the embankment for $\phi = 38^{\circ}$, c = 0. The function ξ_R for the soil in the embankment is given in Figure 2.

CONCLUSIONS

From the calculation using the proposed method we can see that in embankments with loss of density of the ground caused by ε_R, a limiting equilibrium condition may be produced. Similar results have been obtained in model investigations (Litwinowicz, Malcharek and Rosikon, 1975) and from observations of real embankments (Glinko and Litwinowicz, 1977).

The condition for limiting equilibrium in an embankment requires a sufficient reduction in the coefficient of lateral earth pressure associated with a reduction in the horizontal stress component σ_x (Figure 3).

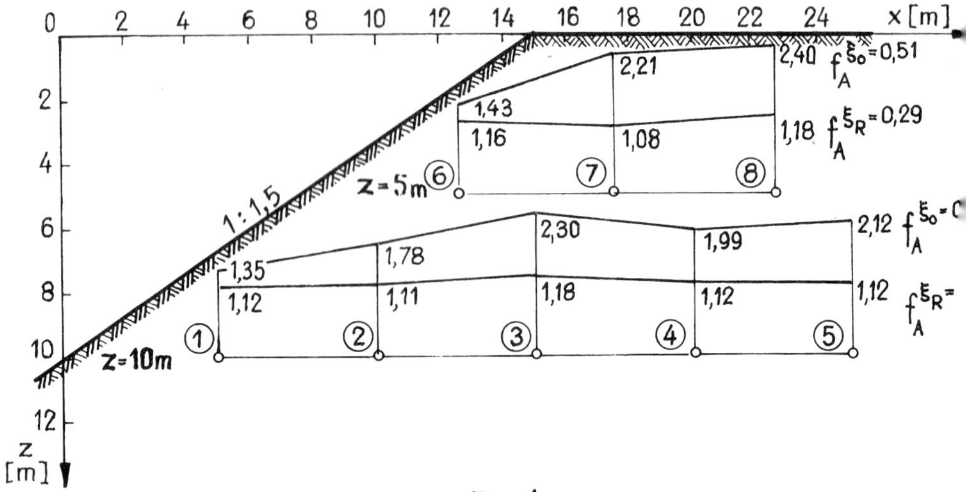

Fig. 1

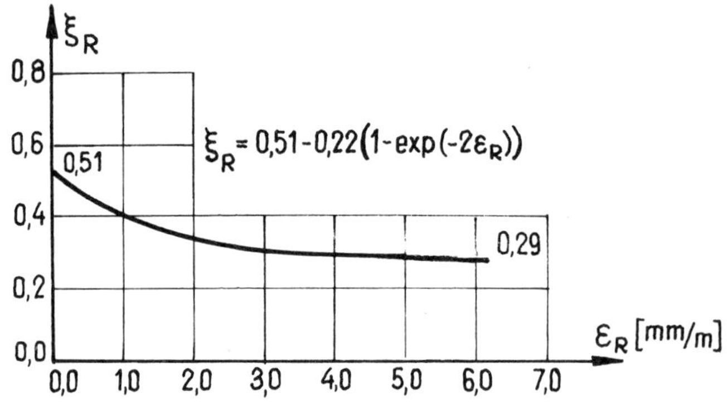

Fig. 2

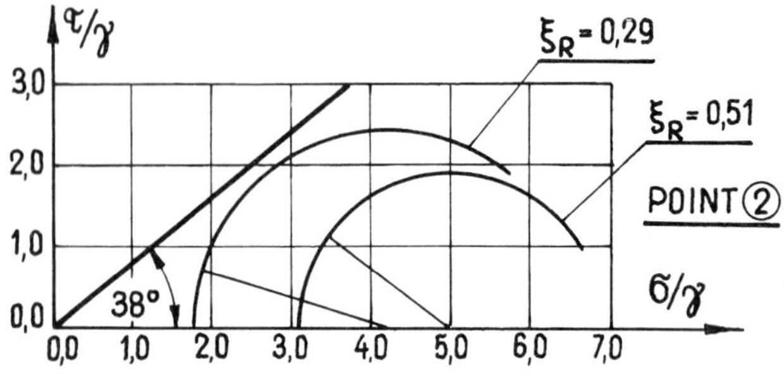

Fig. 3

The proposed method gives a more precise estimation of the stability of embankments subjected to a reduction in density caused by deformation ε_R than other methods (for example Madej, 1978; Szumierz, 1977).

REFERENCES

B.E. Bronsztajn and S.Z. Chadzinow - K opredeleniju izmienienij svojstv gruntov pri ich podrabotkie, Stroitielstvo sooruzenij nad gornymi vyrabotkami,Nedra (1963).

E.M. Dobrow and E.F. Choruzyj - Ocenka obščej ustojčivosti otkosov ziemlianogo polotna s ucetom jego naprjazonnogo sostojanija, Trúdy Sojuzdorinii 3 (1971).

H. Glinko and L. Litwinowicz - Wpływ deformacji górniczych terenu na obwałowania nadpoziomowych osadników przemysłowych, Materiały XXIII Konferencji Naukowej Komitetu Inżynierii Ladowej i Wodnej P A N i Komitetu Nauki PZITB, Krynica (1977).

H. Glinko and J. Kwiatek - Vlijanije gorizontalnogo rozrychlenija grunta na vieličinu bokorogo davlenija, Proceedings of the First Baltic Conference on Soil Mechanics and Foundation Engineering, Gdańsk (1975).

L. Litwinowicz, K. Malcharek and A. Rosikoń - The influence of mining deformations of the subsoil on banks, Proceedings of the First Baltic Conference on Soil Mechanics and Foundation Engineering, Gdańsk (1975).

L. Litwinowicz - Budowle ziemne na terenach górniczych, Materiały Konferencji Naukowo-Technicznej:Budownictwo na terenach górniczych o dużych deformacjach powierzchni, Polska Akademia Nauk Oddział w Katowicach, Katowice (1976).

L. Litwinowicz - Oddziaływanie poziomych odkształceń górniczych terenu na budowle nasypowe w ujęciu mechaniki ośrodków dyskretnych, Materiały Konferencji: Budowle ziemne na terenach górniczych, Polska Akademia Nauk Oddział w Katowicach, Katowice (1977).

J. Madej - Prognozowanie stateczności zapory zbiornika odpadów poflotacyjnych, Materiały Konferencji:Geotechniczne problemy składowania odpadów poflotacyjnych, PAN, KILiW, Sekcja Mechaniki Gruntów i Skał orza Fundamentowania, Lublin (1978).

W. Szumierz - Analiza wpływu górniczych rozpełzań terenu na stateczność skarp podtorza kolejowego, Materiały Konferencji:Budowle ziemne na terenach górniczych, Polska Akademia Nauk Oddział w Katowicach, Katowice (1977).

O.M. Trollope - The systematic arching theory applied to the stability analysis of embankments, Proceedings of the Fourth International Conference on Soil Mechanics and Foundation Engineering (1957).

Wytyczne projektowania i użytkowania powierzchniowych zbiorników wodnych na terenach górniczych, Prace GIG, Katowice (1976).

SESSION II

Chairman: Mr W S Allen, CEng, MIMinE, ARICS
National Coal Board, London

Introduction to Session: Professor J D Geddes

SESSIONAL PAPERS

MINING SUBSIDENCE IN BATHGATE, A TOWN STUDY

P. Carter, MSc, CEng, MICE, MIGeol

Project Engineer, Geotechnical Section
Babtie Shaw and Morton

D. Jarman, BA, MA, MRTPI

Assistant Director, Department of Physical Planning
West Lothian District Council

M. Sneddon

Senior Technician, Geotechnical Section
Babtie Shaw and Morton

INTRODUCTION

Bathgate is an old town of 15,000 population in the heart of the Scottish central belt. Since 1975 it has been the administrative centre of West Lothian. It suffers from problems of poor environment, ageing property and a declining industrial base, more recently alleviated by the arrival of British Leyland. A particular planning problem has been the need to sustain the town centre in the face of competition from Livingston new town nearby. With the imminent opening of the regional shopping centre in Livingston, confidence in Bathgate as a place in which to trade or live was at a very low ebb in the mid-seventies.

An additional blow was dealt by a severe subsidence which commenced on Christmas Eve in 1975 in the Hopetoun Street area of the main shopping centre. During the following year the effects became visible as properties were closed and some demolished as being beyond repair. One entire street block is now being cleared and many other premises may have only a short life remaining. Although there had been a number of subsidence incidents over previous decades the full extent of the problem was not known in 1975. It was known that coal seams outcropped under the Town Centre, but it had been assumed that any workings were confined to the immediate outcrop zones, the location of which were not accurately known.

The 1975 Hopetoun Street subsidence occurred outside the immediate outcrop zone, at a location where the old workings were believed to be at a depth of at least 30m. Initial speculation in Bathgate that this was a local occurrence where particularly ambitious monks had dug unusually deep bell pits gave way to a growing concern that the problem could be more extensive, as the subsidences continued to spread throughout 1976 and 1977.

Early in 1978 the District Council asked Babtie Shaw and Morton to prepare a comprehensive report on the mineral position in the worst affected area, even though much of the area is occupied by private property. The work was funded by the Planning Committee in the general interest of the community in order to establish the true facts and remove the uncertainty which was having a serious effect in blighting a major part of the town centre (Fig.1). Site investigations were concentrated on a number of gap sites that the District Council wished to redevelop.

This study showed that earlier beliefs as to the number and location of worked seams were in error, and that the area underlain by old mine workings is considerably more extensive than had been supposed. It was also shown that subsidence due to old mine workings is not limited to narrow strips along the outcrop zones but may occur over a far wider area of the town centre. Large scale subsidence due to failure of fairly large numbers of stoops at any one time was shown to be a more important risk in Bathgate than localised subsidence due to room failure.

It was appreciated that blanket sterilisation of the area would create a vast gash through the heart of the town, and attention was therefore concentrated on developing suitable structures for the development sites. It was also appreciated that some of the newer structures already located in the affected area were built on foundations inappropriate to an area of mining subsidence. Attention was therefore given to assessing the degree of risk involved, both to the structure concerned and its users, and to remedial measures where considered necessary. Grouting of the old mine workings beneath one block of flats has recently been completed and further grouting and monitoring work is planned.

This paper deals with Bathgate's history of mining and mining subsidence; the investigations which have been carried out, and the proposals evolved for designing new structures and making existing structures safe, in the zones of potential mining subsidence.

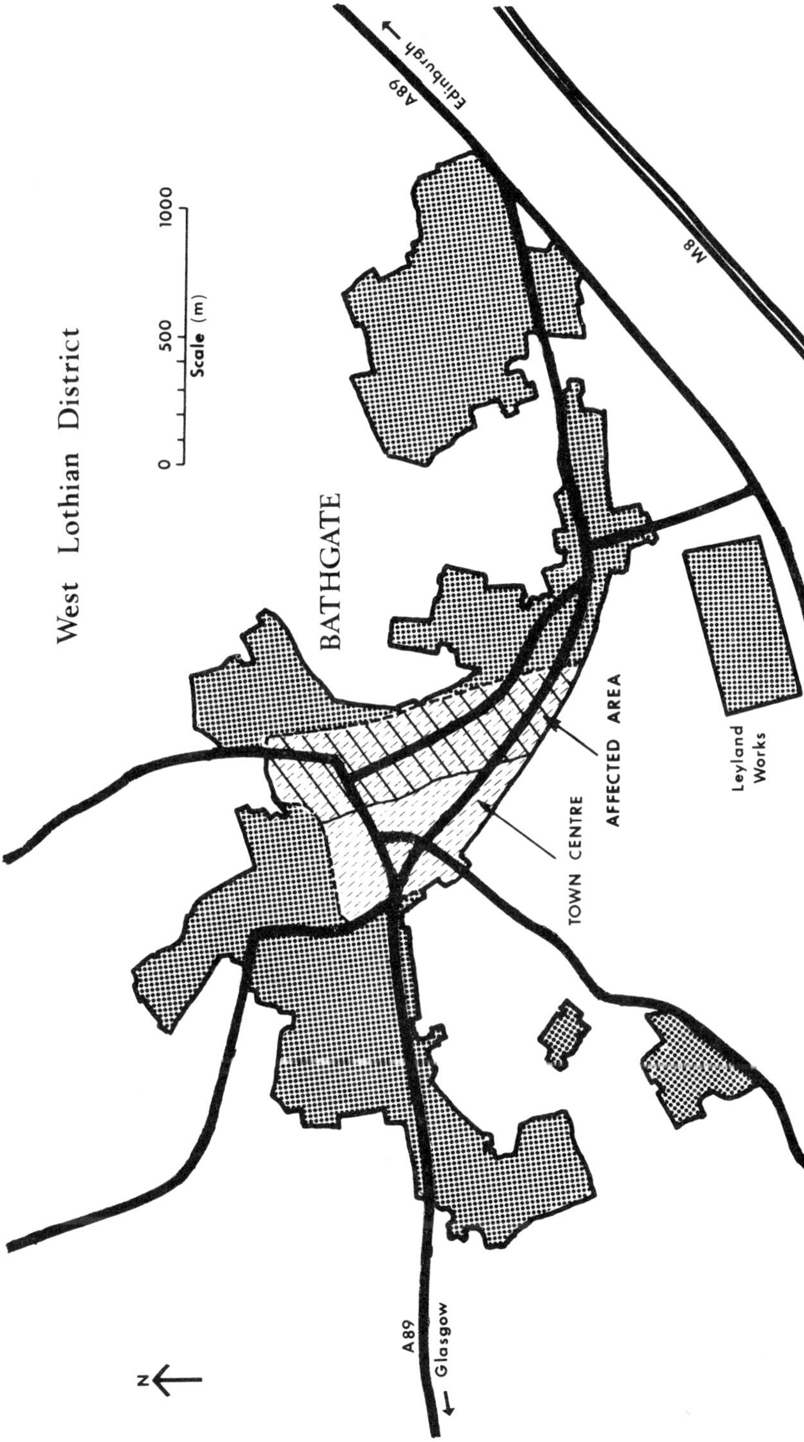

LOCATION PLAN

FIGURE 1

LOCAL GEOLOGY

The affected area of Bathgate is situated on limestone coal group strata which strike north-south (Fig.2) and dip westwards at a gradient of about 1 in 2.5 (Fig.3). Underlying thick contemporaneous basalts form the Bathgate Hills to the immediate east of the affected area. Overlying thinner basalts form the stable part of the town centre to the immediate west.

The Limestone Coal Group is just over 100m thick in Bathgate and contains six named coal seams and some blackband ironstones (Fig. 4). Three seams, the Balbardie Gas, Jewel and Wilsontown Main Coals, were known to have been worked in the general Bathgate area. Only the Balbardie and Main Coals were originally believed to have been worked in the affected area but the investigations found workings in all three seams. Earlier locations for the seam outcrops were found to be in error, due mainly to the total lack of exposure and mis-correlation of the scanty original mining and borehole evidence. Correlation is a particularly hazardous operation in the Limestone Coal Group of the Bathgate area, due to the lack of distinctive marker horizons. The investigations accurately located the worked seams and pushed the assumed sub-drift outcrop of the Main Coal eastwards, thus extending the zone of undermining into areas previously thought to be clear of the lowest worked seam.

The drift cover in the area is generally stiff boulder clay 7 - 10m in thickness, with some areas of peat and granular soils.

MINING HISTORY

An initial view of past mining activities in the district was obtained from information provided by statutory bodies such as the Institute of Geological Sciences and the N.C.B. Abandonment Plans Office. The information obtained was however mainly of nineteenth century origin and dealt with the more recent deeper workings. Some clues as to the extent of earlier coal mining were obtained from many other non-statutory sources, such as the local library, offices of local architects and engineers, the District Council's own archives and the nearby office of the N.C.B. at Polkemmet Colliery. The latter proved to be a "mine" of useful information. An important mine plan dating back two centuries was also obtained from the Scottish Records Office. The information gathered gave positive records of old mine workings in less than 30% of the area now known to be undermined.

Mining started at least three centuries ago in Bathgate, as shown by the reference to the occupation of "colheuer" in the Kirk Session Records for 1677 (Bisset 1906). Local outcrop mining carried on sporadically for the next hundred years, as shown on old plans of the area (Lesslie 1764). This catered for the limited local household fuel market in competition with wood and

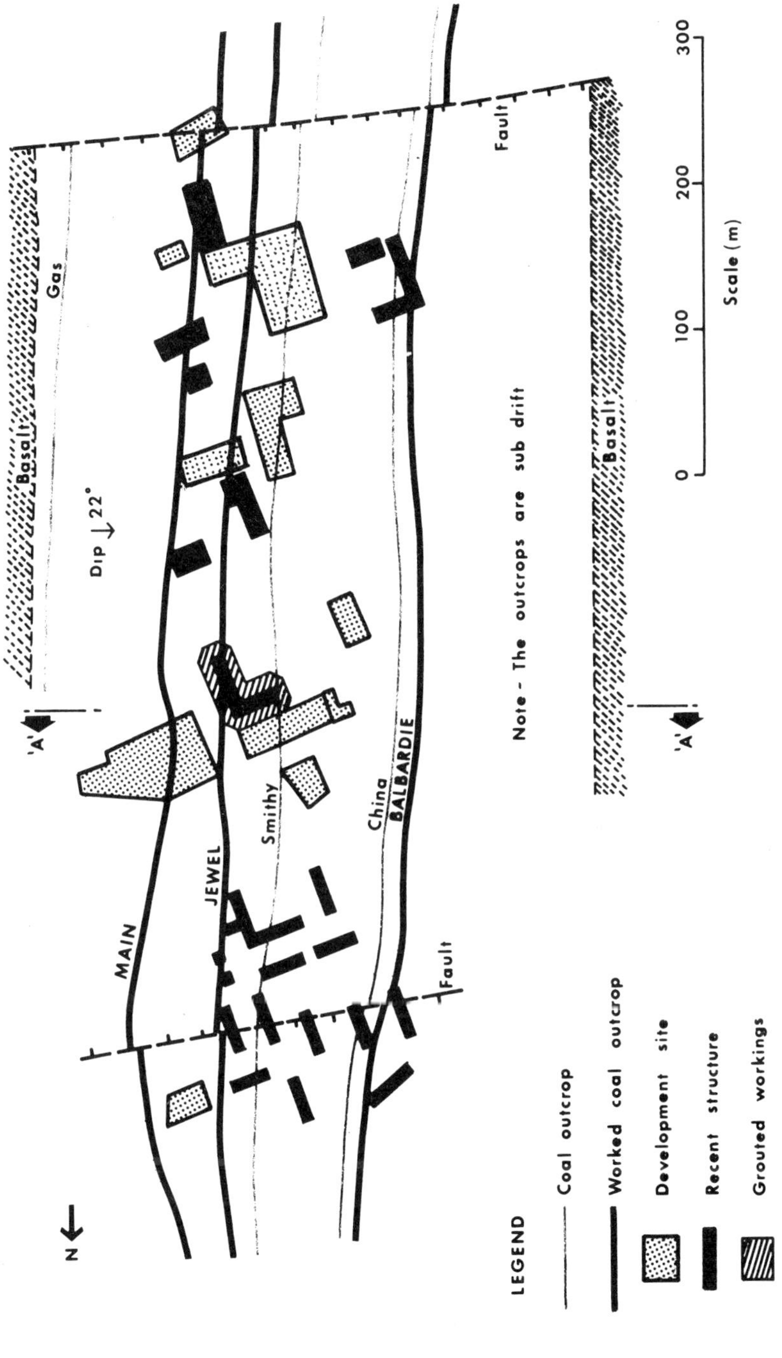

GEOLOGICAL PLAN OF AFFECTED AREA

FIGURE 2

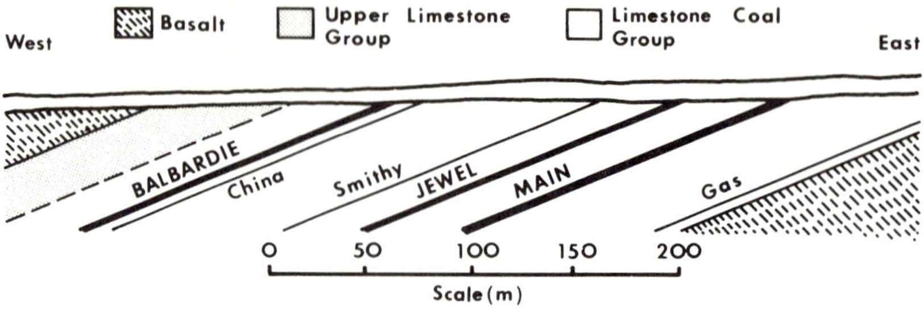

GEOLOGICAL SECTION -'A A'

FIGURE 3

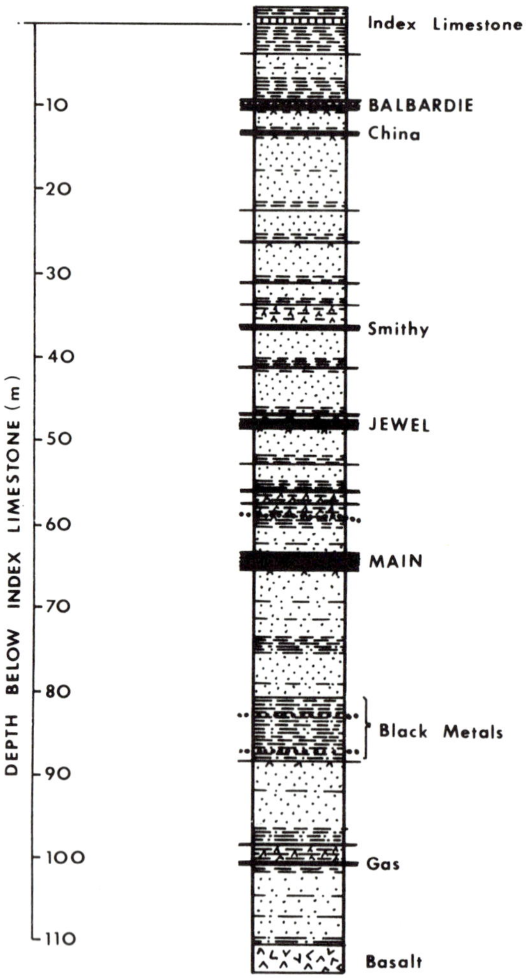

STRATIGRAPHICAL SECTION OF LIMESTONE COAL GROUP AS PROVED BY DRILLING

FIGURE 4

plentiful supplies of peat. In the mid-eighteenth century, mining became more extensive with a substantial area of coal being worked by means of a day level draining to the Bathgate Water. The locations of these old workings are only approximately known.

A rapid advance of Scottish coal mining took place in the eighteenth century, due to increased industrial use of coal and the introduction of the Newcommen Engine. Bathgate had one of the earliest Scottish engines (Duckham 1970) installed by the Earl of Hopetoun in the 1770's. This is shown on an old mine plan (Lesslie 1777) as a 'fire engine' draining an engine pit 31 fathoms (57 m) deep. Coal was raised to the surface by way of a Horse Gin Pit 21 fathoms (37 m) deep, and a Stair Pit (Fig.5). The stair pit would have been used by women bearers carrying creels full of coal from the workings below to the coal hill above. Bald (1812) estimated that the women bearers had to climb "the equivalent of Ben Lawyers (4000') every day loaded".

The sinking of the engine pit meant that coal could be worked to much greater depths than had been achieved by the old day level. For the first few years the coal was worked by leaving stoops 7 m long along the strike and 2.7 m wide down the dip and taking out rooms 3.7 m wide, giving a 66% extraction rate. In 1777 the mine was visited by John Grieve, then one of Scotland's small band of "viewers" or mining consultants. He recommended that the stoops and rooms should both be made 3.7 m square, giving 75% extraction and, in his opinion, a more stable working. This recommendation was very likely followed for the next few decades.

By the late eighteenth century mining had brought about a 50% increase in the population of Bathgate (Old Statistical Account 1791), so much so that the seams within the town were soon worked out near the outcrops. In the early nineteenth century new mines were opened on the outskirts (New Statistical Account 1844). A plan of Bathgate in 1833 still shows the Engine Pit but other shafts shown are already referred to as "old pits".

The development of oil shale distillation and iron foundries later in the nineteenth century led to a further upsurge in mining activity (Caddell 1925). Shaft depths in the vicinity of the town increased progressively to 100 m and then 300 m below ground level (Dron 1902). Coal was once again mined from directly beneath the town as the seams were followed up-dip from the new deep shafts (Fig. 6). This time however large stoops of 25 m square were left with room size being limited to 3.7 m, giving an extraction rate of only 25%. These dimensions were the conditions of a lease between the coal company and the Hopetoun Estates, made to protect the expanding town.

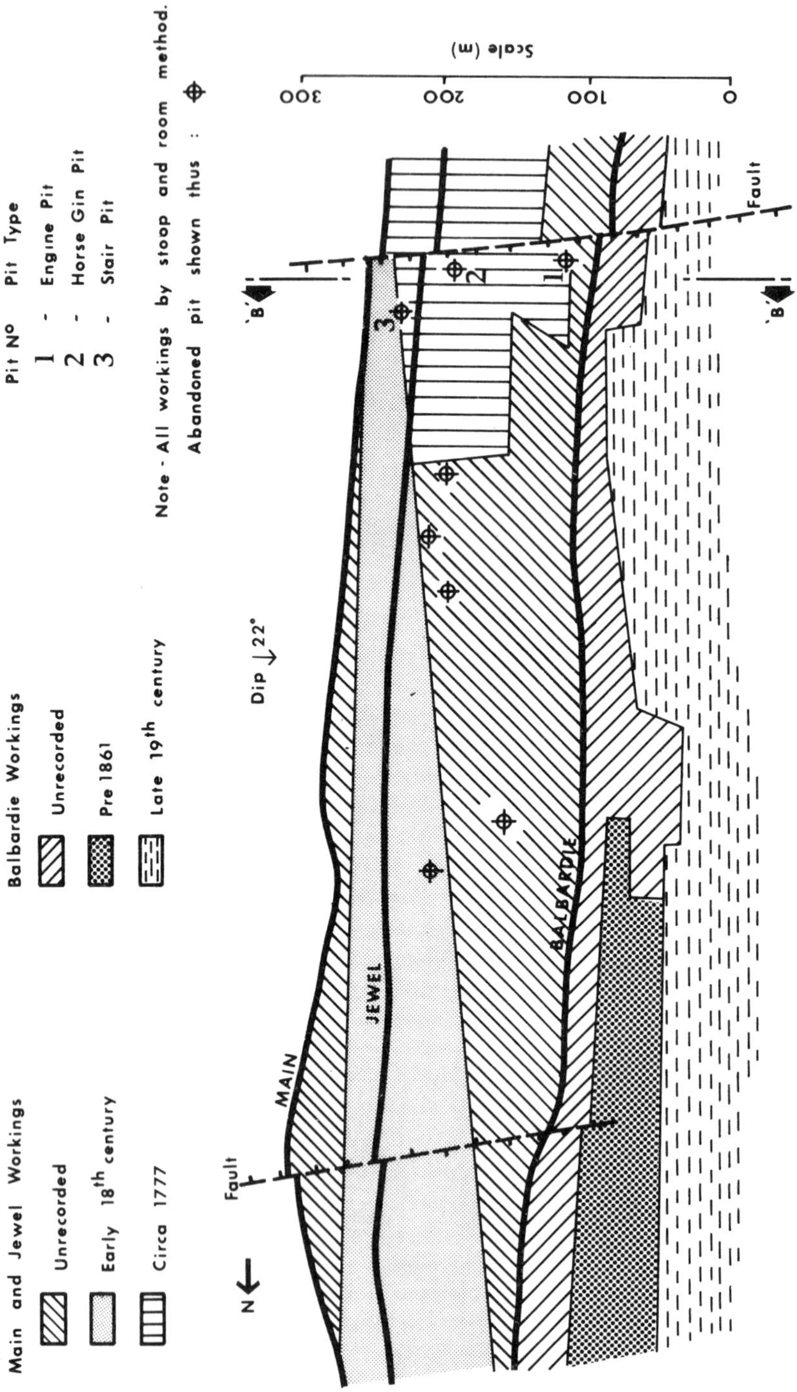

FIGURE 5

MINING INFORMATION PLAN

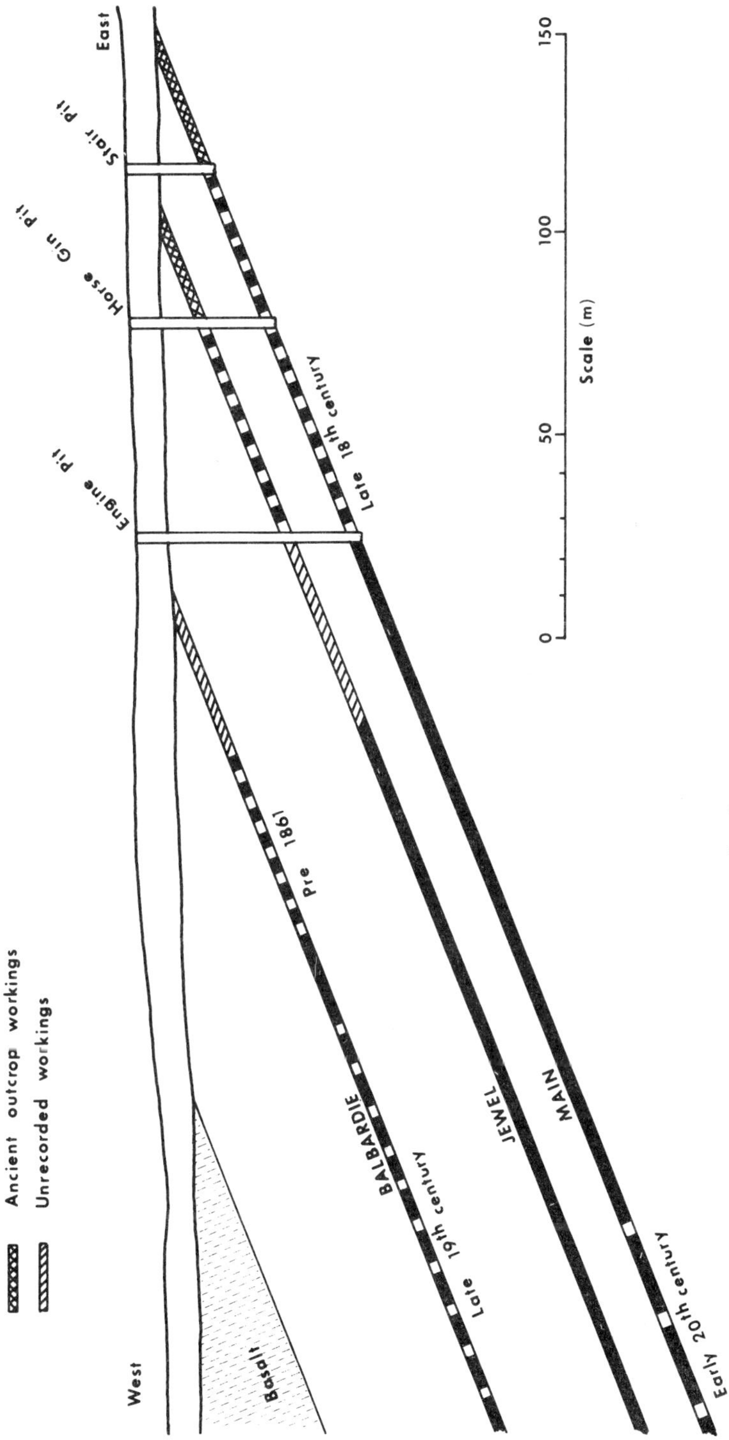

MINING INFORMATION : SECTION 'B-B'

FIGURE 6

Coal mining beneath Bathgate ceased in the 1930's and most of the coal beneath the town has been exploited, although some reserves are present in places between the more recent workings and the eighteenth century workings. Mining is however still carried out from the nearby Polkemmet mine.

Subsidence problems in the town are related to the eighteenth century workings which, as shown above, were carried out at shallow depths using small stoop sizes and a very high extraction rate.

SUBSIDENCE HISTORY

Mining subsidence may have started in Bathgate almost as soon as the mining itself began. Evidence of this is given by an old feu disposition of 1799 between Lord Hopetoun's trustees and a weaver called James Anderson, for a plot of land in Hopetoun Street. The disposition states that the plot "is situated upon the old coal waste" and that if James Anderson erected "houses or any other buildings thereon the said Earl and his heirs and successors shall not be liable for any damage that may happen to them from the coal operations below ground." This foresight was justified nearly 200 years later, when the plot was involved in the recent major Hopetoun Street subsidences.

There is a good deal of evidence of mining subsidence this century. A major subsidence occurred in June 1923 in Hopetoun Street (Fig. 9) which was described as measuring 100 m by 50 m and causing big vents across the streets, broken gas mains and very severe cracking of buildings. The subsidence was said to have been preceded by "strange noises and rumblings" and to have resembled an earthquake (Scotsman 1923). A sequence of levels on a benchmark in the vicinity between 1855 and 1979 shows a total settlement of 0.5 m, almost all of which occurred at some time between 1914 and 1954.

Several other subsidences have occurred and indeed some of the gap sites which the West Lothian District Council wish to develop have resulted from demolition of badly cracked buildings. One building that suffered severe subsidence was the old Police Station, and the story is told locally, with some glee if not complete veracity, that the policemen had to climb out of the windows in the morning.

Not all settlement in Bathgate can be ascribed to mining subsidence. Some old stone buildings, probably bound with lime mortar, have distorted to considerable angles without apparent structural distress, due to consolidation of buried peat deposits.

The subsidences for which we have the most complete records are those which commenced in Upper Hopetoun Street on Christmas Eve 1975 (Fig. 7 and 8). A local resident said she heard the

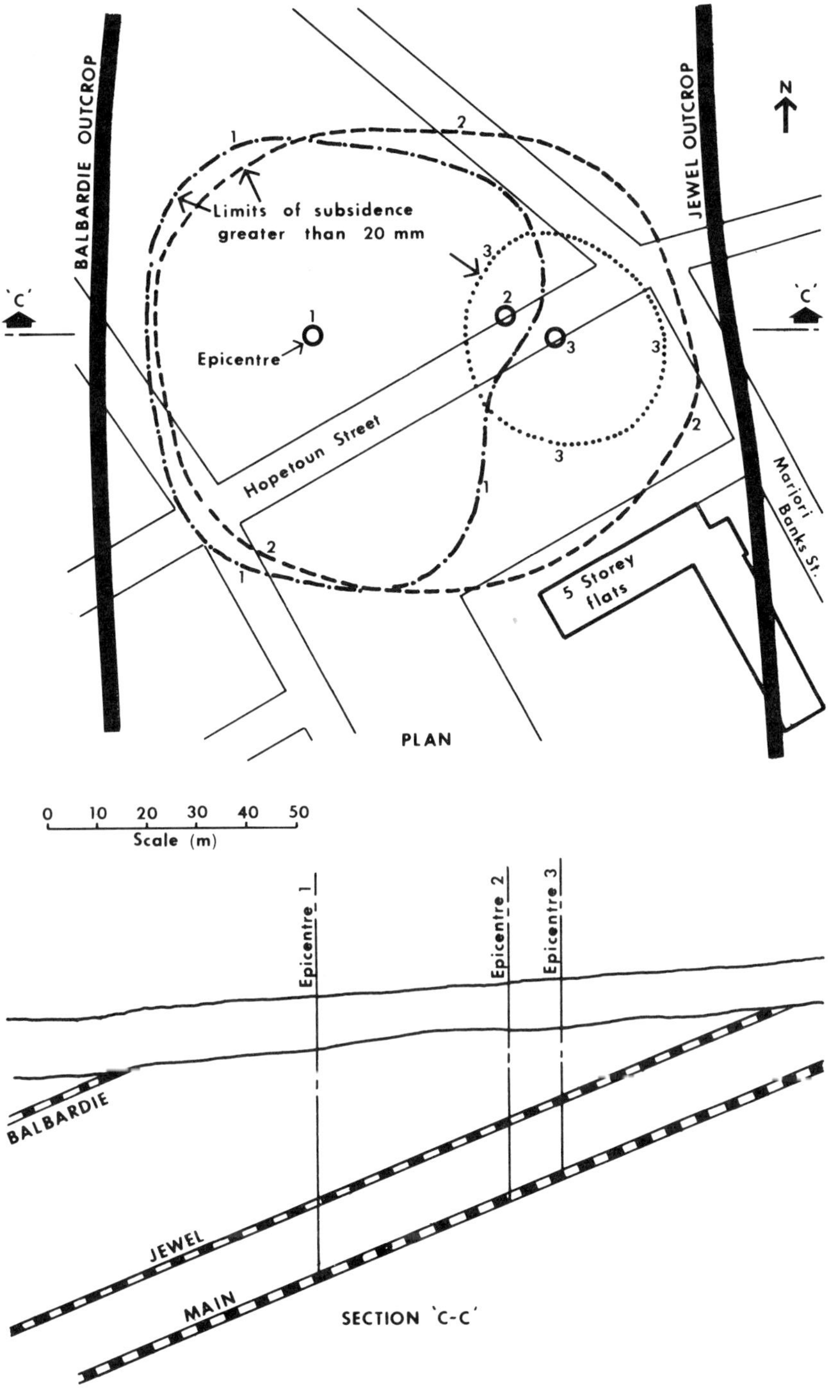

HOPETOUN STREET SUBSIDENCES

FIGURE 7

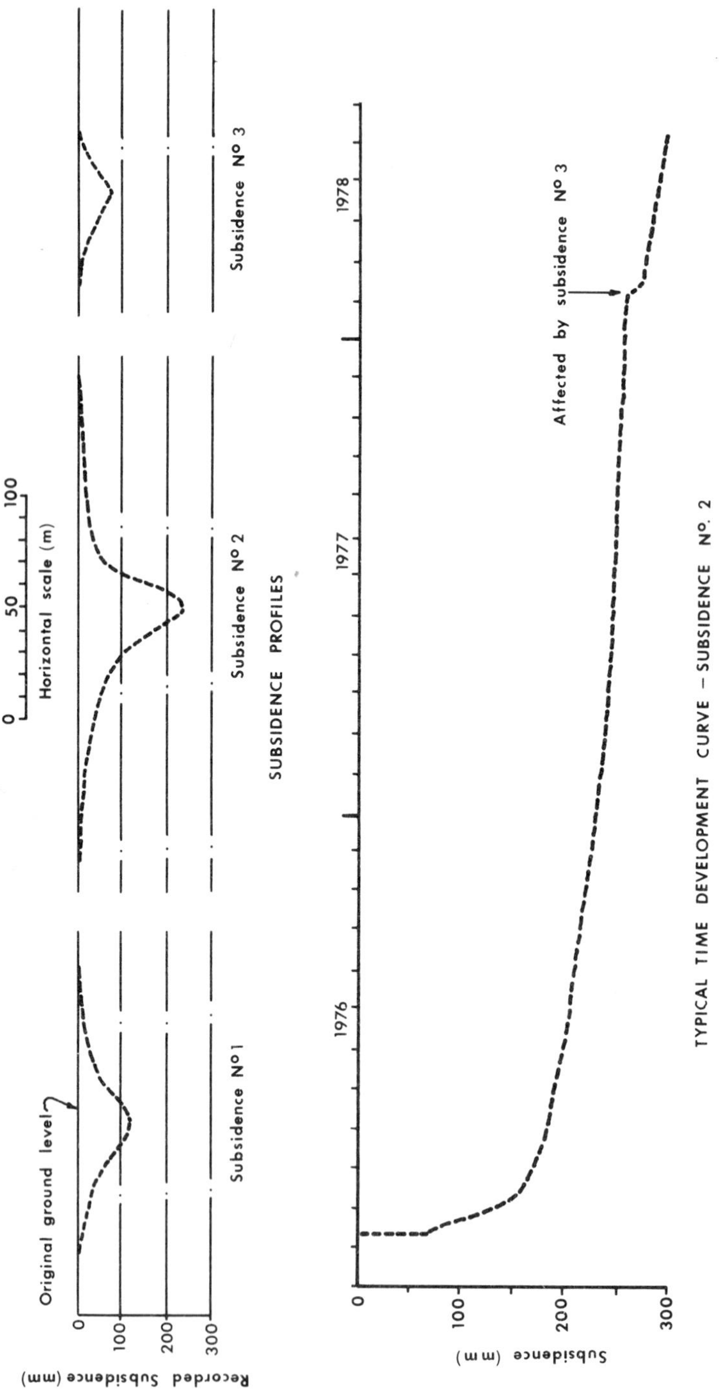

HOPETOUN STREET SUBSIDENCES

FIGURE 8

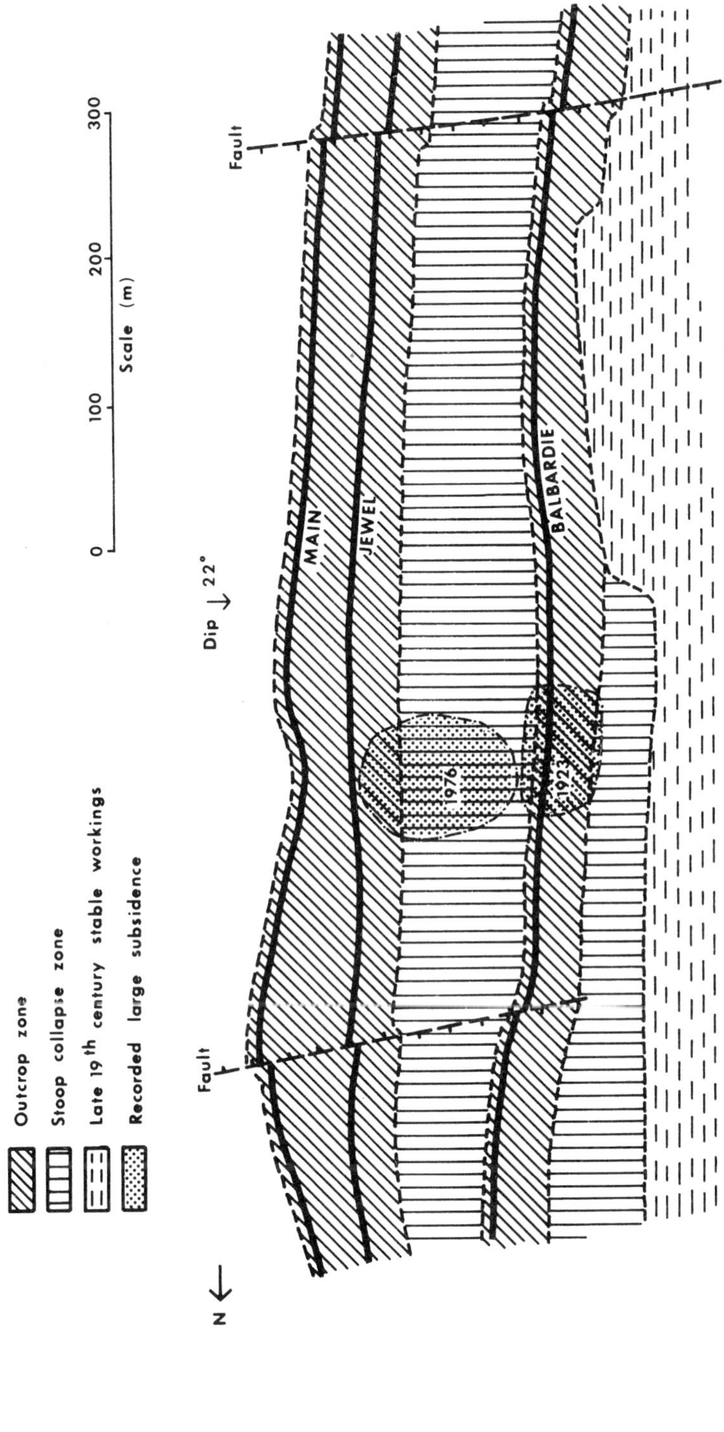

FIGURE 9

POTENTIAL SUBSIDENCE ZONE PLAN

first earth movement early on Christmas morning and that "it was like an earthquake" (Lothian Courier 1976). Cracking of buildings and distortion of windows and door frames was very severe. After a period as the subsidence progressed some buildings were evacuated and eventually demolished. Others required safety measures such as removal of heavy chimneys.

The National Coal Board began monitoring settlements within two weeks of the first sign of subsidence and have continued to take levels over the subsequent few years. Altogether three subsidences occurred in Hopetoun Street over a three year period, all of them commencing in winter periods.

The maximum recorded subsidence exceeds 300 mm to date due to overlap of subsidence zones, and the total volume of ground lost exceeds 650 cu.m. The subsidences take the form of broad troughs with a deep central portion (Fig. 8) and affect a surface area of more than 7,500 sq.m. (Table I).

Time development curves for subsidence No. 2 are shown on Fig. 8. These show that about 30% of the subsidence took place immediately and 50% had occurred after the first two weeks. After one year however subsidence had not finally finished, although probably 90 - 95% had taken place.

SITE INVESTIGATIONS

A total of 52 rotary cored boreholes have been drilled on the re-development sites and in the vicinity of recently built structures in the affected area. The drilling was carried out by Hydracrat Limited. All the boreholes were fully cored in order to provide full stratigraphical information and hence avoid miss-correllation of seams, coring also reveals such telltale signs as bedding plane separation, oxidation of coal etc., which may indicate the presence of workings even where the coal is found to be intact. A summary of seam conditions found in the boreholes is given in Table II.

Voids were generally found to be 0.5 m high overlying waste material although voids in excess of 2.5 m high were encountered occasionally. Workings were encountered right up to the sub-drift outcrops of the seams. Some voids were surveyed with a down-hole television camera, which showed that these extended to at least 6 m laterally, this distance being the limit of the light source. Sandstone forms a fairly high proportion of the strata between the coal seams (Fig. 4) and in particular occurs close to the roofs and floors of the Jewel and Main Coal workings.

Twenty three soft ground holes were bored by Whatlings Limited to investigate the superficial soils on the redevelopment sites. Boulder clay was found to be the predominant soil

EPICENTRE	1	2	3
Depth to old Workings (m)	42	26	20
Rock Cover to old Workings (m)	34	18	12
Maximum Measured settlement (mm)	115	235	82
Area of subsidence greater than 20 mm (m^2)	5,000	7,500	1,200
Volume of subsidence (m^3)	275	331	49
Maximum slope over 6 m (ratio)	1:250	1:63	1:225
Maximum strain over 6 m (mm/m)	3	7	4
Commencement of subsidence	Dec.75	Feb.76	Dec.77

TABLE I - HOPETOUN STREET SUBSIDENCE VALUES

	Average Thickness (m)	Percentage of Boreholes Encountering			
		Voids	Waste	Oxidised Coal	Fresh Coal
Seam					
Balbardie	1.0	0	40	40	20
Jewel	1.5	50	13	12	25
Main	2.0	42	21	25	12

TABLE II SEAM CONDITIONS ENCOUNTERED IN THE BOREHOLES

Depth to old Workings (m)	Maximum Settlement (m)	Maximum Ground Slope	*Max. Ground Strain (mm/m)
20	0.4	1:27	9
50	0.2	1:85	4
90	0.09	1:250	2

* Average for 10 metre wide structure

TABLE III CALCULATED MAXIMUM SETTLEMENT - JEWEL COAL SEAM

type although some granular soils and peat were also found. The thickness of the superficial soils was generally found to be in the order of 7 - 10 m, decreasing in places to 2 m and increasing locally to 15 m.

MODES OF COLLAPSE

Two forms of collapse are possible in areas of old mine workings, these being room collapse and stoop collapse. Room collapse is generally considered to be by far the most likely cause of subsidence due to old mine workings in the British coal fields. Stoop collapse is generally thought to be a rare exception. Accordingly, the rules of thumb that have gained currency for predicting safe depths to old mine workings, such as a minimum required rock cover of 5 to 10 times seam thickness depending upon local conditions, have been based on the mode of collapse involved in room failures. In Bathgate the rock cover at the main centres of subsidence is significantly greater than 10 times the seam thickness, and simple room collapse cannot account for the magnitude of subsidence that took place.

We consider that a combination of adverse factors was responsible for the Hopetoun Street subsidence, these being:-

- high rate of extraction, 65 - 75%
- age of the workings, two centuries or more
- steep dip, 1 vertical to 2½ horizontal
- strong roof and floor strata, generally sandstones
- proximity of workings in two seams, Main and Jewel

This combination of factors has resulted in small stoops being left to carry large loads over a long period of time. Protection of the stoops by natural infilling of the rooms with collapsed material has been limited by the presence of competent sandstones just above roof level, and by the down-slope movement of any material that has caved in. Natural room infilling due to floor heave, or stoop settlement due to punching has been prevented by the presence of competent sandstones just below floor level. The unprotected stoops have therefore been left open to gradual weakening by the combined forces of oxidation, erosion by ground water flowing down the slope, and spalling due to locally high stresses. It is significant that nearly half the boreholes encountered voids at seam level, and that the coal was found to be oxidised where the boreholes passed through stoops.

We consider it to be very likely that an area of stoops in the Jewel workings weakened in this manner collapsed on Christmas Eve 1975 causing the Hopetoun Street subsidence No. 1. This would then throw additional stress onto the up-dip stoops which collapsed two months later, resulting in subsidence No. 2.

Conditions then remained more stable until the winter of 1977/78 when the remaining workings up to the outcrop failed, leading to subsidence No. 3. (Fig. 10).

Alternatively, downwash of roof collapse material due to groundwater flow and slumping down the steep slope may have led, with time, to the creation of "super-voids" locally (Fig. 10). Roof collapse voids could be created in this manner that would be larger than the original rooms, and which could engulf one or two surrounding stoops. Migration of such voids to rock-head level could then occur, causing consequent large scale subsidence at ground level.

It is quite possible that a combination of these two effects took place, particularly, in view of the proximity of workings in the Main and Jewel seams. Migration of roof collapse voids upwards from the Main coal workings could well undermine stoops in the overlying Jewel coal workings leading to large subsidences at ground level. A few such migrated voids from the Main coal have been found immediately below the Jewel pavement during the grouting operation at Marjoribanks Street which is described below.

Whatever the mode of collapse assumed, it is noteworthy that major subsidences of fairly large surface area occurred in an area where rock cover to the old workings is greater than that normally considered "safe". Caution should therefore be used when considering sites where a particularly adverse combination of conditions exists.

Collapse of old mine shafts is an additional potential hazard in the affected area and many of the old shafts have not been located. Shaft collapse does not however appear to have been a problem locally, possibly due to the predominant covering of stiff boulder clay.

POTENTIAL SUBSIDENCE ZONES

Two broad subsidence zones have been ascribed to each seam, the 'outcrop zone" and the 'stoop collapse' zone (Fig. 9).

In the 'outcrop zone' subsidence may occur due to both room collapse and stoop collapse. The limit for surface subsidence due to room collapse has been taken as a rock cover of not less than seven times the seam thickness. Room collapse voids which migrate above rockhead level will cause conical depressions of limited lateral extent at ground level. The volume of material concerned in any one collapse will be relatively small but high values of differential settlement will occur locally. Deep plump holes are however unlikely to develop in most of Bathgate, due to the nature and thickness of the superficial soils. Stoop collapse may also occur in the

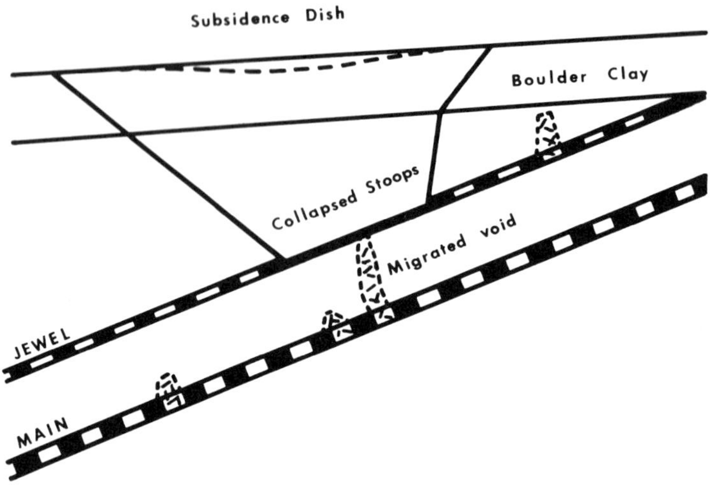

a; Subsidence due to stoop collapse.

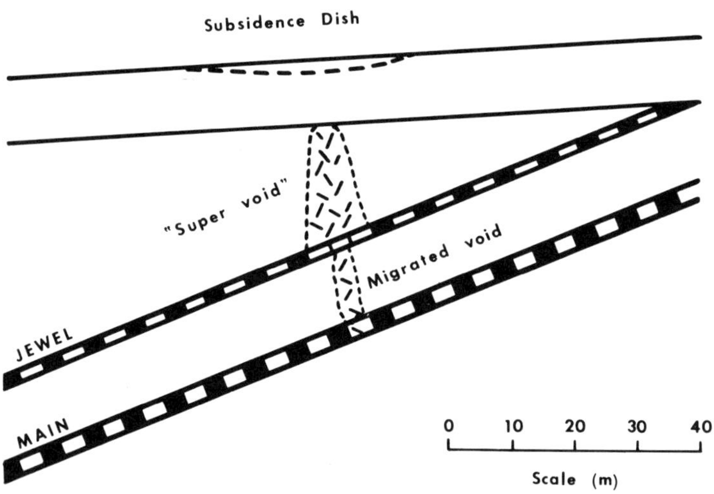

b) Subsidence due to room collapse.

POSSIBLE MODES OF COLLAPSE FIGURE 10

'outcrop zone' due to erosion and oxidation of stoops and will result in widespread depressions with very high values of differential and total settlement.

The 'outcrop zones' for the Main and Jewel workings merge together (Fig. 9) due to the proximity of these workings.

The 'stoop collapse' zones extend back from the outcrop zones to the area of more modern mine workings. No limit of depth has been given to these zones although the surface effects obviously decrease with increase in depth. Potential subsidence profiles in this zone have been based on the methods given in the Subsidence Engineers Handbook (N.C.B. 1975) for longwall workings. A 600 square metre area of collapsing old stoop and room workings has been assumed to be represented by a similar area of long wall workings, with the seam thickness taken to be only 30% of the full seam thickness. This reduction has been derived from summing up the following assumed factors:-

Percentage of coal left in stoops	30%
Bulking of mine waste and roof collapse material	20%
Bedding plane separation and arching	20%

These factors have been estimated from the study of old records and the local geology, and are of necessity very approximate.

The calculated settlements given in Table III are rather higher than the actual values recorded in the Hopetoun Street subsidence No. 2, suggesting that the assumed seam "thickness" of 30% true thickness may be on the high side.

PLANNING NEW STRUCTURES

Mining subsidence was not considered a risk in planning for the majority of Bathgate during the 1960's. The large area of potential mining subsidence is only now apparent. The large scale Lothians Regional Survey and Plan of 1966 has a plan of minerally-suspect areas which shows the affected area of the town centre as being clear. Reports of 1964 71 on the redevelopment of the town centre make no reference to any history or future liklihood of subsidence. Several sites which were redeveloped by the council were bored and some precautions were taken, but no areas were sterilised until 1973, when a 23 metre band over an outcrop at Mid Street was kept free of all building on N.C.B. advice extended to 60 m by the Mineral Valuer. The current study has shown that a 275 metres wide strip is liable to mining subsidence to a greater or lesser extent.

The District Council has therefore been presented with a planning problem with important implications for the development of the town. Only two alternatives are available, either to convert present and future gap sites into public parks and car parks, or to redevelop the sites. The Council has decided to promote redevelopment because:-

(1) They own a number of gap sites in the area which constitute a mis-use of valuable town centre land and create a depressing and deteriorating physical environment.

(2) The psychological effect of an increasing number of empty sites, even if grassed or tarmacadamed, is blighting a much wider area of the town than is actually affected and is undermining confidence in the whole of central Bathgate.

By redeveloping the gap sites the Council hope to stitch together the fabric of the town visually and support the town by providing for fringe commercial uses, service industries and housing.

Five basic ways of engineering around the problem of mining subsidence have been studies, these being:-

(1) Removing the risk of settlement by grouting up the old workings.

(2) Supporting the structures on piles carried through the workings to sound bearing strata.

(3) Construction of fully flexible structures which conform wholly to subsidence profiles when movements occur.

(4) Construction of partially flexible structures where differential settlements are expected to be small.

(5) Construction of rigid structures designed to with stand the tilting action of any settlement without distortion.

Grouting up the old mine workings is an expensive solution or the kind of development envisaged, particularly where two orked seams are present. This method is therefore only onsidered to be appropriate for fairly high-rise or high value evelopment such as major new public buildings or as a remedial ethod where structures are already present and at risk.

Pile foundations are also considered to be inappropriate or the same reasons. They are also rather suspect in view of he high dip of the seams and possibility of the transference f shear forces or drag-down effects to the pile shafts.

Voids of up to 3 m high were encountered, with combined voids/rubble heights of up to 5 m. A few voids were found to have migrated upwards from the Main coal workings to near the floor of the Jewel seam. The percentage of holes that encountered old workings was 50% in the Jewel seam and 70% in the Main seam. Some 1700 tonnes of grout were injected into the Jewel seam and 3900 tonnes into the Main seam, representing about 25% and 40% of the seam volumes treated respectively. These figures may be compared with the discussion of subsidence prediction above, where it was considered that maximum subsidence should not exceed 30% of the seam height. The grouting contract was carried out by Messrs. Colcrete Limited, took four months to complete, and cost about £110,000 which represents approximately 20% of the replacement value of the building being stabilised.

CONCLUSION

Bathgate's "earthquakes" have shown that major mining subsidence can occur in areas where the coal seams are at a depth normally considered "safe", given an unfavourable combination of factors. In such areas the 'rules of thumb' of 5 to 10 times seam thickness for predicting safe depths for old mine workings are not applicable.

Failure to recognise and deal with this situation where it exists in urban areas can eventually lead to various problems for town development. These include blighting of localities, erection of potentially unsafe structures, and the need for expensive remedial action.

An assessment of the degree of risk involved for any area of potential stoop collapse will include investigation of local mining history, establishment of local geology and study of local subsidence. Once this assessment has been made, appropriate engineering measures may be taken to allow development to take place and avoid sterilisation of the areas concerned.

ACKNOWLEDGEMENTS

We wish to thank the following for their assistance in the work described in this paper, I. Forsyth and M. Brown of the Institute of Geological Sciences, A. Anderson and G. Archiebald of the National Coal Board, I. McElhinney of the Lothian Regional Council, J. Mungall of the Lothian Health Board, D. Bruce of Colcrete Limited, S. Learmonth of Hydracrat Limited, W. Miller of the West Lothian District Council, A. McGilp, D. Cochrane, B. Colford and R. Zirmer of Babtie Shaw and Morton. We also wish to acknowledge the permission of the West Lothian District Council and consulting engineers Babtie Shaw and Morton for permission to publish this paper.

REFERENCES

Anon - The (Old) Statistical Accounts of Scotland, (1974).
Anon - The New Statistical Account of Scotland, (1844).
Anon - The Scotsman, 18th June, (1923).
Anon - Lothian Courier, 16th January, (1976).
A.M.Bisset - History of Bathgate and District, Bathgate (1906).
R. Bald - A General View of the Coal Trade of Scotland (1808).
H.M. Cadell - The Rocks of West Lothian, Edinburgh, (1925).
R.W. Dron - The Coalfields of Scotland, Edinburgh, (1902).
B.F. Duckham - A History of the Scottish Coal Industry, David & Charles, Newton Abbot, (1970).
J.Lesslie - Plan of Bathgate, Scottish Records Office, Reference H.O. 6818, (1764).
J.Lesslie - Map of Bathgate Lands and Coal, Scottish Records Office, Reference R.H.P. 6713 (1777).
N.C.B. - Subsidence Engineers Handbook, Mining Department (1975).
R.J.Piggot & P.Eynon - Ground Movements arising from the presence of Shallow abandoned mine workings, Proceedings of the Conference on Large Ground Movements & Structures, U W I S T, Cardiff (1977) 749-780.

GROUND OVER ABANDONED COAL WORKINGS AT BAGILLT

S. Adams, C Eng , F I Struct E , M Cons E

Partner with W.G. Curtin & Partners in charge of Mold Office, Clwyd

INTRODUCTION

In 1978/79 a study was made of the ground at Bagillt, on Deeside. This paper outlines how the study was approached and what it found. It concludes with general advice on developing such an area and comments on how to arrive at the likely surcharge on normal building costs.

More detailed information is contained in the published report :-

> "Survey and Treatment of Abandoned, Shallow Coal Mine Workings" a report on a research and development programme at Bagillt for Delyn Borough Council and the Welsh Development Agency by W.G. Curtin & Partners, Consulting Engineers.

EARLY HISTORY

Figure 1 locates Bagillt and shows the main areas of coal and lead mining. The general drift material is boulder clays, although these become much more sandy in the Bagillt area. The juxtaposition of lead, limestone, clay and sand with access to the sea meant that this area of North Wales was very busy in the 19th century mining coal and lead; winning limestone and kilning for lime; producing refractories; extracting sand; and exporting many of the products. Well into the 20th century the sailing schooners of North Wales were commonplace throughout the U.K. and Europe.

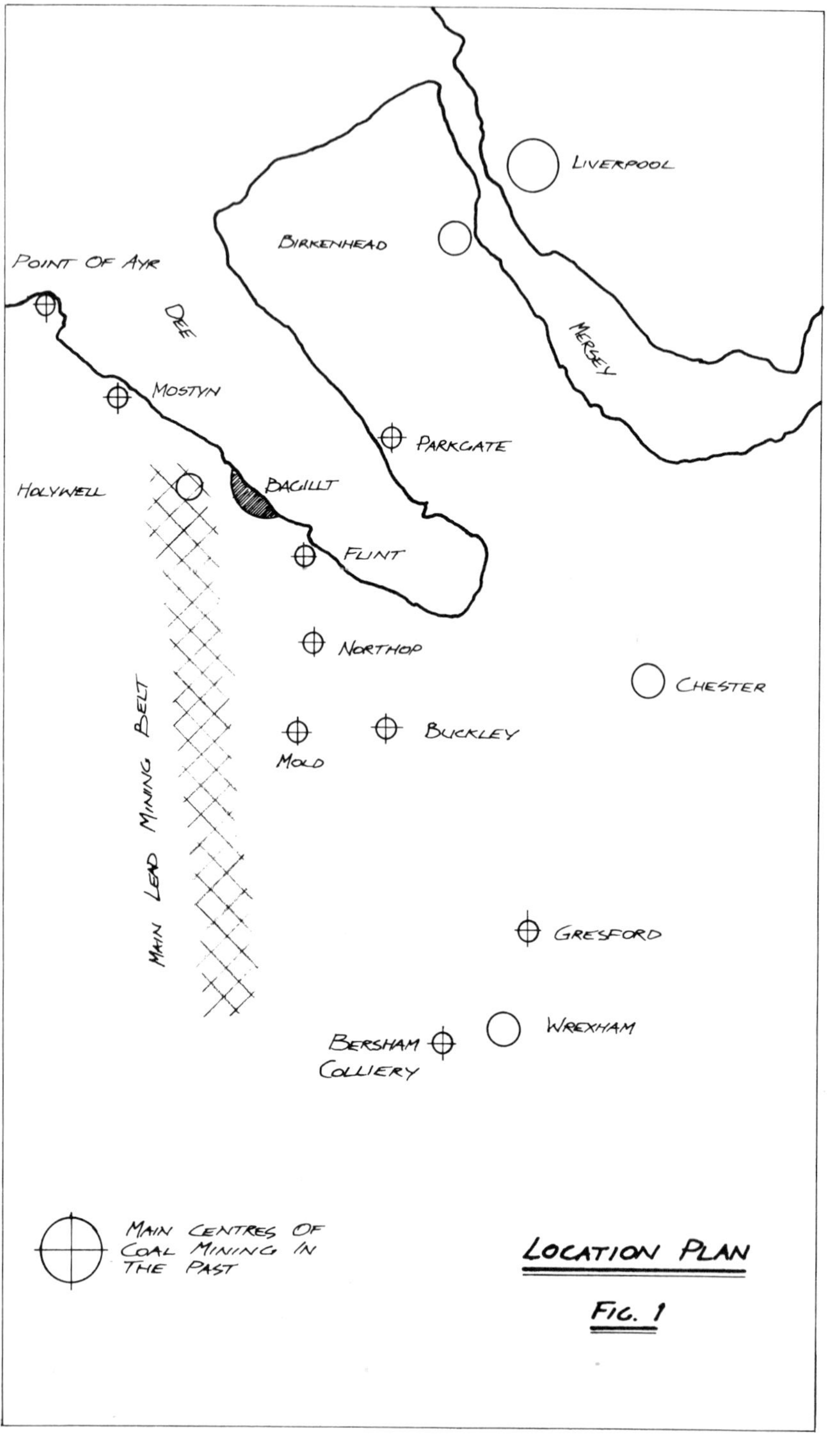

LOCATION PLAN

FIG. 1

In the late 19th century Bagillt started to fall into decline as first the lead then the top coal seams were worked out. Today the only active coal mines are at Point of Ayr, working deep coal beneath the estuary; and at Bersham near Wrexham.

The opening up of the massive Oxford clay beds has virtually wiped out the many small brickworks in North Wales. The area declined.

Figure 2 shows the generalised sequence of coal seams in the area. The 5 yard (Main) and 3 yard (Lower Bench) outcrop in the borough and were fully worked out before 1872, when it became a statutory obligation to lodge plans of abandoned workings. In consequence records of shafts and workings were poor.

Figure 3 shows a composite "section" drawn from some of the shaft records in 1849.

RECENT HISTORY

During the late 1960's and early 1970's the former Flint Borough Council acquired land for an ambitious plan of redevelopment. Investigation of a site for a block of flats at Highfield Estate led to the discovery of a coal shaft and that project was abandoned. Drilling in the Pentre area confirmed how widespread were the workings.

At this time several collapses in this and other counties re-emphasised to all local authorities that old mine workings can cause loss of support at the surface a very long time after their abandonment, and in an unpredictable manner. Conversely pressure built up to use this sort of derelict land. Without much more investigation any sensible redevelopment of this derelict area was impossible.

The scale of investigation was outside the resources of the Borough Council who considered that the problem was, of its very nature, a national one and, as such, should be helped from a National source. The Welsh Development Agency agreed to consider the problem.

In 1975 W.G. Curtin and Partners were asked to write outlining proposals for a study of the feasibility of developing Bagillt. These were submitted in August 1975.

In 1977 authority was given to proceed on the lines of the 1975 proposals. Very briefly these had suggested :-

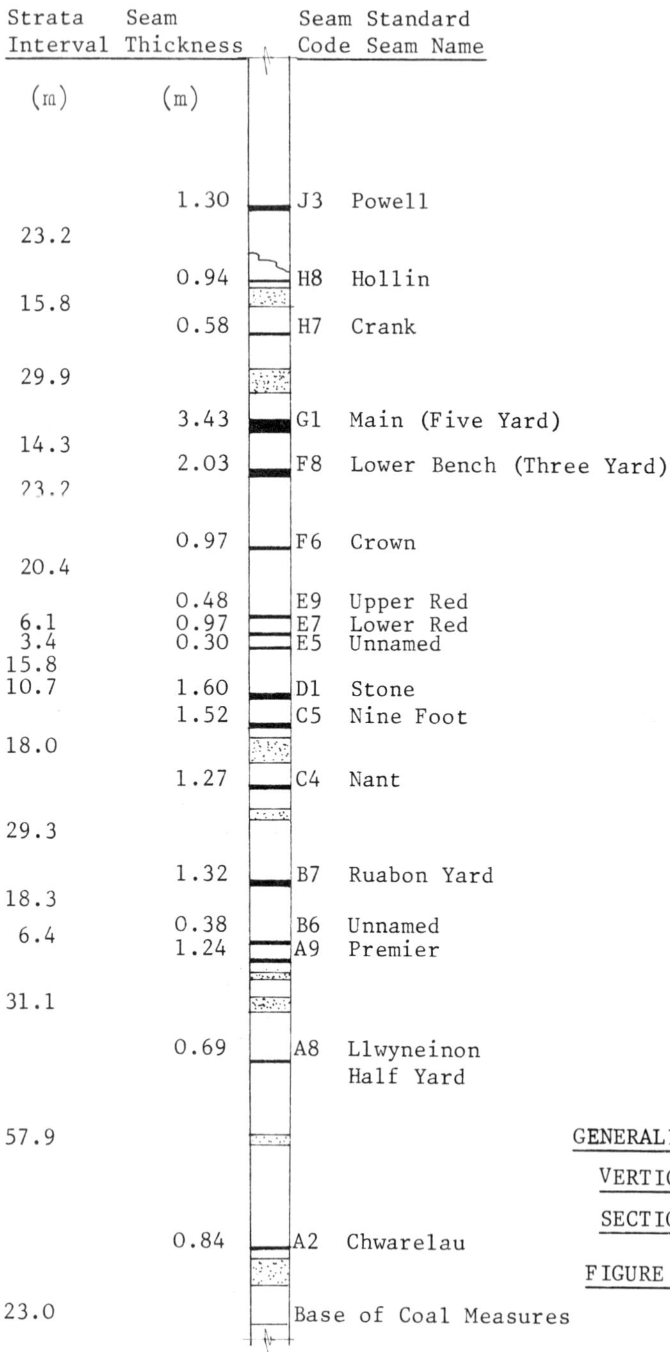

GENERALISED VERTICAL SECTION

FIGURE 2

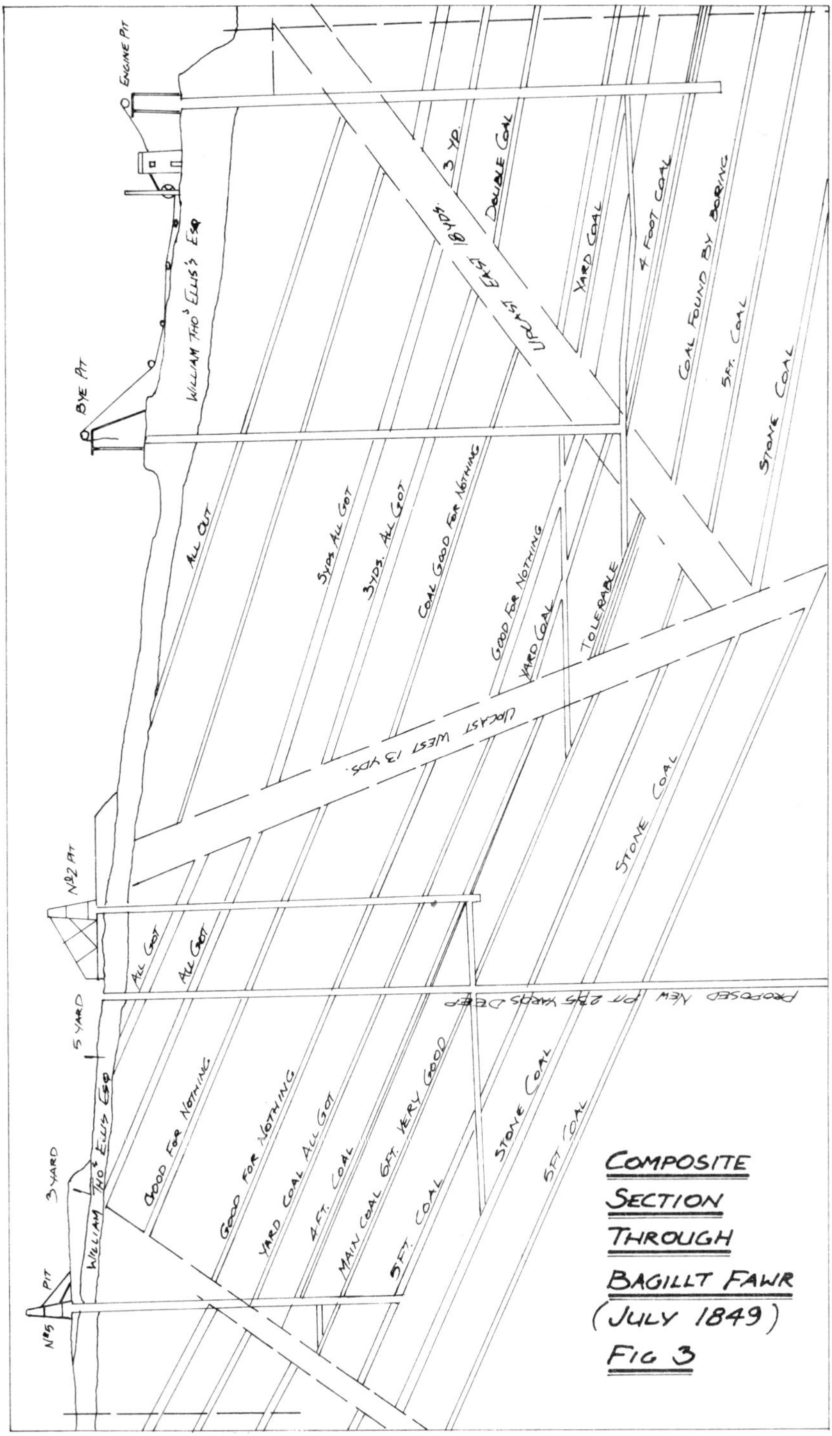

COMPOSITE SECTION THROUGH BAGILLT FAWR (JULY 1849) FIG 3

(1) A search of all likely sources of documentary information.

(2) A drilling programme.

(3) Trial grouting of a pilot site so that likely treatment costs could be extrapolated.

In addition the W.D.A. asked that a number of geophysical techniques be used and compared to assess their value in such an investigation. That particular aspect will not be repeated in this paper.

INITIAL INFORMATION

Figure 4 shows a simplified picture of what was known from the N C B records on shafts and shallow workings. From this the following were anticipated :-

(1) Several seams underlie the area at "shallow" depth. The most important are the 5 yard and 3 yard.

(2) Seams dip to the N.E. at 1 in 3 to 1 in 5.

(3) Seams are likely to outcrop roughly on the long axis of the Borough, i.e. parallel with the High Street, Railway and Coastline.

(4) The 5 yard seam probably outcrops below High Street at the southern end, and further up the hill (to the S.W.) in the northern end.

(5) The underlying 3 yard outcrops on a parallel line to the S.W., the actual displacement depending on the seam interval, the dip and the slope of rockhead.

(6) Such thick seams at shallow depths will have been worked very early on by pillar and stall methods or even bellpits; therefore many shafts could be expected along the backs of the outcrops because of the limitations of haulage and ventilation in primitive workings.

(7) Insufficient shafts are shown in the N.W. and S.E. ends to support the workings. There will be more shafts, pits and adits in the areas of known working.

(8) The central area has shafts where no workings are recorded. Again this is incompatible. There will be workings in all the main seams throughout the district and far more access to them than is recorded.

(9) Faulting will be frequent. Most faults will run parallel with the dip, i.e. roughly S.E/N.W.

DESK STUDY

This aspect of the work threw up some information on

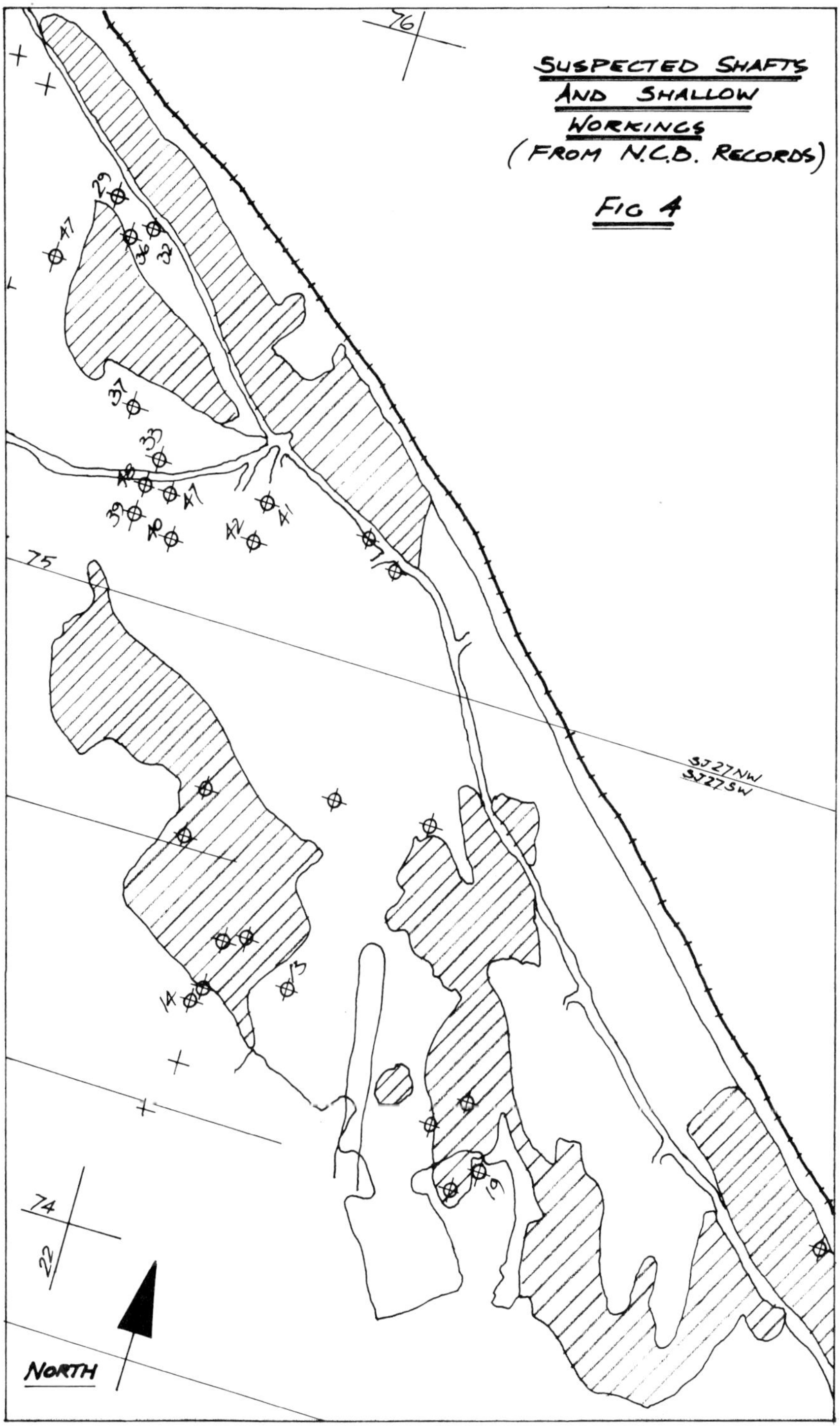
SUSPECTED SHAFTS
AND SHALLOW
WORKINGS
(FROM N.C.B. RECORDS)
FIG 4
76
75
74
22
SJ27NW
SJ27SW
NORTH
29
47
36
32
37
33
40
17
39
42
41
14
13
19

shafts but little on the workings themselves. Many potential sources were approached including :-

(1) N C B Deep Mines.

(2) N C B Opencast Executive.

(3) County Archivist.

(4) Clwyd County Council.

(5) British Rail.

(6) Institute of Geological Sciences, Leeds.

(7) British Museum.

(8) O.S. maps for 1870, 1909, 1970, etc.

(9) Geological Maps and Memoirs.

(10) Safety and Health Commission.

(11) Mine Owners' Legatees, e.g. Mostyn Estates.

(12) Site Investigation Contractors.

(13) Aerial photographs from 1948 and 1959.

(14) The general public via a newspaper appeal.

(15) Delyn Borough Council records of boreholes in the Gadlys Lane area.

From the information obtained a start could be made on drawing up the geology of the area. The ensuing site drilling aimed to build on this to get as accurate a picture as the cost limits would allow. To get meaningful results the Gadlys Lane area was covered in more detail than elsewhere as this was central to tentative redevelopment plans, had most land in the L A ownership and included two streets of council houses (Nant-y-Glyn and Pen-y-Glyn) which were badly cracked.

SITE DRILLING

Figure 5 shows the positions of holes bored by the NKC Partnership and Colcrete Limited. Those by NKC were for the general investigation. Those by Colcrete were for a separate commission to look at specific infil sites along High Street and necessitated Odex bit drilling because of coarse gravel in the drift.

Figure 6 illustrates two of the sections drawn from the drilling logs. They are up Nant-y-Glyn and Pen-y-Glyn and show :-

(1) External cracks in houses.

(2) Seams located.

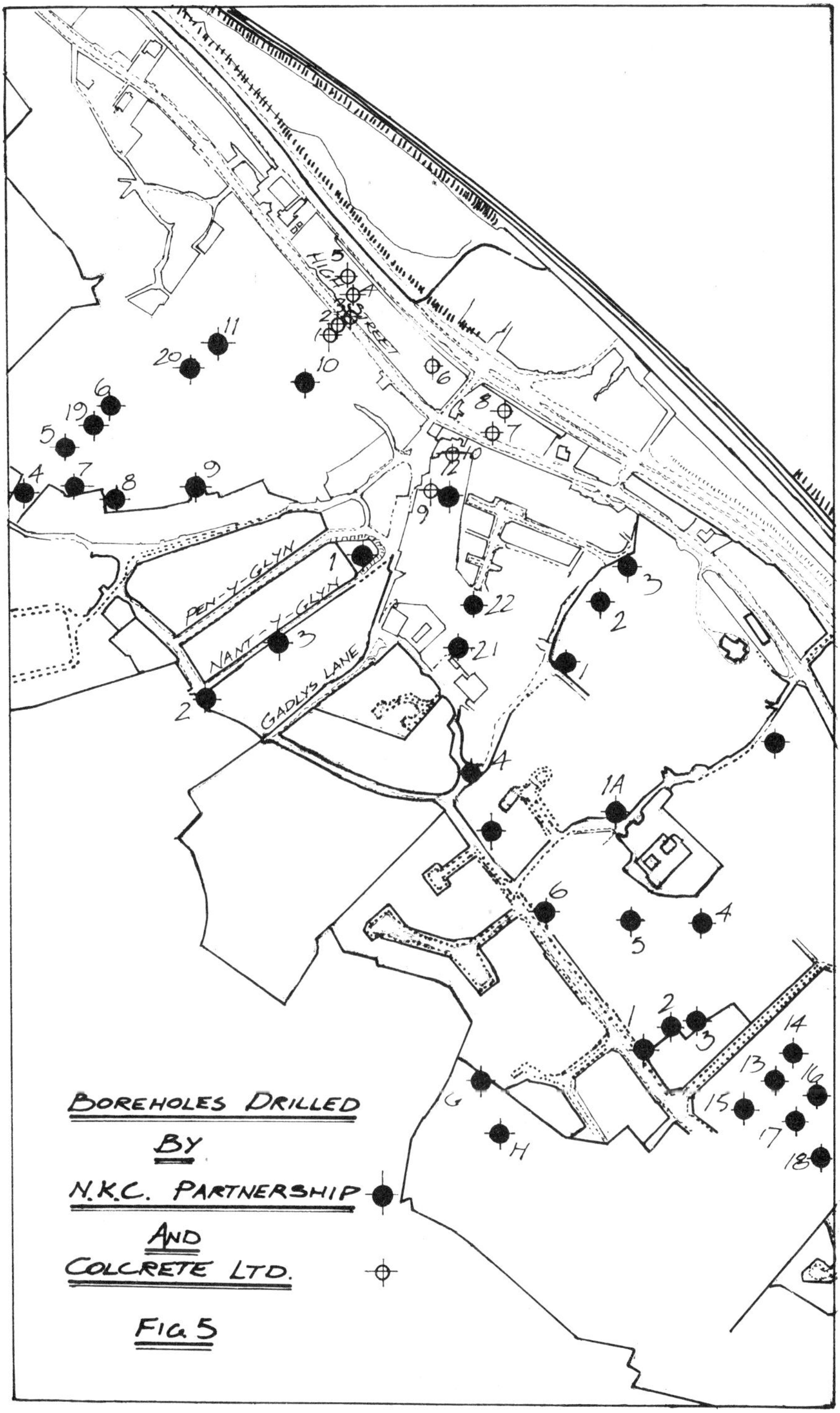
HIGH STREET
PEN-Y-GLYN
NANT-Y-GLYN
GADLYS LANE
BOREHOLES DRILLED
BY
N.K.C. PARTNERSHIP
AND
COLCRETE LTD.

FIG. 5

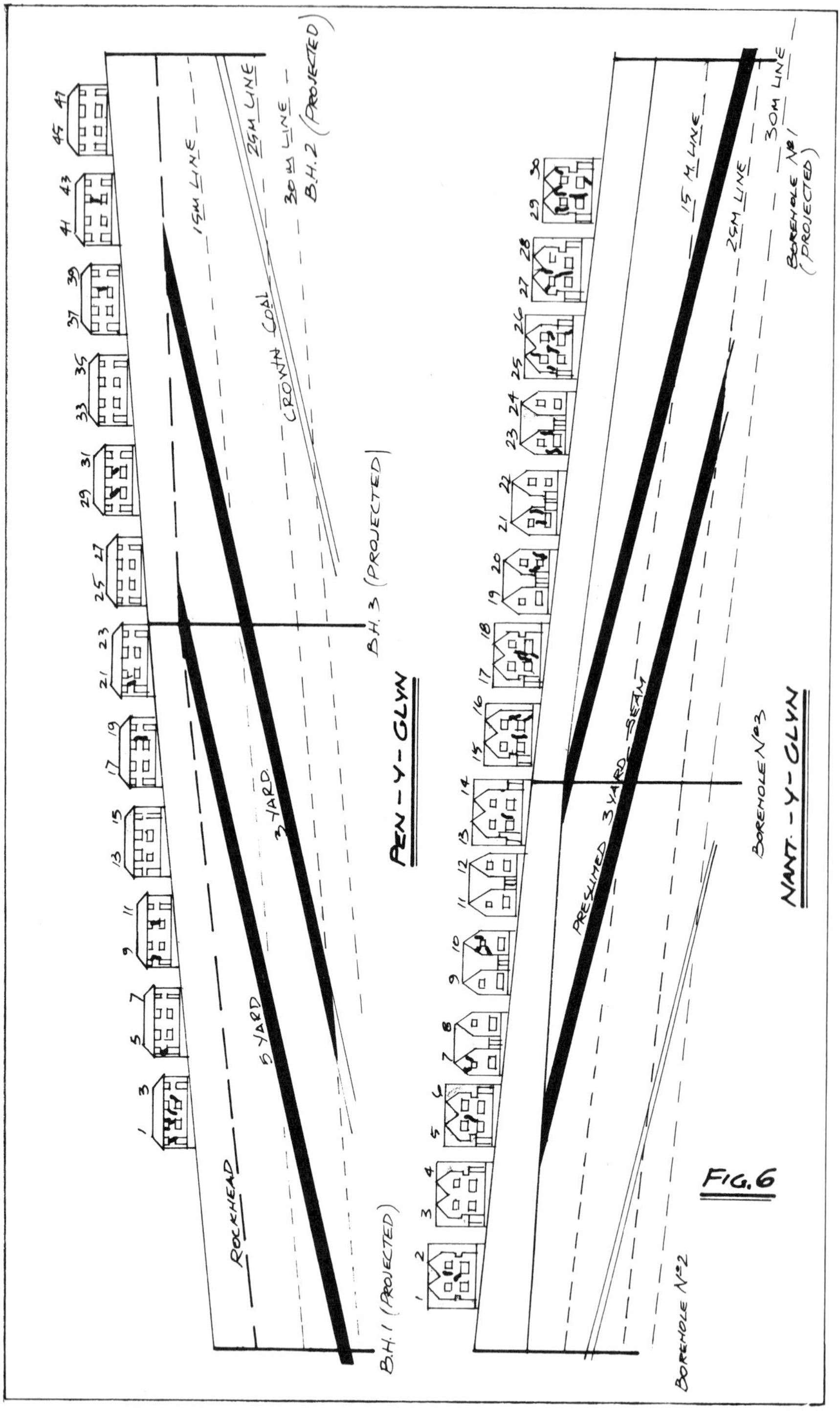
PEN-Y-GLYN
ROCKHEAD
5 YARD
3 YARD
CROWN COAL
15M LINE
29M LINE
30M LINE
B.H.1 (PROJECTED)
B.H.2 (PROJECTED)
B.H.3 (PROJECTED)
NANT-Y-GLYN
PRESUMED 3 YARD SEAM
15 M. LINE
29M LINE
30M LINE
BOREHOLE No 1 (PROJECTED)
BOREHOLE No 2
BOREHOLE No 3

FIG. 6

(3) Depths to which each seam would be considered shallow on current N C B guidelines, i.e. portions of seams which would need to be explored for and treated before new development would normally be sanctioned.

There is a correlation between damage and the workings. The only undamaged property is beyond the outcrop of the 3 yard seam. Extensive damage begins once the 3 yard has sufficient rock cover to give a safe roof for working. It becomes very extensive in the corresponding location on the 5 yard seam. Here the 3 yard is still "shallow". The houses are 40 to 60 years old, but tenants noted movement occurring 12 years previously, well over 100 years after the seams were worked.

The section suggests the possibility of normal seam collapse being compounded by slope instability. Costings did not allow this aspect to be fully investigated but the classic ingredients are there :-

(1) A lubricated, sloping plane of weakness (the worked and water-logged seams).

(2) Lack of weight on the toe because the ground falls away very steeply at the bottom of the streets.

A major slip would not appear likely but, at the very least, cavitation rising to the underside of a wet, sandy, drift will induce lateral as well as vertical movements in the drift in closing the gap.

CONDITION OF SEAMS

Seams were found to have been generally worked, by pillar and stall, and to be largely collapsed and water-logged. The largest identifiable underground void was approximately ½m deep.

SHAFT EXPLORATION

Excavations were carried out to locate the shafts within the Nant-y-Clyn/Pen-y-Clyn estate. Shafts 39 and 40 exist approximately where the records suggest. No. 40 had a small, unreinforced concrete cap just below ground and, below this, is open for at least 160 ft. Excavations around the suspected position of Shaft No. 49 showed no evidence of a shaft or the spoil which could be expected from one.

Other shafts were trenched for on open land. No. 48 was found 22m from the "recorded position". Shafts 26, 27 and 42 were 18m, 20m and 6m from the anticipated positions respectively.

TRIAL GROUTING

Figure 7 shows one area where trial grouting was carried out.

Two holes hit solid coal. Four were assumed to pass through solid coal at the seam horizon although water flush had been lost by then. Twelve holes hit collapsed workings but no significant cavities were encountered. Grout take was quite low but no rogue takes were recorded. It would not be safe to extrapolate likely grout takes from this alone even though general experience in the North Wales area does suggest a low average. For example, on another site 124 holes averaged 1T (dry) each. Individual infil holes, however, did have takes up to 38T.

There is probably a shaft close to this pilot site as suggested by spoil met in the boring. The lie of the spoil suggests it is to the south of hole No. 10.

If open roadways are hit on treating any specific site the average grout take can quickly rise to 5T (dry weight) per hole.

SUMMARY OF FINDINGS

Figure 8 shows part of the derived geology for the north and west areas of the Borough. The south and east end continues the pattern with similar slight offsets in the outcrops due to faulting. From this and the sections, Figure 9 outlines which seams would need treatment in that same area under current guide lines. Table I shows the calculation of critical depth for the top seams.

TABLE I

Seam	Code No.	Effective Coal Thickness	(10 x seam thickness) Plus Drift	Crit. Depth from G.L.
Hollin	H8	1m	10 + 5 = 15m $<$ 30m	15m
5 yd	G1	$3\frac{1}{2}$m	35 + 5 = 40 $>$ 30m	30m
3 yd	F8	2m	20 + 5 = 25 $<$ 30m	25m
Crown	F6	1m	10 + 5 = 15 $<$ 30m	15m

In general the findings are a more detailed confirmation of the initial broad assumptions on what would be found. In terms

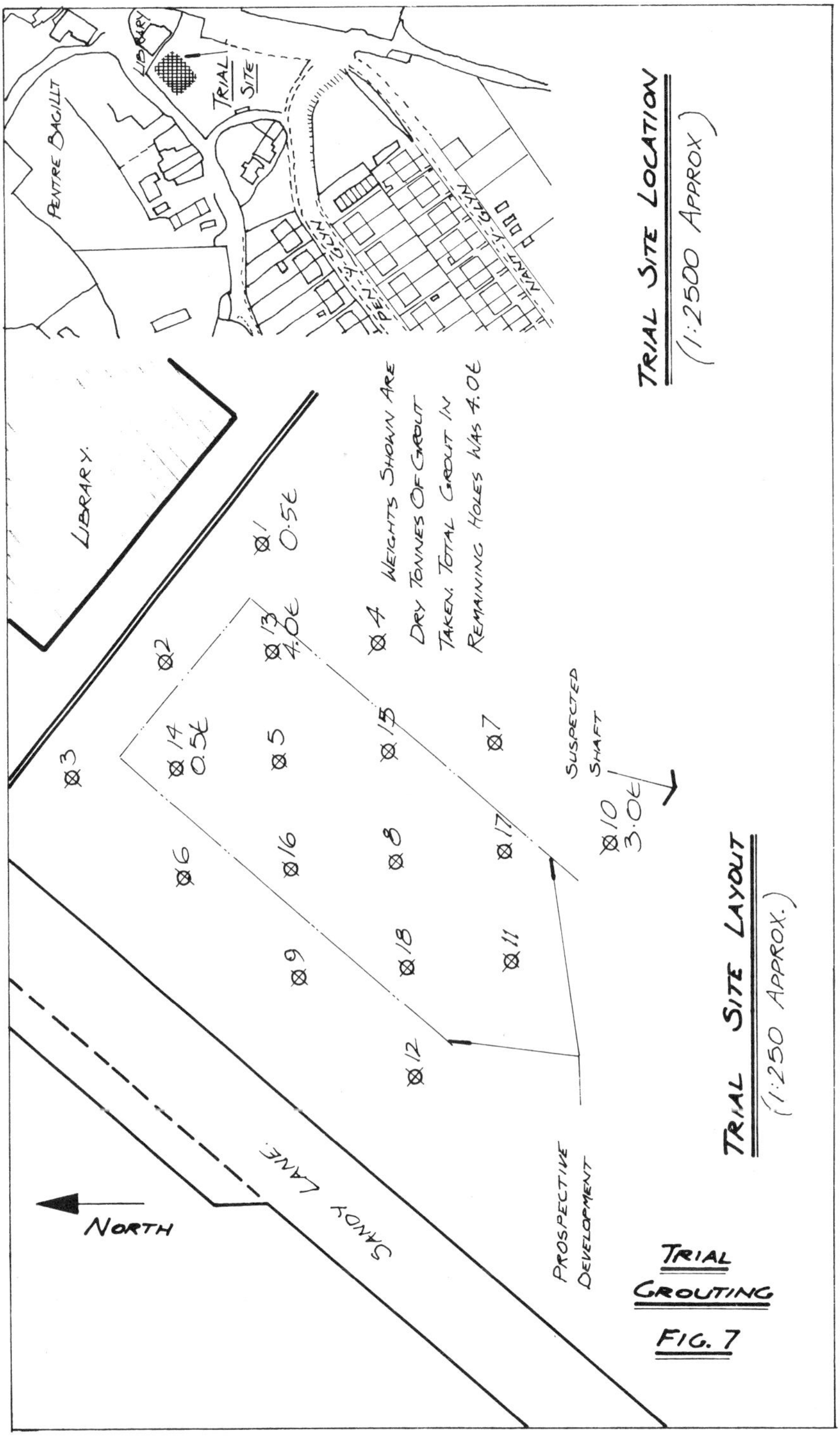

Trial Grouting
Fig. 7

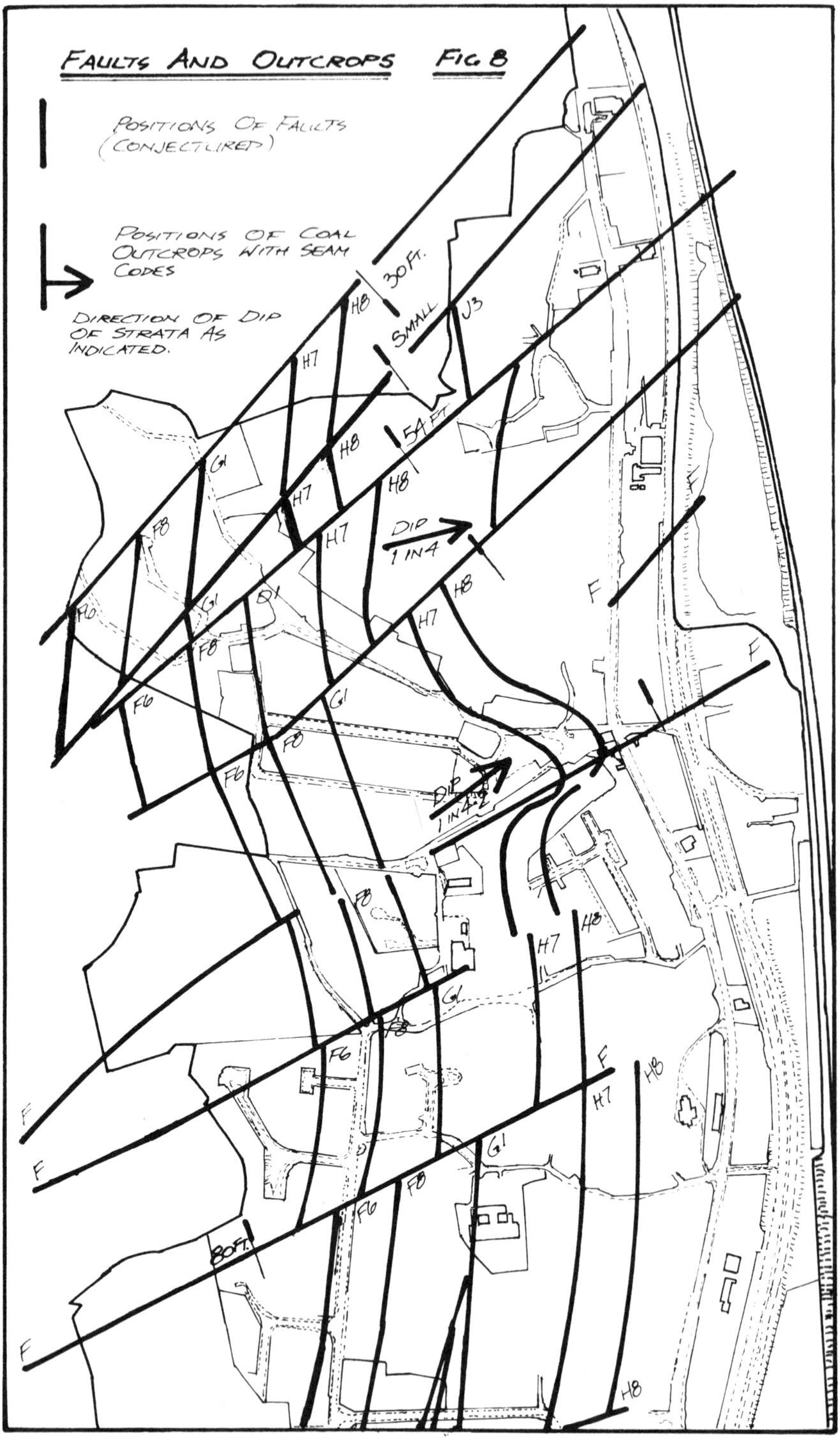
FAULTS AND OUTCROPS FIG. 8
POSITIONS OF FAULTS (CONJECTURED)
POSITIONS OF COAL OUTCROPS WITH SEAM CODES
DIRECTION OF DIP OF STRATA AS INDICATED.
30 FT.
SMALL
54 FT.
DIP 1 IN 4
DIP 1 IN 4.2
80 FT.
H8
H7
J3
G1
F8
F6
F0
F

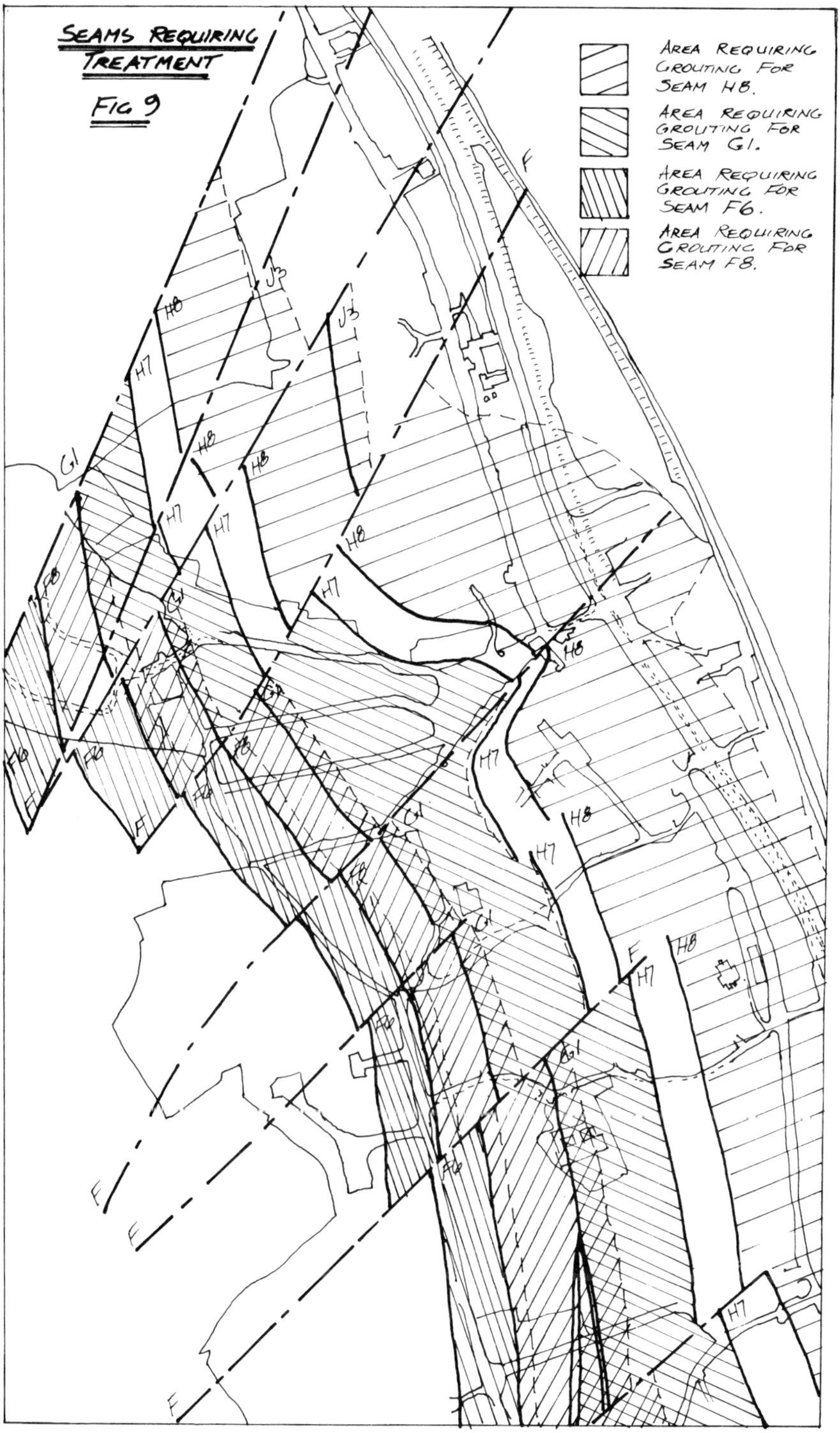
SEAMS REQUIRING TREATMENT
FIG 9
AREA REQUIRING GROUTING FOR SEAM H8.
AREA REQUIRING GROUTING FOR SEAM G1.
AREA REQUIRING GROUTING FOR SEAM F6.
AREA REQUIRING GROUTING FOR SEAM F8.

of damage to existing property the council houses at Nant-y-Glyn/Pen-y-Glyn are the only ones in Bagillt with evidence of any significant damage. Because there was local concern, all residents were publicly invited by individual letter, as part of the investigation, to report any suspicious cracking or other evidence of possible subsidence to the Local Authority. Only 6 properties were referred through for inspection. None had any damage which was not more sensibly explained by other more immediate causes, e.g. tilting of a garage where a neighbour had cut away the adjacent land. One house had sinking in the garden which appeared to be possible evidence of a shaft.

During inspections on the ground a shaft which appeared to be funnelling was found in a children's playground. Several shafts are known to be near existing houses and there is evidence of a shaft on the 1948 aerial survey on land which has subsequently been built over.

COSTS

It is possible now to predict with reasonable accuracy for large areas of Bagillt what seams are likely to be met at what depth and, hence, to judge which seams are "shallow" and need treatment. The same data allows sensible prediction of the depth of grout holes and, hence, the cost of each hole.

It is a little more difficult to go from there to derive the likely total cost of pre-treating any site. A reasonable estimate of grout take per hole might be 4T (dry weight). Cost per unit is patently lower if the plan area of each unit is small and units are close together so that, on a long axis, treatment of one house benefits its neighbour. This would suggest that large bungalows should be discounted and articulated terraces, or courtyard developments or, at the very least, semis should be used. (It is, of course, possible to turn this argument almost on its head and opt for very expensive houses which can swallow the cost of abnormal foundations. Although the area is derelict it is potentially very beautiful with unobstructed views over the Dee Estuary).

For any specific site,costings could be upset if plots had to be sterilised because of shafts and if those shafts had to be made safe. Whilst there is a reasonable starting point here many totally unrecorded shafts can be expected and these will tend to come to light as the ground is opened up and wet spots or spoil found.

At today's prices it would be reasonable to anticipate a figure of up to £2000 above building costs as a likely surcharge. This is a massive burden but is a true reflection of the problem of inheriting a sequence of shallow, worked seams, including one of the thickest in Britain.

GENERAL CONCLUSIONS

(1) Most new houses in Bagillt are likely to need prior ground treatment to give the degree of protection that we, as a society, demand. Then two or three storey structures could be built on light rafts.

(2) The amount of land near the centre not requiring treatment is likely to be confined to a few narrow windows which do not fit into any sensible development plan.

(3) Because of the rapid variation in conditions through topography, faulting, depth of drift, etc. the summary (Figure 9) may only be taken as a guide. All potential developers will need to confirm their local geology by drilling.

(4) Much of the drift is sandy and wet and holes may need lining at least to rockhead. Towards the bottom of Gadlys Lane and on to the High Street normal rotary percussion will be difficult because of the nature of the drift and an underreaming bit will be necessary.

(5) All developers and building inspectors must be aware of the likelihood of finding unrecorded shafts. Many will be unlined and backfilled with spoil, small coal or drift material. There are also unfilled, bricklined or poorly capped shafts in the area.

(6) In treating workings care will be needed in case drainage levels are blocked, e.g. the active Milwr tunnel bringing water from lead mines near Holywell.

(7) Care is also needed in treating shafts which form part of a linked system as treatment of one shaft may cause instability elsewhere, e.g. through removal of hydraulic support.

(8) Cavitation is still working its way to the surface and could affect existing houses. However, the broad lack of evidence of any significant, widespread damage to existing property suggests that there is no action worth taking in general terms to prevent potential damage. Damage should be treated if and when it arises in individual cases.

SUBSIDENCE OVER ABANDONED MINES IN THE PITTSBURGH COALBED

Robert W. Bruhn, Project Engineer
GAI Consultants, Inc., Monroeville, Pennsylvania

Malcolm O. Magnuson, Supervisory Mining Engineer
PMSRC, U.S. Bureau of Mines, Bruceton, Pennsylvania

Richard E. Gray, Vice President and Manager of Technical Services
GAI Consultants, Inc., Monroeville, Pennsylvania

INTRODUCTION

The Pittsburgh Coal, long one of the United States's most valuable mineral resources, was named in the 1840's by H. D. Rogers of the First Geological Survey of Pennsylvania for the city where it was first extensively mined (Eavenson, 1938). The mineable sector of the coalbed, which has an areal extent of 6000 square miles and underlies portions of Pennsylvania, Maryland, Ohio and West Virginia, has yielded more than eight billion tons of bituminous coal over the past 200 years (Ashley, 1938; Eavenson, 1938; Keystone, 1978). This represents almost 35 percent of the total bituminous coal production of the Appalachian region and more than 20 percent of the total coal production of the United States (McCulloch et. al., 1975).

Much of the mining practised decades ago was less efficient than today. Deterioration and failure of coal pillars, roof and floor of these mines results in subsidence of the ground surface. Much of the subsidence is probably unnoticed, particularly in areas of sparse population. However, each year in western Pennsylvania about 45 subsidence incidents associated with abandoned mines in the Pittsburgh Coal are of sufficient importance due to property damage or the creation of hazardous conditions to be reported by newspapers or to be investigated by state agencies concerned with mine subsidence abatement. The repair of damaged homes and property, elimination of hazardous conditions, and implementation of measures to abate on-going subsidence movements associated with these incidents presently cost the taxpayers of

Pennsylvania about $450,000 annually (Molinski, 1977). While this amount is small on a per capita basis, individual subsidence incidents still present a substantial financial and psychological hardship to those whose homes or property have been damaged. As communities spread over areas mined long ago, the number of reported incidents will undoubtedly increase.

This paper discusses subsidence above abandoned mines in the Pittsburgh Coal based on data from 354 documented incidents that took place in western Pennsylvania and West Virginia in the period 1955 to 1976. The data have been compiled from records of the Pennsylvania Office of Resources Management, Ebensburg; the Pennsylvania Mine Subsidence Insurance Fund, Ebensburg and Pittsburgh; the U.S. Bureau of Mines, Bruceton; Greater Pittsburgh newspapers; personal observations; and other sources. Data prior to 1955 are few since no state agencies had the responsibility to deal with subsidence. Past mining practice is related to subsidence and available information regarding the geometry of subsidence features, and their time of occurrence is summarized. Practical inferences are drawn, and several cases are cited where activities by man have apparently hastened the onset of subsidence above abandoned mines.

REGIONAL SETTING

The Pittsburgh Coal lies in a shallow elongate basin that extends from Pittsburgh, Pennsylvania, 200 miles southwestward to Huntington, West Virginia (Figure 1). Overburden consists primarily of alternating nearly flat-lying sequences of shale, claystone, sandstone and limestone of Pennsylvanian and Permian age that reach a cumulative thickness of 1500 feet near the southwest corner of Pennsylvania. The topography is mature and the rock strata are commonly mantled by residual soil and colluvium whose combined thickness is typically 10 to 15 feet.

MINING

Nearly all of the Pittsburgh Coal has been mined by room and pillar methods. Mines have ranged from little more than burrows dug into the hillside to extensive operations occupying thousands of acres.

The ready access to the coal offered by outcrops in the northern part of the coalfield led to the early development of western Pennsylvania as a center of mining and industry.

The first reported mining dates to the late 1750's when British troops stationed at Fort Pitt, near what is now central Pittsburgh, mined coal from the hillside overlooking the fort. As the demand for coal increased, mining gradually spread away from the city to where the coal is several hundred feet below ground surface.

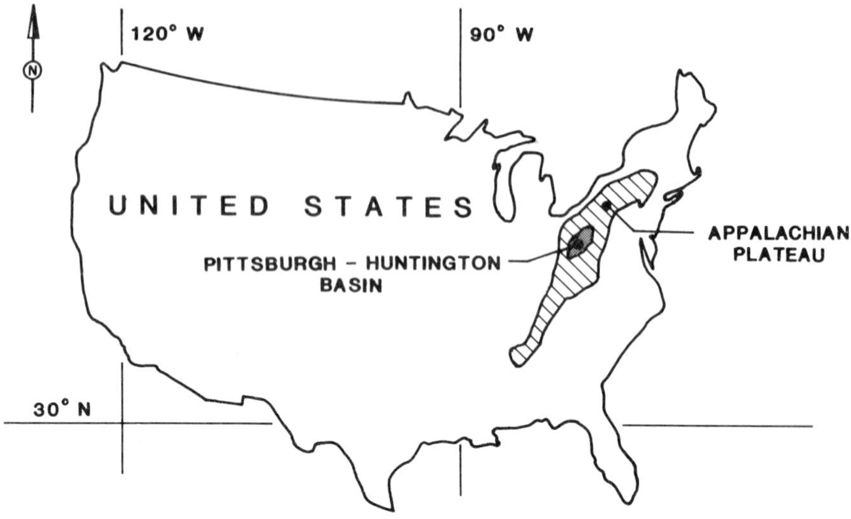

FIGURE 1 - LOCATION OF THE PITTSBURGH - HUNTINGTON BASIN AND THE APPALACHIAN PLATEAU

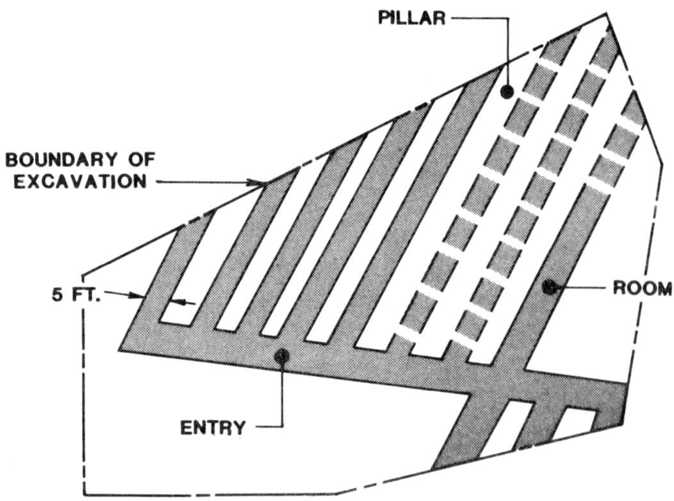

FIGURE 2 - PLAN OF UNDERGROUND MINE NEAR CENTRAL PITTSBURGH OPERATED SOMETIME PRIOR TO 1870. EXPOSED IN EXCAVATION.

During the 18th and 19th centuries, mines were small, hand operations under shallow cover. Coal was cheap and the spacing, size and regularity of pillars were somewhat arbitrary. Coal pillars were left in place as a matter of convenience and safety to the miners. Increased production by the mid- to late- 19th century brought mechanization and ventilation requirements to mines that necessitated a systematic arrangement of pillars, but still resulted in considerable coal being left underground. An excavation for a new building located two miles east of Pittsburgh's central business district exposed a mine operated some time prior to 1870. The open excavation revealed an orderly pattern of pillars and openings, with rooms five to six feet wide alternating with pillars of equal width. Such a pattern did not permit extraction much in excess of 50 percent (Figure 2).

At some time in the 1880's and 1890's, extraction was increased by widening rooms to 20 to 24 feet and pillars to 10 to 18 feet, increasing coal recovery by 10 to 20 percent. These dimensions were characteristic until the advent of the block system of mining several decades later.

In the latter part of the 19th century, total extraction mining was initiated to achieve greater production of the coal that was becoming increasingly valued for its coking properties by the steel industry and was first implemented in partial extraction mines of the day. The distinction from partial extraction mines was that the long narrow pillars left between rooms during first mining were now extracted in a second stage of mining. Subsidence of the ground surface in a properly executed operation took place contemporaneously with pillar extraction.

By the 1920's, block systems of mining came into favor, wherein square pillars 60 to 100 feet on a side were separated by narrow rooms and entries, reducing roof deterioration, roof falls and support problems during pillar extraction (Paul and Plein, 1935). In the period 1948 to 1952, most remaining mines of the old pattern were converted to the block system as continuous mining machines were introduced on a large scale (Black, 1977).

Shortcomings of Total Extraction Methods

Although present total extraction methods using the block system are not always successful in recovering coal pillars, they are considerably more successful than the long narrow pillar systems of several decades ago.

Figure 3 depicts a portion of an early total extraction mine in western Pennsylvania operated just after the turn of the century. Pillar extraction was conducted between 1909 and 1910. In the portion of the mine shown, about 78 percent of the

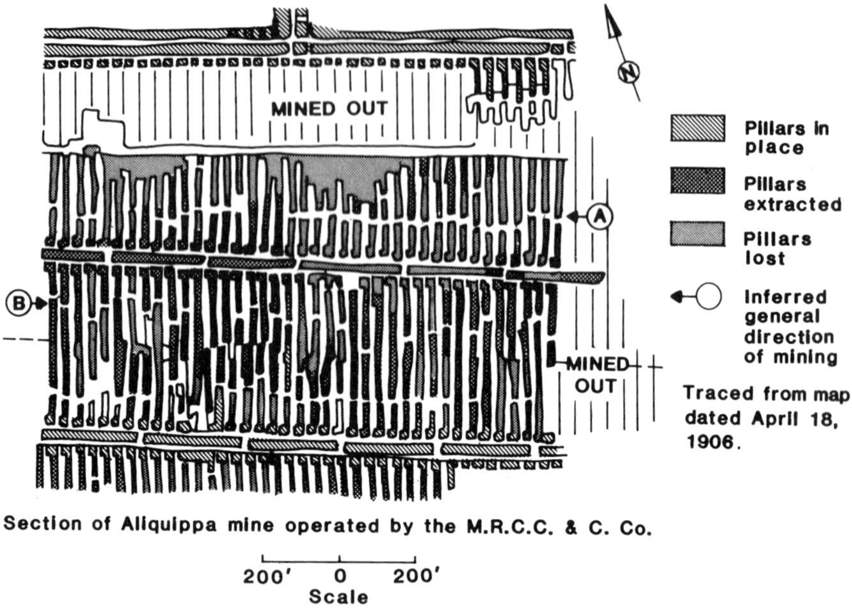

FIGURE 3 – AN EARLY TOTAL EXTRACTION MINE DEVELOPED USING LONG PILLAR LINES, ALLEGHENY COUNTY, PENNSYLVANIA

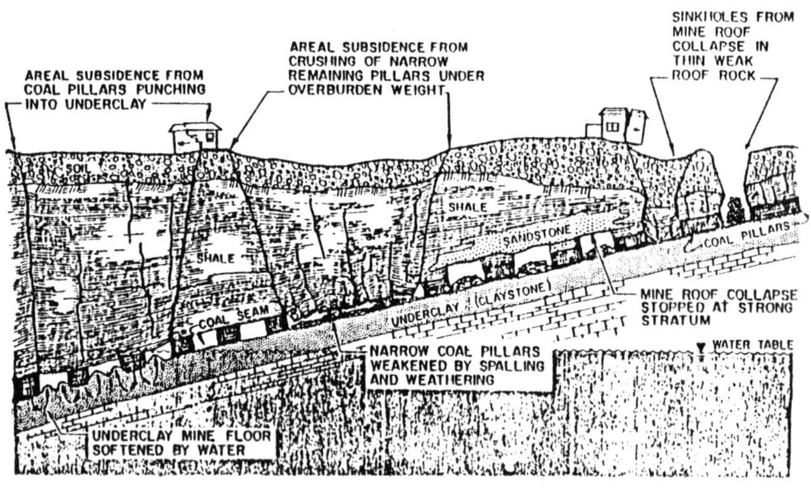

FIGURE 4 – MODES OF SUBSIDENCE

coal was extracted, including some pillars; and about 22 percent was lost. Roof falls, squeezes, or other complications brought about by factors no longer evident may have been responsible for the inability to extract the pillars. As a result, the potential for subsidence remains a real possibility above this mine many years after its abandonment.

Such old mines which sometimes lie at depths of 25 feet or less are gradually succumbing to gravity, resulting in an irregular, but continuing series of subsidence incidents scattered over several hundred square miles of mined-out coalbed.

TYPES OF SUBSIDENCE

Subsidence features above abandoned mines in the Pittsburgh Coal Region are categorized as either sinkholes or troughs (Figure 4).

A sinkhole is a depression in the ground surface that occurs from collapse of the overburden into a mine opening (a room or an entry). Boundaries between ground surface and walls of the sinkhole are often abrupt, and because diameter generally increases with depth, the sinkhole in profile may resemble an open bottle. Erosion of soil at the sinkhole's periphery may increase the diameter near ground surface to create an hourglass profile.

A trough is a shallow, often broad, dish-shaped depression that develops when the overburden sags downward into a mine opening in response to roof collapse, the crushing of mine pillars or the punching of pillars into the mine floor.

Although concrete or wood covers over abandoned mine shafts sometimes collapse with dramatic results, these collapses are considered structural failures rather than subsidence and are not discussed further.

Table I summarizes data from 354 incidents of subsidence above abandoned mines in the Pittsburgh Coal Region of western Pennsylvania that took place in the period 1955 to 1976. The data are drawn from subsidence cases occurring in a populated region that posed sufficient hazard or caused sufficient damage to structures to be reported to state authorities. Data distributions might change if all subsidence cases were included. Although complete details are not available for each subsidence incident, enough information is known to establish a general impression of the nature of subsidence over abandoned mines in the region.

INFERENCES DRAWN FROM SUBSIDENCE CASES

Several concepts potentially important to engineers,

TABLE I

CHARACTERISTICS OF SUBSIDENCE FEATURES ASSOCIATED WITH ABANDONED MINES IN THE PITTSBURGH COAL REGION*

	Sinkholes	Troughs
1. Mode of development	Roof collapse	Roof collapse, Pillar crushing, Pillar punching
2. Geometry		
a. Characteristic Profile:	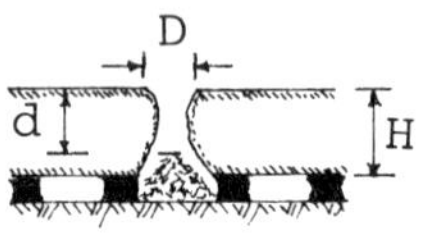	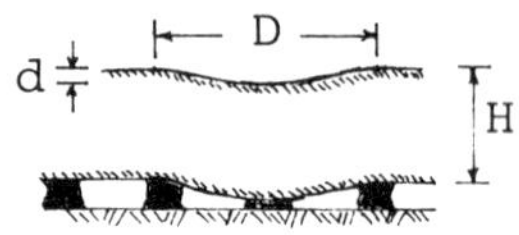
b. Diameter, D:		
Range	1.5 - 45 ft[°]	30 - 1600 ft[Δ]
Mean	9 ft [170][+]	300 ft [15]
80%	≤ 12 ft	≤ 300 ft
c. Depth, d:		
Range	1.5 - 45 ft	0.2 - 3 ft
Mean	14 ft [170]	<3 ft [15]
80%	≤ 18 ft	≤ 3 ft
3. Occurrence		
a. Overburden thickness, H:		
Range	2.5 - 150 ft	35 - 450 ft
Mean	45 ft [212]	120 ft [15]
80%	≤ 70 ft	≤ 150 ft
b. Distance from outcrop	75% of occurrences within 500 ft[X]	No apparent relationship

[*] Height of typical mine opening is 5.5 to 6 feet.

[+] Number in brackets denotes size of sample.

[°] Sinkholes are roughly circular in plan, i.e., $D_{max} = D_{min}$.

[Δ] Nominal trough diameter $D = (D_{max} + D_{min})/2$.

D_{max}/D_{min}: Range = 1 - 3.2; Mean = 1.45 [15].

Relationship of trough diameter D to overburden thickness H,

D/H: Range = 0.4 - 4.7; Mean = 2.0 [15]; 80% lie between 0.5 and 2.5; and 50% between 1 and 2.

[X] Hillslope inclinations commonly range between 10 and 20 degrees.

developers and others concerned with the planning, design and construction of buildings above abandoned mines in the Pittsburgh Coal Region are apparent from a review of the several hundred subsidence cases documented thus far:

1 *Unless total extraction has been achieved, there is no identifiable height above an abandoned mine that ensures a site total freedom from subsidence, nor necessarily a reduction in severity of damage. An increased interval above mine level, however, seems generally associated with a reduced frequency of subsidence.*

Subsidence has been observed above abandoned mines where overburden ranges in thickness from a few feet to as much as 450 feet. Provided appropriate conditions exist at mine level, there is little reason to expect, based on present concepts of subsidence mechanics, that subsidence by pillar crushing or pillar punching will not some day be reported above an abandoned mine where the overburden thickness is significantly greater than 450 feet. The 450-feet interval is presumed to represent no more than a temporary upper limit.

A subsidence incident documented at a site where cover ranged from 230 to 450 feet is noteworthy in that it covered 40 acres and damaged more than 40 structures. It represents the single most extensive subsidence incident recorded above an abandoned mine in the Pittsburgh Coal Region and supports the view that a substantial thickness of overburden does not necessarily ensure safety from subsidence.

Reported incidents of subsidence and damaged structures tend to be fewer as overburden thickness increases (Figures 5 and 6), supporting the contention that the likelihood of subsidence tends to reduce with increased interval above mine level. Sites located 60 or more feet above mine level avoid the majority of sinkholes, which constitute about 95 percent of all reported subsidence incidents and are responsible for more than half the reported incidents of damage to structures.

2 *Unless total extraction has been achieved, there is no identifiable time after mining when the threat of subsidence is clearly past, and there appears little assurance that subsidence will be limited to a single episode.*

At some time after mining, complete collapse of abandoned entries and rooms is to be expected as a result of natural causes and activities by man. Until that point is reached, the ground surface overlying a mine may experience a variable frequency of subsidence incidents: perhaps only a few incidents immediately after mining, while the rock surrounding the openings is still relatively sound; increasing

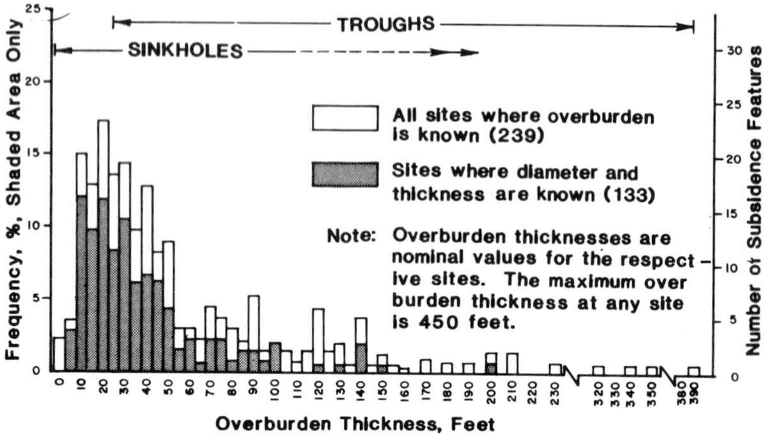

FIGURE 5 – OVERBURDEN THICKNESS AT SUBSIDENCE SITES

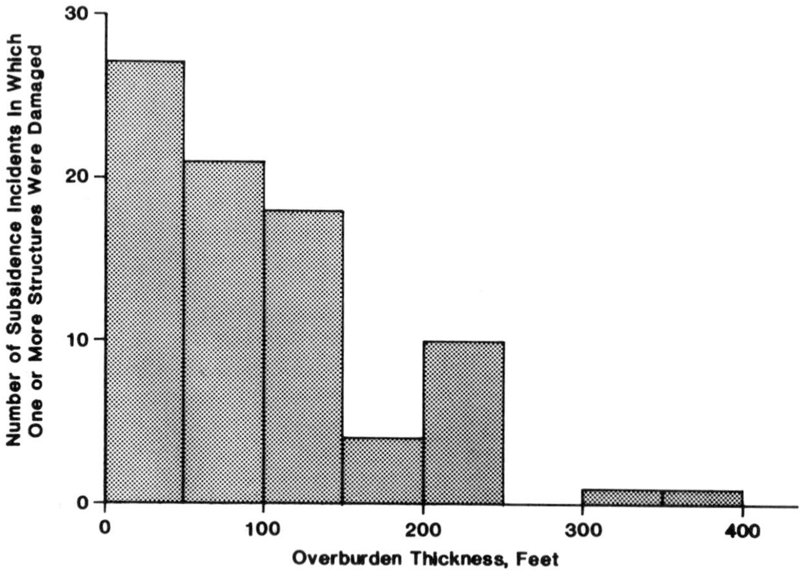

FIGURE 6 – OVERBURDEN THICKNESS AT SUBSIDENCE SITES WHERE ONE OR MORE STRUCTURES HAVE BEEN DAMAGED numbers of incidents for an extended period of time as progressive deterioration and failure of the rock surrounding the openings becomes more pronounced; and later, a diminishing number of incidents as the void spaces at mine level become fewer and fewer. This sequence is shown conceptually in Figure 7.

The several hundred subsidence incidents documented in the Pittsburgh Coal Region over the past twenty years are associated primarily with mining that had been conducted 30 to 200 years previously (Figure 8). Inasmuch as drilling investigations to mine level typically reveal open mine voids in locations surrounding the subsided area, it seems probable these incidents correspond to some point rather early in interval AB of Figure 7. Many hundreds, if not thousands, of years must probably elapse before the complete collapse condition represented by Point B is attained.

This implies that the possibility of future subsidence at a site cannot be ruled out merely because subsidence has not been recognized in the first 50 or 100 years after mining. If abandoned mine openings beneath a site have not been designed for long-term stability, the potential for subsidence remains until the openings collapse, or until they are stabilized by backfilling, grout columns, or some other means (Gray et al, 1974). Precisely when collapse might take place in the absence of stabilization is not predictable.

Even after subsidence has taken place at a particular site the possibility of future subsidence may remain. Figure 9 shows a community that has experienced subsidence on at least four occasions over the past fifty years. Subsidence appears first to have taken place in the 1920's while mining was in progress. After abandonment of the mine, a long period of quiescence followed until 1964, when three homes were damaged by subsidence in an area measuring 300 by 400 feet that overlapped part of the 1921 subsidence area. This was followed in mid-1972 by subsidence of seven homes 400 feet east of the 1964-area and in late 1972 by subsidence of six more homes in an area that partially coincided with the southerly part of the 1964-area. Subsidence stopped for the next few months but resumed in April of 1973, affecting an area 450 by 800 feet and damaging 14 more homes. At nearly the same time, two other homes 1,100 feet east were similarly damaged by subsidence.

Multiple episodes of subsidence have been documented at thirty-three other sites in the region, each episode representing a report of damage to a structure or the appearance of a hazardous condition. Figure 10 shows data from nineteen of the sites. Data from the remainder are nearly identical. Multiple subsidence episodes of lesser significance may have

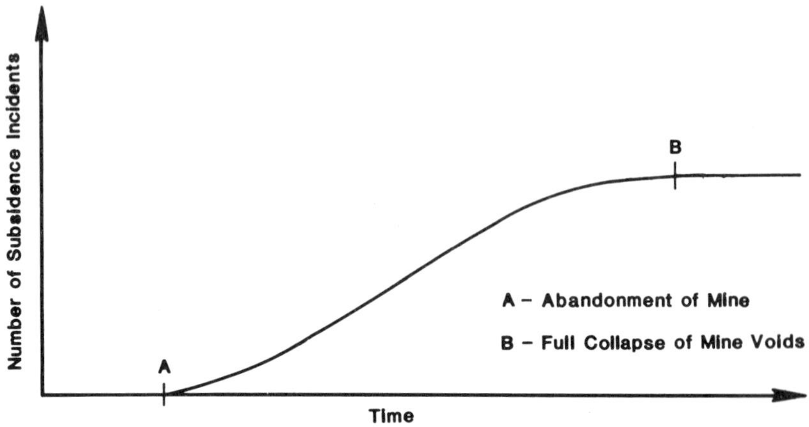

FIGURE 7– CONCEPTUAL REPRESENTATION OF THE POSSIBLE INCIDENCE OF SUBSIDENCE ABOVE AN ABANDONED MINE CONTAINING OPEN VOIDS

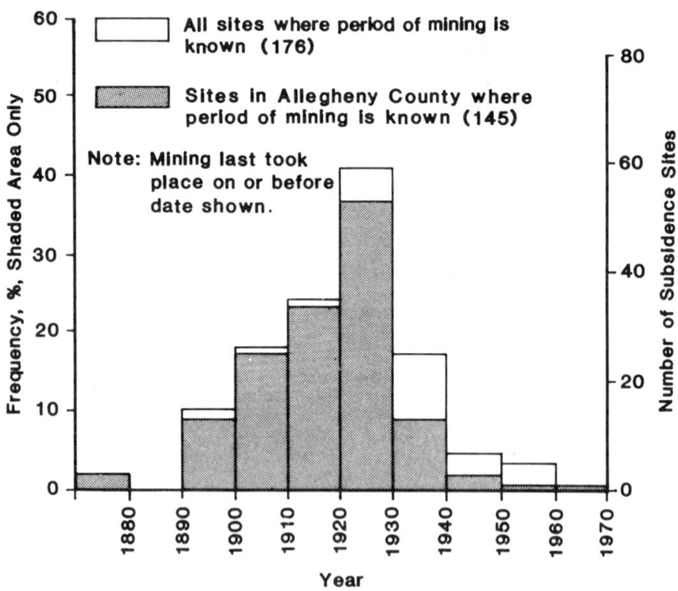

FIGURE 8 – DISTRIBUTION OF PERIODS OF MINING BENEATH SITES THAT LATER EXPERIENCED SUBSIDENCE

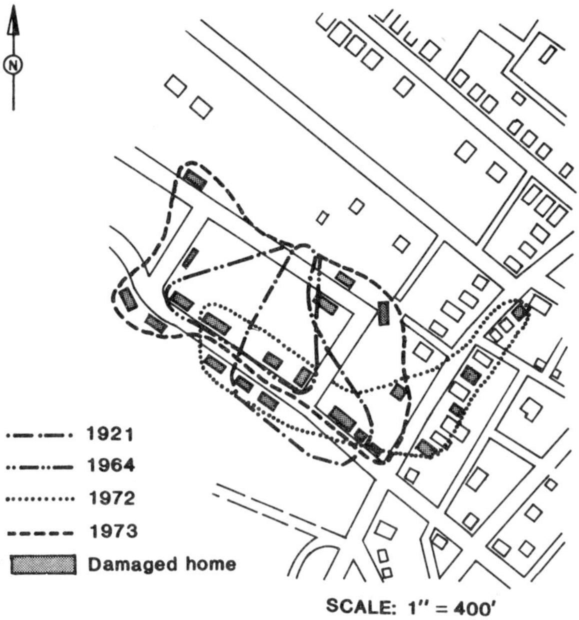

FIGURE 9 – COMMUNITY THAT HAS EXPERIENCED MULTIPLE INCIDENTS OF ABANDONED MINE RELATED SUBSIDENCE

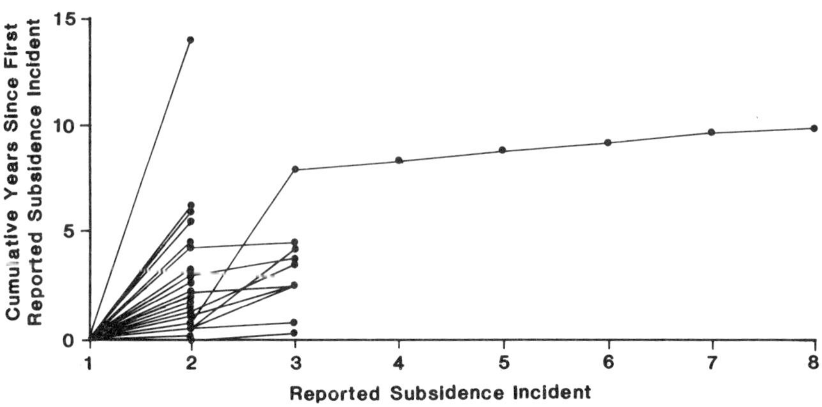

FIGURE 10 – INTERVALS OF RECURRENT SUBSIDENCE AT 19 SITES OVERLYING ABANDONED MINES IN THE PITTSBURGH COAL REGION

taken place at these or other sites and not been recorded.

Inasmuch as multiple episodes of subsidence above abandoned mines are not uncommon, a subsidence occurrence at a site development is perhaps prudently interpreted as one of a continuing series of subsidence episodes that may have begun sometime earlier and will continue sporadically until such time in the future as collapse of underground mine openings is complete. It should be evident that site surveillance programs of a few months duration or, in fact, indefinite duration, cannot provide definitive evidence that a site overlying a mine with open voids will experience no future subsidence.

3 *Although the specific cause of subsidence above an abandoned mine is not often determined, it is apparent that in some cases activities by man have hastened the onset of subsidence if not initiated it.*

The chief concern following subsidence is the protection of lives and property and the mitigation of hazardous conditions. Detailed subsurface investigations to ascertain the factors responsible for subsidence are generally conducted only in rare cases. In the absence of investigatory information, one commonly presumes that natural deterioration of mine roof, pillars or floor are responsible for the initiation of subsidence, and in many cases this is probably correct. However, it has become evident in several cases that activities by man following mining sometimes initiate subsidence or at least hasten the onset.

(a) In one small community, sewage is disposed of by flushing it into an old mine through wells installed from individual homes. When a new sewage disposal hole intercepted a coal pillar at mine level, the contractor chose to split the pillar using explosives, rather than drill another hole. Soon after he did so the ground surface subsided, damaging three homes located 220 feet above the mine.

(b) The increased price of coal prompted a mining company to dewater one of its old mines to extract coal that had been too costly to mine earlier. While dewatering was underway, several homes located 200 feet above an adjoining abandoned mine subsided. As a consequence, the dewatering program was discontinued and the proposed mining program was aborted.

(c) Subsidence at another site is attributed to pillar extraction in an adjoining portion of the mine. The site had been undermined earlier, but the mined-out area was not of critical dimensions and therefore experienced additional subsidence when adjacent mining produced a wider mined-out area. Several homes were damaged.

(d) Urbanization has accelerated sinkhole development in certain areas. The terrain around Pittsburgh is hilly and residential lots are commonly created by cutting benches into the hillslopes. Ten or twenty feet of material may be removed at the rear of a lot. This not only decreases the thickness of overburden above the mine and lessens the time required for caving to reach ground surface, but also sometimes permits caving to reach ground surface in areas where it could not before. Excavation, too, sometimes alters site drainage and increases percolation of surface water to mine level. Indiscriminate discharge of water from rain gutter downspouts has been linked to several cases of sinkhole development. In addition, backyard rubbish fires at outcrop have sometimes spread to abandoned mines, burning away remaining pillars and subsiding ground surface and homes.

The cases presented above indicate that activities by man subsequent to first mining can sometimes be the immediate cause of subsidence, and can hasten what natural deterioration would require a much longer time to accomplish. The obvious implication is that all development and mining activities at a prospective site should be evaluated for possible impacts on the stability of abandoned mines and on structures at ground surface that might be subjected to subsidence if the abandoned mines should be adversely affected.

CONCLUSIONS

Most areas that experience subsidence above abandoned mines in the Pittsburgh Coal Region are rather small compared to the total mined area. Small localized details at mine level - perhaps a fractured roof zone at an intersection, an inordinate thickness of underclay in the mine floor or one or two weak coal pillars, probably dictate where subsidence will take place and when. Even a very carefully executed subsurface investigation cannot ordinarily delineate such small-scale potentially critical mine conditions that might later lead to subsidence. There are no means available as yet to predict exactly when or where subsidence might take place, and so subsidence must generally be expected anywhere in the region unless it can be proved that the area has not been mined, that long-term pillar support has been provided or that the mine voids are fully collapsed.

While it is doubtful that one will ever be able to predict with precision the time of occurrence or the extent of subsidence at a particular site, it may be possible to establish correlations between subsidence damage to structures, thickness of overburden and other parameters. The ability to make such determinations would materially assist the engineering profession in more economic design of structures above mined-out areas and in establishing the most economical subsidence prevention measures for particular site situations. Development of correlations and criteria depend on more refined data than have been collected todate. The collection of such data is encouraged.

ACKNOWLEDGEMENTS

The work described herein is abstracted from a study conducted for the U.S. Bureau of Mines under Contract J0366047 entitled, "Study and Analysis of Surface Subsidence Over the Mined Pittsburgh Coalbed", part of the Bureau's program of Environmental Protection of Surface Areas Near Underground Mining (Magnuson and Cox, 1975). The final report is available for reference at Bureau of Mines libraries and is available from NTIS (PB281522/AS). The views and conclusions are those of the authors and should not be interpreted as necessarily representing the official policies or recommendations of the Bureau of Mines.

REFERENCES

G.H. Ashley - Production and reserves of the Pittsburgh Coalbed, The American Institute of Mining and Metallurgical Engineers Transactions 30 (1938) 56-64.

W. Black - Personal Communication, (1977).

H.N. Eavenson - The Pittsburgh Coalbed: Its Early History and Development, The American Institute of Mining and Metallurgical Engineers, New York (1938).

R.E. Gray, J.C. Gamble, R.J. McLaren and D.J. Rogers - State of the Art of Subsidence Control, Report ARC-73-111-2550, prepared by General Analytics, Inc., for the Appalachian Regional Commission, Washington, D.C. (1974), (NTIS Number PB24275/A5).

Keystone Coal Industry Manual, McGraw Hill, Inc., New York, (1978).

C.M. McCulloch, W.P. Diamond, W.P. Bench and M. Deul - Selected Geologic Factors Affecting Mining of the Pittsburgh Coalbed, U.S. Bureau of Mines Report of Investigations 8093, Washington, D.C. (1975).

M.O. Magnuson and R. Cox - Environmental protection of surface areas near underground mining sites, Coal Age 80 (1975) 135-138.

A.E. Molinski - Personal Communication, (1977).

W.J. Paul and L.N. Plein - Methods of development and pillar extraction in mining the Pittsburgh Coalbed in Pennsylvania, West Virginia and Ohio, U.S. Bureau of Mines Information Circular 6872, Washington, D.C. (1935).

THE EFFECT OF OLD UNDERGROUND WORKINGS ON THE DEFORMATION OF A BUILDING

E.A. Sorochan, doct of tech sciences, professor
A.V. Vronsky, cand of tech sciences
G.M. Troitzky, senior engineer

Research Institute of Bases and Underground Structures
Moscow, U.S.S.R

A 5-storey block of flats was erected at a site situated near to the underground workings of a coal deposit. The workings at a depth of 28-30 m dated from 1948. The absolute level of the bed of the seam is 177.0 m and its thickness is 1.6-3.6 m. At the site, mining subsidence was believed to be ended so the building was erected without special structural measures. The building was situated on the right bank of a small valley, about 80 m away from its water course (absolute level 195.0 m). The whole site slopes towards the valley. At a distance of 100 - 150 m up the slope some depressions were found which collected seasonal water.

The surface of the site was levelled with cuts of 2 to 3 m and fills of up to 2.5 m. The absolute level is in the range 207-208 m. In the NE direction, the surface falls at a slope angle of 10°-12°. The surface displays cuts, undulations, artificial excavations and saucer-shaped depressions. Some of the depressions are believed to result from slope displacement and mining subsidence. In 1970-1971 an embankment was constructed at the foot of the slope. The height of the embankment at the SE part of the site was 1 to 2 m, increasing to 3 m towards the NW.

The afore-mentioned building has an incomplete internal structural frame. The outside walls consist of large panels. The overall dimensions of the building are 94.0 x 11.6 m (Figures 1a, 1b).

The strip foundations for the outside walls and the staircases were assembled from prefabricated concrete elements 1.2 to 1.4 m wide. The loading on the walls at the base level is 208-268 kN/m, and 315kN/m on the staircases. The columns of the framework are founded on separate pillar-type foundations 2.4 x 2.4 m at the base and support loads of 1140 kN. The depth to the underside of the foundation bases is in the range 1.38-1.67 m measured from the level of the undisturbed soil, and 2.17-4.08 m from the embankment level. The basement is at -2.4 m, measured from groundfloor level. The actual contact pressure of the foundation is 0.19-0.22 MPa, compared to an admissible value of 0.23 MPa according to Constructional Design Norms.

The underground structures were constructed in the summer of 1970. The erection of the superstructure began in January and ended in March 1971. In April, deformation of the structure was noticed and instrumental observation of the settlement of the building was started. At the beginning of May, 7 cm of settlement was measured at position A-1. The building tilted towards the valley. Vertical and inclined cracks appeared on the surfaces of the plinths, on walls between stations A-20 and A-22 and on crosswalls between rows 19 and 20. A horizontal joint between the plinth and the wall elements opened there. The cracks had a tendency to narrow towards the top of the building. In mid-May the settlements reached 10.5 cm and by the end of the month had increased to 19.6 cm. The NE plinth was curved, moving 6 cm towards the valley.

Special geological investigations carried out during June to July 1971 showed that the NW part of the building was being affected by mining subsidence. Filling of the underground workings and strengthening of the structure were planned. The works were completed in 1972. As a result the settlements stabilized and no new damage to the structure was found. In December repair work began. The geodetic observations were continued to 1975. During that time additional settlement along the SW side of the building increased by 15-30 mm, and along the NE side by 14-68 mm. No additional structural damage took place.

In April 1978 new cracks appeared in the structure. Observations were renewed and additional soil investigations were made. The settlements continued. New lateral displacements of the building towards the valley were recorded. The results of those investigations indicated a significant deterioration in the physical properties of the soils in the NW part of the site and very complex hydrogeological conditions. Due to the continuing deformation, the building was judged to be unsuitable for further repair.

GEOLOGICAL CONDITIONS

A geological section through the site is presented in Figure 1a. The strata are indicated as follows:-

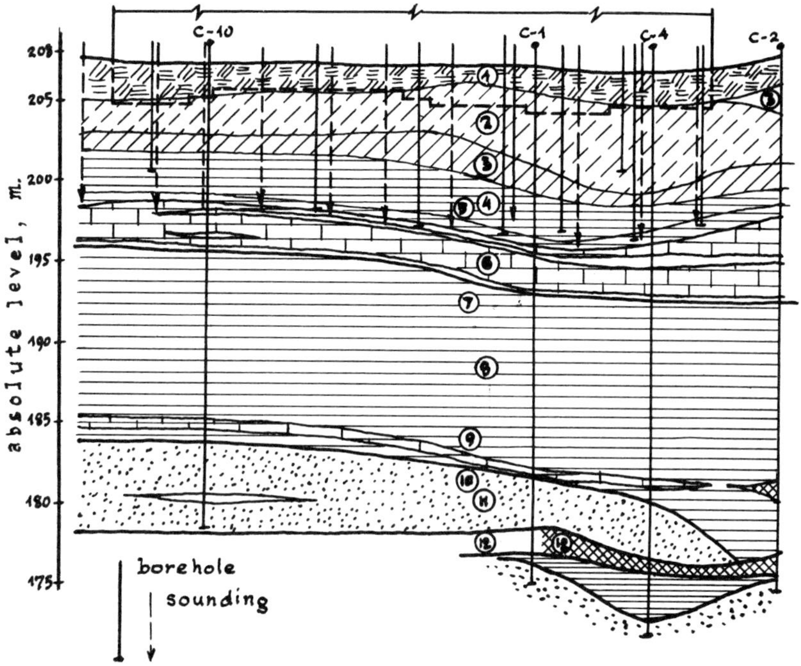

FIGURE 1(a) Soil profile

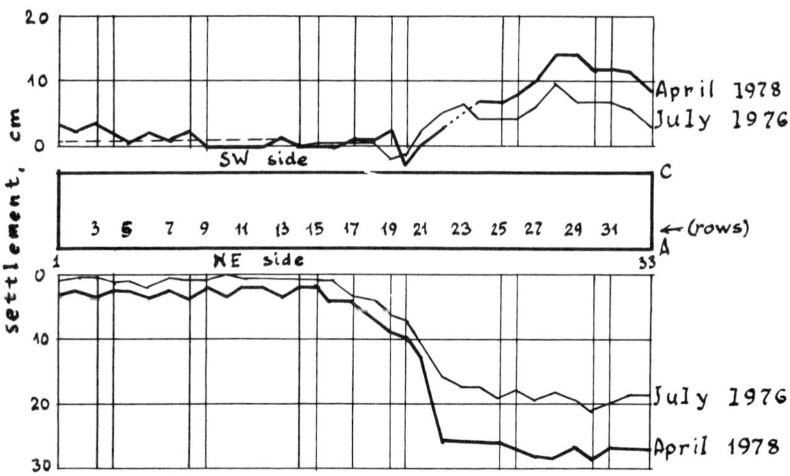

FIGURE 1(b) Plan of the building and settlements along the walls

1 embankment.

Quaternary deposits

2 sand clay, brown, from stiff to plastic consistency,
3 clay, red-brown to brown, stiff consistency, containing thin layers (2-3 cm) of fully saturated silty sand.

Carboniferous deposits

4 clay, yellow-grey to dark-grey, solid consistency,
5 clay, yellow-brown, solid consistency,
6 limestone,
7 brown coal,
8 black clay,
9 limestone,
10 clay,
11 sand,
12 brown coal, worked-out layer,
13 zone of collapsed roof.

Two layers of water were disclosed. The lower ground water is held within the clay (3) of the Quaternary deposits. It is underlain by the clay layer (4) of the Carboniferous deposits. The depth of the underground stream is 2 m at the SE and up to 6 m at the NW of the site. Because of some thin layers of sand within the Carboniferous clay (3),the latter cannot be regarded as fully impervious. The upper water layer is mainly fed by direct rainfall. As already noted, seasonal rainfall accumulates in depressions at the top of the slope.

Because of the position of the building at right angles to the flow of the groundwater a pressure head is produced on the SW side of the building. In April 1978 the surface of the ground water at the SW was at an absolute level of 208.9-209.0 m and water appeared at the soil surface. At the NE (towards the valley) it was at 203.5-204.4 m. The difference in the levels was 5.5 m. The lowest levels occur in February and September, the latter usually being about 1 m lower. The higher levels occur towards the end of April and May. At the SW of the building the changes in levels during the year are as much as 3 m; for the NE only 1 m. Insitu investigations of the coefficient of filtration of the sand clay (2) was 0.08 m/day at the SW and 0.28 m/day at the NE. From the time that the building was erected the level of the ground water increased, and its gradient down the slope was 25°-30°. At the SW of the building it was only 2°-2.5^{o}. Due to the fluctuations in level during the year, parts of the foundations emerge from the ground water, other parts stay submerged. In some parts of the site, for instance in a position at the North of the building, no water was found. At another location, where two close borings were made

(13.5 m between the borings) the observed difference in water levels was about 10m, and along the SW side of the building, down the slope, water emerged at the ground surface.

It was suggested that such complicated hydrological conditions could only result from the underground workings. According to archival sources two main drifts were believed to be located at a distance of 2 to 5 m. The NW part of the building (rows 22-33) is affected by the underground workings, in the zone of continuous mining subsidence. It was found that a local sagging of the dark-grey clay (4) and the limestone (6) layers mirrors the subsidence in the locality. A highly disrupted stratum about 8-10 m thickness was found by deep borings, this being the collapsed roof of the workings. Thus, it was believed that mining subsidence brought about the main structural changes in the soils.

Movements of the slope towards the valley bottom complicated the geological and hydrogeological conditions of the site. Special investigations have shown that although the slope is now stable, it is predisposed to movement and its stability may be only be guaranteed by drainage measures. However, at the time of construction and repairs to the building the ground water conditions brought about a deterioration in the soil properties and in the later stages it could have caused instability of the slope.

Supplementary geological investigations helped to define the differences in the physical properties of the soils for the NW and the SE parts of the site (Table I). Different properties were found for soils in the SW and the NE parts of the site i.e. down and up the slope. Thus the moisture content was 0.282 at the NE and 0.257 at the SW; the void ratio was 0.829 at the NE and 0.762 at the SW. The amount of clay particles in the NE is 7 per cent higher than in the SW. This can be explained by the movement of clay particles by the flow of water. At the same time, the modulus of soil deformation measured at a depth of 4-7 m was halved.

ANALYSIS OF OBSERVATIONS

The results from the instrumental observations of the settlements are presented in Figure 1b. In July 1971 the building showed large uneven settlements. The SE part (rows 1-6) remained almost level. Between rows 25 and 33 the differential movements were relatively insignificant. The greatest differentials were registered between rows 20 and 21 (S/L=5.2/260=0.02). At that location, the walls effectively lost their bearing capacity. The pattern of settlements along the SW wall copied those of the NE side but to a reduced scale. The differences in settlements between the NE and SW walls increased approximately linearly from the 16th to the 25th row reaching a maximum of 14 cm. In addition to significantly

TABLE I PHYSICAL PROPERTIES OF SOILS

Properties	Mean values					
	Sand clay (2)		Clay (3)		Clay (4)	
	SE	NW	SE	NW	SE	NW
Water content w	0.275	0.257	0.214	0.227	0.202	0.247
Liquid limit w_L	0.375	0.330	0.352	0.370	0.410	0.490
Plastic limit w_P	0.216	0.190	0.205	0.200	0.230	0.240
Plasticity index I_P	0.16	0.14	0.15	0.17	0.18	0.25
Liquidity index I_L	0.38	0.45	0.06	0.16	0.0	0.03
Specific gravity γ_s (g/cm^3)	2.72	2.70	2.70	2.69	2.70	2.71
Density γ (g/cm^3)	1.92	1.97	2.02	1.98	2.00	2.00
Dry density γ_d (g/cm^3)	1.51	1.57	1.66	1.62	1.67	1.61
Void ratio e	0.809	0.720	0.624	0.670	0.614	0.688
Degree of saturation G	0.92	0.97	0.93	0.91	0.93	0.96
Coeff. of permeability k_f (m/day)	0.9×10^{-5}	1.0×10^{-5}	-	1.5×10^{-5}	-	-
Cohesion c (MPa)	0.025	0.021	0.014	0.022	0.037	0.037
Angle of internal friction ϕ (deg)	21	25	17	22	16	16
Modulus of deformation E (MPa)	9.0	9.0*	10.0	10.0	12.0	24.0

* For a depth of 4 m or less; from 4 to 7 m, $E = 4.5$ MPa

unequal settlements along the SW and the NE walls, a tilting of the NW part of the building towards the valley (rows 16-33) was recorded. The latter produced torsion of the building. Lateral displacements of the cornice at the 33rd row reached 12 cm. Up to April 1978 settlements were continuing (Figure 1b). The greatest relative settlement (A-30) reached a value of 26 cm; the difference in settlement between stations A-22 and C-22 reached 23 cm. The greatest distortion (S/L) was 0.04 at A-21, A-22 . It was therefore decided that the NW part of the building was completely unsuitable for further repair.

New observations of the lateral displacements of the plinth were made in April 1978:

- the measured displacements of the cornice and of the plinth were more or less the same;
- in the middle of the building the displacements of the cornice were greater than those of the plinth; in this zone, tilting was greatest;
- the recorded tilting and lateral displacements at station A-33 and their changes with time indicated increasing torsion of the building;
- taking the slope phenomenon into account, the lateral displacements of the plinth were relatively high and averaged 4.6 cm/year, which should be interpreted as of considerable significance.

The settlement/time curve for 4 stations (Figure 2) along the NE wall (the numbers refer to the row in which they are located) indicate that:

- the greatest increase in settlements corresponds to the first half of 1971;
- the settlements of the SE part of the building are relatively small and are stabilizing;
- the settlements for the NW part are significant, no stabilization with time has been registered. (During 7 years of observations the average rate of settlements was as much as 17 cm/year and was accompanied by new cracks in 1978).

CAUSES OF THE BUILDING DEFORMATIONS

Influence of mining subsidence

Comparing the position of the depressions in the settlement curves (Figure 1a) with the position of the collapsed roof over the workings (Figure 3) it may be concluded that the most probable cause of movements was the collapse of the roof over the drifts. The main cause of this occurrence may be assumed to be the loading of the soil surface by the weight of the building.

A further effect was a deterioration in the

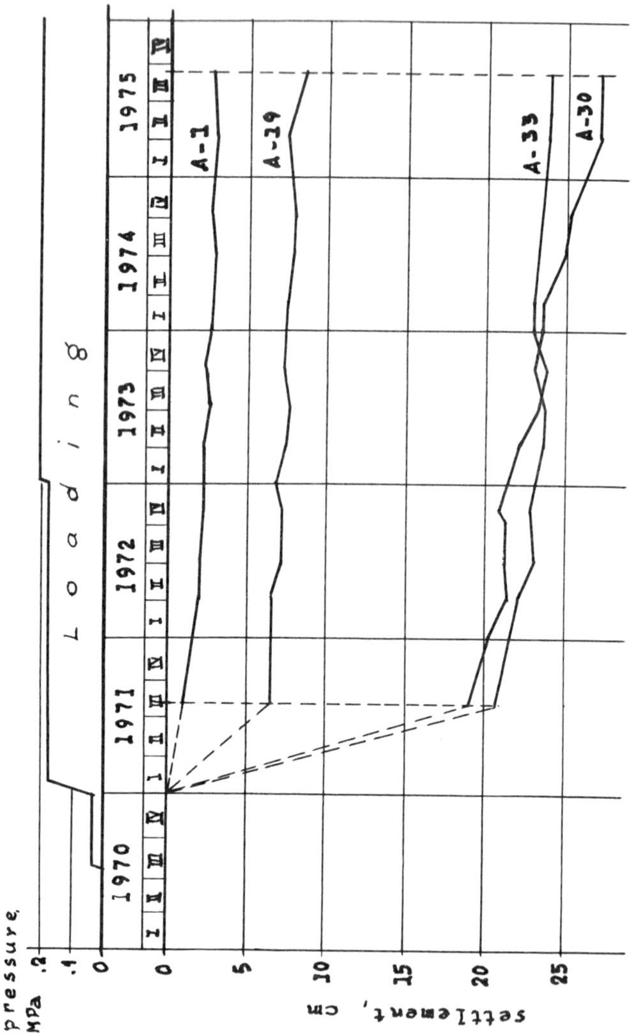

FIGURE 2 Settlements of four stations

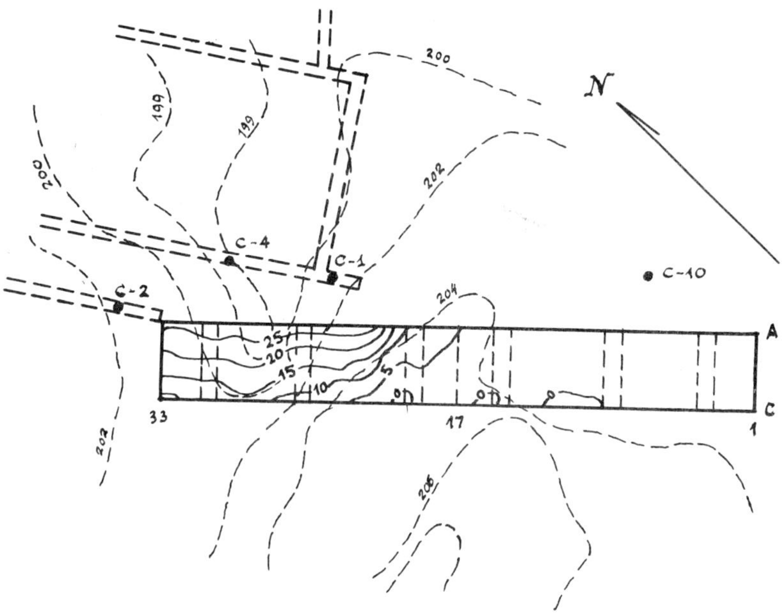

FIGURE 3 Site plan

hydrogeological conditions of the site which produced unequal changes in the soil properties. Complications due to slope phenomena might also be partly responsible.

It should be noted that the coal was worked in 1948 at a time when control of workings was not regulated, and this was confirmed by the state of the surface of the site.

Calculations of the possible settlements of the soils due to roof collapse over the drifts, within the area of the building, show these to be of the order of 30 to 35 cm i.e. approximately as measured. It should be noted that the settlements have probably not yet ended. This is corroborated by the presence of loose rocks at a depth of 26.0-27.3 m and of undecomposed organic debris in the workings. The full value of the settlements is thus likely to be higher.

Influence of slope phenomenon

The changes in the hydrological conditions at the site during 1970-1978 produced a variable deterioration in soil properties. It cannot be supposed that such changes will not promote further incidences of soil deformation due to mining subsidence. The observed lateral displacements of the building in April-May 1978 confirmed the presence of probable slope movements. Probably these displacements are due to extremely slow movements of the slope. As mentioned, it is believed that soil deformations have not yet stabilized and additional deformations may be stimulated by slope movements.

Influence of hydrogeological changes and of soil deterioration

Due to changes in the hydrogeological conditions and in the structure of the soils in the vicinity of the underground workings, large changes in the physical properties of the soils were registered. The changes in the hydrogeological conditions developing during 1972-1978. The modulus of soil deformation at a depth of 4-7 m was halved in value. However the great unevenness in the settlements cannot be explained by those changes. According to calculation the actual settlements could not be explained by the increase in height of the underground water, nor by changes in its levels, particularly since it was higher on the SW side and lower on the NE side.

CONCLUSIONS

1 Due to the presence of underground workings (and the hydrogeological and geological conditions of the site, the latter should be regarded as highly complex. The NW part of the building described, is being affected by mining subsidence and slope movements.

2 The collapse of the roof of the workings is to be seen

as the main cause of the developed uneven settlements of the site. The movements developed after the site was additionally loaded by the weights of the building and embankment.

As contributory factors the following may be noted:
(a) the continuing disturbance to the soils due to mining subsidence;
(b) changes in the hydrogeological conditions due to the creation of an artificial barrier to the flow of groundwater, produced by the substructures of the building which led to the deterioration in soil properties;
(c) due to the occurrences mentioned, the stability conditions of the slope have also deteriorated.

3 The uneven settlements of the surface of the site led to inadmissible deformations of the NW part of the building. Thus the greatest tilt of the building (row 22) reached a value of 0.02 (exceeding the admissible value by a factor of 10). The distortion of the large wall panels (between A-21 and A-22) reached a value of 0.04 (30 times the admissible value).

4 Settlements and lateral movements are still taking place. Their highest values are developing at rates of 1.7 and 12.0 cm/year. Cracking of the structure has been renewed.

The greatest danger is due to the lateral displacements of the building down the slope. The latter may even be linked to an extremely slow failure of the slope.

As a result of the large and continuing movements of the site the building was deemed to be unsuitable for further repairs and was transferred to a unit to be used for experimental investigations.

SETTLEMENT BEHAVIOR OF FOOTING ABOVE A VOID

M.C. Wang, PhD, PE

Associate Professor, Department of Civil Engineering
The Pennsylvania State University

R.L. Baus, PhD, PE

Assistant Professor, College of Engineering
University of South Carolina

INTRODUCTION

Structural damage due to mining or tunnelling under existing buildings has been a major problem in geotechnical engineering. Recently, the expansion of urban sprawl to areas of prior mining and tunnelling activity due to population growth adds additional concern to geotechnical engineers regarding foundation stability.

Very few studies on the stability of foundations above a mined or tunnelled void are presently available. This paper presents the settlement behavior of a footing situated above a void. Analysis of the settlement behavior was made using the finite element method. Model footing tests were performed to validate the results obtained. The variables considered in the analysis were the size, location and shape of the void.

FOUNDATION SOIL

The foundation soil was a silty clay which had a Liquid Limit of 37%, a Plastic Limit of 19%, a specific gravity of 2.67, and a classification of CL according to the Unified Soil Classification System. The soil was compacted using the static compaction method, to a dry density of 1602 kg/m^3 (100 pcf) with a water content of 17%. The optimum water content for standard Proctor compaction effort was approximately 18%.

The stress-strain and strength behaviors of the compacted soil in compression were determined using triaxial compression tests. The test specimens had a length of 71.5 mm (2.82 in) and a diameter of 3.3 mm (1.40 in). The deformation rate used was 1.3 mm/min (0.05 in/min). Because the stress-strain curves were highly non-linear, the deformation moduli in compression were obtained using the method proposed by Simons (1957). According to this method, the compression modulus is equal to the ratio of one-half of the axial stress at failure, to the axial strain at failure.

The stress-strain and strength behaviors in tension were determined using indirect tension tests. The test specimens were 102 mm (4.0 in) in diameter by 114 mm (4.5 in) in length. Loads were applied at a deformation rate of 1.3 mm/min (0.05 in/min). The vertical loads and horizontal diametral deformation were monitored using an electronic load cell and a linear potentiometer, respectively. The modulus in tension was calculated from the slope of the initial portion of the stress-strain curve. The strengths and moduli in tension and compression are summarized in Table I. Included in the Table are Poisson's ratio, the initial earth pressure coefficient (i.e. the ratio of the horizontal to the vertical earth pressure) and the wet unit weight of the test soil. The Poisson's ratio and initial earth pressure coefficient used were 0.28 and 0.60, respectively; they were estimated from the available literature (Bowles, 1974; Winterkorn and Fang, 1975). The strength, modulus, Poisson's ratio, pressure coefficient, and unit weight of the model steel plate footing (the model footing tests are described later) are also tabulated in Table I. These values were used in the theoretical analysis which follows.

THEORETICAL ANALYSIS

The finite element method of analysis was performed to investigate the settlement behavior of a continuous footing subjected to static loading. The footing is situated above a void which is either rectangular or circular in shape. As shown in Figures 1(a) and 1(b), B, Df, D and W designate the footing width, depth of foundation, depth to the void, and void width (or diameter of the circular void), respectively. In the analysis, the foundation soil mass was idealized as an assemblage of a number of discrete structural elements interconnected at a finite number of joints or nodal points. The sizes of the elements were chosen to vary in accordance with the anticipated stress gradients. Figure 2 shows a typical finite element mesh for a 5 cm (2 in) footing with a 15.2 cm (6 in) wide void. The void is 7.6 cm (3 in) deep and the top of the void is 38 cm (15 in) below the bottom of the footing. The depth of foundation is 6.4 cm (2.5 in).

In the finite element analysis, the foundation soil was treated as an elastic-perfectly plastic material. Before yielding, i.e. within the elastic range, the stress-strain relationship of the soil was described by Hooke's law. After yielding, the plastic strain rate was computed using the assoc-

TABLE I MATERIAL PROPERTIES OF TEST SOIL AND MODEL FOOTING MATERIAL (STEEL PLATE)

	TEST SOIL	STEEL PLATE
Unit Cohesion, C, (KN/m^2)	65.6	1.24×10^5
Angle of Internal Friction (°)	13.5	0
Modulus in Compression (KN/m^2)	4670	200×10^6
Modulus in Tension (KN/m^2)	10380	200×10^6
Tensile Strength (KN/m^2)	14	2.48×10^5
Poisson's Ratio	0.28	0.30
Initial Pressure Coefficient	0.60	0.43
Unit Weight (Kg/m^3)	1878	7850

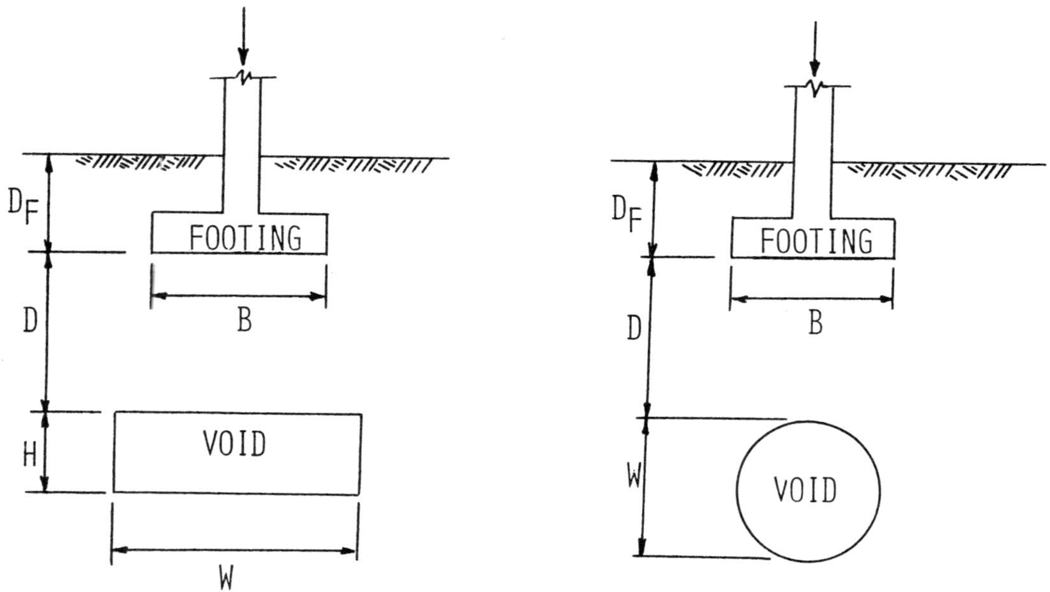

A. RECTANGULAR VOID

B. CIRCULAR VOID

Figure 1 Continuous Spread Footing and Underground Void

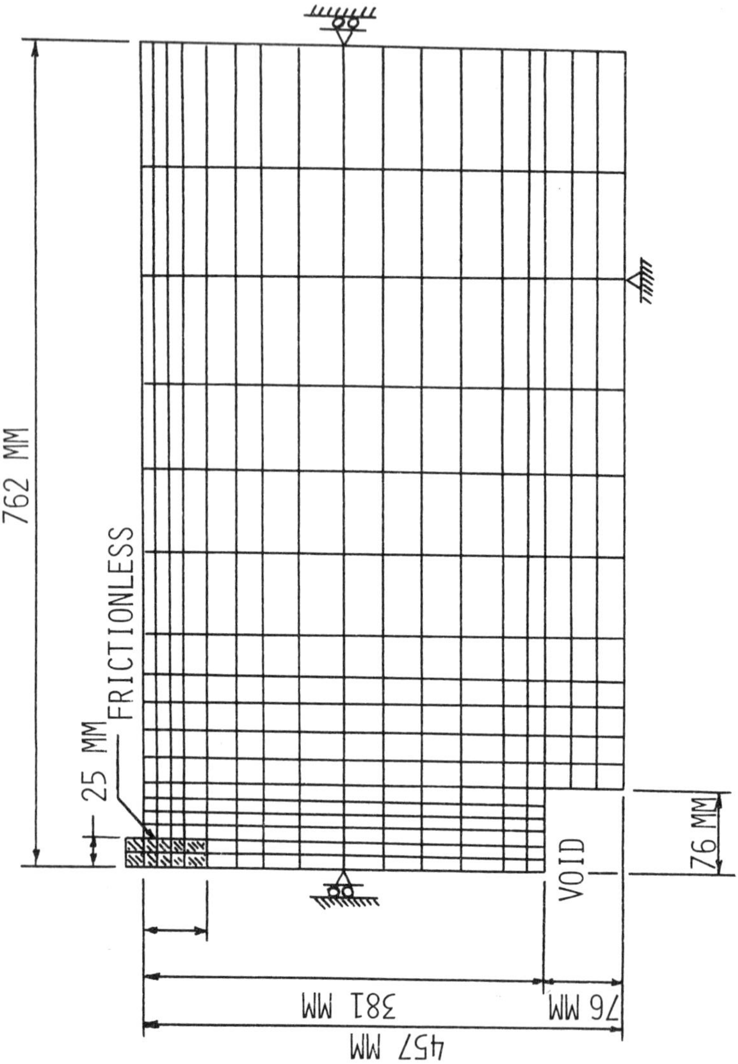

Figure 2 Finite Element Mesh

iated flow rule. Based on these constitutive laws, a computer program was developed by Chang et al (1974). This was adapted for analyzing footing behavior with various boundary conditions. The computer analysis was performed using the IBM 370 Model 168 computer at The Pennsylvania State University.

In the computer analysis, loading increments of 5% of the total load to be applied to the footing were used so that the non-linear settlement behavior could be better accounted for. Three iterations at each load increment were found sufficient to provide an acceptable level of refinement in the final displacement values. Results of a sensitivity analysis indicate that within the range of values studied, the modulus in tension and the initial earth pressure coefficient have a negligible effect on computed stress, strain and footing displacement. The input value of Poisson's ratio was found to have a moderate effect; for one set of boundary conditions, changing Poisson's ratio from 0.25 to 0.40 yielded a decrease in displacement of approximately 15 to 20 percent.

Using the material properties in Table I, the settlement behavior of a continuous footing above an underground void was analyzed for a total of 11 boundary conditions. Some of these were used in the model footing testing which is described below to validate the results of the theoretical analysis.

EXPERIMENTAL INVESTIGATION

Model footings were tested in a plexiglass-sided tank which was approximately 1.2 m (4 ft) high, 1.7 m (5.5 ft) long and 14 cm (5.5 in) wide. The test tank was constructed of plywood and plexi-glass heavily reinforced with 76 mm (3 in) steel channels. The model footings were 13 mm (0.5 in) steel plates with three widths: 51 mm (2 in), 76 mm (3 in) and 127 mm (5 in). The footings were welded to the loading device to minimize possible rotation during testing.

The test soil was compacted in 76 mm (3 in) layers to a prescribed density using a steel loading head which covered the entire length and width of the tank. A thin line of lightly colored sand was placed between the layers, along the plexiglass side to aid viewing of the slip line fields. Voids of various widths were centered under the test footing at the bottom of the tank. The voids were formed by placing a removable, lubricated, wooden block at the bottom of the tank. Void widths of 25 mm (1 in), 152 mm (6 in), 305 mm (12 in) and 584 mm (23 in) were used. The no-void condition was also tested.

A Universal testing machine was used to apply load to the model footing using a constant rate of displacement. The load applied was read directly from the testing machine and the footing displacement was monitored with a 0.025 mm (0.001 in) dial gage. The load was applied at the same constant displacement rate as that used in the strength tests. A total of 47 model footing tests were performed.

COMPARISON OF RESULTS

The results of the finite element analyses and model footing tests are compared in Figures 3 and 4. Figure 3 shows the footing pressure-displacement curves for a 51 mm (2 in) footing with a 152 mm (6 in) void located at approximately 381 mm (15 in) below the bottom of the footing. The moisture content and dry density are approximately 17% and 1602 kg/m^3 (100 pcf) respectively. A very good agreement is seen between the two sets of results. Figure 4 illustrates the pressure-displacement curves for a 51 mm (2 in) footing with a 152 mm (6 in) void at a depth of 533 mm (21 in) from the footing base. It is seen that the curve obtained from the finite element analysis resembles and is bracketed between the two curves of the model tests. The water content for one of the two curves was greater than and the other smaller than that of the finite element curve. This Figure reveals the effect of water content on footing displacement behavior.

The close agreement between the results of finite element analysis and model testing justifies the use of finite element procedures to investigate the settlement behavior of continuous footings with various boundary conditions. The results of the investigation are presented for various conditions including footing size, depth of foundation, and location, size and shape of the void; the void size and location are expressed in terms of the footing size.

SETTLEMENT BEHAVIOR

Figure 5 illustrates the effect of D/B ratio on the settlement behavior of the footing for a constant W/B ratio of 3.0 . It is seen that, as would be expected, under the same footing pressure, settlement increases with decreasing D/B ratio, because of less soil support at smaller D/B ratios. When the D/B ratio remains constant, the void width increases with an increase in W/B ratio. As a consequence under the same footing pressure, the footing settles more at greater W/B ratios as demonstrated by Figure 6.

For a given D/B ratio, footing pressures required for a constant settlement vary considerably with both W/B and D/B. Figure 7 illustrates the variation of the required pressure for three levels of settlement (i.e. 1, 2 and 3 cm) and three levels of W/B ratio (1.0, 3.0 and 10.0). Note that for clarity, data points are given only for the condition of 1 cm of settlement with W/B = 3.0 . The Figure indicates that the required footing pressure increases with an increase in D/B ratio. In other words, for a given footing width, as the depth to the void increases, the footing pressure also increases. The rate of

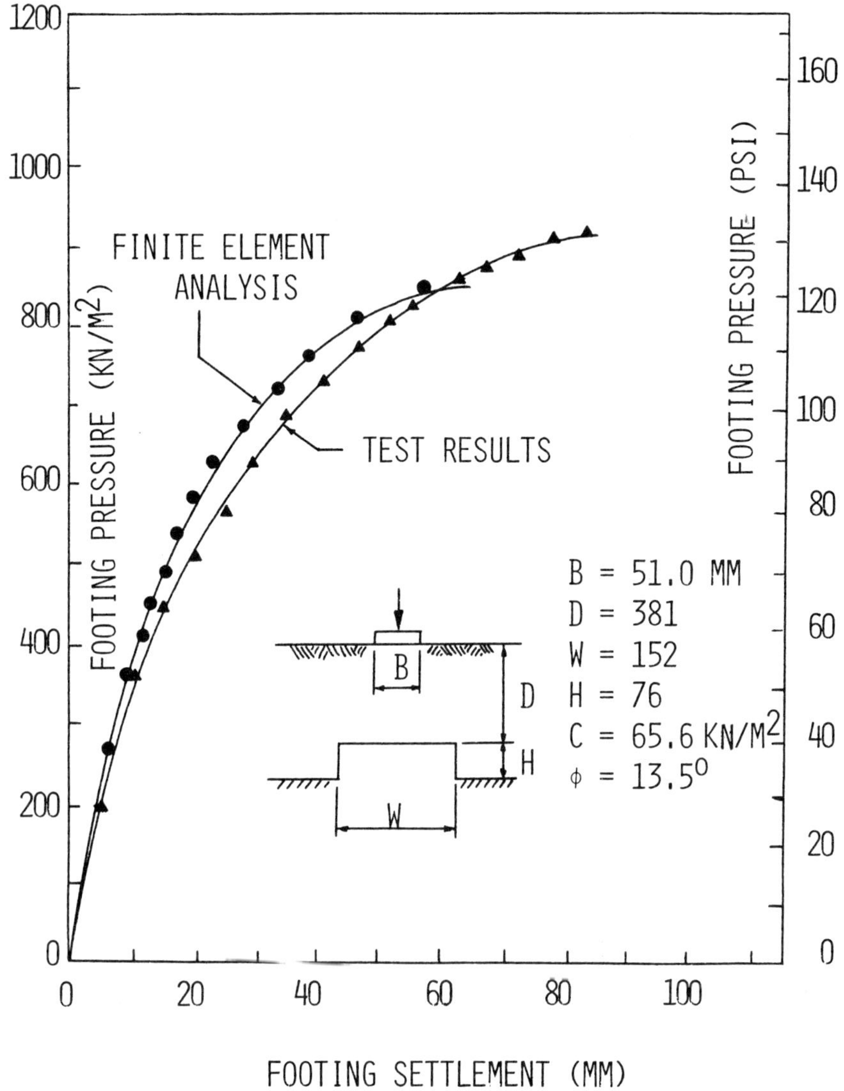

Figure 3 Comparison of Results of Finite Element Analysis and Model Footing Tests

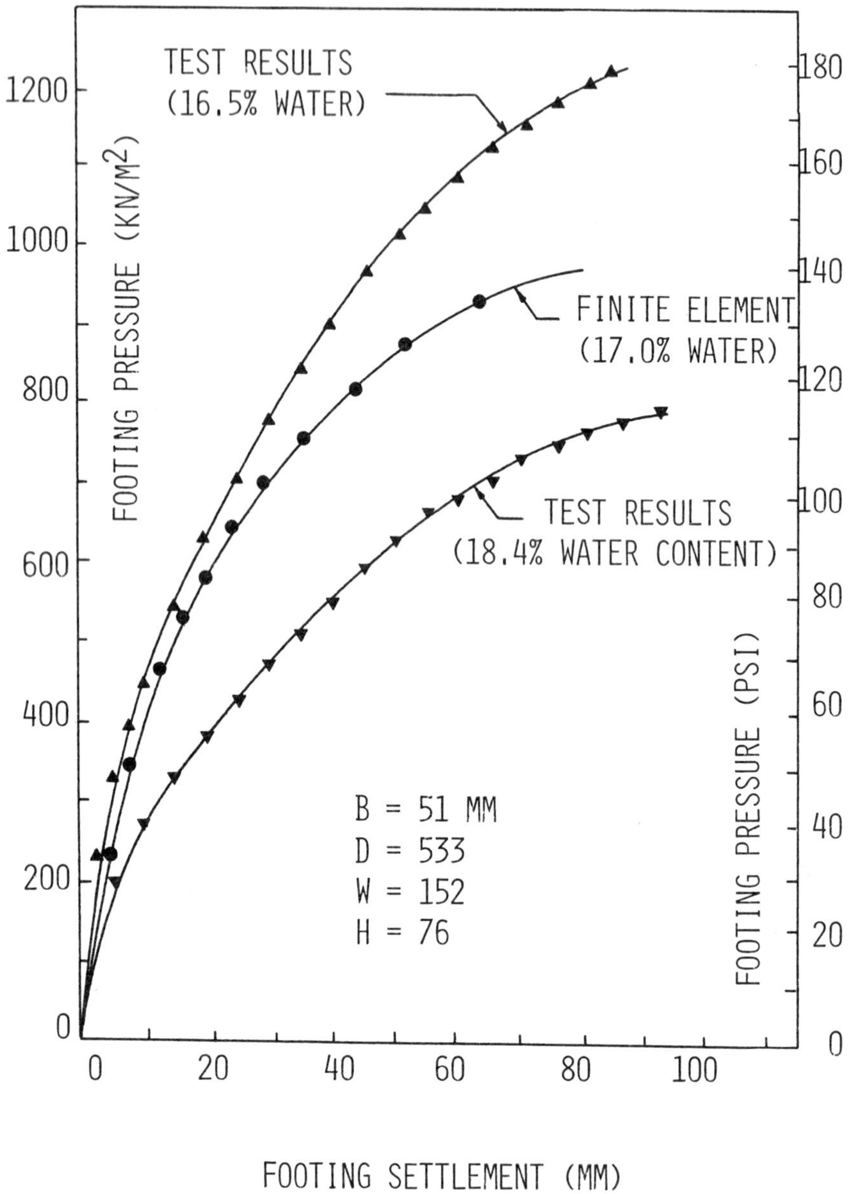

Figure 4 Comparison of Finite Element Analysis with Footing Tests for Different Water Contents

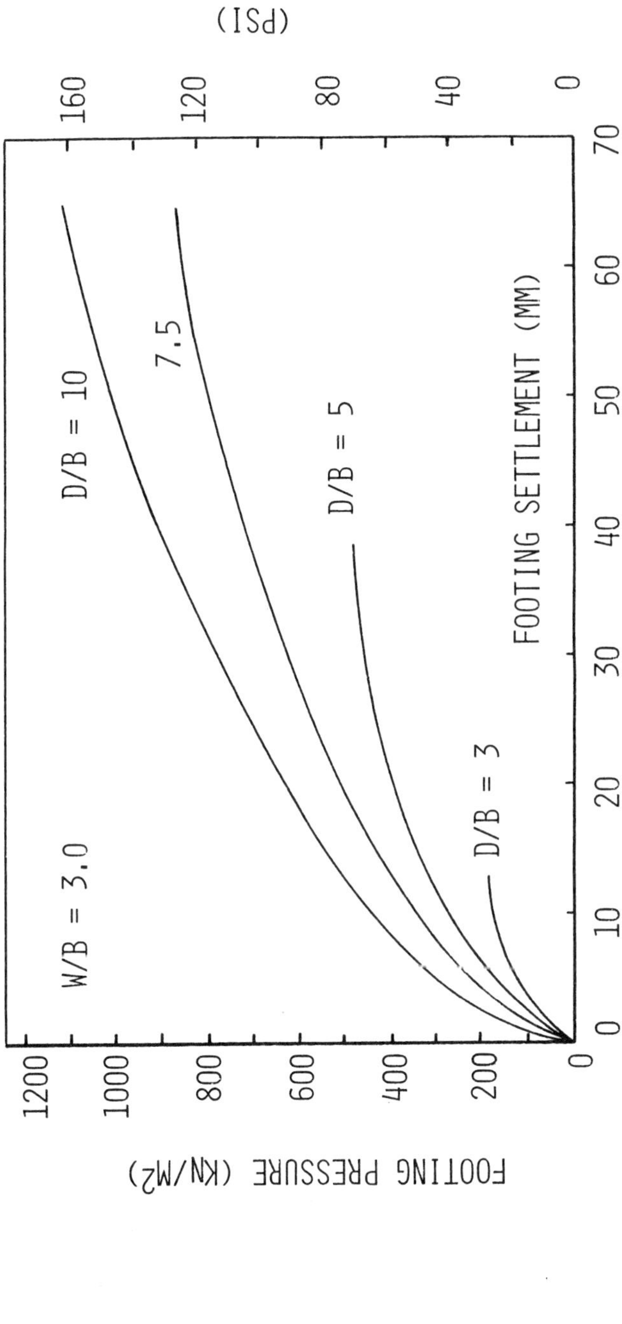

Figure 5 Variation of Footing Displacement With D/B Ratio

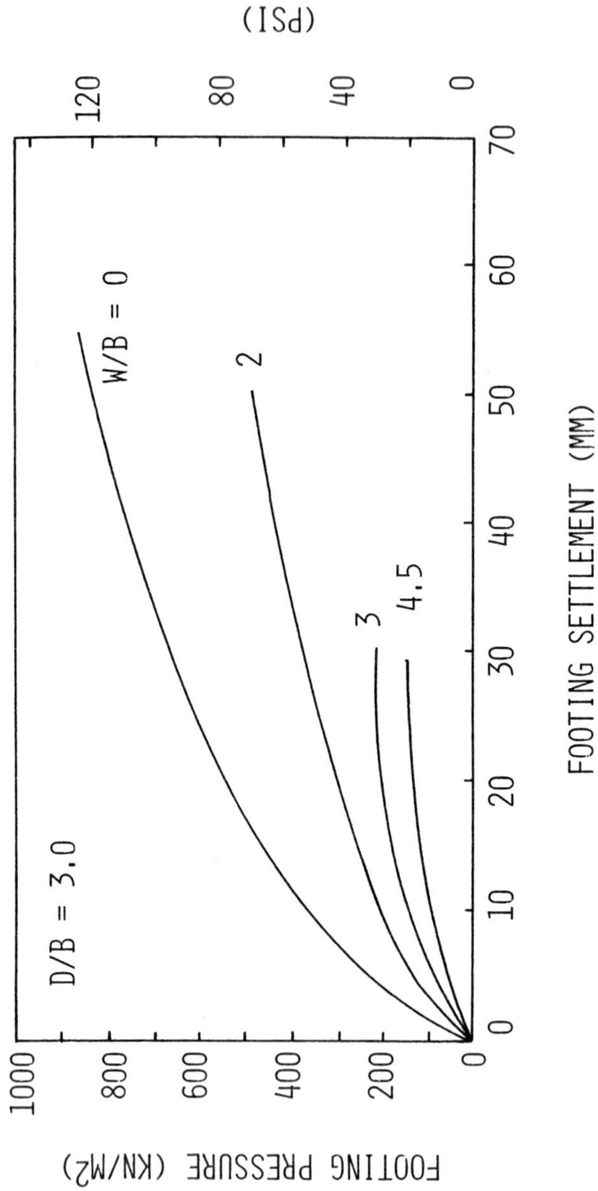

Figure 6 Variation of Footing Displacement with W/B Ratio

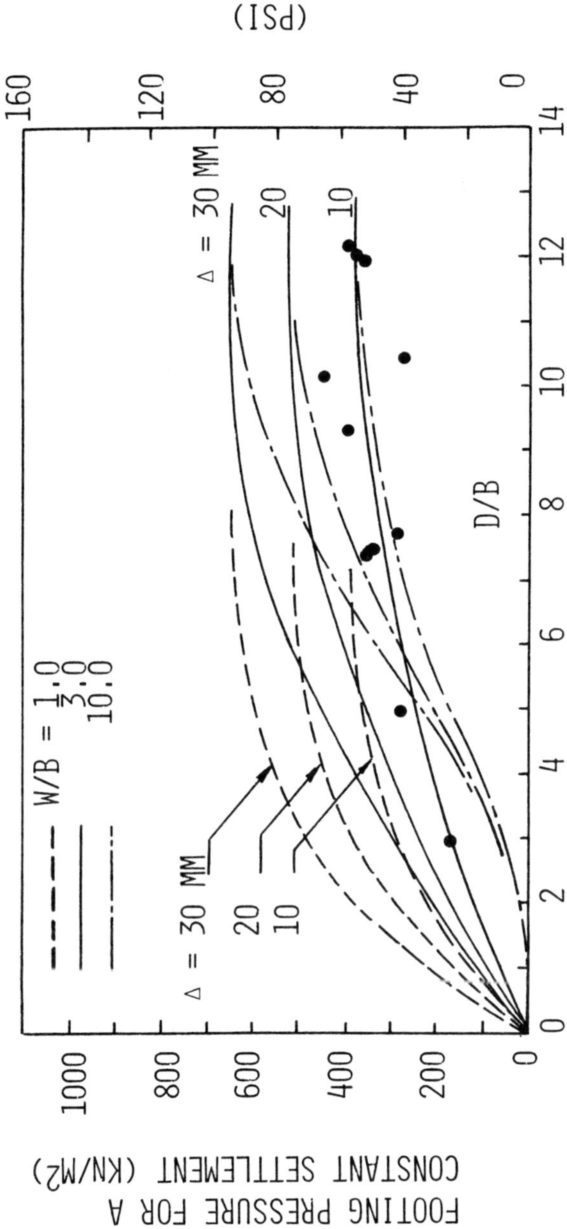

Figure 7 Effect of Void Size and Location on Footing Pressure

pressure increase varies considerably with different values of W/B. For a given W/B ratio, however, no significant variation of the rate of footing pressure change with the level of settlement is seen.

The footing pressures in Figure 7 are expressed in terms of the percentage of footing pressure for the no-void condition and are given in Figure 8. Since the rate of footing pressure increase with D/B ratio does not vary appreciably among the three levels of footing settlement, only one curve is given for each level of W/B ratio. According to these curves, the effect of an underground void on the footing settlement diminishes as the D/B ratio increases. The effect becomes insignificant when the D/B ratio reaches approximately 7, 11 and 13 for W/B ratios of 1, 3 and 10 respectively.

The variation of footing pressure required to cause a constant settlement as a function of the size and location of the void is also given in Figure 9. This Figure demonstrates that the effect of void width on footing pressure is greater when the void is nearer the bottom of the footing. This is as would be expected because at smaller depths to the void, the soil mass immediately underneath the footing is thin, so that the total shearing resistance which can be mobilized is low.

Figure 10 shows the footing pressure versus displacement relationships for a circular and a rectangular void. The footing width is 51 mm (2 in), void width or diameter is 152 mm (6 in), depth to void is 381 mm (15 in). The Figure indicates that the void shape has a negligible effect on the settlement behavior, at least for the boundary conditions analyzed.

PRACTICAL APPLICATION

Because void shape has no appreciable effect on the settlement behavior, the results presented in Figure 9, which was developed for rectangular voids, can be used as a first approximation for voids of any shape. Figure 9 provides a basis for determining the footing pressure at which the footing settlement will not exceed a certain value. For instance, for a footing located above a void for which D/B and W/B ratios equal, say, 2 and 3 respectively, the maximum footing pressure which produces 2 cm settlement is approximately 25 percent of the pressure causing the same amount of footing settlement when no voids exist under the footing. Note that each curve in Figure 9 represents average values of pressure for three levels of settlement, namely 1, 2 and 3 cm. Therefore, the footing pressure determined from Figure 9 will be valid for any level of settlement up to at least 3 cm.

SUMMARY AND CONCLUSIONS

The settlement behavior of continuous footings with an

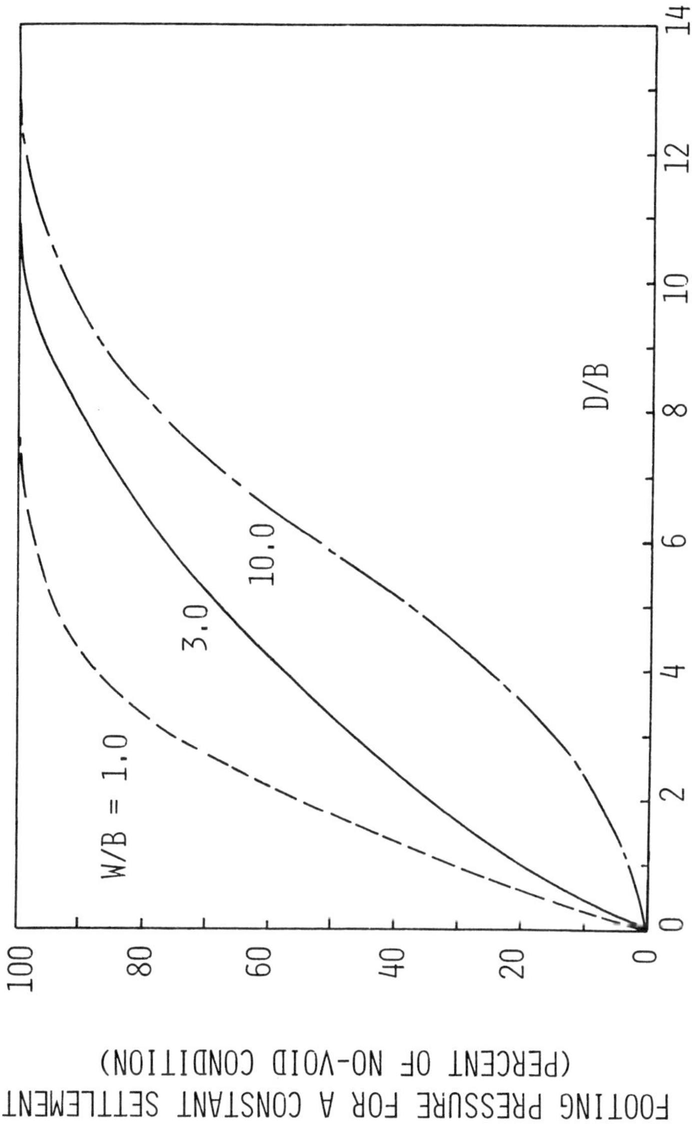

Figure 8 Effect of Void Size and Location on Footing Pressure
Expressed in Terms of No-Void Condition

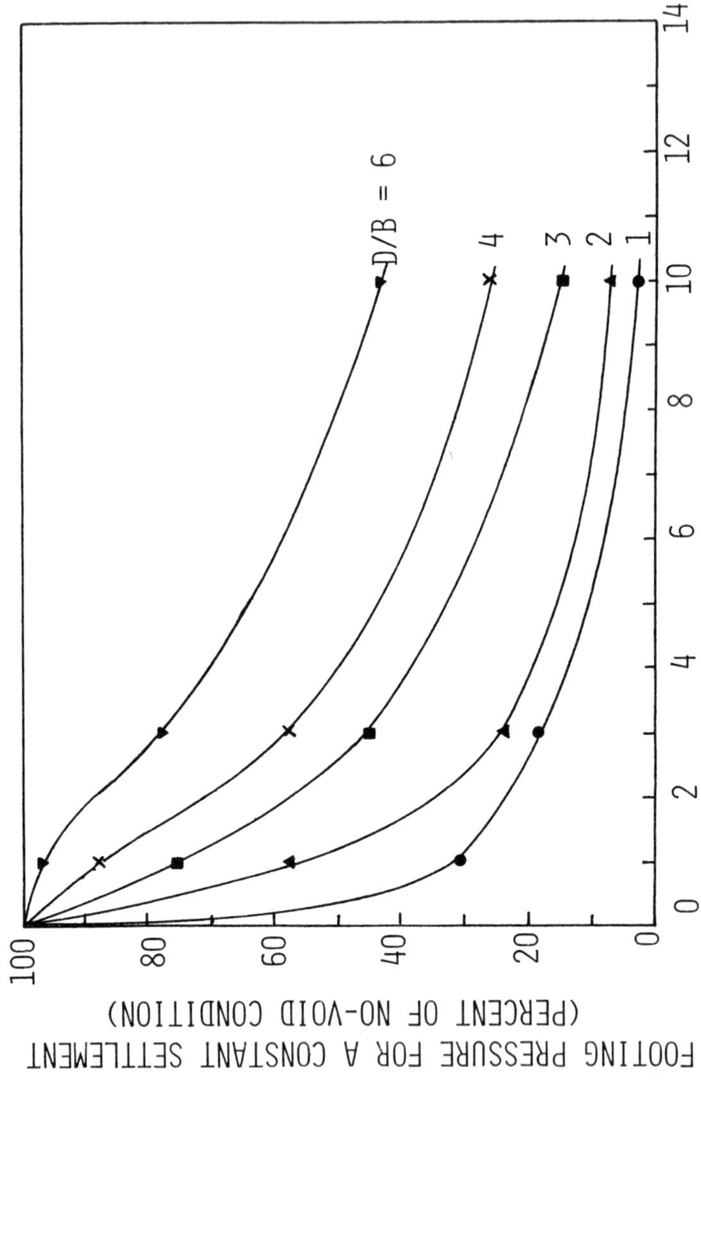

Figure 9 Effect of Void Size and Location on Footing Pressure in Terms of No-Void Condition

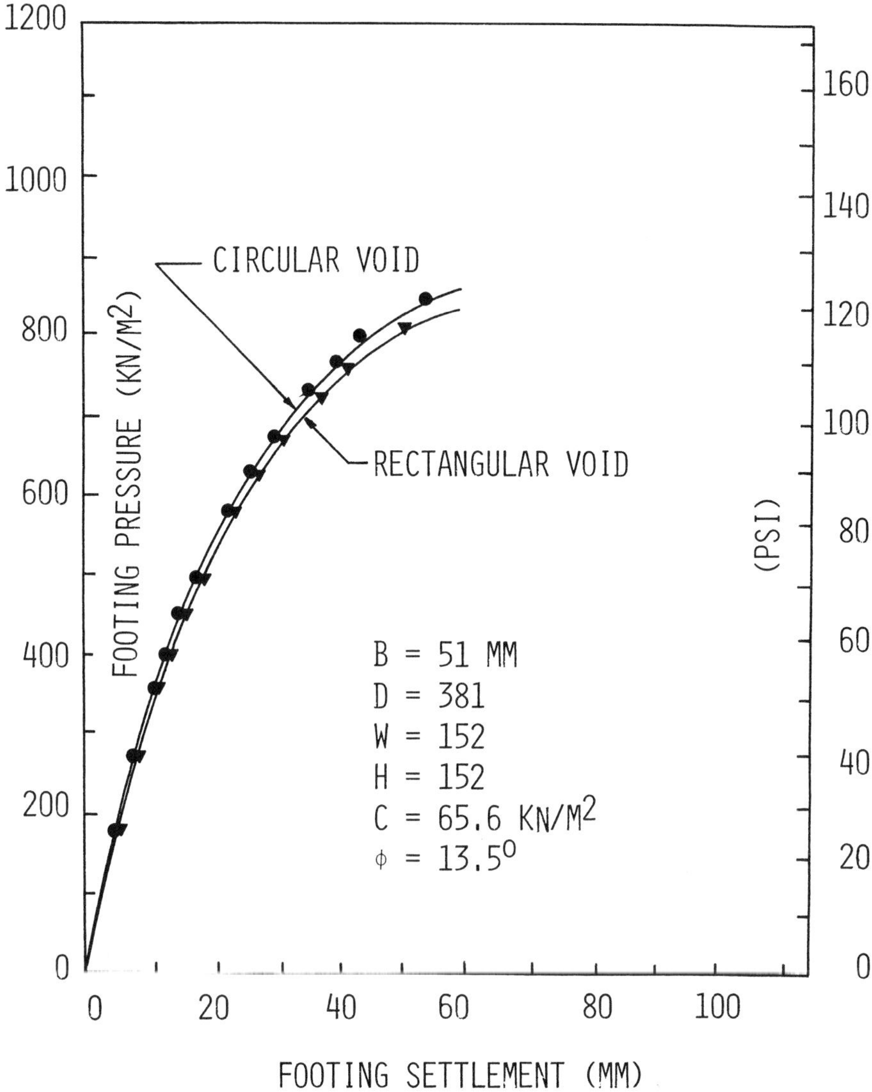

Figure 10 Effect of Void Shape on Settlement

underground void has been investigated, using the finite element method of analysis and model footing tests. In the finite element analyses, the foundation soil was treated as an elastic-perfectly plastic material. The results of the theoretical analyses agree well with the model tests.

The results of the study indicate that the settlement behavior of a footing depends greatly on the size and location of the underground void. The footing pressures required to cause a certain level of settlement are expressed in terms of the pressure for no-void condition and are summarized in Figure 9. This provides a basis for designing continuous footings located above an underground void.

ACKNOWLEDGMENTS

The research reported here was sponsored in part by the Central Fund for Research, The Pennsylvania State University. The theoretical analysis was performed at the computation center of The Pennsylvania State University. These financial supports are gratefully acknowledged. The authors wish to express their sincere appreciation to Dr. C.Y. Chang, Woodward-Lundgren & Associates, for his guidance during the course of adapting his finite element program.

REFERENCES

J.E. Bowles - Analytical and Computer Methods in Foundation Engineering, McGraw-Hill Book Inc. (1974) 51.

C.Y. Chang, K. Nair and R.D. Singh - Finite element methods for the nonlinear and time-dependent analyses of geotechnical problems, Proceedings of the Conference on Analysis and Design in Geotechnical Engineering, ASCE, University of Texas, Austin, 1 (1974) 269-301.

H. Winterkorn and H.S. Fang - Foundation Engineering Handbook, van Nostrand Reinhold (1975) 116.

N. Simons - Settlement studies of two structures in Norway, Proceedings of the Fourth International Conference on Soil Mechanics and Foundation Engineering, London, 2 (1957) 431-436.

THE USE OF ORDNANCE SURVEY BENCH MARKS FOR THE STUDY OF LARGE-SCALE MINING SUBSIDENCE

Richard Edward Hughes, B A , F G S

Geotechnical Archaeologist, Ove Arup and Partners

INTRODUCTION

The Ordnance Survey grew out of 18th century military surveying and was formally instituted by Act of Parliament in 1841.

The potential for civil purposes was readily appreciated and early examples of use are for civil engineering works associated with improving sanitary conditions in growing towns of the early Victorian period and for planning railway routes between towns. The value for mapping geological features such as rock outcrops was also quickly realised.

It is proposed that the same maps provide unique information concerning the collapse of mine workings as expressed by the degree, extent and pattern of ground subsidence. The method is based on the use of Bench Mark Levels taken from the many editions and revisions of the Ordnance Survey maps. How these levels are obtained, interpreted and used will be discussed in the following pages. While apparently not applied to the mining industry before, similar techniques were used as far back as 1932 for the study of "The Subsidence of London" (Longford,1932) and this work is still quoted (Wilson and Crace,1942 and Bell, 1975).

Ordnance Survey maps are the major source of information for the accurate study of topographic changes that have taken place for nearly 150 years. A recent paper on geotechnical mapping discusses their importance for the location of surface geological and mining features (Dearman,Morey,Coffey,Scott and Wheeler,1977). This was because it was as late as 1872 that it became a statutory duty for mine owners to send boundary plans of their abandoned mine workings to the Secretary of State to be preserved in the Mining Record Office. Only in 1887 was it made compulsory to show details of mining technique and their relationship to surface topography (see Appendix A).

For many years, on a sometimes regular but more often a superficial basis, observations were made by coal company surveyors on the vertical lowering of the land surface over mined areas (Heriot-Watt University,1966). It was only when the National Coal Board was created in 1947 that systematic scientific research into mining subsidence was started. For example, Orchard and Ogden commenced field monitoring in North Staffordshire in 1948 (Orchard and Ogden,1959). Because records of the earlier coal company observations and the more recent research by the N.C.B. are not available for public consultation Bench Mark surveys provide the only readily available source of information. An example in the North Staffordshire area Ordnance Survey levelling is recorded on maps from 1876 onwards.

CASE HISTORIES

To demonstrate how bench mark surveys are carried out, the value of the readings and their limitations, 5 case histories are presented in Appendix B.

The five case histories have been chosen to demonstrate the use and shortcomings of such bench mark surveys. Case history 1 records an investigation in Alfreton, Derbyshire where different amounts and rates of mining subsidence can be seen around the town centre. Case history 2 demonstrates the relationship of surface settlement patterns and geology/mining features at Hanley, Stoke-on-Trent. Case history 3 shows the difficulty of using such surveys in rural areas and case history 4 shows that the Ordnance Survey information can be wrong. Case history 5 illustrates how bench mark information, post dating the subsidence, can be misleading.

THE HISTORY OF BENCH MARKS

"A Bench Mark is defined as a mark, the value of which relative to Ordnance Datum has been determined by spirit levelling" (Harley,1975). Bench marks are of various types depending upon their function. Fundamental Bench Marks, set in stable rock, are at 40 km intervals along lines of the main survey triangulation system (geodetic survey) across the country. Flush Brackets are at 1½ km intervals along the triangulation lines and these are metal plates cemented into the faces of buildings. The country is divided by this triangulation system into areas averaging 400 sq. km within which there is a system of Cut Bench Marks. These are the commonest form of mark that is used by the author for examination of mining subsidence.

Cut Bench Marks are positioned on buildings and other relatively solid structures so that their position, and hence measured level, remain stable over long periods. However, in areas of mining subsidence,shrinking soil and landslides, the measured value is accurate only at the time of survey.

As a result of urban expansion, the reshaping of the topography by redevelopment and ground movements, the Ordnance Survey has on many occasions relevelled and established new Bench Marks. Generally moorland and mountain areas are re-levelled every 40 years and the remaining areas every 20 years.

As far as it can be ascertained all maps before 1925 show bench marks established during the 1st geodetic survey of Britain and are related to Ordnance Datum located at Liverpool. A more precise datum at Newlyn, Cornwall was established and used for the second geodetic survey during which the fundamental levelling programme was done in England and Wales between 1912 and 1921, and in Scotland between 1936 and 1952. A third triangulation of England and Wales was done between 1951 and 1956 and Scotland between 1956 and 1958.

Bench marks are provided at a density appropriate to the nature of the country. In recent times the density has been reduced to not more than 5 per square km in rural areas, and up to 40 per sq. km in urban areas. In some areas the existing density of bench marks still exceeds the density given above and consequently some marks are not used when relevelling takes place. Sometimes the marks are levelled but not shown on the published maps, but are registered on the Ordnance Survey Bench Mark lists. During Case Study 2 it was noted that the density of bench marks on the 1st Edition maps greatly exceeded the figures quoted above and are still more than the current density.

It is assumed that readers are familiar with the presentation of Bench Mark Levels and Spot Heights on Ordnance Survey maps, but may not be familiar with the manner in which the Ordnance Survey provided tabulated data and an example is shown in Table 1.

It is possible to obtain the same values by referring to old editions of Ordnance Survey maps at 1:10,560 and 1:2,500 scale held by national, city and town libraries. It should be noted that bench mark values are not shown on the recent 1:10,000 scale maps.

CALCULATING GROUND MOVEMENTS FROM BENCH MARKS

It is generally appreciated that bench mark values shown on the most recent maps are important for relating existing structures, active construction projects and site topography to Ordnance Datum. However, as previously discussed, a bench mark may have last been levelled more than 30 years ago. Therefore unless familiar with the history and geology of an area, it is wrong to assume a bench mark has maintained its last

TABLE 1

National Grid Reference		Description	Altitude in feet above				
			Newlyn Datum			Liverpool Datum	
			1965	1957	1947-49	1885	1872-73
ST 0789							
0706	8963	Wall High Street	277.95	277.94	277.92	-	-
09	48	52 Phillip Street	369.66	-	369.72	-	-
11	71	Rivet Steps Court House Street	248.24	248.24	-	-	-
11	71	High Street Northwest Side	-	-	-	249.2	249.9
16	54	Wall 37 Graig Street	324.09	324.09	324.10	-	-
16	76	Urinal Station Terrace/ High Street	-	218.34	218.32	-	-
18	81	Railway Bridge	208.66	208.61	-	-	-
25	63	Rivet Steps Graig Terrace	-	290.88	290.91	-	-
31	74	Abattoir	203.59	203.53	203.52	-	-
38	69	Wall N side Rickards Street	259.29	259.24	259.21	-	-
50	69	Wall N side Rickards Street	267.40	267.34	-	-	-

Note: 0.4 feet should be subtracted from altitudes based on Liverpool Datum to refer them to Newlyn Datum.

: The cost for this work was approximately 80p. per bench mark. Since September 1974 the price for this documentation has increased and for example in April 1978 a price of £3. 80p. per bench mark was charged.

determined value.

The correct value of a bench mark can only be verified by surveying back to a Fundamental Bench Mark or by prediction from the mark's previous values. This latter method is made possible by the Ordnance Survey's policy of, whenever possible, relevelling the same mark rather than establishing a new one.

For each mark the amount of movement can be calculated and by plotting the level values against time the patterns and rate of movement are disclosed. This is demonstrated in Case Study 1. Even where a Bench Mark is destroyed and a new one established nearby, they can in conjunction still be used for demonstrating movement trends. If there is a sufficient density of bench marks, usually in urban centres where little rebuilding has taken place, it is possible to produce plans showing contours of subsidence and this is shown in Case Study 2.

THE USE OF BENCH MARK HISTORIES FOR INVESTIGATING MINING SUBSIDENCE

Bench Mark Histories are of great value during desk studies of coal mining areas and also areas where other minerals such as fireclay, clayband ironstone, salt and gypsum have been extracted. The histories can also be useful in areas such as East Anglia, where peaty soils have been drained, and large settlement of the surface has occurred.

Before the industrial period, coal seams were first worked at their outcrop and working was generally restricted to extracting material less than 12m below ground level. Subsidence would therefore have been limited to a narrow zone following the outcrop of each seam. During the early industrial period and up to the mid-19th century coal was extracted by 'pillar and stall' method. At first the remnant pillars, left to support the roof, were irregular in size and arrangment but as mining became more systematic the pillars and roads were excavated to a regular grid.

It is pillar and stall mining at shallow depths that gives rise to subsidence and instability problems long after extraction ceases. This is because the length of time before the collapse of unsupported short spans of the roof rocks is unpredictable. Thus the location and extent of collapse and the resultant subsidence at the surface cannot be readily estimated.

Extraction of coal and other minerals by long wall workings was introduced in the mid 19th century. By the long wall method collapse takes place immediately behind the working face and most of the surface subsidence generally takes place within a short time after extraction. The maximum vertical movements can be as much as 90% of the extracted thickness although pack-

-ing in which the void is partly refilled by waste material can reduce subsidence to about 30%.

CONCLUSION

The use of Bench Mark data provides a simple, rapid and relatively cheap method of calculating the actual amount of ground lowering during the last 130 years. It is a particularly effective method for determining the pattern and rate of subsidence caused by mining. The methods described above can be used by geotechnical and mining engineers who need to understand the local ground conditions prior to site investigation. The data can also supplement historical information required for Engineering Geological mapping.

The value of a Bench Mark Study is :-

(a) To confirm whether or not mining subsidence has taken place.
(b) To indicate the presence of unpredicted old workings.
(c) To indicate anomalies that need detailed investigation.
(d) To provide data for calculating ground strains and hence account for structural distress to buildings.
(e) To examine the stability of land in reclamation areas.
(f) To account for changes in above and below ground drainage patterns.

The case histories show that :-

(1) If the information is good and can be corroborated, a valuable account of subsidence is gained.
(2) The data can be wrong.
(3) The data may fail to show up the presence of old mine workings.

Therefore the information obtained from Bench Mark histories should always be used in conjunction with other information and never accepted at face value.

APPENDIX A

Background to the legislation regarding coal mine abandonment plans

An Act of 1859 required the coal owner to keep a plan of workings showing the system of ventilation for inspection by the government mines inspector. Since there were so few inspectors it was probably only the richer and more progressive owner who would produce such plans and keep them up to date. The Act of 1859 allowed mine owners an escape clause with regard to the keeping of plans. If an inspector found that no plan existed at a mine, he could not penalize the mine owner beyond insisting that a plan be made within a reasonable time of his visit. This clause was not removed until the Act of 1872. The Act also stated that an inspector was forbidden to make a copy of the whole,or any part of a,plan.

An Act of 1860 extended the requirements of the 1859 Act to include all mines of ironstone or the coal measures, but not mines of limestone. It was not until 1872 that mines of clay ironstones, unconnected with collieries, were included.

The legislation of 1872 required mine owners to make plans of the boundaries of their abandoned workings to enable future mine owners of neighbouring leases to work and plan their mines free from the danger of flooding from unrecorded workings. Therefore, a tracing of the boundaries shown on the existing working plan at the time of abandonment would satisfy the inspector.

In 1873 it became the statutory duty of the owners of mines to send plans of the boundaries of their abandonment mine to the Secretary of State to be preserved in the Mining Record Office.

The Act of 1887, recognising the practice of copying the boundary from the existing working plan, insisted that the copy be done 'accurately' by a competent draughtsman. The Act also allowed the existing working plan to be offered as an abandonment plan. This arrangement seldom occurred, as it left the mine owner without a plan of his mine. The Act also demanded that the abandonment plan should show the boundaries of the workings correlated to the surface, the general direction and rate of dip of the workings, and the depth of the shaft (not necessarily the same depth as the seam). Indirectly, this caused the mine building to be recorded on the plan. The buildings were the surface features most surveyed for the purpose of correlation.

The Act of 1896 amending the 1887 Act improved the accuracy and further enlarged the scope of both the abandonment plan and the working plan. Plans had to be certified by a surveyor before being lodged with the Secretary of State. All headings in advance of the boundary, pillars of mineral remaining unworked, and the position, direction and extent of every known fault or dislocation of the seam with its vertical throw, had to be shown.

Following an appeal for the deposition of old working plans with the Mines Inspectors in 1925, they were, along with the abandonment plans, catalogued and published.

APPENDIX B

Alfreton, Derbyshire. Case History 1 (see Figure 1)

The development of two sites in the centre of Alfreton, Derbyshire necessitated comprehensive desk studies including Bench Mark level histories. The sites are underlain by the productive Middle Coal Measures.

Ordnance Survey maps show the town to be surrounded by mines and the N.C.B. have confirmed that coal extraction has taken place below the sites.

From the old Ordnance Survey maps it has been possible to record ground movements, going back to 1876, at 15 points around the town centre and these are shown in Figure 1. These bench marks confirmed that ground movements had taken place at the sites.

Without reference to mine plans it is possible to note several features associated with the extraction process. The large movements recorded at points 1, 2 and 3 appear to be coming to an end. Their origin may be associated either with the robbing of pillars in the Highfields Colliery (abandoned prior to 1898) or with extraction from the Standwick Collieries which the maps show to be expanding in size between 1876 and 1962.

It is possible from the levels recorded at points 4, 5, 9 and possibly 6, to infer that no recent mining has taken place under Alfreton town centre. Point 11 suggests a similar situation under the railway further to the east. Point 13 and perhaps 12 show movements associated with mining are still taking place. Movements at points 6 and 15 are consistent with long wall workings or a rapid collapse of the roof in stall and pillar workings. At points 4 and 10 the original bench marks were destroyed and at a later date new ones established nearby. In such an instance it is still possible to use the information to demonstrate settlement trends, by estimating the amount of settlement during the intervening period between successive readings.

Although it was concluded that the centre of Alfreton was unlikely to be affected by subsidence, other nearby areas are clearly affected and the pattern of which is different at different locations.

Hanley, Stoke-on-Trent. Case Study 2 (see Figures 2 and 3)

This case history shows the relationship between geological features, main working patterns and surface settlement.

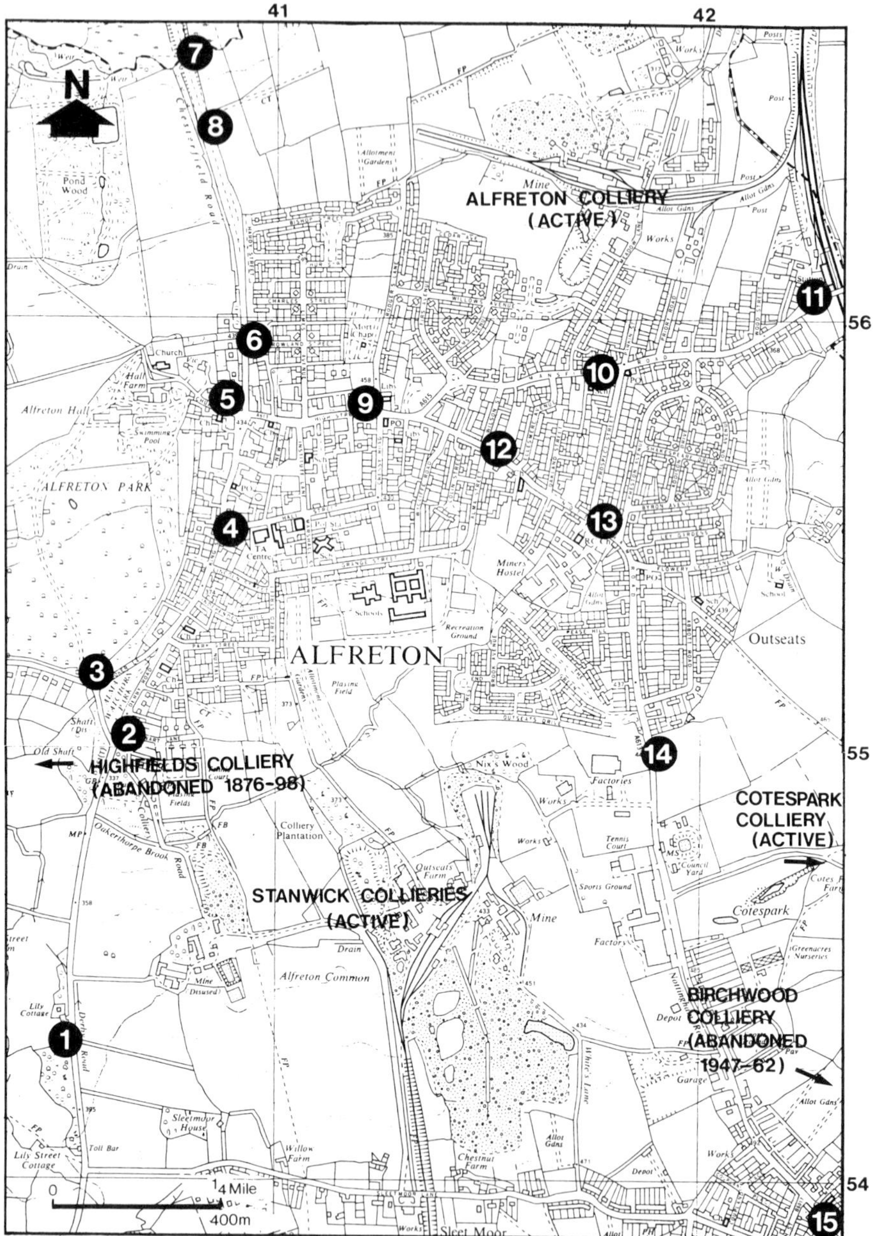

This map is reproduced from the Ordnance Survey Map with the sanction of the Controller of H.M. Stationery Office

LIVERPOOL DATUM

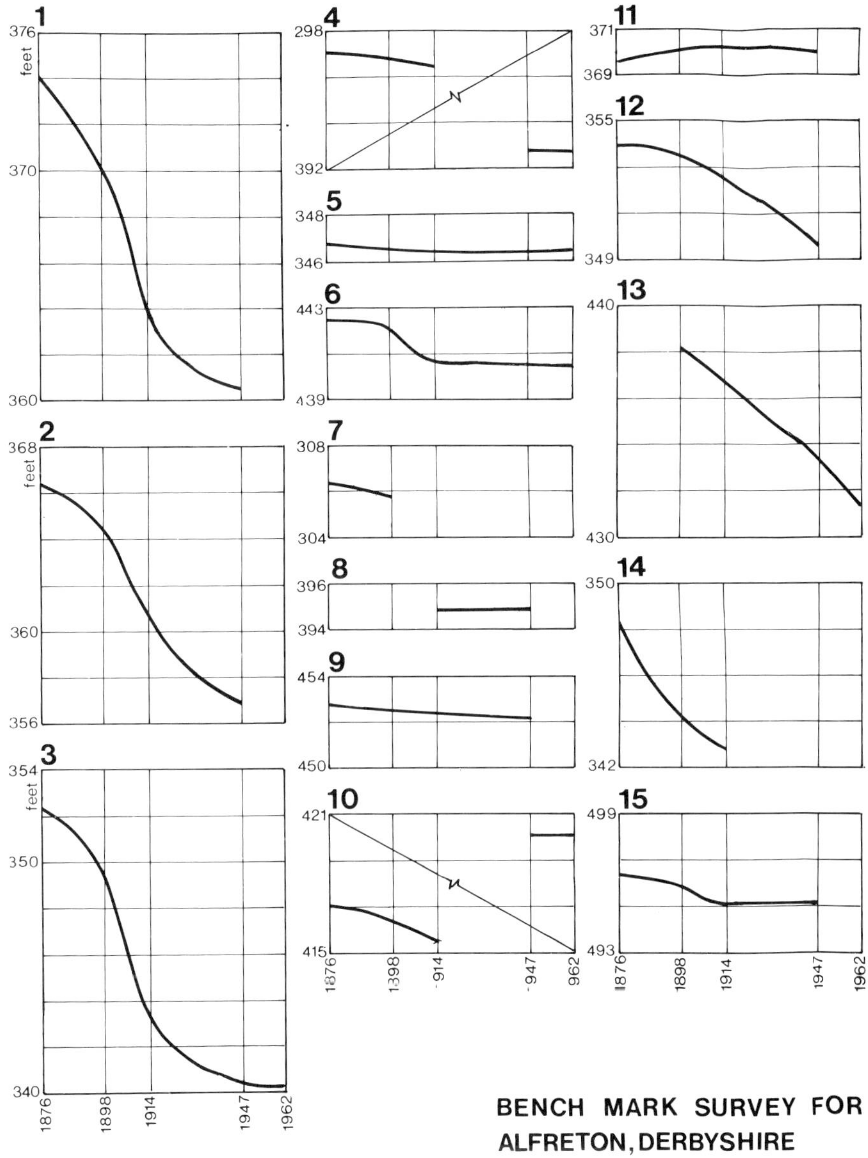

BENCH MARK SURVEY FOR ALFRETON, DERBYSHIRE

FIGURE 1

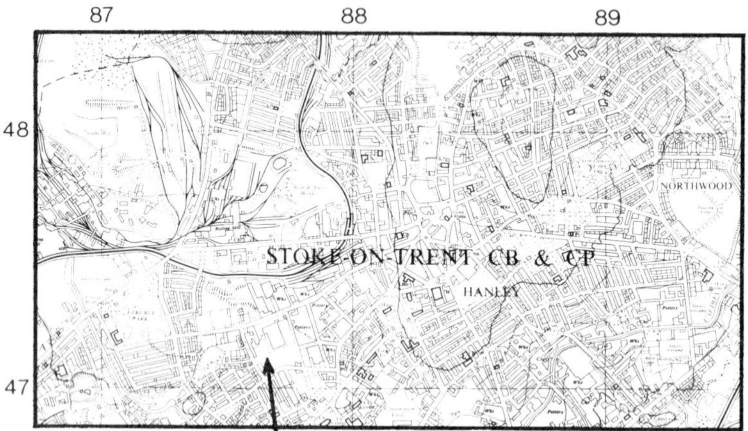

SUN STREET MINE

TOPOGRAPHIC SURVEY OF HANLEY
BASED ON O.S. 1:10 560 SURVEY

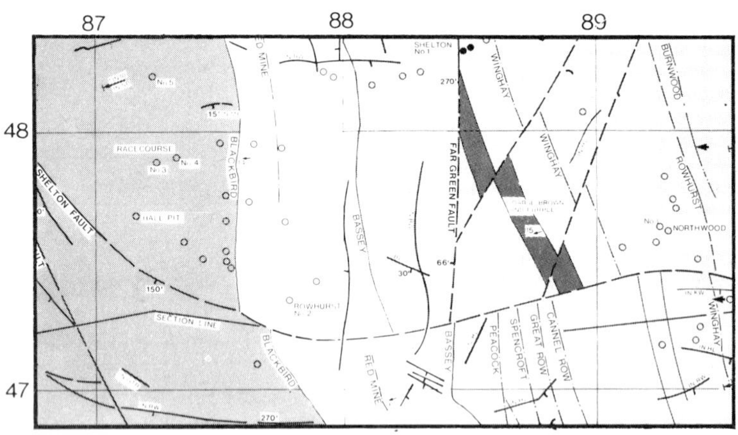

GEOLOGICAL MAP OF HANLEY
BASED ON I.G.S. 1:10 560 SURVEY

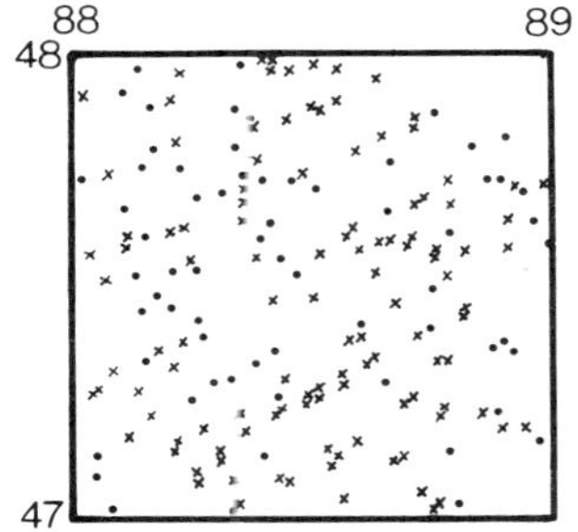

TYPICAL DISTRIBUTION OF LEVEL RECORDS FROM WHICH SUBSIDENCE CONTOUR PLANS OF HANLEY HAVE BEEN CONSTRUCTED

• = B.M. ESTABLISHED IN 1876 AND NOT RE SURVEYED IN 1923

× = LEVEL RECORDS FROM WHICH SUBSIDENCE WAS CALCULATED

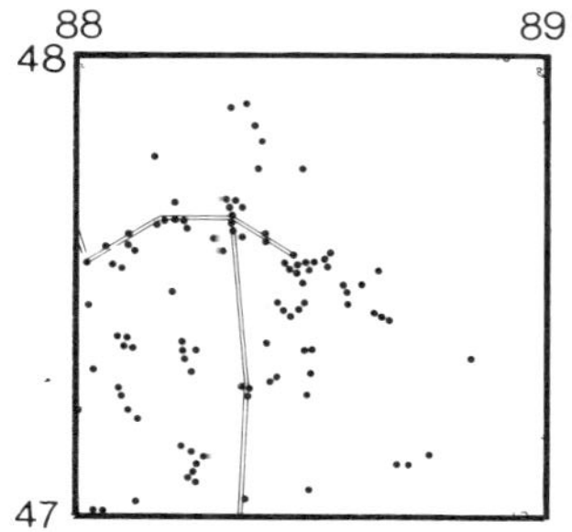

DISTRIBUTION OF MINE SHAFTS AND UNDERGROUND DRAINAGE CHANNELS IN HANLEY

HANLEY

0 — 1000m

FIGURE 2

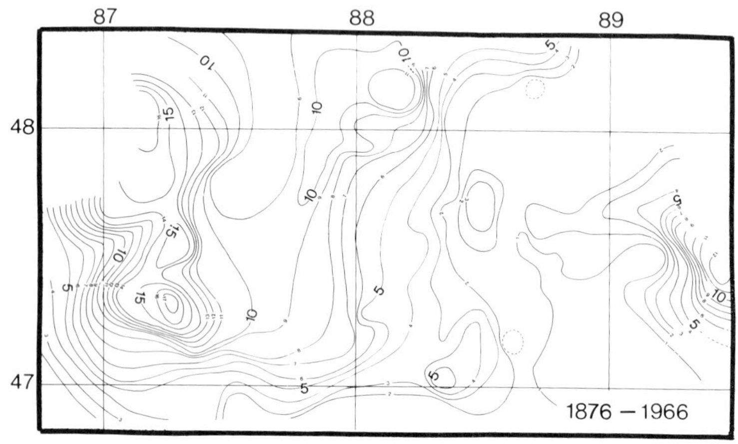
87
88
89
48
47
1876 – 1966

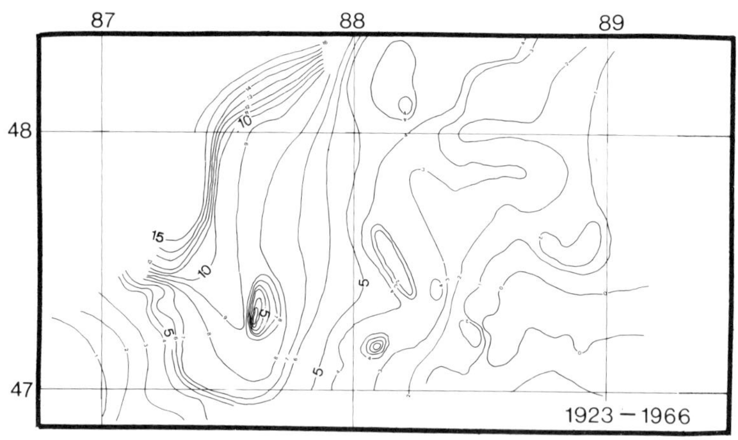
87
88
89
48
47
1923 – 1966

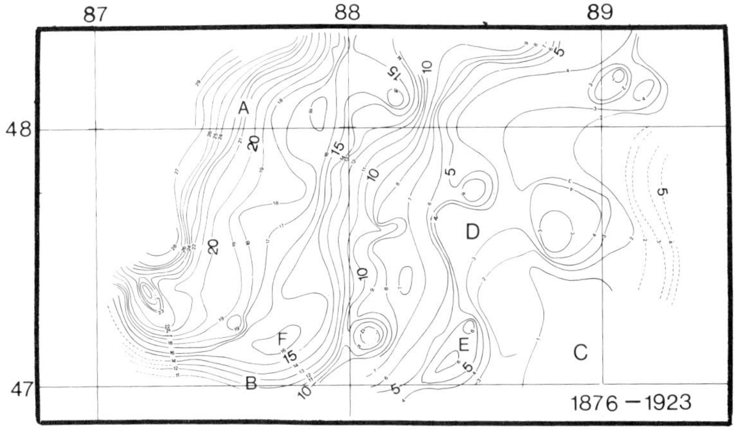

0 1000m

A Major trend follow strike of strata
B Major trend follow fault
C Left unworked
D Concentrations of shafts
E Deep point anomaly
F Sun Stcot Mine

SUBSIDENCE CONTOURS
HANLEY – STOKE ON TRENT

FIGURE 3

The study area in Hanley covers approximately 570 hectares, which is the area covered by 3 half sheets of the 25 inch to 1 mile Ordnance Survey County Series of 1876. Hanley has shown typical development from a hamlet to an industrial town. Mining originally occurred around the nucleus of settlement which subsequently spread over the mined areas leaving behind problems of old mine shafts and unstable ground.

Hanley is situated on the eastern flank of the Potteries Syncline, the geological structure that forms the basin of the Potteries coalfield. Rocks of the Upper and Middle Coal Measures outcrop within the study area. The dip of the strata varies between 8° - 15° to the west. Coal, iron and fireclay have all been extracted by mining. Records relating to the extraction of coal go back to at least AD. 1297. Most mining, however, took place in the late 19th and 20th centuries with the majority of seams extracted lying west of the town centre. At the western limit of the study area it is estimated that a minimum of 25 seams have been worked with approximately 24m (80 feet) of products having been removed.

Prior to examing the Bench Mark data it was evident that a great amount of subsidence had taken place. For example, Josiah Wedgewood's Etruria Works, opened in 1769 and fronting onto the canal, was originally built with its ground floor at water level, but is now 4m below it. The canal banks have continually been raised to compensate for subsidence. It is also worth noting that this is an area that is subject to occasional earth tremors. These are thought to be a result of movement concentrated along fault planes caused by collapse of mine workings.

The area has been levelled on three occasions from which it has been possible to produce subsidence contour maps for the periods 1876-1932, 1923-1966 and 1876-1977 (see Figure 3). The density of Bench Marks originally varied from 30 per square km in the rural areas to more than 140 per square km in the centre of Hanley. After the 1st Edition maps the frequency of Bench Mark points was reduced by about 20% in urban and 40% in rural areas. Approximately 10% new Bench Marks were established at the two re-surveys to fulfil a new need, principally in the new suburbs.

With reference to any other records these subsidence plans clearly demonstrate the extent to which the ground has subsided. Between 1876 and 1966 more than 29 feet (8.83m) of subsidence has taken place on the west side of the study area while under the centre of Hanley movements less than 0.3m are recorded. From these plans differential movements on a site can be estimated.

Because there are many mine abandonment plans for this area

it has been possible to correlate the subsidence with the extraction pattern, and hence make comment on the sensitivity of the Bench Mark data. All three subsidence plans illustrate two types of features which for convenience are termed Major Trends and Anomalies. Major trends are shown by the contours of subsidence parallel to the strike of the rock strata (with greater values down dip). Others follow the orientation of fault lines, the result of mining up to and not beyond the fault plane (see Figure 3 points A and B).

Anomalies appear as contours forming features intercepting the general pattern and recognised as High and Deep points. High point anomalies result from areas not mined perhaps due to adverse geological structures and from areas left unworked, so that the ground does not subside (point C). Within the study area more than 340 shafts have known positions and high point anomalies coincide with clusters of shafts, hence with extensive shaft pillars (point D). The contour plans appear to be too insensitive to isolate single mine shafts. The size and distribution of such features as underground drainage courses and coal barriers, with tendency to run along the strike makes them too insignificant to be observed on the subsidence plans.

Deep point anomalies have resulted from areas of more intensive extraction, for example where six seams have been worked beneath an area generally mined in only one or two (point E). Extraction from small shallow mines that would produce surface subsidence over a small area is usually associated with 'early' and 'rural' mining. These mines are normally too old to be recorded on plans and to be observed on the subsidence plans. However, Sun Street Mine, a small complex that closed down in 1877, is an exception as it coincides with a deep point anomaly (point F). It has not been possible to locate anomalies that result from the collapse of earlier pillar and stall workings.

This case history has not attempted to show the rate of movement at the various survey points; reports of recent damage to structures in the area suggest that subsidence is still continuing. The differential settlements between various points of the area will have given rise to strains which may have adversely changed the pattern of groundwater level and undergroundwater flows.

<u>Broxstowe, North West Nottingham. Case Study 3 (see Figure 4)</u>

This desk study case has been included,although unsuccessful to demonstrate the problems in using Bench Marks in rural areas.

The site is positioned approximately 6.5 km of Nottingham city centre. It lies immediately south of Assarts Farm in a traditionally rural area. The fields are now becoming

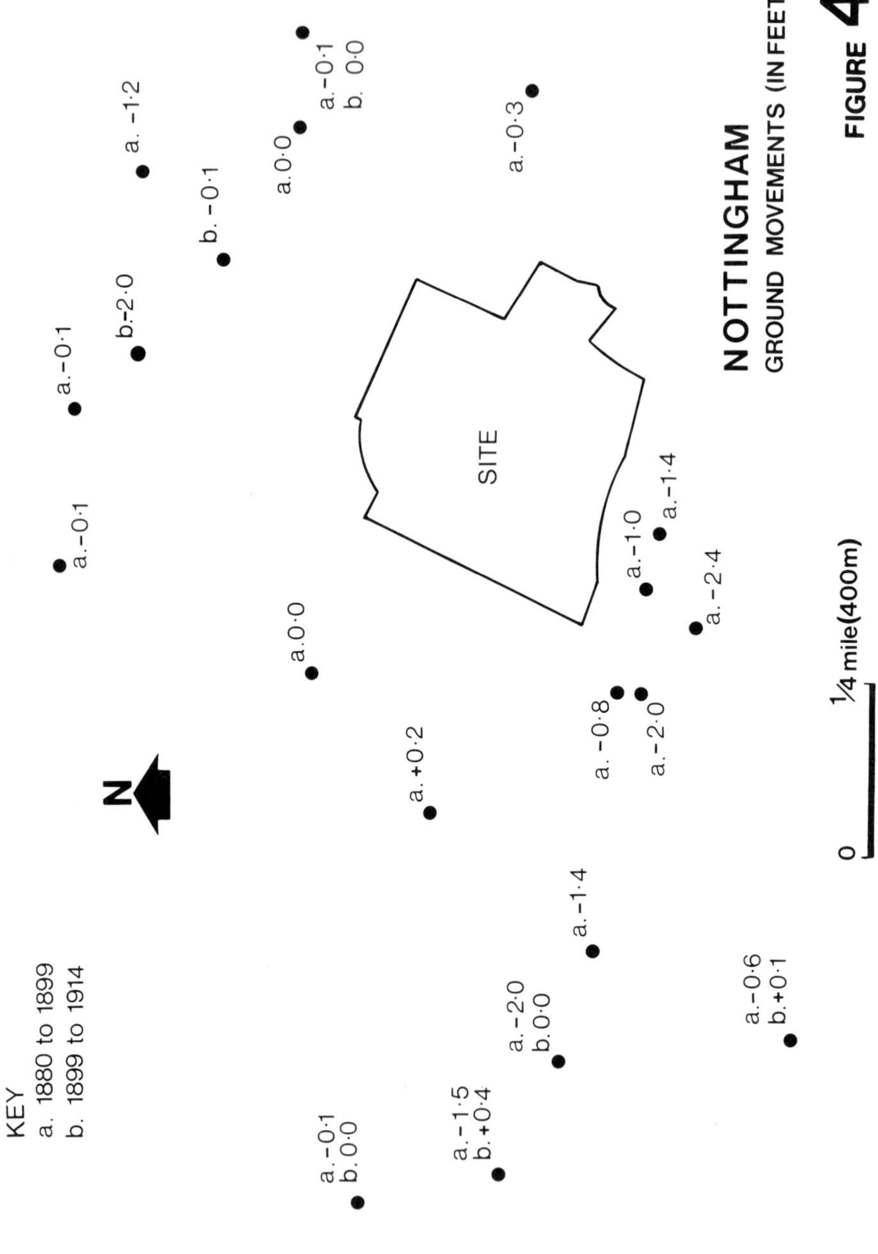

FIGURE 4
NOTTINGHAM
GROUND MOVEMENTS (IN FEET)

engulfed by housing estates.

Beneath the site is approximately 9.14m of massive and tabular red/yellow magnesian limestones. This is followed by 4.57m of blue to light brown shales. Below this are shales, sandstones and coals with associated seatearths of the Middle Coal Measures.

This area of Nottingham has a long history of coal mining, but it is considered that west of Nottingham no pits were sunk through the magnesian limestone until the 1840's

Plans of abandoned workings of unknown date show that workings have taken place in three seams below the site : Top Hard 91.4m; Deep Soft 248.3m and Deep Hard 253m below ground level. It is possible that two shallower unnamed seams have also been worked at 23.7m and 46m below ground level.

There is evidence on the site of old mining activities and this includes old shafts, a tramway and engine house. Aerial photography also indicates shallow workings and associated unstable ground that may be connected with unrecorded workings.

The first edition Ordnance Survey map of 1879 - 1882 shows within the study area 54 Bench Marks including some Spot Heights (along roads, a tramway and field tracks) though none of these were within the site boundary. On the 1899 second edition map only 18 level points survived to enable subsidence to be calculated from readings (a) shown on Figure 4. Fifteen new points were established at this time. Of the 33 points available for relevelling in 1913-1914 only 7 were read and are indicated by (b) on Figure 4. These 7 Bench Marks are no closer than 400m to the site and are considered to be too remote to provide reliable information. The study was therefore abandoned.

<u>Case Study 4. Bank Hill, Airdrie. (see Figure 5)</u>

This case history is reported to show that Ordnance Survey information cannot always be relied upon, and should always be cross checked when the opportunity arises.

The site is located 1.4 km south of Airdrie town centre and adjacent to the dismantled Caledonian Railway, Airdrie and Newhouse Branch.

The site occupies the eastern flank of a N-S boulder clay ridge that is some 15-23m thick. Beneath this are coal measures consisting of weak mudstone, siltstones and sandstones. Eight named coal seams and a major ironstone band are present within the upper 40m of the solid strata.

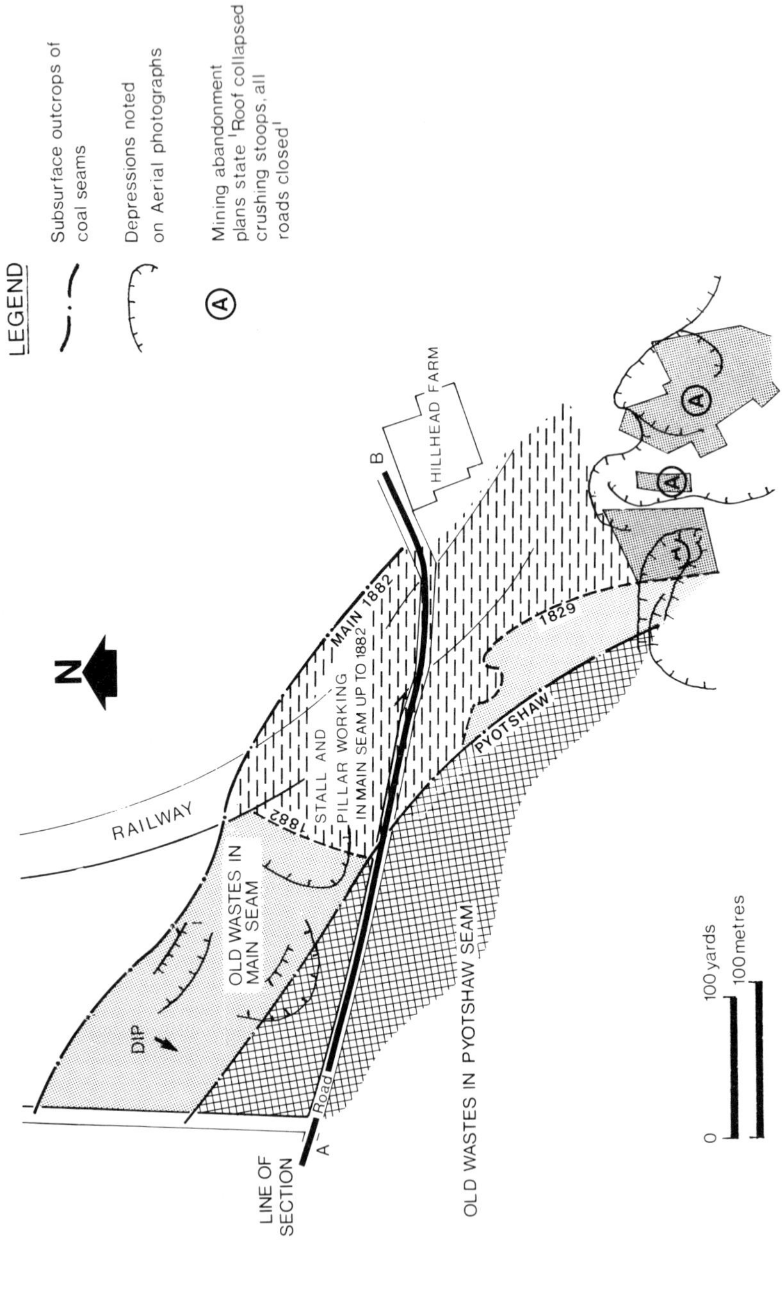
LEGEND
Subsurface outcrops of coal seams
Depressions noted on Aerial photographs
A Mining abandonment plans state 'Roof collapsed crushing stoops, all roads closed'
N
HILLHEAD FARM
B
MAIN 1882
STALL AND PILLAR WORKING IN MAIN SEAM UP TO 1882
1829
PYOTSHAW
RAILWAY
1882
OLD WASTES IN MAIN SEAM
DIP
OLD WASTES IN PYOTSHAW SEAM
Road
A
LINE OF SECTION
0
100 yards
100 metres

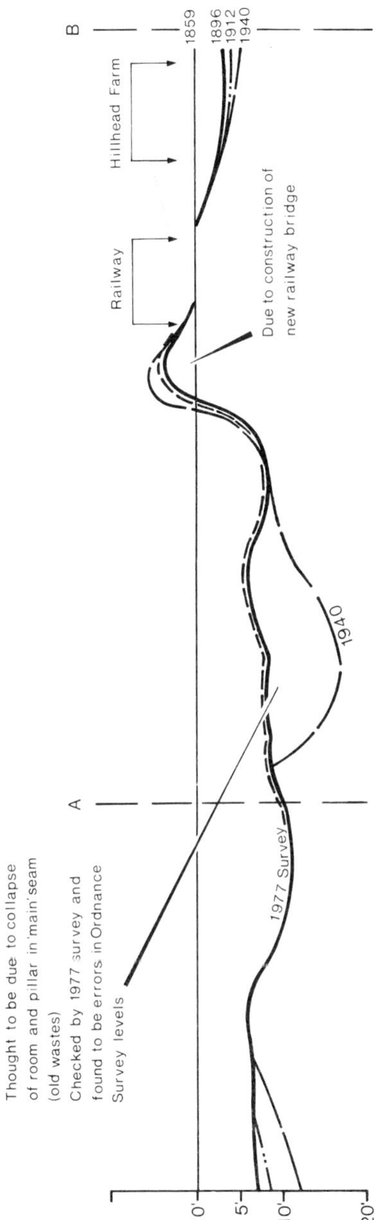

ARDRIE
FIGURE 5

Below the site four coal seams outcrop below the boulder clay and dip 2-5° SW. These are the Pyotshaw, Main, Hump and Virtuewell. The oldest known pit is located ½km to the south and was actively working in 1791. By 1829 there was a small mine on the south boundary of the site certainly working the ironstone band at a 45 fathom (82m) depth and possibly the coal seams above it. An abandonment plan of 1882 shows workings in the Main seam below the site by stall and pillar method, but also records workings in this seam, and the Pyotshaw, up to their sub surface outcrop in 1829.

It is known from mine abandonment plans that south of the site roof collapses have occurred crushing remnant pillars in the Main seam. Aerial photographs show surface depressions coinciding with the known collapses and indicate similar features occur at the south end of the site. It was hoped that a Bench Mark study would confirm and show the extent of this post-abandonment collapse. The survey was carried out using both Bench Marks and Spot Heights, the latter being used with caution since they are positioned on road surfaces and are recorded to the nearest foot. Figure 6 shows the profile of subsidence along the E-W road that forms the south boundary of the site.

The Bench Marks show 7 - 11 feet (2.1 - 3.3m) of subsidence took place along the whole road between 1859 and 1896. It was first thought that the subsidence between 1912 and 1940 showed 8 feet (2.4m) of movement along the eastern half of the road adjacent to an observed surface depression. However, doubt existed about the reliability of the 1940 levels and a check was made by surveying the road from a recently surveyed bench mark. This showed that the road surface was at the same level as in 1896 and indicated that the 1940 levels were in error.

Case Study 5. Burnley. (see Figure 6)

In this case history, the settlement of the ground from the survey records has been small, but site investigation has shown the site to be underlain by partially collapsed old workings.

The site is located in the centre of Burnley and is bounded on the west and south sides by the River Calder in a channel, the Manchester Road, and on the north and east sides by Yorke, Bath and Victoria Street, see Figure 6.

The Town Hall is the principal building on the site and is a four storey sandstone masonry structure, with a semi-basement, constructed in 1864. The Mechanics Institute (now a Social Club) is of a similar style and date.

Early maps of the town (dating from 1848 and 1857) show the site to be located within the marshy valley flood plain of

The River Calder and showed no indication of mine workings.

The 1:10560 scale geological map indicates the site to be underlain by superficial drift deposits - mainly Glacial Tills and the upper strata of the Middle Coal Measures which are composed of sandstone and mudstones with a number of coal seams. The seams existing immediately beneath the site are the Shell mine, Burnley four foot and Maiden mine. Beneath these a thickness of some 50-55m of barren coal measures are encountered before reaching the Lower Yard mine.

Evidence of shallow mining in the immediate vicinity is suggested by an historical account of several small collieries which existed close to the site at the beginning of the 1800's. Due to the age of the workings no specific documentary evidence either from the N.C.B. or local mining literature exists for these workings. It is therefore almost certain that shallow workings, if any exist beneath the site, would have been of the pillar and stall type and would have been abandoned prior to 1850.

Evidence of workings was encountered by the site investigation boreholes, positioned as shown on Figure 6 but the layout of the workings, their extent and amount of roof/pillar collapse could not be precisely determined. A section drawn through the site on line A-A is shown in Figure 6.

From a study of the early editions of the Ordnance Survey maps the bench mark on the Social Centre shows the settlement of the building since the 1880's has been approximately 50mm. A possible explanation for the small settlement of the building is that the mine workings collapsed prior to the establishment of the Ordnance datum and have remained stable ever since.

In view of the poor quality of the ground at the levels of the old workings, they are to be consolidated by grouting, since the proposed development could otherwise initiate further settlement.

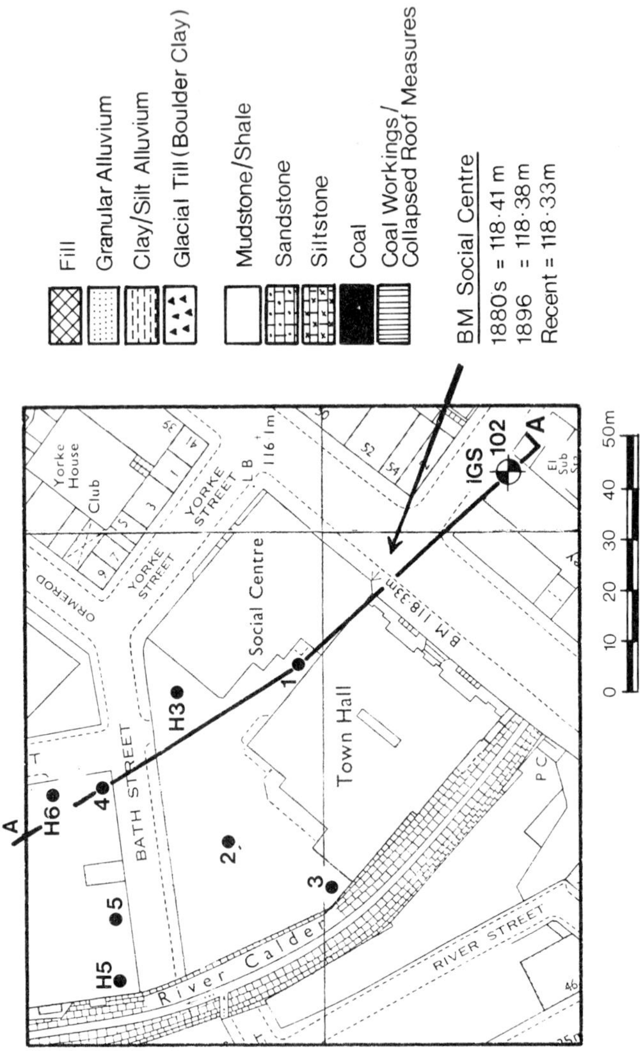

Fill
Granular Alluvium
Clay/Silt Alluvium
Glacial Till (Boulder Clay)
Mudstone/Shale
Sandstone
Siltstone
Coal
Coal Workings / Collapsed Roof Measures
BM Social Centre
1880s = 118·41 m
1896 = 118·38 m
Recent = 118·33m
Yorke House Club
YORKE STREET
ORMEROD
Social Centre
Town Hall
BATH STREET
River Calder
RIVER STREET
BM 118·33m
IGS 102
El Sub Sta
P C
A
H6
H5
H3
1
2
3
4
5
0
10
20
30
40
50m

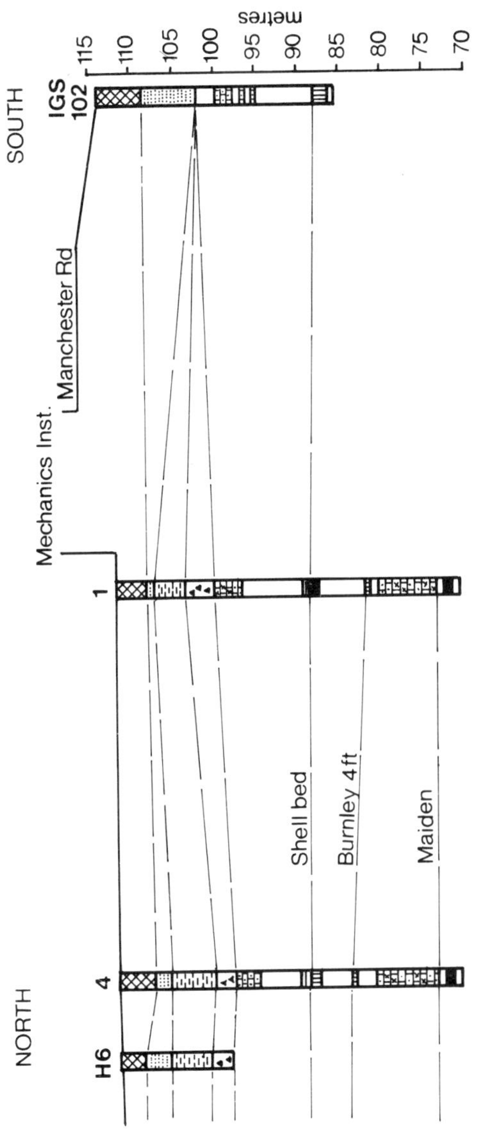

SECTION A – A

FIGURE 6 BURNLEY

REFERENCES

F.G. Bell - Ground conditions in mining areas, Chapter in Methods of Treatment of Unstable Ground, Newnes-Butterworths, London (1976).

W.R. Dearman, M.S. Money, J.R. Coffey, P. Scott and M. Wheeler - Engineering geological mapping of the Tyne and Wear conurbation, North-East England, Quarterly Journal of Engineering Geology 10 (1977) 145-168.

Heriot-Watt University - Principles of Mining Subsidence, Departments of Mining and Civil Engineering, Course on Mining Subsidence, September (1966).

J.B. Harley - Ordnance Survey Maps, A Descriptive Manual, Ordnance Survey Southampton.

T.E. Longford - The subsidence of London, Ordnance Survey Occasional Paper No. 4, HMSO (1932).

R.J. Orchard and H. Ogden - Ground movements in North Staffordshire, North Staffordshire Institution of Mining Engineers 119 (1959) 259-272.

G. Wilson and H. Crace - The settlement of London due to underdrainage of the London Clay, Journal of the Institution of Civil Engineers 19.

PRELIMINARY SITE INVESTIGATION IN OLD COAL MINING AREAS - PROBLEMS OF CORRELATING COAL SEAM NAMES

Malcolm V. Symons, B Sc , Ph D , F G S

Department of Civil Engineering and Building Technology
University of Wales Institute of Science and Technology

SUMMARY

Information on the occurrence of old, shallow mine workings, of vital importance to the planning and design of civil engineering structures, is held in archive sources. It can be difficult to interpret certain of this information, however, because it refers to the past mining activity in terms of the coal seam names in use at the time the mining was carried out. This information will be of practical benefit to the construction industry only if the old coal seam names are correctly correlated with present known seams.

This paper examines the problems involved in such a correlation for an area where coal was mined between the 16th and mid-20th centuries. The practical benefits to be gained from this correlation will be assessed, and it will be advocated that a systematic search be carried out for coal seam name correlations for all of Britain's old coalfields.

INTRODUCTION

Abandoned, shallow mine workings constitute a serious hazard to the safety of structures in old coal mining areas. Existing structures, stable for many years, may suddenly exhibit signs of distress due to the collapse of underlying old workings (Littlejohn 1979, Bruhn, Magnuson and Gray 1980, Carter, Jarman and Sneddon 1980) and, in particular, new structures, which can impose increased loadings on the ground, are in jeopardy if constructed over undetected and untreated shallow workings (Malkin and Wood 1972, Bell 1975, Taylor 1975, Thorburn and Reid 1977).

Whenever construction is proposed in an area known or suspected to have been a once-active coalfield it would, therefore, seem merely a common-sense procedure to examine, at the preliminary site investigation stage, all information sources which can prove, or imply, the presence of old shallow workings. The knowledge gained can then be used as the basis for the subsequent physical site investigation which should include boreholes into the Coal Measures rock.

This procedure is seldom adopted, mainly because of limitations of time and money, and normally only primary source information (National Coal Board Abandonment plans, Ordnance and Geological Survey plans, aerial photographs and any other readily available data) is consulted. If this does not reveal the presence of old, shallow workings, the physical site investigation usually consists only of shell and auger boreholes or trial pits which do not penetrate beyond the interface between the overlying superficial deposits and the Coal Measures rock. This normal procedure has serious limitations because the primary source information invariably post-dates 1872 and tends to provide details of a later and deeper phase of coal mining, which took place after the earlier and shallower mining had ceased. It can, quite erroneously, indicate a complete absence of shallow workings in areas where all the seams have been worked to their outcrops (Symons 1977).

It is, therefore, essential that archive sources containing information on the earlier mining be consulted and, in view of the mounting body of evidence regarding the importance of this information (Fookes 1967, Taylor 1969, 1971 Price 1971, Dumbleton and West 1972, 1976, Dept. of Trade and Industry 1973, Symons 1977, RICS 1978) there is little doubt that, in future, planners and engineers will have to refer to it whenever development is proposed in an old coal mining area.

When the archive sources are consulted two distinct types of information, in terms of practical application, will emerge:-

(1) Direct information, usually in the form of plans and sections, which allows shallow workings to be quite accurately located on present-day Ordnance Survey plans. This information can be easily abstracted, is immediately understandable, and allows engineering decisions to be made.

(2) Indirect information, usually in the form of documents, leases, letters, rentals, mining agents' notebooks, newspapers etc., which proves or strongly implies the existence of shallow workings, but does not allow their accurate location on plan. This information is more

difficult to abstract and needs considerable interpretation before it can be put to practical use. Nevertheless, it is essential information because it provides the planner and engineer with the knowledge, or positive suspicion, that shallow workings exist under a proposed development area.

A paper to the previous conference on Large Ground Movements and Structures, UWIST (Symons 1977) recommended that archive sources should be searched for information they hold on past coal mining activity, but expressed reservation concerning the abstraction of indirect information because of the problems involved in interpreting it. Subsequent research has shown that the main obstacle to understanding the indirect information (and sometimes the direct information) is the fact that all the old documents, leases etc. referred to the coal mining in terms of the coal seam names in use at the time they were written. Unfortunately, because the early coal viewers and mining agents rarely understood the geological structure of their coalfields, the same seam was known by many different names at quite close locations and, conversely, different seams were mistakenly identified with one another. To complicate the issue, names also altered with the passage of time. Unless the old coal seam names can be correlated with present known seams the important indirect information in the archive sources will remain of limited use to the construction industry.

The problems likely to be encountered in attempting to correlate old coal seam names will be described by summarising the findings of a case study carried out in an old coal mining area.

CASE STUDY

All known information relating to coal seam names has been examined for an area of 5200 hectares (approximately 20 square miles) centred around the town of Llanelli, Dyfed, South Wales. The boundaries of the case study area are: northern - Ordnance Survey grid line SN04; southern - the land/sea boundary; eastern - the Plas Isaf Fault which represented a natural discontinuity in coal workings and coal seam outcrops; western - the Moreb Fault which similarly represented a discontinuity in workings and outcrops (Figure 1). Coal was mined from more than 20 seams between the early-16th century and 1953 with peak output being reached in the mid-1860s. Many collieries had been worked-out and closed before the 1872 Coal Mines Regulation Act stipulated that working plans had to be deposited with the Mines Inspectorate on abandonment. As a result, many records, in the form of plans and written statements, have ended up in scattered archive collections and have not passed into the custodianship of the National Coal Board.

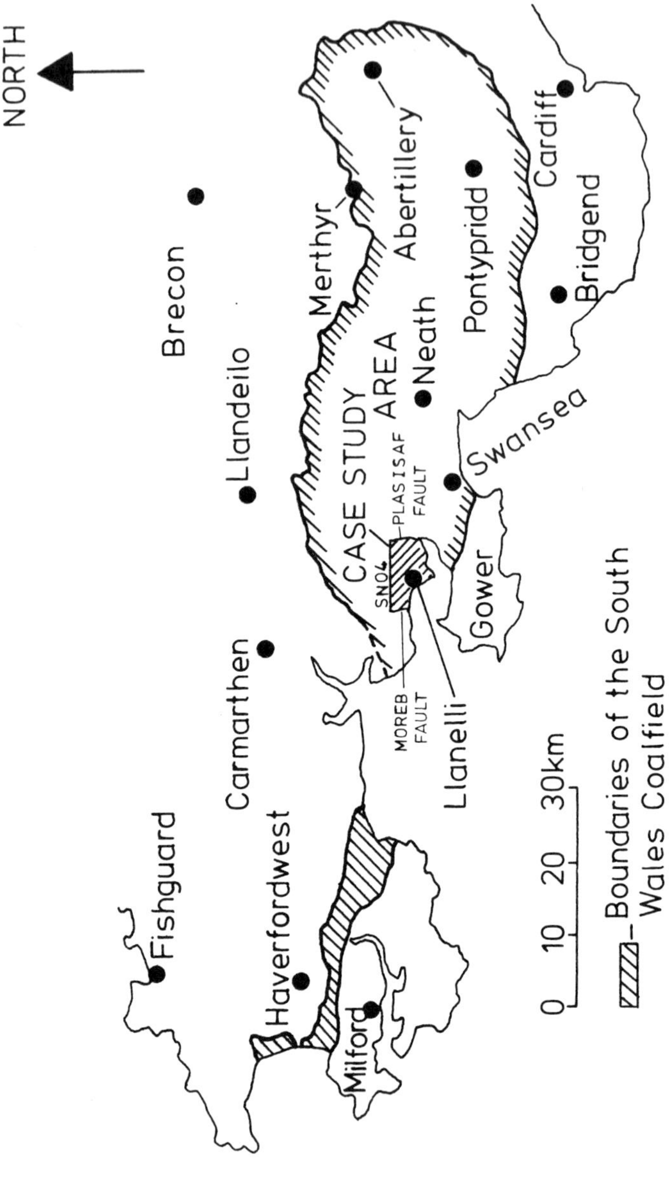

FIGURE 1

LOCATION OF CASE STUDY AREA

RESULTS OF THE CASE STUDY

(a) The geological structure in terms of present knowledge

Coal mining took place entirely in the Upper Coal (or Pennant) Measures, which consist principally of a succession of sandstones and shales with more than 20 coal seams occurring at different horizons within these rocks. These seams were exploited to a maximum depth of about 400 metres below ground surface level. Extraction took place from seams in the Grovesend, Swansea and Hughes Beds of the Upper Pennant Measures and from the Brithdir and Rhondda Beds of the Lower Pennant Measures. The main exploited seams varied in thickness between 0.3m and 2.7m.

The geological structure of the case study area is shown in Figure 2. The coal seam outcrops are represented by dashed lines and the main faults by heavy full lines. The non-standard representation of the faults is deliberate because, in the context of old coal mining, a fault represented a barrier and was seldom worked through. The main faults, therefore, divided the case study area into separate coalfields, each of which was exploited independently of the others. It can be seen that the geological structure is predominantly synclinal with many of the seams having a northern and southern outcrop. The syncline is asymmetric and the coal seams dip more steeply from their southern outcrops. The synclinal axis has a pronounced southwards inclination with depth and occupies a slightly different position, in plan view, for each coal seam. The structure is more complex to the south of the main east/west Carnarfon Fault, being governed by an east/west anticline, and is not known with any certainty to the present day.

For the purposes of this paper, the old coalfield bounded by the Moreb Fault to the west and the Stradey Fault to the east is termed region 1. Similarly, regions 2, 3 and 4 are bounded by the Stradey and Box, Box and Gors and Gors and Plas Isaf faults respectively. The main east/west Carnarfon Fault also acted as a barrier to workings and the entire area south of it is termed region 5. Regions 2, 3 and 4 are further divided into sub-regions north and south of the synclinal axis, designated by suffixes N and S respectively (Figure 2).

Quaternary deposits are present above the Upper Coal Measures rock over much of the area, varying in thickness from a few centimetres to over 30 metres. Alluvial deposits laid down by rivers, streams and estuarine waters predominate near the shore, with glacial deposits in the form of boulder clay and sand and gravel lying mainly inland. The Upper Coal

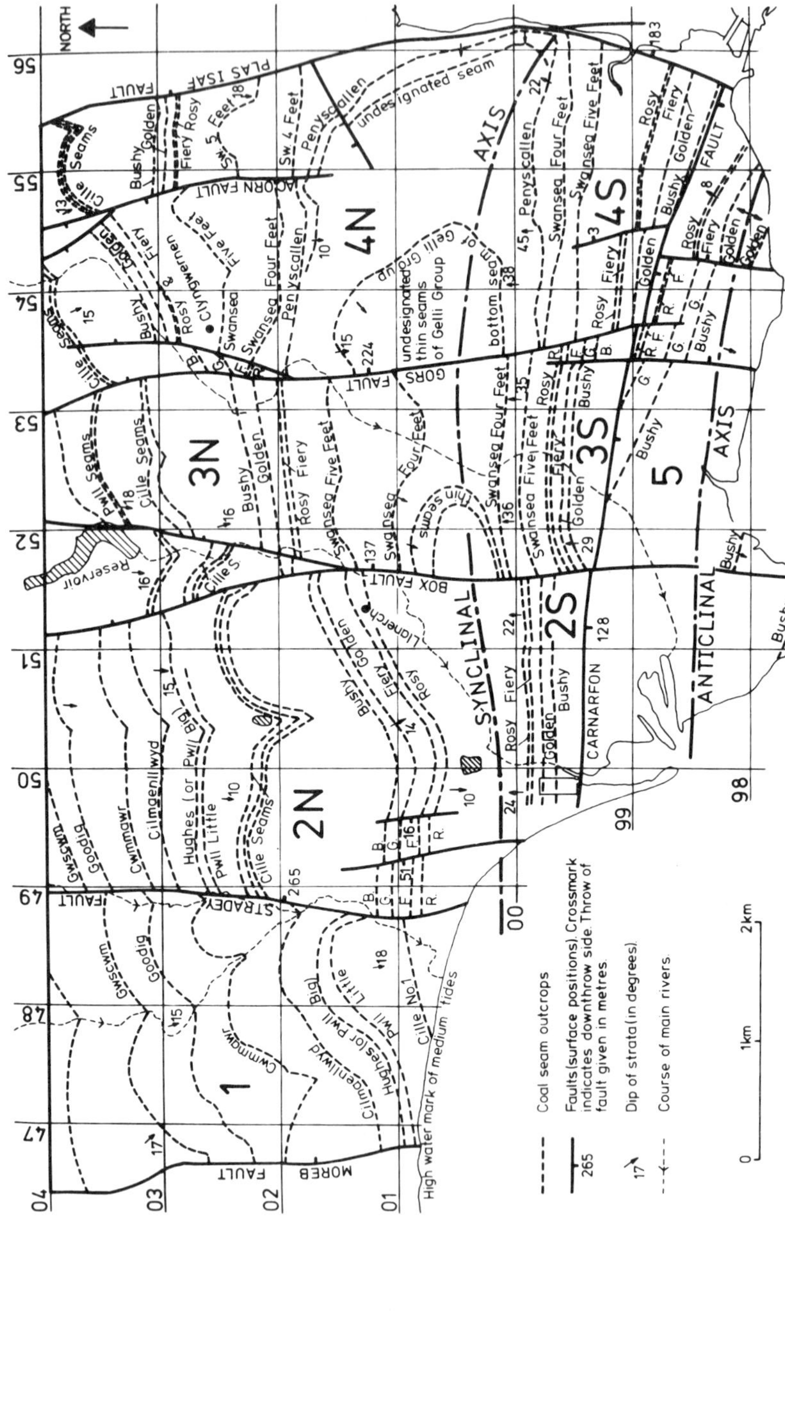

FIGURE 2

GEOLOGICAL STRUCTURE OF CASE STUDY AREA

Measures rock outcrops over a significant proportion of the area.

(b) Coal seam names in terms of present knowledge

The sequence of coal seams exploited in the case study area is given in Table I in terms of the names allocated to them on present-day 1:10560 Geological Survey plans (SN40SE, SN50SW, SN50SE, SS59NW, SS59NE). These names reflect the standard terminology introduced when the Geological Survey published their first 1:10560 plans of the area (Carmarthenshire 54SW, 54SE, 55SW, 58NW, 58NE, 58SW, 58SE, 59NW, 59SW) together with their associated Memoirs (Strahan 1907a, b; Strahan, Cantrill, Dixon and Thomas 1907, 1909) in the first decade of the 20th century.

(c) Coal seam names used in the past

All the coal seam names known to have been used in the case study area are given in Table II in relation to their period and region of use. These names are correlated with present known seams in the sequence adopted in Table I. It can be seen that considerable confusion existed until the early 20th century, with some of the seams having been known by more than 20 different names over the years.

(d) Reasons for the allocation of different coal seam names

Leases of coal were being granted in the case study area by the 16th century and, from the very beginning, recognition of individual seams was important in order that the terms of the leases could be observed. Knowledge of the structure of the coalfield grew slowly, however, and the first instance seen of the allocation of a name to a seam was in 1729. After that time all known and newly discovered seams were allocated names.

In order to appreciate why so many different names were allocated, it was found necessary to gain an understanding of the manner in which the complementary factors of coal mining and geological knowledge developed in the case study area (Symons 1979). Even a summary of these developments would occupy many pages and cannot be presented in a paper of this limited length. A full explanation for the allocation of each of the coal seam names in Table II cannot, therefore, be given. Instead, the main reasons governing coal seam name allocation in the case study area will be examined in order that an appreciation may be gained of the difficulties involved in correlation.

The reasons encountered for the allocation of a name to a coal seam are given in Table III.

TABLE I

COAL SEAM SEQUENCE IN THE CASE STUDY AREA

Coal seam names	Thickness	Notes
Gelli Group of seams	thin to 0.8m	5 or more seams
Un-named seam(s) Penyscallen Un-named seams	thin 0.6m approx. thin	1 or 2 seams 1 or 2 seams
Swansea Four Feet (or Llanelly Six Feet)	0.6m to 2.7m	
Un-named seam(s)	thin	1 or 2 seams
Swansea Five Feet (or Llanelly Four Feet)	0.9m to 2.7m	
Un-named seam	thin	
Upper Swansea Six Feet (or Rosy) Lower Swansea Six Feet (or Fiery)	0.3m to 0.9m 0.9m to 1.2m	Although generally 12-21m apart, these seams merged at certain locations
Swansea Three Feet (or Golden)	0.6m to 0.9m	
Swansea Two Feet (or Bushy)	0.3m to 0.6m	
Un-named seam(s) Cille Nos. 1,2 and 3	thin thin to 0.6m	1 or 2 seams 3 seams
Pwll Little	0.3m to 0.6m	
Hughes (or Pwll Big)	0.6 to 1.2m	
Cilmaenllwyd	0.3m to 0.6m	
Cwmmawr	0.3m to 0.9m	
Un-named seam(s)	thin	1 or 2 seams
Goodig	thin	
Gwscwm	thin to 0.9m	

TABLE II

COAL SEAM NAMES USED IN THE PAST

Present Name(s)	Past Name(s)	Known use	Region	Basis
Gelli Group of seams	Gelly-Gilly	1803-1905	4N/4S	A(i)
	Gelly Little	1824-1926	"	A(i)(iii)
	Gellygele	c1825	"	A(i)
	Gellicilau	1836	"	A(i)
	Bryn	1836	"	A(i)
	Gellywhiad	1836	"	A(i)
	Small	1836-88	"	A(iii)
	Little	1836-39	"	A(iii)
	Brynbach	1836-49	"	A(i)
	Gelly	1863-1912	"	A(i)
	Gelly Rider	1907	"	A(i)(iii)
	Upper Gelly	1905-25	"	A(i)(iii)
	Lower Gelly	1926	"	A(i)(iii)
	Gelly Fach	1926-28	"	A(i)(iii)
	Gelly Fawr	1928	"	A(i)(iii)
	Loughor Little	1953	"	E
	Loughor Little Rider	1953	"	E
Un-named seam(s), Penyscallen, un-named seam(s)	Little	1805-22	3N	A(iii)
	Genwen Small	1812	4S	A(i)(iii)
	Small	1825	3N	A(iii)
	Penprys	1825	4N	B(iii)
	Rosy	1835-6	4S	B(i)
	Box Little	1850	3N	A(i)(iii)
	Upper	1858	3N	A(iii)
	Lower	1859	3N	A(iii)
	Little	1888-94	4S	A(iii)
	Clement's Nos 1,2 and 3	1890-1907	3N/3S	A(ii)
	Quarry	1891	4S	A(i)
	Lower Genwen	1894	4S	A(i)(iii)

Present Name(s)	Past Name(s)	Known use	Region	Basis
Un-named seam(s), Penyscallen, un-named seam(s) (continued)	Little No. 2 or Upper	1894	4S	A(iii)
	Penyscallen	1907 to present	4N/4S	E
	Ar-y-cwar	1907	4S	A(iii)
	Dan-y-cwar	1907	4S	A(iii)
	Loughor Little	1907	4S	E(error)
	Little	1908	4N	A(iii)
	Stinking	1943	4N	A(iii)
	Dirty or Little	1917	4S	A(iii)
Swansea Four Feet (or Llanelly Six Feet)	Penyfigwm	1762-1808	3N	A(i)
	Rosey	1762-1806	"	B(i)
	Great	1762-1896	"	A(iii)
	Box	1808	"	D(i)
	Six Feet	1833-1930	3N/3S	A(iii)
	Box Great	1836	"	D(i)
	Nine Feet	1836-60	"	A(iii)
	Big	1840-1925	"	A(iii)
	Llanelly Six Feet	1901 to present	"	E
	Box Big	1907-27	"	A(iii)
	Warde's Fiery	1907	"	B(iv)
	Swansea Four Feet	1907 to present	"	E
	Fiery	1772-1836	4S	B(i)
	Old Penyfan or Daven Pill Fiery	1807	"	D(i)
	Daven Pill Burry River Fiery	1808	"	D(i)
	Old Fiery	1809	"	D(i)
	True Fiery	1813	"	D(i)
	Six Feet	1842-1923	"	D(ii)

Present Name(s)	Past Name(s)	Known use	Region	Basis
Swansea Four Feet (or Llanelly Six Feet) (continued)	Great	1842-1900	"	D(ii)
	Warde's Fiery	1865	4S	B(iv)
	Nine Feet	1880	"	D(ii)
	Big	1897-1908	"	D(ii)
	Llanelly Six Feet	1907 to present	"	E
	Swansea Four Feet	1907 to present	"	E
	Penprys	1803-58	4N	A(i)
	Glanmurrog	1803	"	A(i)
	Glanmwrwg	1824	"	A(i)
	Glanmwrwg Great	1825	"	A(i)(iii)
	St. David's	1836-62	"	D(i)
	Llangennech	1836	"	A(i)
	Fiery	1847	"	D(ii)
	Six Feet	1858-1902	"	D(ii)
	Llanelly Big	1881	"	A(i)(iii)
	Clyngwernen	1895-1907	"	B(iv)
	Big	1903-09	"	D(ii)
	Llanelly Six Feet	1907 to present	"	E
	Swansea Four Feet	1907 to present	"	E
Un-named seam(s)	Golden	1836	4N	B(i)
Swansea Five Feet (or Llanelly Four Feet)	Stradey	1767	3N	B(i)
	Llanerch	1805	"	B(iii)
	Fiery	1805-35	"	B(iii)
	Penygare	1813-27	"	A(i)
	Warde's Fiery	1823-27	"	A(ii)D(i)
	Box	1836-54	"	C
	Four Feet	1840-1930	3N/3S	A(iii)
	Upper	1860-76	"	A(iii)

Present Name(s)	Past Name(s)	Known use	Region	Basis
Swansea Five Feet (or Llanelly Four Feet) (continued)	Llandafen	1860	3N/3S	A(i)
	Box Lower	1880	"	A(i)(iii)
	Box Five	1907	"	A(i)(iii)
	Llanelly Four Feet	1907 to present	"	E
	Swansea Five Feet	1907 to present	"	E
	Marsh	1755-61	4S	A(i)
	Great	1772-1812	"	A(iii)
	Golden	1772-1808	"	A(iii)
	Gwndwn Mawr Golden	1807	"	D(i)
	Old Gundunmaur	1808	"	D(i)
	Old Golden	1812	"	D(i)
	Great Fiery	1835	"	B(ii)
	Cwmfelin	1835-36	"	A(i)
	Fiery	1836	"	B(ii)
	Nine Feet	1861	"	B(ii)
	Genwen	1880	"	A(i)
	Genwen Great	1907	"	A(i)(iii)
	Genwen Big	1907	"	A(i)(iii)
	Llanelly Four Feet	1907 to present	"	E
	Swansea Five Feet	1907 to present	"	E
	Big	1908	"	B(iv)
	Four Feet	1927	"	D(ii)
	Alt Cole	1773-74	4N	A(i)
	Clyngwernen	1803-96	"	A(i)
	Cornhwrdd	1836	"	A(i)
	Llangennech Four Feet	1877	"	A(i)(iii)
	Four Feet	1896-1908	"	D(ii)
Un-named seam	Little	1892-1916	3S	A(iii)
Upper Swansea Six Feet (or Rosy)	Thomas David's	1754-58	2N/2S	A(ii)
	Ca-Main	1754-58	2N	A(i)
	Rosey, Rosy or Rose	1762 to present	2,3,4	A(iii)

Present Name(s)	Past Name(s)	Known use	Region	Basis
Upper Swansea Six Feet (or Rosy) (continued)	Penyfigwm	1776-1806	2N	B(i)
	Stinking	1788	2N	A(iii)
	Carnarfon Upper	1795-1836	5	A(i)(iii)
	Great	1806	2N	B(i)
	Little	1836	2N/2S	B(i)
	Binea Upper	1852	5	A(i)(iii)
	Four Feet	1866	5	A(iii)
	Upper Swansea Six Feet	1907 to present	2-5	E
Lower Swansea Six Feet (or Fiery)	Fiery	1749 to present	2-5	A(iii)
	Llwynwilog	1754	3S	A(i)
	Ca-Plump	1758	2N	A(i)
	Carnarfon Lower	1795-1836	5	A(i)(iii)
	Bowen's Real Old Fiery	1808	2S	D(i)
	Old Fiery	1811	2N	D(i)
	Lady Mansel's or Raby's Old Fiery	1819	2N	D(i)
	Rosy	1836	2N/2S	B(i)
	Binea Lower	1852	5	A(i)(iii)
	Little	1866	5	A(iii)
	Lower Swansea Six Feet	1907 to present	2-5	E
Swansea Three Feet (or Golden)	Bwysva	1729	2S	A(i)
	Boisva	1754	2S	A(i)
	Parkycrydd	1758-1805	2N	A(i)
	Killyveig	1758-1805	2N	A(i)
	Country Coal and Dray Pitt	1762-1776	2N	A(iii)

Present Name(s)	Past Name(s)	Known use	Region	Basis
Swansea Three Feet (or Golden) (continued)	Engine	1762-1776	2N	A(iii)
	Old Golden	1794	2N/2S	B(i)
	Golden	1794 to present	2-5	B(i),D(ii)
	Trosserch	1825	4N	A(i)
	Fiery	1836	2N/2S	B(i) or B(ii)
	Penbryn or Small	1852	5	A(i)(iii)
	Penbryn Little	1867	5	A(i)(iii)
	Swansea Three Feet	1907 to present	2-5	E
Swansea Two Feet (or Bushy)	Biggin or Hill	1754	2S	A(i)
	Cwm	1762-99	2N	A(i)
	Caereithin	1794-1836	2N	A(i)
	Bushy	1806 to present	2-5	A(iii)
	Lower	1819	2N/2S	A(iii)
	Golden	1836	2S	B(i)
	Llanlliedi	1835	3N	A(i)
	Carnarvon	1854	5	A(i)
	Swansea Two Feet	1907 to present	2-5	E
Cille Nos. 1,2 and 3	Brickyard Seam	1822	2N	A(i)
	David Charles'	1870	2N	A(ii)
	Cille Nos. 1,2 and 3	1872 to present	1-5	A(i)
Pwll Little	Black Rock	1832-1952	1	A(i)
	Pwll Little	1907 to present	1,2N, 3N	A(i)(iii)
	Clay	1907	1	A(iii)
	Rock	1910	2N	D(ii)
	Black	1918	2N	D(ii)

Present Name(s)	Past Name(s)	Known use	Region	Basis
Hughes (or Pwll Big)	Pool	1765-1855	1	A(i)
	Pwll	1836-1909	1,2N, 3N	A(i)
	Bettytor	1907	1	A(iii)
	Cille No. 4(?)	1907	2N	A(i)
	Hughes	1907 to present	1,2N, 3N	E
	Pwll Big	1919 to present	1,2N, 3N	A(i)(iii)
Cilmaenllwyd	Stradey	1762-1836	1	A(i)
	Cwtta	1822	2N	A(i)
	Cilmaenllwyd	1836 to present	1,2N	A(i)
	Weasel	1907-33	1	A(iii)
	Cille No.5(?)	1907	2N	A(i)
	Shenkin	1907	1	A(ii) or A(iii)
Cwmmawr	Hengoed	1822	2N	A(i)
	Cwmmawr	1875 to present	1,2N	A(i)
	Rees	1888-1907	1,2N	A(ii)
	Cille No.6(?)	1907-52	2N	A(i)
Un-named seam	Pantylliedy Fawr	1822	2N	A(i)
Goodig	Goodig	1907 to present	1,2N	A(i)
Gwscwm	Trevenna or Tre-fanau	1822-1923	1,2N	A(i)
	Gwscwm	1859 to present	1,2N	D(ii)

TABLE III

REASONS FOR THE ALLOCATION OF COAL SEAM NAMES

(A) - REGIONAL AND DESCRIPTIVE NAMING based on (i) the district, farm or landshare where the seam was discovered (ii) the person exploiting the seam (iii) colloquial terminology
(B) - NON-DELIBERATE MIS-IDENTIFICATION WITH A PREVIOUSLY-NAMED, BUT DIFFERENT, SEAM due to (i) inability to appreciate the effects of faulting on seam occurrences (ii) equating seams of equal thickness (iii) visual extrapolation of a seam outcrop, beyond the point of working, to the distant outcrop of a known seam (iv) incorrect transmission of oral knowledge over the years
(C) - DELIBERATE MIS-IDENTIFICATION WITH A DIFFERENT, BUT SUPERIOR, SEAM IN ORDER TO PROMOTE SALES
(D) - LOCALISED RE-NAMING in order to (i) promote new interest in the export markets (ii) correlate with the same seam in another part of the coalfield
(E) - RE-NAMING BY THE GEOLOGICAL SURVEY OR NATIONAL COAL BOARD FOR CORRELATIVE PURPOSES IN THE SOUTH WALES COALFIELD

(A) Regional and descriptive naming

Names were being allocated to the coal seams in the case study area by 1729 at the latest but, for more than a hundred years after that time, the area's geological structure was barely understood. The early coal viewers and agents did not employ lithological and palaeontological techniques to identify beds and seams and their understanding of stratigraphy was imperfect. It is hardly surprising they could not correlate newly-discovered seams with already-known and named seams and, instead, they allocated a new name almost every time a coal seam was encountered at a new location. Regional and descriptive naming, therefore, was the basis for the area's subsequent coal seam name pattern.

There were three main approaches to regional and descriptive naming:-

(i) Naming after the district, farm or landshare where the seam was discovered - this was, by far, the most common approach and the first known allocation of a name to a seam in the case study area was on this basis; the Swansea Three Feet Vein in region 2S being called the Bwysva Vein, in 1729, after the landshare of the same name across which the seam outcropped. The extent to which regional naming pre-dominated in the early days, and has often carried through to the present time, can be seen by reference to Table II in which the successive names of Gelly, Bryn, Genwen, Penprys, Box, Penyfigwm, Penyfan, Daven, Glanmurrog or Glanmwrwg, Llangennech, Clyngwernen, Stradey, Llanerch, Pinygare or Penygare, Llandafen, Gwndwn, Cwmfelin, Alt, Cornhwrdd, Ca-Main, Carnarfon, Binea, Llwynwilog, Ca-Plump, Bwysva or Boisva, Parkycrydd, Killyveig Trosserch, Penbryn, Biggin, Caereithin, Llanlliedi, Cille, Black Rock, Pool or Pwll, Cwtta, Cilmaenllwyd, Hengoed, Cwmmawr, Pantylliedy, Goodig, Trevenna or Tre-fanau, Gwscwm are all names of districts, farms or landshares in the case study area.

(ii) Naming after the person exploiting the seam - this was a less common method, the successive names in Table II of Clement, Warde, Thomas David, Bowen, Lady Mansel, Raby, David Charles, Rees, representing this form of naming.

(iii) Naming on colloquial terminology - this was a common method but it is difficult, in some cases, to provide an explanation for the choice of a particular name. The allocation of Little, Small, Great, Big, Upper, Lower, Stinking, Nine Feet, Six Feet etc. are self-explanatory, but the true origins of Rosy, Fiery, Golden, Bushy, Bettytor and Weasel have been lost with the passage of time and can only be guessed at today.

(B) Non-deliberate mis-identification with a previously named, but different seam

The case study area experienced its real industrialisation from the middle of the 18th century when capitalists invested heavily into the local coal industry. Their coal viewers and agents made repeated attempts to correlate the coal seams but they were working intuitively, without modern geological knowledge, and also held fundamental misconceptions regarding the effects of faulting. It was inevitable that their attempts at correlation would result in widespread mis-identification. This mis-identification carried through unchecked until the 1840s when the newly-formed Ordnance Geological Survey systemised and rationalised the locally held knowledge. Errors were perpetuated after that time, though, and have persisted into the 20th century.

There were four main reasons why seams were accidentally mis-identified with one another:-

(i) Inability to appreciate the effects of faulting - the early coal viewers regarded faults merely as zones of disturbed ground which halted workings in a given seam and, in the case study area, referred to them as "bars". They did not appreciate that seams could be displaced by up to hundreds of metres vertically by faulting and worked on the assumption that the sequence of seams, from ground surface downwards, was identical in all parts of the coalfield (Symons 1979). This led to widespread mid-identification, which was further compounded by the fact that the true sequence of seams was not known everywhere because not all seams had been located. For example, in 1772 the Fiery Vein in region 2S and the Swansea Four Feet Vein in region 4S were both given the same name (Fiery) on the assumption that they were one and the same, because they were the second known and worked seams below ground level in their particular areas. Again, in 1805, a legal case, disputing the right of an industrialist to work a given seam, was based entirely on the premise that the Rosy Vein in region 2N and the Swansea Four Feet Vein in region 3N were the same, because both were the topmost known and worked seams in their areas (the undesignated thin seams between 3N and 3S not having been located and worked at that time) - See Figure 2. Other examples of this form of mis-identification exist for the case study area.

(ii) Equating seams of similar thickness - this type of accidental mis-identification was not common but did sometimes occur. This was particularly the case with the Swansea Four Feet and Swansea Five Feet Veins, which both reached the exceptional thickness of 2.7m in certain areas, and were regularly confused with one another.

(iii) <u>Visual extrapolation of a seam outcrop, beyond the point of working, to the distant outcrop of a known seam</u> - this mis-identification usually occurred when seam outcrops were visually extrapolated over a fault. In 1805, the outcrop of the Swansea Five Feet Vein in region 3N was visually extrapolated, at ground surface level, over the Box Fault, to the known outcrop of the Fiery Vein in region 2N at a location called Llanerch. As a result the Swansea Five Feet Vein was mistakenly identified as the "Llanerch or Fiery Vein". This type of mis-identification also took place between faults, plan evidence of 1825 showing that the south-eastward orientated outcrop of the Swansea Four Feet Vein, immediately east of the Gors Fault in region 4N, was extrapolated to the outcrop of the Penyscallen Vein west of the Acorn Fault in the same region; both seams being termed the Penprys Vein after the farm where the Swansea Four Feet Vein outcropped (Figure 2).

(iv) <u>Incorrect transmission of oral knowledge over the years</u> - the reluctance of the early coal viewers to commit their knowledge to writing led to a widespread word-of-mouth transmission of coal seam names. Errors inevitably crept in with the passage of years and, in some cases, have remained to the present day. It is known that the Swansea Five Feet Vein in region 4N was known as the Clyngwernen Vein, after the farm where it outcrops, as early as 1803, and retained this name locally for most of the 19th century. Incorrect transmission of oral knowledge led to use of this name for the Swansea Four Feet Vein, before the end of the 19th century. The Geological Survey, in accepting the opinion of local mining engineers, perpetuated this error in the 1907 Memoir covering the area (Strahan 1907a), with the result that it is incorrectly documented as an alternative name for the Swansea Four Feet Vein to the present day. In a similar manner, Warde's Fiery Vein, a name originally allocated to the Swansea Five Feet Vein in region 3N in 1823, became mistakenly transferred to the Swansea Four Feet Vein with the passage of time and has also been erroneously documented in the 1907 Memoir.

(C) <u>Deliberate mis-identification with a different, but superior, seam in order to promote sales</u>

Coal output in the case study area was directly related to export trade demand until the last quarter of the 19th century and, for most of the time, the ships' masters were free-traders, not committed to any contract, who would only come to buy the local coal if they knew that there was a ready and profitable market for it. In this respect, coal from a named seam which had proved to be in demand in both home and overseas markets was always given preference. When such a seam became worked-out in a colliery, the coal-owner was often tempted to call one of his other partly-worked, and possibly inferior, seams by the same name in order to retain his share of the export market. This occurred in the mid-1830s at the Box Colliery in region 3N. The Swansea Four Feet Vein had been extensively exploited, under the name Box Vein, at this colliery since 1808 and was fast becoming exhausted. The Swansea Five Feet Vein, known as Warde's Fiery Vein at the colliery, had proved to be an inferior seam and had not sold well but, by 1836, it too was known as the Box Vein and was sold, under this name, to the ships' masters.

(D) Localised re-naming

There were two reasons for the localised re-naming of a seam:-

(i) To promote new interest in the export markets - this was quite distinct from reason (C) and usually coincided with the opening of a new colliery or the expansion of an existing concern. On the opening of the Box Colliery in region 3N, in 1808, the Swansea Four Feet Vein (previously known as the Penyfigwm or Great Vein at this location) was re-named the Box Vein. Similarly, soon after the opening of the St. David's Colliery in region 4N, in 1832, the Swansea Four Feet Vein (previously known as the Penprys or Glanmwrwg Vein at this location) was re-named the St. David's Vein.

If the seam at a new or expanded colliery had previously been in particular demand, the coal-owners would often capitalise on this fact by allocating a new name which was a minor variation on the old name, but which would be likely to catch the interest of the discriminating ships' masters. Old Penyfan or Daven Pill Fiery Vein, Old Fiery Vein, True Fiery Vein, Bowen's Real Old Fiery Vein, all variations on the original name of Fiery Vein, are examples of this re-naming for advertising purposes.

(ii) To correlate with the same seam in another part of the coalfield - this corrective re-naming took place as knowledge of the geological structure of the area increased and correlation became possible. It really began in the late 1830s/mid 1840s when, firstly William Logan and, later, Henry De la Beche of the newly-formed Ordnance Geological Survey mapped the area (Bassett 1967, 1969). They rationalised the existing geological knowledge and corrected many of the misconceptions held up to that time. An example of this corrective re-naming is provided by the Swansea Four Feet Vein which was correctly correlated throughout the case study area, between 1833 and 1858, under the name Six Feet Vein.

(E) Re-naming by the Geological Survey or National Coal Board for correlative purposes in the South Wales Coalfield

The geological Memoirs covering the case study area were

published in 1907 and 1909, the survey having been carried out during the previous few years. The Geological Survey was involved in correlation work for all the South Wales Coalfield and used modern stratigraphic and palaeontological techniques. Seams were identified and correlated more accurately than ever before and the re-naming which took place, at that time, is in general use today. Basically, the most important exploited seams were re-named in accordance with the terminology adopted in the Swansea area of the South Wales Coalfield, with an alternative local name retained for identification purposes. Examples of this are all the seams with the word Swansea in their titles, the Penyscallen Vein and the Hughes Vein. Thinner, less-exploited and, hence, less-well known seams were standardised in terms of their contemporary local names, although this has tended to change since 1907/1909.

The National Coal Board has also standardised seam names since 1947, an example of this being the use of the name Loughor Little Vein for the previously used Gelli terminology in the 4N/4S region.

THE PRACTICAL BENEFITS TO BE GAINED FROM CORRELATING OLD COAL SEAM NAMES WITH PRESENT KNOWN SEAMS

When construction is proposed in an old coal mining area the engineer should, ideally, at the preliminary site investigation stage, refer to all extant information on the following aspects of the past mining activity:-

(1) the extent of shallow workings
(2) the method of coal extraction adopted
(3) the extent of pillar robbing, if partial extraction methods had been used.

(Information should also be obtained on the presence of old shafts but discussion on this point is not relevant to the theme of this paper).

This information would allow the engineer to estimate the probable extent of upward void migration due to roof collapse of old workings, and also the likelihood of pillar collapse due to time-dependent crushing or punching into underlying seat earth.

Because shallow mining was often an early phase of mining, the information relating to it is usually contained in archive sources pre-dating 1872. Similarly, information on pillar robbing before, and often after, 1872 is usually held in archives and invariably refers to the workings in terms of the seam names in contemporary usage. If the site investigation engineer does search the archives he is regularly faced with the frustrating situation of unearthing documents which refer to shallow workings or pillar robbing in the area under investigation, without being able to identify which of the seams on the Geological Survey plans are being referred to.

This problem has continually been encountered in the case study area and the following examples illustrate the practical benefits that accrue, if coal seam name correlations are available.

Shallow workings - A mining report of 1772 specifically stated that the Golden or Great Vein had been worked out at shallow depth in the eastern half of region 4S, but the report did not allow location of the workings on plan. Interpretation of the available archive information relating to the boundaries specified in the original coal leases, known old mining sites, farm names etc., allows the Golden or Great Vein to be identified as the Swansea Five Feet Vein in this region (Figure 2 and Table II). No other information exists to even suggest that the seam has been worked at shallow depth in this region. A preliminary site investigation based on primary source information only would erroneously declare this area free from shallow workings.

In 1850, a lease of the Box Little Vein and the "little vein lying above the same" was granted to a William Clement for as long as it took to work out the coal, or for 14 years, whichever was the longer. The lease was surrendered within 12 years implying that the coal had been worked out by that time. Later archive information retrospectively referred to the fact that the Clement's Nos 1, 2 and 3 Veins had been extensively worked and the Box Little Vein worked out. No plan of these workings has survived, but the land boundaries specified in the 1850 lease show that they would have lain immediately east of the Box Fault, between the northern and southern outcrops of the Swansea Four Feet Vein, in region 3N/3S (Figure 2). Interpretation of available evidence shows that the seams named correspond to the Penyscallen Vein and the thin un-named seams above and below it, whose theoretical outcrops are tentatively shown on present editions of the Geological Survey plans. Development is proposed in this area at the present time and there is no doubt that parts of it are underlain by uncharted, old shallow workings. A preliminary site investigation based on primary information sources will clear the area of shallow workings and, if reference is made to archive sources, it is unlikely that the significance of the statements quoted above will be recognised, because no published correlation exists for the Box Little Vein and the Clement's Veins.

Pillar robbing - An archive letter, dated 1836 stated that pillar robbing was being carried out in shallow workings in the 0.8m thick Gellywhiad or Small Vein. Details given showed that the initial pillar and stall extraction had removed two-thirds of the coal with the subsequent robbing reducing each pillar by half, giving a final extraction of about 83%. No plan of the shallow workings has survived but it can be concluded that it would be dangerous to build over them. The Gellywhiad or Small Vein can be correlated with one of the lower seams comprising the Gelli Group, immediately east of the Gors Fault in region 4N (Figure 2). Development was proposed in this area in 1978 and, in the light of the evidence summarised above, the decision was taken, at the preliminary site investigation stage, that sub-surface grouting would have to be employed before any construction work could proceed.

THE FEASIBILITY OF OBTAINING OLD COAL SEAM NAME CORRELATIONS FOR BRITAIN'S OLD COALMINING AREAS

Whereas some of the seam name correlations listed in Table II were obtained by straightforward reference to primary source information, the majority were the product of many hours research into archive collections situated throughout England and Wales. These collections varied from well-scheduled and well-stored to unscheduled and disordered, but even the best collections did not possess a sufficiently detailed schedule to allow a researcher interested in coal seam names to pin-point the relevant documents. Consequently, all coal mining information in each collection had to be scrutinised, and Table II is the product of the examination of tens of thousands of documents. In order to correctly interpret the information obtained, it was also found necessary to gain an understanding of the case study area in the historical mining and geological contexts. There is no doubt that the same problems of correlation exist for Britain's other old coalfields. South Wales, and the Swansea/Llanelli area in particular, was the first coal mining area to be examined in detail by the newly formed Ordnance Geological Survey in the 1840s (Bassett 1969)and, in view of this, it is possible that correlation problems in the case study area are slightly simpler than for other areas.

It can be seen that the time which has to be outlaid to obtain coal seam name correlations for an area of similar size and character to the case study area must normally be measured in years. A full-time skilled researcher, concentrating on the one area, could gain the basic information from the archives in a matter of months, however, although correct interpretation of this information would take much longer to achieve.

In terms of normal preliminary site investigation procedure it is obviously impractical, in terms of time and cost, to attempt to obtain old coal seam name correlations in order that archive information relating to shallow workings and pillar robbing can be interpreted. The only possible way in which these correlations could be achieved for Britain's old coalfields is by the initiation of a long-term, systematic search of both primary and archive sources for coal seam names, with the results of the search being made available to all interested parties.

CONCLUSIONS

(1) Information on the presence and condition of uncharted, old shallow workings, of practical relevance to the construction industry, is held in archive sources.

(2) Because this information refers to the past mining activity in terms of the seam name in use at the time the mining was carried out, it is usually difficult to decide which of the present seams shown on modern Geological Survey plans is being referred to.

(3) Correlation of the old coal seam names with present known seams is essential, if this information is ever to become of use to the construction industry.

(4) It is normally impractical to attempt to obtain this correlation, in the course of a normal preliminary site investigation, because of the excessive time and cost involved.

(5) The knowledge that most of the information relating to the presence and condition of uncharted, shallow workings lies in archives, and is only partially understood because of the lack of correlation of old coal seam names, makes it imperative that the greatest care be taken when planning and designing structures in old coal mining areas. If preliminary site investigations, based on primary information sources only, do not reveal old, shallow workings it must never be assumed that this is actually the case. Whenever it is known, or suspected, that a coal seam lies at shallow depth under a proposed development area, it is essential that the physical site investigation employs rotary boreholes, or an equivalent technique, to check for old workings and to assess their extent and condition. This will require a fundamental change in attitude by the construction industry with the acceptance of higher site investigation costs in old coal mining areas.

RECOMMENDATION

It is recommended that a programme of research be initiated to correlate old coal seam names with present known seams for all of Britain's old coalfields. This would entail extensive research into archive sources by skilled personnel and, if this single research into a limited area of 5200 hectares is typical of Britain's old coalfields, then a large and costly operation on a scale requiring central Governmental backing is required. The results of the research, if collated and made readily available to all interested parties, would be of the greatest practical benefit to the construction and mining industries.

ACKNOWLEDGEMENTS

The work reported in this paper was carried out as part of a research programme jointly sponsored by the University of Wales Institute of Science and Technology and Llanelli Borough Council, Dyfed.

Acknowledgement is given to the following for help provided in the form of supplied information or technical discussion:- Dr. D.A. Bassett, Director, National Museum of Wales; Messrs. G. Lewis (Area Surveyor), V. Jones and C. Hopkins, all of the South Wales Area Survey Department of the National Coal Board; Messrs. W.A.M. Jones and D. Pennington of the Valuation Office of the Mineral Valuer, Wales; Mr. H.A. Prescott, Llanelli Borough Librarian; Miss M. Patch, Dyfed County Archivist.

REFERENCES

D.A. Bassett - A Source-Book of Geological, Geomorphological and Soil Maps for Wales and the Welsh Borders (1800-1966), National Museum of Wales, Cardiff (1967) Vl + 239.

D.A. Bassett - Wales and the Geological Map, Bulletin of the National Museum of Wales 3 (1969) 10-25.

F.G. Bell (Editor) - Site Investigations in Areas of Mining Subsidence, Newnes-Butterworth, London (1975) 1-24.

R.W. Bruhn, M.O. Magnuson and R.E. Gray - Subsidence over abandoned mines in the Pittsburgh Coalbed, 2nd Conference on Ground Movements and Structures, UWIST, Cardiff (1980).

P. Carter, D. Jarman and M. Sneddon - Mining subsidence in Bathgate, a town study, 2nd Conference on Ground Movements and Structures, UWIST, Cardiff (1980).

Department of Trade and Industry - Inrush at Lofthouse Colliery, Yorkshire, HMSO (1973).

M.J. Dumbleton and G. West - Preliminary sources of site information for roads in Britain, Quarterly Journal of Engineering Geology 5 (1972) 15-18.

M.J. Dumbleton and G. West - Preliminary sources of information for site investigations in Britain, Department of the Environment RRL Report LR403, (1976).

P.G. Fookes - Planning and stages of site investigation, Engineering Geology 2 (1967) 81-106.

G.S. Littlejohn - Surface stability in areas underlain by old coal workings, Ground Engineering 12 (1979) 22-30.

A.B. Malkin and J.C. Wood - Subsidence problems in route design and construction, Quarterly Journal of Engineering Geology 5 (1972) 179-194.

D.G. Price - Engineering Geology in the Urban Environment, Quarterly Journal of Engineering Geology 4 (1971) 191-208.

RICS - The Problem of Disused Mine Shafts, The Royal Institution of Chartered Surveyors Minerals Division (1978).

A. Strahan - The Geology of the South Wales Coalfield, Part VIII, The Country around Swansea, Memoir of the Geological Survey (1907a).

A. Strahan - The Geology of the South Wales Coalfield, Part IX, West Gower and the Country around Pembrey, Memoir of the Geological Survey (1907b).

A. Strahan, T.C. Cantrill, E.E.L. Dixon and H.H. Thomas - The Geology of the South Wales Coalfield, Part VII, The Country around Ammanford, Memoir of the Geological Survey (1907).

A. Strahan, T.C. Cantrill, E.E.L. Dixon and H.H. Thomas - The Geology of the South Wales Coalfield, Part X, The Country around Carmarthen, Memoir of the Geological Survey (1909).

M.V. Symons - Sources of information for preliminary site investigation in old coal mining areas, Conference on Large Ground Movements and Structures, UWIST, Cardiff, Pentech Press (1977) 119-135.

M.V. Symons - Coal Mining in the Llanelli Area, Volume 1, 16th Century to 1829, Llanelli Borough Council (1979) XX + 371.

R.K. Taylor - Site Investigations in coalfields - the problem of shallow mine workings, Quarterly Journal of Engineering Geology 1 (1969) 115-133.

R.K. Taylor - The functions of the engineering geologist in urban development, Quarterly Journal of Engineering Geology 4 (1971) 221-234.

R.K. Taylor - Characteristics of shallow coal-mine workings and their Implications in Urban Redevelopment Areas, Site Investigations in Areas of Mining Subsidence, Newnes-Butterworth, London (1975) 125-148.

S. Thorburn and W.M. Reid - Incipient failure and demolition of two storey dwellings due to large ground movements, Conference on Large Ground Movements and Structures, UWIST, Cardiff, Pentech Press (1977) 87-99.

SESSION III

Chairman: Mr J A Derrington, BSc(Eng), DIC, FEng,
FIStructE, FICE, FCI Arb
President The Institution of Structural Engineers

Introduction to Session: Mr L Slowikowski, BA, CEng,
FIStructE, MConsE
Victor Slowikowski, James Blackshaw
and Partners

SESSIONAL PAPERS

THE DESIGN OF RIGID FRAME STEEL STRUCTURES IN MINING AREAS

B.L. Davies, C Eng , B Sc (Eng), A C G I , M I C E

Chief Engineer - Bridges and Structures,
South Yorkshire County Council

INTRODUCTION

The South Yorkshire County Council in seeking to promote a policy of industrial development has been undertaking the construction of light industrial units for leasing by small firms. The contracts for these have been on a design-and-construct basis and it has been the responsibility of the Bridges and Structures Division of the South Yorkshire County Department of Engineering to produce structural design specifications and to check the designs as submitted for structural adequacy.

Several of the sites for development are in areas where there are workable resources of coal and the consequent probability of these being extracted during the life of the structure.

Whilst there is much information to enable estimates to be made of surface movements due to coal extraction, very little guidance is available as to how non-articulated structures should be proportioned to ensure adequacy. This is particularly true with the multi-bay steel pitched roof portal frame which is generally the Contractor's preferred submission for the size of building required. This paper sets out to

(i) describe criteria which have been evolved for the above situation,

and

(ii) examine the implications of the above criteria on a specific Contractor's submission.

CURRENT PRACTICE IN GREAT BRITAIN

At the time of writing, the deemed-to-satisfy code for Building Regulation purposes is British Standard 449. The basic design philosophy is elastic analysis of the structure with limiting permissible stresses. As an alternative for fully rigid structures the design may be based on the principles of plastic design so as to provide an adequate load factor and with deflections under working loads not in excess of the limits implied in the British Standard.

No guidance is given in the standard on load factor values but it is generally accepted that 1.7 for dead and live loading and 1.4 for wind effects are adequate.

A maximum deflection of beams 1/325 of the span is stated but this may be exceeded where it will not impair the strength and efficiency of the structure, lead to damage of finishings or be unsightly. A specific waiver is given in the case of sheeting rails and for the purlins the figure may be increased to suit the characteristics of the roof covering.

There are no specific requirements for dealing with large ground movements.

DRAFT CODES

The British Standards Institution Technical Committee B/20 is at present revising the above Standard and a draft-for-comment has been issued.

Limit state philosophy is used in this document the designer having to ensure that the structure complies with

(i) safety limits concerning the ability to carry load, and
(ii) serviceability limits concerning normal use and durability.

The major safety limits are strength, stability, excessive deformation and fracture whilst for serviceability, deflection is the limiting criterion. Partial load factors are given for the safety limits for dead, imposed and wind loading, a further partial factor being included in determining the adequacy of section.

For rigid frame structures the plastic design method may be used for the safety limits with elastic analysis to determine deflections. Deflection limits are defined more precisely being in the case of roof beams not carrying a plaster finish 1/200 of the span.

The European recommendations adopt an identical philosophy to the British draft but no deflection limits are stated.

I.S.O. Standard 4356 deals with the question of acceptable deflection limits and considers for roof beams maximum values between 1/250 and 1/125 of the span depending upon the degree of flexibility of the roofing material.

With the exception of the I.S.O. Standard no reference at all is made to criteria which can be adopted for dealing with large ground movements.

APPLICATION TO RIGID FRAME PORTAL STRUCTURES

The application of the elastic approach of B.S. 449 is considered to be very conservative for portal frame structures and plastic theory is used almost universally in Great Britain.

The most significant mining effect on normally proportioned portal frame structures is usually that of horizontal ground strains. These can be eliminated by the utilisation of the floor slab acting as a tie or strut, this member being placed on a mining slip layer and being adequately reinforced for the tensile strains caused by ground movements. When, however, the structure is of sizeable plan proportions, movement joints in the slab will be required and horizontal ground strains will then be transmitted into the superstructure.

Occasionally with shallow workings differential vertical movements can be a sizeable effect.

J.F. Baker has shown that displacement of supports has no influence on the ultimate strength of the structure. Neglecting strain hardening and secondary effects this is correct but designers have inferred from this that no account need be taken of mining movements when proportioning a rigid frame portal structure using plastic theory. The writer does not accept the validity of this inference.

It is implied in the plastic approach that, under unfactored loads, there is little or no permanent deformation of the structure. An elastic analysis of a framework proportioned using theory and using working loads and predicted ground strains will show that stresses well in excess of those permitted by B.S. 449 and, depending on the proportioning of the structure and the magnitude of the ground movements, in excess of yield may be developed.

It is contended that a stress should be specified as a serviceability limit state to ensure there is no permanent damage to the main structural framework.

Clause 5.7.1 of the British Standard B/20 draft code suggests a design strength of 0.93 x guaranteed minimum yield stress for the material provided that this is not in excess of 0.73 x ultimate stress. Tabular values are given for the normal structural grades of steel in British Standard 4360.

Taking into account the degree of unpredictability of mining movements the writer considers it prudent to reduce this figure by a further amount to give a stress limitation of 0.87 x minimum guaranteed yield stress.

With regards to permissible deflections, with pitched roofs at 1 in 5 a deflection limit of 1/200 of span has been given under dead and working loads. Deflections in excess of this may occur under mining movements with the resulting possibility of minor damage to roofing or cladding. Bearing in mind that this is predictable and can be closely monitored, the damage can be the subject of a claim against the Coal Board, and as it is superficial ie. not part of the main structure, this damage is considered acceptable.

The criteria which have been specified, therefore, are summarised as follows:-

(i) Limit State Collapse
Design of structure using plastic analysis with minimum load factors of 1.7 for dead and imposed loads and 1.4 for wind loading.

(ii) Limit State Local Damage

(a) Deflections of roof beams limited l/200 of span under unfactored dead and imposed loads.

(b) Stress in main steel framework not to exceed 0.87 x yield stress with unfactored loads and predicted mining movements.

APPLICATION TO 7 BAY CONTINUOUS PORTAL STRUCTURE

The structure considered consisted of seven 18m. bays, there being 5 frames at 6.5m. centres. The height to eaves was 5.4m. with a further 0.95m. to the ridge (Plate 1).

Member sizes were proportioned using plastic theory and gave a load factor of 1.92 under dead and imposed loading with Grade 43 steel. Wind loading was not critical.

Mid span roof beam deflection under dead and imposed loading had a maximum value of l/283 of the span.

The foundation floor slab was cast monolithically across the building with transverse joints at 19.2m. maximum centres. A mining slip layer was provided underneath the slab, the latter being reinforced to cater for ground strain frictional forces.

The National Coal Board's current proposals are for extraction in the Cortonwood Silkstone seam, 3 ft. thick at 690 yards depth and Figs. 1 and 2 show the predicted subsidence contours for the faces which affect the site. From these it will be seen that the maximum ground strain is 0.3mm/m. on the completion of face X21 which reduces to a lower figure when X22 is extracted.

The most critical stress levels are in the end bays and these are summarised in Fig. 3. It will be seen that at no section has the stress level exceeded 217 N/mm^2, 0.87 of the guaranteed yield stress for grade 43 steel.

CONCLUSIONS

Although the predicted ground strains in this example are comparatively low, the exercise has shown that the structure should perform satisfactorily under working conditions with the super-imposition of the mining wave. Close monitoring of the structure

Plate 1 : 7-bay portal structure investigated.

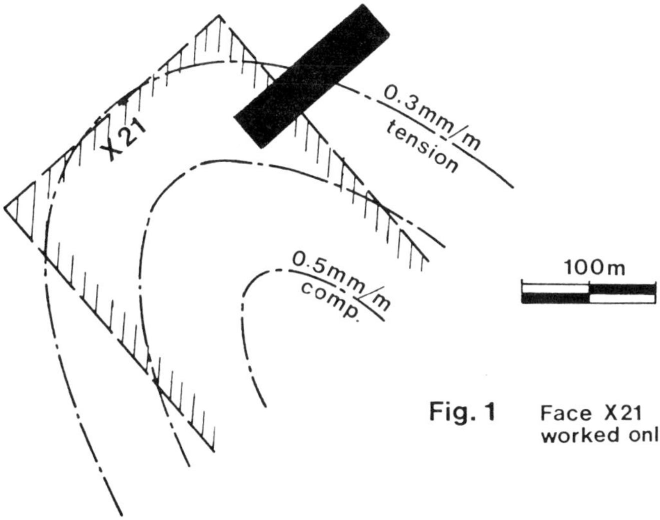

Fig. 1 Face X21 worked only

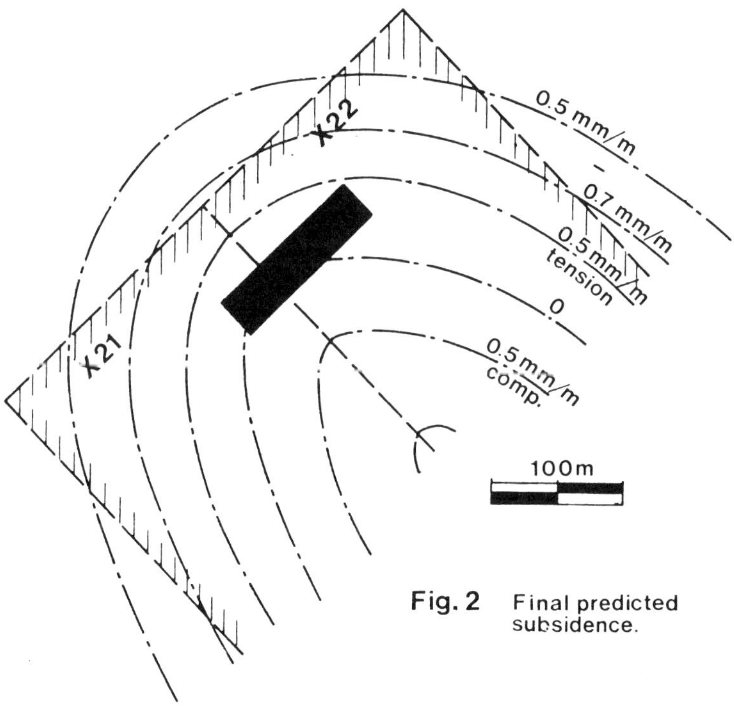

Fig. 2 Final predicted subsidence.

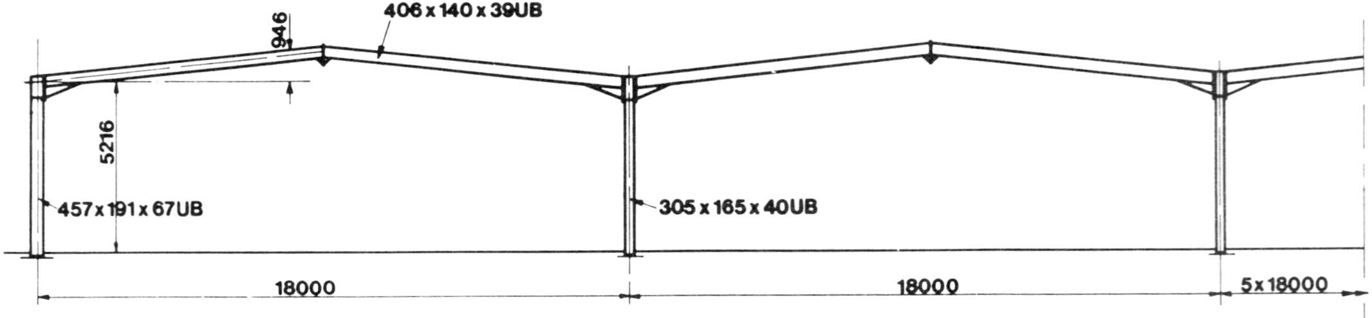

TYPICAL END AND INTERNAL BAY

MAXIMUM STRESS LEVELS
(0.2mm/m average ground strain)

RAFTER BEAM	215.9N/mm^2
EXTERNAL STANCHION	184.5N/mm^2
VALLEY STANCHION	96.4N/mm^2

Fig. 3 Frame analysed with maximum stress levels.

during the progress of mining will be undertaken to compare the predicted with actual behaviour.

Should the mining movement predictions have been much in excess of those used in the analysis then the criteria given would have indicated that other precautions would have required to have been incorporated in the structure. Alternatives which could be considered are

(i) use same section for end bays but with higher grade steel,

(ii) reduce the section sizes by use of higher grade steel. Displacement stresses would have been reduced with a more flexible structure but with the penalty of higher deflections,

or

(iii) provide some articulation in the centre to reduce the structure to effectively 2 three bay portals.

Consideration could also be given to a reduction of wind and/or roof live loading to take into account the reduced possibility of these maximum effects occurring during the limited duration of the passage of the mining waves. This would be particularly relevant if residual effects were small on completion of the coal extraction.

ACKNOWLEDGEMENTS

The author would wish to express his thanks to Mr. J.F. Kirkham, C Eng , M I Mun E , County Engineer to the South Yorkshire County Council for permission to present this paper.

REFERENCES

J.F. Baker - <u>The Steel Skeleton - Vol. 2</u>
Cambridge University Press (1956).

British Standard 449 : 1959 - <u>The Use of Structural Steel in Building</u>
British Standards Institution (1959).

British Standards Institution Document 77/13908 DC - Draft Standard Specification for the Structural Use of Steelwork in Building - Part 1 : Simple Construction and Continuous Construction. British Standards Institution.

European Recommendations for Steel Construction. European Convention for Constructional Steelwork (1978).

I.S.O. Standard 4356 - Bases for the Design of Structures Deformations of Buildings at the Serviceability Limit States. International Standards Organisation 1977.

THE STATE OF STRESS IN A BUILDING CONSTRUCTED OF LARGE PRECAST WALL PANELS SUBJECTED TO MINING SUBSIDENCE

Marian Kawulok, Mgr inz

Institute of Building Technology - Gliwice Division, Poland

INTRODUCTION

A complex investigation has been carried out on the effect of coal extraction on a five-storey block constructed of large precast wall panels, which included measurements of the deformation of the terrain around the building as well as of the resultant strains in the structure. A general analysis of the results of the investigation was presented by the author at the first LGMAS Conference in Cardiff in 1977.

A subsidence basin has developed along the long axis of the building, characterised by compressive horizontal movements of the ground and a concave shape to its settlement curve. Because of the way in which the movement joints were packed with woodwool, the four sections of the building behaved in practice as one unit, under the influence of compressive horizontal displacements in accordance with the scheme shown in Figure 1.

The investigation gave the fullest information on the strains in the external longitudinal walls at the stairwells, obtained from strain measurements on the reinforcement in the horizontal bracing provided at each floor level. This allowed a determination of the edge loads on the walls, which in turn made it possible to separate the effects of the vertical and horizontal displacements of the ground on the structure, although only their combined effects had been measured.

The scale of the study allows only the basic assumptions and the design procedure to be given in this paper which ends with a discussion of the results of the loading analysis, defining the state of stress in the structure.

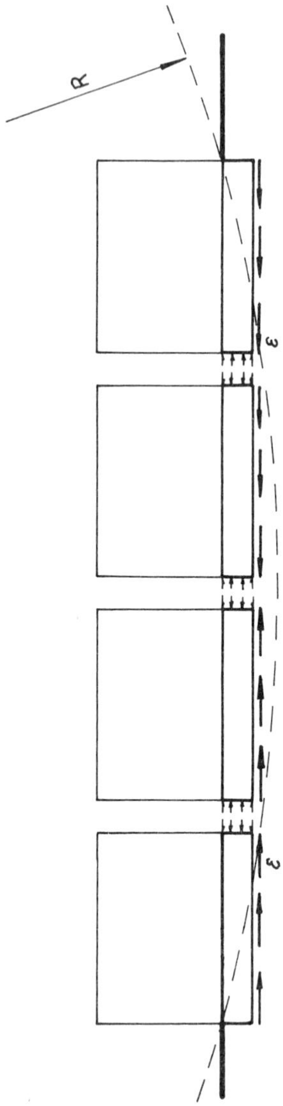

Fig. 1

ASSUMPTIONS AND THE ANALYSIS OF THE LOADS

The analysed investigation deals with a complex process which in general is represented by three fundamental categories of effects:

- those taking place in the ground, forcibly deformed as a result of mining operations;
- those arising on the interface between the ground and the building during the process of deformation;
- those induced in the structure by the previous two.

Only an approximate description of the effects is possible on the basis of the author's investigation, principally due to the great complexity of the examined building.

Measurement of the strains in the structure as well as other technical data derived from the investigation were used to provide a basis for the subsequent analysis. A general analysis of the survey data for the ground surface and the building justified the assumption that the structural deformations were solely the effects of loads induced by the ground deformation. Figure 2 shows the logical scheme for the solution of the postulated problem of load identification.

BASIC MODEL FOR STATICAL ANALYSIS

For each of the external longitudinal walls of the building it was possible to determine the deformations at 9 positions, "k". With the notation adopted in Figure 3 these are:

3 relative vertical displacements of the wall (Δ_1, Δ_2, Δ_3), at cross wall positions and with the straight line joining the external points on the wall as datum,

strains in the longitudinal reinforcement in the footing (Δ_4),

strains in the bracing beam at the basement ceiling level (Δ_5),

strains in the bracing beams at each floor level (1-4) of the superstructure (Δ_6 - Δ_9).

Edge loads p_1 - p_{12} were introduced for a side wall considered in isolation (Figure 3). For the wall of an internal block (see Figure 1) they denote:

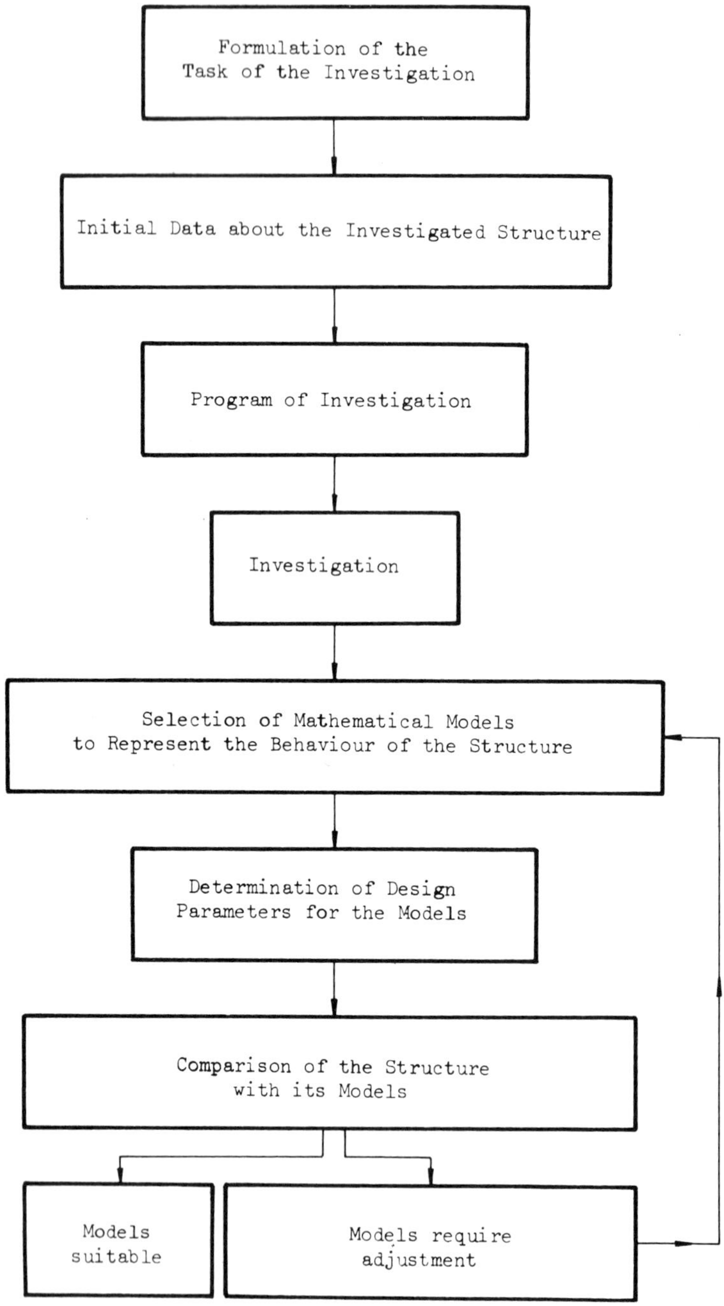
Formulation of the
Task of the Investigation
Initial Data about the Investigated Structure
Program of Investigation
Investigation
Selection of Mathematical Models
to Represent the Behaviour of the Structure
Determination of Design
Parameters for the Models
Comparison of the Structure
with its Models
Models
suitable
Models require
adjustment

Fig. 2

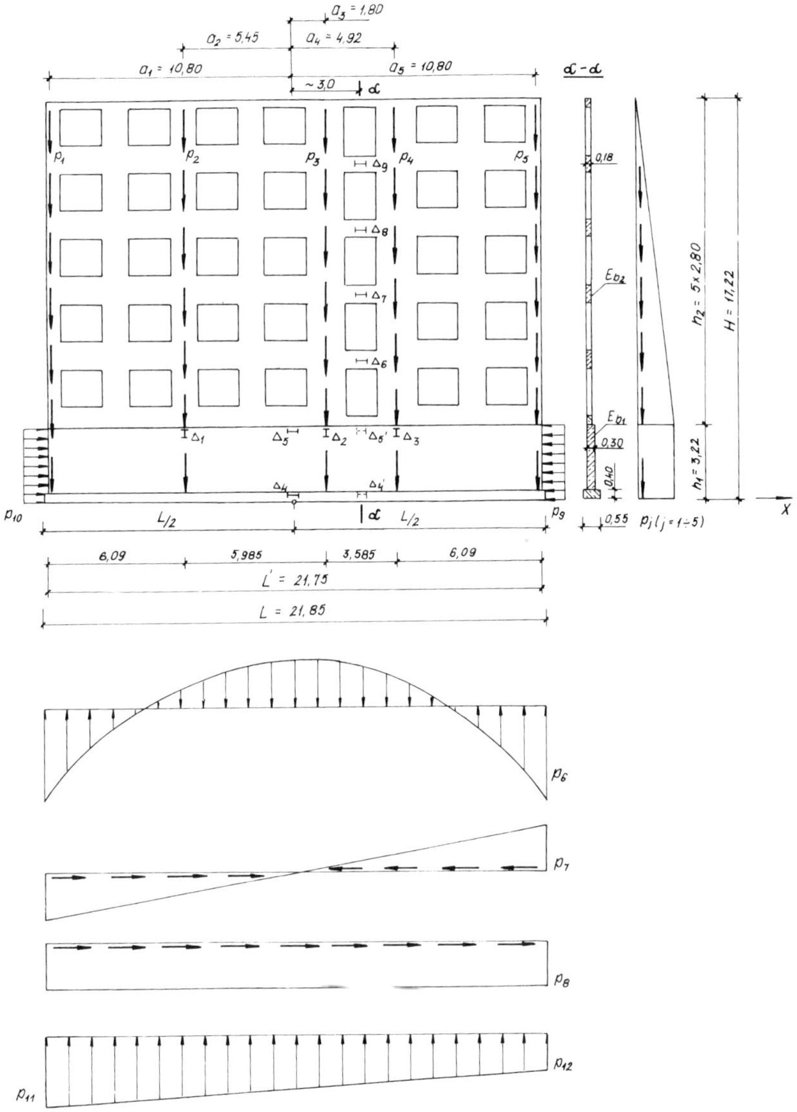

$a_3 = 1,80$
$a_2 = 5,45$
$a_4 = 4,92$
$a_1 = 10,80$
$a_5 = 10,80$
~ 3,0
α - α
p_1 p_2 p_3 p_4 p_5
Δ_9 Δ_8 Δ_7 Δ_6
Δ_1 Δ_5 Δ_2 Δ_5' Δ_3
Δ_4 Δ_4'
0,18
E_{b_2}
E_{b_1}
0,30
0,40
0,55
$h_2 = 5 \times 2,80$
$H = 17,22$
$h_1 = 3,22$
X
p_{10} L/2 L/2 p_9
$p_j (j = 1 \div 5)$
6,09 5,985 3,585 6,09
L' = 21,75
L = 21,85
p_6
p_7
p_8
p_{11} p_{12}

Fig. 3

p_1 - p_5 edge loads caused by ground curvature replacing the effect of the cross walls,

p_6 vertical, self-balancing ground reaction on the wall footing due to the curvature of the ground,

p_7, p_8 horizontal forces in the foundations due to symmetric (p_7) and antisymmetric (p_8) horizontal displacements of the ground, with respect to the centre of the wall,

p_9, p_{10} reactions of adjacent sections at the basement level, caused by horizontal movement,

p_{11}, p_{12} vertical ground reactions on the wall footing induced by the weight of the structure.

This system of edge loads is based on some simplifying assumptions of which the two most important are:

- Monolithic behaviour of the in-situ constructed basements with the precast wall panels of the superstructure, disregarding the specific conditions of interaction between the two parts.

- Insignificance of the edge forces between the floors and the wall.

For this statical system the problem of analysis is tantamount to a calculation of the edge loads p_1 - p_{12} for different states of deformation of the structure, defined by the data obtained from a given cycle of the survey of the building (values Δ_1 - Δ_9). Thus formulated, the problem with 12 unknowns and 12 known values (9 measured deformations and 3 equations of equilibrium) would have 0 degrees of freedom, defined as the difference between the number of unknown and known values. To obtain the required number of degrees of freedom with reference to the models of loading described later, certain relationships had to be introduced between some of the loads.

ANALYTICAL MODELS

Models of Loading Conditions

Determination of the edge loads requires a solution of two fundamental problems, namely an analysis of the effects of the vertical and horizontal displacements of the ground on the given building. Solution of the problem for the vertical ground movements follows Budzianowski's approach (1964) which considers the interaction between a rigid structure and deformable ground to be in accordance with Winkler's model. A determination of the reaction on the foundations allows us to establish the loads $p_1 - p_6$ as functions of the parameters of the Winkler model and of the subsidence profile.

The behaviour of the ground was modelled elastoplastically to permit analyses of the effects of the horizontal ground movements based on Wasilkowski's solution (1966). In this manner, the edge loads $p_7 - p_9$ were defined as functions of the properties of the ground.

The two models of ground behaviour were treated as independent of each other. The vertical movements were assumed to affect only forces $p_1 - p_6$ and the horizontal movements, forces $p_7 - p_{10}$. Loads p_{11} and p_{12} were derived from the equilibrium of the forces $p_1 - p_5$ and of the vertical forces introduced by p_8.

MODEL OF THE STRUCTURE

The statical analysis of the wall was based on a model combining a rigid frame, according to Borowiec (1975), and a model of an edge panel wall (Kosicyn, 1971). The rigid frame model is particularly suited to the analysis of monolithic walls with openings. The Kosicyn model of the wall takes into account the behaviour of a structure consisting of large panels.

FORMULATION OF CRITERIA FOR LOAD ANALYSIS

For the schematic model of a wall shown in Figure 3 it is necessary to establish values of the edge loads $p_1 - p_{12}$ which provide the best agreement between the results of the calculations, according to the given assumptions, and the survey data for:

- vertical displacements and strains in the reinforcement at the positions k = 1 to 9 (where Δ_k and δ_k denote the measured and calculated values respectively)

- the magnitude of the resultant horizontal force acting at the section $\alpha - \alpha$ (Figure 3)

- the position of the resultant force in the height of the wall.

The method of least squares (Kawulok,1977) was used to derive the necessary design parameters. The results of the survey were used in delimiting the magnitude of the parameters used in the statical models.

DISCUSSION OF THE RESULTS OF THE ANALYSIS

The process of analysis covered 9 independent states of loading of the investigated building, described in each case by a set of strain data for the structure. This led to the determination of parameters for the statical models used in the calculations. Positive results were only obtained with regard to intermediate states of deformation of the structure when the combined statical model of the building represented relatively correctly the state of the deformation in the prefabricated walls. The results for the remaining states of deformation did not conform with the survey data. In part, this can be accounted for by the specific behaviour of the precast panel structure characterised by a stiffness varying with an increase in the vertical deformations $\Delta_1 - \Delta_3$ and in part by the fact that the ground stresses exceeded the elastic limit. The results confirmed that it is possible to resolve the complex state of interaction of a structure with the ground into two independent models of the vertical and horizontal ground movements.

THE STATE OF DEFORMATION OF THE STRUCTURE

Separation of the Effect of Vertical and Horizontal Ground Movements

The measured strains in the structure Δ_k resulted from the combined effects of the vertical and horizontal movements of the ground. The system of analysis enabled these effects to be separated by assigning appropriate values to the horizontal movements ($\delta_{k,\varepsilon}$) and to the vertical movements ($\delta_{k,R}$). The nature of the distribution of the vertical deformations (k=1,2,3) is shown on the graphs in Figure 4 and that of the horizontal deformations (k= 4 - 9) in Figure 5.

VERTICAL DEFORMATIONS

Vertical deformations of the structure ($\delta_1, \delta_2, \delta_3$) were mostly due to the vertical ground movements, the effect of horizontal ground movements being small but noticeable. The compressive horizontal movement created a tendency for the structure to assume a convex deflected shape which was in opposition to the actual curvature. This action was countered by the effect of the vertical ground movements, which eventually led to a resultant concave shape of the deflection curve in accordance with the profile of the subsidence basin (Figure 1).

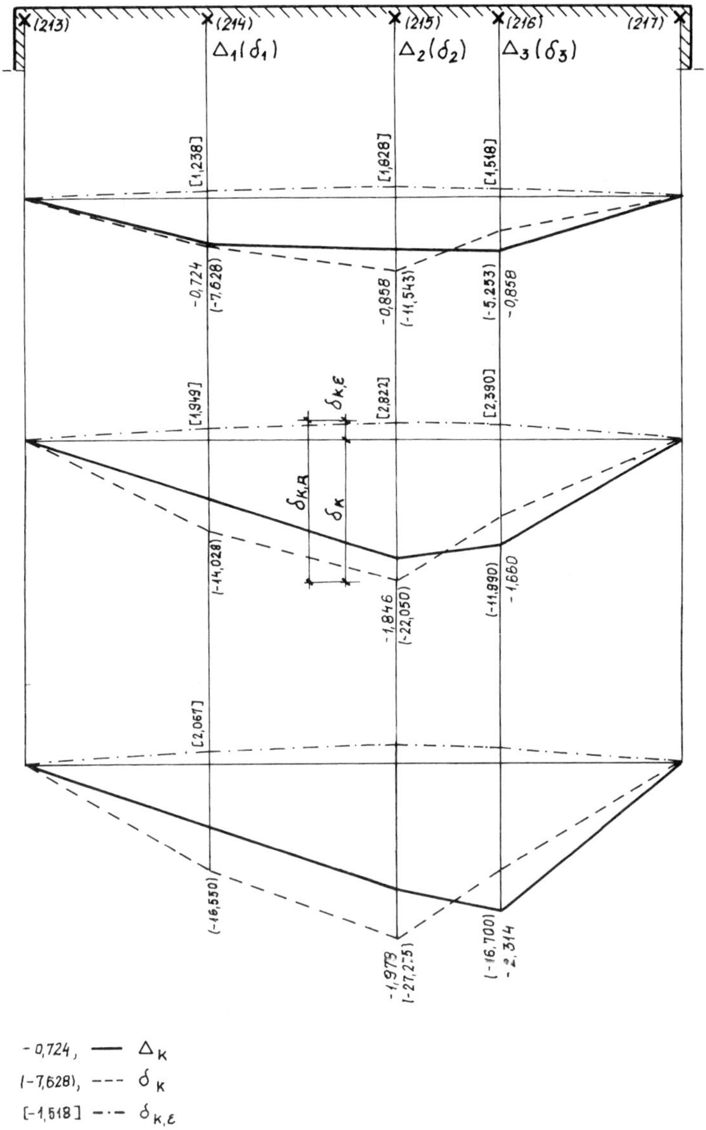
(213)
(214)
(215)
(216)
(217)
Δ1(δ1)
Δ2(δ2)
Δ3(δ3)
[1,238]
[1,828]
[1,518]
-0,724
(-7,628)
-0,858
(-11,543)
(-5,253)
-0,858
[1,949]
[2,822]
[2,390]
δK,ε
δK,R
δK
(-14,028)
-1,846
(-22,050)
(-11,890)
-1,660
[2,067]
(-16,550)
-1,973
(-27,275)
(-16,700)
-2,314
- 0,724, — ΔK
(-7,628), --- δK
[-1,518] -·- δK,ε

Fig. 4

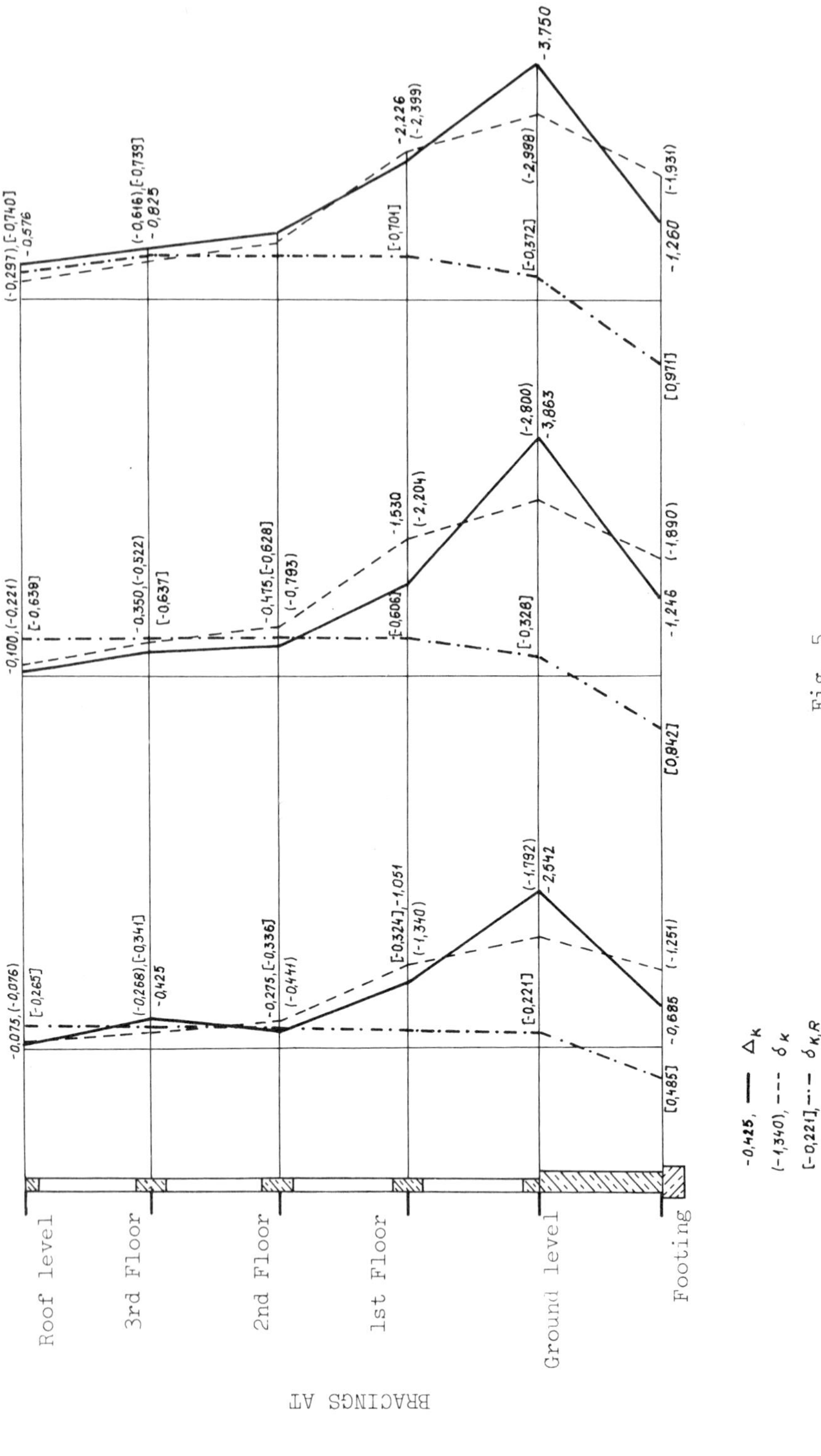
BRACINGS AT
Roof level
3rd Floor
2nd Floor
1st Floor
Ground level
Footing
-0,075,(-0,076)
[-0,265]
(-0,268),[-0,341]
-0,425
-0,275,[-0,336]
(-0,441)
[-0,324],-1,051
(-1,340)
[-0,221]
(-1,792)
-2,542
(-1,251)
-0,685
[0,485]
-0,100,(-0,221)
[-0,638]
-0,350,(-0,522)
[-0,637]
-0,475,[-0,628]
(-0,783)
[-0,606]
-1,530
(-2,204)
[-0,328]
(-2,800)
-3,863
(-1,890)
-1,246
[0,842]
(-0,297),[-0,740]
-0,576
(-0,616),[-0,739]
-0,825
[-0,704]
-2,226
(-2,399)
[-0,372]
(-2,998)
-3,750
(-1,931)
-1,260
[0,971]
-0,425, —— Δ_K
(-1,340), --- δ_K
[-0,221],-·- δ_{K,R}

Fig. 5

HORIZONTAL DEFORMATIONS AT VARIOUS LEVELS

Horizontal deformations of the wall at various levels $\delta_4 - \delta_9$ were produced by a superposition of the effects of the horizontal and vertical movements of the ground. The horizontal displacements, had a decisive influence on the magnitude and distribution of the deformation of the building so that even with the concave curvature of the building, the foundations were subjected to compressive strains. As can be seen from the graphs of Figure 5, the effect of horizontal displacements was most important in the zone from the foundations to the bracing at the first floor level. At other levels (k=7,8) the vertical displacements dictated the state of deformation of the structure. At the highest level (k=9) there arose strains due to the horizontal displacements which tended to reduce the compressive strains caused by the vertical movement of the ground.

CONCLUSIONS

1 The case analysed in this paper was somewhat untypical, especially in relation to the effects of ground movements, because of the way in which the expansion joints were constructed. As a consequence, large horizontal forces were developed in the structure. Quite apart from this, the possibility of interaction between the different elements of the building over the full height of the basement created particular conditions of strain in the structure, corresponding to the combined effects of forces p_7, p_8 and p_9. In this situation the decisive influence on the magnitude and distribution of strains in the building, caused by the horizontal ground displacements, was that of the largest of the loads, p_9.

2 The nature of the expansion joints had no appreciable effect on the behaviour of the building under the influence of the vertical ground movements. For this particular case, especially significant is the distribution, over the height of the wall, of the strains in the structure caused by the vertical ground movements ($\delta_{k,R}$ - Figure 5). The derived strain distribution leads to a distribution of forces in accordance with Figure 6. This approximation is based on the assumptions:

- that the neutral axis is located at 0.1 L, where L is the length of the building and equals in this cause 21.85 m.

- that the lever arm of the internal forces is H=0.4 L

- that a good approximation of the distribution of normal forces in the superstructure can be obtained by assuming a uniform distribution of normal stresses at that level.

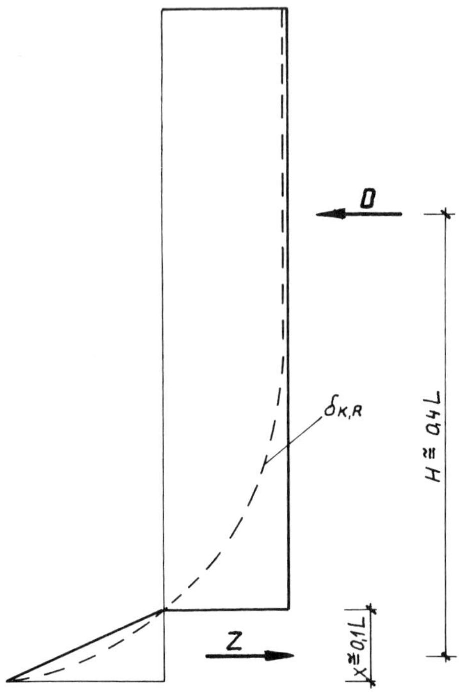

Fig. 6

3 The rheological deformations of the structure are the most important factor in approximating the statical models of analysis to the real structure. The simplified solution of this problem suggests that they account for 50% of the measured deformations.

REFERENCES

Z. Borowiec - Stress and deflection analysis in the perforated walls, Reports of the II International Symposium on Bearing Walls, Warsaw (1975) 129-140.

Z. Budzianowski - Działanie wygiętego podłoza na sztywną budowlę znajdującą się w obszarze eksploatacji górniczej, Inżynieria i Budownictwo 6 i 7 (1964) 199-206, 235-239. (Effect of ground deformed by coal extraction on a rigid building).

M. Kawulok - Investigation of a precast concrete panel structure subjected to mining subsidence, Proceedings of the Conference on Large Ground Movements and Structures. University of Wales Institute of Science and Technology, Cardiff (1977) 606-620.

B.A. Kosicyn - Staticzeski rasczet krupnopanelnych i karkasnych zdanij, Izdatielstwo Literatury po Stroitelstwu, Moskwa (1971) 216. (Statical calculation of buildings consisting of large precast concrete panels and skeletons).

F. Wasilkowski - Wpływ pełzania podłoża na fundamenty budowli posadowionych na terenach górniczych, Inżynieria i Budownictwo 7 i 10 (1966). (The effect on foundations of the creep of the ground caused by mining subsidence).

INVESTIGATIONS OF PILED FOUNDATIONS FOR BUILDINGS IN AREAS OF MINING SUBSIDENCE

P.E. Klezhev,
R.A. Muller,
S.E. Shagalov, Masters of technical sciences,
eng. Gadimba I.M

All Union Research Institute of Surveying, U.S.S.R

To prove the possibility and expediency of building in areas to be undermined in the Karaganda basin, as required by modern town-planning needs, the construction of two five-storey experimental residential buildings on piled foundations was carried out at the site of the "50 years of October Revolution" mine. One of them is a 56 flat, frameless, large-panel building No.14 built as a typical project of the 1-464A-17B series and the other is a 68 flat, brick building of the 1-438AB-33K series (Figure I).

One can find deposits of Quaternary, Neogene and Jurassic periods at the given site. The strata of Quaternary period are found at depths of up to 7-8 m and are represented by running-and running, laminated loams and sandy clays with low bearing capacity. The strata of Neogene and Jurassic periods are discovered at depths of more than 7 m and are represented by sandy clays and clays of semi-solid consistency.

The series of coal seams, which are being developed at the basin, run under the experimental buildings. At the present time the "50 years of October Revolution" mine has extracted the upper seam K_{18} under the experimental buildings. The seam has the following mining-geological conditions: the average working depth is 410 m; the seam thickness is 1.4 m; the manner of roof control is complete caving.

THE LARGE-PANEL BUILDING

To protect the large-panel building from damage due to

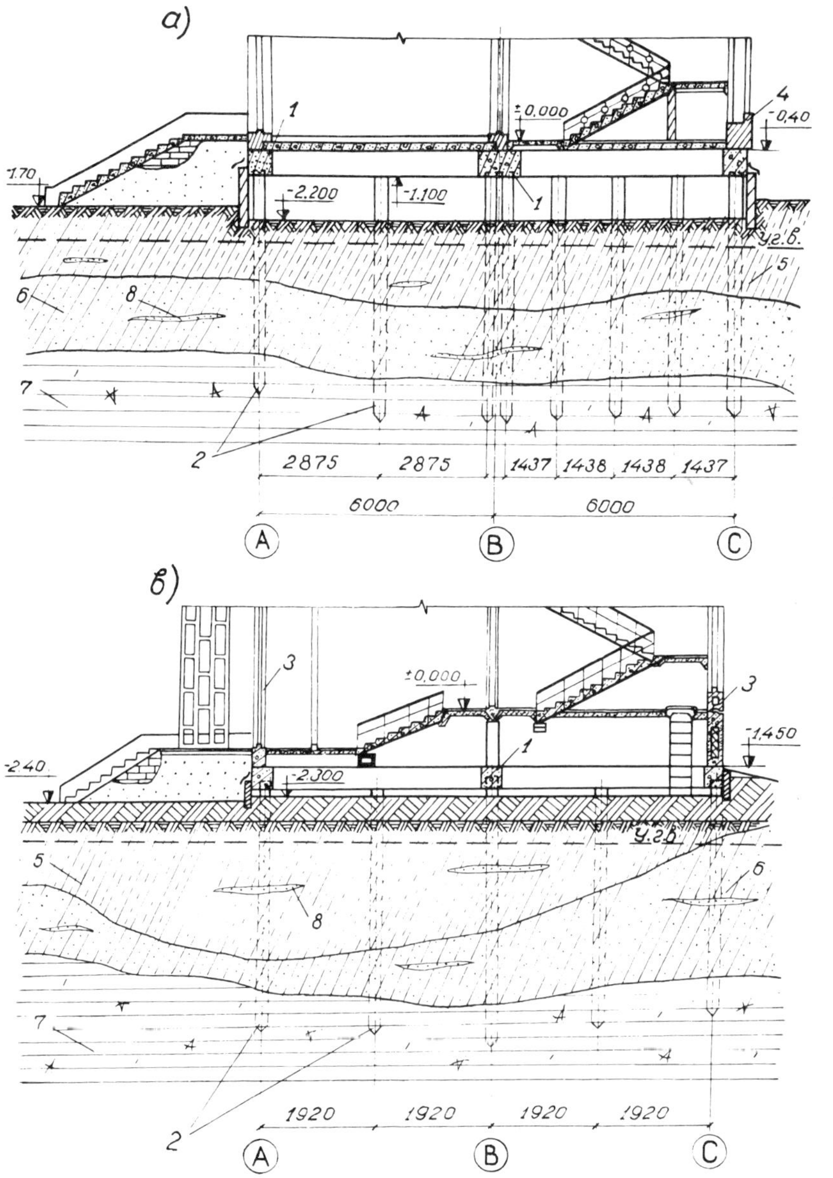

Fig.1 Details of the foundation-basement parts of the (a) brick, and (b) large-panel experimental buildings

1- foundation mat 2- cast-in-situ reinforced concrete piles 3- layered wall panel 4- brickwork 5- sand loams 6- sandy clays 7- clays 8- sand lenses

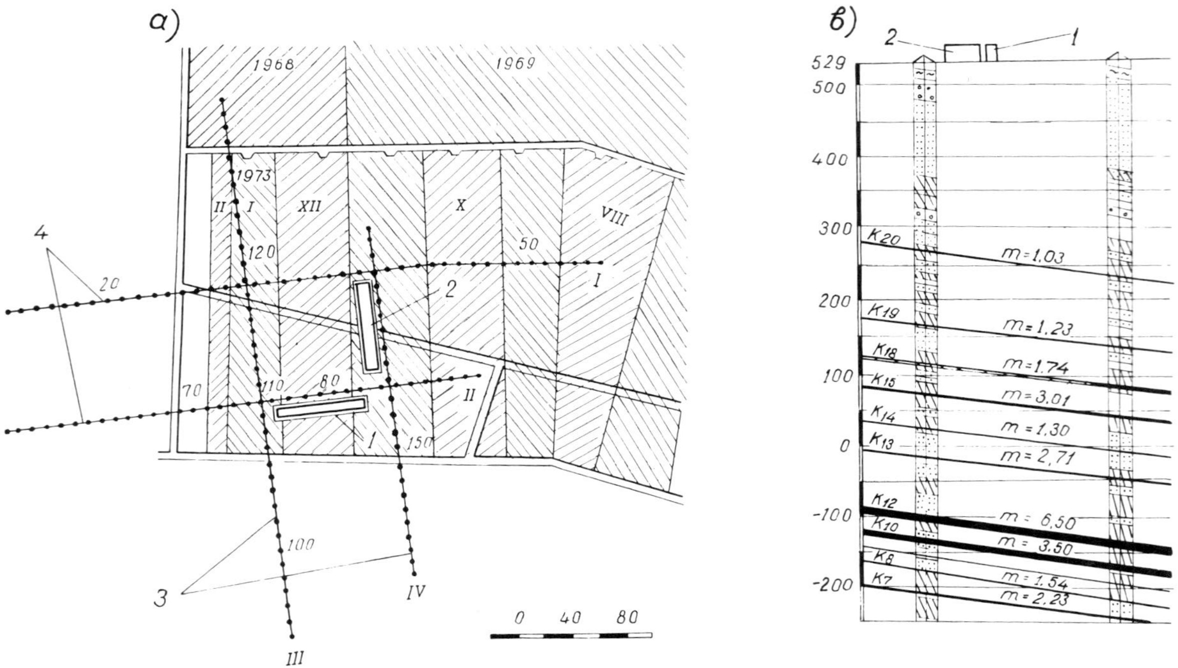

Fig. 2 (a) Plan of development of seam K_{18} ; (b) Geological section through the site of the experimental buildings

1- large-panel building No.14 2- brick building No.15
3- lines of ground observation stations along the seam's dip
4- lines of ground observation stations along the seam's strike

mining it was planned as three independent units with the plan dimensions: 11.52 x 14.2 m (units Numbers 1 and 2) and 11.52 x 28.4 m (unit Number 3). The subdivision of the building was achieved by means of movement joints and the erection of twin walls with independent foundations.

In design concept, the building is constructed as a frameless structure with transverse bearing walls and panel floors supported around their perimeters. Three-dimensional rigidity and stability of the units is provided by longitudinal and transverse reinforced concrete wall panels in one room. They are connected with each other and with the floor panels by steel loop ties into the vertical and horizontal stiffening diaphragms. The outside wall panels are triple-layered, their thickness being 30 cm; the inside wall panels are single-layer, with a thickness of 12.0 cm. The outside and inside plinth wall panels differ from the standard ones in their height, and the reinforcement provided is 15-20 per cent greater.

To further the objectives of the experiment, the sub-structures of the individual units of the building are provided in different forms of construction, which differ mainly in the manner in which the piles are joined to the foundation mat.

Thus the substructure of unit Number 1 was planned with a stiff connection between the piles and the foundation mat. Rigidity was provided by building the heads of the piles, with bare longitudinal reinforcement, into the foundation mat to a depth of at least 30 cm; as far as the upper foundation mat reinforcement. Such fixity between the foundation mat and the piles provides for the transmission of stress to the foundation mat itself. The stress arises from undermining, due to the vertical and horizontal displacements of the foundation soil.

The substructures of units Numbers 2 and 3 were designed on the basis of a flexible construction scheme, with pinned pile connections to the foundation mat through DS-50 dynamometers, which are set on the head of each pile by means of a special metal frame. Such a form of construction not only provides complete hinges at the junctions of the piles and the foundation mat, but makes it possible to measure the vertical loads transmitted from the superstructure to each pile. At the same time it allows an investigation of the redistribution of these loads between the piles as the building is undermined.

The piles in all units of the large panel building are of in-situ reinforced concrete, with a section of 30 x 30 cm and a length of 10.0 m. The pile reinforcement is provided on the basis of calculations of the bending stresses due to the translation of the foundation soil and is equal to 4 Ø 18All.

The number of piles depends on a calculation of the pile's

maximum bearing capacity in the ground, with the additional stresses arising from flexure of the foundation and with expected changes in the physico-mechanical properties of the ground also being taken into account.

The foundation mat in each of the units is placed clear of the soil and provides a monolithic reinforced concrete frame of high rigidity in the horizontal plane. The frame has a section of 560 x 650 mm, with reinforcement of 10 Ø 20A11. The influence of horizontal ground translation is also taken into account. The joints of the foundation mats which have to cater for variable stresses in the horizontal direction, are additionally reinforced by curved bars and haunching.

THE BRICK BUILDING

The main protection of the brick experimental building from the influence of vertical and horizontal ground displacements is provided by its sub-division into 4 independent units with the following plan dimensions: outer units, 12.0 x 18.8 m; inner units, 12.0 x 14.8 m. The structural arrangement of all the units is the provision of three longitudinal bearing walls.

The outer walls consist of inner and outer leaves of clay brick. All the walls are reinforced at the upper line of the damp-proof course of each floor with close-reinforced concrete braces, having sections of 250 x 250 mm on the outside walls and 380 x 220 mm on the inside walls and reinforcement of 4 Ø 18A11. In addition, all the junctions between adjoining and intersecting walls and partitions are reinforced with fabric of Ø 5-6 mm, which is placed in the horizontal joints between every 3-4 rows of brickwork.

The foundation basement of each unit is designed according to a structural arrangement defined by the manner in which the piles are connected to the foundation mat. In unit Number 1, the piles and foundation mat are pinned together with dynamometers type DS-50, in a similar way to those in units Numbers 2 and 3 of the panel building.

In unit Number 2, the piles are also pinned to the foundation mat. However, in this instance, the pile head carries a precast reinforced concrete cap which is in turn rigidly connected to the mat, using a construction similar to that already described.

In unit Number 3, the piles are cast rigidly into the foundation mat. The junctions of the mat are dealt with in a similar way to those of unit Number 1 of the panel building.

In unit Number 4, the piles make contact with the

foundation mat through sliding joints which are provided on the smooth surface of the monolithic reinforced concrete caps. The piles and the caps are rigidly joined. The sliding joint is made of three-layered ruberoid, powdered mica being placed between the layers. The pile sections are the same as in the panel building, namely 30 x 30 cm, with a length of 10.0 m; but the pile reinforcement is increased to 4 Ø 250A11. From the results of static tests the pile bearing capacity was found to be 400-470 kN, but after undermining it will be reduced to 350-400 kN. The maximum planned load on the piles, taking into account the additional load resulting from the mining operations is about 320 kN.

The foundation mat is provided clear of the ground. It takes the form of a closed frame in each compartment, with cross-sectional dimensions of 600 x 500 mm under the outside and inside transverse walls and 1000 x 500 mm under the inside longitudinal wall.

Reinforcement of the foundation mat under the outside walls was provided by 13 Ø 20A11 and 5 Ø 10A11 and under the inside longitudinal wall by 14 Ø 20A11 and 10 Ø 10A11. High-resistance strain gauges were fixed on the longitudinal reinforcement in the foundation mat, in the reinforced concrete braces at the floor levels and in the joints between the panels. As a further contribution to the experiment, during the process of building the apartments, 19 experimental piles were equipped with low resistance strip extensometers to measure the total stresses at different positions along their lengths. They were also fitted with contact pressure vibrating wire devices (SDKS-3) to determine the lateral ground pressures produced on them by displacements of the soil.

Tension piles were placed in the foundations of two units of different length in the panel building and of one unit in the brick building. They were positioned at various distances from the centres of the units to determine the effect of pile position on the stresses produced in the piles by mining.

OBSERVATIONS

To allow investigation of the stressed and deformed state of the piled foundations of the experimental buildings and their interactions with the moving ground, while the latter was being undermined, a comprehensive set of observation stations was set up. These assisted visual observations, geological surveying and various measurements on the buildings. In addition, to evaluate the influence of undermining on the strength and stiffness characteristics of the ground two comprehensive engineering - geological investigations were carried out by field and laboratory methods. The latter were performed before and during undermining, on 17 experimental sites in the immediate proximity of the experimental buildings and along ground station observation lines sited in zones of differing ground movement patterns.

Observations of the stressed-deformed state of the experimental buildings and their interaction with the ground were started in August 1969 and continued to 1979 while the piled foundations were built. This period covered construction, the initial building settlements to their stabilization, and the period of active mining movements to the termination of the first phase of undermining.

RESULTS OF OBSERVATIONS

Table I shows the results of the instrument observations. It presents data on the movements of the ground surface, in the strike direction and across the strike of the coal seams. They were measured during undermining within the boundaries of each building.

The settlement of each unit as a function of time was practically identical with a maximum rate of up to 9.8 mm in 24 hours. As a result of the undermining, all units of the experimental buildings experienced equal tilts towards the workings. The inclination had reached 4.3×10^{-3} in buildings Nos 14 and 15 when the last readings were taken. At the same time, settlement of the buildings resulted in further penetration of the piled foundations; the maximum settlements being 26 mm (brick building) and 15 mm (panel building). The latter process continued until the ground displacements ceased.

The horizontal strains at the ground surface, averaged along the length of a unit, were relatively large (Table I). As measured they were equal to those which had been estimated. Nevertheless, the measured horizontal strains proved to be about two-thirds of the corresponding strains on the adjoining sites beyond the limits of the buildings. The horizontal strains beyond the building limits are unusual; the strains in the line of the pile foundations are 5-6 times those measured along the line of the ground bench marks, which are 6.0 m from the buildings.

Apart from the strain measurements and the ground surface displacements, observations on the experimental buildings revealed many other features. Extensional and changing loads on the pile heads, compression of the pile shafts and the changing lateral pressures acting on them during erection of the buildings and during loading, before and after undermining, have been observed. Tensile stresses in the piles, the foundation mats, the reinforced concrete braces and the joints between the panels have been observed as well.

Load measurements using the dynamometers (DS-50) made it possible to determine the weights of two units of the panel

TABLE 1 MAXIMUM MOVEMENTS AND STRAINS OF THE GROUND AND STRUCTURES

	Type of deformation		Experimental building No.14	Experimental building No.15
Strike direction	Settlement, mm	building	716	803
		ground	709	797
	Slope, $1x10^{-3}$	building	4.3	4.0
		ground	4.5	3.6
	Radius of curvature, km	building	4.3	2.3
		ground	2.4	1.0
	Horizontal strains, $1x10^{-3}$ (tension)	building	-2.6	-1.8
		ground	-2.6	-6.8
Normal to Strike direction	Settlement, mm	building	716	803
		ground	709	797
	Slope, $1x10^{-3}$	building	+2.8	+2.3
		ground	+3.6	+3.3
	Radius of curvature, km	building	2.7	4.0
		ground	1.3	2.4
	Horizontal strains, $1x10^{-3}$ (tension)	building	-1.8	-0.9
		ground	-7.6	-4.0

building and one unit of the brick building and to monitor the changes in the longitudinal stresses in each unit during the period of the observations. After erection of the buildings and their occupation, the disparities in the measured weights of the units were not more than 15 per cent (they were less each springtime and at the beginning of summer). At other times of the year, the coincidence between the measured and planned weights of the units was practically exact (within the limits of precision of similar projects).

The maximum vertical load on the piles, measured by DS-50 dynamometers, was 250-300 kN during construction. The diagram representing the longitudinal stresses in the piles is of irregular shape near the walls of the building.

During undermining, as a result of the changing geometry of the ground surface and its inclination, the longitudinal stresses increased in some piles and decreased in others. There appeared to be no direct relationship between the load redistribution between the piles and distortion of the building during undermining. Nevertheless, it was discovered that the inclination of the ground surface greatly influenced the stress changes. It resulted in increasing loads on the piles nearest to the goaf and in decreasing loads on the piles furthest from it. The maximum load increase on some piles reached 180 kN and the maximum reduction was 190 kN.

Measurements of the loads on the piles and their penetration into the ground made it possible to calculate the soil modulus for each pile and to monitor its variation during the six years of observations. The completed calculations showed that the soil modulus of the experimental units was sharply reduced (by 40 to 50 per cent i.e. about halved) during the time of the initial and intensive ground movements and it was related to the magnitude and sign of the ground surface deformations. A continuous reduction in the moduli was mainly due to changes in the stress-strain condition of the ground mass and has the effect of generally reducing effects on undermined buildings.

The data collected corroborated the results of the engineering/geological investigations. From the latter, the dynamic sounding index which characterises the density and compressibility was reduced by 40-50 per cent. The stress-strain modulus of the ground was reduced (according to laboratory tests) by 20 per cent and the horizontal modulus (from compression tests) was reduced by 45-55 per cent. The unit cohesion and the ground resistance to displacement were considerably reduced (up to 30-40 per cent). All these measurements prove that considerable reserves of planned loads exist on the foundations of buildings being undermined.

The measurements of lateral earth pressure, with SDKS vibrating wire devices, showed that the distribution of pressure on the pile shafts during driving was irregular along the length and sharply increased with depth, reaching in some cases 0.50-0.52 MPa. When the piles "were rested" for 15-20 days, reduction of the stresses was observed and the lateral pressures reduced by 20-25 per cent.

The horizontal surface strains produced by mining resulted in changes in the lateral pressures on the piles. The latter can be defined in terms of the difference in pressure on the opposite sides of a pile shaft taking the initial observations as datum. Measured in this way the maximum lateral pressure due to mining was found to be applied to the upper part of a pile as had been expected, and reached 0.28 MPa. The lateral pressure reduced to zero at the bottom of the pile i.e. the ground translation during undermining exerts a maximum pressure at the top of the pile where the pile is fixed to the foundation mat.

An analysis of the data from all the experimental piles showed that the ground exerts a pressure on a pile's lateral surface when the latter is displaced towards the goaf, loading the surfaces of the piles which face the ground movement on each side of the central axis of the unit. Therefore there was no clear-cut proof of the existing belief that the value of the lateral pressure is a function of the pile's distance from the centre of the group.

The stressed-deformed state of the piles was determined by measuring the fibre stresses and bending moments along their lengths with the help of strain measuring devices. The recorded data showed that a variable bending moment, up to a value of 28.0×10^5 Nm, developed in the piles at installation.

Following this, during the time of the piles "rest", and when the buildings were under construction the values of the bending moments along the whole lengths of the piles fell and only increased again sharply during the active stage of ground movement. Despite this, the maximum value of the bending moment caused by the mining activities only reached 19.0×10^5 Nm; and taking the initial stressed state of the piles into account it was 29.0×10^5 Nm, that is one-half of the ultimate bending moment.

The results of the observations with the measuring devices gave an insight into the dynamics of stress increases in the foundation mats and the reinforcing braces as the buildings were constructed. They also gave an idea of the character of the stress changes in the reinforcement of the buildings at undermining. Thus, the stress increase caused by mining reached 80-130 Mpa in the foundation mats of the panel and brick buildings. The stress increase in the reinforcement of the floor braces of the brick building was 70-80 MPa. The stress reduction over the height of a building was five-fold between the 2nd and 5th floors.

CONCLUSIONS

The values of the stresses developed in the piles, the main load carrying members and in other elements of the experimental buildings showed that the initial undermining was carried out without any noticeable damage, thus proving the validity of the construction decisions and the effectiveness of piled foundations for building in mining areas.

The favourable results from the tests on the buildings with piled foundations justifies their use in large-scale construction in mining areas with complicated engineering-geological conditions.

NON-LINEAR FOUNDATION BEHAVIOUR IN THE ANALYSIS OF FRAMELESS BUILDINGS UNDER THE ACTION OF FOUNDATION DISPLACEMENTS

S.N. Klepikov, Prof

Head of Department in NIISK, Kiev, USSR

F.N. Borodatcheva, MSc

Senior Scientist in NIISK, Kiev, USSR

I.V. Matveev, Eng

Engineer in NIISK, Kiev, USSR

A special feature of the design of buildings for mining areas and for other complex engineering-geological conditions is the necessity to consider not only conventional loads but also the influence of differential movements of the footings. For the approximate evaluation of the stressed state of frameless buildings, a number of calculation methods have been developed (Code of Practice, 1977; Lishak, 1977; Kosizyn, 1971).

In the present paper, a more rigorous method is proposed for the three-dimensional design of multi-storied frameless buildings under the action of external loads and horizontal displacements of their footings, taking into account non-linear behaviour. The essence of the method is a conversion of the modules of the building into a system of cross beams whose properties are selected so that the forces and displacements in the beams and in the corresponding walls differ by insignificant amounts. The design philosophy for the footings is based on a model with a variable stiffness coefficient, according to Klepikov (1977). Such a footing has the following characteristics: vertical contact pressures are proportional to the vertical displacements and the horizontal forces are proportional to the horizontal displacements. The displacement characteristics of the footings are determined simultaneously by the use of two stiffness coefficients $K_{сж}$ (compression) and $K_{сд}$ (shift).

To commence, the individual walls of the building are considered as flat plate systems with conventional loads acting on them and they are calculated by the finite element method. The stiffness coefficients of the connections between the plate elements are determined using the recommendations given in the Manual (1974).

Numerical design is carried out using the program "Super-76", intended for calculations with the "Minsk-32" computer. This program uses the finite element method based on displacements.

From the calculation, the support reactions, the forces in the members at joints and the vertical and horizontal displacements of nodes are determined. On the basis of this calculation it is possible to find the values of the normal and transverse forces and the values of the bending moments in the cross-sections of the wall at the junctions of the panels. Then, knowing the support reactions and the applied load, it is possible to calculate general values of the normal and transverse forces and bending moments in an ordinary beam of the same cross-section.

The following conditions must be satisfied absolutely:

$$N_i^c = N_i^b \quad ; \quad Q_i^c = Q_i^b \quad ; \quad M_i^c(z) = M_i^b(z) \tag{1}$$

The first two equations (1) must be fulfilled automatically, to ensure that the design of the wall by the finite element method is correct.

In the third equation, or more strictly in $M_i^b(z)$, the values of h_i are involved (Figure 1).

Equations $M_i^b(z)$ take the form:

$$M_0^b(z_0) = H_0(h_0 + z_0)$$

$$M_1^b(z_1) = H_0(h_0 + z_1) - V_0 a$$

$$M_i^b(z_i) = H_0(h_0 + z_i) + H_1(h_1 + z_i) - V_0 2a - V_1 a$$

etc.

Equating in sequence:

$$M_0^c(z_0) = M_0^b(z_0)$$

$$M_1^c(z_1) = M_1^b(z_1) \quad , \; .... \; ,$$

$$M_i^c(z_i) = M_i^b(z_i) \quad , \; \text{etc.} \; ,$$

we determine the values h_0, h_1, ... , h_i . With the values h_i the bending moments at any distance from the beam soffit can be calculated.

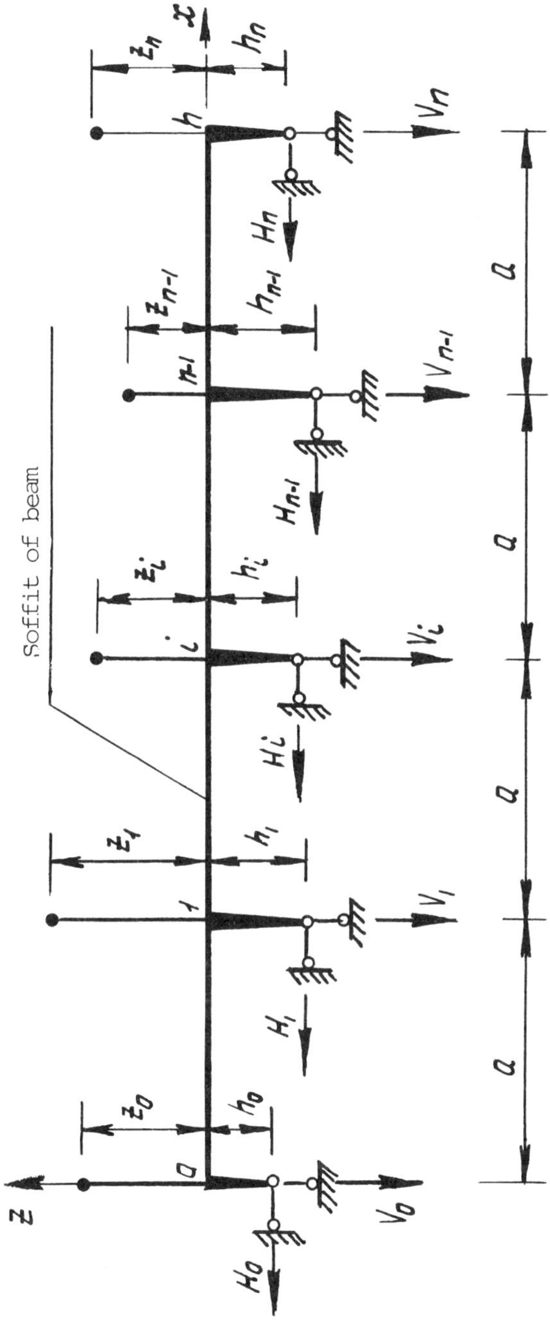

Figure 1 Beam layout

In determining the values of h_i by this method we attempt to equate the bending moments in the corresponding sections of the wall and the beam.

It is then necessary to select the bending stiffness EJ and the tensile stiffness EF of the beam by applying the condition that the vertical and horizontal displacements of the wall and the beam are equal.

We make use of the following relationships:

$$EJ = \frac{M}{z''} \quad ; \qquad EF = \frac{N}{x'} \tag{2}$$

For a given distance from the bottom of the wall, the diagrams of vertical and horizontal displacements are constructed. The values follow from the finite element computations for the wall. The diagrams of bending moments and of normal forces are also drawn for the same distance from the bottom of the beam.

By numerical differentiation one then finds the derivatives z'' and x'. Following this and using equations (2), one determines the stiffnesses EJ and EF.

To illustrate the proposed method of transforming the modules of the building into a system of cross-beams we give an example of an actual nine-storied apartment block made from large panels. It was intended to build this block in a mining area. Loading is provided by the dead weight and the vertical and horizontal displacements of the ground surface due to the underground workings. Measures of the deformation of the ground surface are taken as $R = 7$ km and $\varepsilon = 5 \times 10^{-3}$.

Figure 2(a) shows the stiffness (EJ and EF) diagrams calculated by the proposed method. As indicated by trial calculations, one can replace these diagrams by step functions as shown in Figure 2(b). With this simplification, the diagrams of displacements and of forces in the beam differ very little from the more precise diagrams.

Using this method a calculation is carried out for each wall. We then consider the problem of selecting values for the parameters of the beam analog for the walls of the building. These parameters determine the stiffness EJ and EF and also the height of the support h_i. They are selected so that the force and displacement diagrams of the beam differ very little one from another.

After values of the parameters are selected, separately for each wall, we can proceed to replace the modules of the building by the system of cross-beams.

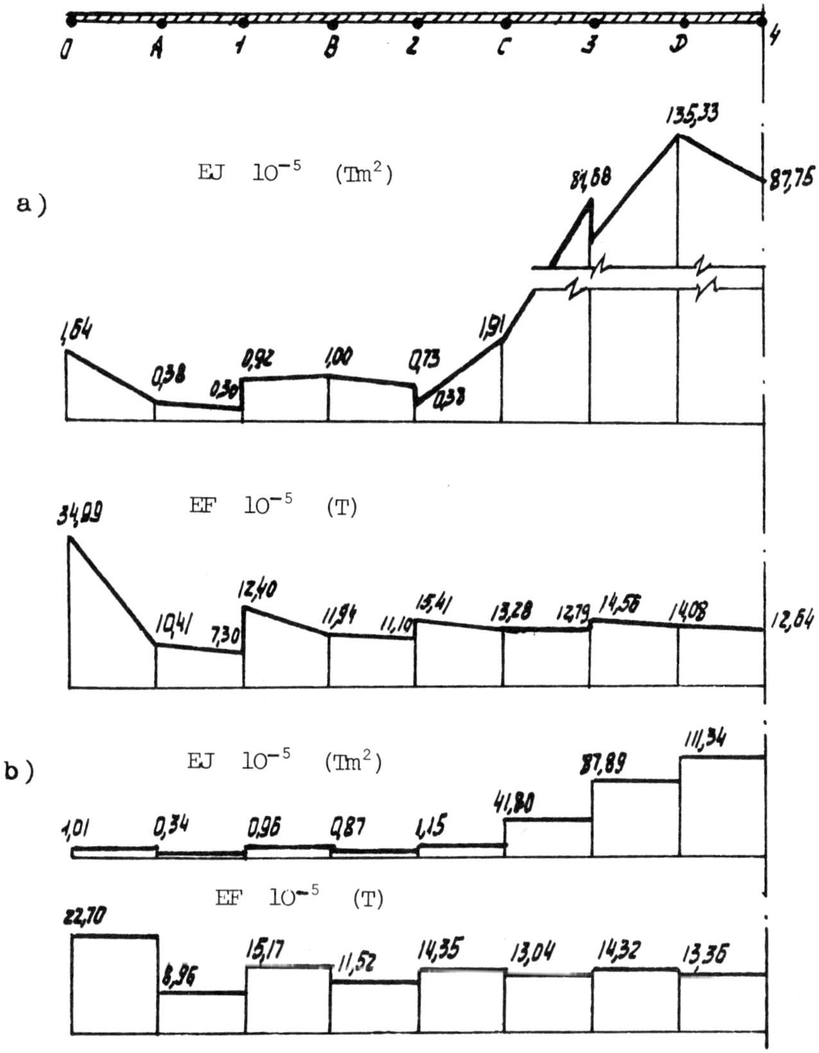

Figure 2 Stiffnesses EJ and EF for applied loads and for footing displacements

The values of the parameters of the beam equivalents of the wall were calculated in relation to a datum chosen at a distance z = 1.2 m from the bottom of the wall. This permitted the beams to be joined into a grid in the datum plane 1.2 m from the bottom of the wall.

Figure 3 shows a numbered layout of the building as a system of cross-beams. The plane XOY is selected as the datum plane. The origin of coordinates is taken at the centre of the beam equivalent to the gable-end wall. The geometrical sizes of the grid system coincide exactly with the outer longitudinal walls along sections 1-1 and 3-3; along 2-2 with the inner longitudinal wall; along 4-4 with the gable-end wall; and along 5-5 and 6-6 with the inner transverse walls. The junctions of the inner walls with the outer walls are taken as stiff.

To account for the action of the soil support during the displacement of every beam of the system from its plane, a horizontal prop is added to each of the junctions. The junctions bring together the stiff supports and the soil props. The stiffnesses of these props equal the stiffness of the beam in its plane. We also assume that in the horizontal plane the soil possesses the same properties in all directions.

The grid system is loaded with the same loads as the separate beams.

As a result of the design of the grid system of beams by the program "Super-76", the vertical and horizontal displacements of the system, the bending moments and the normal and transverse forces at specified sections of the system are obtained. Figures 4 to 7 (dotted lines) illustrate these displacements and forces.

Non-linear displacements of the footings are allowed for by applying corrections to the values of the stiffness coefficients at the individual bases. The algorithm for determining the stiffness coefficients of a non-linear elastic footing is as follows:

(1) The calculation is started by solving the linear-elastic problem. Thus all loads and actions in the system are allowed for. From this calculation the forces in the construction and the vertical and horizontal displacements of the footing $z^{(1)}$ and $x^{(1)}$ are determined (where the number in the brackets means the number of the approximation of the calculation).

(2) In the second approximation, we first calculate the values of the stiffness coefficients by the formulae (Klepikov, 1977):

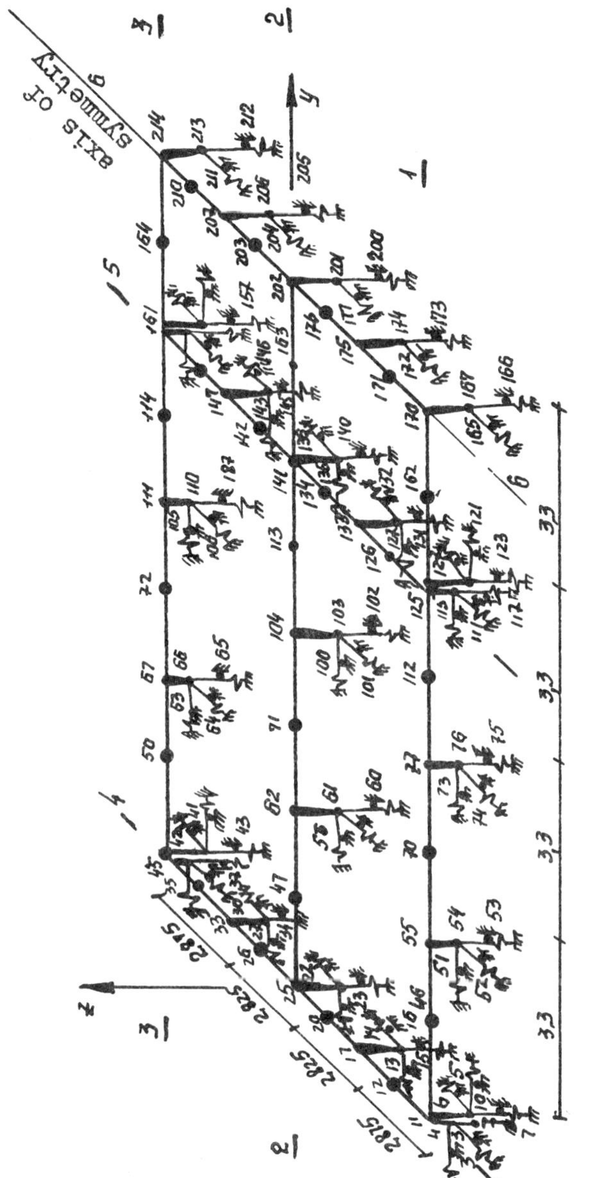

Figure 3 Layout of building as a system of cross beams on elastic supports

Y (mm)

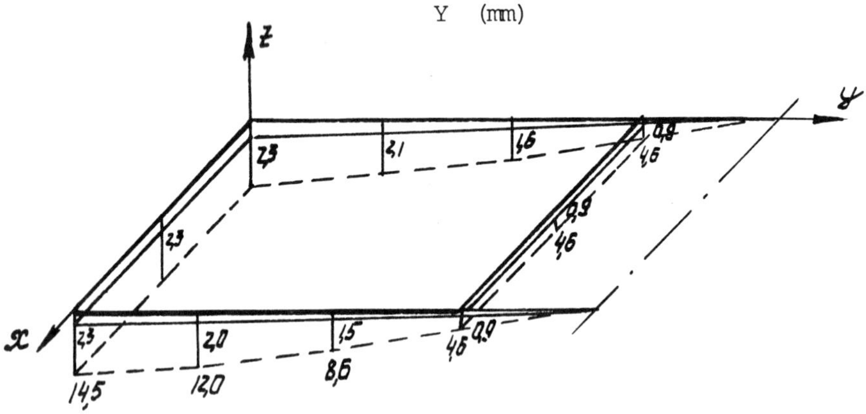

Z (mm)

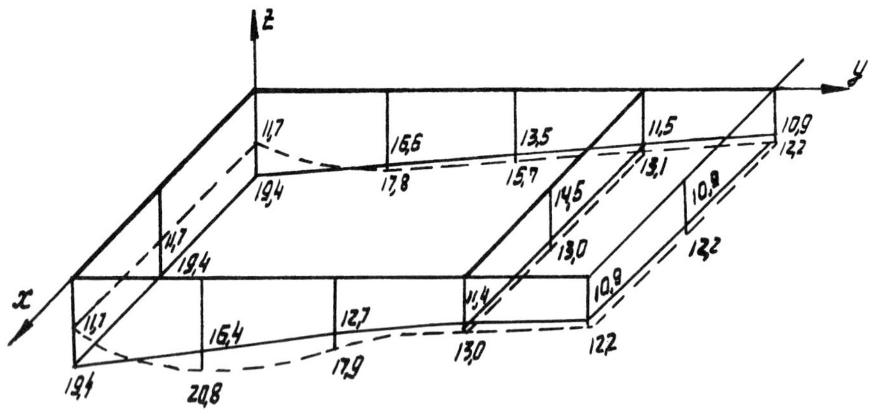

Figure 4 Displacement diagrams

P_Y (T)

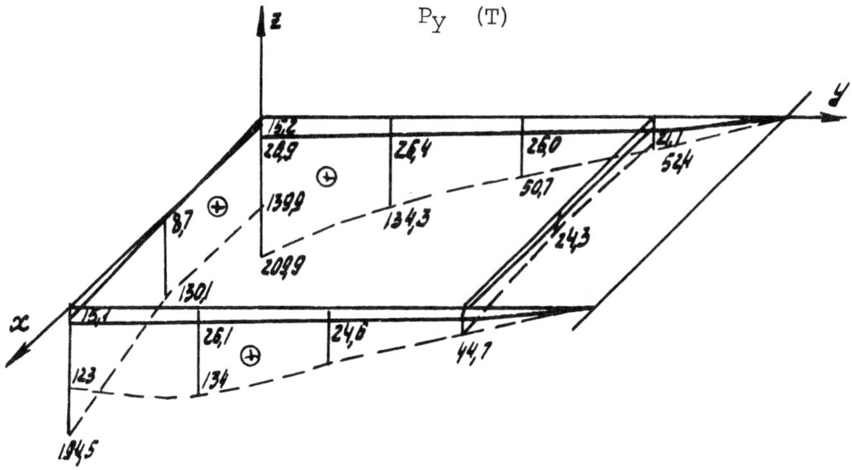

P_Z (T)

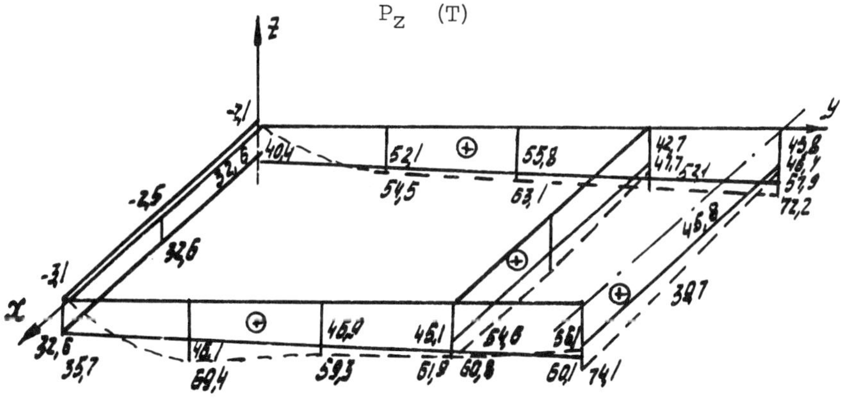

Figure 5 Reaction diagrams

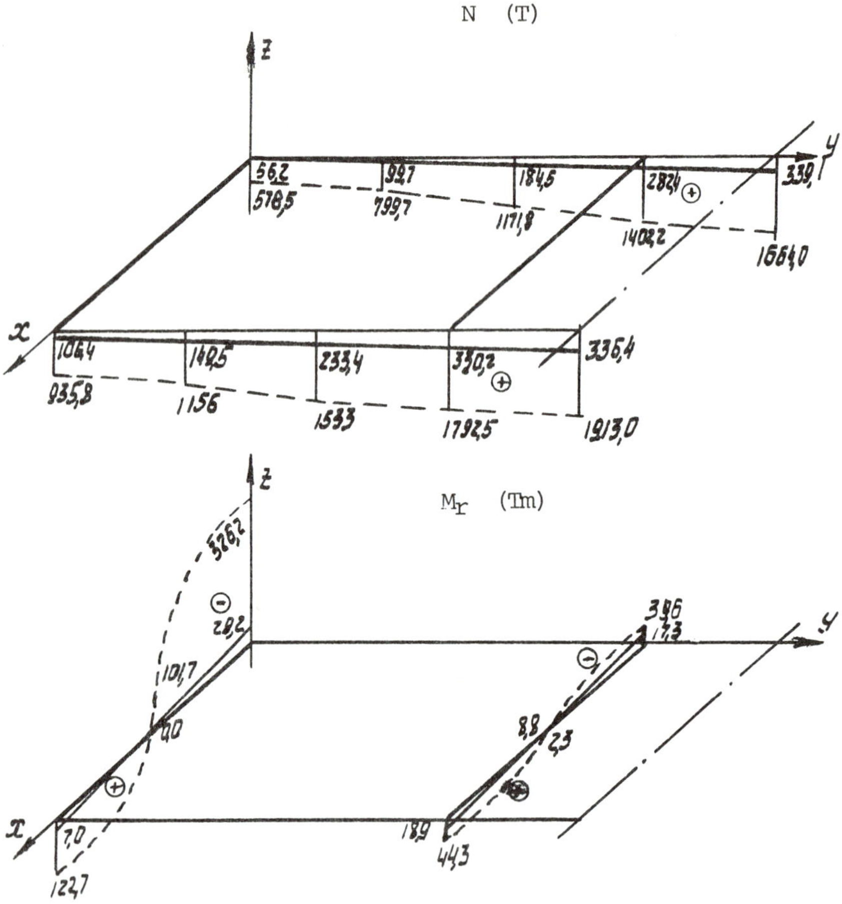

Figure 6 Forces N and M_r

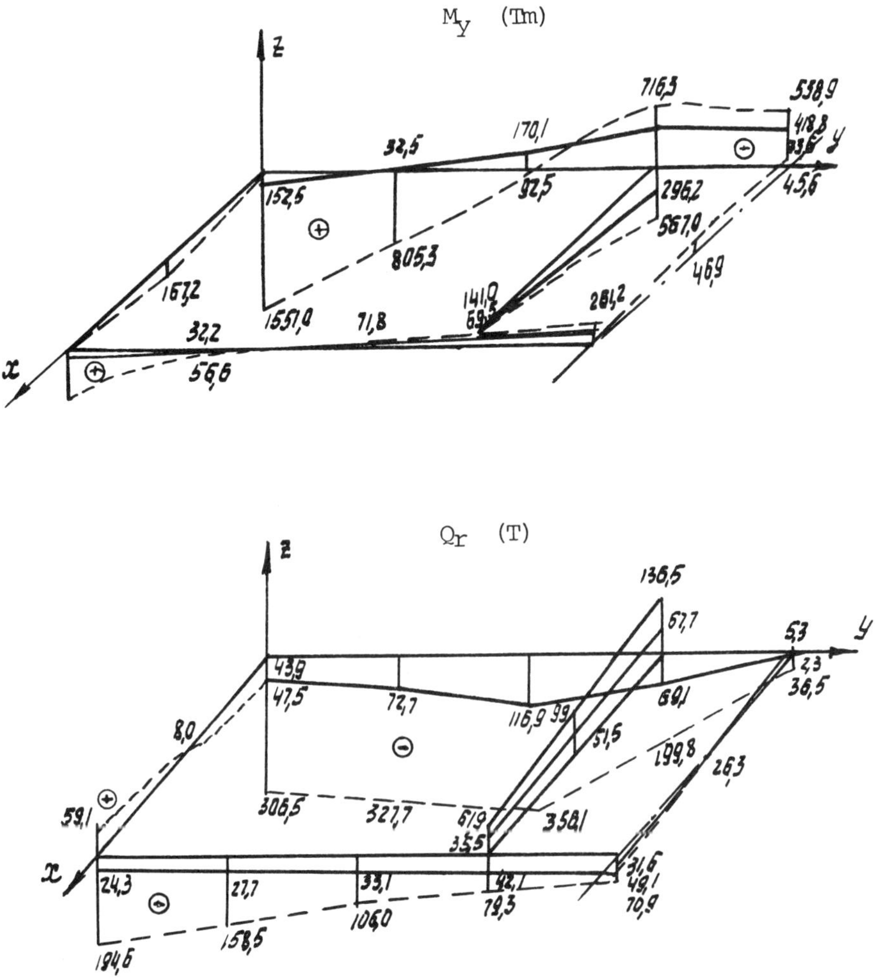

Figure 7 Forces M and Q_Z

$$K^{(d+1)}_{сж(i)} = \frac{R_i}{z_i^{(d)} + S_i}$$

$$K^{(d+1)}_{cg(i)} = \frac{\Phi_i}{x_i^{(d)} + \dfrac{0.8\Phi}{K^{(d)}_{cg(i)}}} \qquad (3)$$

$$S_i = z_i \left[\frac{R_i}{\bar{P}_i} - 1 \right]$$

in which, $\bar{P}_i$ is the average uniformly distributed contact pressure on the footing; R_i is the ultimate bearing pressure that corresponds to a loss of bearing capacity of the footing; Φ_i is the bearing capacity of the footing under the action of a tangential (horizontal) load. This bearing capacity is determined as in Klepikov (1977); i is the number of the footing.

In the formulae for the stiffness coefficients (Equations 3) the displacement values $z^{(1)}$ and $x^{(1)}$, obtained from the first cycle of calculations, are substituted. With the values $K^{(2)}_{сж}$ and $K^{(2)}_{cg}$ from the second cycle of the calculations we again solve the linear-elastic problem with the same loads and actions as in the first cycle.

(3) The calculation in the third and following approximations is carried out in the same way.

The calculation is repeated until the results in a given approximation are close to the corresponding results in the preceding approximation. Experience with the calculations shows that the process always converges and the rate of convergence is rapid; by the third or fourth approximation we obtain acceptable accuracy of results. Illustrations of the displacements and forces obtained in a non-linear calculation are also given in Figures 4 to 7 (continuous lines).

Taking non-linear behaviour of the footings into account the results of the calculation show that, in relation to the results for linear behaviour, the horizontal reactions of the footings are considerably reduced and the vertical reactions are redistributed between the beams of the grid system. As a result the horizontal deformations are considerably less and the vertical reactions are more uniform and the stresses are reduced.

REFERENCES

Руководство по расчёту и проектированию зданий и сооружений на подрабатываемых территориях. М., 1977.
(Guidance for the calculation and design of buildings in mining areas).

В.И. Лишак - Расчёт бескаркасных зданий с применением ЗВМ. Стройиздат, М., 1977.
(Calculation of frameless buildings with the aid of a computer).

Б.А. Косицын - Статический расчёт крупнопанельных и каркасных зданий. Стройиздат, М., 1971.
(Statical calculations for large-panel and frame buildings).

С.Н. Клепиков - Методические рекомендации по определению коэффициентов жёсткости оснований зданий и сооружений. Киев, 1977.
(Methodical recommendations for the determination of the stiffness coefficients of building footings).

Пособие по расчёту крупнопанельных зданий. Выпуск 1. Характеристика жёсткости стен, элементов и соединений крупнопанельных зданий. М., 1974.
(Manual for the calculation of large-panel buildings. Part 1: Stiffness characteristics of walls and other elements of the building).

SUBGRADE RESTRAINT AND SHEARING FORCE EFFECTS DUE TO MOVING GROUND

J.D. Geddes, BSc, PhD, CEng, FICE, FASCE, MIHE, FGS

Professor of Civil Engineering
University of Wales Institute of Science and Technology, Cardiff

INTRODUCTION

When structures are supported at the ground surface and the soil is subjected to horizontal movements produced by causes other than those of structural loadings on the foundations, there is a natural tendency for the structure to be translated horizontally with the soil from its original position. If the movements are of constant value over the surface of contact with the foundations, and movement of the foundations is not opposed, the structure will move as a rigid body over a distance equal to that of the ground translation. If the motion is slow, no forces will be imposed on the structure via its foundations. When motion of the foundations is opposed by the introduction of structural props, or when the movements are not uniform over the surface of contact, the subgrade exerts shearing forces on the underside of the foundations. This subgrade reaction may produce considerable forces. These forces, applied in the form of a shear, are also obviously capable of producing significant bending moments on a superstructure, due to the eccentricity at which they act. In an earlier paper (Geddes, 1977A) this phenomenon was described at greater length than will be attempted here. For the present purpose, only those aspects required for an understanding of the contents of this paper will be repeated.

Horizontal movements of the kind referred to, may be produced by several causes, of which nearby excavations, tunnelling and coal mining operations are obvious examples. In each of these instances, the horizontal movements at ground surface level are at a maximum near the edge of the excavation, or near the edge of the underground cavity, and they diminish with distance from these positions until at a finite distance they become practically zero. It is therefore clear that any structure within the zone of movement will experience shearing forces due to subgrade restraint. The ground movements may be placed in one of two categories; first those which increase in value at a given point with time, to a maximum and then remain constant, and in these cases there may be symmetry of the distribution about an axis; and secondly, those which undergo a cycle, increasing to a maximum at a point and then diminishing again with time, to perhaps a zero value. The latter category of movement is that generated along the axis of an advancing underground mining extraction, often described in mining literature as a dynamic wave (Figure 1). The values of the movements differ from point to point and the difference in the values at two adjacent points, expressed as a proportion of the original distance between the points, is commonly referred to as the horizontal ground strain. Figure 2, drawn using data from the NCB Handbook (1975), shows a typical dynamic strain curve for a coal seam of 210 m width and 1.5 m thickness, at a depth of 150 m below ground surface. It is in this form that mining data is usually presented, but as will be seen later, curves of the type of Figure 1 are of more direct use to those seeking an understanding of structural behaviour. Figure 1 has been derived from Figure 2 by numerical integration, since the latter curve represents the first differential of the former i.e. $e = \frac{ds}{dx}$, where e is the horizontal strain, s is the horizontal ground displacement, and x is the horizontal co-ordinate of the point at which the former are measured.

In this paper a numerical calculation process is advanced, which permits an examination of the forces produced by subgrade restraint in a surface-supported structure as it is exposed to the cycle of displacements of Figure 1. For this purpose, the structure is idealised and the treatment simplified, the main objectives being to demonstrate how the forces may be established and to show that approaches which equate ground strains (in size and/or sign i.e. tensile or compressive) with the same response in structural behaviour may be misleading. An understanding of the relationships requires a consideration of the ground movements (as the primary quantity i.e. not the ground strains) and subgrade restraint behaviour at the interface of the structural foundation and the soil. In the analysis, the effect of vertical ground movements (which generally accompany horizontal ones) is eliminated by restricting consideration to the case of a structural foundation which is completely flexible in bending, whereby the value of the vertical support reaction at each point does not vary during the process to be described.

SUBGRADE RESTRAINT

If a foundation slab on the horizontal surface of a soil is displaced relative to the soil (or vice-versa) the subgrade restraint typically develops in the way shown in Figure 3, from which it is seen that a finite displacement is needed before a maximum shearing force is produced. The curve may be approxim-

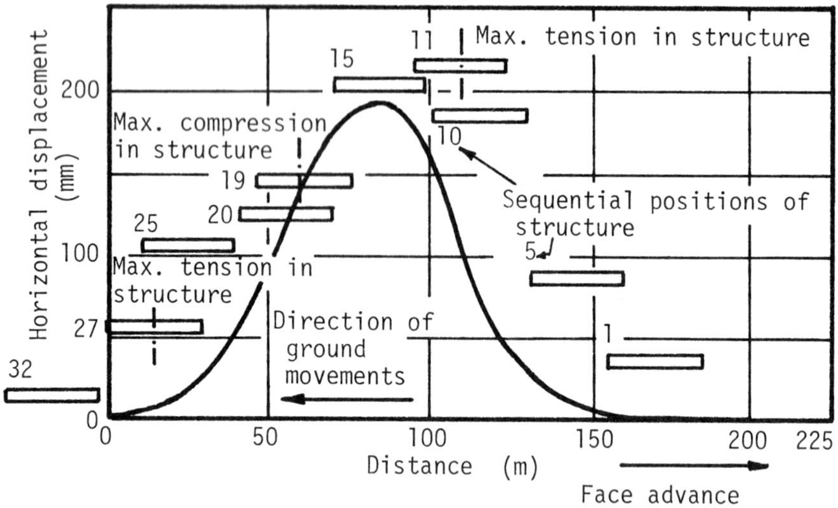

FIGURE 1 Horizontal ground movements

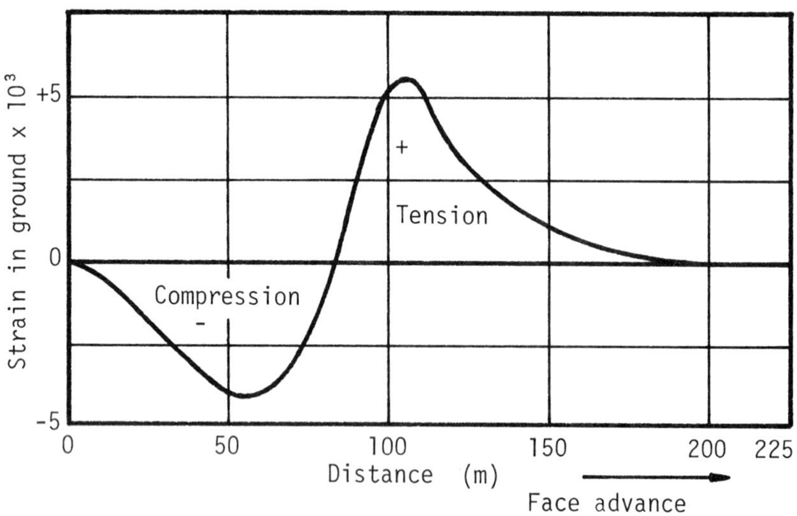

FIGURE 2 Dynamic ground strain curve

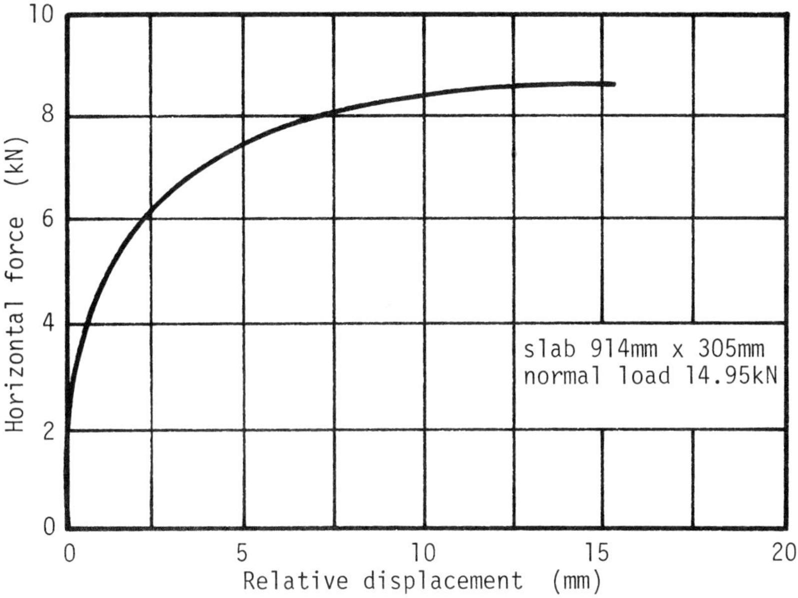

FIGURE 3 Subgrade restraint

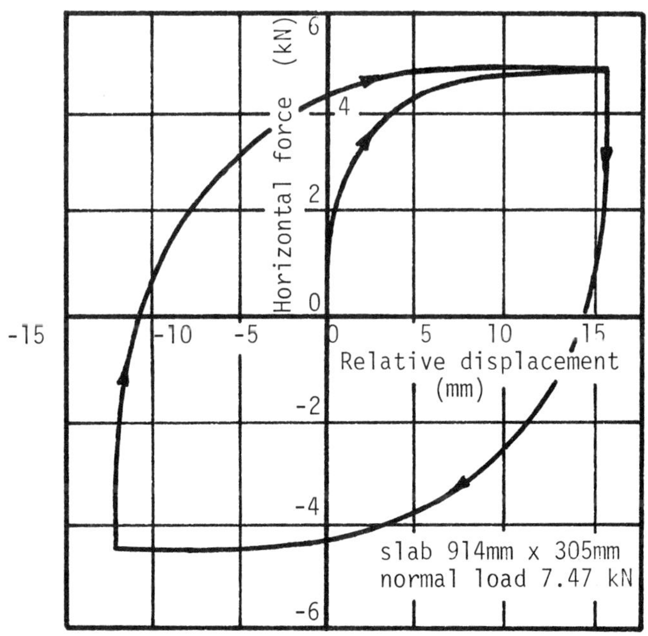

FIGURE 4 Hysteresis effect

ated by two straight lines, which intersect at what may be defined as a critical relative displacement or slip. If the latter is exceeded, no further increase of restraint takes place. If the direction of relative displacement is reversed, the original curve is not retraced and the shearing force-displacement curve takes the form shown in Figure 4. A subsequent reversal produces the hysteresis-type loop, also shown in the Figure. As a simplification, these curves may also be replaced by straight-line approximations (see Figure 6). These factors of non-reversability and critical displacement are of great significance to an understanding of why ground movements and strains are not mirrored or equalled by structural displacements and strains, irrespective of structural rigidity.

METHOD OF ANALYSIS

Consider the simple linear structure shown in Figure 5, which rests on the horizontal surface of a soil. As already mentioned, it is assumed to be infinitely flexible in bending and perfectly rigid in terms of its reaction to horizontal forces applied along and parallel to its length. These criteria are similar, for example, to those applying in the case of CLASP-type structures (see Bell, 1977; Geddes, 1977B), where the foundation slabs rest at ground level, are provided with flexure joints and consist of reinforced-concrete slabs (essentially non-deforming in compression and of limited extensional capability). It is further assumed that the model structure is subjected to a uniform vertical contact pressure along its length and that the shearing stress-displacement relationship at a point on its interface with the ground is of the form shown in Figure 6. It is also assumed that the structure remains intact during the movement process.

The structure is first sub-divided into n - 1 elements (Figure 5), with the boundaries between the elements numbered 2 to n - 1. The ground movements at the different positions are g_1 to g_n and the structural movements are s_1 to $s_n = s$ (constant in this example, since the structure has been assumed to be non-deforming).

For the element i to i+1 ; its length is $L_{i,i+1}$, its breadth (in plan) is $B_{i,i+1}$ and the shearing stress acting on its lower boundary varies from f_i to f_{i+1} , where, for example, according to the principles stated earlier (Figure 6) :

(1)

$f_i = \alpha_i \, (g_i - s_i)$, when $(g_i - s_i)$ is less than or equal to the critical value,

$f_i = f_{i\ max}$, when $(g_i - s_i)$ is greater than the critical value (path B of the curve, Figure 6),

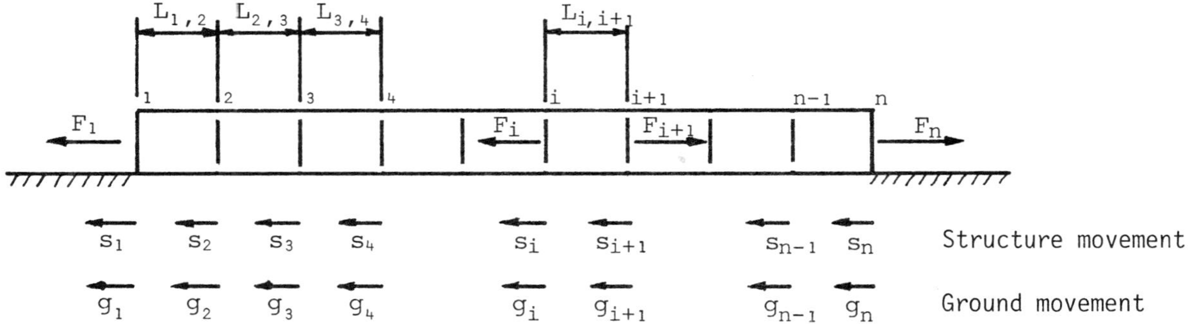

FIGURE 5 Model of structure considered

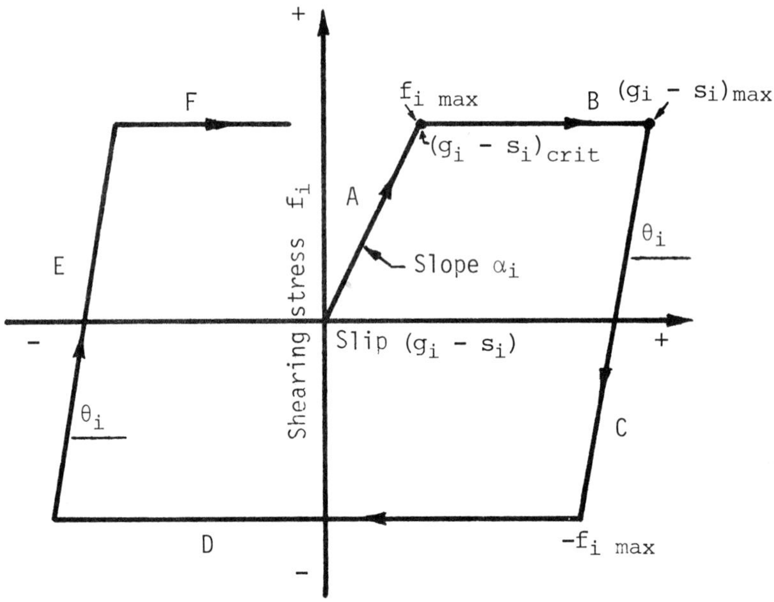

FIGURE 6 Assumed shearing stress-relative displacement relationship

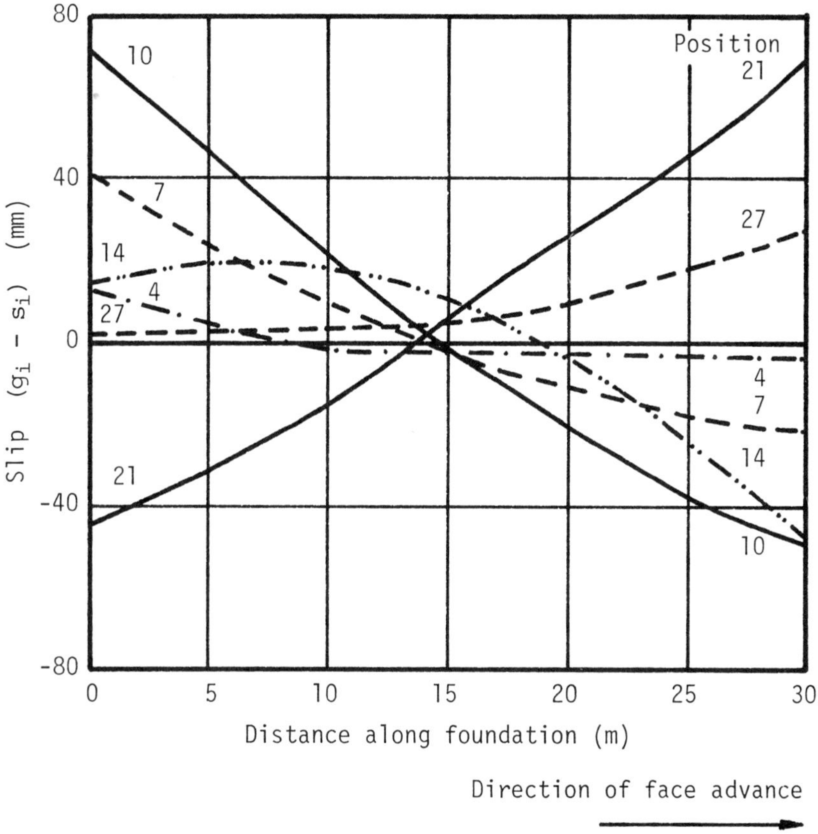

FIGURE 7 Relative movements at interface

$$f_i = f_{i\,max} - \tan\theta_i\{(g_i - s_i)_{max} - (g_i - s_i)\} \qquad (1)$$

> α_i, θ_i being constants defining the curves, subject to a lower limit of $-f_{i\,max}$, when relative displacements, greater than the critical value, are reducing (as on path C of the curve, Figure 6).

Similar relationships can be derived for other parts of the hysteresis curve (paths D, E and F).

There are n Equations (1) with $1 \leqslant i \leqslant n$.

If the shearing stress is assumed to vary linearly between the values f_i and f_{i+1} over the length of the element (i, i + 1), equilibrium requires that the forces at the boundaries of the elements, F_i and F_{i+1} are related by :

$$F_i + 0.5(f_i + f_{i+1})\, B_{i,i+1}.L_{i,i+1} = F_{i+1}$$

or

$$F_i - F_{i+1} = \beta_{i,i+1}(f_i + f_{i+1}) \qquad (2)$$

where $\beta_{i,i+1} = -\,0.5(B_{i,i+1}.L_{i,i+1})$ (a geometrical constant)

There are n - 1 Equations (2) with $1 \leqslant i \leqslant n - 1$.

In general, the forces F_i and F_n , at the ends of the structure, will be known. Hence the number of available equations is :

$$n + (n - 1) + 2 = 2n + 1$$

and the number of quantities to be determined is :

$$n\ (F_i) + n\ (f_i) + 1\ (s_i = s) = 2n + 1$$

A solution may therefore be obtained.

From Equations (1) and (2), <u>with all relative displacements less than or equal to the critical</u> $(g_i - s_i)$:

$$F_i - F_{i+1} = \beta_{i,i+1}\{\alpha_i(g_i - s) + \alpha_{i+1}(g_{i+1} - s)\} \qquad (3)$$

and

$$F_i - F_{i+1} + \beta_{i,i+1}(\alpha_i + \alpha_{i+1})s = \beta_{i,i+1}\{\alpha_i g_i + \alpha_{i+1}g_{i+1}\} \qquad (4)$$

Assuming that $F_1 = 0$

$$-F_2 + \beta_{1,2}(\alpha_1 + \alpha_2)s = \beta_{1,2}(\alpha_1 g_1 + \alpha_2 g_2) \qquad (5)$$

and assuming that $F_n = 0$

$$F_{n-1} + \beta_{n-1,n}\,(\alpha_{n-1} + \alpha_n)\; s = \beta_{n-1,n}\,(\alpha_{n-1}g_{n-1} + \alpha_n g_n) \qquad (6)$$

The system of Equations may be written in matrix form as :

$$\begin{bmatrix} -1 & & & & & & & & \beta_{1,2}(\alpha_1+\alpha_2) \\ 1 & -1 & & & & & & & \beta_{2,3}(\alpha_2+\alpha_3) \\ & 1 & -1 & & & & & & \beta_{3,4}(\alpha_3+\alpha_4) \\ & & 1 & -1 & & & & & \beta_{4,5}(\alpha_4+\alpha_5) \\ & & & & \cdot & & & & \\ & & & & 1 & -1 & & & \beta_{i-1,i}(\alpha_{i-1}+\alpha_i) \\ & & & & & & & & \\ & & & & & & 1 & -1 & \beta_{n-2,n-1}(\alpha_{n-2}+\alpha_{n-1}) \\ & & & & & & & 1 & \beta_{n-1,n}(\alpha_{n-1}+\alpha_n) \end{bmatrix} \begin{Bmatrix} F_2 \\ F_3 \\ F_4 \\ F_5 \\ \\ F_i \\ \\ F_{n-1} \\ S \end{Bmatrix}$$

$$= \begin{Bmatrix} \beta_{1,2}(\alpha_1 g_1 + \alpha_2 g_2) \\ \beta_{2,3}(\alpha_2 g_2 + \alpha_3 g_3) \\ \beta_{3,4}(\alpha_3 g_3 + \alpha_4 g_4) \\ \beta_{4,5}(\alpha_4 g_4 + \alpha_5 g_5) \\ \\ \beta_{i-1,i}\,(\alpha_{i-1}g_{i-1} + \alpha_i g_i) \\ \cdot \\ \beta_{n-2,n-1}(\alpha_{n-2}g_{n-2} + \alpha_{n-1}g_{n-1}) \\ \beta_{n-1,n}(\alpha_{n-1}g_{n-1} + \alpha_n g_n) \end{Bmatrix} \qquad (7)$$

When <u>a critical displacement is first exceeded at point i</u>

α_i is deleted from the expressions on the left-hand side of Equations (7) and f_{imax} replaces the expressions α_i g_i on the right-hand side of Equations (7).

When the third condition of Equation (1) is to be satisfied, further modifications must be made to the left- and right-hand matrices. The procedure is continued for other points on the hysteresis curves.

It is thus seen that the numerical solution of a problem consists of setting up the Equations (7) with the initial set of values of the ground movements g_i and solving for all F_i and s. Back substitution then shows if the critical value of displacement is exceeded anywhere. If it is not, the calculation proceeds with the next set of values, g_i. If it is exceeded at one or more points, the relevant alterations are made to the matrices and recalculations are performed until the assumptions inherent in a calculation are satisfied. As the calculations proceed for successive sets of values of g_i, additional modifications are made to the matrices in accord with the hysteresis criteria. The history of relative movements at each point is therefore important in determining the next stage of the calculation. In practice not more than two iterations are usually needed to bring about a satisfactory solution of the equations.

EXAMPLE OF CALCULATION

The process described has been applied to a typical problem, with the following parameters :

Length of structure 30 m
Number of elements 30
Breadth of elements 1 m
Lengths of elements 1.75 m and 0.25 m, alternating, (with first i.e. left-hand element 1.75 m)
Maximum shearing stress 8 kN/m^2
Critical relative displacement for each element 15 mm
End forces $F_1 = F_{30} = 0$ kN
Displacement required to move from maximum positive shearing stress to maximum negative shearing stress 10 mm i.e. $\tan\theta_i = 1600$ kN/m^2/m .

The pattern of ground movements assumed was that already introduced in Figure 1, and the calculations were made for 32 separate positions of the structure relative to the displacement profile, each representing a face advance of 6 m. The calculation commenced with the structure just clear of moving ground and was continued until ground movements under the structure ceased, with the complete passage of the dynamic wave. Several positions of the structure are marked on Figure 1. Some of the results are shown in a series of diagrams.

Figure 7 illustrates the way in which the distribution of relative movements ($g_i - s_i$) (slip) varies along the structure, as a function of position on the ground movement curve. The pattern changes from small initial values (position 4) to initial maxima (position 10), these then decrease to maximum values of reversed sign (position 21) and then diminish again to residual values (position 27), when the ground movement curve has passed through.

Figures 8(a) and 8(b) show distributions of slip as a function of time (i.e. position of the structure on the displacement curve) assuming a uniform rate of mining face advance, for four stations on the foundation. These each display two peaks (of opposite sign) followed by a return to a residual value. The two ends of the structure, in general, display relative movements of opposite sign.

The force developed at the centre of the structure (not always the maximum value at any stage) is shown in Figure 9, as a function of the position of the structure on the movement curve. This clearly shows that the single subsidence wave of extensional strain followed by a compressive strain, produces a tension phase, followed by compression, followed by tension in the structure and demonstrates the error in considering ground strains to be mirrored by structural strains of the same sign. It also means that some current design philosophies may need re-examination. The positions of the structure at maximum loading conditions are also shown on Figure 1.

Figure 10 shows how the shear stress values at the interface vary with time, for two stations. This also demonstrates the existence of a three-stage process. Figure 11 shows how the shearing stress at station 1 is conditioned by its prior history.

Comparisons may be made with Figures 12, 13(a) and 13(b) and 14, reproduced from an earlier paper (Geddes, 1977A) which present data measured in a full-scale site experiment on a linear structure subjected to a dynamic sequence of mining movements. Although the latter relate to different site and mining conditions, they clearly confirm the general patterns of the relationships derived from the analysis described in this paper.

CONCLUSIONS

The method of analysis presented in this paper is capable of describing the behaviour of a structure, of the form defined, when subjected to horizontal ground movements. It shows that over-concentration on the values of ground strains (in terms of magnitude and particularly sign) can mislead in the interpretation of structural effects. As demonstrated, a more rational

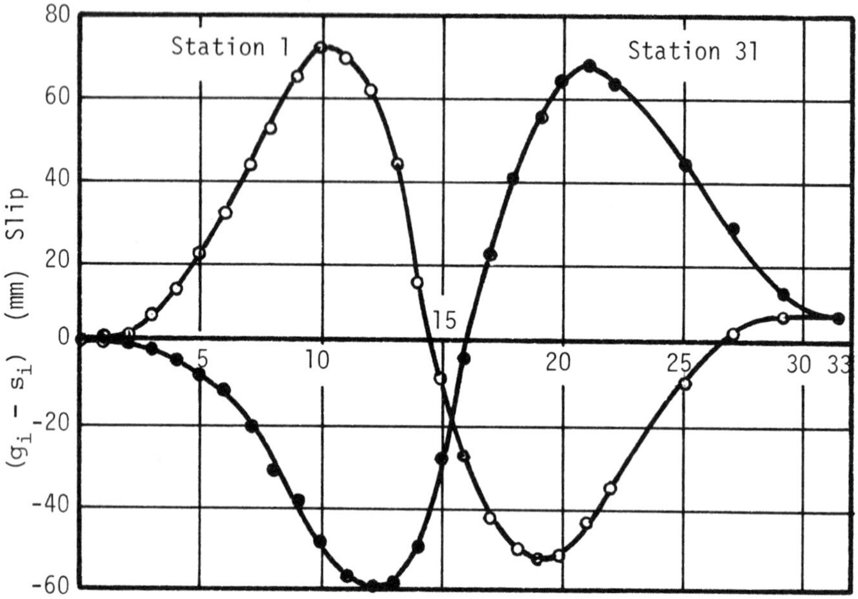

FIGURE 8(a) Relative movements at two stations versus time

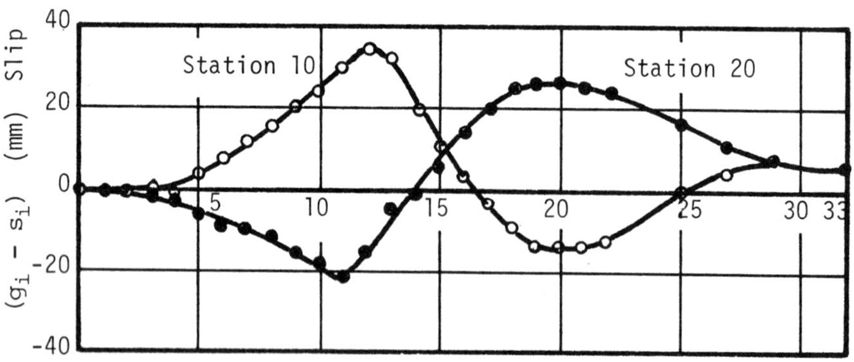

FIGURE 8(b) Relative movements at two stations versus time

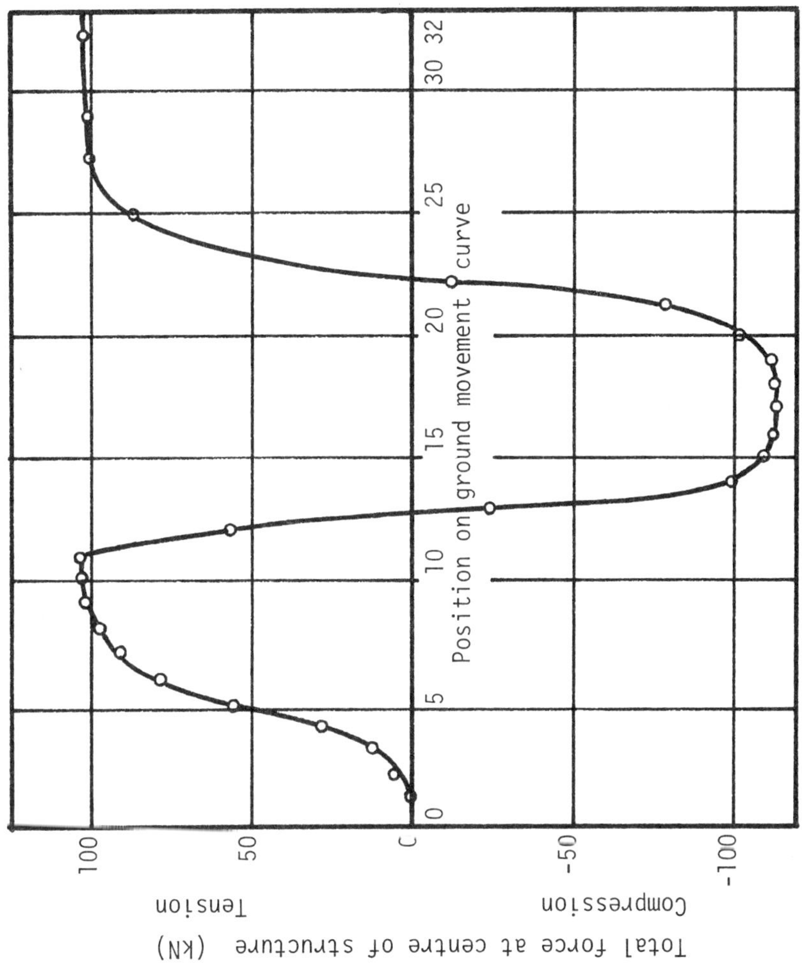

FIGURE 9 Force developed at centre of structure

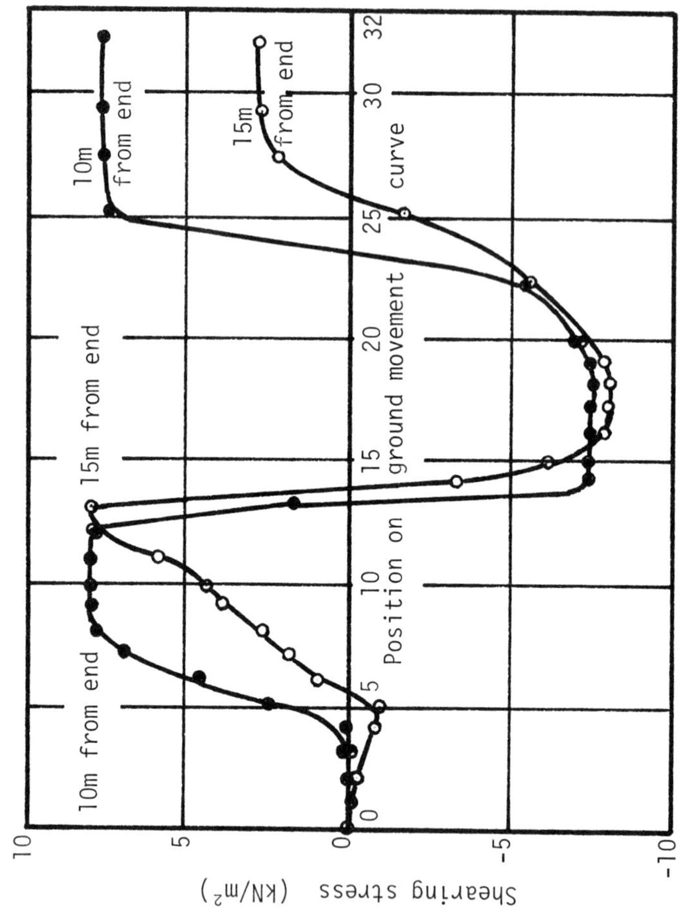

FIGURE 10 Shear stress developed at two stations

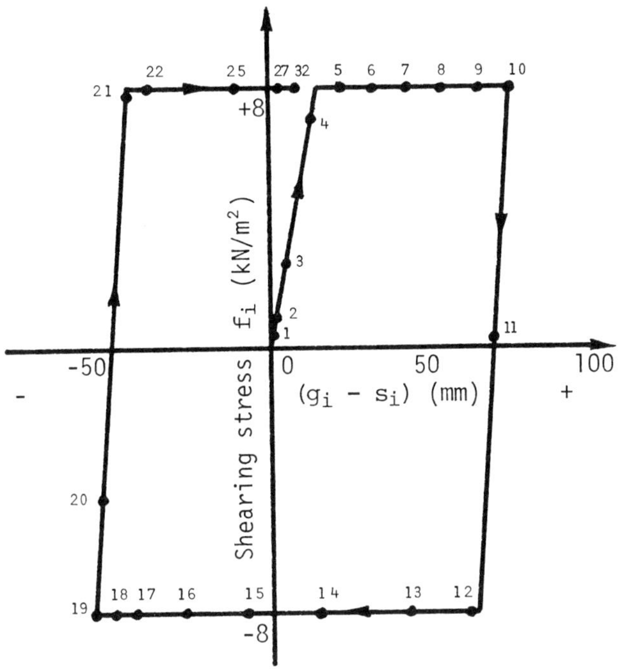

FIGURE 11 Shearing stress at end of structure (station 1) during passage of ground movêments

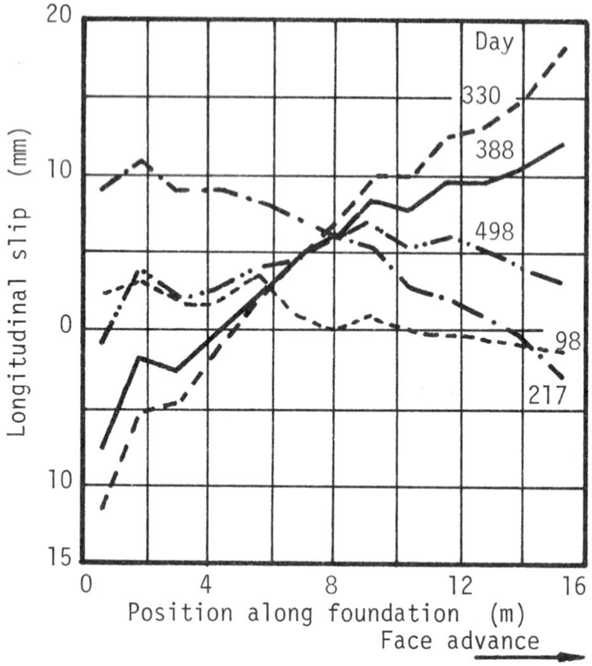

FIGURE 12 Relative movements at interface - field experiment

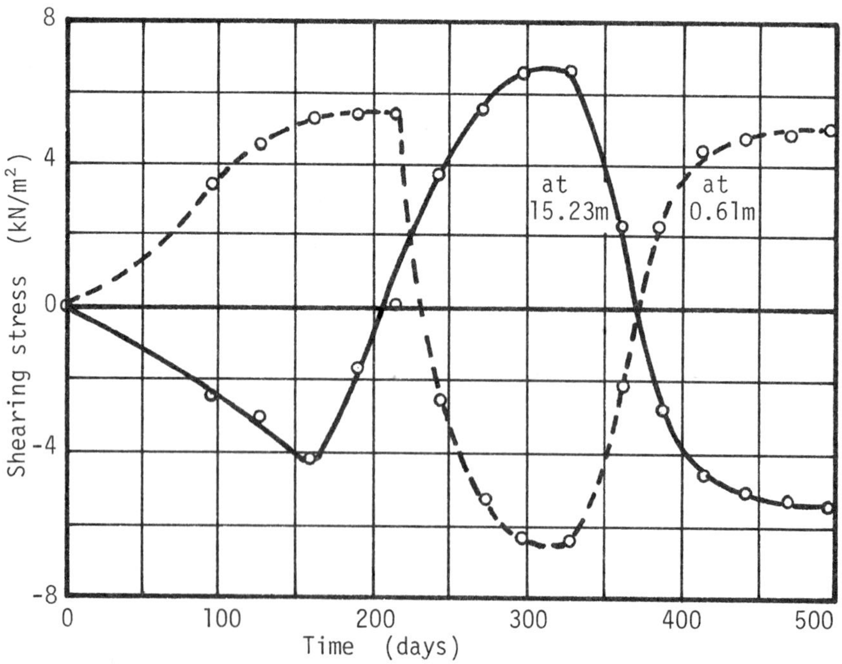

FIGURE 14 Shear stress developed at two stations - field expt.

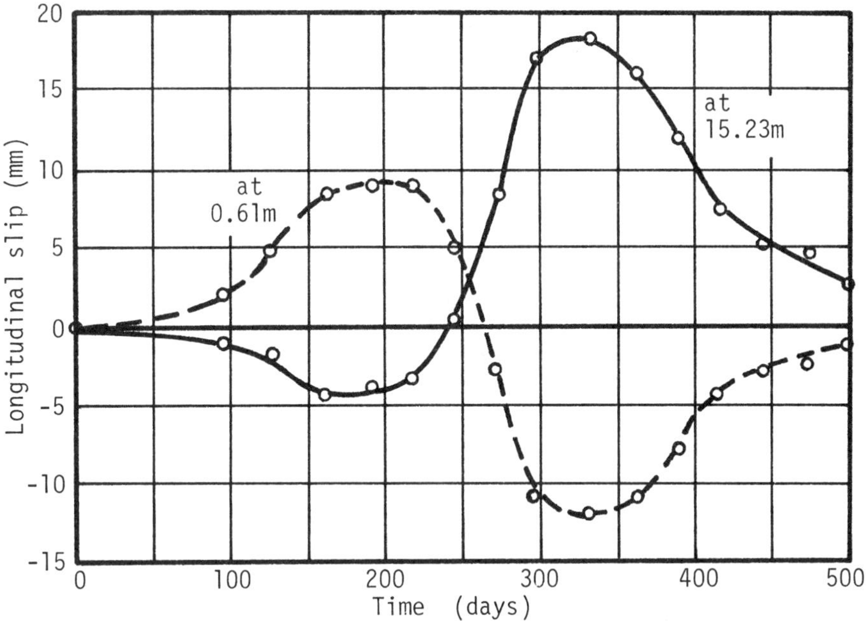

FIGURE 13(a)

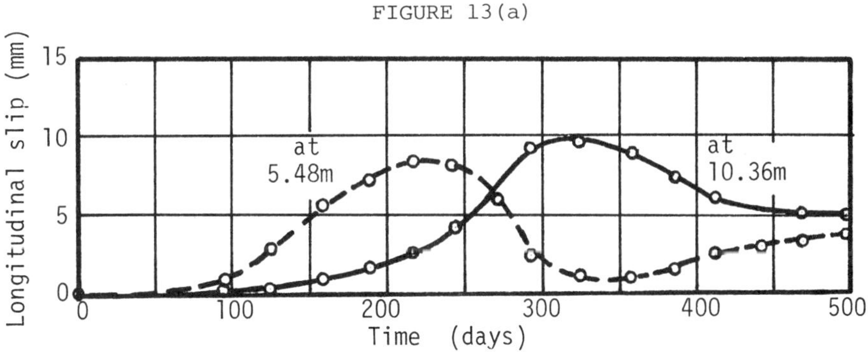

FIGURE 13(b)

FIGURE 13 Relative movements at four stations versus time - field experiment approach is to make use of a knowledge of the sequence of ground movements together with a model of subgrade restraint behaviour. When this is done, the way in which a structure moves relative to the soil and the way in which forces are developed on it are seen to yield results in conflict with some conventional thinking. The method outlined is being extended to deal with a wider range of situations and more complex structural behaviour.

ACKNOWLEDGEMENT

The author gratefully acknowledges the help received from Dr.J.P. Baty, in the preparation of computer programs.

REFERENCES

S.E. Bell - Successful design for mining subsidence, Proceedings of the Conference on Large Ground Movements and Structures, Cardiff, Pentech Press (1977) 562-578.

J.D. Geddes (A) - The effect of horizontal ground movements on structures, Proceedings of the Conference on Large Ground Movements and Structures, Cardiff, Pentech Press (1977) 623-646.

J.D. Geddes (B) - The behaviour of a CLASP-system school subjected to mining movements, Proceedings of the Conference on Large Ground Movements and Structures, Cardiff, Pentech Press (1977) 579-596.

National Coal Board - Subsidence Engineers' Handbook, 2nd Edition (1975) pp 111.

MODELLING STOCHASTIC BEHAVIOUR USING THE FRICTION TABLE WITH EXAMPLES OF CRACKED BRICKWORK AND SUBSIDENCE

D.W. Cox, Ph D , M Sc , B Sc , M I C E , F G S

Lecturer, Polytechnic of Central London

The friction table has been widely applied to demonstrate the behaviour of rock slopes. The same methods can be applied to other fissured or discontinuous media such as brickwork or rock in a subsidence zone. The paper demonstrates the effect of different ground deflections on crack patterns in brickwork. The secondary effect of window and door openings on the pattern are also shown. A stereographic method of crack measurement is detailed.

With regard to subsidence the effects of thickness and depth of seam, pillar size and spacing or change in strata may be indicated by the model. Of particular interest is the change in behaviour when a fault or continuous fissure is introduced.

The progressive development of a settlement crack or roof collapse can be arrested and examined at any stage.

INTRODUCTION

Strains in granular particulate matter such as soils may be calculated from a knowledge of elastic and shear properties. However if the material is composed of interlocking but uncemented blocks then the behaviour is more difficult to predict especially at small strains.

The base friction model or friction table may be used in this context. The device consists of a continuous sandpaper belt driven by rollers which moves beneath a flat block model laid horizontally on the belt. The block model may represent a vertical cross section through or along a tunnel, a cutting slope, or the face of a masonry structure. The block model is supported at the lower edge by a cross member representing stable ground at depth. Details are shown in Fig. 1.

The down drag due to the dynamic friction between the belt and the back of the blocks represents the force of gravity. The sides of the block model may be constrained by various methods depending on the required boundary conditions. The top of the model represents the free ground surface but may be surcharged if required. Deflections may be inserted at the base of the model and the resultant distortion in the model recorded as it develops. The device is detailed by R.E. Goodman 1972, 1977 and has been used to demonstrate the behaviour of rock slopes at Imperial College for a number of years. The exact origin of the device remains obscure. The previous use of the device has most frequently been in the analysis of rock slopes and also tunnels. The results must be regarded as qualitative rather than quantitative for reasons which are explained later.

BASE FRICTION MODEL BEHAVIOUR

Gravitational Creep Mechanism

If a block or series of blocks sit on a rock slope at an angle less than the angle of repose, there will be a very slow movement down the slope due to cyclic movements such as temperature expansion and contraction which occur preferentially in the most favourable direction (i.e. downhill). Vibration will have a similar effect. The friction table models these movements and accentuates them. Fig. 2 shows the rate of movement for blocks on an inclined plane on the table at varying angles below the static angle of repose. As can be seen there is no stable angle. While the creep of one block is of little significance a group of blocks may magnify the behaviour by accumulation and this is what the friction table indicates qualitatively but not quantitatively.

Toppling Mechanisms

A column built of blocks will topple if the base is tilted so that the centre of gravity moves outside the base area. However, a vertical column of blocks on a horizontal base may theoretically be built to infinite height. In practice, surface defects in the blocks cause the column to curve over, wobble or ultimately collapse at relatively small height. Where two blocks are in contact there will be three main points or areas of contact between the surfaces since three points define a plane.

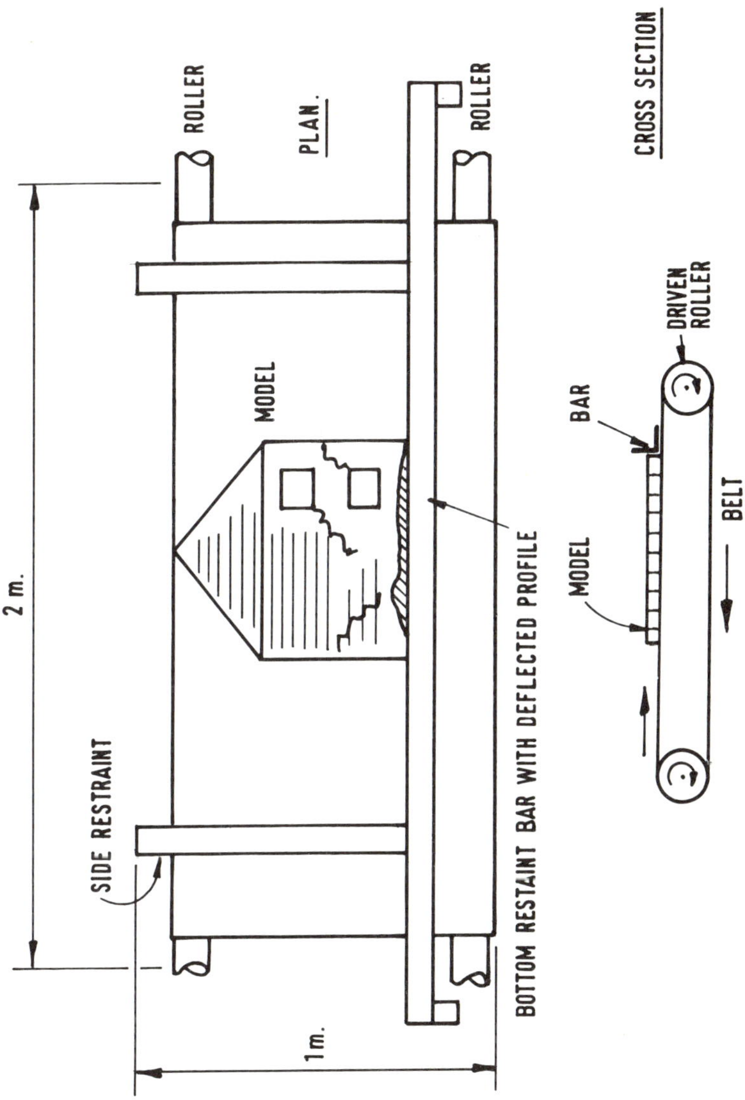

FIG.1

THE FRICTION TABLE

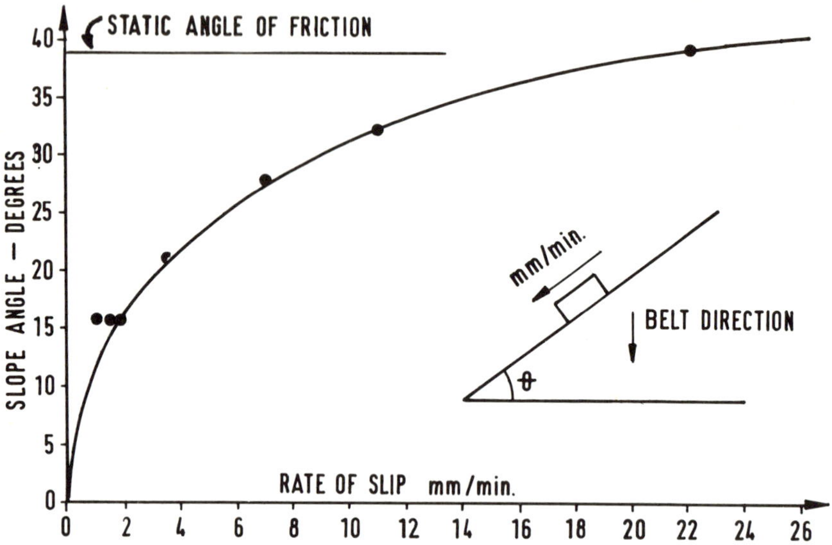

GRAPH SHOWING RATE OF SLIP VARYING WITH SLOPE ANGLE FIG. 2

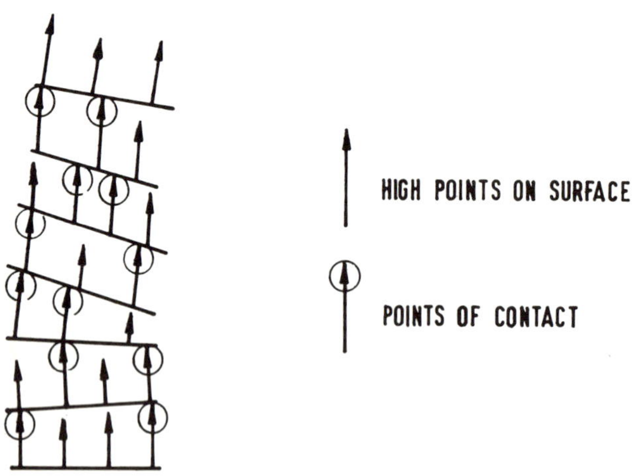

FIG. 3.

However, many other points on the planes are in close proximity, though not quite touching, and, if disturbed, the blocks can rotate or wobble about axes between any two areas of contact. When sufficient rotations accumulate in the same direction the column topples. Toppling can therefore be attributed to a combination of base tilting and random tilting due to surface irregularities which accumulate up the column.

The effect of base tilting may be calculated or determined from a static model. The combined effects of base and random tilting may be modelled qualitatively on the friction table.

The photographs in Plate 1 show a vertical column in the process of toppling.

In two dimensions each block can be considered as a beam pivoting about one or more central points of contact, as a 'see-saw'. A column is formed by a stack of such 'see-saws'.

This is shown diagramatically in Fig. 3. A static model may be built up until the ultimate block sets off a chain reaction of pivoting down the column and complete collapse follows. The height will vary depending on the sequence of blocks.

To model this a series of columns of varying heights were assembled on the friction table and the time to collapse recorded for each column. The results are shown in Fig. 4.

The action of the sandpaper belt is to cause random tilting of the individual 'see-saws' until a sequence which causes cumulative failure develops. The shorter the column the less probable this is and the longer it takes. As can be seen there is a lower bound stable height of about 20 blocks. A static model would reach a considerably greater average height of, say, 100 blocks.

It may be noted in passing that the design of the traditional brick with a central cavity or 'frog' has probably developed as a means of restraining toppling by keeping the points of contact at the perimeter of the block.

A block model of a rock slope combining the effects of both base tilting and random tilting due to surface irregularities is shown in Plate 2.

<u>Arching Mechanisms</u>

Plate 3 shows long horizontal arches developed in various vertically jointed model block systems. The arches are stable because they are under compression and gravitational creep can only lead to further compression of the arch.

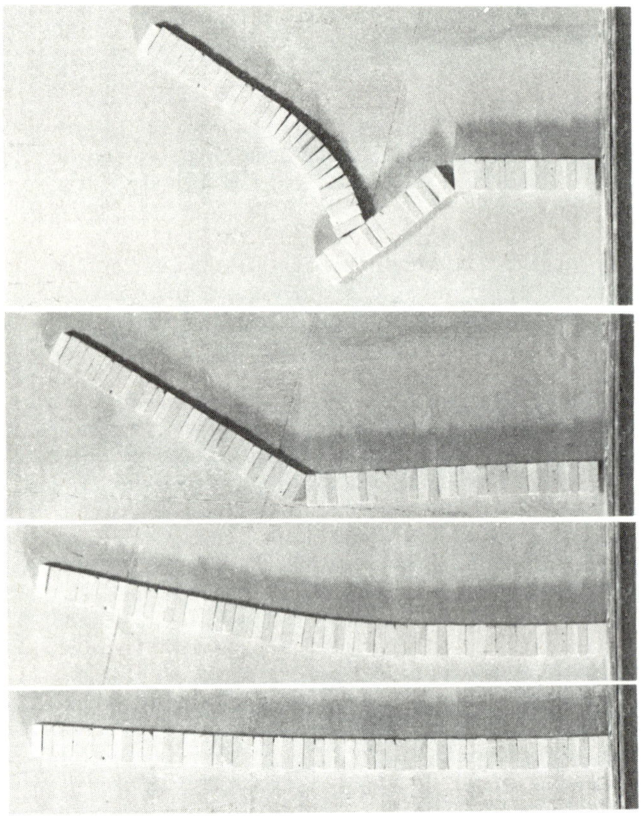

PLATE 1 VERTICAL COLUMN TOPPLING

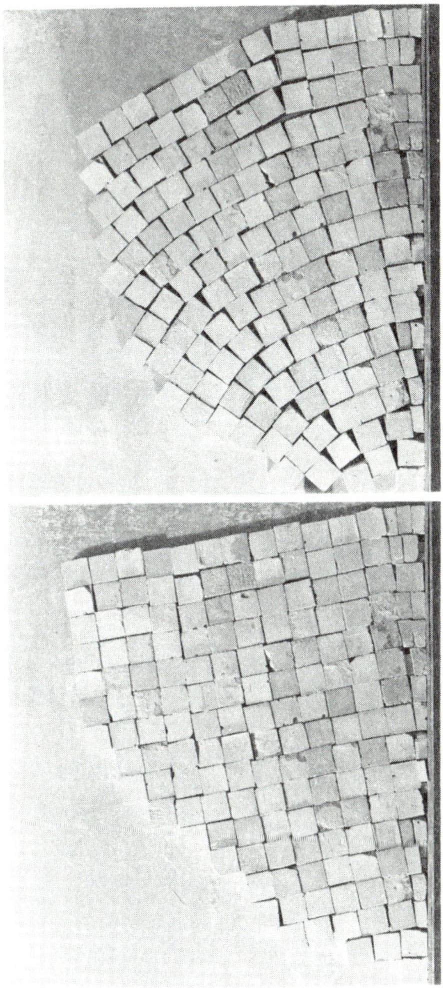

PLATE 2 ROCK SLOPE TOPPLING

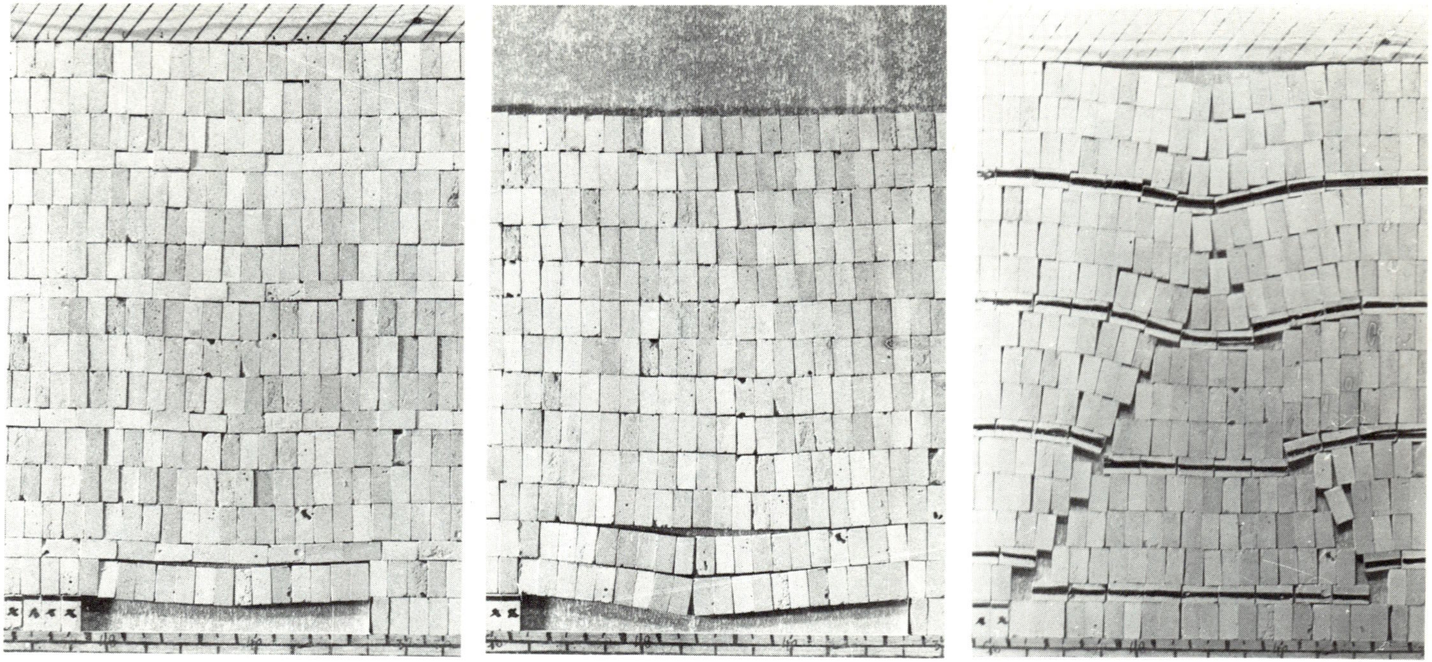

(a) Tight Joints (b) Tight Joints (c) Open Joints

PLATE 3 LONG HORIZONTAL ARCHES

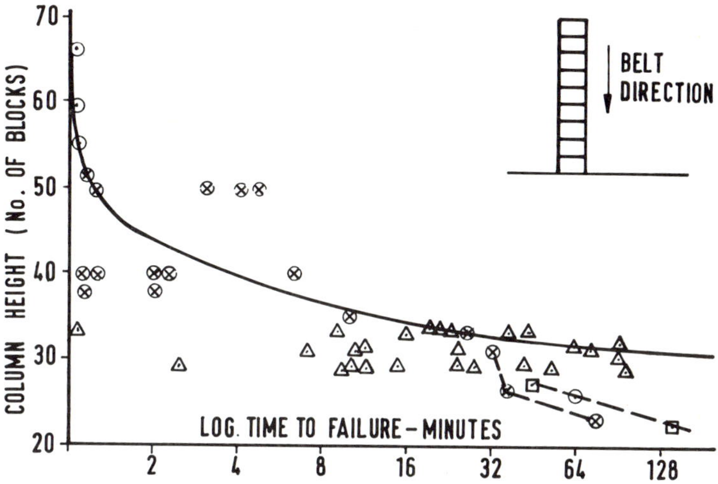

TIME TO FAILURE FOR COLUMNS OF VARIOUS HEIGHTS FIG. 4

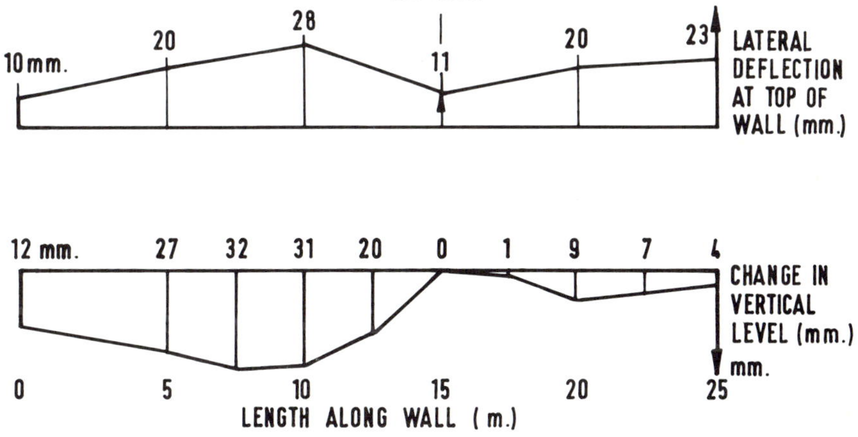

VERTICAL & LATERAL DEFLECTIONS OF BRICKWALL 3m. HIGH. FIG. 5

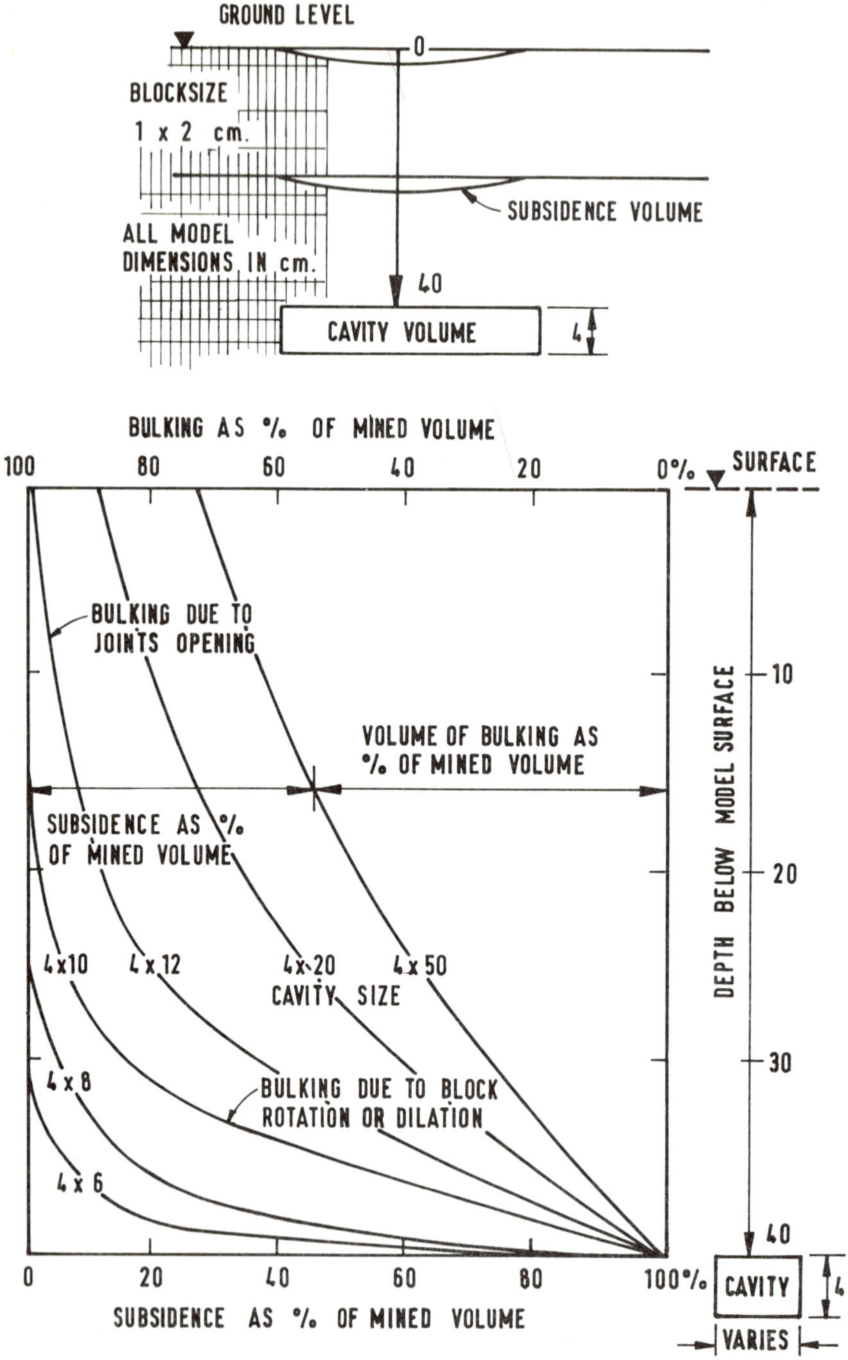

FIG. 6 DISTRIBUTION OF SUBSIDENCE / DEPTH

Ward, 1979, has presented a quantitative analysis of the behaviour of such blocks.

The base friction model can be used to demonstrate the effect of open joints. These enable the block structure to compress laterally and this restricts the formation of arches. The models suggest that the lateral compressibility of a rock strata (in terms of the volume of open vertical joints) will be a good guide to the extent of arching which can be assumed in a tunnel roof.

CRACK PATTERNS IN BRICKWORK

Cracks in brickwork can be caused by vertical deflections of the foundations which require the brick work above to hog or sag and this movement can be modelled on the friction table. Other cracks are due to shrinkage of certain types of brick or longitudinal expansion (extension) of the sub soil and these cracks can also be modelled. One other type of crack is due to torsion or twisting of the brickwork. This cannot be modelled in two dimensions on the friction table, but it should be noted that in practice the ability of brickwork to bow sideways or twist in torsion often enables an otherwise rigid structure to accept relatively large deflections in the vertical plane without cracking.

Hogging (heave) and Sagging (settlement) Cracks

Plate 4 shows the effects of hogging and sagging as found in the model (uncemented) walls.

Brickwork is very strong in compression and very weak in tension. Bending is likely to be accommodated by tension cracks in the tension zone with little or no compression elsewhere. This is the behaviour of the model.

In the sagging (settlement) case tension cracks are partly suppressed by the friction against the sub soil which does not extend and the brickwork tends to arch over the settlement profile. An exception occurs where the d.p.c. acts as a slip membrane as shown in Plate 5. Settlement cracks are also suppressed by the ability of the top of the wall to bow out sideways, thus shortening the length of the upper compression zone, and permitting bending. Measurements of vertical deflection as compared with lateral deflection of the top of a wall are shown in Fig. 5.

In the hogging (ground heave) case, tension cracks are much more severe. This is because there is little arching and the lower 'compression flange' is restrained from bowing out by the ground. This behaviour is found in practice and a detailed study was made by Cheney (1974) and reported by

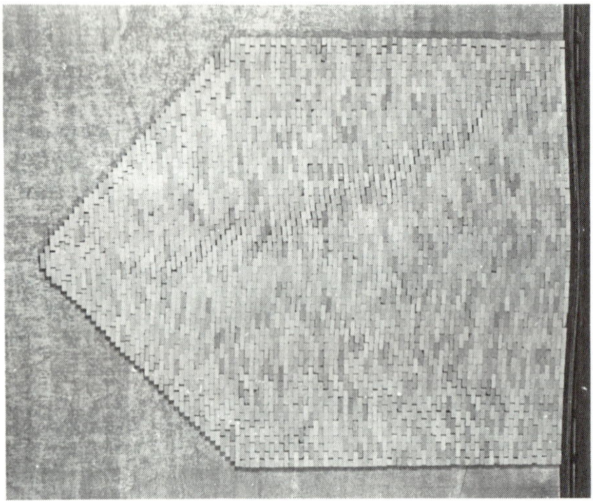

Sagging

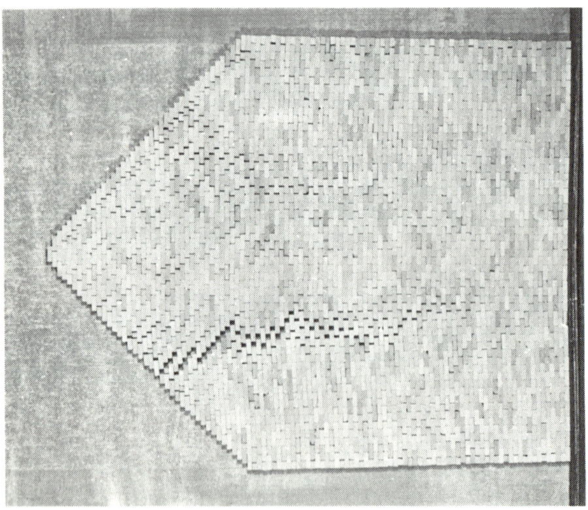

Hogging

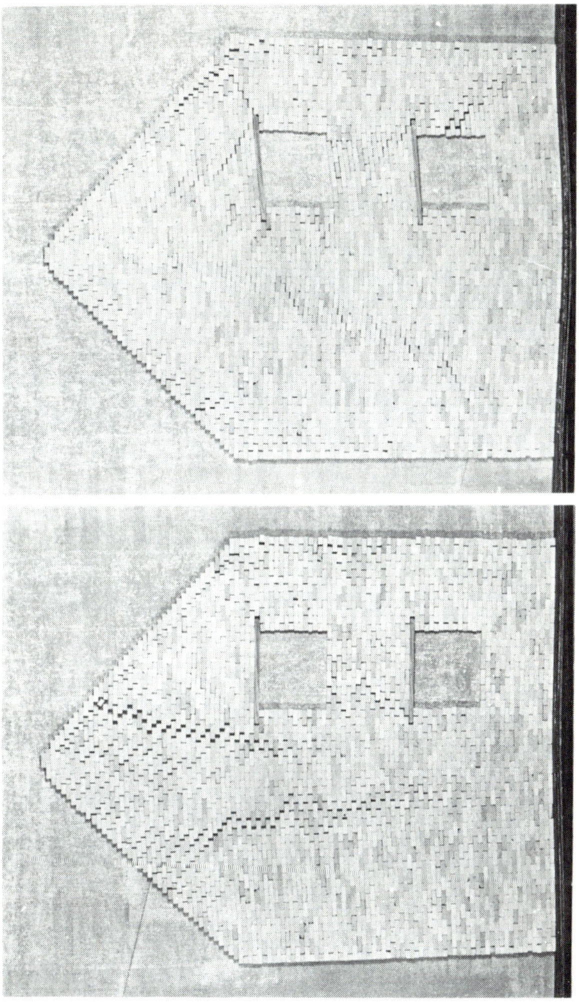

PLATE 4 CRACKS IN MODEL BRICKWORK

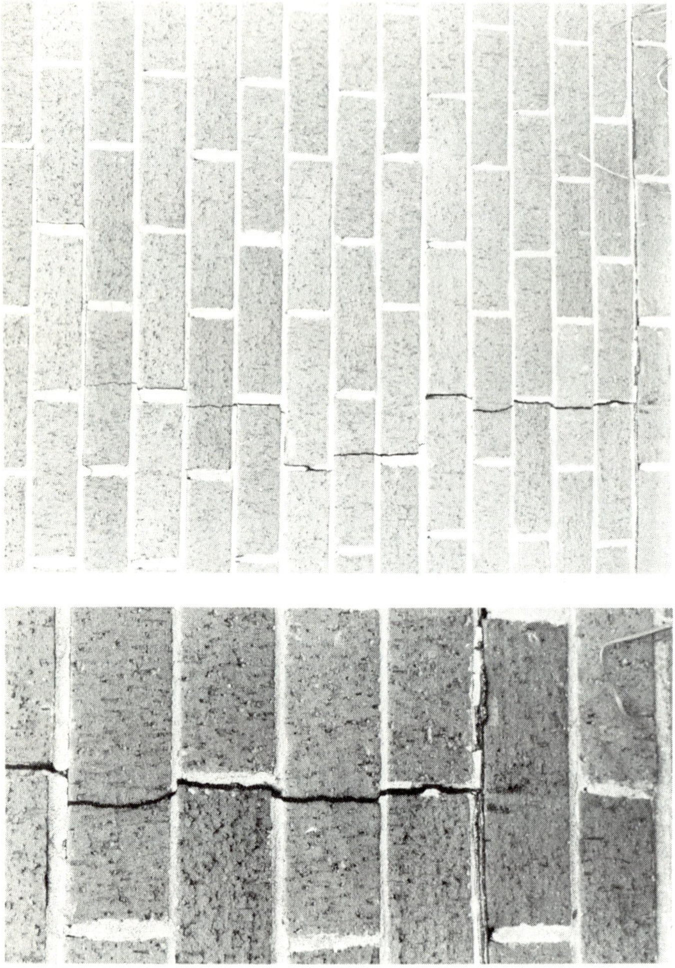

PLATE 5 SLIP ON D.P.C.

Burland (1977). The effect of openings is shown in Plate 4.

STEREOSCOPIC CRACK MEASUREMENT

The crack patterns shown in Plate 4 are exaggerated in order to show up the patterns. The crack width, when scaled up, is some ten times that which would give concern. Since the cracks occur in a number of directions there is some difficulty in complicated models in establishing absolute movements unless the position of each brick is recorded relative to a reference frame. The problem of recognising small cracks and determining the relative movements was solved by using time lapse photography.

As a model was deflected successive photographs were taken. The photographs are viewed stereoscopically using a stereo comparator. An example of this is shown in Plate 6(a) - the initial condition of a wall; Plate 6(b) - the deflected condition: and Plate 6 (c) - a reconstruction in three dimensions of what is seen through the stereo comparator when viewing Plate 6(a) and (b) stereoscopically.

The relative movement to the left or right of a brick causes the brick to appear to advance or retreat from the viewer. Cracks between groups of bricks are visible as sudden changes in the level or distance from the viewer. Since the camera remains fixed and most distortions cancel out, accurate measurements can be made using a parallax bar or stereo comparator.

The method has been used successfully by the Polytechnic to record small changes in crack widths in actual structures. An account of somewhat similar work is shortly to be published by the Transport and Road Research Laboratory, 1979/80.

The main use of the friction table combined with stereographic techniques is to model actual structures, deflect them by scaled amounts and then record the extent and pattern of cracking and compare this with the structure. Susceptible areas for cracking can also be predetermined.

SUBSIDENCE OF UNDERGROUND CAVITIES : THE SUBSIDENCE WAVE

The base friction model may be used to indicate the behaviour of a surface subsidence 'wave' by mining from the base of the model and recording collapse at various levels up to the surface. Plate 7 indicates a mine working from the right towards the left. At the surface a brick wall has been constructed to show the effects of the subsidence wave on structures. Initially the wall is subjected to a hogging deflection, Plate 7(b), then to extension, Plate 7(c)

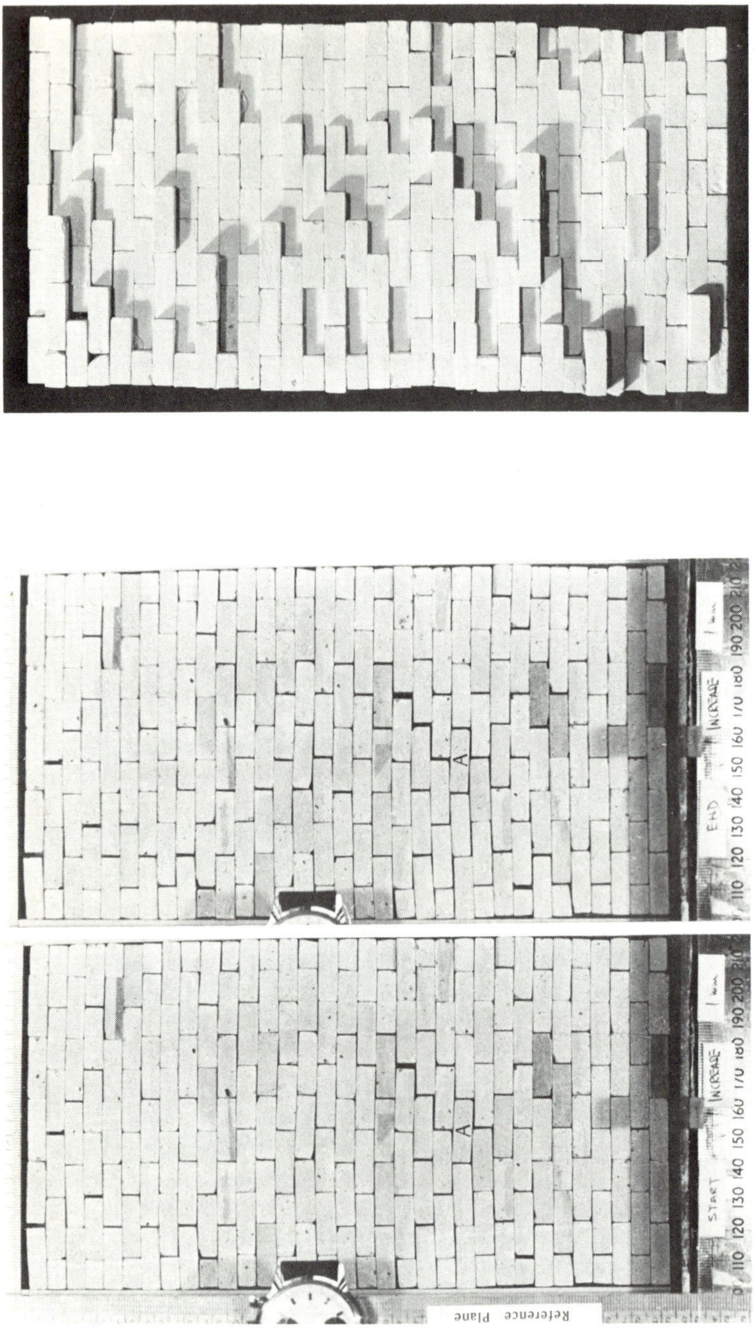

(a) Initial condition

(b) Final condition

(c) Reconstruction of view through stereo comparator

PLATE 6 STEREO COMPARISON

and finally to compression, Plate 7(d). The deflected surface profile may be measured from the model and compared with field records. In this way it is hoped to develop the technique as a means of predicting ground movements quantitatively.

Extent and Distribution of Bulking and Settlement

The eventual collapse of a roof may be sudden as in loose strata, or consist of the progressive collapse of a series of arches where the strata are relatively incompressible. The amount and distribution of bulking and settlement which occurs will vary depending on the mechanism as is shown in Plate 3(b) and (c). It will also depend on such factors as the width and height of excavation which is allowed to collapse, and block shape and size. Fig. 6 shows the settlement and bulking distribution as a proportion of the total volume mined. Plate 8 shows typical arrangements for the experiments. The effects of pillar size and spacing, depth of mine and strata dip can be investigated similarly.

Effect of Faulting

Additional subsidence damage is often reported where faults outcrop and this is sometimes due to a change in excavation pattern as the fault is intersected.

The models show a change in subsidence behaviour if the excavation proceeds uniformly through or below a fault zone. The strata above an excavation are put into tension by the subsidence wave as shown in Plate 7. Competent strata can resist some tension assuming most vertical fissures or joints to be discontinuous. Where a fault or continuous fissure occurs there is a concentration of tensile strain resulting in a severe local movement at the surface.

This is illustrated by Plate 9(a), (b) & (c), showing a fault zone in a stratum with discontinuous vertical joints: A surface cavity opens at the fault outcrop as the mine passes below, and then closes during the compression wave. The effect is exaggerated by the arrangement of the blocks but similar effects occur in models of other strata.

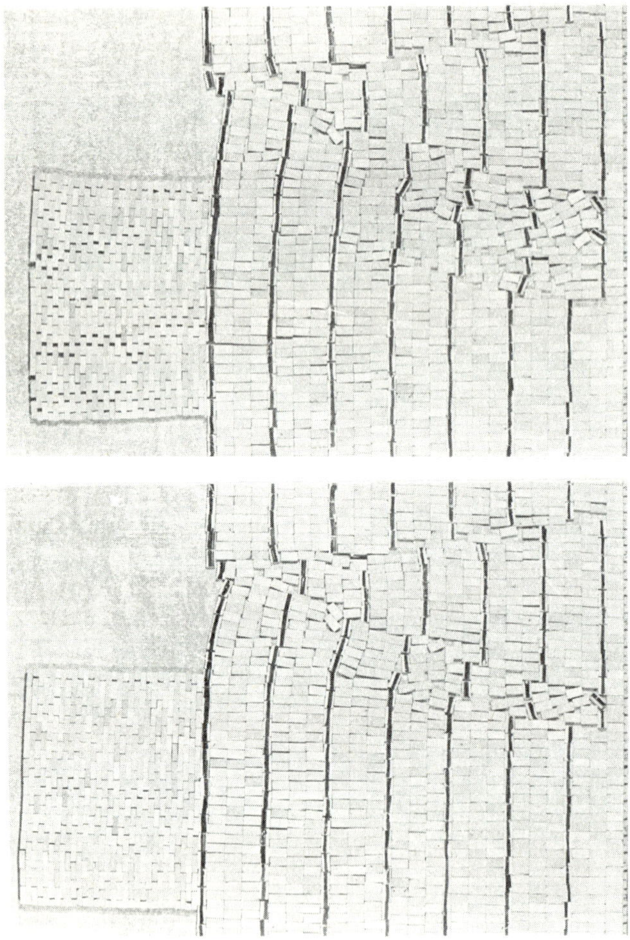

(b)

(a)

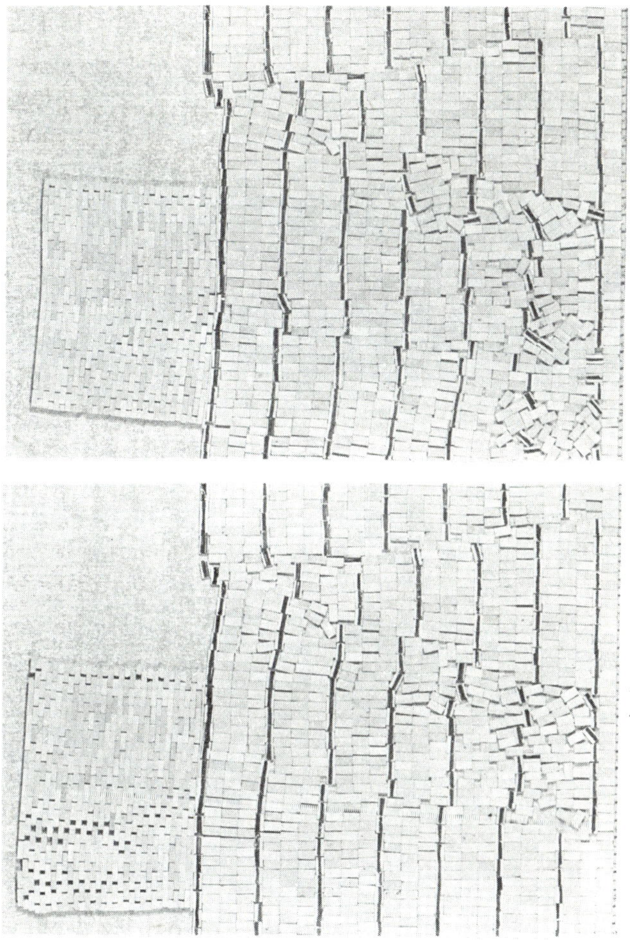

(c) (d)

PLATE 7 SUBSIDENCE WAVE

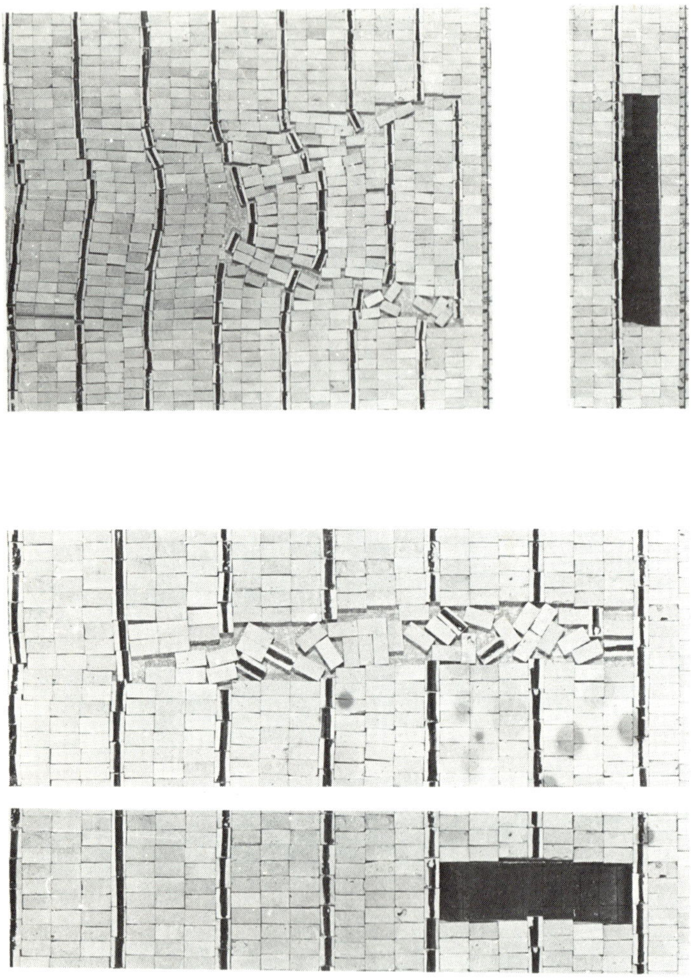

PLATE 8 DISTRIBUTION OF SUBSIDENCE

(a) (b) (c)

PLATE 9 EFFECT OF FAULT ZONES

ACKNOWLEDGEMENTS

The writer would like to acknowledge the kind help and contribution of Messrs. Vaghela, Yeo & Christofi, project students in 1976, 77 & 78, S. Douglas and E.C. Francis of PCL, and W. Fisher of F. Graham & Partners.

REFERENCES

Cheney & Burford - Damage to three storey office block on clay, and also Discussion, Proceedings of Conference on Settlement of Structures, Cambridge (1974) Pentech Press.

Burland, Broms & de Mello - Report on Session 2, 9th International Conference on Soil Mechanics and Foundation Engineering, Tokyo 1977.

Goodman & Ergulvani - Applications of Models to rock excavations, Bulletin of Association of Engineering Geologists, 1X 1972.

Goodman - Methods of Geological Engineering, West Publishing Co., New York & St. Paul Minnesota 1977.

Transport and Road Research Laboratory. Report on Photogrametric measurements of settlement during tunnelling near Kings Cross, London, to be published in Ground Engineering 1979/80.

Ward - Ground supports for tunnels in weak rocks, Geotechnique 28 - 1977, 133-171.

SESSION IV

Chairman: Professor R Ciesielski, DrScEng
Technical University of Krakow, Poland

Introduction to Session: Professor P B Attewell, BSc, PhD, DEng, CEng, MICE
Department of Geological Sciences, University of Durham

SESSIONAL PAPERS

ESTIMATION OF GROUND SUBSIDENCE OVER URBAN TUNNELS

C. Sagaseta, Dr Ing de Caminos, C y P

Professor of Geotecnia y Cimientos, Universidad de Santander Spain

J.F. Moya, Ing de Caminos, C y P

Laboratorio del Transporte y Mecanica del Suelo, Madrid, Spain

C.S. Oteo, Dr Ing de Caminos, C y P

Head of Department, Laboratorio del Transporte y Mecanica del Suelo, Madrid, Spain

INTRODUCTION

When a tunnel is excavated through a soil mass, an alteration of the initial stress state is produced. The achievement of a new equilibrium is produced by means of ground deformations around the excavation. Generally speaking, the greater these movements are for a given soil, the lower the stresses are in the tunnel lining. However, in urban tunnels, the existence of nearby buildings and structures implies a need to keep these ground movements below allowable limits to avoid structural or functional damage.

The solution of this problem must be undertaken both in the design stage (selection of layout, design of cross-section, tunnel diameter, etcetera) and in the construction stage (rate of advance, excavation method, support system, special procedures such as grouting or ground freezing, etcetera).

The width of the zone affected by subsidence phenomena can be very important; up to 15 to 20 times the tunnel diameter, as has been shown by instrumentation of actual tunnels (Tinajero and Vieitez, 1972; Attewell and Farmer, 1974).

The prediction of the magnitude and distribution of ground movements needs a careful study, taking into account the soil type, its stress-strain properties and its time-dependence, the actual geometry of the problem, the construction process (soil-tunnel interaction, rate of advance, etcetera). A complete analysis would be very difficult and the resulting predictions not necessarily good, because the actual movements depend on many details of the construction process which cannot be considered in a theoretical analysis. For the design stage, it is desirable to have a simple method to predict ground subsidence. This method should include the main factors controlling the soil movements, i.e. soil deformability, lining and support stiffness, tunnel diameter and depth, with an overall allowance for the construction method. The results of the calculations must give the magnitude of the ground movements and their distribution, at least at the ground surface. Obviously, the prediction method must be based on subsidence observations in actual tunnels.

In this paper, a set of theoretical results is presented, which can be considered as a first approximation to the solution of the problem. These theoretical predictions are based on finite element calculations, corrected by field measurements. Previous studies leading to prediction methods based on empirical (Peck, 1969) or theoretical analysis (Ghaboussi et al., 1978) have been considered and extended.

DIMENSIONAL ANALYSIS

The primary factors controlling subsidence can be grouped into three main categories: soil behaviour, lining properties and construction procedures. For the analysis of actual cases, these three factors can be taken into account (Sagaseta and Escario, 1974). However, a general analysis whose results can be presented in tables or charts for later use in design, can only include a limited number of parameters. For this reason, the problem has been simplified to the case of a circular unlined tunnel of infinite length excavated in a homogeneous, isotropic, linearly elastic soil with a horizontal surface. The only external force is the soil self-weight. The variables, parameters and unknowns of the problem are:

Variables:

(a) Spatial co-ordinates (x,z)

Parameters:

(b) Depth of tunnel axis (H)
(c) Tunnel diameter (D)
(d) Soil Young's modulus (E)
(e) Soil Poisson's ratio (ν)
(f) Soil unit weight (γ)

Unknowns:

(g) Stresses within the soil mass (σ)
(h) Strains (ε)
(i) Displacements (ρ), both horizontal (u) and vertical $(\delta$

From a simple dimensional analysis, the following equations can be derived:

$$\rho/D = f_\rho \left(H/D,\ \gamma D/E,\ \nu,\ \frac{x}{D},\ \frac{z}{D}\right)$$

$$\varepsilon = f_\varepsilon \left(H/D,\ \gamma D/E,\ \nu,\ \frac{x}{D},\ \frac{z}{D}\right) \qquad (1)$$

$$\sigma/\gamma D = f_\sigma \left(H/D,\ \gamma D/E,\ \nu,\ \frac{x}{D},\ \frac{z}{D}\right)$$

The assumption of linear elasticity leads to:

$$E\rho/\gamma D^2 = f_\rho \left(H/D,\ \nu,\ \frac{x}{D},\ \frac{z}{D}\right)$$

$$E\varepsilon/\gamma D = f_\varepsilon \left(H/D,\ \nu,\ \frac{x}{D},\ \frac{z}{D}\right) \qquad (2)$$

$$\sigma/\gamma D = f_\sigma \left(H/D,\ \nu,\ \frac{x}{D},\ \frac{z}{D}\right)$$

If the soil is considered to behave as an elasto-plastic material, with a Mohr-Coulomb failure condition, a perfectly plastic behaviour and a non-associated flow rule, three additional parameters must be included i.e. the unit cohesion (c), the angle of internal friction (ϕ) and the angle of dilatancy (ν_D), in the formulae:

$$E\rho/\gamma D^2 = f_\rho \left(H/D,\ \nu,\ c/\gamma D,\ \phi,\ \nu_D,\ \frac{x}{D},\ \frac{z}{D}\right)$$

$$E\varepsilon/\gamma D = f_\varepsilon \left(H/D,\ \nu,\ c/\gamma D,\ \phi,\ \nu_D,\ \frac{x}{D},\ \frac{z}{D}\right) \qquad (3)$$

$$\sigma/\gamma D = f_\sigma \left(H/D,\ \nu,\ c/\gamma D,\ \phi,\ \nu_D,\ \frac{x}{D},\ \frac{z}{D}\right)$$

In the following paragraphs, the results of finite element computations of Equations (2) are presented. Equations (3) are solved only for a particular value of the parameters involved.

ANALYSIS OF RESULTS

Maximum displacements

The two parameters of Equations (2) are the relative depth (H/D) and Poisson's ratio (ν). For the relative depth, values of 1.0, 3.0 and 7.5 have been taken, whilst for Poisson's ratio, values of 0.1, 0.2, 0.3 and 0.4 have been considered.

In Figure 1, the first of Equations (2) has been plotted for the ground surface (z = 0). As can be seen, the maximum vertical movement ($E\delta_{max}/\gamma D^2$) and the maximum horizontal one ($Eu_{max}/\gamma D^2$) are almost constant with varying relative depth, its value being dependent on Poisson's ratio only (Figure 2). The curves of Figure 2 can be approximated by the following simplified equations:

$$\frac{E\delta_{max}}{\gamma D^2} = 0.85 - \nu \qquad (4)$$

$$\frac{Eu_{max}}{\gamma D^2} = 0.27 - 0.3\nu \qquad (5)$$

Although the maximum displacements are not dependent on the relative depth, the influence of this factor on the lateral extent of the settlements is very important. The shape of the curves of Figure 1 is very similar to the Gauss normal distribution curve, as proposed by Peck (1969).

Non-linear soil behaviour

The influence of non-linear behaviour of the soil is represented in Figure 3. Only the cases of H/D = 3.0, $\nu = 0.2$, $\phi = 30^\circ$, $\nu_D = 0^\circ$ and 30° have been analysed, with the cohesion factor, $c/\gamma D$, varying between 0.4 and 1.8. As can be seen, for a cohesion factor greater than 1.35 only elastic strains develop. For $c/\gamma D$ between 1.35 and 0.75 the influence of plastic strains is less than 20 percent, whilst for lower values of the cohesion factor, the displacements increase rapidly. The influence of the value of the angle of dilatancy is not very important.

Volume of settlements

Besides the maximum movements, the extent of the subsiding area needs to be determined. This extent can be represented by the volume of settlement, $\Delta 1$, i.e. the area between the original ground surface and the curve representing the surface settlements in a transverse profile. This value has been plotted in Figure 4 as a function of H/D and ν. In the same Figure, the loss of ground in the tunnel section, Δ, has been included. As can be seen, the ground loss is somewhat less than the volume of settlement, $\Delta 1$. In Figure 4, some measured values of Δ and $\Delta 1$ in cohesionless soils (Cording and Hansmire, 1975) have been included. It can be observed that the measured values of Δ and $\Delta 1$ are greater than the calculated ones.

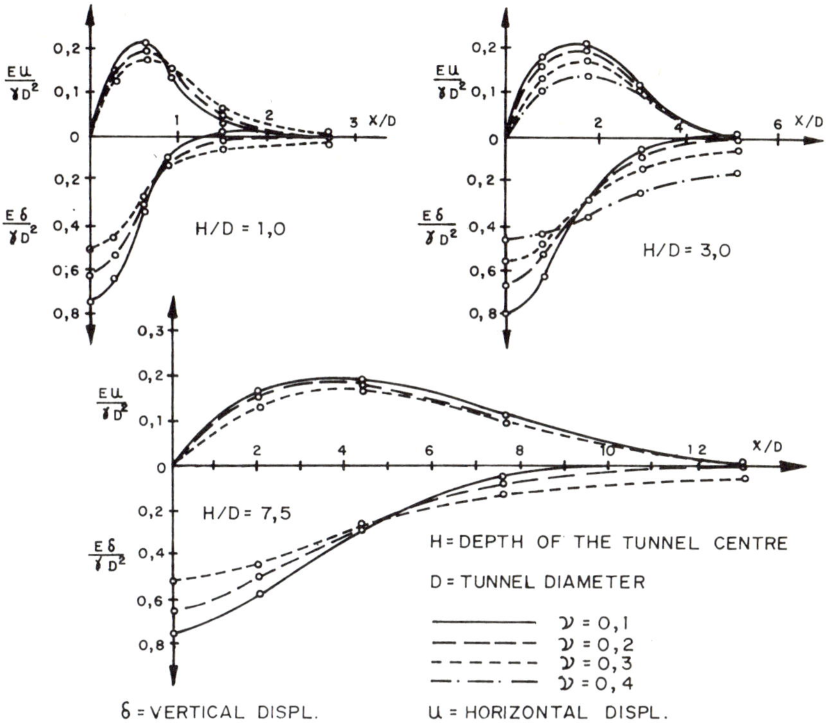

FIG. 1.- NONDIMENSIONAL SUPERFICIAL MOVEMENTS INDUCED BY THE EXCAVATION OF A SINGLE TUNNEL

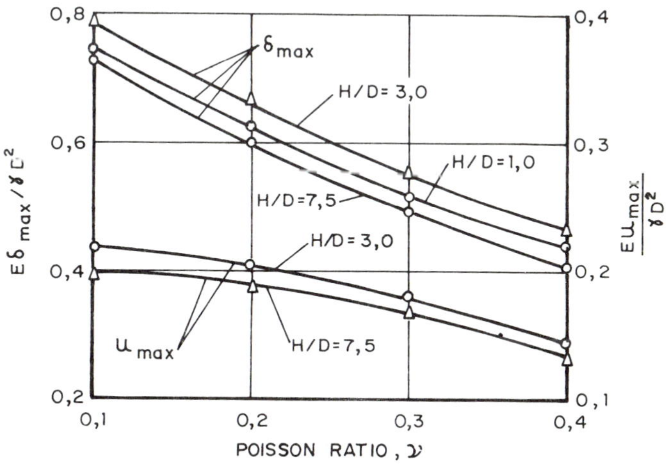

FIG. 2.- NONDIMENSIONAL MAXIMUM MOVEMENTS ON THE TUNNEL CROWN.

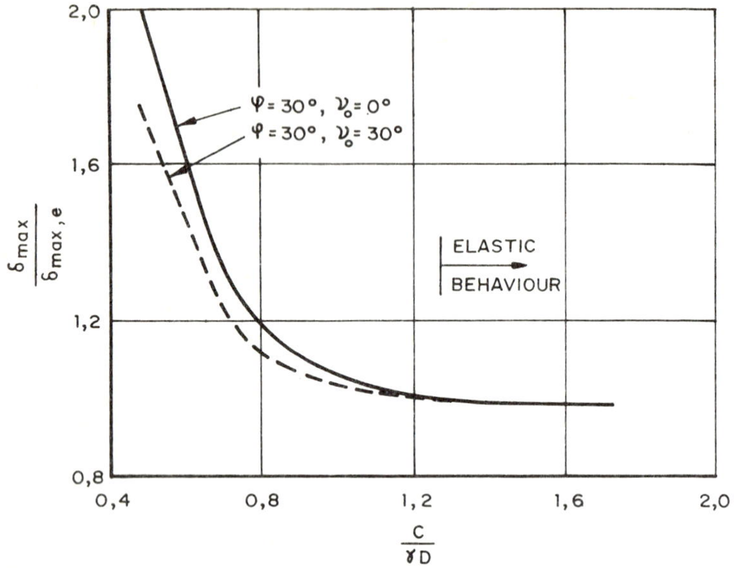

FIG. 3. - SETTLEMENT ON THE CROWN.COHESION INFLUENCE.

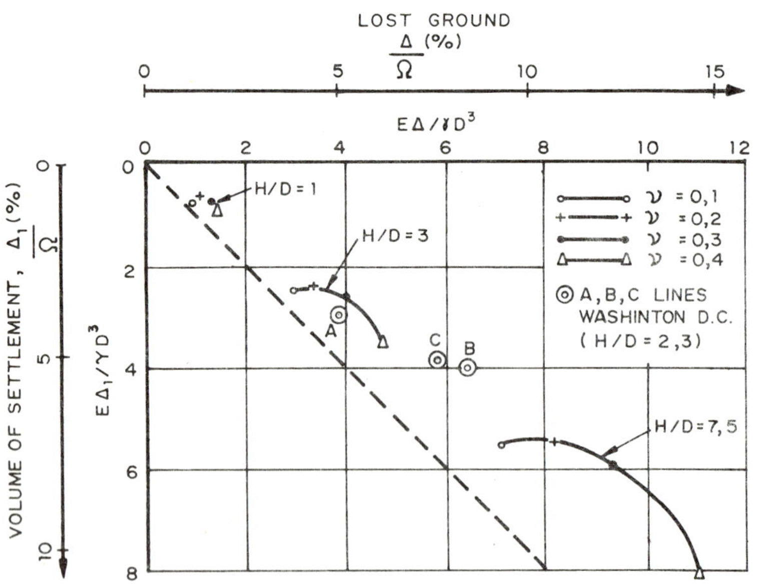

FIG. 4. - RELATIONSHIP BETWEEN THE VOLUME OF SETTLEMENT AND THE LOST GROUND IN THE TUNNEL.

In Figure 5 the settlements of the points located over the crown, related to the surface settlement are shown. The influence of Poisson's ratio and plastic strains are small. In the bottom of Figure 5 these results are compared with field measurements by several authors. Agreement is reasonably good in the upper portion, whilst in the vicinity of the tunnel the measured settlements are lower than the calculated ones, due to the influence of the lining.

It can be concluded from the preceding paragraphs that the simple analysis carried out gives a general picture of the distribution of ground movements which is in reasonable agreement with actual observations. Nevertheless, the absolute values of the displacements are greatly influenced by the existence of the lining, which has not been taken into account in the analysis. This point is further discussed below.

Influence of construction procedure

In a line of the Madrid subway, the settlements have been recorded in several test sections (Oteo and Moya, 1979) (Figure 6); measurements were taken at the ground surface and within the soil over the crown.

In Figure 7 the measured maximum settlements are plotted, together with the above predicted ones. Soil layers affected by the tunnel have been grouped into two types: the so-called "peñuela" and "transition" (hard marly clay) and the "tosco" and "arena de miga" layers (interbedded sandy clay and clayey sand layers). The measured settlements are considerably smaller than those predicted. However, at several points at which the lining erection was delayed, (points 19, 20, 55, 56 and 57), the measured movements are of the same order as the predicted ones.

In Figure 8 the influence of shield advance rate is shown. The case of zero velocity would correspond to the case of an unlined tunnel.

From Figures 7 and 8, the following Equation can be given for the maximum surface settlement:

$$\frac{E\delta_{max}}{\gamma D^2} = \psi(0.85 - \nu) \qquad (6)$$

where the coefficient ψ includes for all the effects of the construction procedure (shield advance rate, shield and lining stiffness, etc.). This coefficient must also depend on the soil properties. For the case of the Madrid subway, this coefficient has been found to vary between 0.4 in the clay layers ("peñuela" and "transition") and 0.5 for the sandy layers ("tosco"

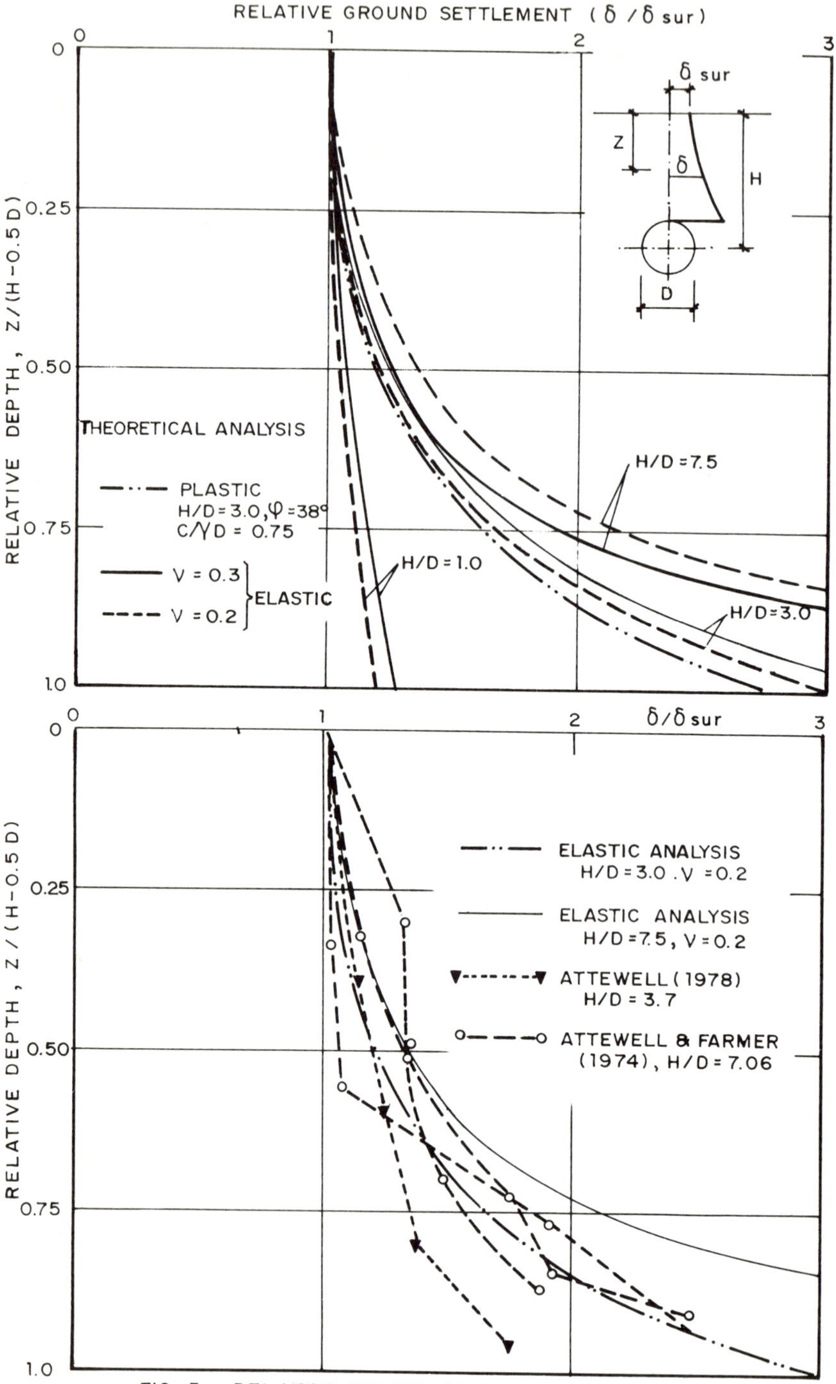

FIG. 5.– RELATIVE GROUND SETTLEMENT ON THE CROWN

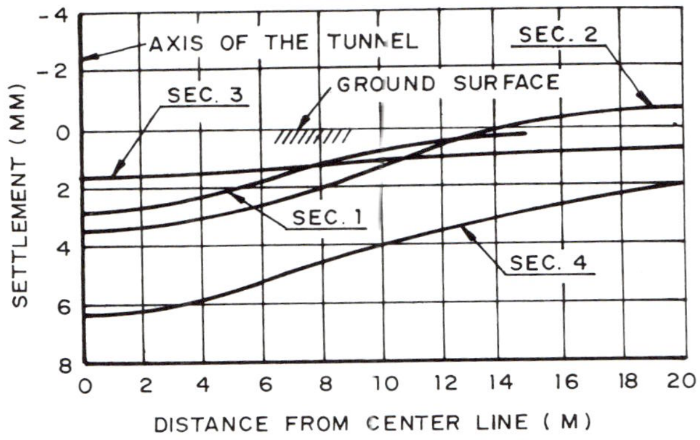

FIG. 6. - MEASURED SETTLEMENT AT MADRID

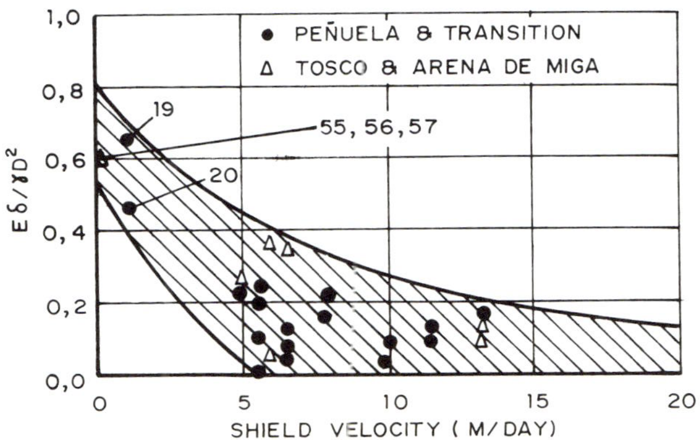

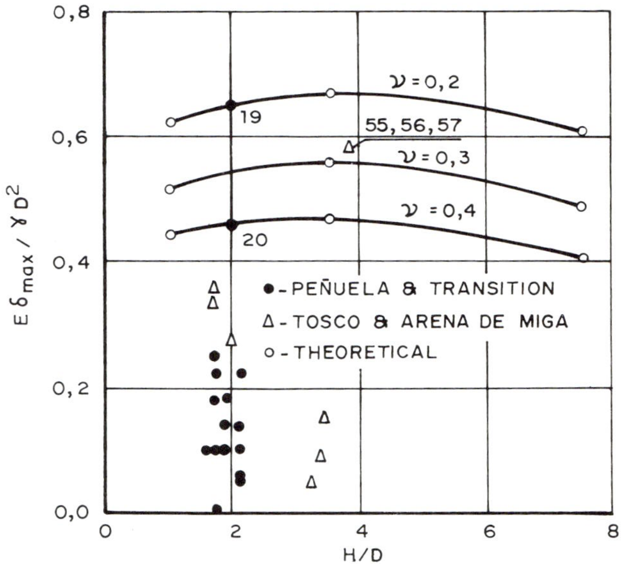

FIG. 7. - COMPARISON BETWEEN THE MEASURED AND CALCULATED SETTLEMENTS ON THE TUNNEL AXIS. (OTEO & MOYA, 1979)

FIG. 8. - NON-DIMENSIONAL MAXIMUM SETTLEMENT AS A FUNCTION OF THE SHIELD VELOCITY (OTEO & MOYA, 1979) and "arena de miga") (Oteo and Moya, 1979).

Prediction of surface settlements

With Equation (6) and a suitable value of ψ, the maximum surface settlement can be calculated.

For the transverse distribution of surface settlements, the curves of Figure 1 can be used or, alternatively, the normal distribution curve proposed by Peck (1969). These curves have been redrawn in Figure 9, but scaling the abscissae, x, with the abscissa of the inflection point, i, instead of the tunnel diameter. It can be seen that the fit is somewhat better with the theoretical curve (ν = 0.2) than with the normal distribution curve, although either of them gives reasonable agreement.

Referring to the abscissa of the inflection point, i, in Figure 10, the suggestions of Peck (1969) have been plotted, together with the theoretical results of Figure 1 (Sagaseta and Oteo, 1974) and some actual observations. In Figure 10, R is the radius of the tunnel. The agreement between the theoretical predictions and actual measurements is very good. It may be noted that for deep tunnels, the theoretical predictions give higher values of i than Peck's curves; however, the measurements shown in Figure 10 seem to support this trend.

The dotted zone in Figure 10 can be defined by:

$$\frac{i}{R} = \eta(1.05 \frac{H}{D} - 0.42) \qquad (7)$$

where η is a random factor between 0.7 and 1.3 (mean 1.0).

With the value of i of Equation (7), the settlement profile is fully defined by use of any of the curves of Figure 9.

Horizontal movements

In Figure 1, the horizontal movements of the ground surface have also been plotted. In two sections of the Madrid subway in which horizontal movements were measured, the recorded values were of the same order as those predicted. However, the available data is insufficient to enable us to make reliable predictions about horizontal movements.

ESTIMATION OF POTENTIAL BUILDING DAMAGE

Once the transverse distribution of surface movements due to tunnelling has been established, it is easy to calculate the appropriate unitary deformations which can be related to the allowable values for buildings. Two of the most widely used

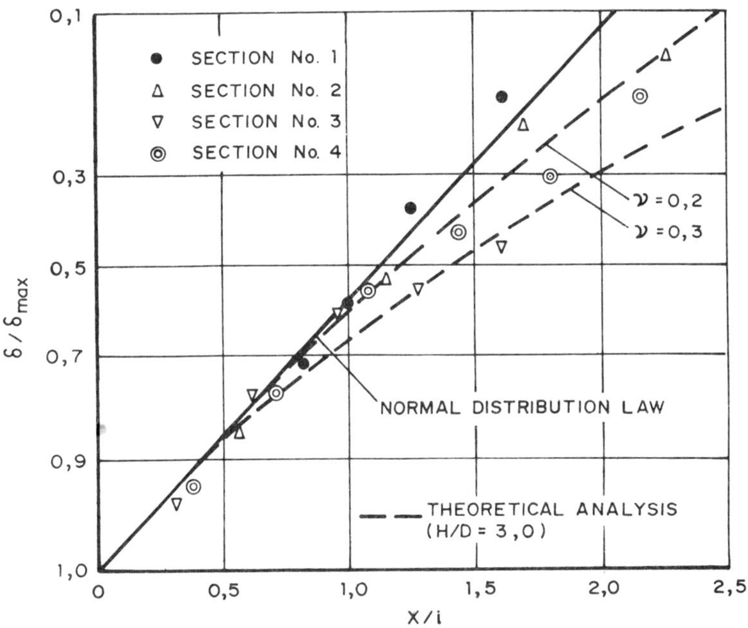

FIG. 9.- COMPARISON BETWEEN THE MEASURED SETTLEMENT AT MADRID AND THE NORMAL DISTRIBUTION LAW AND THE THEORETICAL SOLUTIONS.

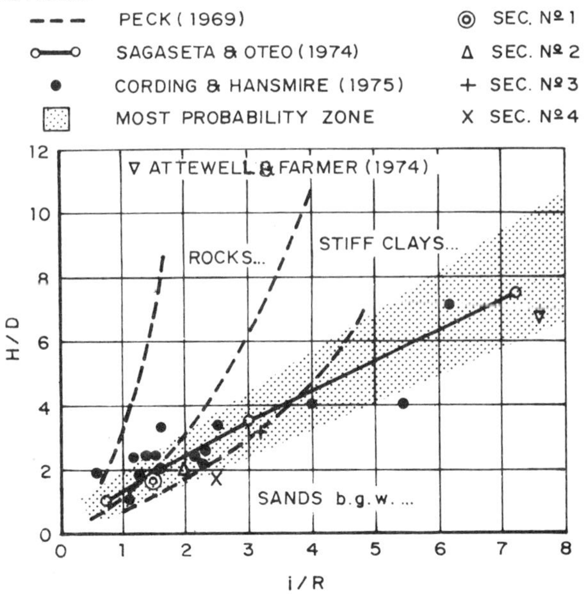

FIG.10.- COMPARISON BETWEEN MEASURED AND CALCULATED SITUATION OF THE INFLECTION POINT.

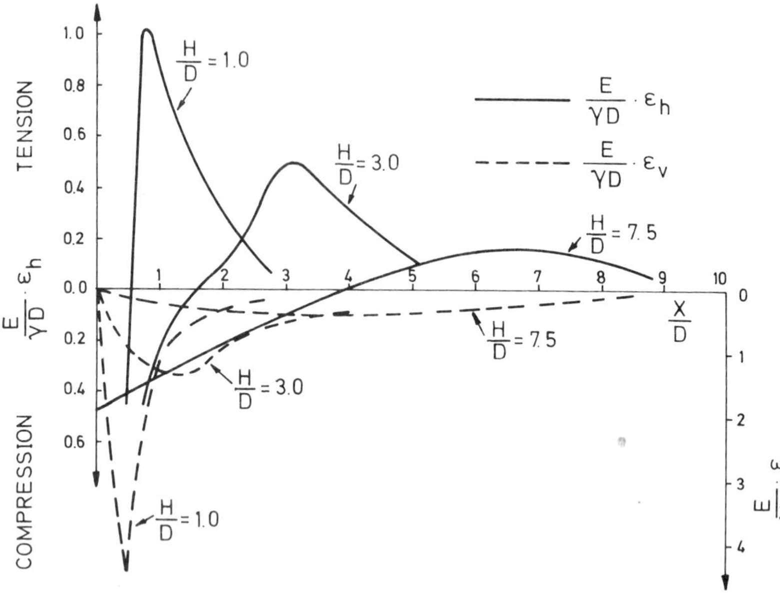

FIG. 11.- HORIZONTAL STRAIN AND ANGULAR DISTORTION AT THE GROUND SURFACE

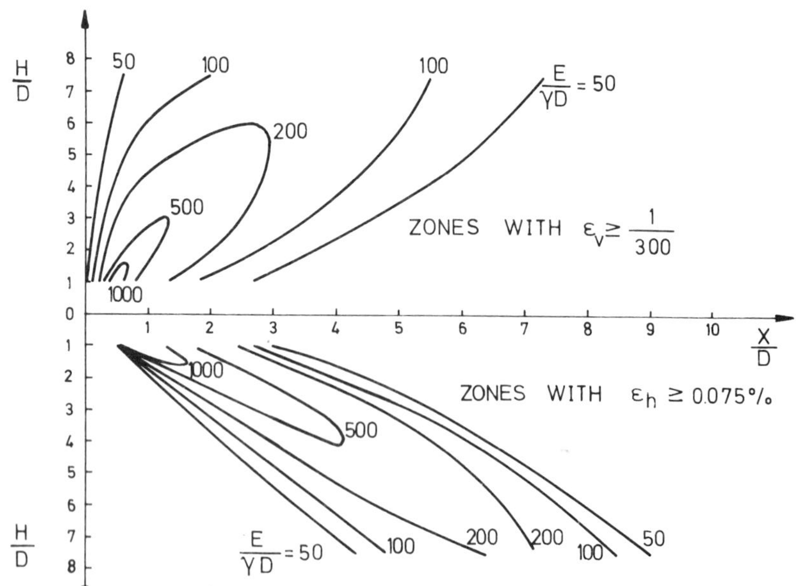

FIG. 12.- LOCATION OF ZONES OF POTENTIAL DAMAGE

deformation parameters are the angular distortion, ε_v, proposed by Skempton and McDonald (1956) and the horizontal elongation, ε_h, of Burland and Wroth (1974). In Figure 11 the distributions of ε_v and ε_h along the transverse profile have been plotted.

A thorough analysis of the areas of potential damage is beyond the scope of this paper. Only as an example, the case of concrete-framed structures with a width to height ratio between 2 and 3 will be considered here. For this particular type of building, the allowable limits to prevent structural damage can be taken as 1/300 for the maximum angular distortion and 0.075 percent for the maximum horizontal elongation.

With these allowable limits, the charts of Figure 12 have been drawn. The region enclosed by each curve, corresponding to a given value of soil deformability ($E/\gamma D$), represents the points at which the predicted angular distortion, ε_v, or horizontal elongation, ε_h, exceed the allowed values. As can be seen, for a given soil and tunnel diameter, if the depth of the tunnel is increased ($E/\gamma D$ constant, H/D increasing), the dangerous zone is widest for intermediate values of H/D. For shallower tunnels, the subsidence area is very narrow, whilst for very deep ones, the area is very wide but the settlements are very uniform, i.e. with small values of ε_v and ε_h.

CONCLUSIONS

Solutions for surface settlements and horizontal movements due to tunnel construction have been given.

With Equations (6) and (7) the maximum settlement and the location of the inflection point can be calculated.

The coefficient ψ includes the overall effect of lining stiffness and construction method. Its value must be determined from field measurements for each soil type and construction method. For the case of the Madrid subway, a value ranging from 0.4 to 0.5, depending on the soil type, has been found to give good agreement between predictions and measurements.

With the transverse distribution of settlements and horizontal movements established, the potential danger for nearby structures can be assessed by means of the usual parameters for allowable settlements. A case has been developed, giving the location and extent of the dangerous zone.

REFERENCES

P.B. Attewell - Ground movements caused by tunnelling in soil, Proceedings of the Conference on Large Ground Movements and Structures, UWIST, Cardiff (1977) 812-948.

P.B. Attewell and I.W. Farmer - Ground deformations resulting from shield tunnelling in London clay, Canadian Geotechnical Journal 11 (1974) 380-395.

J.B. Burland and C.P. Wroth - Settlement of buildings and associated damage, Proceedings of the Conference on Settlement of Structures, University of Cambridge (1974) 611-654.

E.J. Cording and W.H. Hansmire - Displacements around soft ground tunnels, Proceedings of the VIth Panamerican Conference on Soil Mechanics and Foundation Engineering, Buenos Aires (1975) General Report 5, 571-633.

J. Ghaboussi, R.E. Ranken and M. Karshenas - Analysis of subsidence over soft ground tunnels, Proceedings of the International Conference on Evaluation and Prediction of Subsidence, A.S.C.E., Florida (1978) 182-196.

C.S. Oteo and J.F. Moya - Estimation of the soil parameters of Madrid in relation to the tunnel construction, Proceedings of the VIIth European Conference on Soil Mechanics and Foundation Engineering, Brighton (1979) 3, 239-247.

R.B. Peck - Deep excavations and tunnelling in soft ground, State-of-the-Art Report, Proceedings of the VIIth International Conference on Soil Mechanics and Foundation Engineering, Mexico (1969) 225-290.

J.A.J. Salas - Deux cas de subsidence en Espagne, Proceedings of the IVth Congress of the International Society for Rock Mechanics, Montreux (1979).

C. Sagaseta and V. Escario - Influence of construction procedure on ground pressures around a tunnel, Proceedings of the IIIrd Congress of the International Society for Rock Mechanics, Denver (1974).

C. Sagaseta and C.S. Oteo - Análisis Teórico de la subsidencia producida por la construcción de túneles, Comunicaciones del 1er Simposio Nacional sobre Túneles, Madrid (1974) 1, paper 11-10.

C. Sagaseta and C.S. Oteo - Influencia de la plastificación del terreno alrededor de un túnel en los asientos superficiales, Comunicaciones del 1er Simposio Nacional sobre Túneles, Madrid (1974) 2, 76-79

A.W. Skempton and D.H. McDonald - Allowable settlement of buildings, Proceedings of the Institution of Civil Engineers 5 (1956) 727-768.

J. Tinajero and L. Vieitez - Asentamientos en la vecindad de túneles perforados con escudo, Ingeniería, Universidad Autónoma de Mexico (1972) Abril-Junio 209-220.

THE PREDICTION OF THE SETTLEMENT ABOVE SOFT GROUND TUNNELS BY CONSIDERING THE GROUNDWATER RESPONSE WITH THE AID OF FLOW NET CONSTRUCTIONS

A.F. Howland, MSc, DIC, MIGeol, MIMM, CEng, FGS

Senior Geotechnical Engineer, Harrison and Company, Norwich
formerly
Engineering Geology Section, Imperial College, London

INTRODUCTION

When tunnels are driven through soft ground, settlements are known to occur above them. When driven below pre-existing structures the distribution and magnitude of settlement become critical because of the possible detrimental effect on foundations.

The settlements associated with tunnels have been shown to take the form of a trough transverse to the tunnel long axis and approximating to a normal probability curve centred about the tunnel, with the greatest settlement occurring over the centre-line (Fig.1). This distribution is therefore able to produce high differential settlements in structures adjacent to the tunnel and develop both compressive and tensile strains within them. Attewell (1977a) has described a method of establishing the size of the settlement trough based on its mathematical form. It does not directly predict the magnitude of the settlement but rather requires that a number of estimated factors be varied by iteration until a suite of variables is produced which experience suggests is correct for the situation. From this data the strains developed by the settlement may be determined from the size and shape of the trough.

In this paper, the problem is tackled differently. A model is produced whereby the settlement above the tunnel is determined mechanistically. Back-analysis of settlements above two recently reported tunnels in soft ground indicates that the approach will give a good direct prediction of the magnitude and distribution of settlement.

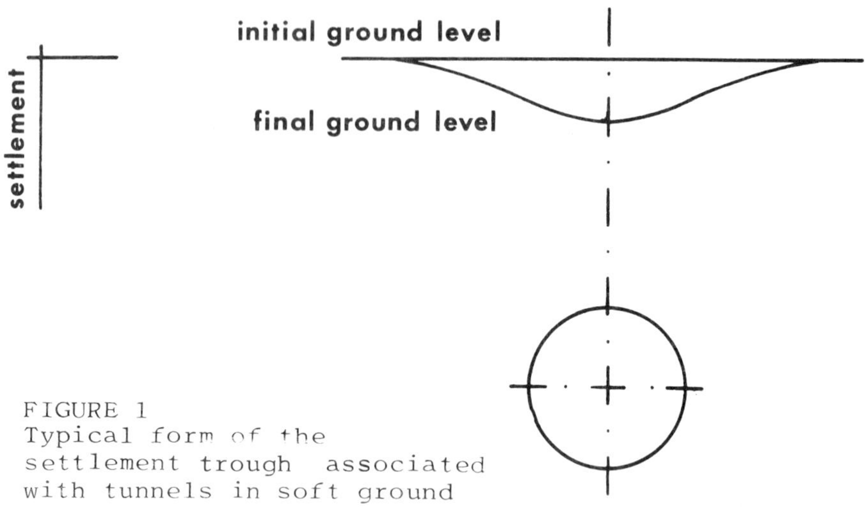

FIGURE 1
Typical form of the settlement trough associated with tunnels in soft ground

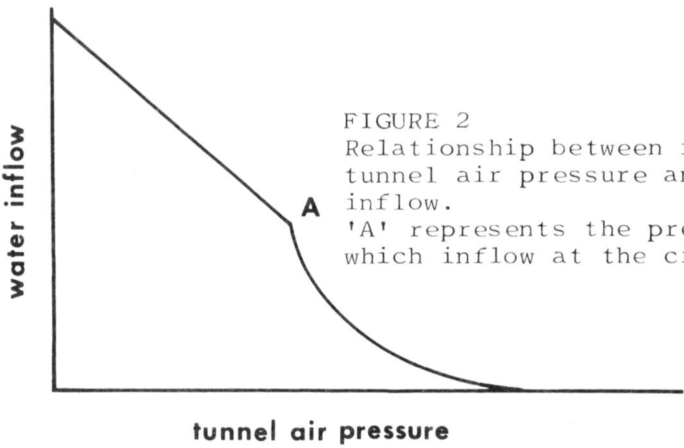

FIGURE 2
Relationship between increasing tunnel air pressure and water inflow.
'A' represents the pressure at which inflow at the crown ceases

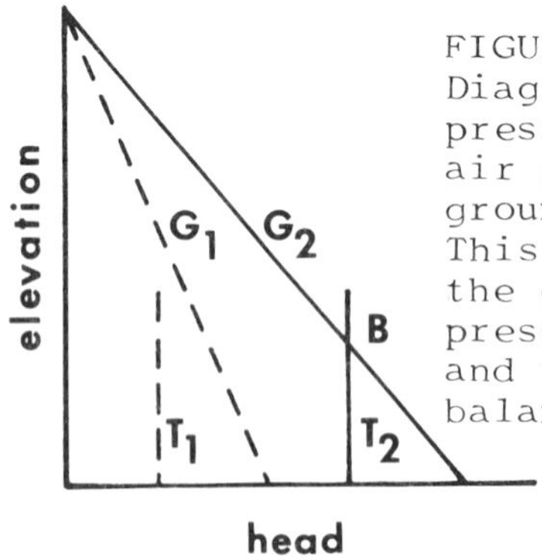

FIGURE 3
Diagrammatic representation of the pressure head distribution of the air in the tunnel T, and of the groundwater G.
This illustrates the response of the groundwater as the tunnel pressure increases from T_1 to T_2 and the development of the balance point, B.

GROUND RESPONSE TO TUNNELLING IN SOFT GROUND

Ground response around tunnels in stiff clay has been considered by Attewell and Farmer (1974a, 1974b) and the approach extended to alluvial clays (Attewell, 1977b) and man-made fill (Dobson et al, 1978). The common approach is that reviewed by Attewell (1977a), based on the surface response of the ground to a tunnel taking the form of a normal probability curve given by:

$$S = S_{max} \cdot \exp\left(\frac{-y^2}{2i^2}\right) \qquad (1)$$

where S is the settlement measured at any point at a distance y from the tunnel centre-line,

S_{max} is the maximum surface settlement and occurs at the centre-line of the tunnel, and

i is the distance to the point of inflection on the settlement trough.

The approach relies on an experience of case histories in similar soil types, to give the likely values for a number of otherwise unknown factors. From these, the likely settlement associated with the tunnel may be established. Little work has been done on the response of groundwater and its role in settlement control, although its effect has been appreciated (Attewell et al, 1978).

When a tunnel is driven through saturated soft ground it acts as a drain and the groundwater responds by flowing towards it. The response can be modelled by flow net construction in a similar way to that used in other groundwater seepage situations. The result of the hydraulic gradient, initiated by this drainage, is to lower the original pore water pressures in the ground in a way fully described by the net. If full saturation is maintained, an increase in effective stress in the ground around the tunnel can be determined, if the hydraulic gradient is quantified and the original pore pressure is known.

The separate groundwater response to a tunnel driven in free or compressed air can be modelled by the appropriate flow net.

The flow net can be constructed for any geometry, given by the ratio of tunnel depth to crown, D, to tunnel radius, r, which will be valid for all sizes of tunnel with the same ratio and for all ground conditions, regardless of soil type or condition, excepting the ratio of vertical to horizontal permeability, for which the net needs to be adjusted according to well-published rules.

The rate of inflow of water, Q, may be determined using

the Darcy Equation:

$$Q = kiA \qquad (2)$$

by determining the hydraulic gradient, i, and substituting k, the ground permeability, and A, the area over which inflow is occurring.

The hydraulic gradient may be found from the flow net, by multiplying the ratio of the number of flow lines, N_f, to the number of equipotential lines, N_e, representing the system, and multiplying by the pressure differential across the system, p, such that:

$$i = \frac{N_f}{N_e} \cdot p \qquad (3)$$

Since the flow net is unaffected by the pressure differential the inflow is inversely proportional to the tunnel pressure (Fig.2). Tunnelling experience shows that at a certain value the tunnel pressure will have increased to a point where the tunnel crown becomes dry when inflow at this point has ceased, while seepage continues at lower elevations. This is diagramatically explained by a head distribution diagram (Fig.3).

As the system is multiphase, consisting of two fluids contained in a solid (air, water and the soil skeleton), the behaviour is controlled by the interaction at the mutual interfaces and the pressures within the fluids. To simplify matters, the head distribution may be considered in terms of pressure head.

The pressure head distribution is governed by the unit weight of the respective fluids. For the groundwater this increases with depth in the usual manner. Any downward drainage affects the distribution by increasing its gradient. The low unit weight of air results in a negligible hydrostatic increase over distances of the scale of the tunnel, therefore, there is no effective increase in pressure at different elevations within the tunnel.

As the tunnel pressure increases from atmospheric, here taken as a datum, the head distribution in the tunnel moves uniformly, by a distance equivalent to the increase in pressure (T_1 to T_2 of Fig.3). The head distribution of the groundwater will also move by the same amount at the lower boundary, but since it is fixed at the upper boundary, the amount of shift will be proportionately less at higher elevations (G_1 to G_2 of Fig.3). At some increase in tunnel pressure the head distribution in the tunnel and in the groundwater will intersect. At the point of intersection the pressure head in the two will be the same so no flow can occur at that elevation on the tunnel surface. Flow to the tunnel will continue below this elevation since the groundwater maintains a higher head than the tunnel.

As the tunnel pressure increases, the point of overlap, or balance point, will first occur at the crown. It is clear that only in exceptional ground conditions would the required tunnel pressure be equal to the hydrostatic head of water above the crown. As the tunnel pressure is increased, the balance point moves down so that while flow continues to the tunnel at lower elevations, the direction of flow would seem to be reversed above. However, the interfacial characteristics of air and water in a solid porous medium are such that a greater tunnel pressure can be balanced than that equal to the groundwater pressure head alone. This is because of the capillarity effects produced by the interaction of the three phases at their interfaces acting into the tunnel.

As the tunnel pressure is raised beyond that required to produce a balance point in the crown, the flow net changes in response to the increased zone of no inflow between the crown and the balance point. The inflow Q falls off exponentially from this point, since it is proportional not only to the pressure differential p, but also to the ratio of N_f to N_e and the reducing area of inflow A (Equations 2 and 3).

CONSOLIDATION ABOVE A TUNNEL INDUCED BY DRAINAGE

By producing the flow net for any tunnel geometry, a distribution of effective stress increase brought about by the pore-pressure reduction can be determined at any point. According to consolidation theory, the ground will settle as a response to an increased effective stress. Since the increase is variable according to the equipotential distribution of the flow net, it follows that the settlement will not be uniform. By taking a vertical line through the flow net adjacent to the tunnel, the increase in effective stress can be substituted into standard consolidation formulae to give a measure of the settlement at that point.

In order to check the hypothesis, the approach has been used on two recently published case histories. Good agreement is found between settlement predicted in this way, both in terms of magnitude and distribution, with that reported in the two cases.

1 Willington Quay

Attewell et al. (1978) provide a detailed case record of the settlement produced by a tunnel forming part of the Northumbrian Authority's Tyneside Sewerage Scheme, situated on the north bank of the River Tyne at Willington Quay. The tunnel was excavated to 4.25 m diameter under compressed air at an axis depth of 13.375 m.

This gives a depth to crown, D, to tunnel radius, r, ratio D/r of 6.29 . If a flow net is constructed for this geometry (Fig.4), the settlement, d, may be determined at any point away from the tunnel centre-line, by substitution into the consolidation formula:

$$d = M_V \cdot H \cdot \Delta p$$

where M_V is the coefficient of consolidation,
H is the vertical distance between two adjacent equipotentials, and
Δp is the increase in effective stress at the mid-point between the two equipotentials, as determined by the equipotentials.

The sum of the settlements between each potential is assumed to give the settlement experienced at the surface at that point (Fig.5).

The drive was completed under compressed air. If the balance point is assumed to be about axis level, a flow net may be constructed for the groundwater response. From this a distribution of settlement may be determined (Fig.6a). The coefficient of compressibility M_V was determined from an e-log p curve for material taken from the face of the tunnel (Farmer, 1977); Δp and H are found directly from the flow net. The settlements calculated in this way are larger than those indicated by the field measurements. However, since the settlement is not instantaneous, the rate of settlement may be taken into account by substitution into the standard formula:

$$t = \frac{T_V \cdot H^2}{C_V} \qquad (4)$$

The coefficient of consolidation has been determined for the ground from the field settlement records (Fig.7 of Attewell et al., 1978), and the drainage path determined from the flow net. Substitution then enables the settlement at any time to be determined.

Attewell et al. (1978) published a distribution of vertical ground movement at 23 days after passage of the shield below the measured station (Attewell et al., 1978, Fig.11),

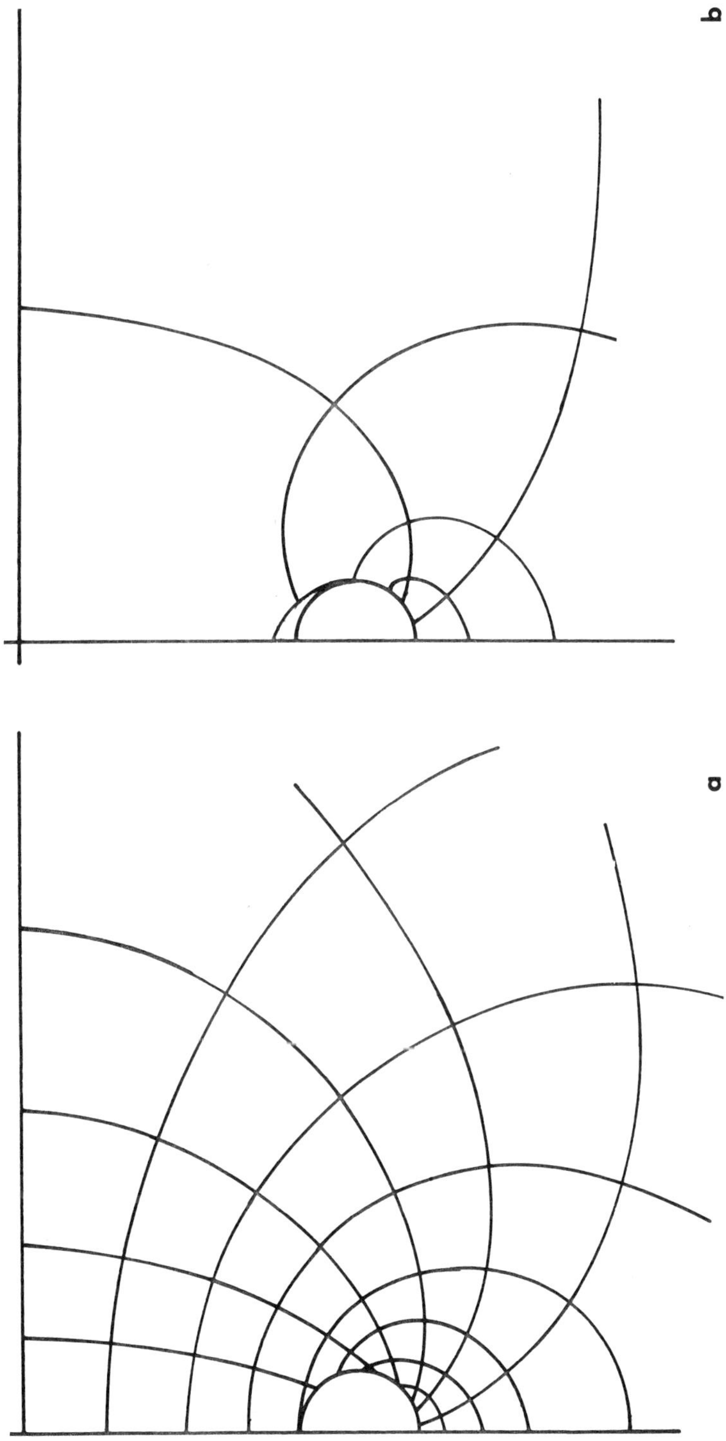

FIGURE 4
Flow net construction modelling the groundwater response to a tunnel at Willington Quay under a) free air and b) compressed air, where the balance point is produced as axis level

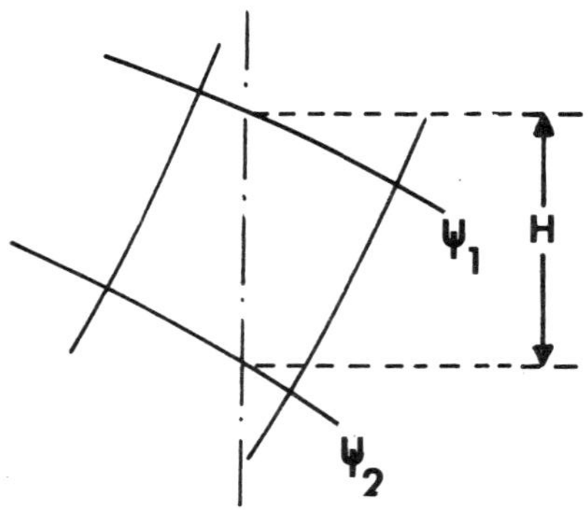

$$d = M_v . H . \frac{\psi_1 + \psi_2}{2}$$

FIGURE 5
Calculation of the settlement d experienced between two equipotentials of a flow net

during which time the compressed air was maintained in the tunnel. The time/settlement records show that settlement began almost 10 days before passage of the face. If a distribution is calculated for a period of 33 days, by substitution into Equation 4, a distribution is found which is remarkably close to the field measurement for the same time (Fig.6a).

After this period, a phase of caulking and high-pressure grouting occurred around the lining. This would have the effect of decreasing C_v by decreasing the effective permeability of the ground, with the result that the rate of settlement would decrease, as borne out by the field records. Sixty-six days after passage of the shield, the tunnel pressure was released. The balance point would move accordingly and the new groundwater response can be modelled by redrawing the flow net. By comparing the two flow nets it is immediately obvious that the release of compressed air will widen and deepen the settlement trough (Fig.4); a fact which has been recorded elsewhere when compressed air is released in a tunnel (Henry, 1974; Glossop and Farmer, 1979).

When the total settlement is calculated, the influence of the grouted lining must be taken into account. Its thickness is insufficient to affect the size of the tunnel opening on the scale of the drawing, but the low permeability barrier set up by the lining will affect the values of the potentials. This is because the lining produces a greater energy loss across it than in an equivalent thickness of the original ground. Since the upper and lower boundary conditions are unchanged, it has the effect of maintaining higher pore pressure in the ground than if the tunnel was effectively unlined. The extreme case would be provided if a totally impermeable lining were to be installed so that no drainage takes place, and no reduction in pore pressures would result.

The scale of the drawing is too small to produce a flow net to take account of the energy loss across the lining. However, there is a relationship between the permeabilities of the two materials and the modified shape of the flow net within the second material (Cedergren, 1967), where the sides of the rectangle of the modified flow are in the same proportion as the two permeabilities. For the purpose of the analysis the lining is considered to have an effective permeability two orders of magnitude lower than the surrounding ground and to influence a 30 cm zone adjacent to the tunnel opening. Thus, the number of equipotentials crossing the system should be increased to 38 and the relevant ones crossing the surrounding ground used to compute the effective stress changes.

The final settlement trough, as determined from the flow net for the tunnel in free air (Fig.6b) can be related to the actual field settlement at 149 days as given by Attewell et al. (1978) in their Figure 12. Since the compressed air was

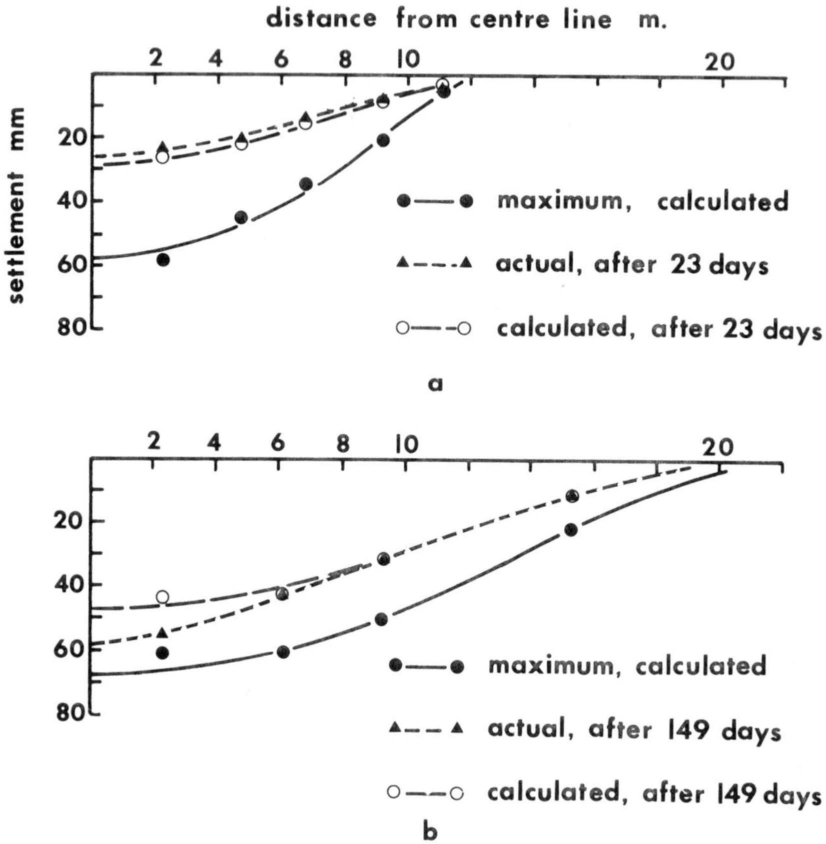

FIGURE 6
The distribution of settlement above the tunnel at Willington Quay with a) compressed air and b) free air in the tunnel. The predicted maximum settlement from the flow net construction, the actual field settlement and the predicted settlement at an equivalent period of time are shown. Partly from Attewell et al 1978.

released 66 days after passage of the face, or 76 days after field settlements began, the predicted settlements for 83 days may be related to the field measurements at 149 days. Good agreement is shown between these values (Fig.6b).

2 Stockton on Tees

The same procedure was followed for a one metre diameter tunnel, driven 7.0 m below ground surface through soft to very soft silty clay at Stockton on Tees (McCaul, 1978).

The tunnel was driven entirely in free air and the settlements were monitored by the Tunnels Division of the Structures Department of the Transport and Road Research Laboratory.

Using the appropriate D/r ratio, the flow net has been constructed and the amount and distribution of settlement determined as outlined above (Fig.7). The value of M_V was assumed to be similar to that determined for the ground at Willington Quay.

The field settlement records indicate that settlement was complete within the period of measurement, with 90 percent occurring within two months (McCaul, 1978), with readings being taken to 200 days. A comparison has been made with the settlement recorded at measurement station D (Fig.6b). Again, good agreement is found between this and the predicted settlement determined from the flow net.

CONCLUSIONS

A prediction of the size and shape of the settlement trough occurring above tunnels driven through soft ground may be determined using flow net construction. The groundwater response to differing tunnel construction conditions is modelled and standard one-dimensional consolidation theory and the changes in effective stress as determined from the flow net are used to calculate the total settlement. The amount of settlement at any time may be determined from the coefficient of consolidation C_V and the drainage path as determined from the flow net. In this way the development of the settlement trough may be established and the critical situation found as it passes below adjacent structures to the tunnel line.

It seems that consolidation settlement may adequately explain the majority of settlement above tunnels in soft ground, while the deepening and enlarging settlement troughs associated with the release of compressed air are shown to be a natural consequence of the changing groundwater response.

The use of low permeability linings in tunnels is shown not to affect the lateral extent of the settlement trough. The lower the permeability of the lining, the smaller will be the maximum settlement experienced above the tunnel.

REFERENCES

P.B. Attewell - Ground movement caused by tunnelling in soil, Proceedings of the Conference on Large Ground Movements and Structures, The University of Wales Institute of Science and Technology, Cardiff, Pentech Press (1977a) 812-948.

P.B. Attewell - Large movements and structural damage caused by tunnelling below the water-table in a silty alluvial soil, Proceedings of the Conference on Large Ground Movements and Structures, The University of Wales Institute of Science and Technology, Cardiff, Pentech Press (1977b) 307-356.

P.B. Attewell and I.W. Farmer - Ground disturbance caused by shield tunnelling in stiff over-consolidated clay, Engineering Geology 8 (1974) 361-381.

P.B. Attewell, I.W. Farmer and N.H. Glossop - Ground deformation caused by tunnelling in a silty alluvial clay, Ground Engineering 11 (1978) 32-41.

R. Cedergren - Seepage, Drainage and Flow Nets, Wiley, New York (1967) xvii + 489.

C. Dobson, I. Cooper, P.B. Attewell and I.M. Spencer - Settlement caused by driving a tunnel through fill, Proceedings of the Symposium on the Engineering Behaviour of Urban and Industrial Fill, Midland Geotechnical Society (1979).

I.W. Farmer - Case histories of settlement above tunnels in clay, Proceedings of the Conference on Large Ground Movements and Structures, The University of Wales Institute of Science and Technology, Cardiff, Pentech Press (1977) 357-371.

N.H. Glossop and I.W. Farmer - Settlement associated with removal of compressed air during tunnelling in alluvial clay, Geotechnique 29 (1979) 67-72.

K. Henry - Grangemouth sewer tunnel, Tunnels and Tunnelling 6 (1974) 25-29.

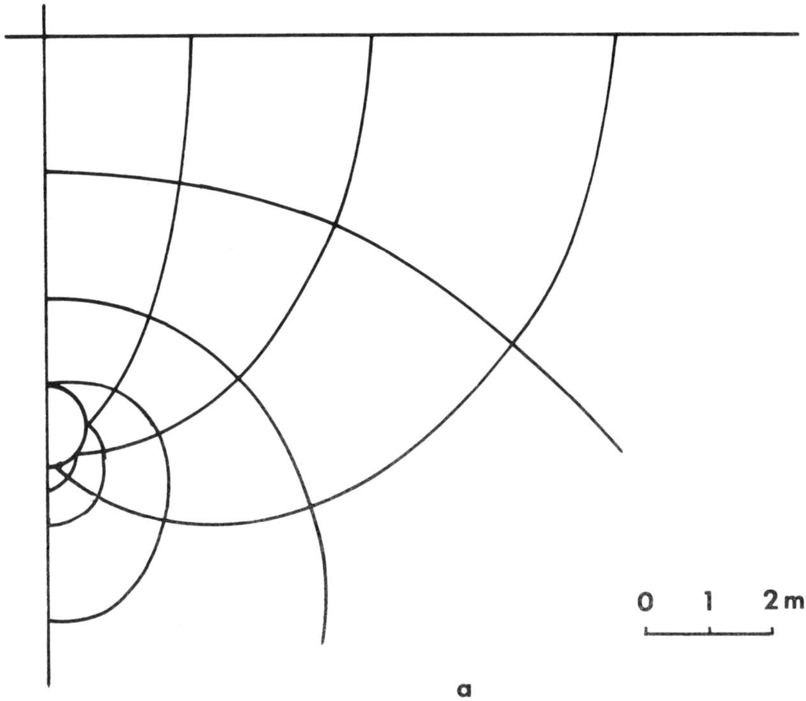

a

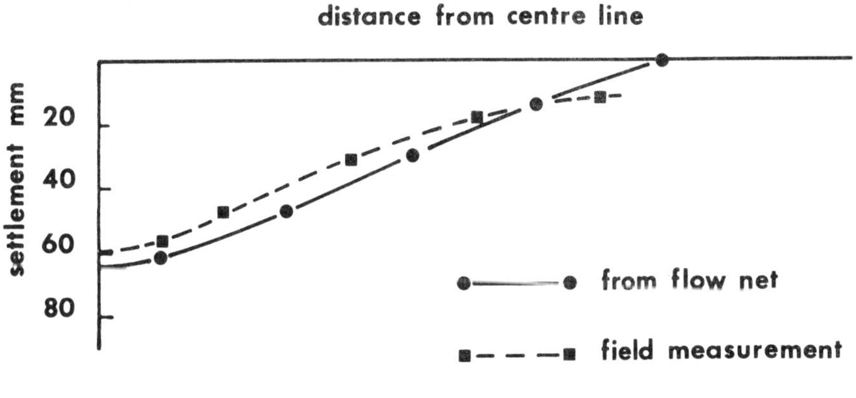

b

FIGURE 7
a) flow net and b) predicted and actual maximum settlements for the tunnel at Stockton on Tees. Partly from McCaul 1978.

C. McCaul - Settlements caused by tunnelling in weak ground at Stockton on Tees, Department of the Environment, Transport and Road Research Laboratory Report SR 383, Crowthorne (1978).

COMPARISON OF SETTLEMENTS RESULTING FROM THREE METHODS OF TUNNELLING IN LOOSE COHESIONLESS SOIL

M.P. O'Reilly ME MICE MInstHE, M.D. Ryley BSc, D.A. Barratt BSc and P.E. Johnson BSc

Tunnels and Underground Pipes Division, Structures Department, Transport and Road Research Laboratory

INTRODUCTION

Tunnelling through cohesionless soils without inducing significant settlements and consequential damage to buildings and services above the tunnel can be difficult. In such conditions compressed air and/or chemical treatment are commonly used to stabilise the ground (West and O'Reilly): the recently developed bentonite and slurry shield processes provide other solutions for the tunneller in these ground conditions (Bartlett et al; Jacob). The construction of three lengths of sewer tunnel at Warrington in loose granular soils enabled ground deformations during bentonite tunnelling to be monitored for a second time in Great Britain and enabled them to be compared with settlements occurring when tunnels were driven by conventional methods.

THE TUNNELLING SCHEMES

General details of the schemes, ground conditions and measurements for the three tunnelling schemes are given in Table I. A one-shot injection of sodium silicate and ethyl acetate using the tube à manchette technique from the road surface was used to stabilise the 110 m length of chemically treated ground on the Acton Grange Sewer. At the fully-stabilised section of the Lumb Brook Sewer sodium silicate was injected from the tunnel face to form an arch of material like weak sandstone 1 m or so thick outside the tunnel above the axis level. For full stabilisation eleven injection lances were driven 5 m ahead of the tunnel and 70 litres of solution were injected at each 0.3 m penetration; in the partially-stabilised sections only 5 or 7 lances were used.

TABLE I

Details of the tunnelling schemes

Name of tunnel	Length m	External/ internal diameter m	Construction method	Ground conditions	Ground water conditions	Details of measurements
Acton Grange Trunk Outfall Sewer	1370	2.87/2.44	0-330 m Conventional full-face tunnelling by machine through full-face sandstone up to 220 m and then a mixed face of chemically-treated sand overlying sandstone 330-1370 m Bentonite tunnelling	Windblown sands and fluvio-glacial sands and gravels overlying Upper Mottled Sandstone of the Bunter series	Ground water at tunnel axis level	Two comprehensively instrumented Sections A and B at Francis Road and Algernon Street. Surface settlements measured at 25 other locations along Ellesmere Road
Lumb Brook Sewer	446	3.6/3.3	Hand excavating of natural and chemically-treated ground from within a shield in free air	Mainly sand and gravel	Ground water lowered below tunnel invert prior to start of driving	Surface settlements measured at four locations at Willow Drive, Hollow Drive and beside the Bridgewater Canal
Mersey Street to Howley Sewer	781	1.93/1.5	Hand excavation within a shield in compressed air at pressures up to 14 psi but normally 10-11 psi	Variable deposits of gravel, sand, silt and clay overlain by made ground	Ground water level $4\frac{1}{2}$ m above tunnel axis	Surface settlements measured in Parr Street.

Tunnel linings all consisted of 600 mm long bolted concrete segments backgrouted after erection and institu concrete placed on completion of tunnelling.

At the two comprehensively instrumented sections on the Acton Grange Sewer fine to medium grained sands predominated and contained bands of silty clay and gravel: glacial erratics, mainly granite and dolerite boulders up to ½ m diameter, were often encountered. Standard Penetration Tests (SPT) made during the contract often gave values below 10 and values of 1 to 5 were not uncommon. Further confirmation of the loose ground conditions (New) was obtained from insitu density determinations at a depth of 4.5 m in a trench for a feeder sewer; these showed an average dry density of 1.57 Mg/m^3 at a moisture content of 13.1 per cent giving a 'relative compaction' of 87 per cent based on the maximum dry density found in the BS Compaction Test (4.5 kg rammer).

On the Lumb Brook Sewer the tunnel at Willow Drive and Hollow Drive was driven through gravel and a mixed face with the crown in stiff sandy clay near the Bridgewater Canal. A few dry density determinations on lumps of the stabilised gravel soil indicated dry densities of about 1.9 Mg/m^3 which for such coarsely graded materials suggest a relative compaction below 90 per cent: the high proportions of gravel sizes in the soil would explain the higher SPT values found in the boreholes.

Although ground conditions varied appreciably over the Mersey Street to Howley Sewer conditions were relatively uniform in Parr Street where the settlement measurements were made. Generally low SPT values ranging from 2-8 were found in the site investigation boreholes down to crown level indicating loose soil conditions above the tunnel.

THE MEASUREMENTS

Vertical settlements of the ground surface were made on all three sewers using ground anchor stations and/or road studs. In addition movements within the ground around the Acton Grange tunnel were made using magnetic rings and reed switch probe at the two comprehensively instrumented Sections A and B. Each section had two inclinometer tubes, one on the centre line 3.5 m long was fitted with 3 magnetic rings and the other, 9.5 m long, located half a metre to the right of the tunnel had 7 magnetic rings (Figs 1 and 2). At least 3 temporary bench marks (TBMs) were provided for each group of measuring points.

Ground water and slurry pressures ahead of the bentonite machine were monitored using pneumatic diaphragm piezometers located at tunnel axis level at Sections A and B. The records

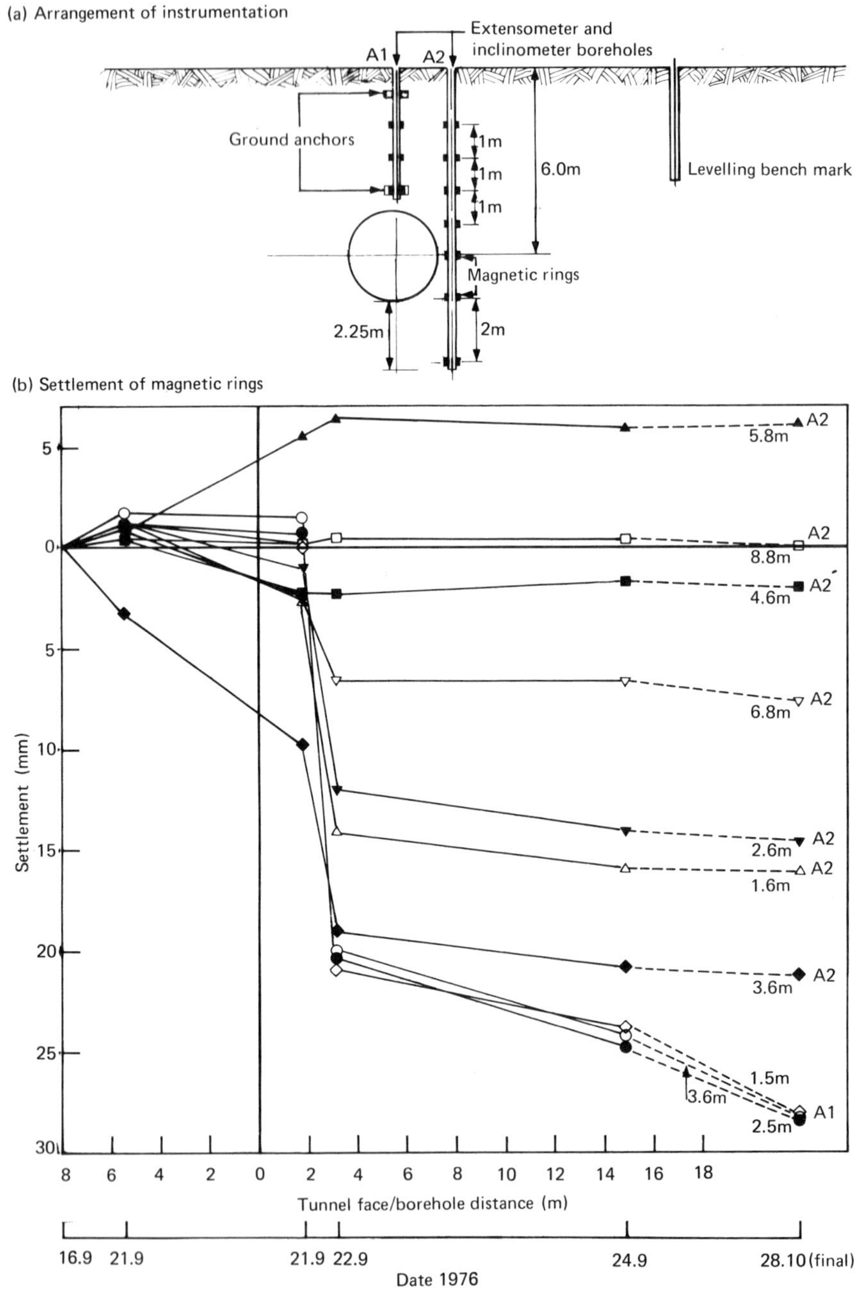

Fig. 1 ACTON GRANGE TRUNK OUTFALL SEWER–INSTRUMENTATION AND SETTLEMENT –SECTION A

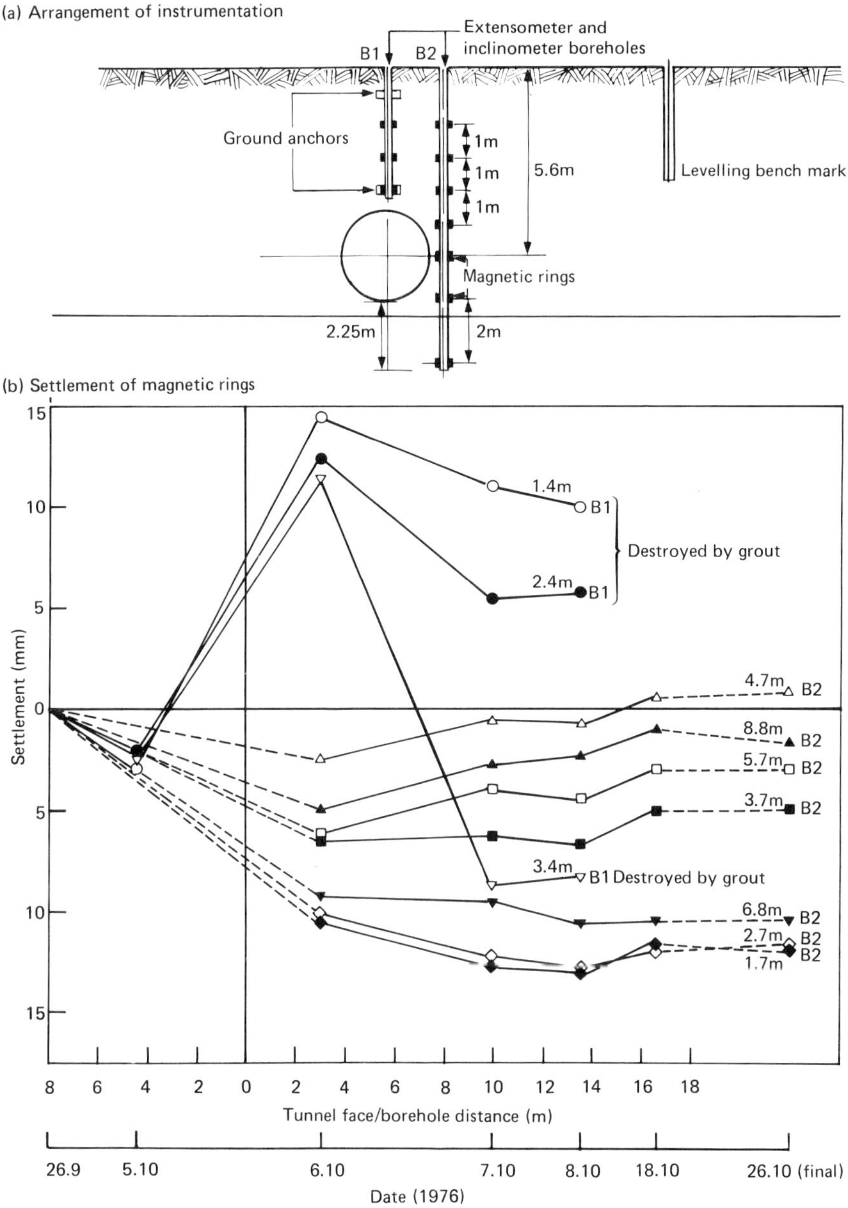

Fig. 2 ACTON GRANGE TRUNK OUTFALL SEWER–INSTRUMENTATION AND SETTLEMENT –SECTION B

obtained as the machine approached are shown in Fig 3.

Horizontal ground movements were measured by inclinometer at Sections A and B; they are outside the scope of this paper and have been described elsewhere (Ryley et al).

THE RESULTS

Of the 27 sets of measurements made on road studs on the Acton Grange Sewer five were over the tunnel in full-face sandstone, four were in situations where the upper portions of the tunnel were in chemically treated sand and the remaining 18 were over tunnels driven by the bentonite tunnelling process. The results are summarised in Table II; the largest settlement was 42 mm measured at a point where tunnelling came to a standstill for a day due to breakdown of the gravel pump on the bentonite shield.

Where one or more settlement values in addition to the maximum settlement on the centre line, S_{max}, are available on a cross-section at right angles to the tunnel an estimate can be made of the area of the settlement trough, A, on the assumption that the settlement profile follows a normal curve; similarly estimates can be made of the distance from the centre line to the inflection point, i, or the standard deviation of trough width. On the above basis the estimated area of the settlement troughs in the full-face sandstone and in the mixed-face sections with chemically treated sand overlying sandstone were 0.02 m^2 and 0.03 m^2, just 0.3 and 0.5 per cent of the tunnel cross-section respectively. For the sections of tunnel driving with bentonite support the corresponding values ranged from 0.043 m^2 to 0.141 m^2 representing 0.7 to 2.2 per cent of the tunnel cross section.

The development of settlement of the road surface on the centre line at Sections A and B with tunnel progress is shown in Fig 4 together with the final settlement profile perpendicular to the tunnel in Fig 4(b). Vertical movements of the magnetic rings in the boreholes are shown in Figs 1 and 2. The settlement recorded in borehole B1 is anomalous as bentonite was forced through the ground and up the inclinometer tubing as the tunnel passed.

The ground settlement results obtained from ground anchors and road studs at 4 locations on the Lumb Brook Sewer tunnel in natural, partially-stabilised and fully-stabilised ground are summarised in Table III together with the measurements on road studs in Parr Street over the Mersey Street to Howley Sewer. The development of settlements with time across and along these sewers are shown in Figs 5 and 6 respectively.

The air pressure in the Mersey Street to Howley Sewer

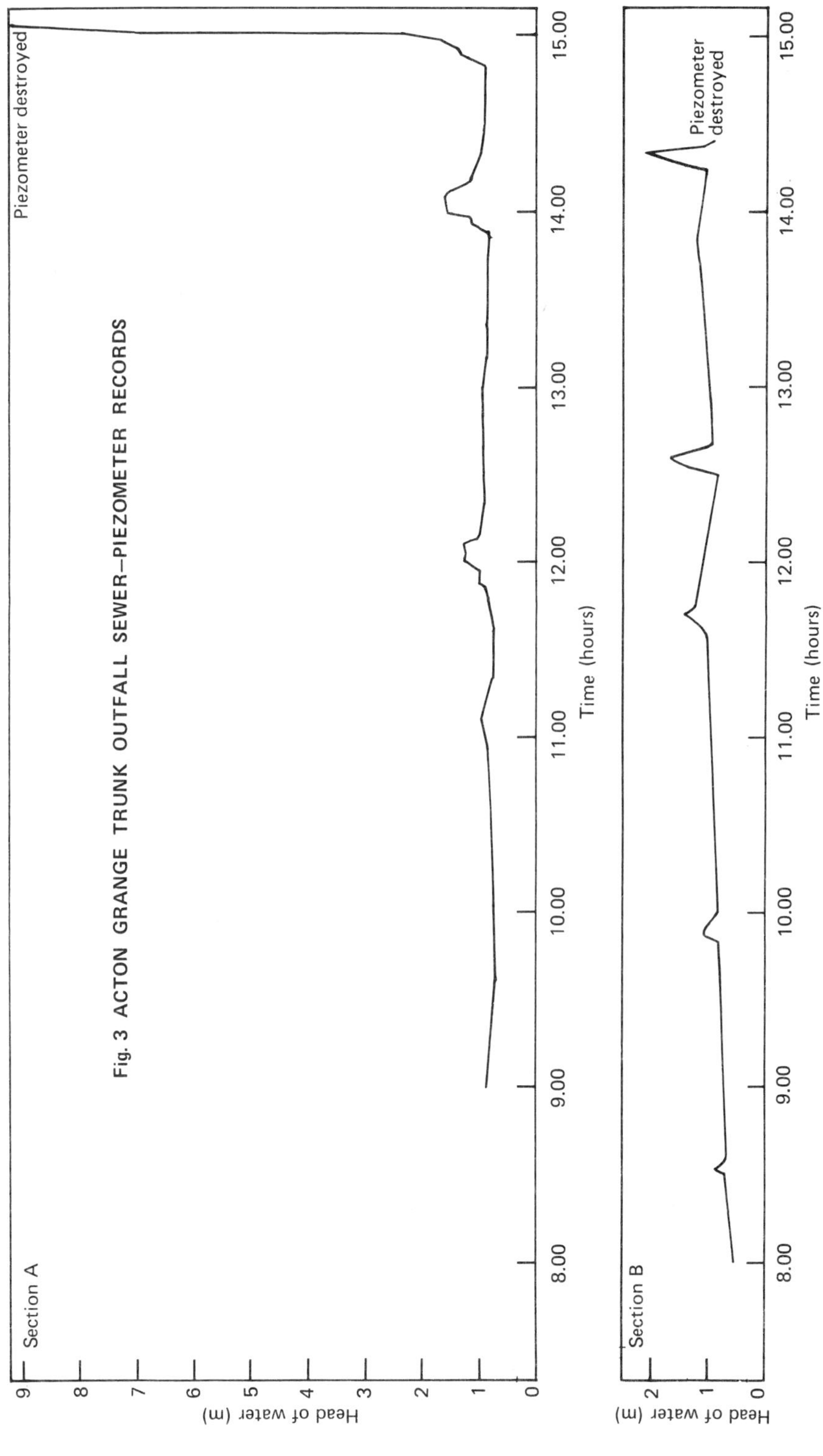

Fig. 3 ACTON GRANGE TRUNK OUTFALL SEWER—PIEZOMETER RECORDS

TABLE II

Summary of ground settlements for Acton Grange Trunk Outfall Sewer

Chainage	Settlement values: 2 m North mm	Settlement values: Centre line mm	Settlement values: 2 m South mm	Best estimate of i m	*Best estimate of area of trough m^2	Percentage of tunnel cross-section	Cover/ diameter	Ground conditions
188	1.8	2.1	2.3	3.60	0.019	0.3	2.1	Full-face sandstone
197	-	1.9	-	-	-	-	2.1	
206	-	1.3	-	-	-	-	2.0	
224	-	1.4	-	-	-	-	2.0	
233	-	1.6	-	-	-	-	2.0	
242	2.0	2.2	1.8	3.81	0.021	0.3	1.9	Mixed face - chemically treated sand - sandstone in the invert
250	-	2.0	-	-	-	-	1.9	
320	2.5	2.8	-	4.13	0.029	0.5	1.8	
325	3.3	4.8	-	2.33	0.028	0.4	1.8	
365	10.2	19.9	-	1.78	0.086	1.3	1.8	Bentonite tunnelling process in natural sand - some sandstone in the invert
370	8.6	14.2	-	1.99	0.071	1.1	1.8	
648	8.7	42.2	-	1.12	0.119	1.8	1.6	
657 (A)	2.3	18.6	10.0	1.78	0.083	1.3	1.6	
721 (B)	3.6	25.3	11.3	1.56	0.099	1.5	1.5	
942	1**	20	-	1.22	0.061	0.9	1.3	Measurements made in connection with the passage of the tunnel beneath services near Northwich Road Bridge (New et al)
947	-	25	-	-	-	-	1.3	
962	-	19	-	-	-	-	1.3	
967	-	24	-	-	-	-	1.3	
1015	-	27	-	-	-	-	1.5	
1020	-	27	-	-	-	-	1.5	
1024	-	20	-	-	-	-	1.5	
1025	-	18	-	-	-	-	1.6	
1030	10**	26	-	2.16	0.141	2.2	1.6	
1035	-	24	-	-	-	-	1.6	
1040	-	15	-	-	-	-	1.6	
1045	3**	8	-	2.14	0.043	0.7	1.6	
1050	-	14	-	-	-	-	1.7	

*Assuming settlement trough follows normal curve

**3 m North

TABLE III

Summary of ground settlements over the Lumb Brook Sewer and Mersey Street to Howley Sewer tunnels

Scheme	Location	Max surface settlement mm	Best estimate of i m	Area of* trough m^2	Percentage of tunnel cross-section	Cover/ diameter	Ground conditions
Lumb Brook Sewer	Willow Drive	78	2.40	0.47	4.6	0.8	Natural sands and gravels
	Between Bridgewater Canal and Greenfield Avenue	19	2.52	0.12	1.2	1.9	Natural sandy clay soil
	Hollow Drive	15 20	1.59 1.79	0.06 0.09	0.6 0.9	1.3 1.3	Partially-stabilised sands and gravels
	Between Hollow Drive and Chester Road	7	2.28	0.04	0.4	1.3	Fully-stabilised sands and gravels
Mersey Street to Howley Sewer	Parr Street	28	3.20	0.23	7.1	3.7	Loose sand below water table

*Estimated - assuming settlement trough follows normal curve

(a) Settlement development profile (centre line)

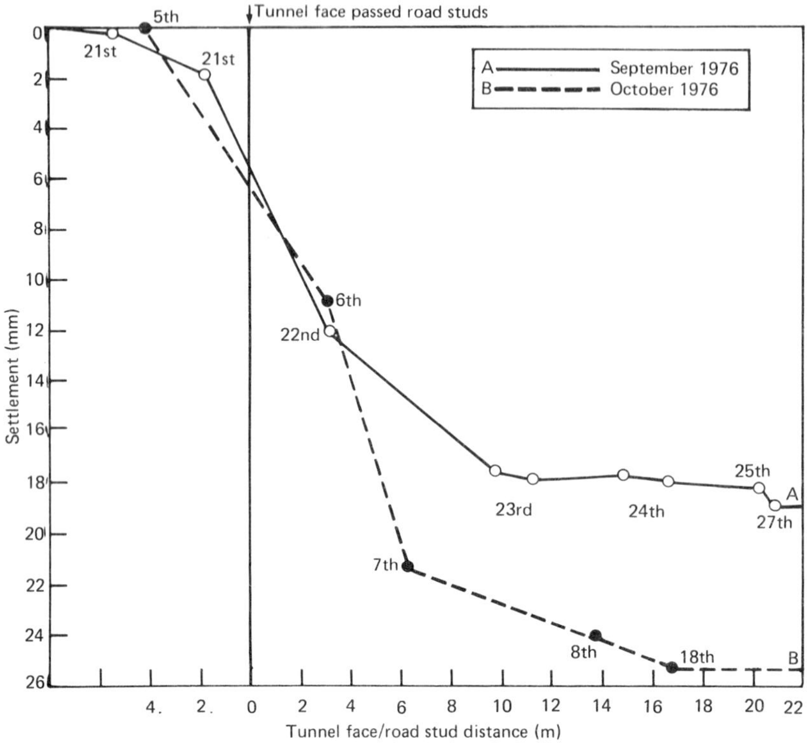

(b) Transverse settlement profile

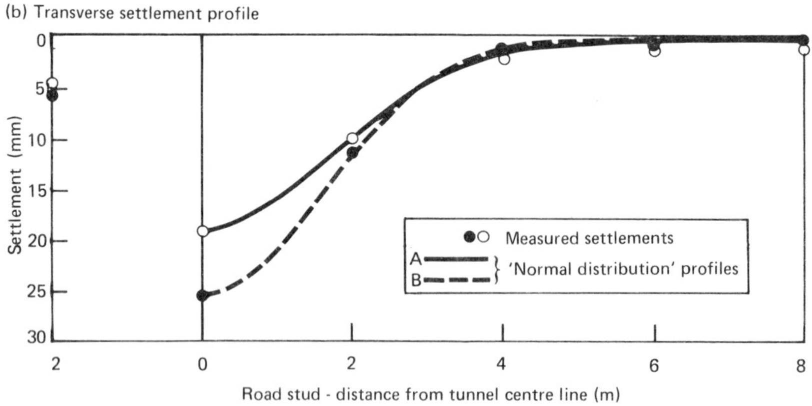

Fig. 4 ACTON GRANGE TRUNK OUTFALL SEWER–SURFACE SETTLEMENT AT SECTIONS A AND B

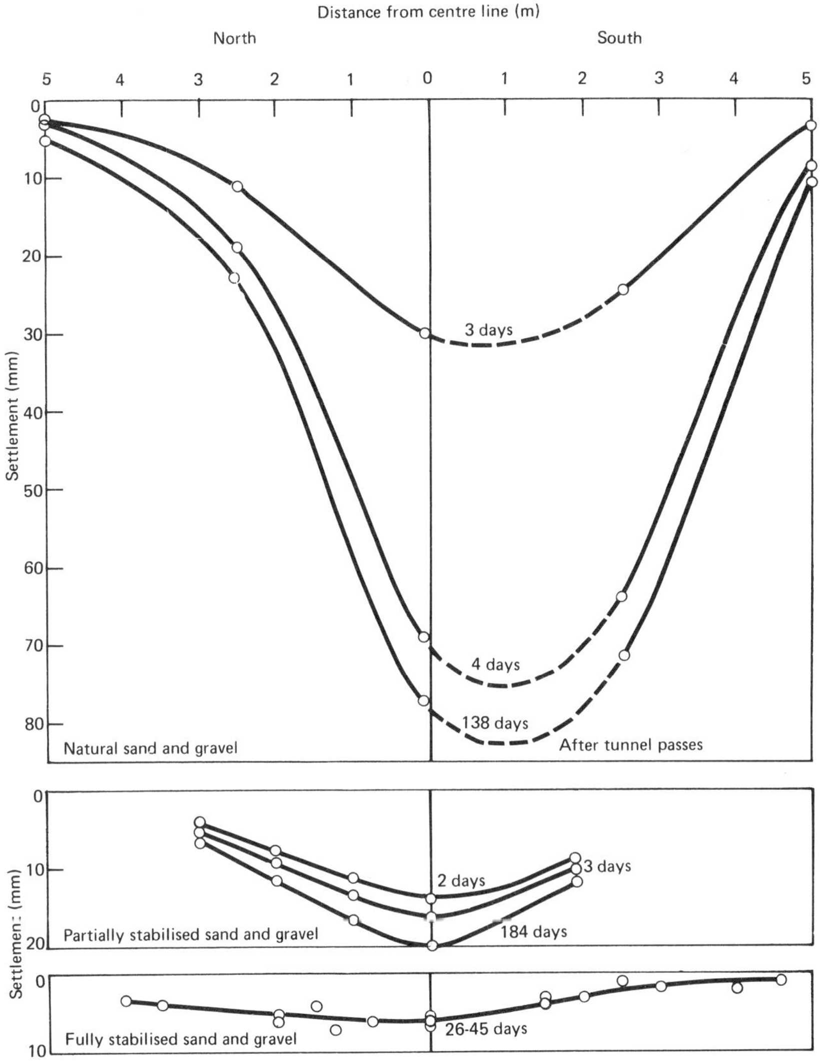

Fig. 5 LUMB BROOK TUNNEL–SETTLEMENT ACROSS THE TUNNEL LINE

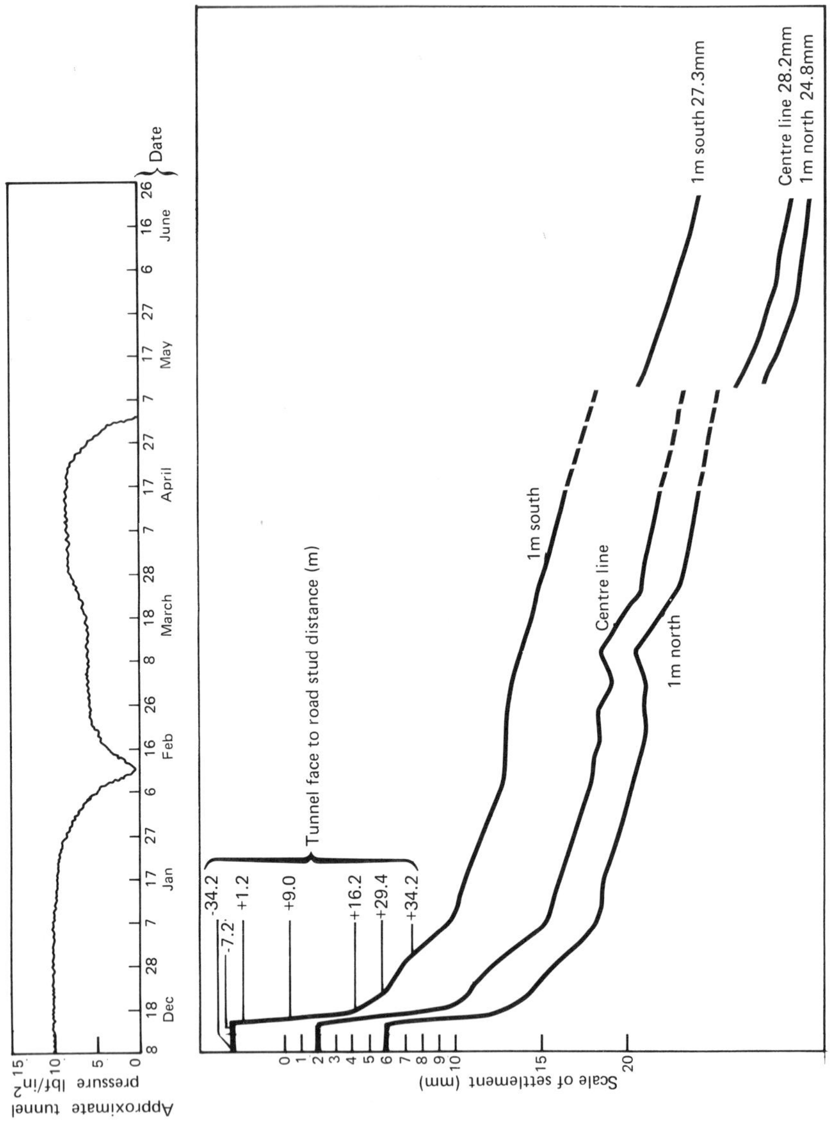

Fig. 6 MERSEY STREET TO HOWLEY SEWER—DEVELOPMENT OF SETTLEMENT

tunnel varied considerably; it was zero on 11 and 12 February 1977 (see Fig 6) and was finally released on 4 May 1977. No appreciable settlement was observed around 11-12 February; although settlement was not being monitored when the air was finally released there would appear to be a 3 mm settlement associated with this operation.

DISCUSSION

The loss of ground during tunnelling is related to the area or more accurately the volume per unit length of the settlement trough above a tunnel. Dilation of the ground above the tunnel may occur in sandy soils during the period of settlement but the expansion is likely to be quite small in loose cohesionless soils such as those at Warrington (Atkinson and Potts; Atkinson and Bransby).

Table II shows that the area of the settlement troughs for the bentonite tunnelling on the Acton Grange Sewer varied from 0.7 to 2.2 per cent of the tunnel cross-section while the values for the portion of the tunnel driven mixed-face in chemically treated sand overlying sandstone were one-half per cent or less. A value of 0.4 per cent was found on the section of the Lumb Brook Sewer driven through fully-stabilised sands and gravels (Table III); where this tunnel was driven through partially-stabilised sands and gravels settlement troughs as a percentage of the tunnel cross-section were somewhat larger at 0.6 and 0.9 per cent. The above values in chemically treated ground do not include any deformation caused by the injection process; limited measurements on the Acton Grange Sewer indicate that significant settlement may be caused during the chemical treatment process (New).

The settlement troughs in the unstabilised sands and gravels on the Lumb Brook Sewer and on the Mersey Street to Howley Sewer were 4.6 and 7.1 per cent respectively of the excavated tunnel cross-section. Both figures are greater by a factor of 2 or more than the settlement troughs found when bentonite tunnelling was used and by a factor of 9 or more when the ground was chemically stabilised. However it must be emphasised that ground water conditions were much more severe on the Mersey Street to Howley Sewer than on the other two tunnels.

While there are few data in the literature which are directly comparable with ground conditions at Warrington the measurements in the section of the Lumb Brook Sewer driven through sandy clay indicate that the method of hand excavation within a shield resulted in ground losses rather similar to those found in London Clay (Barratt and Tyler). Eadie has reported settlement troughs of about 1 per cent of the excavated volume for a tunnel driven under compressed air in somewhat denser sands, SPT range 5-13, near Irvine. Settlement troughs varying from fractions of a per cent to 5.3 per cent are quoted by Peck et al for 9 tunnels in sand while more recently MacPherson et al report troughs of 3.5 to 13.7 per cent of the excavated cross-section for a tunnel driven by machine in medium dense to dense sands and gravels, SPT range 8-52. There are indications, too, (McCaul) that the settlement trough as a percentage of the excavated cross-section is higher for smaller diameter tunnels which can be explained in terms of a constant value of the clearance around the shield providing a void space proportional to the diameter rather than the diameter squared outside the tunnel lining.

Limited data are available from schemes in Japan on the settlement occurring during slurry shield tunnelling which is similar to the bentonite tunnelling process (Yamazaki; Miki et al). At three of the four sites where conditions were similar to those on the Acton Grange Sewer the settlement picture although less comprehensively reported appeared similar to that at Warrington: in the remaining case the settlements reported were considerably lower. In Europe bentonite tunnelling has been used at Hamburg (Jacob) where with a C/D of 4 the settlement trough appears to have been less than 1 per cent of the tunnel cross-section: the process is now being used for construction of the Antwerp Metro and again settlements are said to be low.

The width of the settlement trough defines the zone affected by tunnelling and there will be many cases where the tunnel designer can locate his tunnel so that this zone is clear of buildings and sensitive services. Values of $2\frac{i}{D}$ have been plotted against normalised tunnel depth, $\frac{Z}{D}$, in Fig 7 for the tunnels at Warrington driven through natural sands and gravels: the data correspond well with that given by Peck.

A similar plot of the data for the sections of tunnel in granular soils fully and partially treated with chemicals and also for a section of tunnel in weak rock is shown in Fig 8; for these cases the settlement trough is relatively wide by comparison with those found in the natural soils. The reasons for this behaviour are not clear but in all cases the percentage loss of ground was low, 0.5 per cent or less in the fully treated soil and rock and a maximum of 0.9 per cent for the partially treated materials. Similar trough widths have been reported in Madrid (Oteo and Moya) and London (West et al) where ground losses have been low and the ground relatively stiff.

CONCLUSIONS

All factors considered, it is difficult to conclude other than that the bentonite shield is likely to provide more

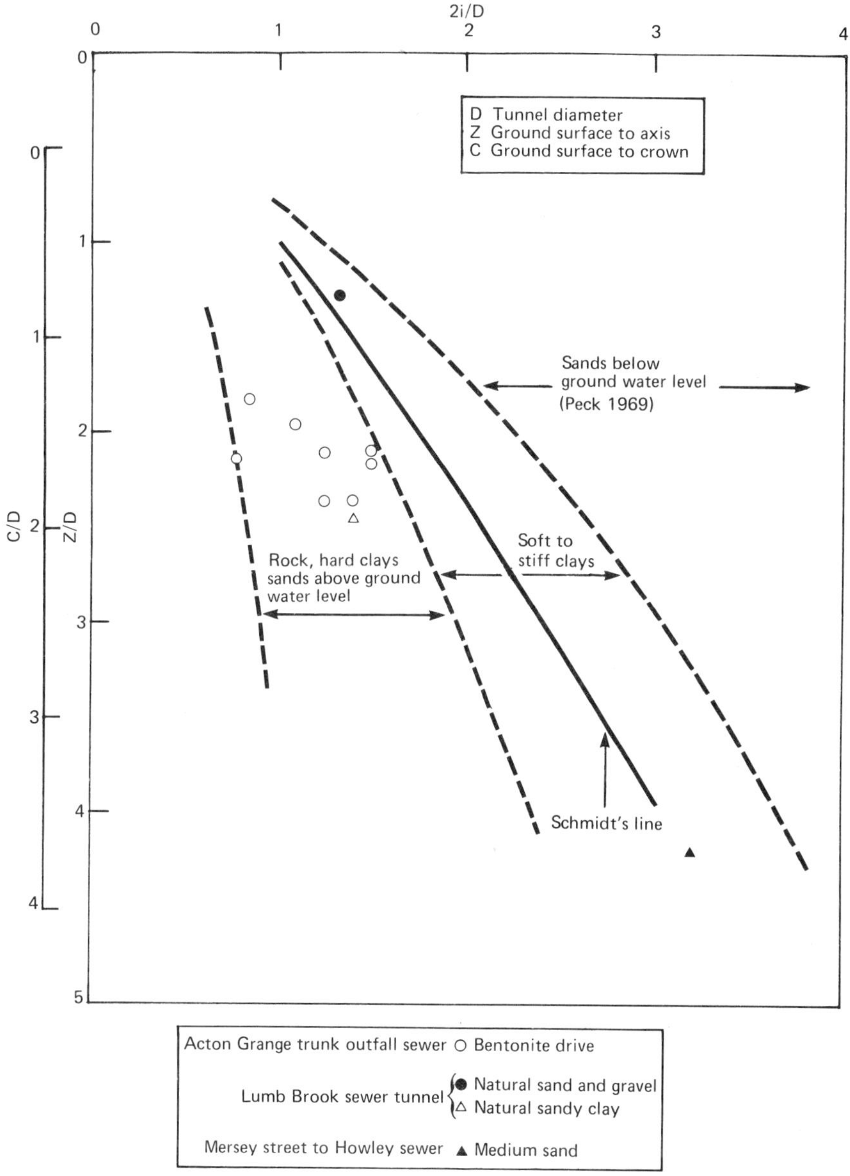

Fig. 7 RELATION BETWEEN 2i/D AND Z/D FOR TUNNELS IN NATURAL SOIL

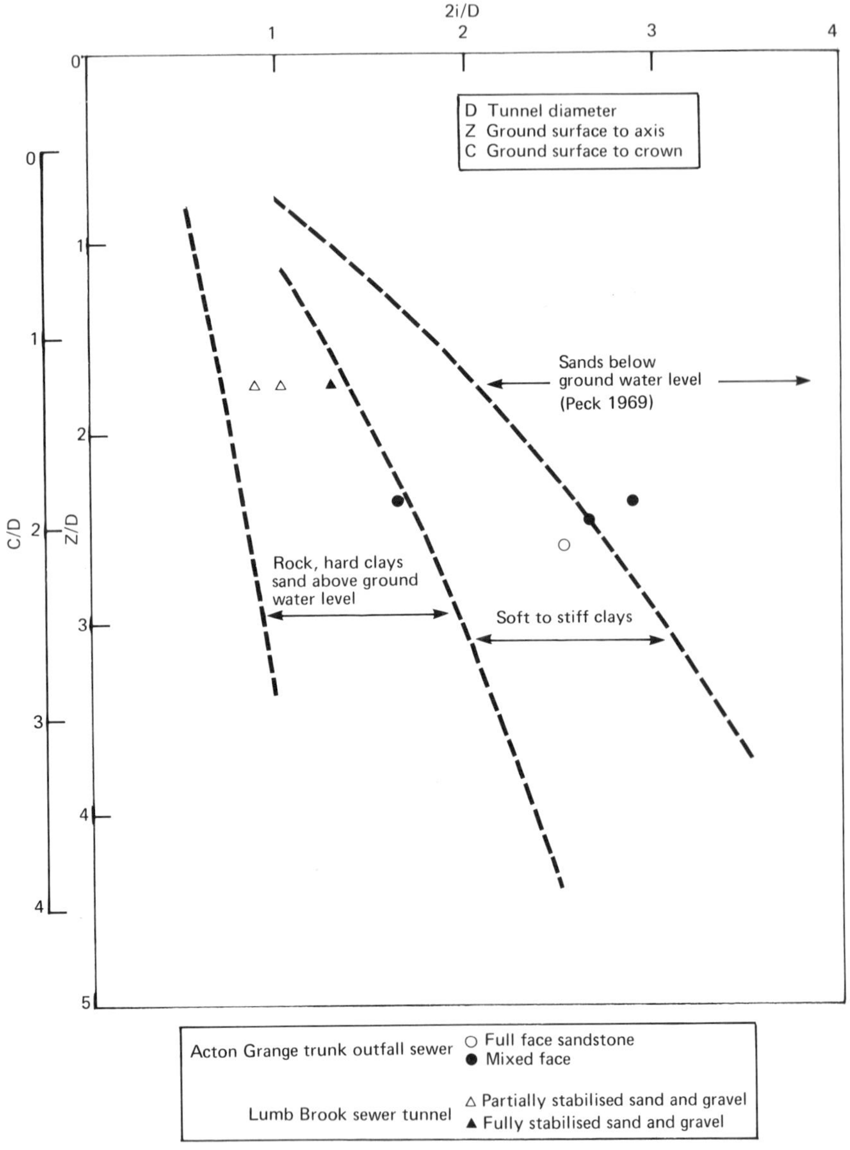

Fig. 8 RELATION BETWEEN 2i/D AND Z/D FOR TUNNELS IN ROCK AND STABILISED GROUND

effective ground support than compressed air in cohesionless soil but it is impossible to quantify the differences with any accuracy at this juncture. Chemical treatment is considerably more effective than either although settlements during the injection process may cause considerable problems.

ACKNOWLEDGEMENT

The work described in this paper forms part of the programme of research of the Transport and Road Research Laboratory and this paper is published by permission of the Director.

The Laboratory wishes to thank the Warrington New Town Development Corporation and the Central Division, North West Water Authority for permission to carry out these investigations, and Ward, Ashcroft and Parkman , consulting engineers, for supplying some of the settlement data at the Lumb Brook Sewer. The assistance of the contractors; Edmund Nuttall Ltd (Acton Grange Sewer), Sheridan Contractors Ltd (Lumb Brook Sewer) and Streeters (Northern) Ltd (Mersey Street to Howley Sewer) is gratefully acknowledged.

REFERENCES

J.H. Atkinson and P.L. Bransby - The mechanics of soils. An introduction to critical state soil mechanics, McGraw-Hill, London (1978).

J.H. Atkinson and D.M. Potts - Subsidence above shallow tunnels in soft ground, Journal of the Geotechnical Engineering Division ASCE 103 (1977) 307-325.

D.A. Barratt and R.G. Tyler - Measurements of ground movement and lining behaviour on the London Underground at Regents Park, Department of the Environment, Transport and Road Research Laboratory Report LR 684, Crowthorne (1976).

J.V. Bartlett, A.R. Biggart and R L Triggs - The bentonite tunnelling machine, Proceedings of Institution of Civil Engineers 54 (1973) 605-624.

H.S. Eadie - Settlements observed above a tunnel in sand, Tunnels and Tunnelling 9 (1977) 93-94.

E. Jacob - The bentonite shield: technology and initial application in Germany, Proceedings of symposium Tunnelling '76, Institution of Mining and Metallurgy, London (1976) 201-207.

C. McCaul - Settlements caused by tunnelling in weak ground at Stockton-on-Tees, Department of the Environment Department of Transport, Transport and Road Research Laboratory Report SR 383, Crowthorne (1978).

H.H. MacPherson, J.W. Critchfield, S.W. Hong and E.J. Cording - Settlements around tunnels in soil: three case histories. U.S. Department of Transportation Report No UMTA-IL-06-0043-78-1, Washington (1978).

G. Miki, T. Saito and H. Yamazaki - The principle and field experiences of a slurry mole method for tunnelling in soft ground, Ninth International Conference on Soil Mechanics and Foundation Engineering, Speciality Session 1, Tokyo (1977) Foundation Publications, London.

B.M. New - The effects of ground vibration during bentonite shield tunnelling at Warrington, Department of the Environment Department of Transport, Transport and Road Research Laboratory Report LR 860, Crowthorne (1978).

C.S. Oteo and J.F. Moya - Estimation of the soil parameters of Madrid in relation to tunnel construction, Proceedings of the Seventh European Conference on Soil Mechanics and Foundation Engineering, Brighton (1979) Vol. 3, British Geological Society, 239-247.

R.B. Peck, D.U. Deere, J.E. Monsees, H.W. Parker and B. Schmidt - Some design considerations in the selection of underground support systems. Report for U.S. Department of Transportation PB-190-443, Technical Information Service, Springfield, Virginia, U.S.A. (1969).

M.D. Ryley, P.E. Johnson, M.P. O'Reilly and D.A. Barratt - Measurement of ground movements around three tunnels at Warrington, Department of the Environment Department of Transport, Transport and Road Research Laboratory Report, Crowthorne (in the press).

G. West, W. Heath and C. McCaul - Measurements of the effects of tunnelling at York Way, London. Paper submitted to Ground Engineering.

G. West and M.P. O'Reilly - Methods of treating the ground, Tunnels and Tunnelling 10 (1979) 25-29.

H. Yamazaki - Problems of the slurry shield method and their counter measures, Proceedings of the Second Austrailian Tunnelling Conference, Melbourne (1976) 113-122.

ADVANCING FACE SIMULATION OF TUNNEL EXCAVATIONS AND LINING PLACEMENT

Ezzat A. Hanafy, M Eng

Research Engineer, Department of Civil Engineering and Engineering Mechanics, McMaster University

John J. Emery, Ph D , P Eng

Associate Professor of Civil Engineering and Engineering Mechanics, McMaster University, and Principal, The Trow Group Limited Consulting Engineers

INTRODUCTION

The plane strain, elastic-creep, finite element method simulation of tunnelling presented at the First Conference (Emery et al, 1978) has been extended through an axisymmetric, elastic-plastic-creep approach to include the representation of the actual three-dimensional advancing face case with lining placement at typical distances (or times) from the face. Tunnel construction in squeezing ground (stress relaxation and/or swelling), and the increasing North American use of precast concrete segmental liners, were the prime motivation for developing an analysis and design package with the following general capabilities: axisymmetric analysis with nonsymmetric radial loadings (full Fourier series representation) and axial loadings due to the initial state of stress; excavation in stages to represent the advancing face condition; elastic-plastic-creep analysis for each stage; lining placement delay; and time-dependent interaction of the final ground-interface-support system. While discontinuities (joints, bedding, faults) and individual liner joints are not incorporated in the axisymmetric simulation, the importance and potential behaviour of these features can be considered in supplementary plane strain simulations (Emery et al, 1978). The reliability of any deformation and stress predictions based on numerical analyses is largely a function of the adequacy of the input (initial state of stress and ground characteristics) and realistic representation of the excavation and construction procedures. The computer has not replaced the significant engineering judgement required at all stages of tunnel design, as "changed conditions" are almost the rule during tunnelling.

As background information on squeezing ground and its influence (Brown and Hoek, 1978; Franklin and Hungr, 1978), tunnels and linings (Craig and Muir Wood, 1978), advancing face simulations (Ranken et al, 1978) and the finite element method in geotechnology (Gudehus, 1977) is readily available, this contribution emphasizes the development of the axisymmetric simulation and its application to typical tunnel designs. A number of application examples are summarized, particularly to indicate the influence of lining placement delay on liner deformations and stresses. For instance, with a simple elastic analysis of a typical tunnel, the stresses are reduced to about 20 percent of the instantaneous values (often still assumed in design) by a liner placement delay of one tunnel diameter from the advancing face.

AXISYMMETRIC ANALYSIS OF TUNNELLING

The plane strain, finite element method of analysis is still widely used to provide deformation and stress input to tunnel designs, as it has the important advantage of simplicity and low cost compared to more appropriate elastic-plastic-creep, three-dimensional finite element method analyses (Ranken et al, 1978, and noting that there are other powerful analysis techniques). It is clear that tunnelling is a three-dimensional, incremental process as shown by the convergence monitoring with face advancement and time in Figure 1 for a large water intake tunnel at depth in horizontally bedded shale with high lateral stresses (Emery et al, 1978). The full simulation of the advancing face and liner installation sequence indicated in Figure 2, followed by progressive ground-interface-lining interaction, can be accommodated by an axisymmetric finite element method approach that has significant computer core and time savings over three-dimensional finite element methods.

The axisymmetric approach involves the representation of nonsymmetric loadings through Fourier series expansions (see for instance Wilson, 1965), but is restricted to isotropic, radially homogeneous, continuous ground and cylindrical or spherical cavities. As the representation of discontinuities (joints, bedding, faults) is critical in many designs, and liners can involve special joints and stress reducers, an elastic-plastic-creep plane strain simulation approach such as that shown in Figure 3 is still used in conjunction with the axisymmetric approach to consider limiting conditions if necessary. This typically involves the use of displacement boundary conditions from the axisymmetric approach, or adjustments based on parametric studies of the type given here.

The coordinate system and Fourier series representation

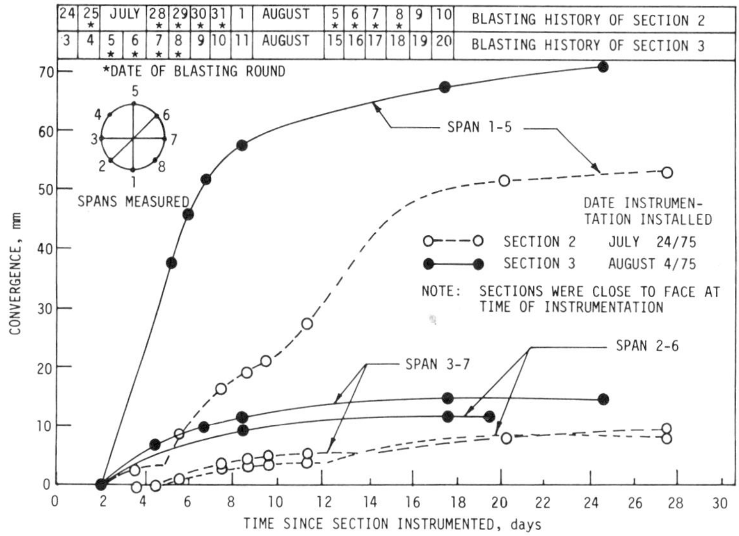

FIGURE 1 MONITORED CONVERGENCE WITH TIME, 4 METRES DIAMETER TUNNEL AT DEPTH OF 61 METRES IN HORIZONTALLY BEDDED SHALE

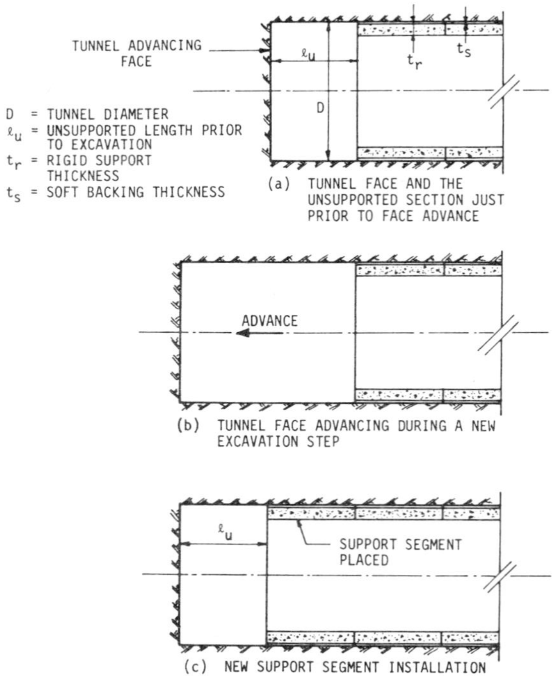

FIGURE 2 TUNNEL EXCAVATION AND PRECAST CONCRETE SEGMENTAL LINER INSTALLATION SEQUENCE

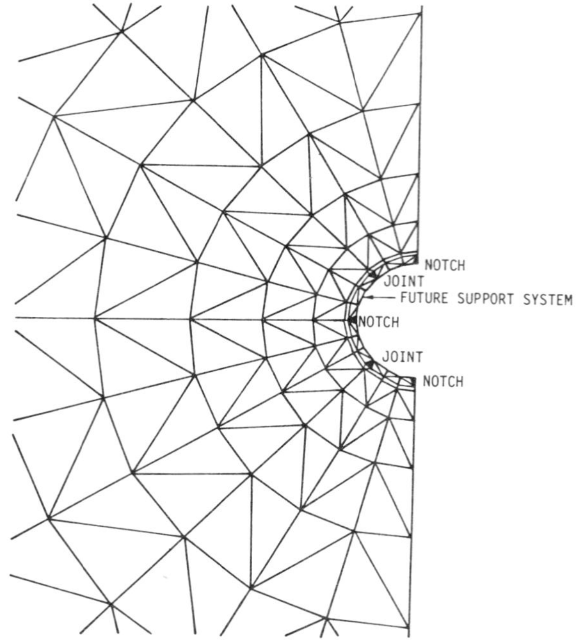

FIGURE 3 FINITE ELEMENT MESH REPRESENTATION OF PRECAST CONCRETE SEGMENTAL LINER WITH SOFT BACKING FOR PLANE STRAIN SIMULATION OF TUNNELLING

of nonsymmetric loadings used in the axisymmetric approach are given in Figure 4 for a shaft orientation. In summary, the axisymmetric elastic-plastic-creep simulation of tunnel excavation and liner placement has the following features:

1. The geometry and materials properties do not vary with the angle of rotation about the shaft or tunnel axis, but can vary along the axis.
2. The primary state of stress is expressed in terms of the vertical, lateral and axial stresses, typically as a ratio of the overburden stress (vertical stress), i.e. K_a = axial effective stress/vertical effective stress and K_ℓ = lateral effective stress/vertical effective stress. It should be noted that K_a and K_ℓ are not constants, but can vary with depth, and nonsymmetric radial loadings are involved when $K_\ell \neq 1$. High K values are indicative of potential squeezing ground conditions (Emery et al, 1978).
3. The excavation process is simulated by a "deactivation" of the stiffness terms for the elements representing the ground removed during the excavation step (advancing face). This is accomplished by reducing their stiffnesses by 1×10^{-6}, and these reduced stiffness elements then remain available for "reactivation" to represent any further construction steps such as tunnel support placement.
4. Tunnel support placement using the reactivation process at a specified distance from the advancing face or a specified time following face advance. The support system can incorporate special features such as soft backings.
5. Elastic-plastic-creep analysis following any change in the stress state, including support system placement. The incremental initial stress method is used for the plastic analysis, and the incremental initial strain method for the creep analysis (Zienkiewicz, 1977 for instance). Both the Von Mises and Drucker-Prager yield criteria with associated flow rule (i.e. elastic-perfectly plastic response to model yielding) have been incorporated (Zienkiewicz, 1977), while the creep relationship must be supplied as a subroutine as required (Emery et al, 1978).

The finite element mesh representation for the axisymmetric simulation of the advancing tunnel face is given in Figure 5. Application of the axisymmetric elastic-plastic-creep simulation will be illustrated through a series of examples in following sections.

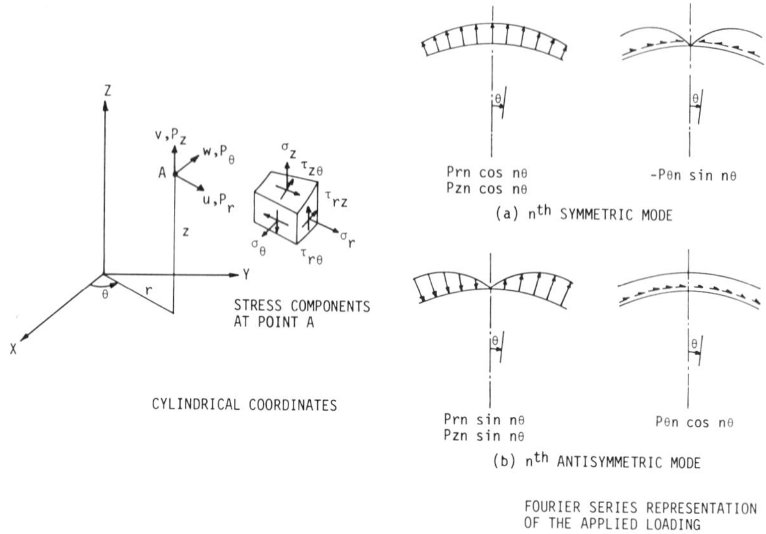

FIGURE 4 COORDINATE SYSTEM AND FOURIER SERIES REPRESENTATION OF NONSYMMETRIC LOADINGS FOR AXISYMMETRIC ANALYSIS OF SHAFTS AND TUNNELS

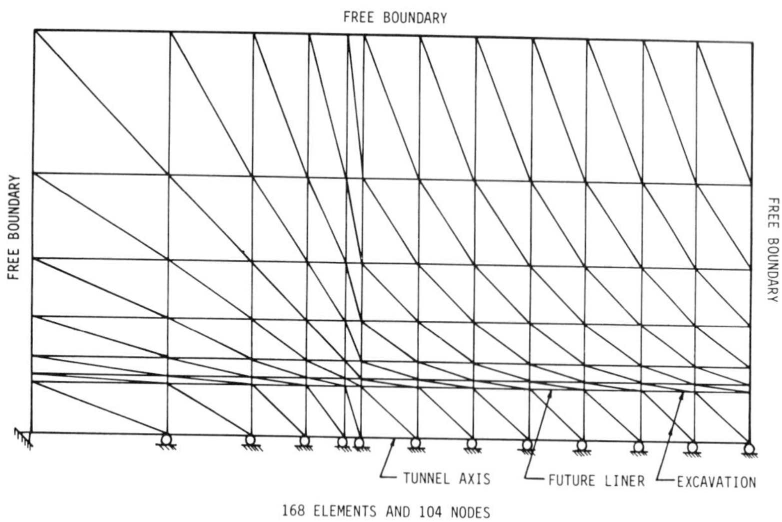

FIGURE 5 FINITE ELEMENT MESH REPRESENTATION OF CYLINDRICAL OPENING FOR AXISYMMETRIC SIMULATION OF ADVANCING TUNNEL FACE

UNLINED TUNNELS

Elastic Response

The effect of tunnel face advance on the elastic radial displacements are shown in Figure 6 for the following rock properties and initial stress assumption: $E = 5 \times 10^3$MPa; $\nu = 0.2$; $K_a = K_\ell = 1$ with a vertical stress of 8MPa (i.e. uniform initial stress state); and the tunnel radius R = 5m. For a face advance (Stage II) of 2R (10m) from the initial position (Stage I) at 5R (25m), the incremental elastic radial displacements are a maximum at about R (5m) from the new face position, and gradually decrease with increasing distance from the new face. At a distance of about 4R, no significant radial displacements occur due to the new advance.

In order to demonstrate more fully the effect of face advancement, the elastic radial displacements at reference point B given in Figure 7 are tabulated in Table I for the same system described above. Point B represents any reference point along the future tunnel wall that the tunnel face will advance past. As shown in Table I, when the tunnel face reaches the point of interest (i.e. point B), an accumulated radial displacement of more than 35 percent of the total displacement will already have occurred. When the tunnel face advances one radius (R) ahead of the reference point, about 80 percent of the total displacement will have accumulated at the point of interest. Clearly the load carried by the support system due to the elastic closure will depend on the distance between the advancing face and location where the liner is placed in contact with the rock.

Time-Dependent Response

In tunnel design cases where time-dependent behaviour of the ground is relevant (squeezing ground or swelling for instance), a creep analysis must be considered in conjunction with the advancing face simulation. Such an analysis is essential to cover design questions such as: the appropriate time (or space) delay before liner placement; the possibility of leaving a gap or using a soft backing to take up some of the closure; and the liner stresses developed with long term creep.

To demonstrate the effect of creep on tunnel inward displacements, the same tunnel described above was considered with the representative power form creep relationship given on Figure 8 (Robertson, 1963). As a uniform initial stress state was assumed, the creep closure is uniform around the opening with values shown in Figure 8 for the case of instantaneous face advance of 7R followed by the creep analysis. For this case, the creep radial displacements are more significant than the elastic response and creep clearly increases the face advance zone of influence from about 4R for the elastic case to

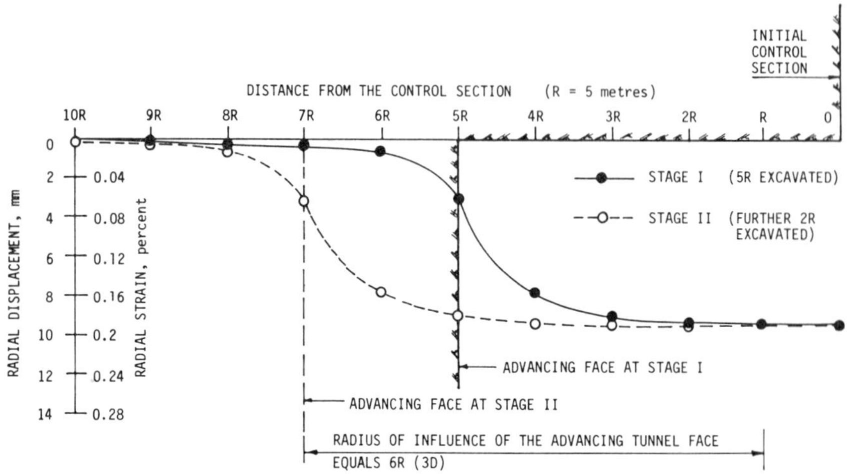

FIGURE 6 INFLUENCE OF ADVANCING FACE ON THE ELASTIC RADIAL DISPLACEMENTS FOR AN UNLINED TUNNEL (FINITE ELEMENT METHOD AXISYMMETRIC SIMULATION)

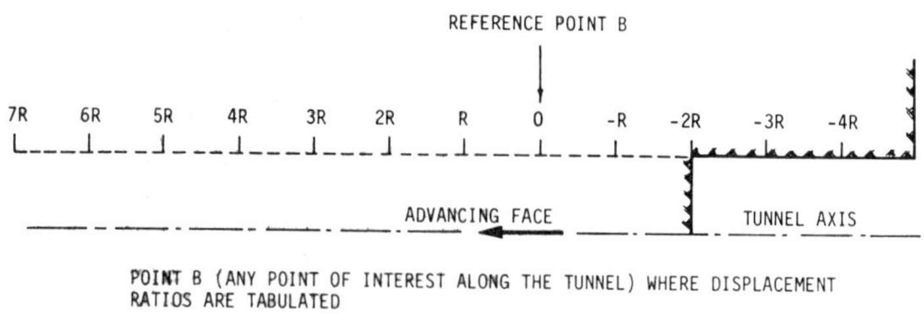

FIGURE 7 REFERENCE POINT FOR RADIAL DISPLACEMENTS

TABLE I

EFFECT OF ADVANCING FACE ON THE ELASTIC RADIAL DISPLACEMENT RATIO

DISTANCE (L) FROM POINT OF INTEREST (B) TO THE ADVANCING FACE (SEE FIGURE 7)	RATIO BETWEEN THE INCREMENTAL AND FINAL RADIAL DISPLACEMENT AT POINT (B)	RATIO BETWEEN THE ACCUMULATIVE AND FINAL RADIAL DISPLACEMENT AT POINT (B)
-4R	0.007	0.007
-3R	0.008	0.015
-2R	0.019	0.034
-R	0.059	0.093
Face at the point of interest (B)	0.284	0.377
R	0.431	0.808
2R	0.130	0.938
3R	0.031	0.969
4R	0.023	0.992
5R	0.003	0.995
6R	0.003	0.998
7R (Approximate distance of influence)	0.002	1.000

Note: R is the tunnel radius. Details on system given in text.

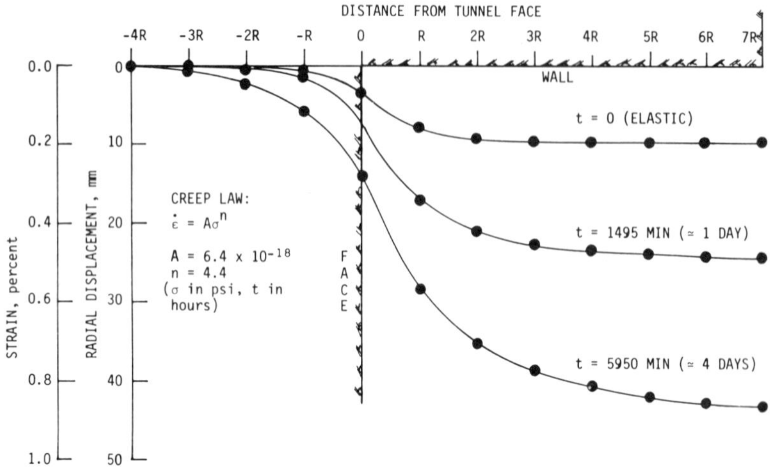

FIGURE 8 TIME-DEPENDENT RADIAL DISPLACEMENTS OF UNLINED TUNNEL WALL (FINITE ELEMENT METHOD AXISYMMETRIC SIMULATION WITH SINGLE STEP ADVANCING FACE OF 7R)

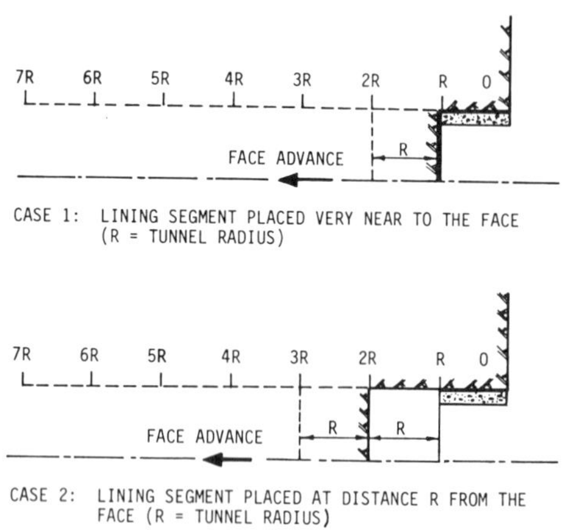

FIGURE 9 TWO CASES TO REPRESENT VARIOUS PRECAST CONCRETE SEGMENTAL LINER PLACEMENT DISTANCES FROM THE ADVANCING FACE

at least 6R after 4 days creep. It should be noted that the examples in this section did not include any liner and this must be considered next.

TUNNEL LINING INSTALLATION

During construction, when the tunnel face is being advanced, the support system is usually installed up to some construction-limited distance from the face. This distance, termed the unsupported length ℓ_u prior to excavation, is shown in Figure 2a. For an excavation step, the tunnel face will advance and reach the new position shown in Figure 2b. Due to this excavation step, there will be a change in the stress field with much of the redistributed load transferred to the tunnel support already in place, particularly the last liner segment installed. A significant portion of the load is also transferred to the unexcavated ground ahead of the face which is not commonly considered during design. This load redistribution will cause a pattern of movements within the ground mass similar to those shown in Figure 6 for the unsupported tunnel. Following this, a new liner segment is installed as shown in Figure 2c and the construction cycle is repeated.

The procedure used to simulate the liner segment placement in the analysis involves reactivation of the elements representing the liner while the stresses and strains within these elements, and the nodal displacements, are set to zero. This is essential to obtain elements free of stress, strains and displacements immediately after installation and before any stress redistribution has occurred. After a new excavation step, the equivalent nodal forces are applied, and the response of both the ground and support system is obtained.

To demonstrate the applicability of the simulation for considering the advancing face situation with lining placement, two elastic cases have been considered to show the effect of the lining installation location on tunnel inward displacements and lining stresses, and to compare these results with those predicted assuming a plane strain condition (Figure 3 without soft backing). Figure 9 shows these two cases where a 600mm thick liner is adopted for the system described in the previous section. In the first case (I) the liner is placed very near to the advancing face, while in the second case (II) the liner is placed at a distance of 1R from the advancing face. The sequence of face advance and lining placement is continued in steps to a distance of 7R from the control section using the mesh shown in Figure 5. The support system modulus of elasticity used of $6 \times E_{rock}$ is in the practical range for precast concrete segmental tunnel liners.

Figure 10 gives the elastic radial displacements along the tunnel wall which are plotted for both cases of support placement as well as for the unsupported tunnel for comparison. It

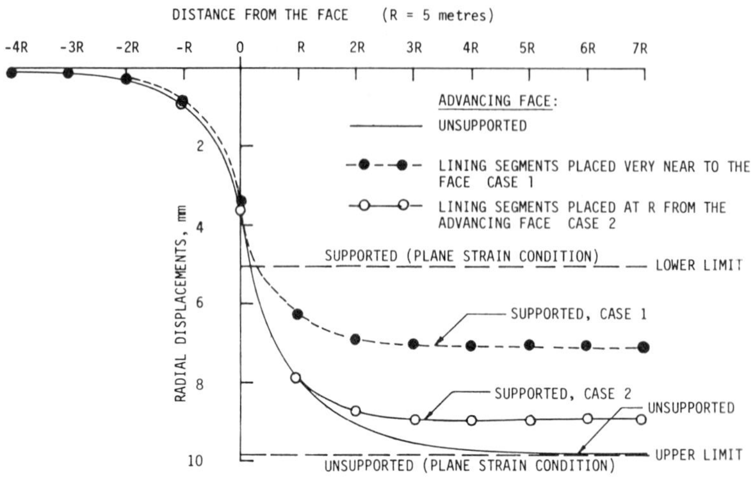

FIGURE 10 INFLUENCE OF ADVANCING FACE AND LINER PLACEMENT DISTANCES ON ELASTIC RADIAL DISPLACEMENTS (FINITE ELEMENT METHOD AXISYMMETRIC AND PLANE STRAIN SIMULATIONS COMPARED)

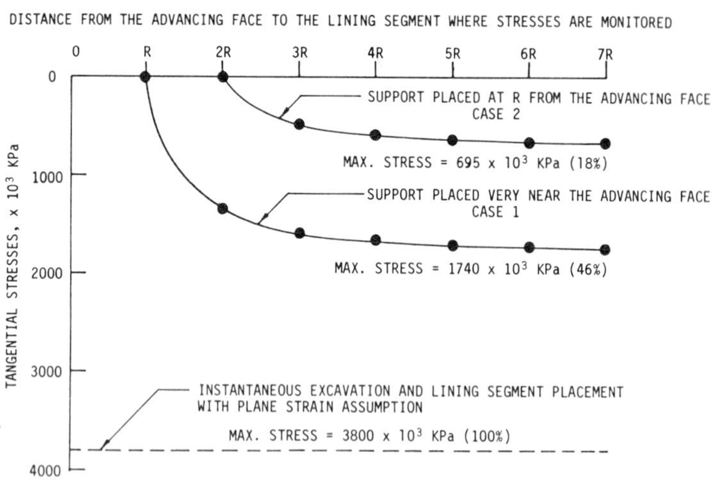

FIGURE 11 INFLUENCE OF ADVANCING FACE AND LINER PLACEMENT DISTANCES ON ELASTIC TANGENTIAL STRESSES (FINITE ELEMENT METHOD AXISYMMETRIC AND PLANE STRAIN SIMULATIONS COMPARED)

can be seen that for Case I the inward displacements have been reduced by about 25 percent from those for the unsupported tunnel, while for Case II, the tunnel inward displacements have been reduced by less than 10 percent. This indicates that if the lining is placed far from the advancing face (i.e. >4R to 6R), it will not give any significant reduction in the tunnel elastic closure due to face advance. The results from a parallel plane strain analysis are also plotted on Figure 10 for both the unsupported and instantaneous supported cases. From this comparison, it can be seen that the plane strain analysis does not simulate the progressive excavation-support placement interaction near the advancing face. This is clearer when comparing the stresses developed in the liner for both cases. Figure 11 shows the support segment stresses as both the tunnel face and lining placement are advancing. The stresses developed in Case I represent about 46 percent of the plane strain case (i.e. instantaneous excavation and support placement), while for Case II, the stresses in the lining segment are less than 20 percent of the plane strain case. It should also be noted that when the face has advanced 6R ahead of the lining segment of interest, the stresses have reached their maximum value and no significant increase occurs after further face advance.

COMPREHENSIVE EXAMPLE

The preliminary design analyses for a 2.67 m outside diameter tunnel to be excavated by a tunnel boring machine at a depth of 93 m through a siltstone formation that may exhibit squeezing will be outlined to indicate the full scope of the axisymmetric elastic-plastic-creep simulation package. The characteristics adopted for the siltstone from the available data and technical literature were: E = 210 MPa; $\nu = 0.4$; $\gamma = 2000\ kg/m^3$; $K_a = K_\ell = 1$ (uniform initial stress condition, but higher K values possible); yield range of 0.7 to 1.0 MPa; Von Mises yield criterion; and Singh and Mitchell's (1968) creep relationship with parameters determined for London Clay (Bishop and Lovenbury, 1969).

$$\dot{\varepsilon} = Ae^{\alpha D}\left(\frac{1}{t}\right)^m$$

with $A = 0.1255 \times 10^{-3}$, $\alpha = 0.4815 \times 10^{-3}$ and $m = 1.03$ (t in days and D in psf).

A precast concrete segmental liner (170 mm thick) with a soft backing (19 mm of Ethafoam) was considered for the preliminary design (see Figure 2 for schematic) with the following properties: $E_{concrete} = 21 \times 10^3$ MPa; $\nu_{concrete} = 0.2$; $E_{Ethafoam} = 1.0$ MPa; and $\nu_{Ethafoam} = 0.45$. A soft backing behind the liner was considered desirable for the following reasons:

1. Allows for some closure associated with squeezing ground

conditions;
2. Provides some tolerance for variations in tunnel diameter;
3. Reduces localized stress concentrations;
4. Provides some waterproofing action; and
5. Should eliminate the need for grouting.

A total overburden weight analysis appeared reasonable for the ground conditions involved as there were impermeable water-barrier materials above the crown. This is a conservative assumption, and some effective weight analyses were also completed for comparison purposes. The liner segments were assumed to be placed 5.5 m from the advancing face (5 hours after the face has passed point where liner segment placed for anticipated boring rate). It should be noted that some preliminary plane strain analyses for K_ℓ = 2 indicated complete collapse with entire yielding, while the K_ℓ = 1 case involved only minor yielding around the opening.

The unlined advancing face case was considered for three conditions with the axisymmetric elastic-plastic-creep simulation:

Case 1.1 Total weight, yield of 1.0 MPa.
Case 1.2 Total weight, yield of 0.7 MPa.
Case 1.3 Effective weight, yield of 0.7 MPa. (hydrostatic stresses must be added to this condition)

The radial deformations of the invert (fairly uniform closure for $K_a = K_\ell = 1$) are summarized in Table II and indicate that the advancing face influences are minimal about 2.7 m from the face. The total weight closures are significantly greater than those for the effective weight case.

The lined advancing face case, in which the liner segment is placed 5.5 m from the face, was considered for two conditions:

Case 2.1 Total weight, yield of 0.7 MPa.
Case 2.2 Effective weight, yield of 0.7 MPa. (hydrostatic stresses must be added to this condition)

The liner compressive stress results are given in Table III. Very low uniform compressive stresses are developed with time (averaged for two triangular elements representing liner thickness) as indicated in Table III. However, as the creep closure has not taken up all of the soft Ethafoam backing thickness after 10 days (Table IV), it would be prudent to carry out longer creep runs to check if the compressive stress increases much more when the soft backing is "used up". It is clear that the delay in liner segment placement and soft backing results in very low liner compressive stress levels.

TABLE II

RADIAL DEFORMATIONS OF INVERT, MM, NO LINER

(one half closure in mm)

		DISTANCE FROM FACE IN METRES				
		2.7[c]	1.8	0.9	0	-0.9[d]
Case 1.1	5 hours[a]	27	24	17	7	2
	10 hours[b]	36	36	29	12	4
Case 1.2	5 hours	27	24	17	8	3
	10 hours	36	37	31	14	5
Case 1.3	5 hours	9	8	6	3	1
	10 hours	11	11	9	5	2

Notes: a. 5 hours to excavate.
b. 5 hours to excavate plus further 5 hours.
c. 3.6m, 4.5m and 5.5m similar.
d. Ahead of face.
e. Yielding was minor for the three conditions and the stresses tended to relax with creep.

TABLE III

LINER COMPRESSIVE STRESSES WITH TIME, kPa

CASE 2.1 TOTAL WEIGHT		CASE 2.2 EFFECTIVE WEIGHT	
TIME (DAYS)	LINER STRESS (kPa)	TIME (DAYS)	LINER STRESS (kPa)
0.199	69	0.13	20
0.34	99	0.35	42
0.57	132	0.70	64
0.94	172	1.26	86
1.54	208	2.16	108
2.50	243	3.62	128
4.04	275	5.96	147
6.55	303	9.73	163
10.0	327	10.0	164

TABLE IV

ADDITIONAL DEFORMATIONS AFTER 10 DAYS, MM

DESCRIPTION	CASE 2.1 TOTAL WEIGHT	CASE 2.2 EFFECTIVE WEIGHT
Inward radial rock movement due to 10 days creep following liner installation	16	10
Inward radial liner movement due to 10 days creep following liner installation	0.2	0.1

CONCLUSIONS

The axisymmetric elastic-plastic-creep simulation allows a wide range of tunnel design factors to be considered, particularly the actual advancing face situation and delays in lining placement. Limitations in the current simulation package are mainly related to the lack of input constitutive information for multi-axial creep and the need to use supplementary plane strain analyses to consider discontinuities.

It would appear that finite element method simulations are now far in advance of our ability to provide ground characteristics. Hopefully it will be possible to have the computing and testing in closer harmony by the time of a Third Conference.

The financial support for this work provided by the Natural Sciences and Engineering Research Council, Canada, is gratefully acknowledged.

REFERENCES

A.W. Bishop and H.T. Lovenbury - Creep characteristics of two undisturbed clays, Proceedings, Seventh International Conference on Soil Mechanics and Foundation Engineering (1969) 1 29-37.

E.T. Brown and E. Hoek - Trends in relationships between measured in-situ stresses and depth, International Journal of Rock Mechanics and Mining Sciences 15 (1978) 211-215.

R.N. Craig and A.M. Muir Wood - A review of tunnel lining practice in the United Kingdom, TRRL Supplementary Report 335, Transport and Road Research Laboratory, Crowthorne (1978) pp. 211 + figures.

J.J. Emery, E.A. Hanafy and J.A. Franklin - Creep movements associated with excavations in rock, Large Ground Movements and Structures, Pentech, London (1978) 387-410.

J.A. Franklin and O. Hungr - Rock stresses in Canada; their relevance to engineering projects, Rock Mechanics Supplement 6 (1978) 24-46.

G. Gudehus (editor) - Finite Elements in Geomechanics, Wiley, London (1977) pp. 573.

R.E. Ranken, J. Ghaboussi and A.J. Hendron, Jr. - Analysis of ground-liner interaction for tunnels, UMTA-IL-0043-78-3, NTIS, Springfield (1978) pp. 441.

E.C. Robertson - Viscoelasticity of rocks, International Conference on the State of Stress in the Earth's Crust,

Elsevier, Amsterdam (1963) 181-220.

A. Singh and J.K. Mitchell - A general stress-strain-time function for soils, Journal of the Soil Mechanics and Foundation Division, ASCE 94 (1968) 21-46.

E.L. Wilson - Structural analysis of axisymmetric solids, AIAA Journal 3 (1965) 2269-2274.

O.C. Zienkiewicz - The Finite Element Method, third edition, McGraw-Hill, London (1977) pp. 787.

STRESS RE-DISTRIBUTION AROUND TUNNELS AND SHAFTS RESULTING IN SMALL GROUND MOVEMENTS

Nyal E. Wilson, Professor of Civil Engineering

McMaster University, Hamilton, CANADA L8S 4L7

ABSTRACT

There is a philosophical difference between the design of temporary structures and the design of permanent structures. Temporary structures, such as access shafts for tunnel boring machines, have a very short life-span and are constantly being inspected by engineers during the course of construction of the main project. In addition, the soil conditions are adequately known as excavation proceeds, rather than deriving soil information by interpolation from adjacent bore-holes.

Basically, the Contractor uses an empirical design which has been refined for the individual soil conditions; the purpose being to develop as economicalatechnique as possible for the temporary work which does not form part of the finished project. Consequently, it is possible to design these temporary structures with small factors of safety, on condition that the structures are regularly inspected and the deformation measured. In the case of access shafts, it is desirable to provide rigidity at the corners of the shaft, with the horizontal walers having sufficient deflection to give the ideal soil/structure interaction. As the soil movements are strictly limited, the earth pressures are minimised and, consequently, there are small ground movements.

The shape effect for cylindrical tunnels and the scale effect for vertical shafts have significant influences on reducing the earth pressures acting on these structures. As the earth pressures are drastically reduced by the re-distribution of stress within the soil mass created by the arching effect, it is a consequence that the surface movements are small.

A shaft, 20 ft x 20 ft x 75 ft deep, was constructed in dense till, with a layer of dense sand near the base of the shaft. The purpose of the shaft was for the installation of a tunnel boring machine for a 13-foot diameter tunnel.

The method of shaft construction was soldier piles and lagging, with many intermediate horizontal walers. The earth pressures on the structure were minimal as the dense till carried the stresses by redistribution.

The trapezoidal loading pattern, as proposed by Peck (1969), for long cuts is related to the depth of the cut. The trapezium is reduced at the base of the shaft from the basic "at-rest" value due to the presence of undisturbed till below the base. Likewise, in the horizontal direction, the stresses are reduced due to the presence of undisturbed dense till on both sides of the shaft. The requiement that the soil is undisturbed and dense must be maintained by providing stability to the soldier piles at the corners of the shaft; this is particularly important for shafts with "holes" for tunnel-boring machines.

The design used "lintel-type" loading in the horizontal plane developed from the reduced pressure diagram given by Peck (1969). The "lintel-type" loading is consistent with the shape effect at the base of the shaft where the distance to undisturbed dense soil is small.

Even after the "eyes" for the tunnel were cut, the lateral soil pressures on the shaft were small as the stability of the corner soldier piles was carefully maintained. The maintenance of corner stability is the "key-stone" to obtaining the minimal earth pressures on the buried structure.

An extensive programme of measurements was organized, with the measurements to be taken both by the Contractor and the Consulting Engineer. The purpose of the measurements being to verify that the design assumptions had been achieved and that the structure would not be overstressed.

Having taken all the precautions about the stability of the piles so that the design could be based on the minimum earth pressures, it was found that the horizontal movements of the buried structure were barely measureable and, consequently, the surface ground movements were insignificant.

Fundamentally, an economical design had been achieved

using the advantages of soil/structure interaction.

PHILOSOPHY OF DESIGN FOR TEMPORARY STRUCTURES

Temporary structures, such as equipment access shafts for tunnel construction, are usually designed by the Contractor, with the design approved by the Consulting Engineer. In Canada, it is not common for the temporary work to be designed by the Consulting Engineer as it reduces the flexibility of the Contractor to use his own procedures on the project. The responsibility for the safety of the entire project remains with the Contractor, while the Consulting Engineer ensures that the safety of adjacent structures is maintained, as well as safety for the temporary work.

Generally, the Contractor specialises in a particular type of construction and has obtained considerable field experience in the temporary construction associated with the main contract. This empirical knowledge is then used to construct temporary work as economically as possible.

Temporary structures have a short time duration and experienced personnel are always on the project to record progress and the behaviour of the structures. Under such constant engineering supervision, it is permissible to have a reduced Factor of Safety for the temporary work, recognising that, in the case of an overstress, it is possible to mobilize men and materials quickly to strengthen the structure. As most engineering materials have a yield, without a rapid decrease in strength above the yield, it is possible to design a buried earth structure where the soil/structure interaction allows the soil to re-distribute the earth pressures and the steel to carry the required loads for stability.

In addition to the constant professional supervision, the temporary structure can be designed and modified as construction proceeds by recognising the actual soil conditions. The preliminary design is usually based on the information from a small bore-hole, whereas the shaft is being constructed inside a full-size excavation which provides the exact soil information.

This philosophy of design precludes any problems with ground-water control which must be tacked on its own merits with a conservative Factor of Safety, as the knowledge of ground-water flows and pressures is seldomly adequately given by a few bore-holes.

PURPOSE OF ACCESS SHAFT

A tunnel-boring-machine was proceeding with the mining of a long 13 foot (4 m) diameter sewer tunnel; the sewer has a finished concrete diameter of 8 feet (2.5 m) and acts as a main trunk sewer on the outskirts of Toronto. The machine, which had already mined a considerable length of tunnel was due for re-conditioning: it was decided to construct an access shaft to remove the machine and install a new machine for the remainder of the project.

The tunnel-boring-machine was mining through a layer of dense sand till at a depth of 67 feet (20 m) below the surface; above the sand till, there is a layer of very dense clay till.

The shaft requirements for access of the tunnel-boring-machine was a clear shaft opening of 16 feet by 16 feet (5 m by 5 m), with the depth to the floor of the shaft of 74 feet (22.5 m); Figures 1 and 2 show the details. The shaft was constructed of soldier piles and timber lagging (Plate A). There was no flow of water in the dense tills as the Contractor had already installed de-watering equipment in preparation for the tunnelling operation.

SOIL CONDITIONS

Considerable geotechnical knowledge is known about the soils around Toronto as there has been extensive deep tunnelling construction. It is known that the glacial tills - both clay tills and sand tills - are extremely dense. There is insignificant flow of water through the clay tills and these can be considered as "impermeable"; however, the sand tills may hold pockets of water, but are not usually connected to a supply of water. The main problem with deep excavations is the occurance of boulders; in fact, eskers and boulder pavements have been encountered in some locations.

The Contractor was advised to take caution that the soils did not become softened by surface run-off. These precautions were taken, together with precautions to avoid heavy surface loading by adjacent construction equipment.

BASIS FOR DESIGN

The earth pressure distributions on deep excavations and tunnels have been extensively documented by Pack (1969). The background for this work was obtained by field measurements recorded during the construction of the Chicago subway in the 1940's, with considerable substantiation of the empirical data

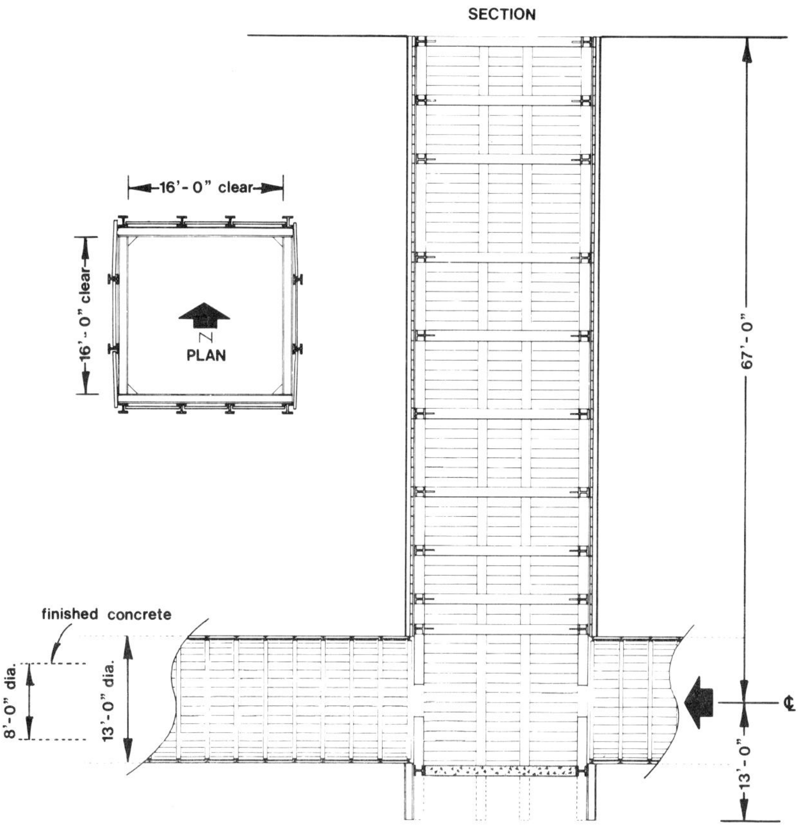

FIGURE 1

Shaft Details

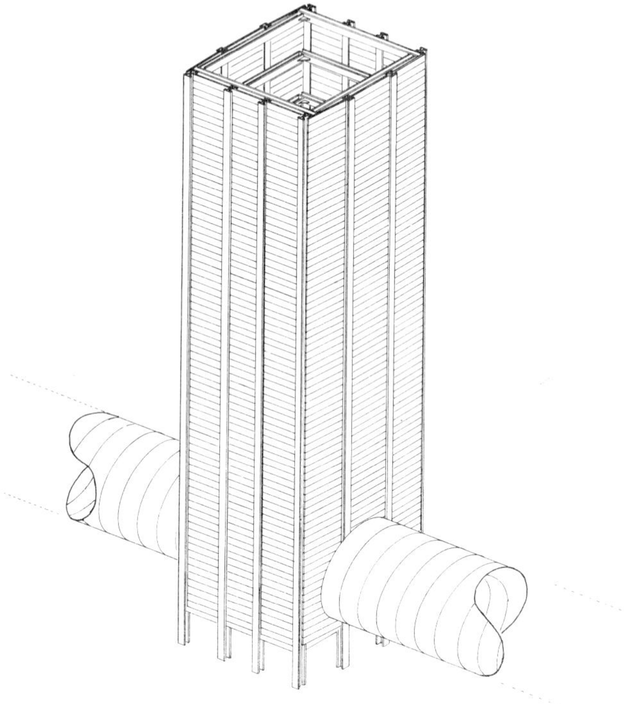

FIGURE 2
Shaft Details

PLATE A

Looking down the shaft during the construction of the floor slab.

in the intervening years.

In the case of the extremely hard clay tills which were encountered for most of the depth of this access shaft, it can be related to the properties of the stiff-fissured clays as discussed by Peck. As the clay till is so dense, it is possible to design, for temporary structures, on the minimum values of the earth pressures proposed by Peck.

Considering stiff clays, it is recommended that the earth pressure diagram has a uniform intensity for the depth of the shaft - with a linear reduction at the top due to surface effects and a linear reduction at the bottom due to the "stiffness" effect at the base (Figure 3). In the case of surcharge loading due to the weight of construction equipment at ground surface close to the shaft, a modification to the upper portion of the earth pressure diagram is necessary.

It is recognised that the basis for design for long cuts considers that the floor of the cut in stiff clays provides stiffness so that arching takes place under the floor of the cut. In empirical design, it is recognised that the basis for design of deep tunnels considers that the soil above and below the tunnel constitutes an arch to reduce the earth pressures on the tunnel lining (deLory and Crawford, 1975; Golder Associates and MacLaren Limited, 1976). The theoretical designs have large margins of safety to take care of unknowns, such as the relative stiffness of the soils. As the engineering properties of the glacial tills are becoming known (Klohn, 1965), it is possible to incorporate these properties into the behaviour of buried structures. Fundamentally, in tunnel design, it is the interaction between the soil and the structure which provides an empirical and economical design.

Previous field measurements on a braced cut in dense sand indicated that the inward movements of the walls of the excavation allowed considerable variation in the earth pressure distribution, even to the final extent when the movements became excessive and the wall failed with very little arching of the soil (Scott, Wilson and Bauer, 1972). In that braced cut, the movements were in the order of 8 inches (0.2 m) due to the combined "compressibility" of the steel struts and the wood blocking or, more particularly, the method of installation of the wood blocking; wood blocking was used - although driven hardwood wedges had been specified - to allow for inaccuracies in the pile-driving operations.

From other field observations, it is recognised that the soil in the floor of the excavation provides passive restraint to the movements of the walls. If this soil should lose its

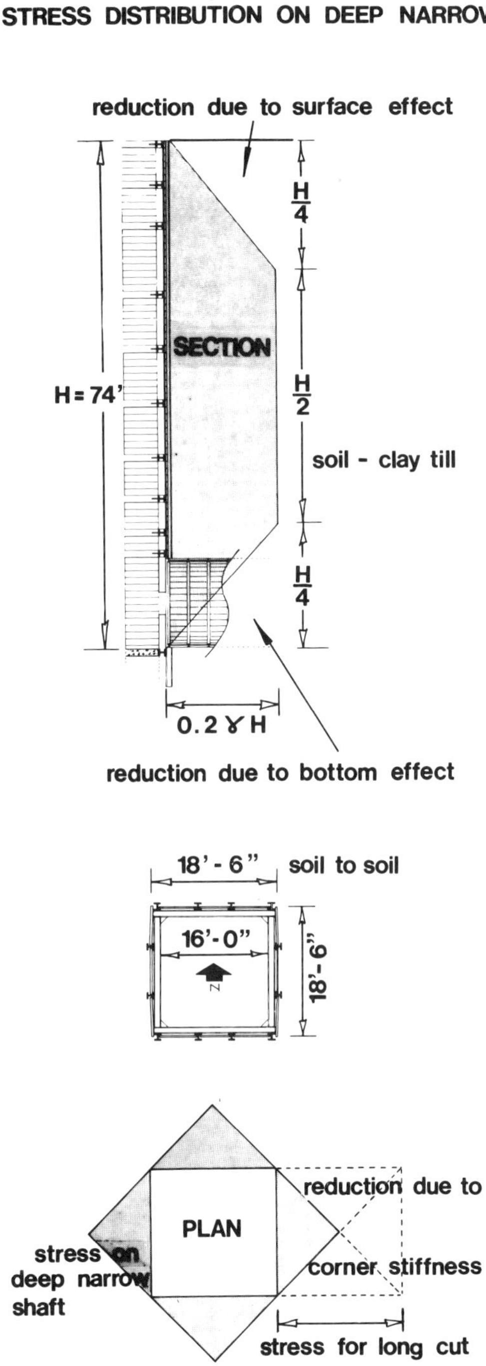

FIGURE 3

Earth Pressure Distribution

integrity, i.e., by water seepage causing the soil to soften or become "quick", then the stability of the entire structure is lost (Wilson, 1979).

STRESS RE-DISTRIBUTION AROUND A NARROW SHAFT

The floor at the bottom of the shaft provides a stiffness to promote arching of the soil in a vertical plane and reduce the earth pressure diagram to zero at that level. In a similar way, on the horizontal plane - analogous to the key-stone of an arch - the edges of the shaft provide a stiffness to promote arching of the soil and reduce the earth pressures to zero at the corners - conditional on the stability of the corner piles being maintained. In other words, the earth pressure distribution on the sides of the shaft in the horizontal plane is similar to "lintel-type" loading in brickwork (Figures 3 and 4).

Consequently, there are two functions for the walers; first, the walers must carry the bending in horizontal span across the shaft and, second, the walers must provide the longitudinal thrust to maintain the stability of the corner soldier piles. Thus, the stresses in the walers is analogous to the stresses in a pre-stressed beam due to a combination of compressive strut loadings and bending moments. As the movement of the soil is strictly limited, the bending stresses on the walers are small. The main criterion for design is deflection, which is related to the movement of soil behind the walers.

As the walers are carrying a combination of compressive strut loads and bending moments, the slenderness of the walers must be considered. This concern was satisfied by installing gusset plates at the corner junctions and by welding the walers to the intermediate soldier piles for the purpose of shortening the "column-length".

CUTTING OF "EYES"

Perhaps the most critical stage of construction is during the cutting of the "eyes" for the passage of the tunnel-boring-machine; there is no realistic design approach and it must be done empirically.

As the shaft was excavated and lagged as a complete box shaft, it was possible to install heavy diagonal braces outside the proposed openings for the "eyes" prior to the removal of the lagging (Figure 5). These diagonal braces do not place additional bending stresses on the corner soldier piles as they have been installed after the box shaft has been

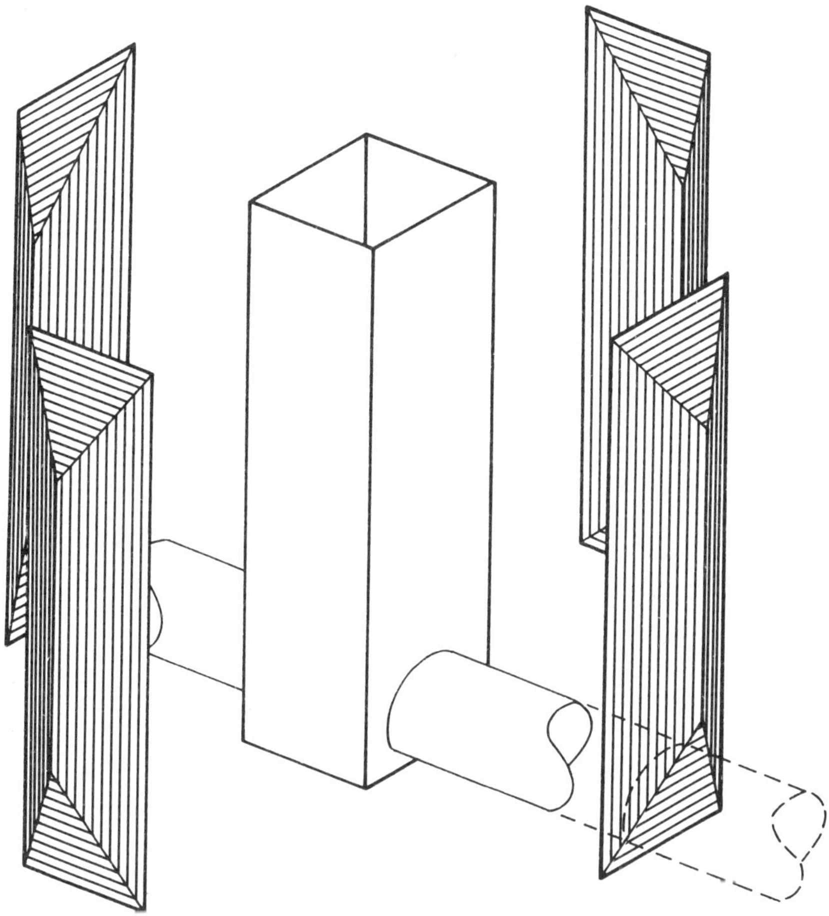

FIGURE 4
Earth Pressure Distribution

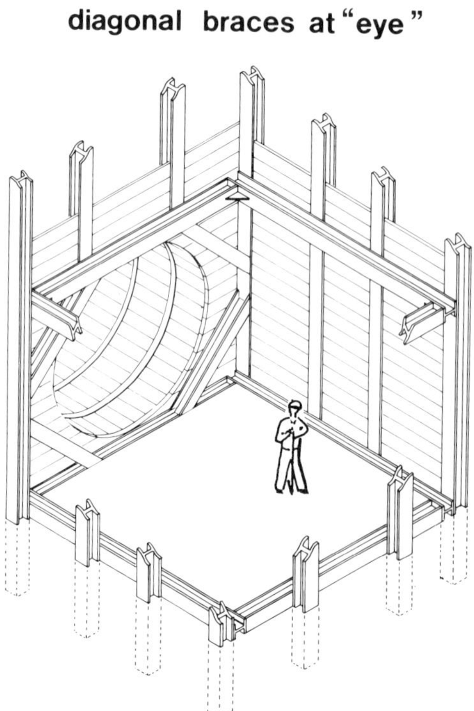

FIGURE 5
Diagonal Braces at "Eyes"

excavated to full depth. The diagonal braces serve the purpose of stabilizing the corner piles as the tunnel "eyes" are being cut. Unfortunately, this tunnelling excavation has to be done by hand for a distance sufficient to install the tunnel-boring-machine. Consequently, the over-break for these hand operations may be greater than the over-break during machine operations, and it occurs at the most crucial stages in construction. In the case of a deep shaft, the stress reduction due to the stiffness of the floor may extend above the level of the "eye" openings (Figure 3).

FIELD MEASUREMENTS

Based on extensive experience which the Contractor had obtained on previous shaft construction, the Contractor was convinced that the earth pressures on the walls of the shaft were low, provided the stability of the piles was maintained.

The concern of the Consulting Engineer for the Owner was safety and the careful control of ground movement. Consequently, on extensive programme of field measurements was developed to record the deflections of the piles from the vertical and the horizontal deflections of the walers. Unfortunately, actual stress measurements were not made.

It was recongised that the deflections that were being measured related to movements which would take place after a particular waler was installed and as excavation proceeded. The assumption was used that the horizontal structural members were unstressed at the time of installation; this assumption is valid as it is based on Peck's (1969) recommendation that the earth pressure at the base of the cut is zero - during construction, the base of the excavation is proceeding downwards. A more significant assumption was made regarding the structural steel members for the walers; it was assumed that the steel, as installed, was perfectly straight when, in fact, previously-used structural steel walers were used for almost all the temporary work.

The actual measurements recorded in the field gave movements less than one eighth of an inch (3 mm) - which can be considered inconsequential. It is anticipated that these movements would increase slightly with time due to stress relaxation within the dense tills; these tills are over-consolidated and the horizontal in-situ stresses may be greater than the over-burden stresses. However, the life span for this shaft was six months approximately and no movement trend could be detected.

CONCLUSIONS

An economical empirical design was made for the temporary structure, based on an earth pressure distribution for long cuts which was modified for deep narrow shafts. As the steel carried strut loadings as well as bending stresses, the deflections were insignificant and, consequently, the ground movements were insignificant.

Provided the stability of the soldier piles was rigidly maintained, the soil/structure interaction reduces the structural requirements and a very economical design can be made. The design was based on the Contractor's extensive experience in deep narrow shafts where it has been shown that the earth pressures are extremely small.

ACKNOWLEDGEMENTS

The excellent cooperation between the Contractor and the Consulting Engineer for the Owner is fully appreciated; this cooperation extended throughout the design and field measurement stages.

The Contractor was S. McNally and Sons, Ltd., Hamilton and Toronto, represented by Mr. P.S. Hobden.

The Consulting Engineer was DeLeuw Cather, Canada, Ltd., represented by Mr. M. O'Brien and Mr. J. Dekker in the design office, and by Mr. J. Washington in the field.

The Owner is the Ontario Ministry of the Environment.

The author developed this design, on behalf of the Contractor, in conjunction with Nyal E. Wilson & Associates, Ltd. of Hamilton, Ontario; this firm specializes in geo-technical/structural research and design.

REFERENCES

deLory, F.A. and Crawford, A.M. (1975) - Measurements and Study of Factors Influencing a Tunnel Lining in Very Dense Clayey Till Soil , Department of Civil Engineering, University of Toronto, Canada.

Golder Associates and MacLaren Limited (1976) - Tunnelling Technology , published by the Ontario Ministry of Transportation and Communications.

Klohn, E.J. (1965) - The Elastic Properties of a Dense Glacial Till Deposit , Canadian Geotechnical Journal Vol. II, No. 2, pp. 116-140.

Peck, R.B. (1969) - Deep Excavations and Tunnelling in Soft Ground , Proceedings of Seventh International Conference on Soil Mechanics and Foundation Engineering, State-of-the-Art Volume, Mexico City, pp. 225-290.

Scott, J.D., Wilson, N.E. and Bauer, G.E. (1972) - Analysis and Performance of a Braced Cut in Sand with Large Deformations, Canadian Geotechnical Journal Volume 9, No. 4, pp. 384-406.

Wilson, N.E. (1979) - Cofferdam Failure at Thunder Bay, report of November 1979 to Counsel for the Underwriters (private communication).

SESSION V

Chairman: Professor F W Williams, MA, PhD, CEng, MIStructE, MICE
Department of Civil Engineering and Building Technology
University of Wales Institute of Science and Technology

SESSIONAL PAPERS

THE EFFECT OF UNSTABLE TERRAIN AND OLD COAL WORKINGS ON THE CHOICE OF THE ROUTE FOR THE PROPOSED CHAPEL-EN-LE-FRITH AND WHALEY BRIDGE BY-PASS

David S Buist, MA,MSc, PhD, MICE, MIHE, CEng, MIGeol, FGS

Specialist in Engineering Geology and Geotechnics

Reginald A Naylor, MIME, AMIMunE, CEng

Coal Mining Specialist (Ex National Coal Board Subsidence Engineer), Department of Transport, Midland Road Construction Unit, PO Box No 6, County Offices, Matlock, Derbyshire DE4 3AG

INTRODUCTION

This proposed dual two lane all-purpose by-pass, a part of the London-Carlisle-Glasgow-Inverness Trunk Road (No 12), will provide an alternative route for through traffic, which at present travels via the existing A6 trunk road (a single carriageway, generally 8 m in width), passing through the towns of Chapel-en-le-Frith and Whaley Bridge or via the A5002 principal road between Buxton and Whaley Bridge (Long Hill). The arrangement is shown on Figure 1, on which the published (northern) route of the Chapel By-pass is also indicated. In addition to the improvement of traffic conditions within these towns the by-pass will remove a high proportion of the vehicles at present using the scenic Long Hill route overlooking the Goyt Reservoir.

Various routes have been investigated for the Chapel By-pass. The preferred one is the so-called 'northern' (published), while the 'southern' is a recognised alternative (Figure 2). Both have been the subject of detailed engineering geological mapping, supported by extensive photogeological and mining studies.

The general area around Chapel-en-le-Frith is characterised by the presence of a number of relatively large landslips, the best known of which is that of Mam Tor (Figure 1), which is active, affecting the A625 (Sheffield) road. Apart from large

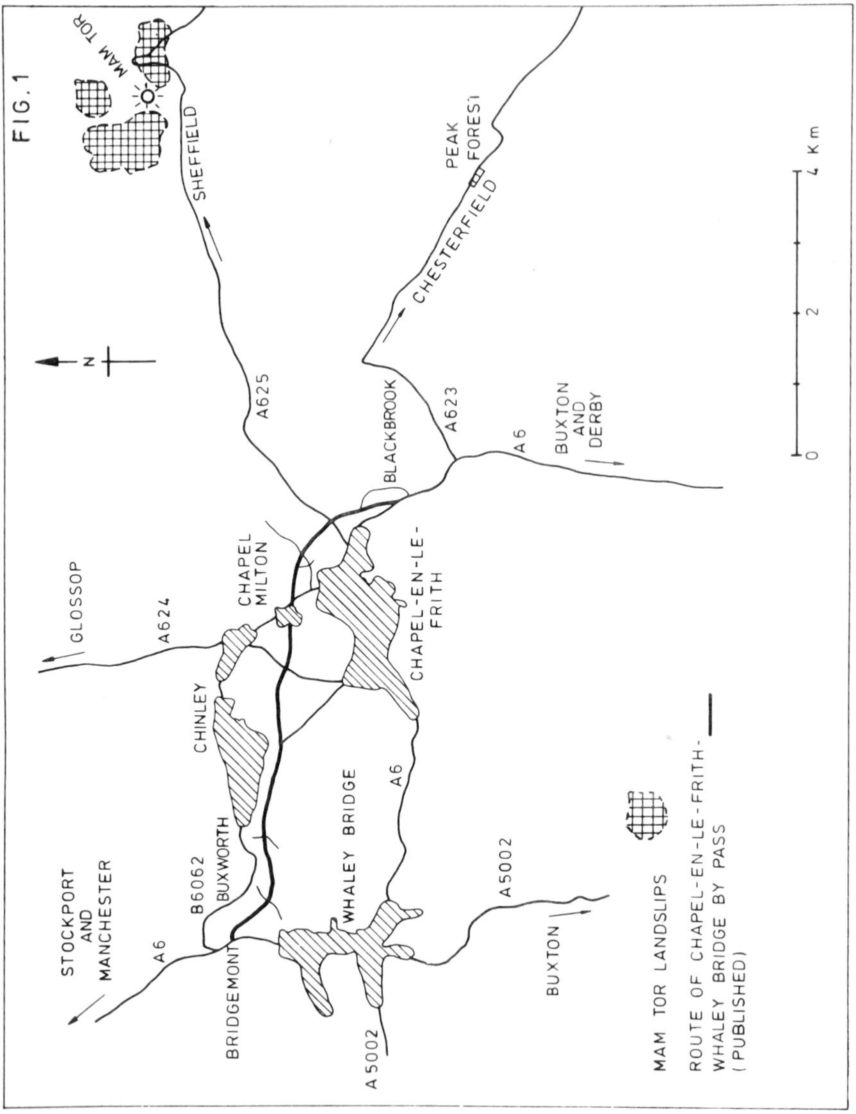
MAM TOR
SHEFFIELD
PEAK FOREST
CHESTERFIELD
N
A625
BLACKBROOK
A623
A6
BUXTON AND DERBY
0
2
4 Km
GLOSSOP
A624
CHAPEL MILTON
CHAPEL-EN-LE-FRITH
CHINLEY
STOCKPORT AND MANCHESTER
A6
B6062
BUXWORTH
WHALEY BRIDGE
A6
A5002
BUXTON
BRIDGEMONT
A5002
MAM TOR LANDSLIPS
ROUTE OF CHAPEL-EN-LE-FRITH-WHALEY BRIDGE BY PASS (PUBLISHED)

FIG.1

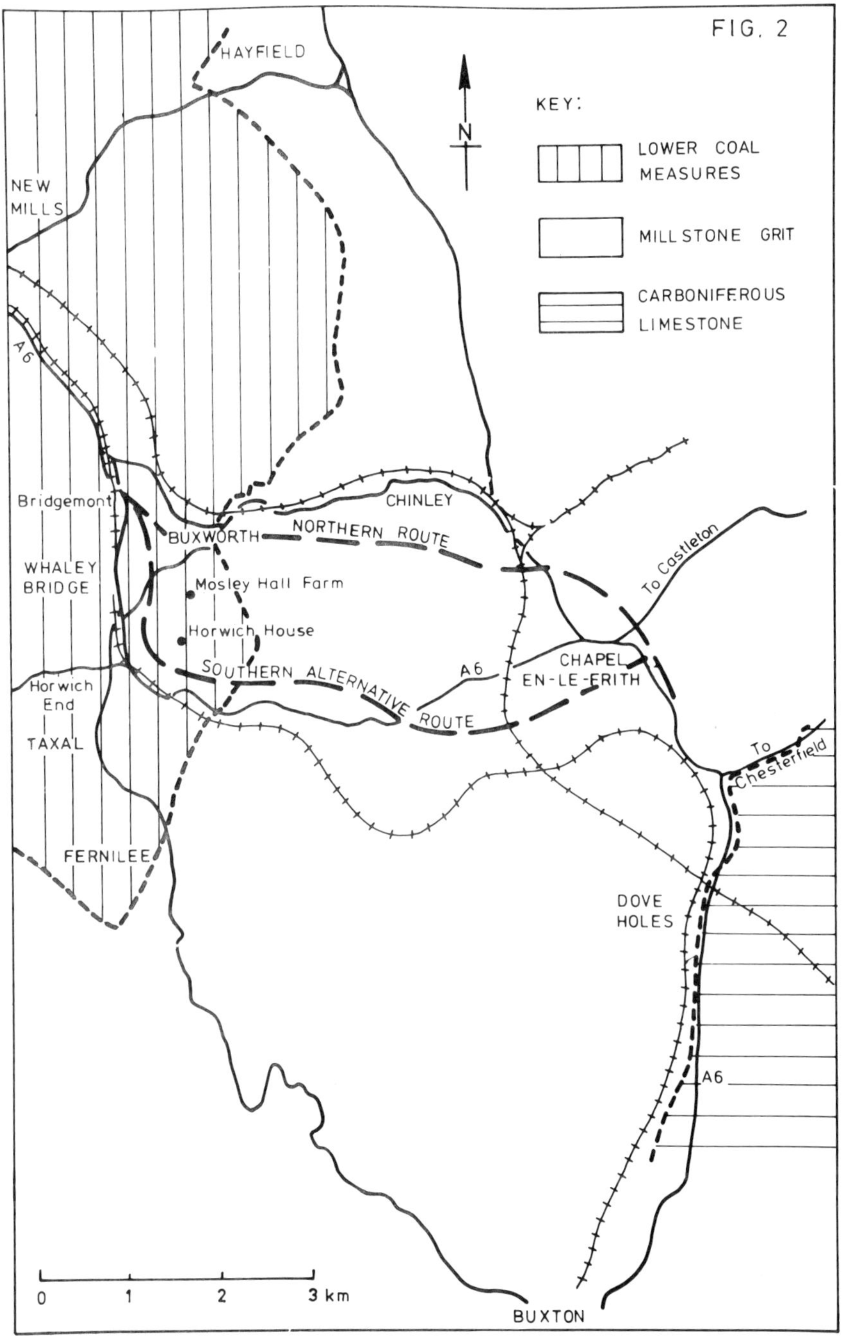
FIG. 2
KEY:
LOWER COAL MEASURES
MILLSTONE GRIT
CARBONIFEROUS LIMESTONE
N
HAYFIELD
NEW MILLS
A6
Bridgemont
BUXWORTH
NORTHERN ROUTE
CHINLEY
WHALEY BRIDGE
Mosley Hall Farm
Horwich House
To Castleton
SOUTHERN ALTERNATIVE ROUTE
A6
CHAPEL EN-LE-FRITH
Horwich End
TAXAL
To Chesterfield
FERNILEE
DOVE HOLES
A6
0 1 2 3 km
BUXTON

landslips, a number of localised unstable areas in materials of glacial origin (till, sand and gravel) can be found in the valley floors of the area.

The horizontal alignment of the original northern route of the Chapel By-pass required modification in order to avoid or reduce impingement on areas which might represent potential engineering hazards. These included minor slipped areas of drift and a relatively large ancient landslip in glacial sand and gravel, which had been partially extracted. Even with this modified line, there still remained the problem of old coal workings at shallow depths in seams in the Bridgemont area north of Whaley Bridge.

There would be potentially more serious geological complications with regard to the southern route. Problem soils include glacial laminated clay. There would be a 22 m cut with a possible sliding wedge failure situation and, perhaps most important of all, very extensive old coal workings (on a much larger scale than those underlying the northern route) in three seams, the most significant of which, the Yard, is up to 1.4 m in thickness. In addition, the Roosdyche (site of special scientific interest) (Johnson, 1963), a presumed sub-glacial overflow channel, would be extensively affected by the alignment of this southern route.

Consideration of the geological and mining factors associated with the two routes suggests that the northern one is preferable. Fortunately, this route also has environmental, economic and other advantages.

GENERAL GEOLOGY OF THE AREA OF NORTHERN AND SOUTHERN ROUTES

Stevenson and Gaunt (1971) have discussed in detail the geology of the Chapel-en-le-Frith area, encompassing that covering both routes. The first author originally surveyed the area of the by-pass in 1952-1957. He re-examined a 250 m zone of interest covering an extended northern route in December 1970 and the results are embodied in an unpublished report (1971), produced for the Midland Road Construction Unit. The brief was to highlight any localities where the geology might lead to constructional difficulties. Evans (1972) briefly discussed the results of this re-survey. One of the present authors (DSB) subsequently produced a comprehensive report on the geology of an extended northern route, while the other author (RAN) compiled a mining report covering both routes in 1978.

A very broad outline of the geology of the area of the two routes is indicated in Figure 2. The solid formations involved are the Millstone Grit (Namurian) and overlying Lower Coal Measures (Westphalian).

A more detailed description of the geology of the northern route is given below. Although the southern route is a recognised alternative, its geology will not be discussed in detail, though reference will be made to mining aspects later in the account.

GEOLOGY OF THE NORTHERN (PUBLISHED) ROUTE

Over the length of the proposed by-pass, Lower Coal Measures rocks (shales, mudstones, sandstones, coals and seat-earths) and strata of Millstone Grit age (shales, mudstones and sandstones) are overlain by superficial deposits ('drift') of Recent and Pleistocene age. A generalised vertical section of the solid geology is given as Figure 3. These latter drift deposits consist of river alluvium and terrace deposits, glacial sand and gravel, boulder clay, also a restricted development of laminated clay. From the proposed Bridgemont interchange (Figure 4), the route of the by-pass encompasses Lower Coal Measures rocks, including the so-called Woodhead Hill rock. This section includes coal seams, largely worked at shallow depths. Near Bothomes Hall, the solid rock is overlain by alluvial and river terrace deposits. Glacial sand and gravel and boulder clay occur along the Black Brook Valley. Near Buxworth, the Millstone Grit Series is met with; this series includes such horizons as the Chatsworth, Roaches, Kinderscout and Shale Grits. Millstone Grit strata pass to the termination of the by-pass at Sheffield Road. 'Boulder clay' (till) overlies the bedrock in most sections of the route between Buxworth and Sheffield Road.

The Chatsworth Grit outcrops in Crist Quarry, where workings have resulted in relatively large areas of waste material being tipped. Kinderscout Grit outcrops south of Chapel Milton, in the existing railway cutting. Shale Grit outcrops in a small quarry near the point where the proposed by-pass crosses Sheffield Road. Areas of alluvium are present at the termination of the by-pass at Blackbrook.

The till, which underlies much of the proposed route, is a stiff, stoney clay with rare large boulders. The deposit occurs mainly as a featureless valley-fill, but erosion has produced oversteepened banks in it, which are in places unstable. In general, thicknesses of about 6 m are normal.

River terrace deposits are of restricted occurrence and are of sand and gravel, in places coarse. The terraces show flat tops lying at up to 4 m above the flood plain of the River Goyt and up to 1-3 m above that of the Black Brook. Alluvium, consisting of silts, sands and fine gravels, is present in both Black Brook and Goyt valleys.

The area of Lower Coal Measures between Bridgemont and

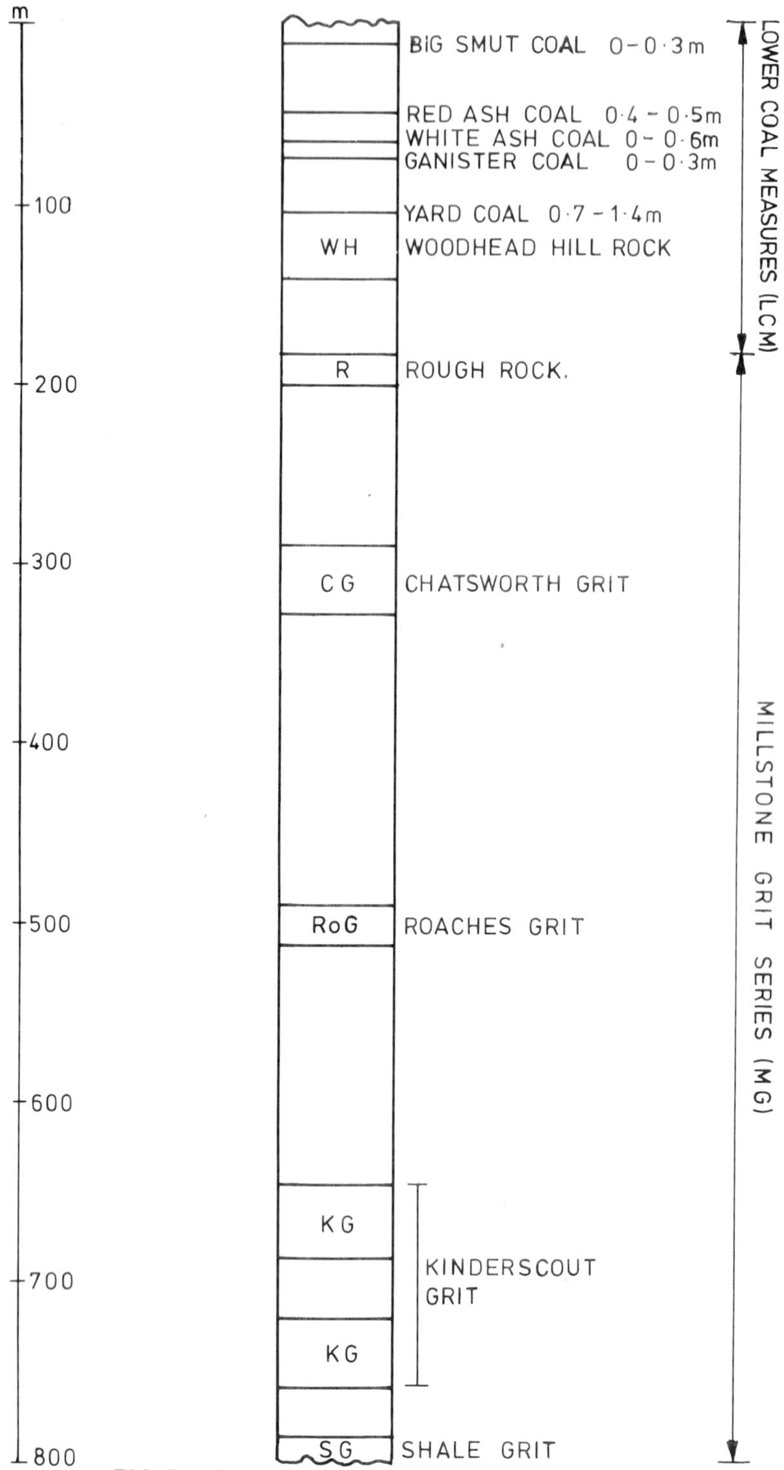
m
100
200
300
400
500
600
700
800
BIG SMUT COAL 0–0·3m
RED ASH COAL 0·4 – 0·5m
WHITE ASH COAL 0 – 0·6m
GANISTER COAL 0 – 0·3m
YARD COAL 0·7 – 1·4m
WH
WOODHEAD HILL ROCK
R
ROUGH ROCK.
CG
CHATSWORTH GRIT
RoG
ROACHES GRIT
KG
KG
KINDERSCOUT GRIT
SG
SHALE GRIT
LOWER COAL MEASURES (LCM)
MILLSTONE GRIT SERIES (MG)

FIGURE 3

Buxworth is affected by a number of faults, some of which have relatively large throws. This area is characterised by a strong folding of the rocks along an approximately north-south synclinal axis, the Goyt Trough. The eastern end of the by-pass shows little structural complexity - faulting occurs and where present, throws are small.

LARGE SCALE LANDSLIPS IN THE GENERAL AREA OF THE BY-PASS

A number of extensive landslipped masses can be found in the area of IGS 1 inch to 1 mile Sheet 99, described in the memoir by Stevenson and Gaunt (1971; their Figure 24). Reference to the latter shows that most of the slips lie to the north of a line drawn through Chapel-Castleton-Hathersage. They occur mainly in valleys within the Millstone Grit outcrop (eg Franks and Johnson, 1964; Johnson, 1965; Lounsbury, 1962). Practically, they can be sub-divided into three types, namely: collapsed scarps, down-dip slides and rotational slumps. Collapsed scarps can be individual or compound, are well fragmented and occur where steep shale slopes occur below massive sandstone scarps, particularly those of Chatsworth or Kinderscout Grit; dislodgement scars are irregular. Down-dip slides are mainly compound types, many of which include areas of unbroken ground and are found where stream incision has produced steep slopes in a similar direction to the tectonic dip; dislodgement scars are typically craggy and wide. Rotational slumps are mainly individual landslips, are well fragmented and most examples occur on steep slopes composed of Shale Grit and/or Mam Tor Beds which have slipped downwards and outwards in front of concave, high but relatively narrow dislodgement scars.

Most of the landslips are inactive but that of Mam Tor (rotational slump type), situated east of Mam Tor itself (Figure 1), is active and very well-known because of its severe and protracted effect on the stability of the A625. The problems result from the steepness of slope of the slipped mass, from the inherently unstable nature of the Edale Shales of which it is largely composed and from the presence of springs at the base of the Mam Tor Beds. The high moisture content of the general area of the slip has been shown by Buist and Penn (1977), using Ektachrome infra-red-photography from a light aircraft, with a hand-held 35 mm camera.

SMALL SCALE LANDSLIPPED AREAS ALONG THE NORTHERN ROUTE OF THE BY-PASS

These would cause potential engineering difficulties to the scheme and the route therefore had to be revised where necessary to avoid them. They are all in material of glacial origin, either till ('boulder clay') or sand and gravel, mainly in the former. Their locations are indicated in Figure 4 and

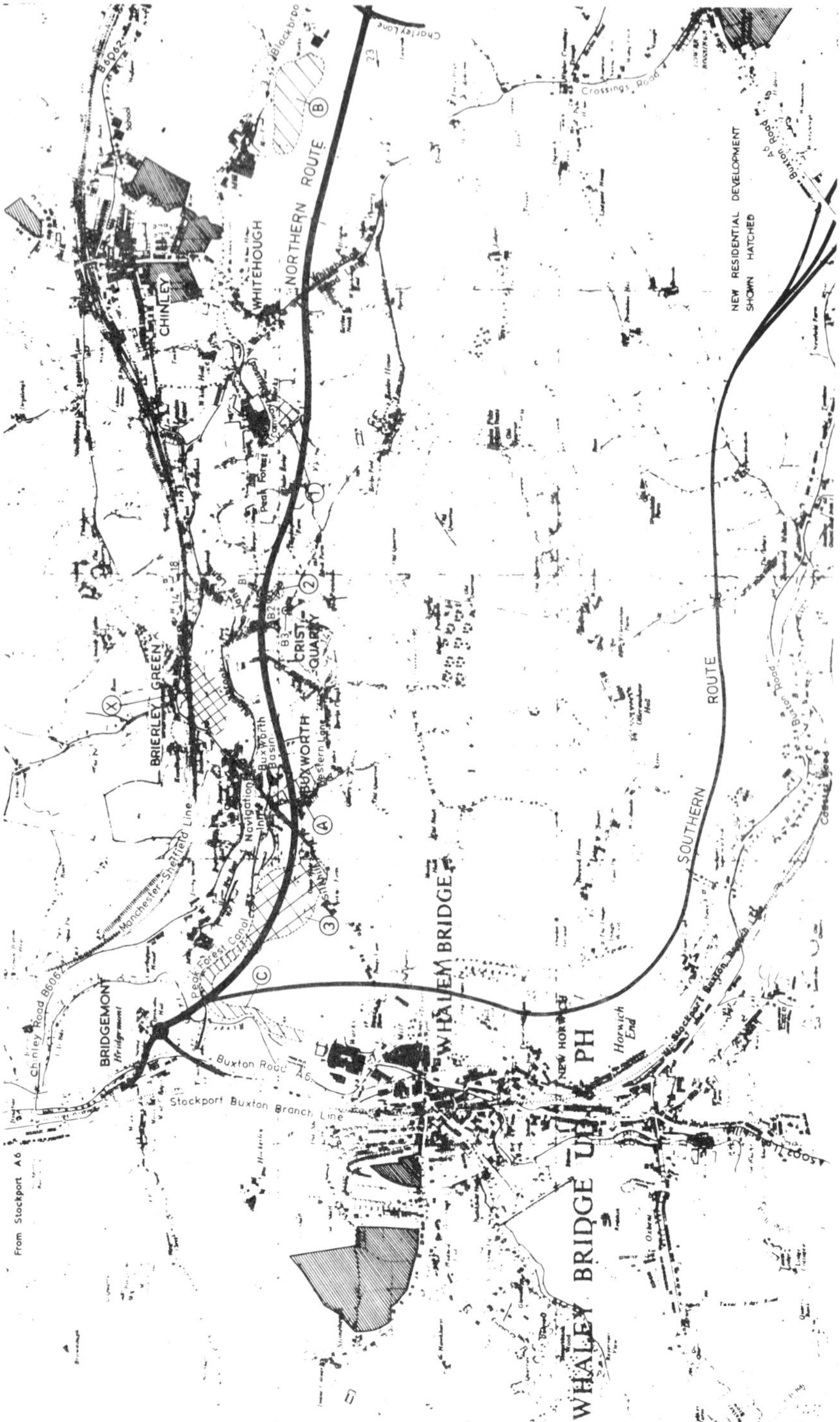

CHINLEY
WHITEHOUGH
NORTHERN ROUTE
SOUTHERN ROUTE
ROUTE
BRIERLEY GREEN
BUXWORTH
Buxworth Basin
Navigation Inn
Western Lane
CRIST QUARRY
Peak Forest
WHALEY BRIDGE
BRIDGEMONT
Peak Forest Canal
Manchester-Sheffield Line
Stockport Buxton Branch Line
Buxton Road A6
Chinley Road B6062
From Stockport A6
Crossings Road
Charley Lane
Horwich End
NEW RESIDENTIAL DEVELOPMENT SHOWN HATCHED

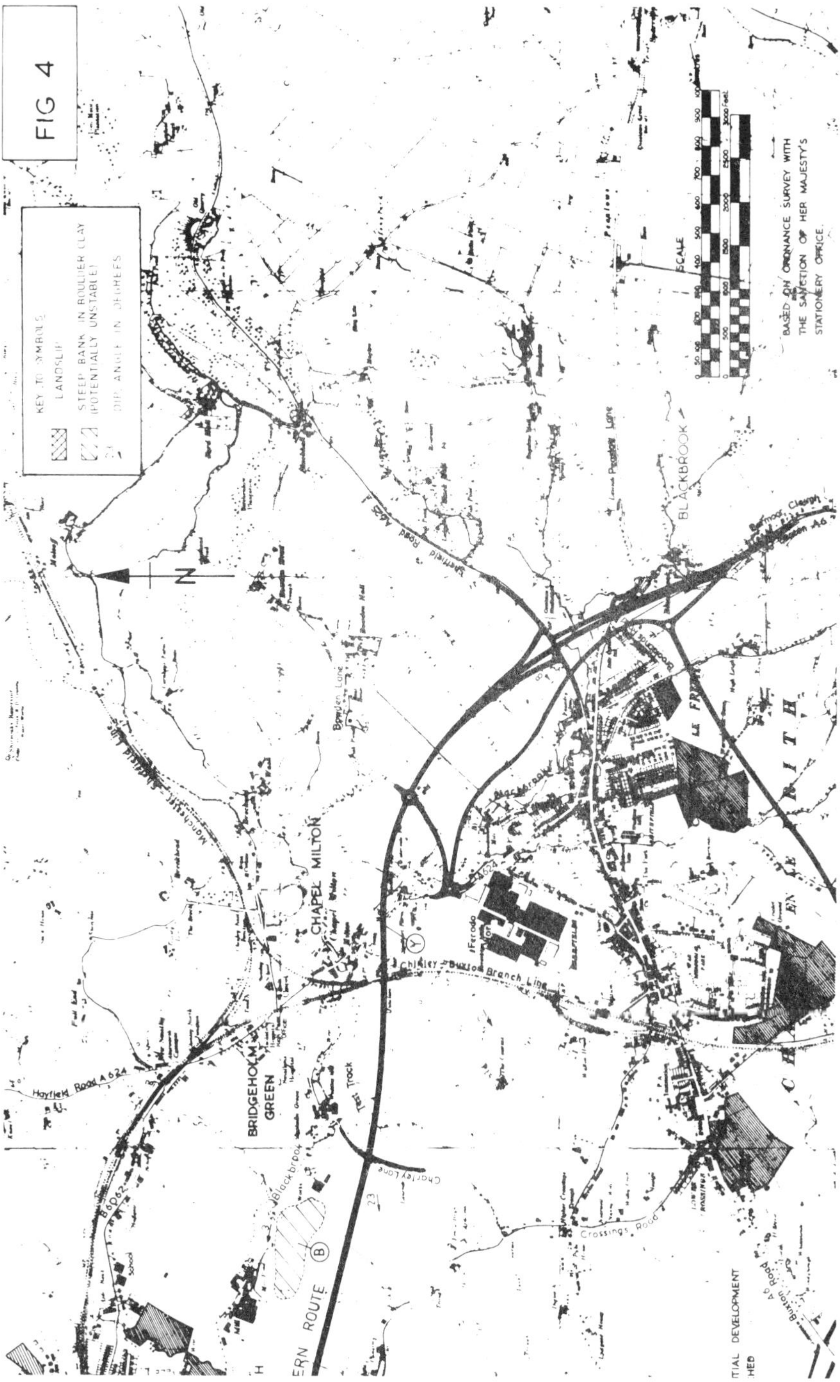
FIG 4
KEY TO SYMBOLS
LANDSLIP
STEEP BANK IN BOULDER CLAY (POTENTIALLY UNSTABLE)
DIP ANGLE IN DEGREES
SCALE
BASED ON ORDNANCE SURVEY WITH THE SANCTION OF HER MAJESTY'S STATIONERY OFFICE.
CHAPEL MILTON
BRIDGEHOLM GREEN
BLACKBROOK
Hayfield Road A 624
Sheffield Road A625
Crossings Road
Chinley - Buxton Branch Line
Bowden Lane
Charley Lane
Test Track
Ferodo
ERN ROUTE
TIAL DEVELOPMENT
HED

they are discussed below, from east to west.

Landslipped area in till, south of Whitehall Works (Position 1 in Figure 4)

The slopes here are 12°; very extensive movement has taken place within a relatively small area (Plate 1). This has been continued in recent times, as occasional trees show 'leaning'. Failure on such flat slopes must be due to a strain - softening effect, leading to progressive failure. Such a result, where the shear strength decreases to the residual value, can happen after many years in overconsolidated clay.

Landslipped area in drift (probable glacial), east of Crist Quarry (Position 2 in Figure 4)

This is interpreted as a past movement of a 'veneer' of surface drift, together with some weathered rock (sand). The drift is probably in part of morainic origin. The unconsolidated material has then slipped on the steep scarp slope of the Chatsworth Grit. The area appears to be currently stable, though cutting into it might cause further movement.

Three boreholes (B1, B2 and B3 in Figure 4) were drilled from near the base of the top of the slope in 1971. B3, drilled outside the slip, with no evidence of movement, passed through 0.3 m of topsoil into sand (with thin bands of grit), then into grit (with a thin clay band). B2 intersected the same thickness of topsoil, thence into sand (with gravel, and grit boulders), completed in shale/mudstone. B1 passed through 0.3 m of topsoil, thence into sand and gravel (occasional grit boulder), completed in grit. The sites of B1 and B2 showed evidence of past subsoil movement, and therefore were within the landslip.

Relatively large landslipped area in glacial sand and gravel, West of Buxworth (Position 3 in Figure 4)

This measures approximately 400 m by 200 m, and has been partially extracted. It is a portion of the mass cited by Jowett and Charlesworth (1929), lying across the valley of the Black Brook. The landslipped area shows extensive evidence of past movements. Apart from this, a section in Buxworth railway cutting shows approximately 16 m of 'clean brown sand with lenses of coarse gravel'. The deposit may have had a deltaic, or, more probably, a morainic origin (Jowett and Charlesworth, op. cit.).

POTENTIAL UNSTABLE AREAS ON OR NEAR THE NORTHERN ROUTE

Three of these were located, shown on Figure 4; they are all composed of glacial till ('boulder clay').

Plate I 12° Slopes in glacial drift (boulder clay), south of Whitehall Works

Area immediately adjacent to Western Lane, Buxworth (Position A in Figure 4)

This is intersected by the northern route, which could not be conveniently moved to avoid it.

Relatively large area west of Bridgeholm Green (Position B in Figure 4)

This is located some 130 m north of the northern route.

Area immediately east of the River Goyt and Canal (Position C in Figure 4)

This is intersected by the northern route, but could not be conveniently moved to avoid it.

PAST EVIDENCE OF GROUND FAILURES IN THE VICINITY OF THE NORTHERN ROUTE

Two areas have been noted, one in the railway area south of Brierley Green (1866) and the other on the A624 in the Chapel Milton area (1970).

Railway failure south of Brierley Green in 1866 (Position X on Figure 4)

This occurrence in the Buxworth area comprised failure of a minor bridge, five-span stone viaduct and a 'slipping' of sixteen acres of land towards the Black Brook (Williams, 1878). The autumn of 1866 marked a period of intensely heavy rainfall which caused widespread railway failures in the United Kingdom. In this Brierley Green example, it appeared that glacial deposits had probably slid on Millstone Grit shales, as a sort of mud flow. The situation was perhaps exacerbated by the presence of nearby, recently worked coal workings - a shaft connected with these may have filled up and overflowed into the glacial area. British Rail were consulted regarding this failure but they had no records for 1866.

Road subsidence on the A624 south-east of Chapel Milton in 1970 (Position Y on Figure 4)

It appeared that movement of a clay wedge on a grit surface had occurred, due to increase in moisture content. There was no direct evidence that the excavation for a new retaining wall had caused the movement directly, though this could have altered drainage conditions. The area within which movement occurred is shown in Plate 2.

Plate II Road subsidence on A624 south-east of Chapel Milton (1970)

PHOTOGEOLOGICAL INTERPRETATION

Stereoscopic examination of aerial photographs of a specially-flown survey of the northern route was carried out. The main object of this was to indicate any areas of unstable ground on and in close vicinity to the line. The previously noted problem areas were confirmed, but no further suspect ground was noted.

Aerial photographs covering the southern route have also been examined with particular reference to unstable ground and any evidence of old coal workings. With regard to the former, none was found, but features relating to the latter (old shafts etc) were noted.

THE EFFECT OF OLD COAL WORKINGS ON THE NORTHERN AND SOUTHERN ROUTES

Brief history of the coal workings

The western halves of the proposed routes overlie a small disused coalfield, sometimes called the 'Goyt Valley Coalfield'. Coal was worked from early 17th century to early this century (60 years ago), and is believed to have been extracted on a small scale in the 16th century. The peak of coal mining was during the second half of last century, when the Buxton Lime Company was a large mining concern.

Coal was used mainly for limeburning, and in the early stage most farms in Derbyshire would have one or two lime kilns for agricultural purposes. In 1650 it is known that fourteen lime kilns were in operation at Doveholes.

The area has hilly terrain and materials were transported by pack-horses up to the 18th century. Turnpike roads in the 18th century enabled the use of carts. A breakthrough was made at the end of the 18th century when Peak Forest Canal was constructed linking Buxworth and Whaley Bridge with Manchester and New Mills. At the same time the Peak Forest Tramway was built connecting Buxworth with Doveholes (one of the main sources of limestone). Two railways were later constructed in the 1860's.

At Buxworth a complex of lime kilns was built (pre 1800) near to the coal sources. According to the 6" IGS geological plan these kilns were sited near the outcrop of the main caol seam, the Yard Coal, overlain by glacial deposits.

This siting may have been a coincidence, as explained later in the account. Other limekilns were built along the canal side.

FIGURE: 5

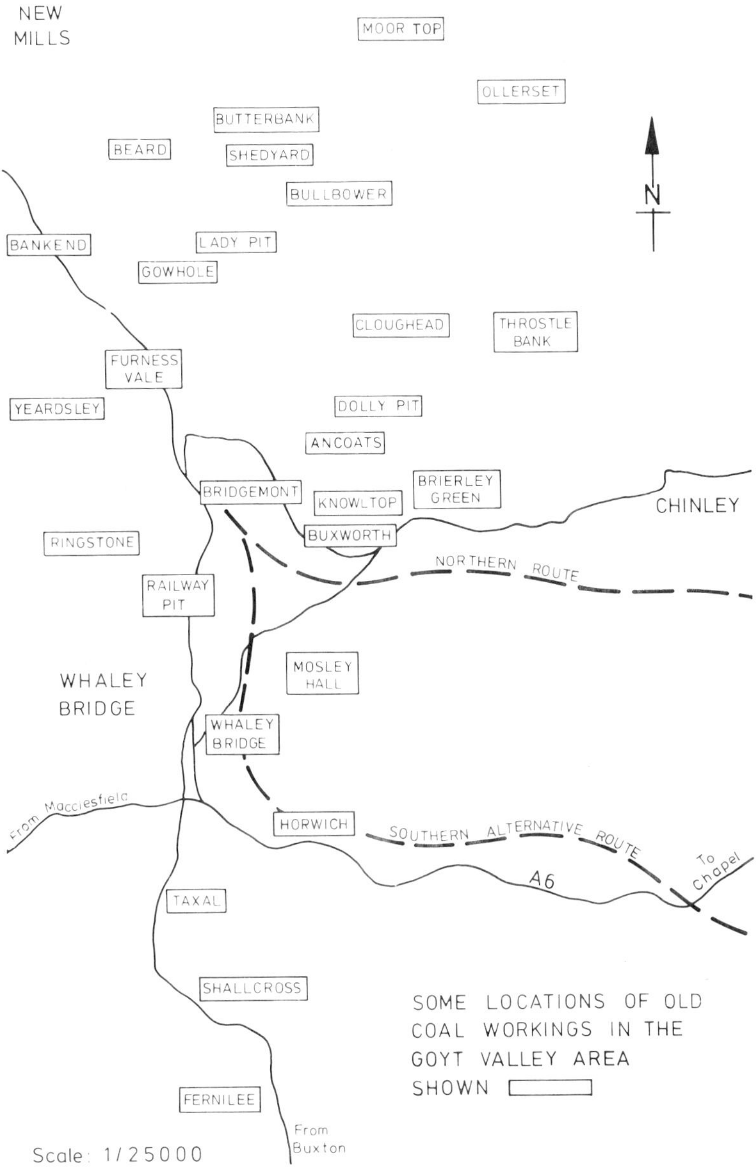

SOME LOCATIONS OF OLD COAL WORKINGS IN THE GOYT VALLEY AREA SHOWN ▭

Some points from the investigation of old coal workings

A coal adit at Mosley Hall Farm (Figure 2) entered an outcrop of the Red Ash coal, but there were no records of shallow coal workings in the Red Ash and White Ash coal seams within a particular area. At Mosley Hall, workings in the lowest seam dated from 1868. It is possible that the Red Ash and White Ash may have been worked at shallow depths before 1868. Shortly after, in 1872, the first statute laws were in force for the keeping of abandoned mine plans.

The 1/2500 first edition Ordnance Survey plan shows a tramway connecting this adit with the Buxworth limekilns. Before 1800, the Yard Coal was worked at several places north of Buxworth and these were sources of coal for limekilns in the Buxworth area.

In the Goyt Valley coalfield, coal was worked from three seams, namely the Red Ash (average coal thickness 18 inches), the White Ash (average coal thickness 16 inches) and the Yard Coal (average thickness 36 inches).

Normally, thin coal seams of 16-18 inch thickness would not be worked. Records do not indicate the working of ancilliary minerals with either of these seams, in the vicinity of the two proposed routes.

After collecting information of old shafts and the old coal workings from the prime sources, four months were spent on historical research, seeking similar information from other sources.

Relative information was gathered indirectly by reading up literature on local history, limestone works, and the building of canals and railways. Leading information may be found by indirect means. After the rather thorough search it was obvious that many old coal workings and old shafts were unrecorded.

Despite the nationwide appeal by the Department of Energy which followed the Lofthouse mine disaster in 1973, there were three cases where plans of old coal workings were retained. The National Coal Board had no record of these particular coal workings.

Sources other than the primary ones need to be examined. For this project only a small amount of additional information was gathered in the research. The NCB Mining Records Officer had covered this area rather thoroughly.

Coal seams and the presence of old workings in the vicinity of the northern route at Buxworth and Bridgemont

The northern route passes over productive coal seams from 400 to 1400 m chainage. A geological fault divides this length at point 'B' as shown in the section, Figure 6. Known coal workings are in the Yard Seam from 440 m to 1090 m chainage (worked 1905-9) at depths varying from approximately 78 to 102 m. The Red Ash and White Ash seams are present over the same length at shallower depths, but no workings have been recorded.

To the east of this fault from 1090 to 1400 m only the Yard Seam is present at shallow depth below the glacial deposits. The limekiln complex of Buxworth was situated over the outcrop of this seam which was mostly used as kiln coal. A thorough search for information of workings in the Yard Seam did not gain any indication of old workings.

The productive coal seams are covered with glacial and river deposits varying from 10 to 25 m in thickness. It is unlikely that coal was worked at shallow depths below thick glacial deposits. In such conditions miners would most likely have met two problems firstly, water continuously draining into their workings and secondly, the glacial deposits, some like quicksand, would easily provide inrushes into very shallow coal workings. It would be an incorrect assumption that no old workings exist in these shallow coal seams.

Coal seams and the presence of old workings in the vicinity of the southern alternative route at Whaley Bridge and Bridgemont

The southern alternative route passes over productive coal seams from 440 m to 2690 m chainage (Figure 6).

In the Red Ash, workings (1867-75) extend from 1700 m to 2570 m at varying depths from 0 to 124 metres. Workings in the White Ash (1868-1880) are from 1670 m to 2520 m at depths varying from 30 to 142 m.

The extensive workings (1868-1910) in the Yard Seam are from 440 m to 2690 m chainage with depths varying from 22 to 181 mètres. Workings in these three seams are from 1867 or 1868. Records of old coal workings do not date any workings before 1867.

Comparison of the problems of the old coal workings along the two routes

Table 1 compares lengths of the two routes over known coal workings. The southern alternative route crosses greater lengths of old coal workings than the northern route (Naylor, 1978 unpublished work).

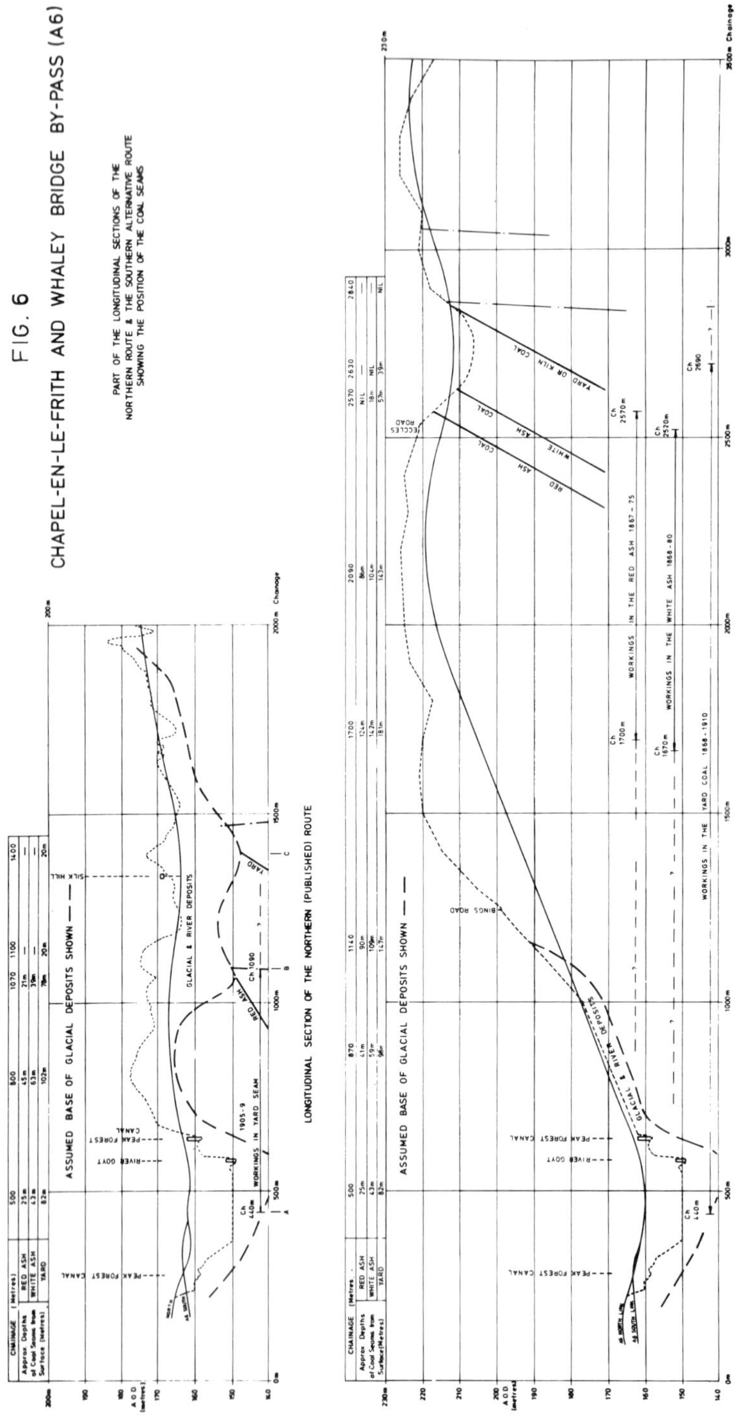

FIG. 6

CHAPEL-EN-LE-FRITH AND WHALEY BRIDGE BY-PASS (A6)

PART OF THE LONGITUDINAL SECTIONS OF THE NORTHERN ROUTE & THE SOUTHERN ALTERNATIVE ROUTE SHOWING THE POSITION OF THE COAL SEAMS

TABLE I Extent of known coal workings below the two routes

	Extent of known workings in metres			
	RED ASH	WHITE ASH	YARD	Total
Northern Route	Nil	Nil	650	650
Southern Alternative Route	870	850	2250	3970

Along the southern route, excavations to depths of 5 metres would be required at the outcrops of each of the three seams. The Red Ash coal would also need to be sealed in the cutting at 2570 m chainage. Two shafts near 1800 m chainage would need to be investigated and later filled in adequately. This route may also encounter several unrecorded old mine shafts.

Along the northern route there would be no need of both the sealing and the extra excavations at coal seam outcrops, because they are covered with glacial deposits varying from 10 to 25 metres thick.

CONCLUSIONS

A number of routes have been proposed for the Chapel-en-le-Frith and Whaley Bridge By-Pass, the principal being the northern (published) route and the recognised alternative southern route. Both these have been examined carefully from the points of view of geology (including photogeology) and mining: socio-economic factors have also been taken into account. Consideration of all these suggests that the northern (published) route is preferable. The present account deals with this route in detail, principally from the geological and mining points of view.

The general area surrounding Chapel-en-le-Frith contains a number of extensive landslips, two of which are active, the best known of which is that of Mam Tor which has caused instability in the A625 road for many years. The northern (published) route is not affected by any large-scale landslips, but its route has been modified to avoid certain small scale slipped areas; even so, other problem areas have unavoidably had to be intersected by the route. These difficult areas are mainly of glacial till ('boulder clay'), with one relatively large former probably morainic glacial sand and gravel. Failure of one of these till areas has occurred with slopes of 10-12°; with still further movement having occurred recently, as evidenced by 'leaning' of trees.

Evidence of ground failures in the vicinity of the northern route was sought and found in the shape of a slip in the A624 in 1970 and an area of movement of 'boulder clay' on a grit surface effecting sixteen acres, including a minor bridge and a fine-span stone viaduct in 1866. The general conclusion is that any slopes in glacial till may become unstable, especially if excavated. The effect of old coal workings on the northern and southern route was examined. It was found that the southern route would cross a much greater length of old coal workings than the northern, published route. There would also be problems in sealing and extra excavation at seam outcrops in the case of the southern route compared with the northern, as the latter would be covered with glacial deposits up to 25 m in thickness.

If additional coal mining information (eg. location of old shafts and plans of old coal workings) is found when searching other sources, it is important that the National Coal Board is notified of the findings. They would appreciate this co-operation as they are responsible for old coal shafts and old coal workings.

ACKNOWLEDGEMENTS

The authors would like to thank Mr A N Brant, BSc(Eng), CEng, FICE, FIHE, DipTE, Director of the Midland Road Construction Unit and Mr C V Underwood, FIMunE, FIHE, Chief Engineer of the Derbyshire Sub-Unit, for permission to publish this paper. Thanks are also accorded to the following:-

Mr. P. Hett, NCB Mining Records Officer, East Midlands.

Mr. I. P. Stevenson, Institute of Geological Sciences, Leeds.

Miss. J. Sinar, Derbyshire County Archivist.

Mr. M. Stakes, Mining Engineer, British Waterways Board, Leeds.

The late Mr. J. Mort (Mining Engineer).

Derbyshire County Library

Bodleian Library, Oxford University.

REFERENCES

D. S. Buist and S. Penn - Low cost aerial photography for terrain analysis, DOE/DTP Construction, 24 (1977), 46-47.

W. B. Evans - Geology of motorways and the role of the Institute of Geological Sciences, Regional Meeting on Engineering Geology, Route Planning, Design and Construction, Leeds, 20-23 September 1971. Quarterly Journal of Engineering Geology 5 (1972) 69-70.

J. W. Franks and R. H. Johnson - Pollen analytical dating of a Derbyshire landslip: the Cown Edge Landslides, Charlesworth. New Phytologist 63 (1964) 209-216.

R. H. Johnson - The Roosdyche, Whaley Bridge: a new appraisal, East Midland Geographer 3 (1963) 155-162.

R. H. Johnson - A study of the Charlesworth landslides near Glossop, north Derbyshire, Transactions of the Institute of British Geographers No 37 (1965) 111-126.

A. Jowett and J. K. Charlesworth - The glacial geology of the Derbyshire dome and the western slopes of the southern Pennines, Quarterly Journal of the Geological Society of London 85 (1929) 307-334.

R. W. Lounsbury - Landslips in the Ashop Valley, Derbyshire, England (Abstract). Special Paper of the Geological Society of America No 68 (1962) 219.

I. P. Stevenson and G. D. Gaunt - Geology of the country around Chapel-en-le-Frith (explanation of one-inch geological sheet 99, new series), Memoir of the Geological Survey of Great Britain (England and Wales) (1971) 444 pp.

M. V. Symons - Sources of information for preliminary site investigation in old coal mining areas. Proc. Conf. Large Ground Movements and Structures, Cardiff (1977) 119-135

F. S. Williams - The Midland Railway: its rise and progress, 4th Edition (1878) 222, 223, 420-422.

APPENDIX

Sources inspected for Coal Mining Information

The prime sources were:-

NCB Mining Records Office, Eastwood, Notts - abandoned coal mine plans, and location of old shafts

Geological plans (6" scale) and memoirs

Institute of Geological Sciences, Leeds

First edition Ordnance Survey 1/2500 and 1/10560 plans

Aerial photographs

The following sources were also inspected for additional coal mining information:-

British Waterways Board

British Rail

ICI (who acquired Buxton Lime Company)

The late Mr. J. Mort (mining engineer) collection of plans of coal workings, coal mining leases, documents and letters

Derbyshire County Library

Derbyshire County Archives - books, and numerous plans of proposed canals and railways

Bodleian Library, Oxford University - inspection of First edition Ordnance Survey 1/2500 plans

Local History Society

Local historians

Derbyshire District Councils

SOME FOOTWALL FAILURE MODES IN SOUTH WALES OPENCAST WORKINGS

G. Walton[1] and H. Coates[2]

[1] Principal, Geoffrey Walton, Consulting Mining and Engineering Geologists, Charlbury, Oxford

[2] Geotechnical Engineer, National Coal Board Opencast Executive, No. 7 South Western Region

INTRODUCTION

This paper considers the stability of footwall slopes in opencast coal workings where the dip of the strata exceeds 15°. Whilst early surface coal mining in Britain was rarely more than 20-25m in depth, new workings have increased in depth and now at times exceed 100m. Footwall slope failures have occurred in many parts of the world and it is important to correctly assess the various constraints to stability especially where coal seams outcrop near major services or structures at the limits of an economic coal deposit. The southern outcrop of the South Wales coalfield provides interesting examples of such constraints and over the last decade it has been possible to examine several cases of footwall collapse and to characterise the modes of failure.

GEOLOGICAL AND GEOTECHNICAL SETTING

The pertinent section of the southern outcrop of the South Wales coalfield is that which lies between Port Talbot and Llantrisant. Figure 1 shows the approximate location of all current and former surface coal mines within the exposed section of the productive Lower and Middle Coal Measures (Westphalian A and B). In this part of the coalfield the dip of the bedding is to the north at gradients ranging from less than 1 in 3 (18°) to more than 1 in 1 (45°). Much of the interburden between the main economic coal seams from the Gellideg to the Two Feet Nine (1) comprises argillaceous strata such as mudstones, siltstones and seatearths rather than sandstones. With the exception of mines where the Gellideg is the pavement seam there is seldom any sandstone in the underburden to the footwall slopes.

Within a number of seatearths persistent intraformational shear zones or clay mylonites (2) are present. These shear zones are sub-parallel to bedding and conform to the northerly dip of the strata. Shearing associated with thrust or lag faulting is commonly found and local, complex repetitions of coal seams occur (1) (see also Figure 4). Jointing is of

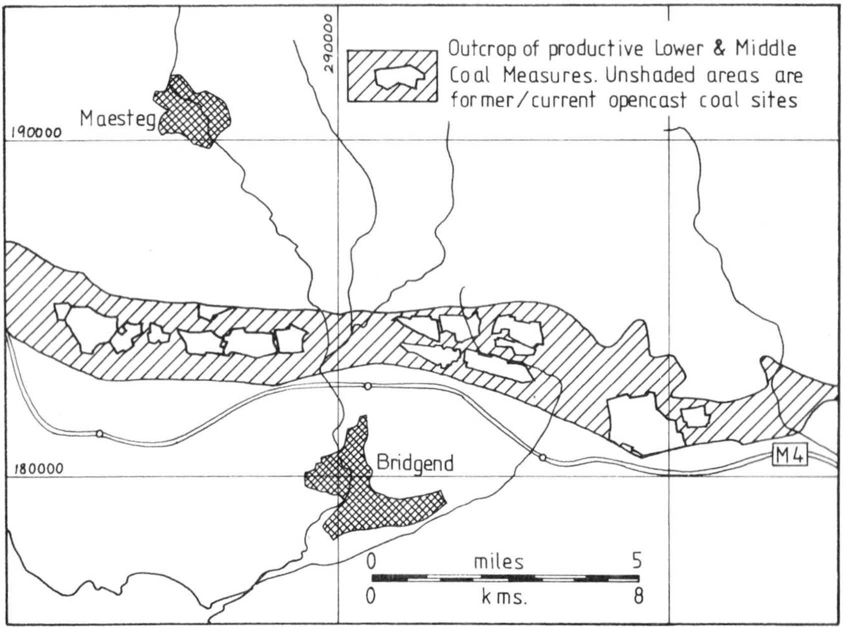

Figure 1 Study area of the southern outcrop of Lower and Middle Coal Measures, S. Wales

three types; two systems of joint sets are approximately perpendicular to bedding and correspond to dip joints (N-S) and oblique joints (NE-SW and NNW-SSE) whilst the third type comprises a shear joint system sub-parallel to the plane of the thrust faults (60/010). The tectonically disturbed zones may be important opencast coal reserves as at times they were not amenable to systematic underground extraction; less complex areas have more commonly been heavily worked. Old mine workings often give rise to joint dilation and a less compact rock mass (3).

The geotechnical constraints on slope stability may be broadly divided into material parameters, structural features and groundwater. Many authors have attempted to characterise the geotechnical properties of coal bearing strata (4)(5). The structural features in the study area are somewhat more complex than most other British coalfields, but the material parameters and groundwater conditions are similar. With a lithology dominated by non-arenaceous materials, rock strength ranges from moderately weak to strong (6). Joint shear strengths are typically 25°-35° for mudstones, but may be more than 45° for siltstones. Lower shear strengths are often found for bedding separation surfaces, whilst intraformational shear zones can have residual shear strengths of 10°-14° and may be present as soft engineering soils.

The major structural controls have been noted above. The intensity of jointing is variable; mudstones are often highly fractured, but the siltstone may have extremely wide spaced joints (6) which yield very large blocks.

The behaviour of groundwater below rockhead is the least predictable of all the parameters associated with stability; flow primarily occurs in joint systems rather than through the rock fabric, and local zones of extension and compression due to subsidence over old workings exert a major control. In general, sandstones, siltstones and coal seams tend to function as aquifers, whilst mudstones, seatearths and especially clay-rich shear zones form aquicludes or aquitards. Overall permeability along the bedding can thus be several orders of magnitude greater than that normal to bedding. Some "bridging" of aquicludes may occur due to overbreak above old workings, or unsealed exploration boreholes. Normally, however, few boreholes penetrate significantly below the intended pavement seam, and in the absence of special control measures, effective drawdown of water pressures beneath the footwall cannot be ensured. Figure 2 illustrates the way in which high water pressures may occur beneath footwall slopes, but not in slopes above horizontal coal seams, even though similar water pressures were found in piezometers before mining.

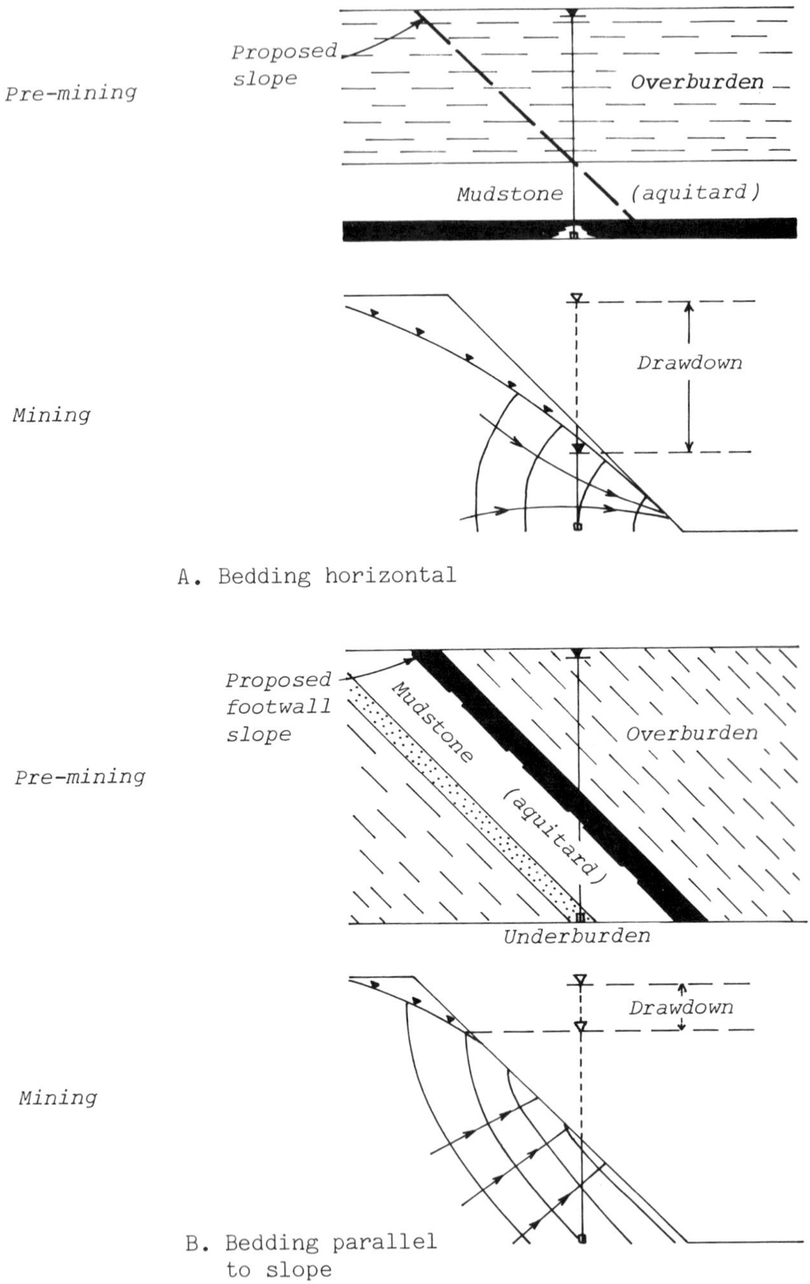

Figure 2 Drawdown effects in mines with horizontal bedding (A) and steep footwalls (B)

SURFACE MINE SETTING

Opencast operations in South Wales have been described previously (7) and general surface mining objectives have been identified by authors classifying different types of opencast mine (8). The Southern outcrop opencast workings are best defined as either steep seam open-pit mines or modified terrace mines. The primary objective in opencast mining is to remove the minimum quantity of overburden for the maximum mineral recovery consistent with safety. Ideally footwall slopes would be excavated parallel to the lowest, or pavement, coal seam. Figure 3 shows diagrammatic cross sections of both narrow and wide mines with excavation confined to the floor of the pavement seam. The layout and development of steep seam mines relies heavily on the siting of the haul road, gradients of which should not greatly exceed 1 in 10; the distance between footwall and hanging-wall crests (Figure 3A) may control the choices available for haul road location, as shown in Figure 3B. Narrow steep-dip mines commonly employ a retreating and reversing ramp technique with the ramp located near the footwall. Wider steep-dip mines may approximate to typical terrace operations with ramps in the spoil and on the advancing face.

It could be inferred from Figure 3B that the narrow steep seam mines expose a larger area of footwall, but this is not always so since wide terrace mines may be deeper.

Instability on the footwall is important for a number of reasons:-

- Coal may be lost or contaminated
- The main haul road may be disrupted, especially in narrow steep-dip operations
- Recession of the crest may extend outside the mine towards existing structures or services, or jeopardize the safety of a spoil dump
- Men and equipment may be at risk and production delayed

Experience over the last three decades has shown that steeper footwall slopes with gradients of >1 in 1.5 (33.7^{o}) commonly collapsed or showed signs of distress when excavations were deeper than 20-25m. Some contractors, therefore, incorporated footwall benching in their design and

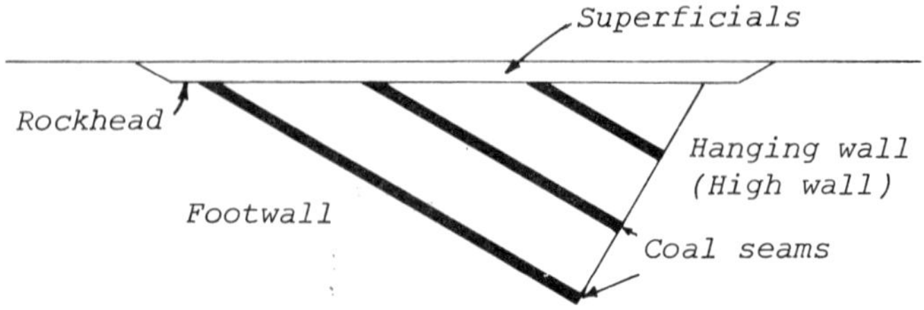

i. Open pit steep seam - Narrow

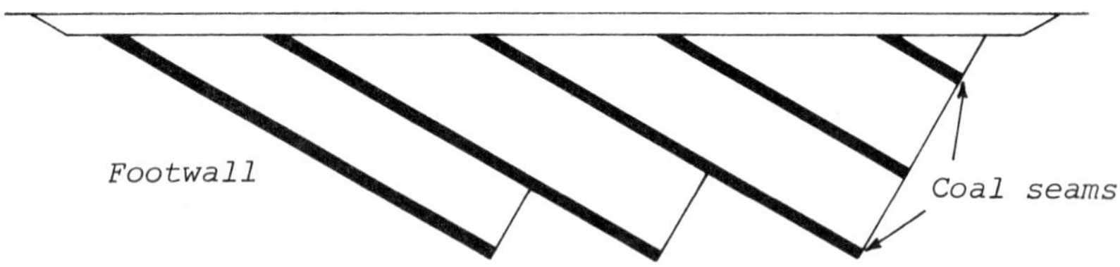

ii. Terrace mine steep seam - Wide

Figure 3A Diagrammatic dip sections of steep seam mines

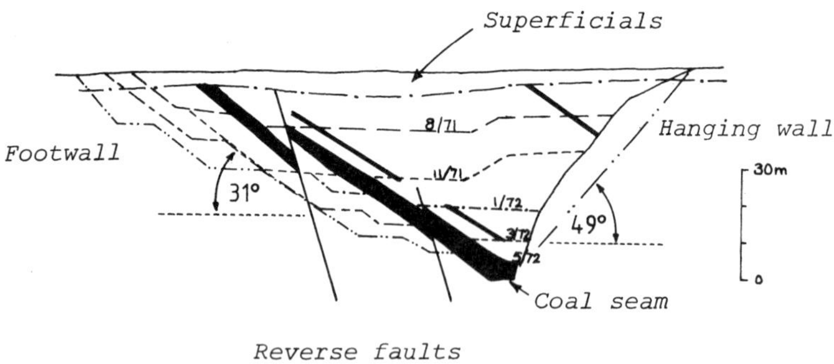

Figure 4 Typical steep seam open pit with footwall benches in underburden

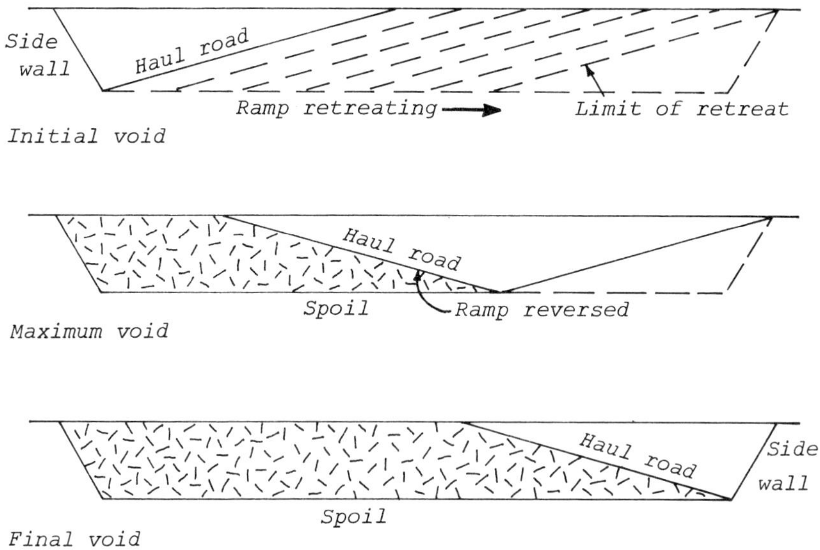

i. Open pit steep seam - Narrow

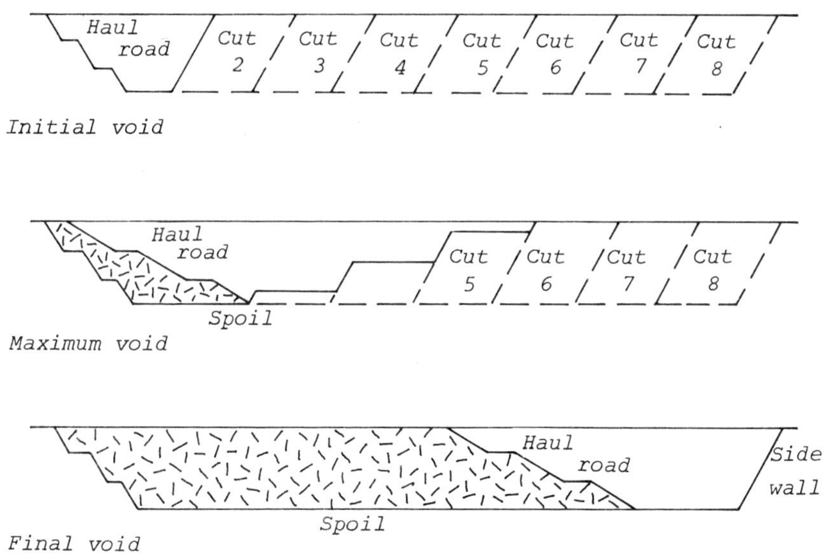

ii. Terrace mine deep seam - Wide

Figure 3B Diagrammatic strike sections of steep seam mines. (Superficials omitted) used this 'under-dig' to locate the haul road. Figure 4 illustrates an example of footwall benching employed in the working of a typical narrow steep-dip deposit and shows the bench slopes and the phasing of the excavation. The spacing and width of footwall benches was empirical and not based on conventional stability analyses.

Footwall slopes inclined at less than 1 in 1.5 to 1 in 2 (26.6^o) rarely gave trouble during the early years of opencast mining, although occasional minor problems occurred when the pavement seam was not totally removed at the outcrop, when small normal strike-faults downthrowing to the north were unexpectedly encountered, or when the dip steepened rapidly at depth. With increasing depth larger and more serious footwall failures occurred on non-undercut footwalls at dips of 1 in 2 to 1 in 3 (18.3^o). The investigation of a large unstable footwall on one mine lead to a better understanding of the modes of footwall failure and a re-appraisal of appropriate analytical methods and site investigation objectives. The findings of this investigation are briefly outlined below.

BRIEF CASE HISTORY

Development

The footwall at the mine was inclined at gradients between 1 in 3.2 (17.4^o) and 1 in 2.7 (20.3^o); the underburden strata is shown in Figure 5. Excavation proceeded to a depth of 30m with no obvious indications of distress within the slope or behind its crest. Cracking, however, appeared up to 20m behind the slope crest when the slope height reached 35m. On deepening the excavation to 40m, pressure ridges developed across the eastern 80m of the 300m long face approximately 10m above the toe of the slope. The eastern section of this ridge continued to deform in a sharp chevron shape, eventually rising 6m above the original mine floor before collapsing, some blocks overturning up-slope. This failure was accompanied by movements at the slope crest, the strata above the pressure ridge subsequently moving down as a sheet and breaking up as it crossed the site of the original buckling. The up-slope sheet or slab apparently moved on a surface parallel to, and approximately 6-7m below, the base of pavement seam. These movements occurred over a period of two months.

To the west the pressure ridges developed as a smoothly curved bulge, consistent with uniform bending rather than hinged or pivoted buckling. Again the bulge rose to 5 or 6m above the original footwall surface, but the whole collapse occurred more rapidly over a period of less than two weeks. Further failures were usually slow moving, but two isolated rapid movements were observed. On one occasion the initial buckling took place during two days, accelerating as the deformation increased and

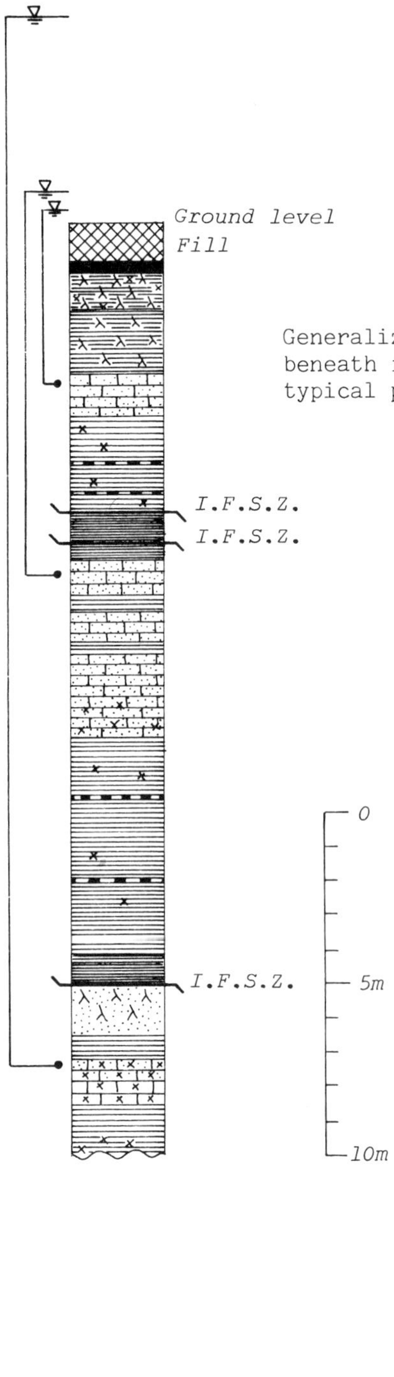

Figure 5
Generalized section of strata beneath footwall slope and typical piezometric levels

eventually failing with a slab slide in which the slope crest slipped from near its original position to the toe of the footwall within a space of five minutes. Another similar collapse occurred in which 4 bulges formed, the up-slope slab sliding down slope very rapidly as the bulges or folds failed. The resulting slip debris assumed the style of a cascade as described in gravity collapse structures in limestone (9).

Further investigations at the edge of a previous collapse showed that both forms of buckling viz. curved bulges and chevron shaped pressure ridges may have resulted from the same underlying mechanism. Within the siltstones which lay 3-4m below the mine pavement, movements occurred along the high angle shear joints such that the rock above the joint was wedging or ploughing underneath the same stratum below the joint. Wedging at, or very near the footwall surface, gave rise to a chevron shaped ridge, whilst a bulge formed if the wedging joint was covered by more than 3m of poorly jointed silty mudstone and seatearth.

Core drilling permitted sampling of the apparent slip surface and the monitoring of water pressures. Boreholes were drilled near a footwall access ramp, west of the main slide area, at an excavated depth of 25-30m. They revealed artesian water pressures as shown in Figure 5 and permitted the sampling of several clay-rich intraformational shear zones. Subsequent shear testing of the 20mm thick shear zones gave drained residual shear strengths of just over 12° ($c_r' = 0$) using reversing techniques. Triaxial strengths were also obtained for materials *above* the principal shear zone which were incorporated in the slip debris further east.

Analysis

At the time of the study (1972-73) there were two widely recognised modes of failure. A slab slide with a curved toe failure (Figure 8A) was consistent with known landslide characteristics, the analytical methods for which were well documented. Similarly the propensity for footwall failure by sliding on low angle 'cross over' joints (Figure 8B) had been investigated with respect to mines in North America and Africa (10). Low angle joints were not present and the shear strength of the intact toe material above the obvious plane of sliding, as found by back analysis, was too low when compared with actual test results.

Three alternative failure modes were therefore investigated:-

- toe disruption due to the ploughing action of the high angle shear joints (Figure 8C)
- buckling at minor eccentricities (Figure 8D)
- footwall heave due to hydrostatic uplift

A preliminary investigation was undertaken using base friction modelling techniques (11) in which plaster bricks simulated rock blocks. Whilst most of the bricks were rectangular, low and high angle joints were specially prepared and the intraformational shear zone was represented by a single block interface coated with P.T.F.E. Eccentricities of varying wavelengths and amplitudes were also incorporated in the models. The findings are shown in Figures 6 and 7. The styles of toe buckling in the models consistently conformed to observed field failure modes, reproducing chevron shaped and gently curved pressure ridges as well as demonstrating more complex movements (Figure 8E). Cross over joint failures were more critical than ploughing action failures for identical joint angles about the bedded plane. The more acute the angle of jointing, the more readily failure resulted. The greater the original eccentricity in bedding the greater the risk of simple buckling failures. High angle dips were necessary for failure in the modelling since water pressures could not be simulated; the role of the structural constraints and the pattern of the block movements at failure are nevertheless apparent.

Various simple analytical methods were investigated to explore further the effects of different geotechnical factors. A modified Euler equation for the critical buckling force (12) was taken to be $\frac{e\pi^2Ed^2}{12L^2}$ where

e = eccentricity factor, E = Young's Modulus, d - thickness of the buckled zone and L = length of the buckled strata beam. This equation had been used to appraise a footwall collapse in Scotland (13). For the South Wales mine it did not appear valid for slides in which the driving force due to the up-slope strata was applied to a rock beam thicker than 1m, given the regular dip of the footwall bedding.

The mechanics of the ploughing action at the shear joints are complex especially when sliding movements follow initial displacement. A limit equilibrium solution was

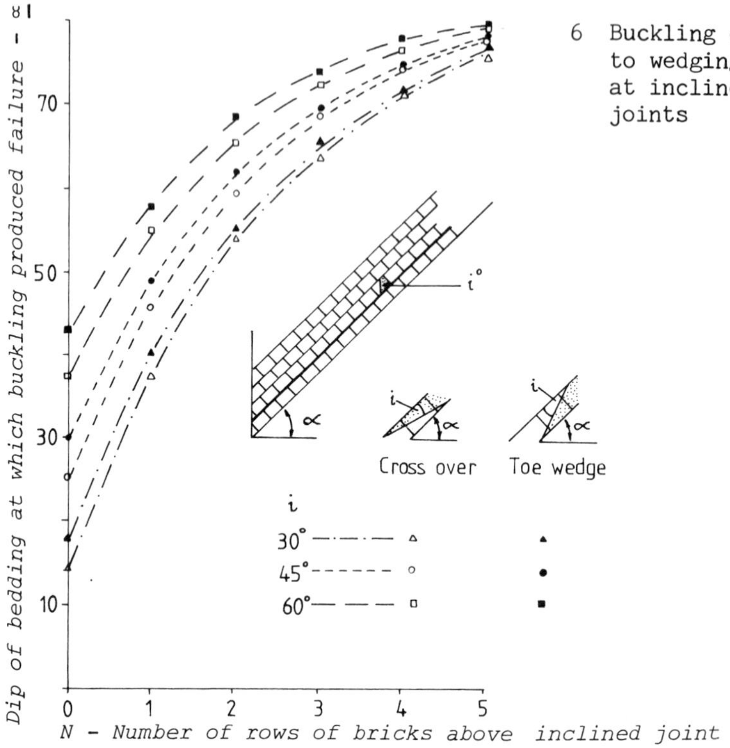

6 Buckling due to wedging at inclined joints

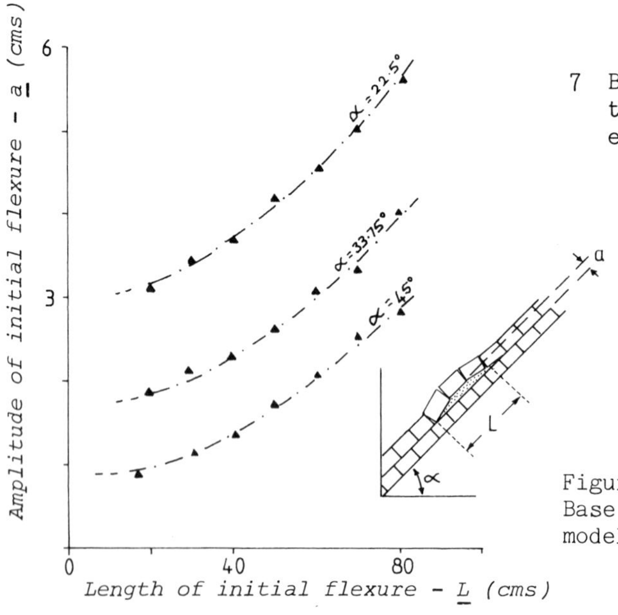

7 Buckling due to structural eccentricity

Figures 6 and 7 Base friction model results

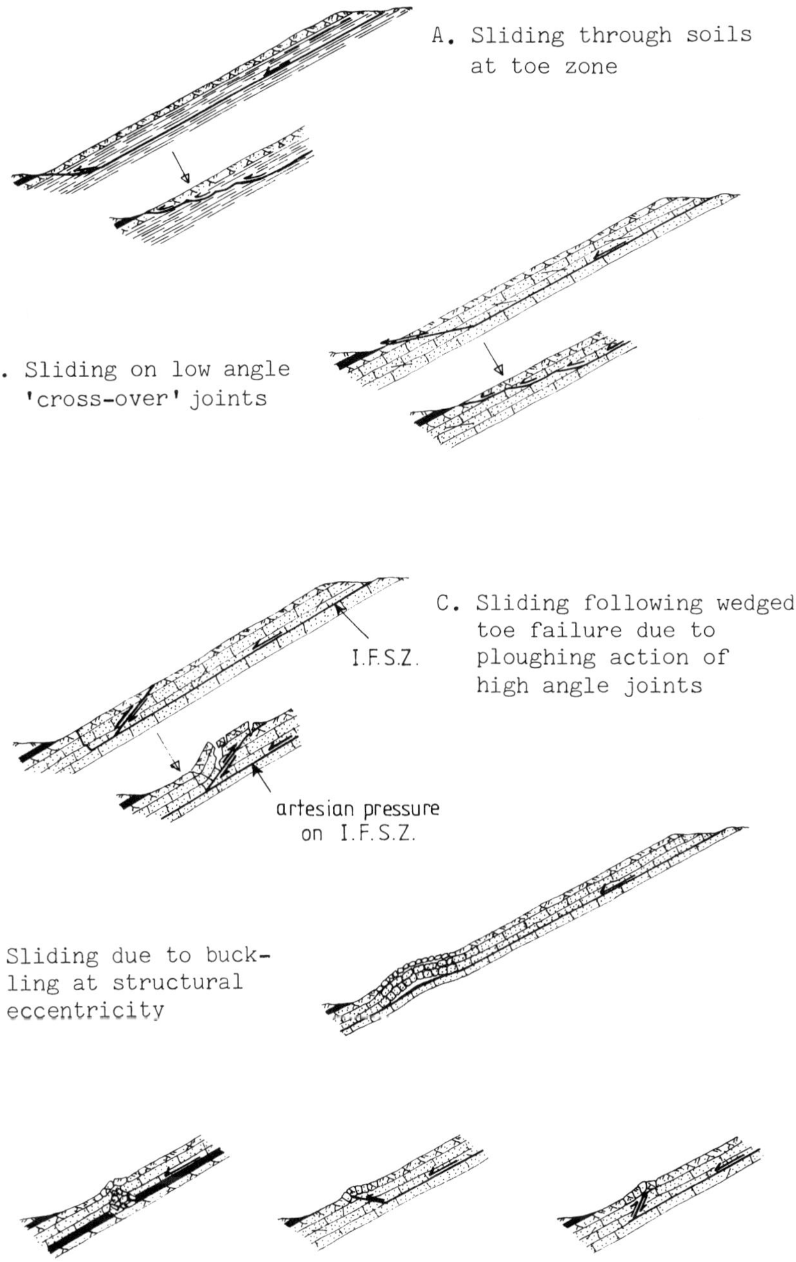

E. Pseudo buckling due to underground mine workings (i) low angle joints (ii) or high angle joints (iii)

Figure 8 Footwall failure modes

used to explore the first small displacement at the shear joint. The solution employed a range of laboratory shear strength values and observed groundwater pressures; potentially unstable conditions were confirmed, for the excavation depths at which pressure ridges had developed.

Hydrostatic uplift may either cause, or contribute to, footwall failure. In horizontally bedded deposits (14) artesian pressures of more than γd are essential for failure in unrestricted pavements (γ = density of the footwall rock, and d = thickness of strata above the confined aquifer). Such pressures were not found at the time of investigation, but could have occurred well before a full depth mine was established. The artesian pressures effectively increased the driving force of the inclined slab resting on the intraformational shear zone (see Figure 8C). It is interesting to note that very little seepage was observed on the footwall, even after collapses. The permeability of the underlying sandstones was moderately low and although the aquifer was probably of limited extent, high pressures could nevertheless develop.

Until recently most stability problems on South Wales footwall slopes were satisfactorily analysed either on the basis of the appropriate failure mode as shown in Figure 8, or as undercut slabs. In one recent failure there was substantial crest displacement, but no discernible bulge feature developed *above* the toe. A slight heave formed in a coal seam beyond the toe of the footwall, but the volume displaced was substantially less than that lost at the crest. The style of the movement accords well with a recent description of the Prandtl prism failure mechanism (15) and it may be reasonable to consider this collapse as akin to a foundation failure. It seems possible that the critical areas for such movements are footwalls across the toe of which there is a zone of highly broken and shattered rock aligned parallel to the strike of bedding; as may be seen (Figure 4) the structural setting in the study area may give rise to this combination of geotechnical constraints.

MINE PLANNING AND PRACTICE

In many steep-seam surface coal mines ground movements are not necessarily critical, and provided the *rate of movement* is not excessive or subject to great acceleration (13) the incipient failure may be unimportant. The planning of footwall slopes near major structures or services is most important and requires a detailed geotechnical investigation; it is imperative that the underburden strata are adequately investigated by drilling *to the full mine depth*, and not merely into the seatearth of the pavement seam. Water pressures should be monitored and discontinuities oriented in cores with respect to bedding dip; monitoring of water pressures and any tension cracks should continue throughout mining.

At the design stage a full range of potential failure modes must be explored. With *gently inclined footwalls* i.e. dip of less than 1 in 2, efficient drainage is almost always adequate to ensure no significant cracking or settlement at the crest. It is essential, however, that drainage is carefully monitored. In the case history outlined above sub-horizontal, slightly rising boreholes were drilled with a blasthole rig to a maximum distance of 12m. Such measures are commonly mentioned in rock mechanics literature as a means of promoting self-drainage. However, in footwalls with dips of 1 in 3, a 12m horizontal borehole penetrates less than 4m of footwall strata. On the footwall in question some drainage occurred, but in the absence of local monitoring this did not apparently achieve adequate depressurization and a further failure ensued. At that stage it was considered necessary to divert a 132kV power line which traversed the length of the footwall crest. Subsequently a further 12m drainage hole was drilled *perpendicular* to the footwall bedding. This penetrated more than 8m of strata beneath that drained by the previous holes: it produced water in large quantities, initially at a high rate of flow. Such drainage holes are less expensive and are now exclusively used in footwalls, since in the absence of face freezing, the pressure relief and drainage is equivalent to that which results from far longer horizontal holes. The performance of drainage holes which are perpendicular to footwall bedding appears to be superior with respect to collapse and silting characteristics, probably as a result of the steady, full-bore flow of water; lining is not usually required. The excavation of drainage slots (10) is not recommended.

More *steeply inclined footwalls* may present problems when planning mine limits near structures. Whilst drainage holes again improve stability, the inclination of weak horizons behind the footwall may promote an excessive disturbing force even after pressure relief. It is moreover possible to analyse deep seated potential failures and find an adequate Factor of Safety, yet have continuing superficial slab failures with the consequent retreat of the slope crest. In these circumstances underdig as in Figure 4 is best planned with slope geometry based on shallow as well as deep seated stability analyses. Bench faces should be parallel to, or flatter than, bedding and if recession continues it may be necessary to allow for the placement of rock fill at the toe of the bench slopes. It is not uncommon to find that the required underdig gives a reduced overall slope 15°-20° flatter than the footwall gradient. These values are appropriate for *preliminary* planning of mine excavations near existing structures and when siting overburden dumps near footwalls inclined at 35°-50°.

Remedial measures are difficult to execute on footwalls. Benching may be attempted, but will rarely prevent failure if movements have already commenced. Similarly slope reduction by grading to a flatter slope will not guarantee success; movements continued on one mine in the study area when the footwall was flattened across the bedding from an initial slope of 1 in 1 to 1 in 2. In the absence of viable drainage measures only toe loading and the abandonment of the underlying pavement coal are feasible. The toe loadings required to eliminate major movement may be small, but the ensuing loss of coal is often substantial. Restraining structures such as anchoring are not usually feasible once movements have started and are rarely cost effective as a routine design solution.

ACKNOWLEDGEMENT

The authors wish to thank T. Atkinson, A. E. Cobb, J. Bray and J. Sullivan for their advice and comments and acknowledge the assistance of the Opencast Executive and its contractors. Part of this work was undertaken whilst the senior author was seconded to work with the Rock Mechanics Reserch Group at Imperial College. The views expressed are entirely those of the authors.

REFERENCES

1 Woodland A.W. and Evans W.B.- Geology of the South Wales Coalfield Pt. IV Pontypridd and Maesteg. 1964 *Mem. Geol. Surv.*

2 Stimpson B. and Walton G.- Clay mylonites in English Coal Measures; their significance in opencast slope stability. In *Proceedings 1st International Congress of International Association of Engineering Geology*, Paris 1970 (Paris: The Association, 1970), vol. 2, 1388-93.

3 Walton G. and Atkinson T.- Some geotechnical considerations in the planning of surface coal mines. *Trans. Inst. Min. Metall.* (Sect. A Min. industry) 87 (1978) A147-171.

4 Bell F.G. ed.- Site Investigations in areas of mining subsidence 1975. London: Newnes-Butterworths.

5 Hassani F.P., Whittaker, B.N. and Scoble M.J.- Strength characteristics of rocks associated with opencast coal mining in the U.K. 1979 In *Proc. 20th Symp. Rock Mechanics* Austin, Texas. (New York: American Society of Civil Engineers 1979) 347-355.

6 The description of rock masses for engineering purposes: report by the Geological Society Engineering Group Working Party. *Q.Jl Enging. Geol.* 10 (1977) 355-88.

7 Whincup G.T.-Some aspects of opencast coal mining in South Wales. *Proc. Sth Wales Inst. Engrs.* 86, (1973) 15-34.

8 Brealey S.C. and Atkinson T.- Opencast mining. *Min. Engr.* 99 (1968) 147-63.

9 Harrison J.V. and Falcon N.L.- Collapse Structures *Geol. Mag.* 71 (1934) 529-539.

10 Brawner C.O., Pentz D.L. and Sharp J.C.- Stability studies of a footwall slope in a layered coal deposit. In Stability of rock slopes Cording E.J. ed. (New York: American Society of Civil Engineers 1972) 329-65. *(Proc. 13th Symp. Rock Mechanics)*

11 Ashby J.- Sliding and Toppling modes of failure in models and jointed rock slopes, M.Sc. Thesis (1972) Imperial College, London.

12 Bray J. pers. comm. 1973.

13 Walton G. and Taylor R.K.- Likely constraints on the stability of excavated slopes due to underground coal workings. In *Proceedings conference on rock engineering*, Newcastle, April 1977 Attewell P.B. ed. (Newcastle upon Tyne: University for British Geotechnical Society, 1978). 329-49.

14 Brealey S.C.- Groundwater control in opencast mining. In Opencast mining quarrying and alluvial mining (London: IMM 1965) 390-415.

15 Kvapil R. and Clews K.M.- An examination of the Prandtl failure mechanism in large dimension slope failures *Trans. Inst. Min. Metall.* (Sect. A Min. Industry) 88 (1979) A1-A5.

HILLSIDE STABILISATION USING ROCK ANCHORS AT GRIMWITH RESERVOIR

H.T. Lovenbury, BSc(Eng), PhD, MICE, CEng

Senior Engineer, Rofe Kennard and Lapworth

B.N. Robinson, BSc, MICE, CEng

Head of Technical Services Dept., Sir Alfred McAlpine & Son (Northern) Ltd

M.R.C. McClenahan, BSc

Contracts Engineer, Ground Anchors Ltd

INTRODUCTION

Grimwith Reservoir, near Grassington in N. Yorkshire, is currently being enlarged by the construction of a 45 metre high embankment dam. The existing earth dam, completed in 1864, will be incorporated in the upstream shoulder of the new structure. The works include a 250 metre long, 5 metre internal diameter, horseshoe-shaped outlet tunnel through the east side of the valley. The downstream 130 metre length of this tunnel was designed to be formed in open cut.

The east hillside consists of alternating coarse sandstones and shales overlying limestone, the beds gently dipping to the N.W., across the line of the tunnel, at about 5 - 10°. The tunnel is founded on the limestone with its sides and roof in shale. This shale stratum is approximately 9 metres thick and is overlain unconformably by gritstone, with a 3 metre thick shale layer some 14 metres above the base of the gritstone. Perched water tables are present above both the shale strata, that over the lower stratum varying between approximately 0.5 and 2.8 metres head prior to excavation at the junction between the open-cut and tunnel sections (Chainage 138). The water pressure in the limestone is sub-artesian.

The gritstone is medium- to coarse-grained, ironstained and fresh to moderately weathered. The jointing is moderately wide to widely spaced, and generally open. The upper shale layer within the gritstone consists of moderately to highly weathered clayey siltstone/mudstone and contains a thin coal band. The lower shale stratum comprises very closely discontinuous mudstone/shale. It is highly to residually weathered over the top 0.5 metres. The limestone is medium to thickly bedded, fossiliferous and variably muddy, with interbedded slightly to highly weathered thin mudstone developments.

The final section of open cut construction passes through the toe of an area of landslipping (Figure 1) which extends for a distance of over 400 metres down the valley. Following a detailed borehole investigation of the east hillside in 1975, it was concluded that this instability was shallow-seated and largely confined to the superficial materials. The side scarp of the feature intersects the culvert excavation approximately 55 metres downstream of the tunnel portal. Beyond this point the gritstone is absent.

The vertical depth of excavation at the upstream end of the culvert section was 18 metres and at the downstream end 8 metres approximately. Excavation in open cut was commenced to slopes of 45° in the shale, up to 60° in the gritstone and $25\text{-}30^{o}$ in both the clay and weathered gritstone overburden, working from the downstream end. Berms of minimum width 2.5 metres were incorporated at the top of both the shale and the gritstone face. At the position of deepest cut the excavation batter intersected the original ground slope at a height of 40 metres above the culvert invert and at a distance of 60 metres from the culvert centreline. This point lay just below the main access road to the site.

The work proceeded with only localised instability in the clay overburden overlying the shale at the downstream end of the culvert. This was overcome by grading back, installing drainage grips and pumping from a temporary shaft.

On July 22nd 1976, with just over half the length of cut substantially complete, the remainder of the excavation had been partly opened up to a depth of approximately 8 metres below the gritstone/shale interface (Figure 2). Deepening of the excavation in the shale was being undertaken when a tension crack up to 80mm wide was observed in the ground above the excavated slope adjacent to the access road. Following this, it was seen that the upper metre of shale below the berm had moved forward some 30mm along a plane sloping at about 8° towards the excavation. This movement occurred quite suddenly and extended over a distance of approximately 50 metres. Tension cracks, up to 30mm wide, also appeared at the top of the gritstone face. A number of springs of water were also issuing from the base of the gritstone from the perched water table over the shale. All work was immediately stopped and the plant withdrawn.

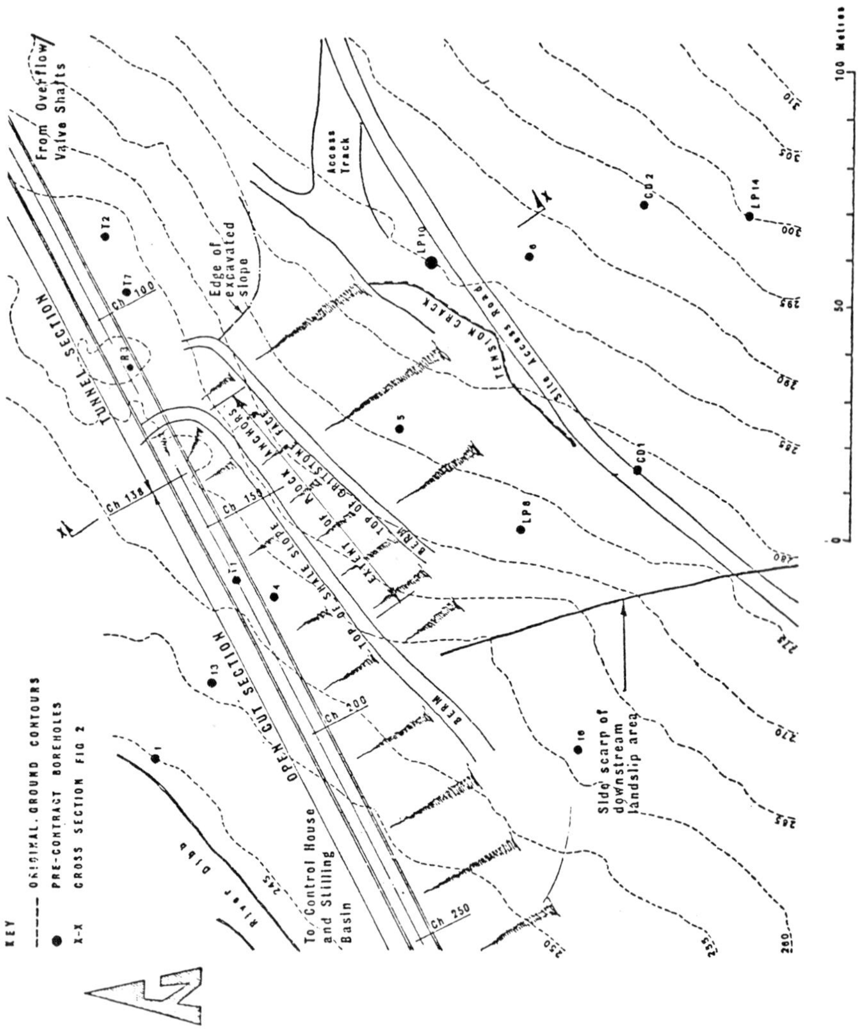
KEY
ORIGINAL GROUND CONTOURS
PRE-CONTRACT BOREHOLES
X-X CROSS SECTION FIG 2
From Overflow Valve Shafts
TUNNEL SECTION
OPEN-CUT SECTION
To Control House and Stilling Basin
River Dibb
Edge of excavated slope
Access Track
TENSION CRACK
Site Access Road
Side scarp of downstream landslip area
BERM
TOP OF SHALE SLOPE
TOP OF GRITSTONE FACE
ANCHORS
EXTENT OF ROCK
Ch 100
Ch 138
Ch 150
Ch 200
Ch 250
0
50
100 Metres

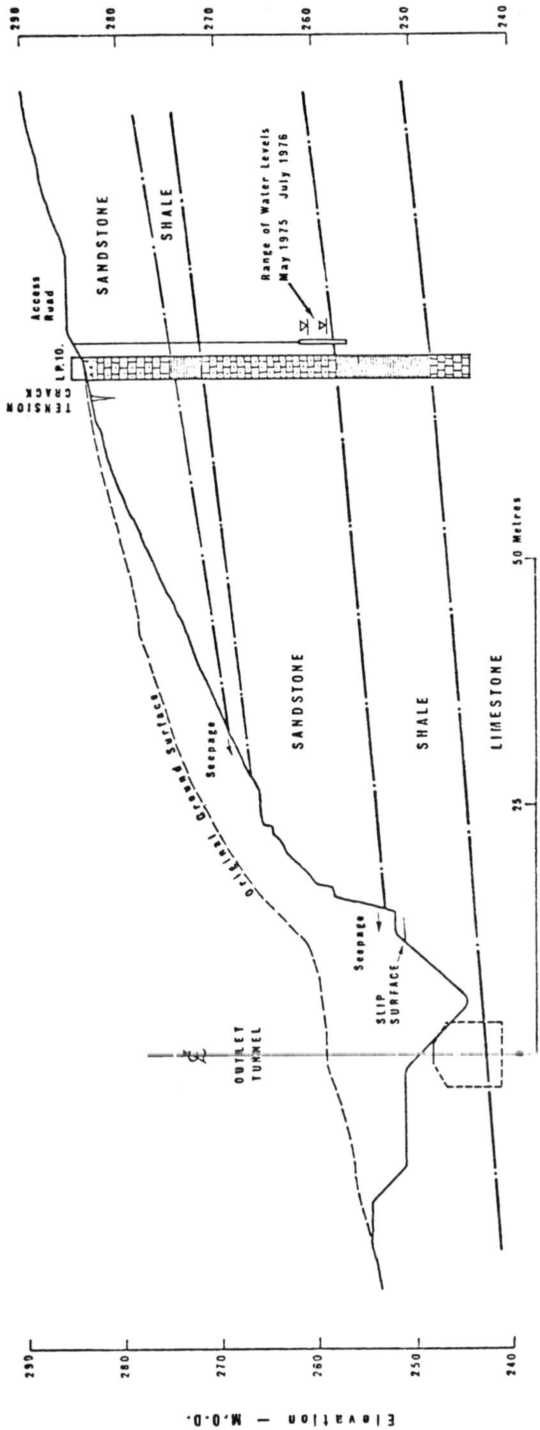

Figure 2 Section through excavation at time of slip

Backfilling of the excavation with imported rockfill and excavated material to the top of the shale in order to arrest the movement commenced on 27th July. By this time, the slip had moved forward approximately 300mm. The slip continued to move, and minor rock falls started to occur, restricting the backfilling operation. The shale berm slowly disappeared and the leading edge of the gritstone face became detached, eventually toppling onto the fill placed in the shale excavation on August 5th.

The fall of the gritstone had some stabilising effect on the slope movement but backfilling against the face was continued to provide further stability and to provide a working platform for subsequent measures.

DESIGN OF REMEDIAL WORKS

The Contractor's first proposal for stabilising the slope involved the installation of a sheet pile or soldier 'H' pile retaining wall at the toe of the slope driven to the shale/limestone interface and restrained by rock anchors. However, this was not considered sufficient without some additional anchoring of the gritstone since it was feared that the slip mass might move forward bodily over the top of the pile wall. It was also felt that the installation of piles through the fill containing the fallen gritstone blocks and rockfill would be difficult and that the vibrations arising from the pile driving operation might endanger the stability of the slip mass.

Consequently, it was decided to rely on rock anchors alone to secure the slope. These would be drilled from the gritstone face through the underlying shale and anchored into sound limestone. A scheme was then prepared by the Contractor for agreement by the Engineer.

In the meantime, two test anchors were installed on site to establish possible load carrying capacities and suitable anchor lengths. Both were drilled from the backfill at the toe of the gritstone face and passed directly into shale, one being founded in the limestone, the other in the shale. The work was carried out by U.A.C. Ltd. The details of the anchors were as follows:-

Anchor No. 1	inclination	30° below horizontal
	drilled through	3m overburden
		18m shale
		10m into limestone

Plate I View showing displacement of upper section of trial pit along slip surface in shale

Anchor No. 2	inclination	11° below horizontal
	drilled through	6m overburden and shale
		15m shale
		into 9m hard shale

This anchor was underreamed with two underream cones.

Each anchor consisted of 7 No. 15.2mm diameter 'Dyform' strands, the total Ultimate Tensile Strength (U.T.S.) of the tendon being 210 tonnes. Anchor No.1 was loaded to a maximum value of 136 tonnes. At this load the bearing pad began to penetrate the underlying fill material and the test was discontinued. Anchor No.2 was loaded to 110 tonnes when penetration of the bearing pad again occurred.

The tests showed that the original working loads of 60 tonnes proposed for the anchors would be perfectly safe, and that the use of anchors with working loads higher than 60 tonnes would be possible, presenting considerable cost savings.

Quotations were therefore obtained for alternative schemes based on 60 tonne and 100 tonne anchors, and a subcontract was awarded to Ground Anchors Limited.

Calculations to determine the anchor loads were then carried out. Since the exact form of failure was not known analyses were carried out assuming different failure surfaces extending between the slip surface observed at the excavation face below the shale/gritstone interface and the tension crack under the access road.

The final calculations were carried out by wedge analysis, based on the assumption that failure occurred by block sliding of the gritstone along a bedding plane within the underlying shale inclined at 8° to the horizontal.

Back analysis showed that the angle of shearing resistance, based on the above assumptions, acting on the plane at failure was of the order of 8.5 degrees, indicating that the movement had probably taken place along a pre-existing shear surface. As it was thought possible that the slip might work back further into the hillside, the restraining forces required were calculated for a slip extending back another 50 metres, taking a factor of safety of 1.2. It was also feared that further slip planes might exist lower down in the shale and so the condition of sliding taking place at the base of the shale was analysed. The angle of shearing resistance of the shale in mass was taken as 25° and that of the interlocked rock as 60°. A perched water table in the gritstone equivalent to an average head of 2 metres over the length of slip plane was assumed for the upper slip.

Due to concern about the possibility of local failure occurring in the shale slope on re-excavation, partly due to the increased vertical loading induced on the shale by the inclined anchors at the toe of the gritstone slope, it was decided to transfer some of the anchors to the shale slope to prevent undermining of the whole anchor installation. This had the further advantage of easing the very close spacing necessitated by installing all the anchors in the gritstone. Sufficient anchors were retained in the gritstone to resist the upper extended slip. The agreed load distribution in tonnes/metre to be provided and based on anchors installed at 35° to the horizontal was:-

Section	Total Load	Load in Gritstone	Load in Shale
Ch. 128-146	320	260	60
Ch. 146-164	275	210	65
Ch. 164-184	230	160	70

The final proposed scheme for satisfying this distribution of loading was for the installation of up to 4 rows of 100 tonne anchors from the gritstone at inclinations of 35° and up to 3 rows of 50 tonne anchors from the shale at inclinations of 30°. The spacing of the 100 tonne anchors varied from 1.5 metres to 2.0 metres and that of the 50 tonne anchors from 1.6 metres to 2.4 metres along the length of the slope. The anchor arrangement is shown in the table below:-

Section	Location	Anchor Load Tonnes	Inclination Degrees	No. of rows	Spacing m
Ch. 128-146	Gritstone	100	35°	4	1.5
	Shale	50	30°	2	1.75
Ch. 146-164	Gritstone	100	35°	4	2.0
Ch. 146-158	Shale	50	30°	2	1.6
Ch. 158-164	Shale	50	30°	3	2.4
Ch. 164-184	Gritstone	100	35°	3	1.8
	Shale	50	30°	3	2.25

The anchors were to be installed one row at a time, re-excavation proceeding to the next level as each row was stressed. Owing to practical difficulties on site many of the 50 tonne shale anchors were actually installed at an inclination of 45°. The spacing of these anchors varied between 1.5 metres and 1.875 metres. The vertical interval between the rows varied between 1.5 and 3 metres.

Consideration was also given to a proposal to install pump wells around the boundary of the slip mass to lower the level of the perched water table in the gritstone, since it was suspected that this could have an influence on the stability of the slip. However it was felt that the work involved in sinking the wells and lowering the water table would be very time consuming and less effective than the rock anchoring solution, so the idea was not pursued.

Undisturbed block samples of the shale containing the shear surface were subsequently tested in the shear box at Manchester University under the direction of Professor P.W. Rowe. The residual strength parameters obtained were $C_{r'} = 16kN/m^2$, $\emptyset_{r'} = 10^{o}$.

The basis of design for the anchors was as follows:-

Anchorages

For the 100 tonne anchors the minimum fixed anchor length in the limestone was to be 8 metres. This represented a rock/grout bond value of 0.40 N/mm^2 in a 100mm dia. hole, which is at the lower end of the range given for massive limestone by Littlejohn and Bruce (1975) . However, a conservative approach was desirable in view of the medium to thick near-horizontal bedding of the limestone, and the difficulty of access and loss of time involved in replacing an anchor.

Several anchors were in fact installed with fixed anchor lengths of 4 metres and tensioned to 125 tonnes, thus indicating that the factor of safety for the 8 metre anchorage was at least 2.5.

The 50 tonne anchors installed from the shale batter had 6 metre fixed anchor lengths.

Tendons

Since the anchors were required for temporary service (up to 3 years) only, the working stress in the tendon steel was specified as 62.5% of U.T.S. The tendons were made up of 15.2mm diameter 'Dyform L.R.' strand with a U.T.S. value of 30 tonnes. Thus the 100 tonne anchors were formed of 6 strands and the 50 tonne anchors 3 strands. Due to the temporary nature of the anchors, no permanent protection against corrosion was provided.

The tendons were made up on site, each individual strand being sheathed in a grease filled plastic sleeve to ensure debonding over the free length. The sleeves were filled using Ground Anchors Ltd's strand greasing machine, which avoids the necessity of having to clean a viscous and sticky grease from proprietary plastic coated strand over the fixed anchor length. The strands were splayed out over the fixed anchor length to increase the bond between the grout and steel. Spacers were placed at 2 metre centres and the strands tied together mid-way between each spacer.

Anchor Heads

To reduce the work which had to be carried out at the site of each anchor installation, it was decided to use precast concrete anchor stressing blocks. These were positioned over the installed tendon and fixed by placing in-situ concrete between the block and the rock face.

The blocks were designed in accordance with the CIRIA guide (Clarke, 1976) to give a net strength of 1.6 x working anchor load. The anchor heads are formed of steel blocks containing tapered holes through which the strands are passed and locked with wedge grips, the load being transmitted to the concrete anchor block via a steel 400 x 400 x 20 load distribution plate.

ANCHOR INSTALLATION

Drilling was initially attempted using open hole/air flushing techniques, as for the trial anchors, but difficulties were experienced in penetrating the shale at its interface with the gritstone due to loss of air circulation and in maintaining an open hole in the shale.

A specially fabricated 137mm dia. high torque resistant drill casing was then tried and after some difficulties a duplex drilling method was evolved using a rock roller bit with water flushing.

For the gritstone anchors the holes were cased through the gritstone into the top of the shale. The holes for the shale anchors were drilled using the rock roller bit, with the upper length of hole cased as a protection against damage from water flushing.

The limestone was drilled open hole, using the roller bit.

Falling head observations of the flush water in the casing indicated that waterproofing of the holes to prevent loss of grout during installation was not required.

Care had to be taken during these operations to prevent contamination of the River Dibb by flushing water and arisings, by leading these to settlement lagoons prior to discharge to the river.

The tendons were brought to the site of the anchors using Ground Anchors Ltd's cable transporter. This is a trailer mounted, hydraulically operated, revolving drum which enables the tendon to be placed in the hole at a controlled rate. The anchors were grouted by tremie pipe, using a 0.40 water/cement ratio rapid hardening cement grout mixed in a high speed colloidal mixer. When full consistency grout emerged from the top of the hole, the casing was slowly withdrawn, the hole being topped up as necessary during this operation. Observation indicated that no significant loss of grout occurred below the gritstone/shale interface, thus ensuring adequate retention over the fixed anchor length. Some grout loss did usually occur in the gritstone. The grout strength was checked by cube testing.

As soon as the strength of the grout had reached 30 N/mm^2 each anchor was tested to 1.25 x working load and locked off at 10% above working load, to counteract the loss in stress on locking off. Check lifting was carried out on all anchors after 24 hours. The average load of the 100 tonne anchors was found to be 105.7 tonnes and that of the 50 tonne anchors 50.5 tonnes. Two of the 100 tonne anchors were locked off at reduced working load due to partial strand pull-out and one anchor was lost completely. Careful monitoring of the position of the anchor blocks during stressing was necessary to enable true extension readings to be obtained. In some instances the blocks moved distances of up to 150mm into the gritstone face. Taped measurements were made at each increment of load to the corners of the block from datum points established some distance away from the anchorage position. The sandstone anchor heads were also subsequently monitored using a 1 second theodolite but no significant displacement (outside the range of scatter) was detected in the 5 month observation period.

A load check on seven 100 tonne anchors in the third row after 3 months indicated a loss in stress of up to 2% in that period. Some minor settlement of the shale anchor heads occurred but this stabilised before stressing was completed.

After completion of the second row of anchors it was found that the gritstone face had in places become quite fractured and loose, even after dressing back, to such an extent that there was a danger of the anchor blocks being undermined by local collapse as re-excavation proceeded. Three anchors were lost due to this cause. As a result, a considerable amount of insitu concrete buttressing, up to 1m thick and incorporating mesh fabric, was constructed to form a suitably coherent mass against which the anchors could be stressed. Since this could only be constructed by under-pinning, as re-excavation proceeded downwards, the work of fixing shutters and placing concrete was laborious and slow. Much of this work was carried out in severe winter conditions. Areas of sounder gritstone were stabilized by rock bolting.

It was felt that a particular weakened area might exist in the upper metre of shale immediately below the gritstone, in the zone where sliding had taken place, and a 1m thick x 2m high reinforced beam was cast at this level through which much of the highest row of 50 tonne anchors was installed. Weep pipes were also installed to drain the perched water table in the gritstone.

To speed up the installation of anchors in the shale, steel grillages, fabricated from 305mm 'H' piles, were substituted for the precast stressing blocks and installed in direct contact with the shale face. Anchor tests were carried out on a grillage to confirm the bearing capacity and a value of 365 kN/m^2 was adopted for the shale. The anchor heads were staggered vertically to ensure a more uniform distribution of load application to the slope. The shale surface was also protected by blinding concrete after installation.

A total of 132 No. 100 tonne anchors (including replacements) and 80 No. 50 tonne anchors were installed. The 100 tonne anchors were placed at angles of between 35° and 37° to the horizontal in order to provide some stagger at the anchorages and vary in length from 23-38 metres. The 50 tonne anchors were initially placed at angles of between 30°-35° to the horizontal but where steel grillage anchor pads were used the inclination was increased to 45°. These anchors vary in length from 10-20m. The arrangement is shown in Figures 3 and 4.

A system for monitoring the displacement of the slip was established as soon as movement was detected. This consisted of three lines of pegs, installed normal to the tunnel centreline, in the overburden slope above the gritstone berm. Interpretation of the observations was complicated by a number of factors, not least that secondary movement developed within the upper shale layer. Nevertheless, the observations show that the movement was effectively halted by the time that stressing of the first row of anchors was completed. The secondary slip was arrested by forming short counterfort drains at its toe.

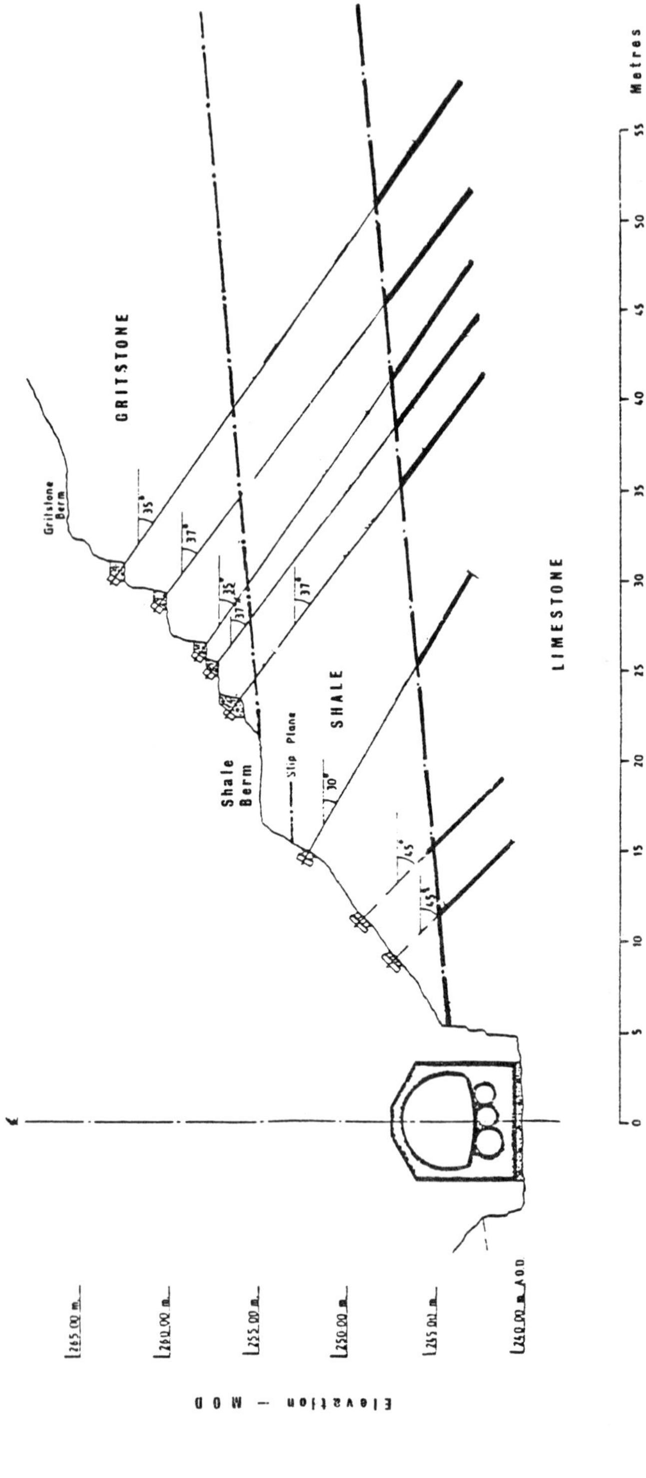

Figure 3 Typical section of completed excavation

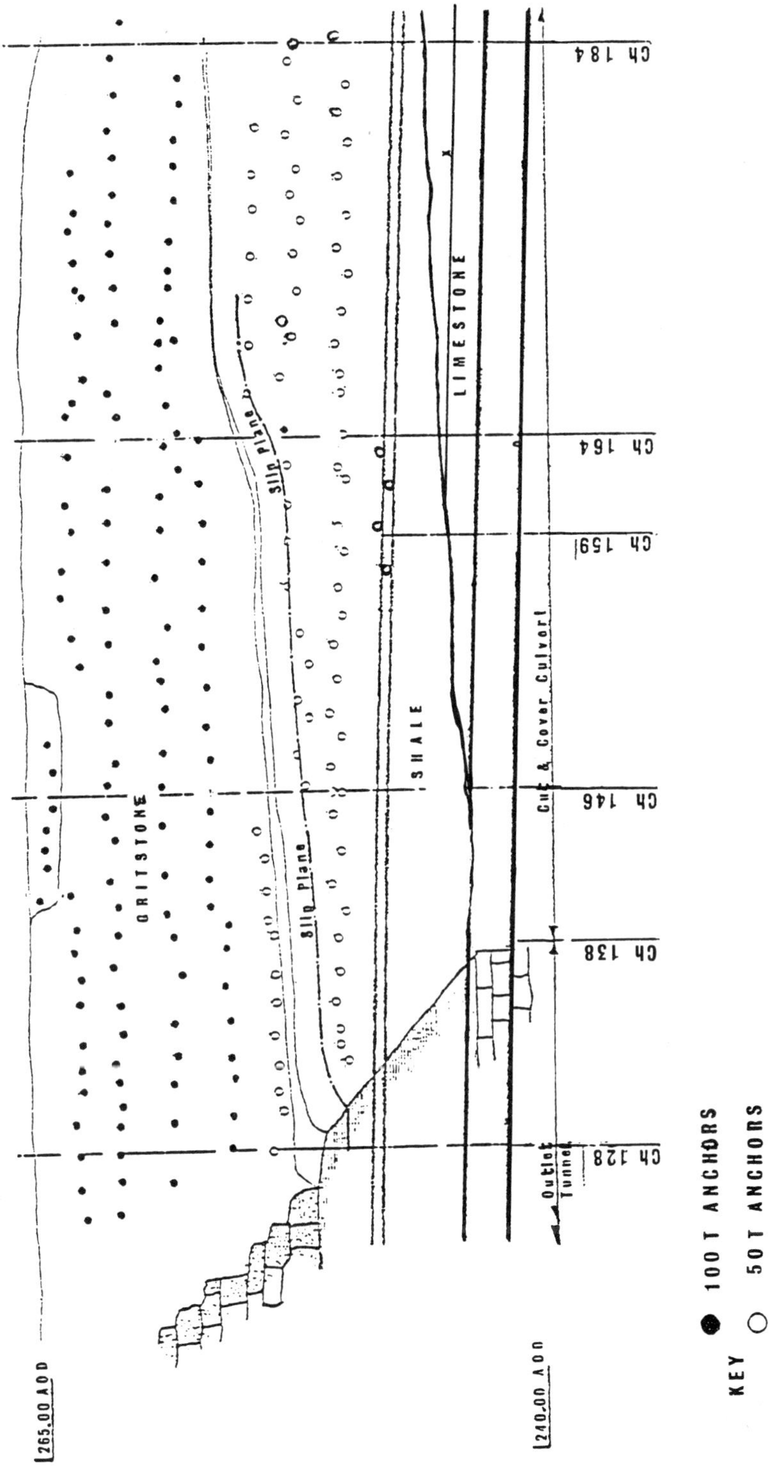

Figure 4 Longitudinal elevation of anchor installation

Plate II View of completed anchor installation

The installation was commenced in September 1976, and took 7 months to complete. The tunnel has now been completed and the excavation largely backfilled. A chimney drain has been incorporated between the back fill and the sandstone face to drain the perched water table in the gritstone. The anchor heads were encased in concrete prior to backfilling. Long term stability will be achieved by forming a berm across the valley at the toe of the downstream shoulder of the dam.

ACKNOWLEDGEMENTS

This paper is published by kind permission of the Division Director, Western Division of the Yorkshire Water Authority. The consulting engineers for the scheme are Rofe, Kennard and Lapworth and the engineering geologists E. Morton and Partner with Professor P.W. Rowe acting as soil mechanics consultant. The main contractor is Sir Alfred McAlpine & Son (Northern) Ltd. and the rock anchors were installed by Ground Anchors Ltd. The authors gratefully acknowledge the assistance of their colleagues in the preparation of this paper.

REFERENCES

J.L. Clarke - A guide to the design of anchor blocks for post tensioned prestressed concrete members, CIRIA Guide 1 (June 1976) pp 50.

G.S. Littlejohn and D.A. Bruce - Rock Anchors - State of the Art, Ground Engineering, (1975-1976) - 8(3), 25-32; 8(4) 41-48; 8(5) 34-45; 8(6), 36-45; 9(2), 20-29; 9(3), 55-60; 9(4), 33-44.

MEASUREMENT OF STRUCTURAL AND GROUND MOVEMENT

R.BALL B Sc (Eng), C Eng, M I E E

Senior Lecturer, Lanchester Polytechnic, Coventry

C.P.LEWIS B Sc, C Eng, M I E E

Principal Lecturer, Lanchester Polytechnic, Coventry

INTRODUCTION

Any structural motion can be considered as comprising a number of superimposed motions. In general a plot of the motion against time has a steady component plus components at a number of frequencies. The instrumentation with which the authors are involved is particularly concerned with the measurement of vibrationary components of structural movement.

CHARACTERISTICS OF VIBRATION MEASURING INSTRUMENTS

Any instrument intended for measuring a time dependent variable will have the following quoted characteristics.:

(a) Sensitivity; i.e. the nominal output signal level produced by a given input variable magnitude.

(b) Sensitivity Frequency Response; a graph of sensitivity magnitude versus frequency.

(c) Phase Response; a graph of the phase difference between the input variable and the output signal versus frequency.

(d) Overload Level; the level at which the instrument ceases to perform correctly. (This may be a function of frequency).

(e) Noise Level; the minimum input variable level that can be discriminated by the instrument.

In addition it will have:

(f) Error; i.e. the actual performance of a particular instrument on a particular day will deviate from the quoted performance.

Other characteristics of importance to the user are:

(g) Power Supply Requirements.

(h) Weight.

(i) Tolerance to overload in use (i.e. damage level) and recovery time.

(j) Ruggedness For Transportation.

(k) Size.

MEASUREMENT TECHNIQUES

Sensed and Displayed Variables

A vibrating structure will produce variations of displacement, velocity and acceleration as well as higher derivative terms. The objective of a measurement task could be to determine the level of any one or more of these variables.

The most suitable variable for observation is the subject of considerable discussion. Acceleration is often quoted, mainly because of ease of measurement, however an acceleration trace is often swamped by noise having very little structural significance. Velocity is said by some authors to have both a structural damage significance and a human sensitivity significance which is virtually independent of frequency over a reasonable range of interest. Displacement records are, however, more visually meaningful. A further discussion is outside the scope of this paper (O'Neil,1978; Kuehn,1978; ISO2631,1975).

Opinion also varies as to the most desirable display, namely a continuous trace or the r.m.s. or peak value.

Since displacement, velocity and acceleration

are related by differentiation then the instrumentation may sense any one of these variables and derive the others. However electronic differentiation of signals implies a frequency response with gain increasing at 20dB per decade of frequency per differentiation hence emphasising high frequency noise components of the input signal or of the input electronic stage.

Integration implies a frequency response which falls at 20dB per decade of frequency per integration thus limiting the high frequency noise response. However any constant input to an integrator (or d.c. error term in the input electronics) will cause the integrator output to increase in proportion to time, and eventually saturate the amplifiers. Consequently very careful electronic techniques are required if a useful instrument is to be produced.

Sensing Acceleration, Velocity And Displacement

Measurement of these variables requires that a reference frame having fixed orientation and origin is defined. Velocity and displacement measurement is simplified if the direction of motion is along one of the chosen co-ordinate directions so that the origin is in line with the motion.

For most ground-mounted structures changes in orientation with respect to an earth-centred rotating reference frame are small and hence co-ordinate directions can be defined in terms of local structural directions. It is more difficult to define a suitable fixed origin since the whole structure and its surroundings are likely to be moving with respect to the earth's centre of mass. Consequently measurements with instruments requiring a fixed datum, such as range finders, dial gauges etc, whatever the sensitivity and static accuracy, are unlikely to be suitable for a vibrating structure. Inertial techniques can be used to overcome some of these difficulties.

Consider the spring-mass, or seismic, system of FIGURE 1; this comprises a mass m attached to a moving point P by means of a spring of stiffness k. Damping of this oscillatory system will be provided by oil or eddy currents. The mathematical treatment of this system is given in standard texts (e.g. Garner,1968). The system will have an undamped natural angular frequency $\omega_0 = \sqrt{k/m}$ rad/s. For vibrations of point P along the axis of this

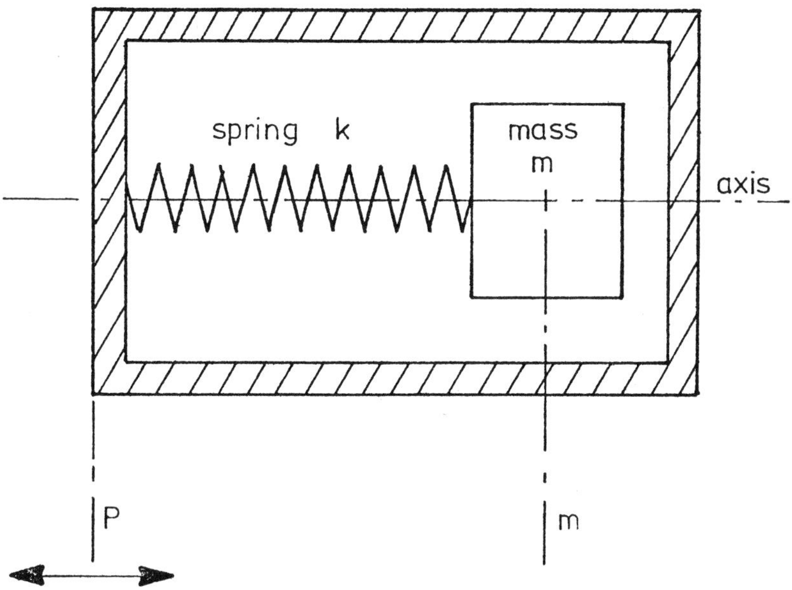

Basic spring-mass system

FIGURE 1

system the average position of the centre of gravity of m will be a fixed distance from the average position of point P. This position of m can be used as a datum for observing the movement of P. It is possible to sense either the relative displacement of m and P by means of the stress or strain in the spring, or the velocity of m relative to P. FIGURE 2 shows response curves when sensing the relative displacement of m and P.

(i) FIGURE 2a applies for constant acceleration of P; for frequencies well below ω_o the sensitivity is independent of frequency. Normally the system would have a damping ratio (ξ) of about 0.7.

(ii) FIGURE 2b applies for constant velocity of P. It can be seen that a heavily damped system ($\xi>2$) will have a sensitivity reasonably independent of frequency.

(iii) FIGURE 2c applies for constant displacement of P; for frequencies well above ω_o the sensitivity is independent of frequency. Normally ξ would be about 0.7.

(iv) If the velocity of m relative to P is sensed (for instance with a permanent magnet and coil) then one other arrangement is in general use which also has the frequency characteristic of FIGURE 2c for constant velocity of P. Again for frequencies well above ω_o the sensitivity is independent of frequency.

Thus we have:

(i) An acceleration sensor suitable for signals below ω_o - this is a conventional accelerometer.

(ii) A velocity sensor suitable for signals around ω_o - this is the basis of some seismometers.

(iii) A displacement sensor suitable for signals above ω_o - this is also the basis of some seismometers.

(iv) A velocity sensor suitable for signals above ω_o - this is the basis of the electromagnetic geophone. The design of seismometers has been previously covered elsewhere (Dennison, 1953; Willmore, 1961).

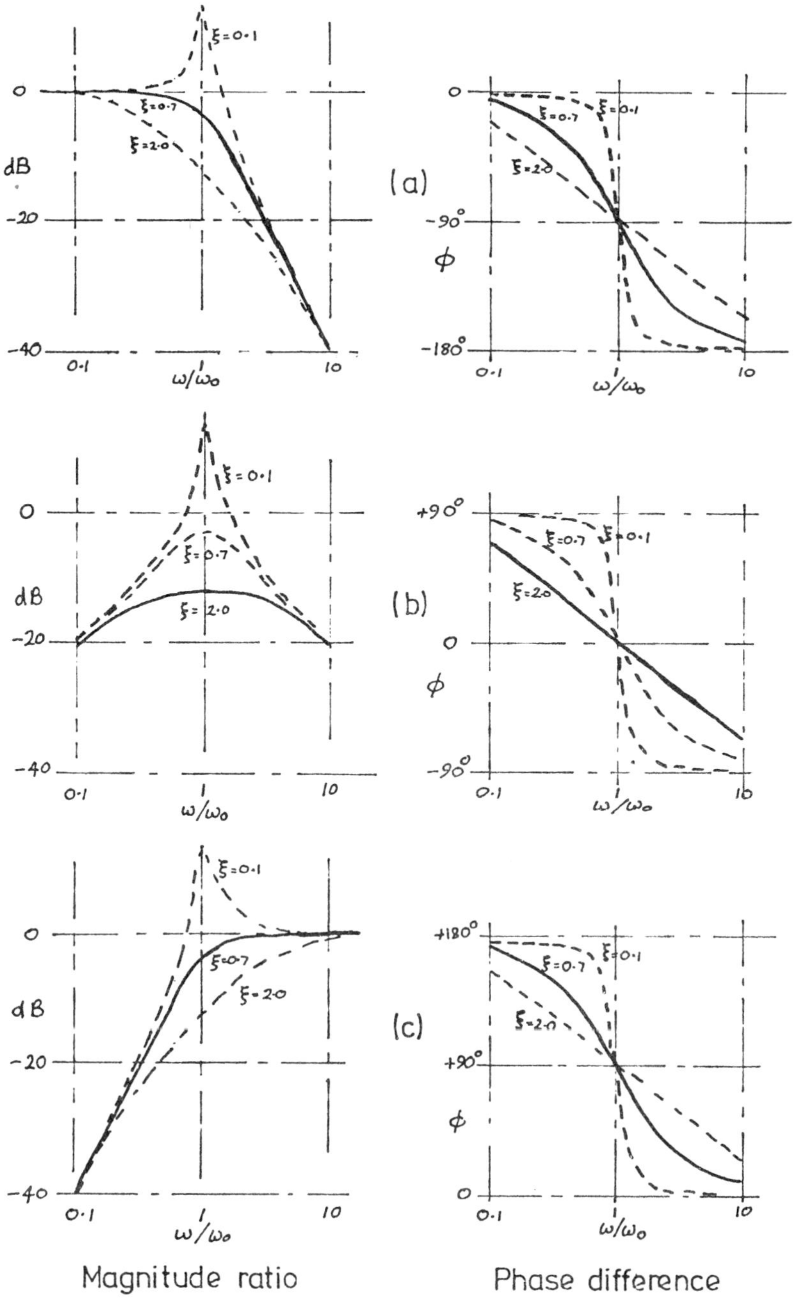

FIGURE 2 Sensor frequency characteristics

THE AUTHORS' INSTRUMENTATION

It was decided to use an accelerometer sensor for several reasons.

(i) Structural vibrations are often at low frequencies, typically below 5Hz and often frequencies below 0.1Hz are of importance (Mainstone,1966; Ward,1966). Geophones with sufficiently low resonant frequencies become very heavy and fragile. Also mechanical oscillation of the transducer due to transient disturbance would continue for a considerable time.

(ii) The range of vibration amplitudes is such that a wide dynamic range is necessary.

(iii) The accelerometer is capable of measuring to very low frequencies and the phase error introduced is negligible.

(iv) Suitable high quality, high sensitivity accelerometers weighing only about 80 grams and having a volume of about $15cm^3$ are available.

The accelerometer used by the authors is normally a Sundstrand Q-flex (PLATE 1). This requires excitation supplies of ± 15Vdc and gives an output of 1.3mA/g. The resolution is approximately 10μg and the resonant frequency about 1kHz.

The accelerometer signal is automatically modified by the instrument to remove any component due to the earth's gravitational field and the signal is amplified and integrated to give signals proportional to structural velocity and displacement. Because of the very low structural accelerations being sensed (for example 1mm peak at 1Hz is approximately $4x10^{-3}$g peak acceleration) the component of acceleration due to gravity must be balanced to a high degree of accuracy to avoid introducing time dependent errors in the output signals.

A typical instrument is shown in PLATE 2. It should be noted that there are no zeroing or setting up controls. This instrument measures velocities from 0.25mm/s to 25mm/s at full scale with a noise level of about 25μm/s peak.

Displacement scales are from 2.5μm to 25mm, although random noise of about 10μm peak in fact precludes use below the 100μm range for most purposes. This is a wide range compared with other instruments (Skipp,1978). Output signals are available for the simultaneous recording of velocity and displacement. Because of complex overload possibilities of an instrument of this type, indicators have been provided to show if the input signal is too great for the instrument or if it is necessary to change range. A weighted velocity range is also provided in accordance with the I.S.O. specification (ISO 2631,1975). The frequency and phase responses of the instrument for both velocity and displacement measurements are shown in FIGURE 3 although for many purposes FIGURE 4 gives a better indication of the range of application. Any vibration level inside the appropriate velocity or displacement line can be measured on an appropriate range.

The instrument is fully self contained with rechargeable batteries and weighs about 1kg. The accelerometer can be connected to the instrument via long cables allowing measurements remote from the recording station.

An additional feature is fast settling following repositioning of the transducer and when first switching on. FIGURE 5 shows the output on the most sensitive velocity range after switching on, indicating that the instrument is ready for use within 1 minute. This has been achieved by non-linear control techniques. FIGURE 6 shows that following a severe disturbance to the transducer (in this case turning it over and thus injecting 2g) the instrument is ready for use within about 1 minute and more rapidly still on a less sensitive range.

TYPICAL MEASUREMENTS

As the prime interest of the authors is the design of instrumentation for vibrational measurement, examples are introduced mainly to demonstrate the capabilities of the instrumentation. These are shown in FIGURE 7 and FIGURE 8.

The instrument has also been used for floor vibration measurements and the measurement of the sway induced in a discharging coal hopper.

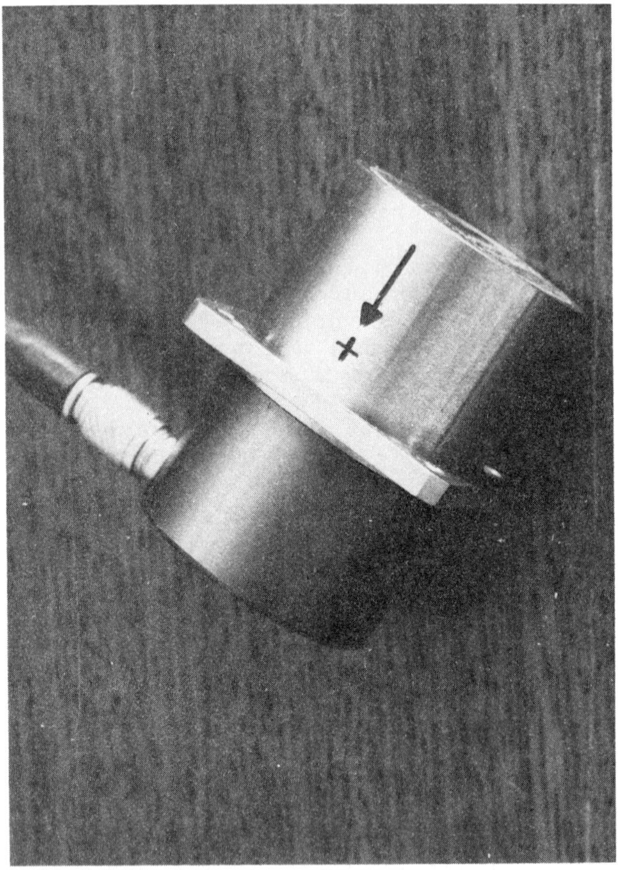

PLATE I

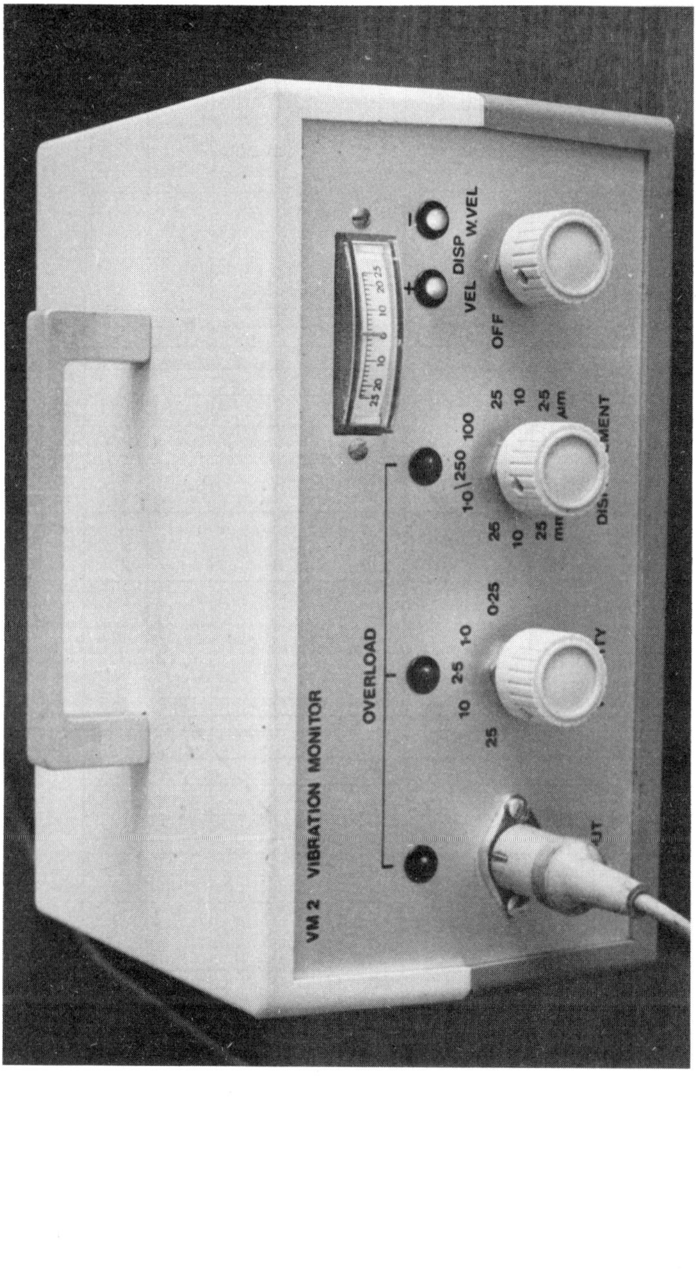

PLATE II

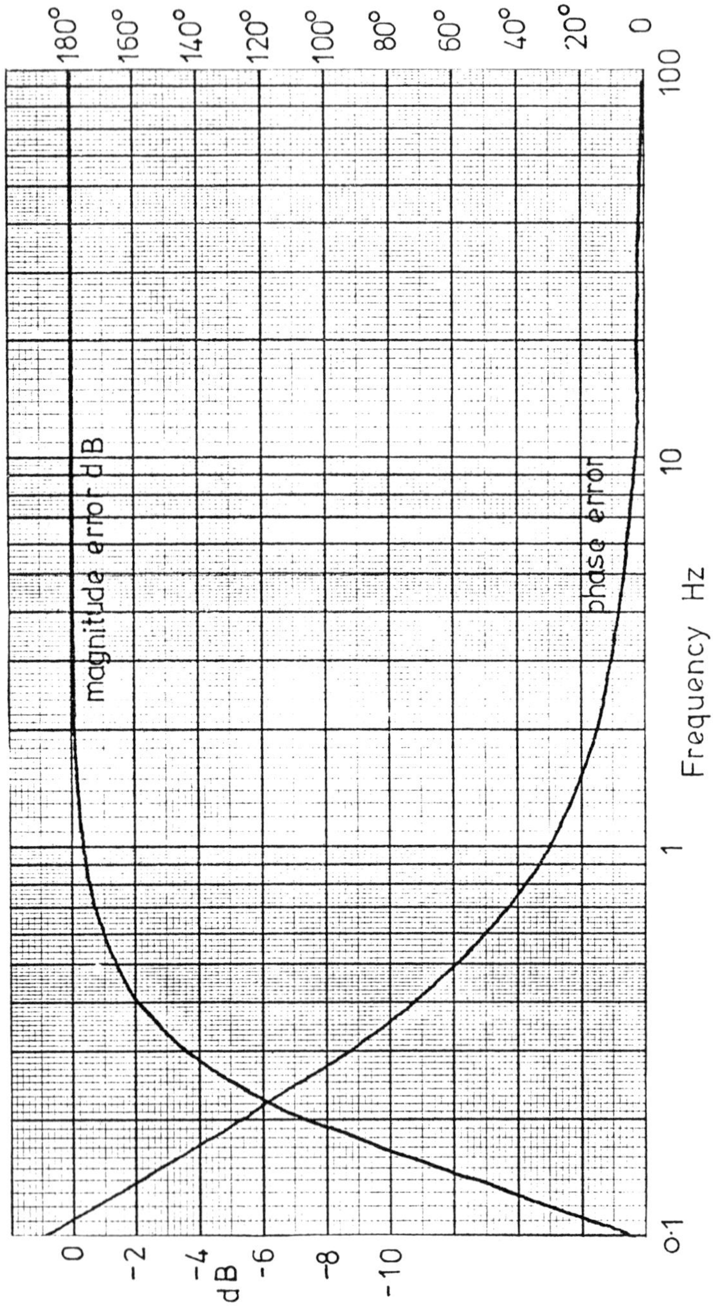

FIGURE 3 Magnitude and phase errors when measuring velocity or displacement

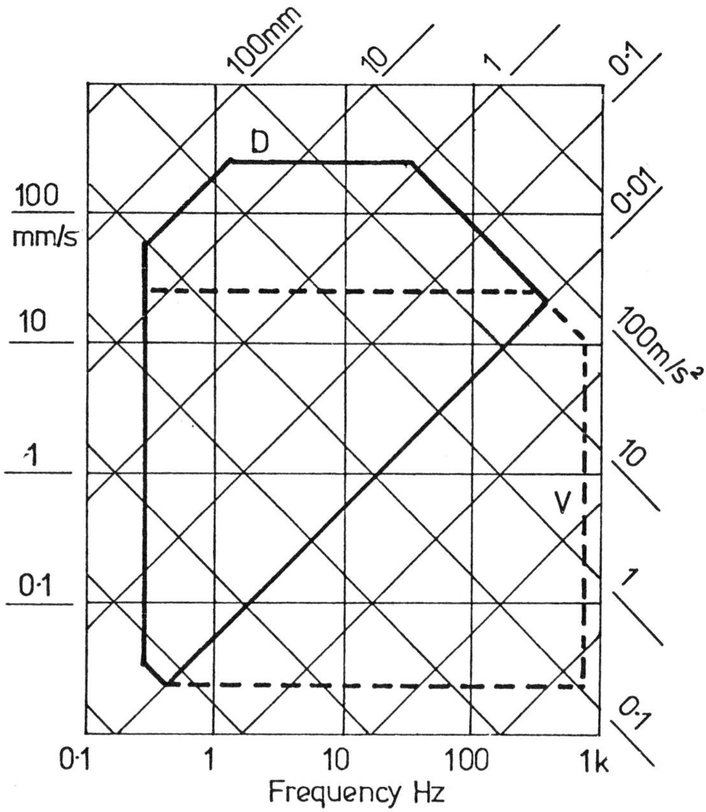

FIGURE 4 Range of application of the instrumentation

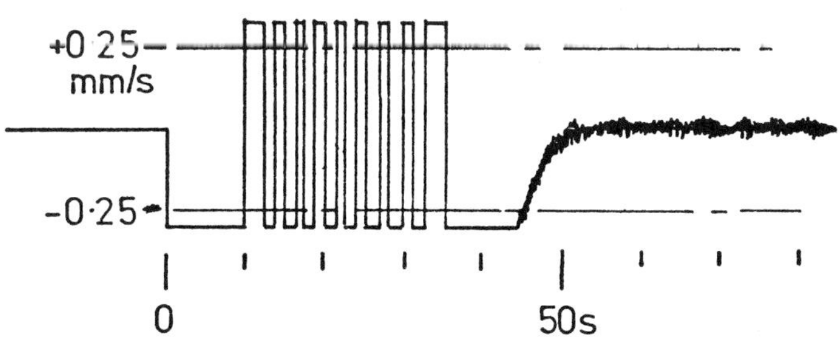

FIGURE 5 Output following switching on, 0·25mm/s range

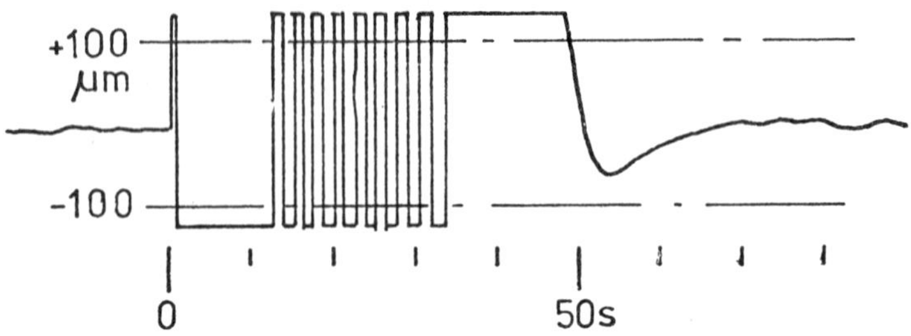

FIGURE 6 Settling after turning transducer over, 100µm range

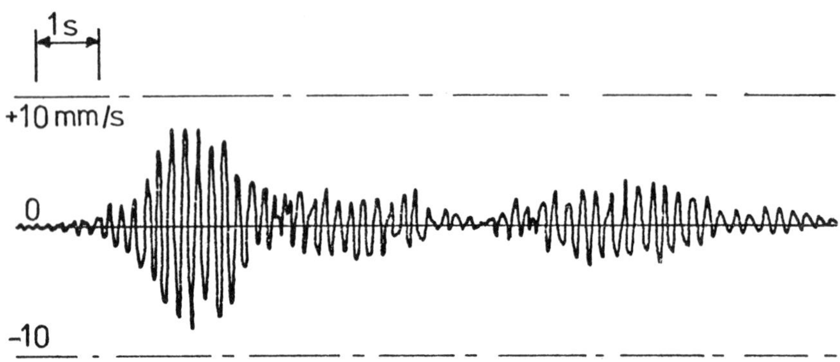

FIGURE 7 Velocity induced in a flyover by a lorry

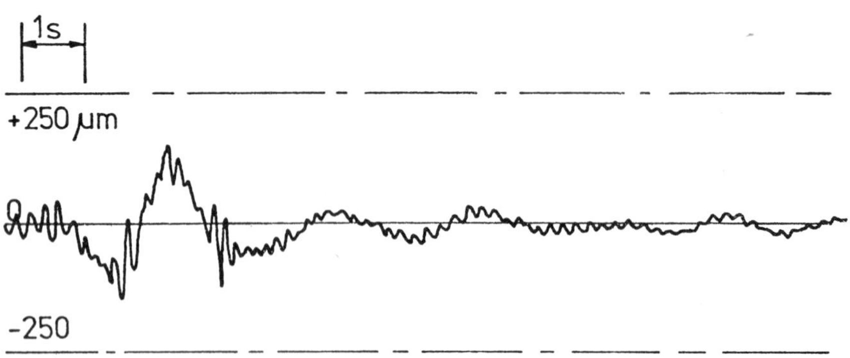

FIGURE 8 Displacement induced in a flyover by a lorry

Some other results of measurements are described by Lewis (1979).

FURTHER WORK

The instrument described is one of a number constructed by the authors and it has been possible to optimise the design for other applications.

One instrument has been designed to measure peak displacements of 10m down to frequencies of 0.1Hz with a noise level of 1mm. This was intended for ship heave measurement. Another instrument capable of measuring steady velocities has been described by Ball (1979).

At present an output display unit is being designed which will perform certain statistical operations on the measurements.

It is intended that this instrument will enable the characteristics of a structure to be determined under 'normal' conditions.

CONCLUSIONS

This paper has described a lightweight portable instrument for the measurement of structural vibration. The instrument has a wide range and produces output signals of velocity and displacement. It is capable of modification for different frequency ranges or sensitivities. Further work is proceeding on techniques for processing and displaying the output of such instrumentation.

ACKNOWLEDGMENTS

The authors wish to thank Mr.S.N.Mustow B.Sc.,C.Eng., County Surveyor for the West Midlands County Council, for permission to make measurements on the Moat Street Flyover, Coventry. We also thank Sundstrand Aerospace for permission to publish the PLATE 1.

REFERENCES

R.Ball and C.P.Lewis - A new technique for measuring lift velocities, Chartered Mechanical Engineer (1979) 56.

A.T.Dennison - The design of electromagnetic geophones, Geophysical Prospecting 1 (1953) 3-28.

K.C.Garner - Introduction to control system performance measurements, Pergammon Press, London (1968) 4-13.

International organisation for standardisation - Vibration and shock limits for occupants in buildings ISO/TC/108/SC4 (1975).

J.H.Kuehn - Ground-borne vibration-instrumentation and method for data acquisition, Proceedings of the Conference on Instrumentation for Ground Vibration and Earthquakes, Institution of Civil Engineers (1978) 59-62.

C.P.Lewis and R.Ball - An instrument for the measurement of structural vibrations, British Journal of Strain Analysis, to be published (1979).

R.J.Mainstone, J.B.Menzies and G.A.Weeks - Dynamic measurements on an experimental helicopter platform and on the Moat Street Flyover, Coventry, Proceedings of the Conference on Stress in Service, Institution of Civil Engineers (1966) Paper 9.

D.B.O'Neill - Vibration problems in civil engineering, Proceedings of the Conference on Instrumentation for Ground Vibration and Earthquakes, Institution of Civil Engineers (1978) 49-54.

B.O.Skipp - Ground vibration instrumentation - a general review, Proceedings of the Conference on Instrumentation for Ground Vibration and Earthquakes, Institution of Civil Engineers (1978) 11-34.

H.S.Ward - Vibration measurements on a modern steel bridge, Engineering Journal 9 (1966) 15-19.

P.L.Willmore - Some properties of heavily damped electromagnetic seismographs, Geophysics, Journal of the Royal Astronomical Society 4 (1961) 389-404.

THE FIELD MEASUREMENT OF THE TRANSMISSION OF VIBRATION FROM RAILWAY TRAFFIC

William Duncan, B Sc , Ph D

Lecturer, University of Strathclyde, Glasgow

SUMMARY

The paper outlines the problem of vibration transmission to structures from a remote source and the effect that the vibration will have on the structure and the people in the structure.

Details are given of an extensive investigation carried out to determine the levels of vibration caused by the passage of trains. The variation of the vibrations with distance from the railway is considered and the response of two houses, one with no damage and one with extensive structural damage, is determined.

TRANSMISSION OF VIBRATIONS

The transmission of vibrations from some source to a remote position is a complex problem depending on a great number of factors. The process of transmission can be separated into several distinct parts. The exciting force (source of vibration) sets up stress waves in the soil and these waves propagate through the soil until they encounter the foundation of some structure. The foundation will then begin to move in response to the ground waves and this in turn will cause a response of the structure, perhaps giving rise to perceptible or even damaging vibrations of the structure.

In this paper, the source of the vibration is a railway line running through the area on which heavy goods trains, and, in particular, oil trains, travel. In this situation there are three ways in which vibrations can be set up. Due to dynamic out-of-balance in the motors and in the wheels of the trains, dynamic forces can be set up between the wheels and the ground. As the train moves along the track the ground reaction at any given position is continuously varying from zero when there is no wheel at that position to a maximum when a wheel is at that position. This varying reaction constitutes a dynamic loading. The third source of vibration occurs when the wheels encounter a discontinuity in the track such as points or a joint in the track.

With modern design and manufacture, the out-of-balance of the mechanical parts has been reduced to a minimum and does not contribute significantly to the problem and so the main causes of the dynamic loading are the variation of surface loading as the wheels pass and the discontinuities in the track. The levels of vibration set up by these dynamic forces would depend on a great many factors such as the axle weights of the train, the spacing of the axles, the speed of the train, the condition of the track ballast, the condition and spacing of the trail-joints, etc.

The dynamic load on the soil will set up stress waves in the soil and those will propagate through the soil mass. Three different types of waves are set up - compression waves, shear waves and surface waves. Each of these waves has a different speed of propagation through the soil and each will reduce in intensity as it gets further from the source. In a perfectly homogeneous semi-infinite soil mass, the surface waves will decrease in proportion to the square root of the distance from the source, while the others will decrease in the ratio of the square of the distance. Thus the characteristics of a measured vibration will change considerably with distance, the surface waves becoming more dominant the further from the source. In a practical situation the soil mass is by no means homogeneous and any changes in density or water content will also alter the characteristics of the wave considerably.

When the ground waves encounter a foundation, the foundation will start to move in response to the ground motion. The actual response of the foundation will vary depending on the type of foundation and the size of the foundation. The response of the structure to the foundation movement will depend on the type of structure. Each structure has its own vibration characteristics and it is possible for reasonance to be set up at certain levels and frequencies of vibration either of the whole structure or of part of the structure.

The vibration set up in the structure can manifest itself in two ways. It can be 'felt' by the occupants of the structure when sitting or standing on a vibrating surface or by the audible rattling of some article in the house. The vibration can also cause damage in the structure. This could be superficial or architectural damage such as cracking of plaster, or if the intensity of vibration is great enough, structural damage could occur.

INFLUENCE OF VIBRATIONS ON PEOPLE

Although much research has been carried out into the problem of human sensitivity to vibrations, as yet no universal agreement on reasonable limits for vibrations in all cases. A comprehensive review of the early work was published by Steffans (1965). At the present time, the only formal guidance as to limiting vibration levels is given in a Draft for Development DD32 (1974) produced by the British Standards Institution. This document which is based on a proposed International Standard ISO 2631 (1978), is primarily concerned with the effects of vibration experienced by people exposed to vibrations in the course of their work such as vehicle drivers or machine operators or as passengers in a vehicle. Three main limits are defined. These are :-

(1) the preservation of working efficiency (fatigue - decreased proficiency boundary);

(2) the preservation of health and safety (exposure limit); and

(3) the preservation of comfort (reduced confort boundary).

The intensities of vibration corresponding to these limits have been found to vary with the time of exposure, the frequency of the vibration and the direction of the vibration relative to the subject and a set of limits in the frequency range 1 - 80 ltz for various times of exposure is given. Figure 1 shows the suggested limits for the reduced comfort boundary for a vibration acting along the spinal direction of the subject.

The use of these limits, particularly the lowest limit, that of the reduced comfort limit, as realistic limits for vibrations in residential properties, particularly at night, is highly questionable. The levels of vibration acceptable to someone travelling as a passenger in a vehicle will almost certainly not be acceptable to someone sitting in the quiet of his home. This fact has been recognised in the proposed German Standard DIN 4150 (1975). In this,proposed limits to vibration are specified for day time and night time situations for different zones of occupancy. These limits are specified in terms of KB values which are defined by the formula

$$KB = v_{rms} \frac{0.18\, f}{\sqrt{1 + (f/f_a)^2}} \qquad (1)$$

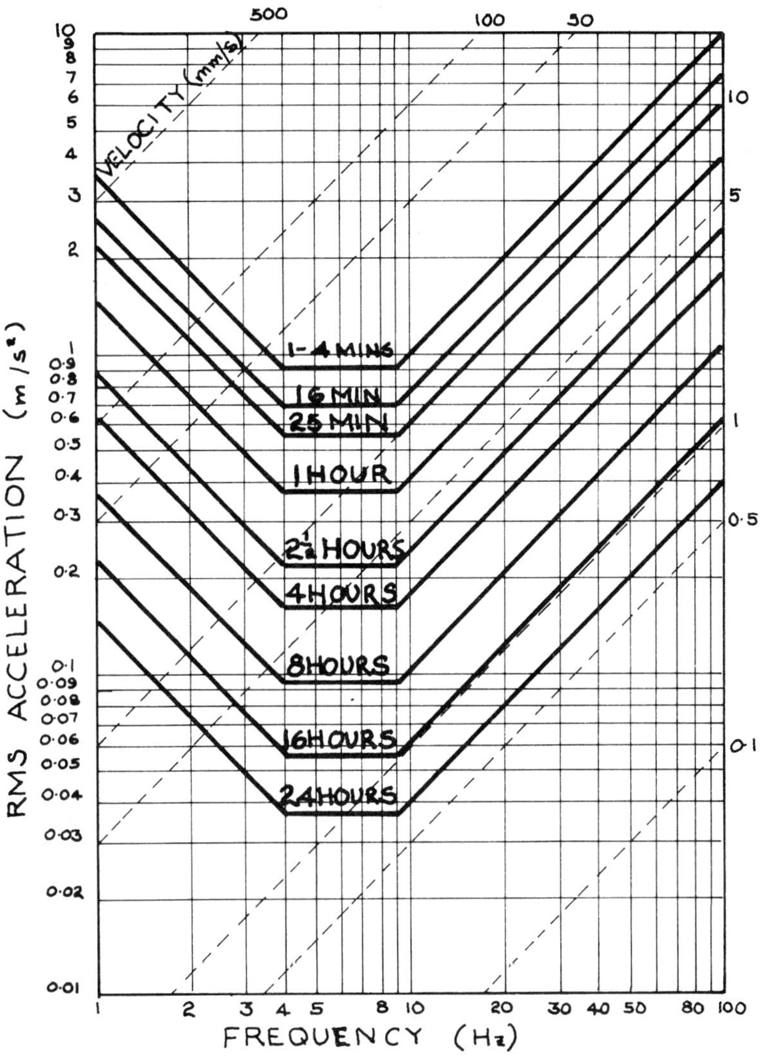

FIGURE I The reduced comfort boundary (DD32)

where v_{rms} is the RMS value of the velocity in mm/sec, f is the frequency of vibration in Hz and f_o is a reference frequency, in this case 5.6 Hz. The suggested limits are given in Table I and they depend on the length and the number of occurrences of the vibration.

The vibrations monitored in this investigation were of the order of 1-2 minutes in total duration, which corresponds to the lowest limit given in DD32 and since they occur only a few times a day they would come into the category of rarely occurring vibrations according to DIN 4150. Figure 4 gives a comparison between these two limits for the case of a domestic house. It is clear that the limits suggested by DD32 are very much higher than those permitted by the proposed rules in DIN 4150, particularly in the case of night time vibrations. The results of earlier work by Reiher and Meister (1931) are also plotted on Figure 2 and it is seen that these limits bear a reasonable relation to those suggested in DIN 4150.

INFLUENCE OF VIBRATIONS IN STRUCTURES

In predicting the effect of vibrations on structures, particularly structures such as domestic houses, again there is a marked lack of information as to reasonable limiting values. Much work has been done in monitoring the effect of blasting and traffic vibrations, but as there is such a wide variety of types of construction and of condition of the structure it has not yet been possible to set down firm guide lines which would prevent damage to the structure, either of an architectural or of a structural nature. Based on this earlier work, many sets of limits have been proposed. One such is due to Dawance and Seguin (1965) and is given in Figure 3 for the various degrees of damage. From this it is seen that there is a strong link between the damage limit and the velocity particularly above a frequency of about 10 Hz. This is reflected in many other suggested limits where the damage risk is measured in terms of the velocity measurement and is independant of frequency. Table II shows the limits suggested by Jorgensen (1973) and Table III contains limits proposed in the proposed German Standard DIN 4150 (1975). Both of these sets of limits refer to structural damage. DIN 4150 has suggested that if the resultant velocity is not larger than 2 min/sec, measured at the foundation, damage of an architectural nature should be excluded if the structure is in good condition and is built to the generally accepted standards for house building. These limits are plotted on Figure 3 for comparison.

All of these examples of limiting vibration levels refer to foundation vibrations or of vibrations of the main supporting walls of the structure. It is quite possible that other parts

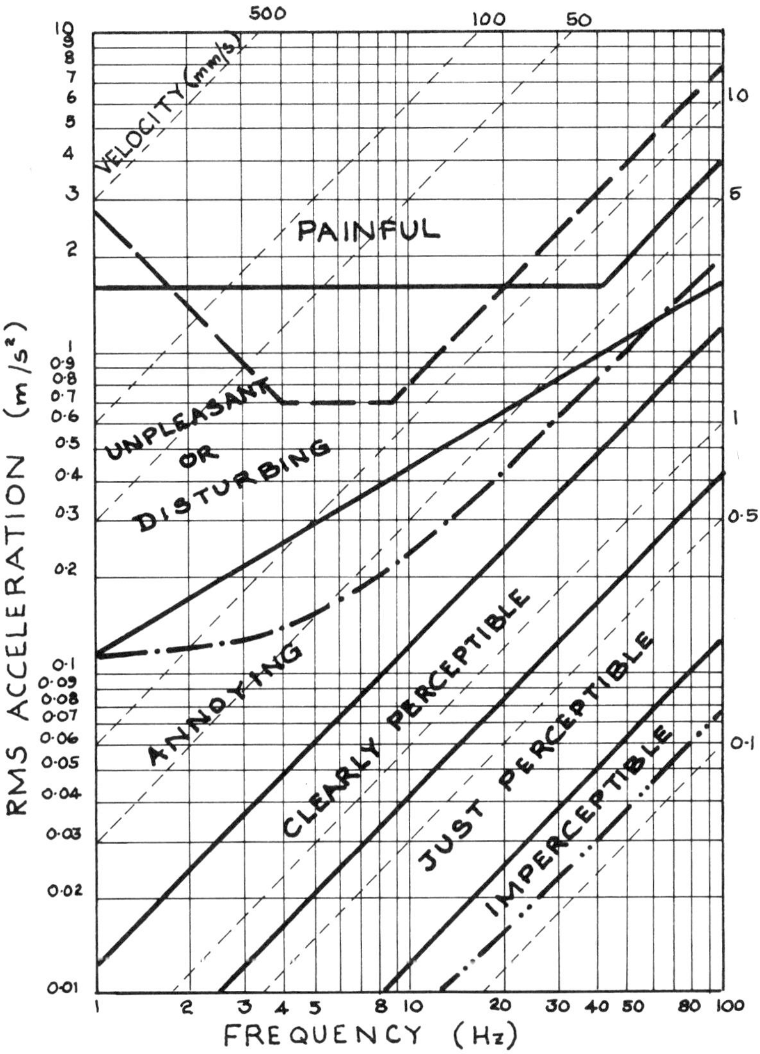

<u>Legend</u>

- ———— Reiher and Meister
- — — — DD32
- —·— Day time Limit, DIN 4150
- —··— Night time Limit, DIN 4150

FIGURE 2 Comparison of the limits of human sensitivities to vibration

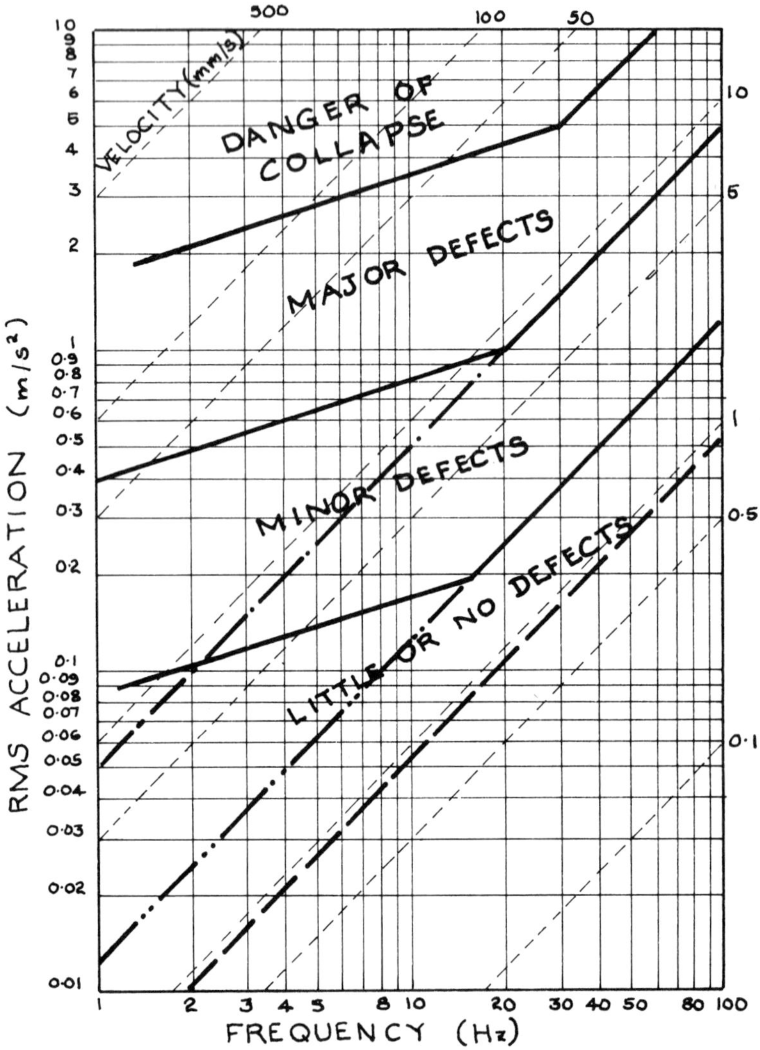

Legend

Dawance and Seguin

No loss in usefulness, DIN 4150.

No architectural damage, DIN 4150.

Continuous or many sisturbances (Jorgensen)

FIGURE 3 Comparison of damage criteria

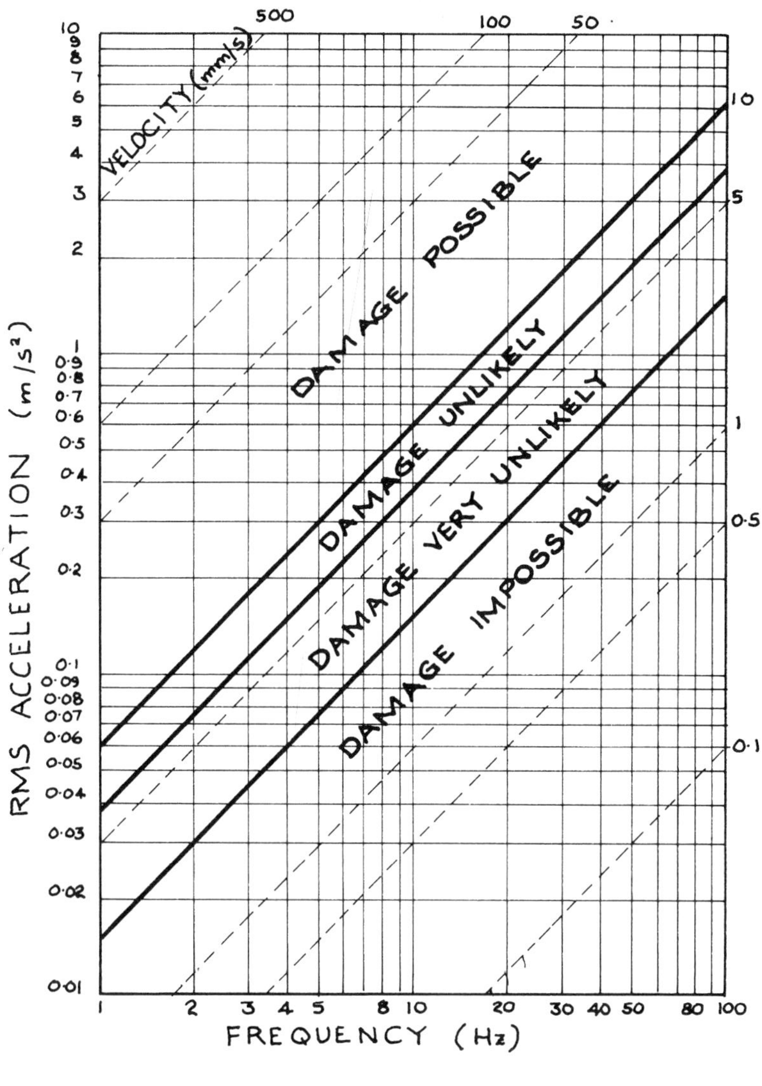

FIGURE 4 Damage criterion for suspended floors subjected to continuous vibration, DIN 4150

Building Zone	TIME	Limiting KB values	
		Continuous and frequently occurring vibrations	Rarely occurring vibrations
Entirely Domestic Housing (WR) Mostly Domestic Housing (WA) Holiday Home Zone (SW) Small Housing Estate (WS)	DAY	0.2	4
	NIGHT	0.15	0.15
Rural Zone (MD) Mixed Zone (M1) Central District (MK)	DAY	0.3	8
	NIGHT	0.2	0.2
Commercial Zone (GE) Valid also for Offices	DAY	0.4	12
	NIGHT	0.3	0.3
Industrial Zone (GI)	DAY	0.6	12
	NIGHT	0.4	0.4

TABLE I Limiting values for the estimation of the effect of vibrations on people according to DIN 4150 (1975)

Band	Kind of Business	Limiting value of velocity
1	Dwelling houses, business premises and other structures of similar construction.	8 mm/sec
2	Well reinforced buildings consisting of heavy components and well stiffened skeletal structures	30 mm/sec
3	Buildings not comforming to Bands 1 and 2 and protected buildings	4 mm/sec

TABLE III Limiting values of velocity on building foundations up to which no significant loss in usefulness will be expected according to DIN 4150 (1975)

Building Class	Type of Building	Safety Limiting Velocity (mm/sec)		
		Single Disturbance	Two Disturbances	Continuous in many Disturbances
1	Ruins, officially preserved and protected buildings	1.3	1.1	0.2 - 0.4
2	Buildings with visible damage and cracks in masonry	2.7	2.2	0.4 - 0.6
3	Buildings in good structural condition without damage. Cracks in plaster are permitted	5.2	4.4	0.8 - 1.0
4	Good and solid industry buildings	6.5 - 25	5 - 20	1.2 - 5.0

TABLE II Safety limits for buildings subjected to disturbances transmitted through foundation according to Jorgensen (1973) of the structure such as suspended floors, could experience much higher levels of vibration without suffering any damage. Figure 4 gives the recommendation given in DIN 4150 for the effect of continuous vibrations on a suspended floor. In the case of short duration and rarely occurring vibrations, the limits would be significantly higher than these perhaps by as much as a factor of 10.

PROGRAM OF FIELD MEASUREMENTS

The investigation described in this paper was mounted as a result of a large number of complaints from residents living near the railway. One main group of complaints was centred on a street running parallel to and at a distance of 80 m from the railway. The houses in this street were for the most part two storey stone built semi-detached or terrace houses. The ground between the street and the track was the site of a former railway goods yard and at the time of the investigation undeveloped. A sample of six houses in this street were visited and the householders interviewed. All complained of discomfort from the vibrations in various parts of the houses, with the normal evidence of audible vibrations of small articles in the houses. As would be expected, the vibrations were said to be worse at night. As a result of these enquiries, it was decided to monitor the vibration levels in a typical house in the street and to measure the vibration at varying distances from the railway to the house. This house is designated House B.

During the course of the investigation, the condition of a house in another part of the town came to light and it was decided to include this house in the investigation. This building was a semi-detached two storey brick building about 40 years old. The external leaf of brickwork had been found to be in very poor condition with severe cracking. No information as to the history of the damage was available. As the house was only 30 m from the railway line, the possibility of the vibrations caused by the railway being a cause of damage had been raised. This house is designated House C.

During the period from July to November 1974, an extensive monitoring programme was carried out. A total of 140 trains were monitored and the vibrations set up at various locations due to the passage of these trains were recorded. For the most part the trains were heavy goods trains or oil trains, but some passenger trains, empty goods trains and single engines were monitored for comparison. In addition to these measurements, recordings were made of vibrations due to traffic, House B, and the effect of normal activities such as walking and shutting doors in both houses.

The measurements fall into three groups:-

Group A - Ground measurements from the railway to House B

Group B - Measurements on House B

Group C - Measurements on House C.

EXPERIMENTAL RESULTS

It is clear from the results that the problem of vibrations set up by the passage of trains is an extremely complex one. Figure 5 shows the vibrations recorded on the ground floor window ledge of House C and it is seen that there is no certain link between the type of train and the severity of the vibrations. This problem is illustrated again in Table IV which shows the vibrations measured on the external wall of House B. The results do not follow any pattern and there is no obvious relationship between the length of the train, the speed of the train, the weight of the train and the level of vibration. In fact there is a randomness, not only in these results but in all the results, which makes detailed conclusions difficult. It is impossible to say that a vibration of greater intensity could not be set up by some other combination of the factors mentioned above, but in view of the number of the trains monitored, it is felt that reasonable bounds can be given to the levels of vibration which can occur in the various positions and that a reasonable estimate could be made of their effect.

Group A Measurements

The measurements taken in Group A are shown in Figure 6. As would be expected, the distance from the point under consideration to the railway has a marked effect on the levels of vibration and on the form of the vibrations. Near the railway the vibrations have a predominant frequency between 50 Hz and 70 Hz. As the distance increases a much lower frequency 5-15 Hz begins to predominate. This change in behaviour is demonstrated in Figure 7 which shows a 1/3 octave analysis of typical recordings made at a distance of 5 m and 80 m from the railway. At 80 m from the track, the velocity due to the low frequency vibrations is of the order of 8 times that due to the high frequency vibrations and consequently is more important.

Group B Measurements

The measurements of Group B are shown in Figure 8. The measurements taken on the ground floor window ledge and on the ground outside the house exhibit the same characteristics as the readings in Group B regarding the two predominant frequencies and it is the low frequency velocities which are the worst.

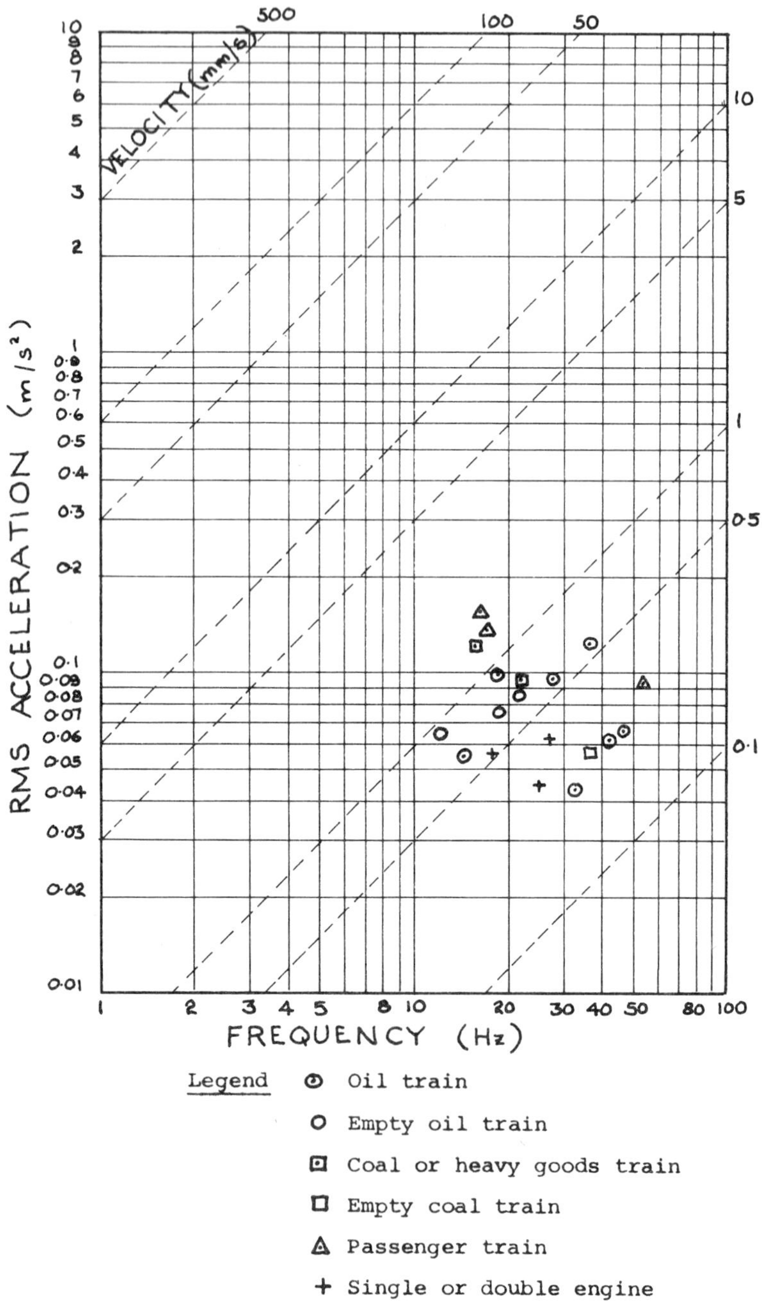

FIGURE 5 Vibrations measured on window ledge at House C showing the effect of train type

Train	Type	Description	Speed (MPH)	Measured Velocity (mm/sec)
4/10/B	Oil	1 Engine, 19 tanks	27	0.82
4/10/D	Oil	1 Engine, 12 tanks	38	0.57
4/10/E	Oil	2 Engines, 32 tanks	18	1.06
1/10/A	Oil	1 Engine, 16 tanks	30	0.9
1/10/B	Oil	2 Engines, 30 tanks	27	0.55
1/10/C	Coal	1 Engine, 16 wagons	25	0.64
11/9/A	Oil	2 Engines, 32 tanks	25	1.65
11/9/B	Oil	1 Engine, 10 tanks	24	0.80
11/9/C	Empty Coal	1 Engine, 42 wagons	46	0.82

TABLE IV Vibrations measured on the ground floor window ledge at House B

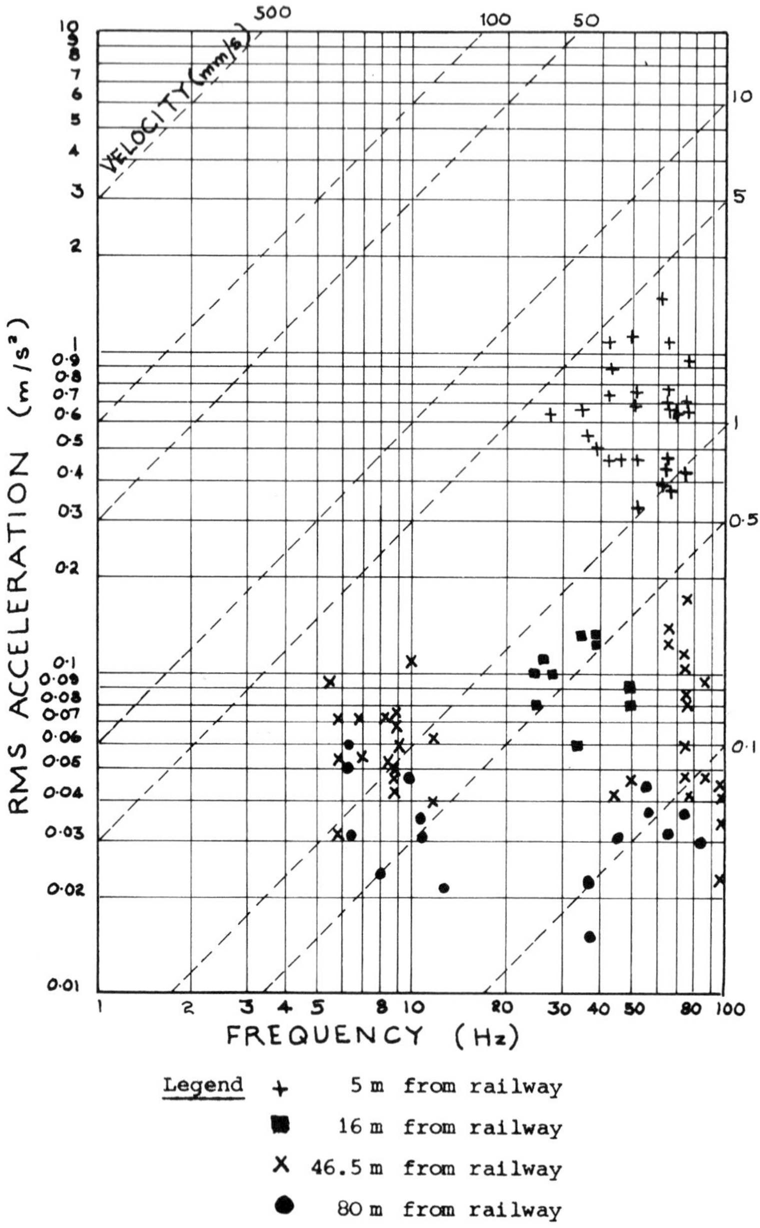

FIGURE 6 Vibrations recorded at different distances from the railway

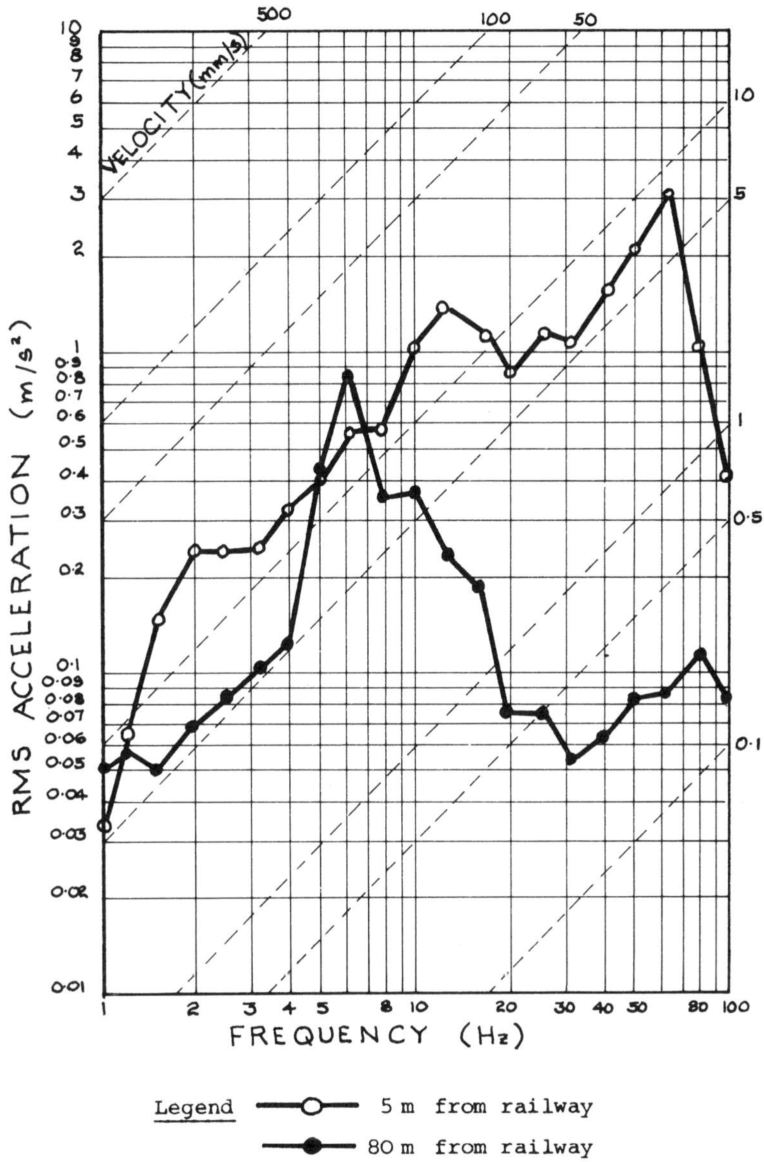

FIGURE 7 1/3 octave analysis of typical records

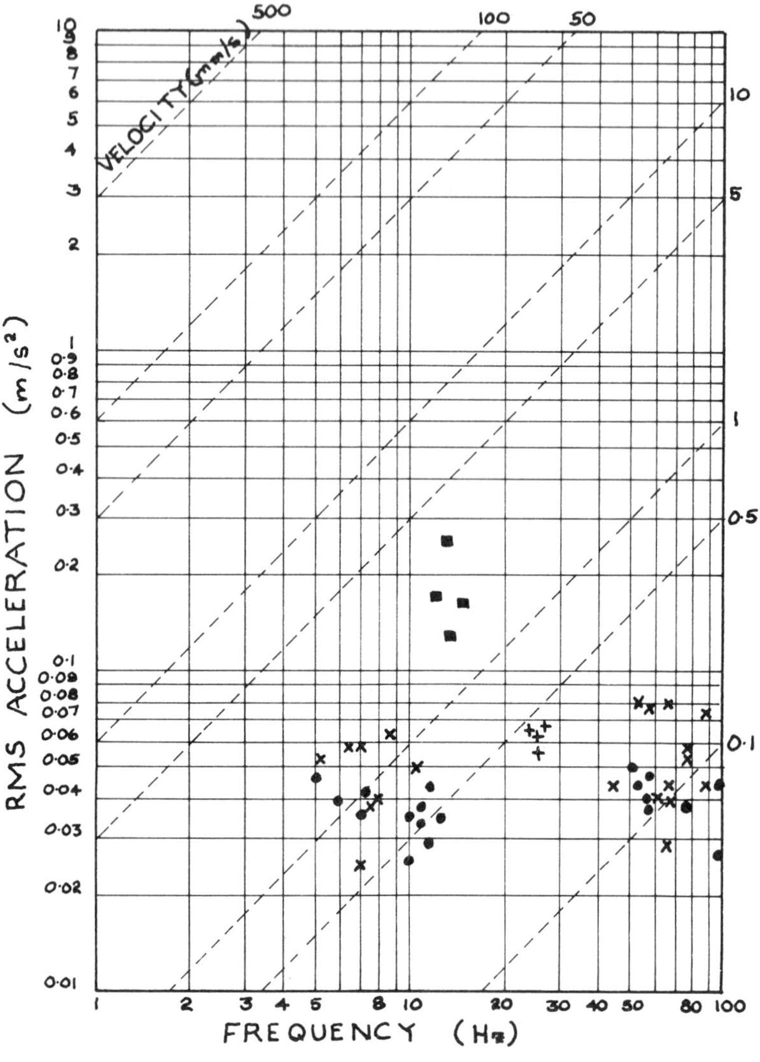

Legend
■ Floor of upper room
\+ Floor of lower room
X Ground floor window ledge
● Ground outside house

FIGURE 8 Vibrations recorded in House B

The maximum velocities recorded indicate that even architectural damage is unlikely.

The levels of vibration measured on the suspended floors are much higher than those measured on the walls. Approximate measurements were made of the natural frequency of the floors and those were found to be 16.9 Hz for the upper and 26.1 Hz for the lower room. The results show that for all the trains recorded, both floors were vibrating very close to their natural frequencies. The levels recorded on the upper floor were worse than those on the lower floor and were of an intensity which would cause annoyance to the occupants.

Group C Measurements

The vibrations of the upper and lower window ledges, Figure 9, exhibit similar characteristics and range of intensity. The frequencies vary from 15 Hz to 50 Hz with no clear pattern emerging as to which type of train is worst or of the effect of speed or length of the trains. The levels of vibration recorded should not have caused any significant damage if the building was in good condition.

The vibrations recorded on the floors of the three rooms, Figure 10, indicate that the floors are all vibrating very near their natural frequency for all the trains monitored. The levels of vibration were such as would cause considerable annoyance to the occupants of the house.

CONCLUSIONS

The results of the investigation demonstrate that there is no simple correlation between the speed of the trains, the length of the trains, the weight of the trains and the intensities of the vibrations set up and that even a relatively small change in one of these parameters can produce a large change in the resulting vibration levels and characteristics. This means that in any investigation, a sufficiently large number of readings must be taken to ensure that a reasonable assessment of the range of values is obtained.

Comparing the levels of vibration recorded in the two houses with the limits given in Figure 2, and bearing in mind the complaints of the occupants, it is seen that the limits due to Reiher and Meister (1931) and DIN 4150 (1975) seem to be reasonable, at least at the lower end of the scale. The limits proposed in DD32 (1974) are unacceptably high and should not be used for this situation. In both houses, the floor vibrations recorded were of an intensity that would cause annoyance to the occupants, particularly at night.

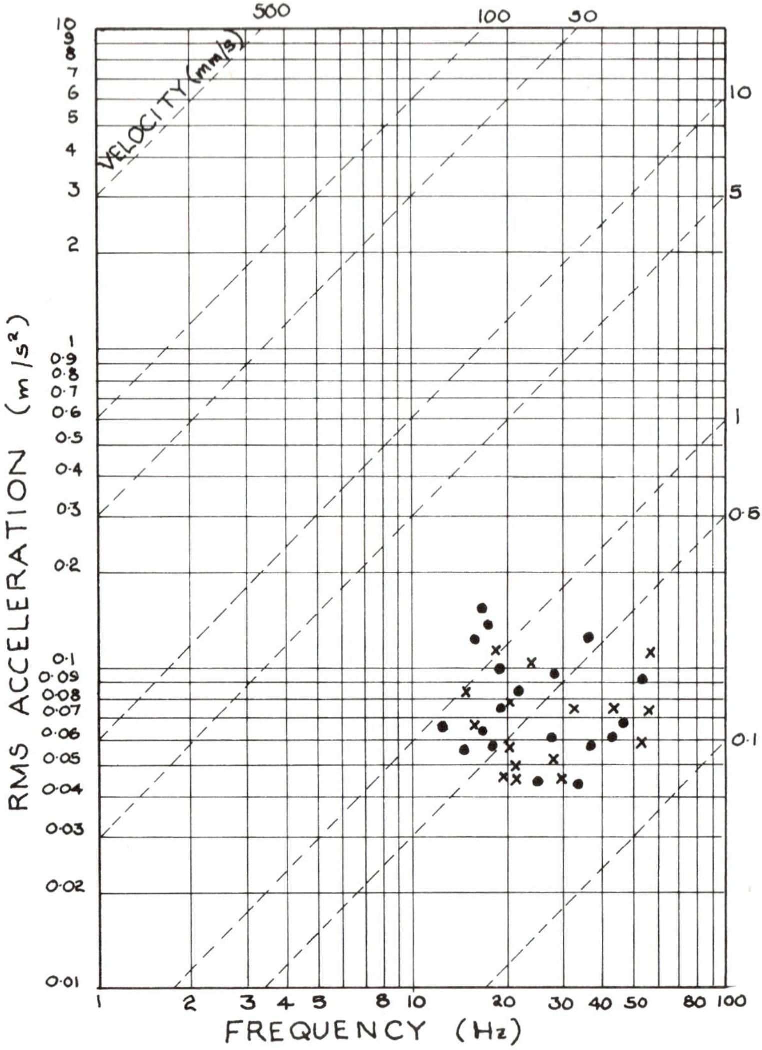

Legend ● Ground floor window ledge

X Upper floor window ledge

FIGURE 9 Vibrations recorded on the wall of House C

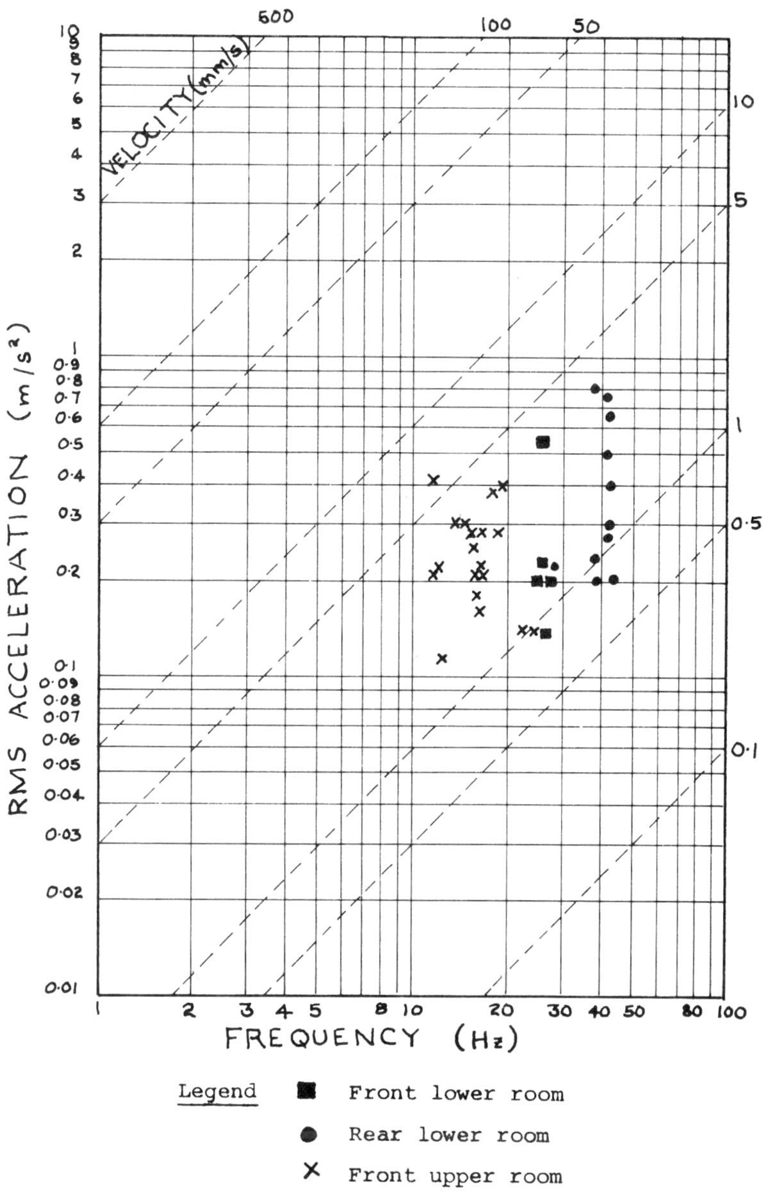

FIGURE 10 Floor vibrations recorded at House C

In assessing the risk of damage to the houses it should be remembered that House B had no signs of any damage while House C had major structural cracking. The vibration levels in both cases were similar and were below the levels set in DIN 150 (1975) and Dawance and Seguin (1965) for architectural damage. On the basis of this, no damage should have occurred at House C if it had been in good structural condition. Subsequent investigation indicated that the mortar used in this house was of a very poor quality and that the foundations were incorrectly constructed. It is possible that the vibrations recorded could have aggravated the condition of the already poor structure and triggered off the damage, but this is conjecture.

REFERENCES

G. Dawance and M. Seguin - Effects of vibrations in buildings, Proceedings of the Symposium of the International Society for Earthquake Engineering, London, (1965) 31-33.

DD32 - Guide to the evaluation of human exposure to whole body vibrations, British Standards Institution, (1974).

DIN 4150 - Vibrations in Buildings Deutsches Institut fur Normuny (1975).

ISO 2631 - Guide for the evaluation of human exposure to whole body vibrations, International Standards Organisation (1978).

A.R. Jorgensen - Effect of vibrations on structures. Journal of Military Engineering, Sweden, 1973.

J.H. Reiher and Meister - Human sensitivity to vibrations Forsch auf den Geb des Ing. 2,11 (1931) 381-6.

R.J. Steffans - Some aspects of structural vibrations. Proceedings of the Symposium of the International Society for Earthquake Engineering, London (1965) 1-30.

EFFECT OF PARASEISMIC VIBRATIONS ON BUILDINGS

Roman Ciesielski
Technical University of Cracow

INTRODUCTION

The problem of the effect of seismic action on buildings has been comparatively well resolved by engineering science. Experimental techniques are very advanced and their use has had a substantial influence on verification of the theory. Earth tremors are communicated directly by means of longitudinal and transverse waves and further by surface longitudinal waves (Rayleigh) and surface transverse waves (Love). The waves are transmitted through the ground to the buildings and cause their movement and set up inertial forces resulting in additional internal forces which may lead to damage or even destruction of buildings. Tremors induced in a building by the ground may be due to other than natural causes. They can be caused by man's activities, either partially or completely under his control. Their effects have been termed paraseismic. In many practical cases they assume a great and constantly growing importance because of the increase in activities causing vibrations and increase in their intensities. In the case of paraseismic vibrations, the question is whether and to what extent it is possible to take advantage of the known methods of analysis of seismic effects. To answer this, it is necessary to consider the nature of the paraseismic movements, their propagation through the ground and the reactions induced by them in structures. The first part of this paper concentrates on this aspect, emphasising the differences between paraseismic movements and earth tremors. The second part gives an approximate method of assessment of the degree of damage such effects may have on buildings. Figure 1 shows the block diagram used in the analysis.

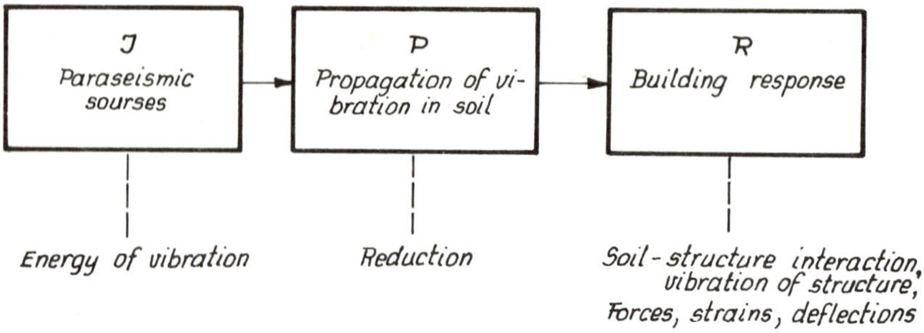

Fig. 1
The flow diagram for the investigation of paraseismic effects of structures

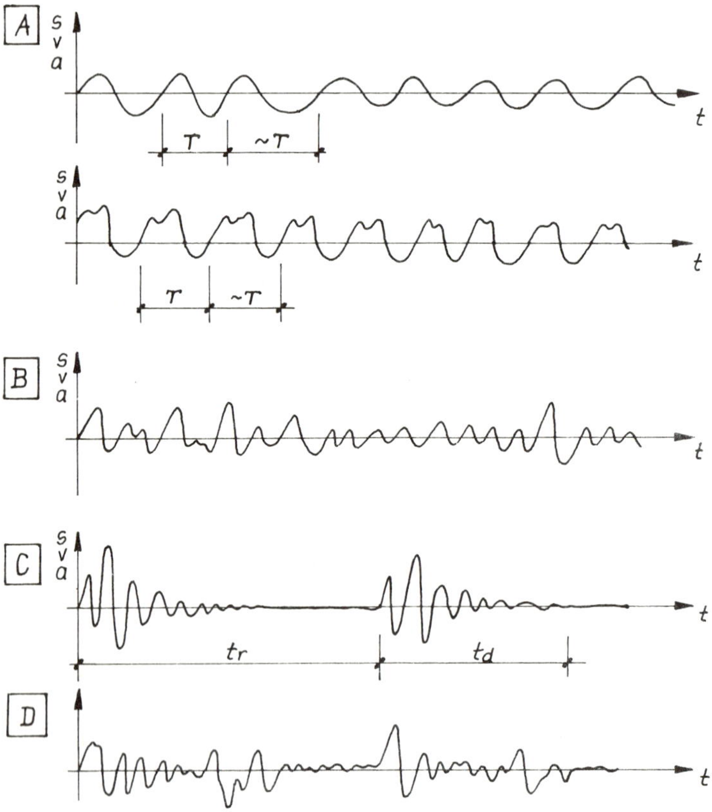

Fig. 2 4 characteristic types of vibration

NATURE OF THE SOURCES OF PARASEISMIC FORCES

Frequent sources of paraseismic forces are machines or plant resting directly on the ground or on special foundations, machines moving over ground or on roads or rails, driving of piles and sheet piling, dropping of heavy loads, explosions, quarry blasting, collapse of shallow workings. In consequence of such activities vibrations are produced in the ground which are transmitted to buildings. They can be divided into four groups in accordance with Figure 2.

A. Long lasting or harmonic or quasi harmonic vibrations.
B. Long lasting vibrations of a standing wave, or similar type of varying parameters.
C. Single or sporadic tremors arising at intervals greater than the damping time, ($t_r > t_d$).
D. Irregular vibrations and tremors of unpredictable characteristics.

The dynamic characteristics of the sources of paraseismic movements are represented by spectra of vibrations. These are normally spectra of acceleration but may also be those of velocity or displacement. Such spectra show the total response of a chosen object such as an oscillator with one degree of freedom of dynamic movement - Figure 3 - vibrating as a result of the movement of the ground on which it is installed. Generally engineers are concerned with horizontal movements but vertical movements can be considered in a similar manner. For the system with one degree of freedom and the Kelvin-Voigt damping model (assuming small damping force) there is a well-known equation (14), (16),

$$m\ddot{y}(t) + r_v \dot{y}(t) + cy(t) = p\ t \qquad (1)$$

After transformation and introduction of simplifying relationships

$$\begin{aligned} p(t) &= m\ddot{u}(\tau) \\ \omega_d &= \omega\sqrt{1-D^2} \approx \omega \\ D &= \Delta / 2\pi \end{aligned} \qquad (2)$$

we obtain:

$$dy(t) = \frac{m\ddot{u}(\tau)}{m\ \omega}\left\{\exp\left[-D\omega(t-\tau)\right]\sin\omega\ (t-\tau)\right\}d\tau \qquad (3)$$

$$y(t) = -\frac{1}{\omega}\int_0^t \ddot{u}(\tau)\ \exp\left[-D\omega(t-\tau)\right]\sin\omega(t-\tau)d\tau \qquad (4)$$

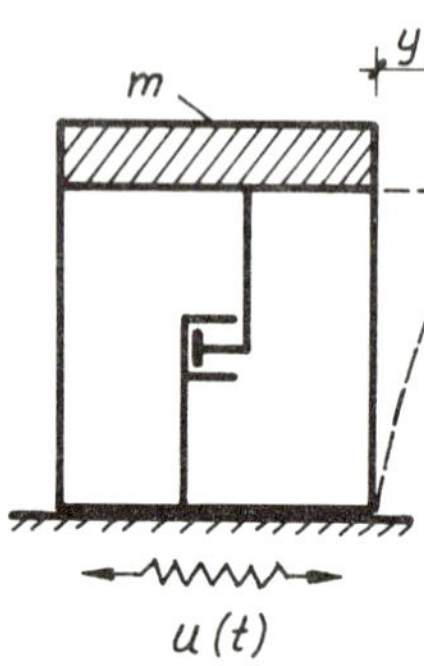

Fig. 3
Model of a system with one degree of freedom of movement with damping

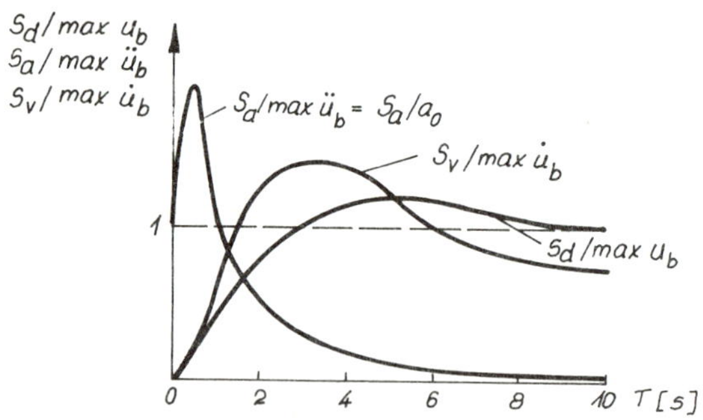

Fig. 4
Examples of smoothed out spectra of vibrations S_y, S_v, S_a

Differentiation of equation (4) gives us equations for speed and acceleration

$$\dot{y}(t) = -\int_0^t \ddot{u}(\tau)\ \exp\left[-D\omega(t-\tau)\right]\cos\omega(t-\tau)d\tau \qquad (5)$$

$$\ddot{y}(t) = \omega\int_0^t \ddot{u}(t)\ \exp\left[-D\omega(t-\tau)\right]\sin\omega(t-\tau)d\tau - \ddot{u}(t) \qquad (6)$$

The summation of the maximum displacements of the system is called the spectrum of displacements

$$S_y = \max\left[y(t)\right] = \frac{1}{\omega}\max\left\{\int_0^t \ddot{u}(\tau)\ \exp\left[-D\omega(t-\tau)\right]\sin\omega(t-\tau)d\tau\right\} \qquad (7)$$

Similarly it is possible to determine the spectra of velocity and acceleration as maxima of S_v and S_a, which can be represented in a general form by

$$\begin{aligned} S_y &= \max\left|y(\omega,D,t)\right| \\ S_v &= \max\left|\dot{y}(\omega,D,t)\right| \\ S_a &= \max\left|\ddot{y}(\omega,D,t) + \ddot{u}(t)\right| \end{aligned} \qquad (8)$$

Values obtained in this way give, depending on the type of oscillator, a graph of the spectrum of vibrations (displacements S_y, velocity S_v or acceleration S_a). As an approximation it is possible to make use of the relations (15)

$$\begin{aligned} S_v &= \omega S_y = \frac{1}{\omega} S_a \\ S_a &= \omega^2 S_y = \omega S_v \end{aligned} \qquad (9)$$

It is usual to present spectra of vibrations smoothed out and in relation to the maximum displacement, velocity or acceleration, for instance the gravity g Figure 4. There are many known spectra of earthquakes, e.g. the Californian earthquake at El Centro, Figure 5 a,b,c, (21). An interesting comparison of smoothed out acceleration spectra for several earthquakes is made in Figure 6. Therefore the basic problem is the determination of the spectra of vibrations for paraseismic forces. This can be done according to the described methods on the basis of measurement of vibrations of the ground (8), (17). From our own measurements of vibrations (8), certain basic rules were established for several types of source of vibration. For others, hypothetical rules were adopted in the form of smoothed out spectra.

Generally, four types of spectra were distinguished - Figure 7.

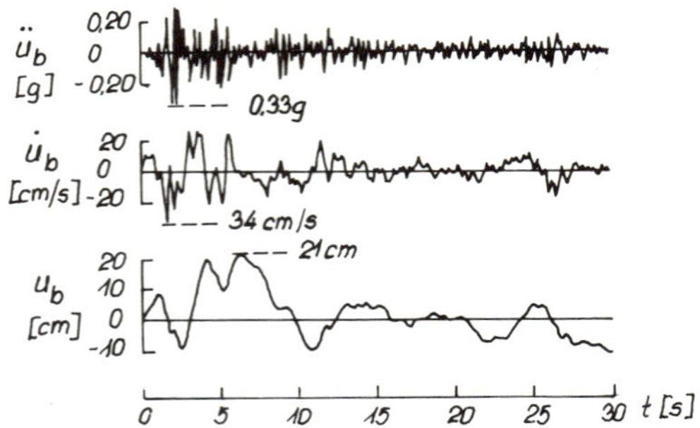

Fig. 5a. [21]

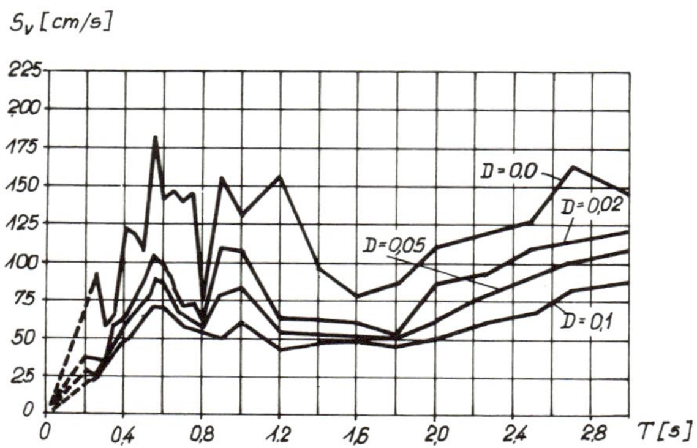

Fig. 5b. [21]

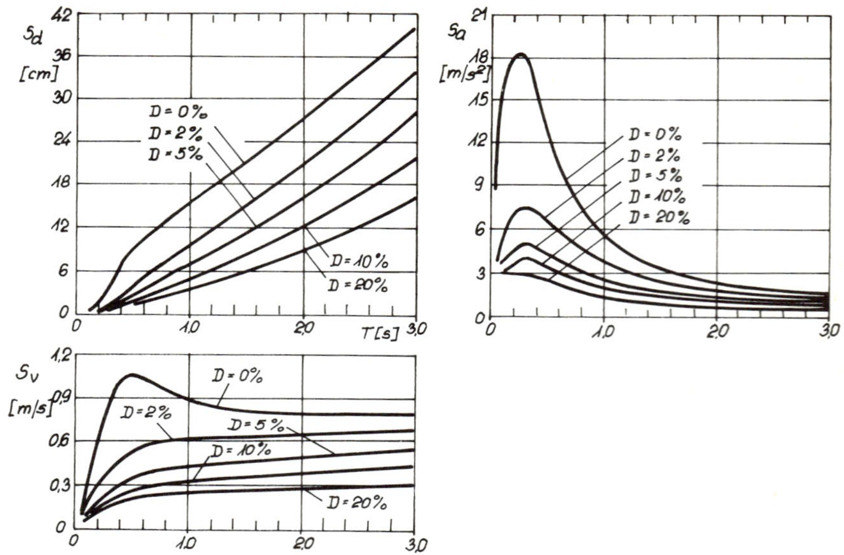

Fig. 5c. [21]

Fig. 5

The El Centro, California earthquake, 1940 (21)

a) graph of tremors, displacement, velocitý, acceleration

b) spectrum of velocity for different damping values of D

c) smoothed out spectra of vibrations S_d, S_v, S_a

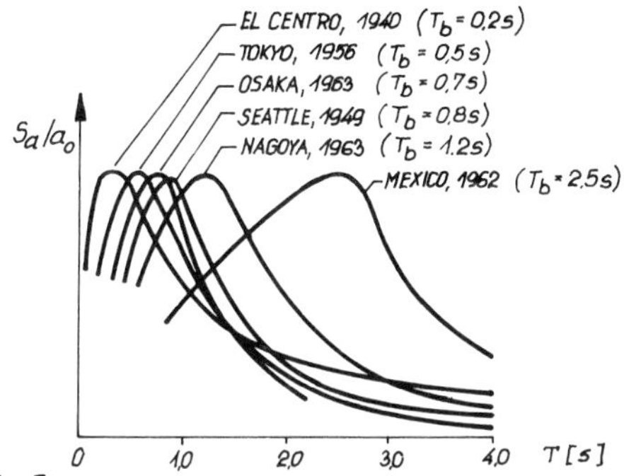

Fig. 6 [13]

Examples of smoothed out spectra of acceleration of several earthquakes. Maxima a_o dependent on T_b, characteristic for the given ground conditions

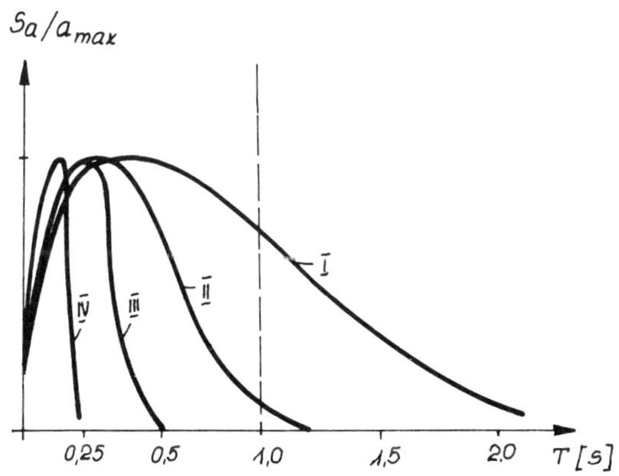

Fig. 7

4 types of spectra of acceleration for paraseismic vibrations I, II, III, IV

I - wide spectrum of frequencies e.g. 0.1 - 100 Hz i.e. T = 0.01 s - 10 s.

II - medium wide spectra e.g. 1 - 50 Hz.

III - narrow compact spectra e.g. 5 - 25 Hz.

IV - very narrow spectra approximating single frequency e.g. 2 and 5 Hz.

Single frequency spectra do not arise during the transmission of tremors through the ground, because the latter acts as a filter. Examples of the original record of ground tremors and its analysis (analysed by Bruel and Kjaer) are shown in Figure 8. It is necessary here to stress the effect of the damping action as defined by the critical damping coefficient, termed Lehr's damping 'D' (1), (15). This factor determines the final shape and magnitude of the spectrum ordinates used in calculations.

The principles used in the case of natural earth tremors may be considered important with respect to paraseismic tremors of types c and d. On the other hand, with vibrations of the standing wave form of types a or b, damping action must be more carefully differentiated with respect to the building under consideration, because in the latter case it is also necessary to take fatigue into account. Table 1 gives suggested typical damping values for various types of construction, taking into account structural vibrations only in the elastic region. The building norms for most countries (13), (15) and (19) give spectrographs of dynamic coefficients for determination of seismic forces from equation

$$S = f\left[(\alpha, \beta, \eta)\ Q\right] \qquad (10)$$

where α denotes vibrations of the ground

β " the interaction between vibrations of the ground and of the structure

η denotes dynamic properties of the structure

Q " the weight of the structure or of its part

Spectrographs for several seismic standards are given in Figure 9.

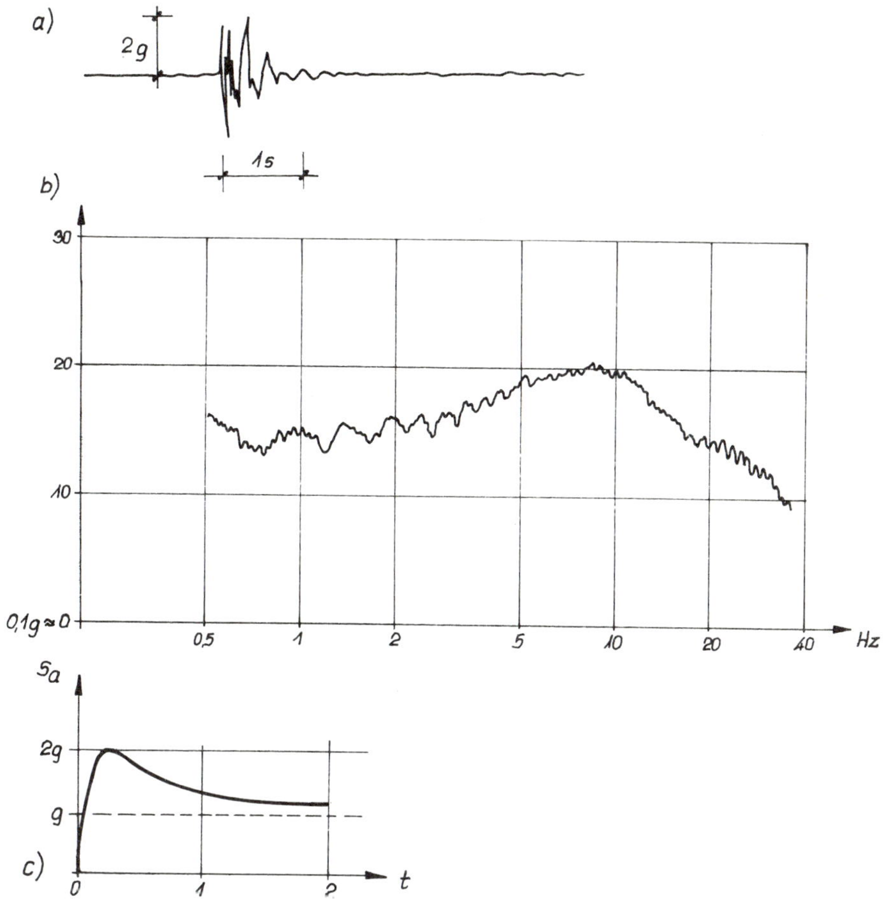

Fig. 8 Examples of measurement of ground vibrations under impact tests

a) graph of acceleration a
b) spectral analysis (a - n)
c) smoothed out spectra of vibrations (a - T)

Tab. 1 Damping Coefficients for Vibrations of Normal Buildings

	Constructional Material	Subsidiary Materials	Type of Structure	D (%)	Remarks
1	Load bearing brick walls	Masonry blocks Lightweight concrete	Dwelling Houses Buildings of general public use. Other buildings (working in shear)	1.8 - 3.5 1.5 - 3.0	In the case of damage assume greater values
2	Load bearing brick walls	-	Industrial buildings	1.2 - 2.5	
3	Concrete	Concrete	Monolithic buildings for dwellings and general public use. Other monolithic structures. Prefabricated buildings with load bearing walls.	1.2 - 2.5 1.0 - 2.0 1.5 - 2.8	

4	Concrete	Masonry, bricks and blocks, Light concrete	Framed or skeletal structures for public use	1.8 - 3.0	Without solid infill panels reduce by 0.5
			Industrial	1.0 - 2.0	
5	Steel	Masonry, bricks and blocks, Light-weight concrete	Skeletal structures without infill panels		
			Bolted connections	0.8 - 1.0	
			Welded connections	0.4 - 0.7	
			Ditto with infill panels		
			Bolted connections	1.0 - 1.3	
			Welded connections	0.6 - 1.0	
6	Timber	Lightweight panels	Dwellings and farm structures	1.0 - 3.0	Depending on the state and type of structure and of connections.

NB: The values of 'D' given in the table do not take into account damping due to ground or arising in transfer of vibrations from ground to structures, which in some cases may increase the values by up to 50%.

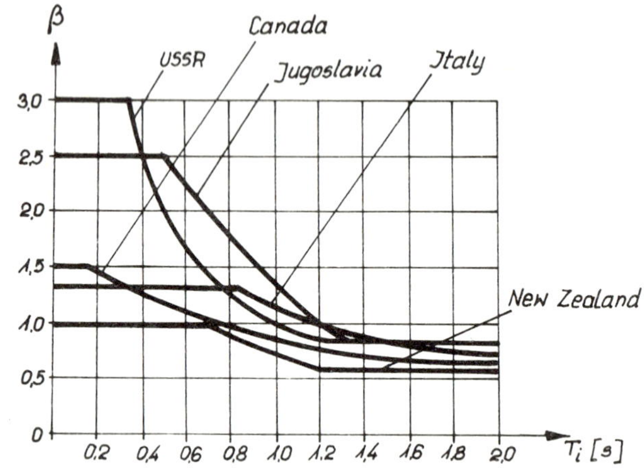

Fig. 9 Examples from building standards of spectral graphs for the calculation of structures for seismic effects

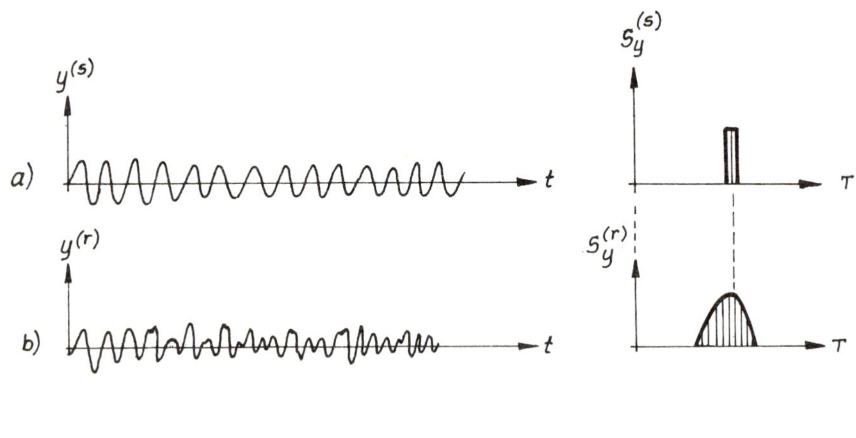

Fig. 11 Graphs of ground vibrations and their spectra

a) at the point of origin (quasi harmonic)
b) at a certain distance away

PROPAGATION OF PARASEISMIC TREMORS THROUGH THE GROUND

Due to the location of the source of vibrations near the earth's surface and due to their, as a rule, short path of travel, the propagation of paraseismic tremors differs from those caused naturally. Hence it is necessary to sum up vibrations in which case the decisive role is played by longitudinal surface waves (Rayleigh). Assuming that the ratio of the velocity of longitudinal waves to that of the longitudinal surface waves is on average 1.1, only after a distance exceeding 200 m it is possible to detect the difference in the time of arrival of the two waves. In practice this distance is about 500 m. Figure 10 shows a graph of vibrations of the foundation of an earth dam caused by quarry blasting approximately 6 km. away. The difference in the time of arrival in the two types of waves can be seen clearly. Figure 10b shows also vibrations due to blasting at another quarry about 2 km away, showing that the surface waves are decisive. Their horizontal transmission through the ground causes effects similar to absorption, which are self-evident, and also to filtration and distribution and 'reduction' or 'augmentation' of particular frequencies (7). Harmonic tremors after travel through the ground show, for example, a spread of the spectrum and increase in the amplitude of frequencies similar to characteristic frequencies of the ground - Figure 11. An important element of the diagnosis of the effect of paraseismic vibrations on buildings is the dynamic interaction between the ground and the structure (9), (11), (18), (20). In spite of a considerable amount of work on the subject, there appears to be a lack of information which would enable us to distinguish the factors causing the quantitative and qualitative change in the vibrations of the building in relation to those transmitted through the ground. The case is complicated by simultaneous, although not very large, change in a phase of the vibrations transmitted to the structure at various points and in different ways, (compression, friction). Figure 12, derived from our own experiments, illustrates this (6).

DYNAMIC CALCULATIONS FOR THE PARASEISMIC LOADING OF BUILDINGS

Methods of calculation are virtually identical with those used in seismic loading, which are: Time History Analysis (THA) and Response Modal Analysis (RMA) see (15), (16) and (21). An important part of the calculations is an adoption of a physical model of the structure which is appropriate to the dynamic model.

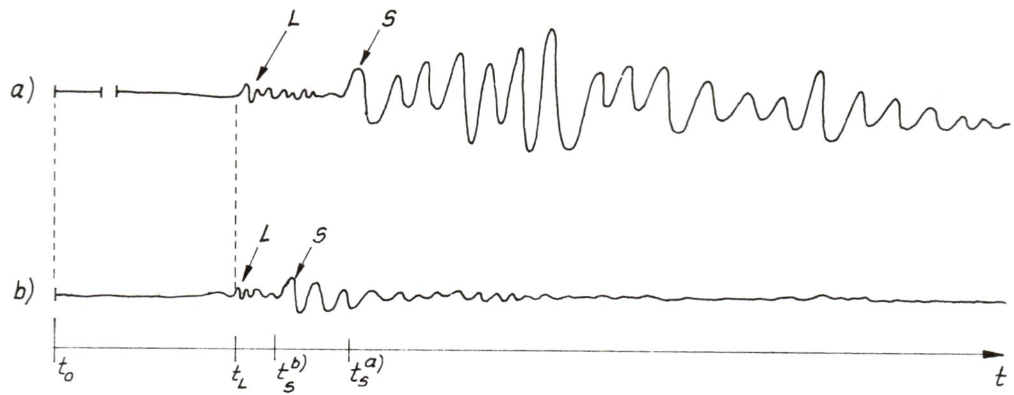

Fig. 10 Vibrographs of horizontal vibrations of the ground under an earth dam

a) due to a strong quarry blast approximately 6 km away

b) due to weak blast approximately 2 km away

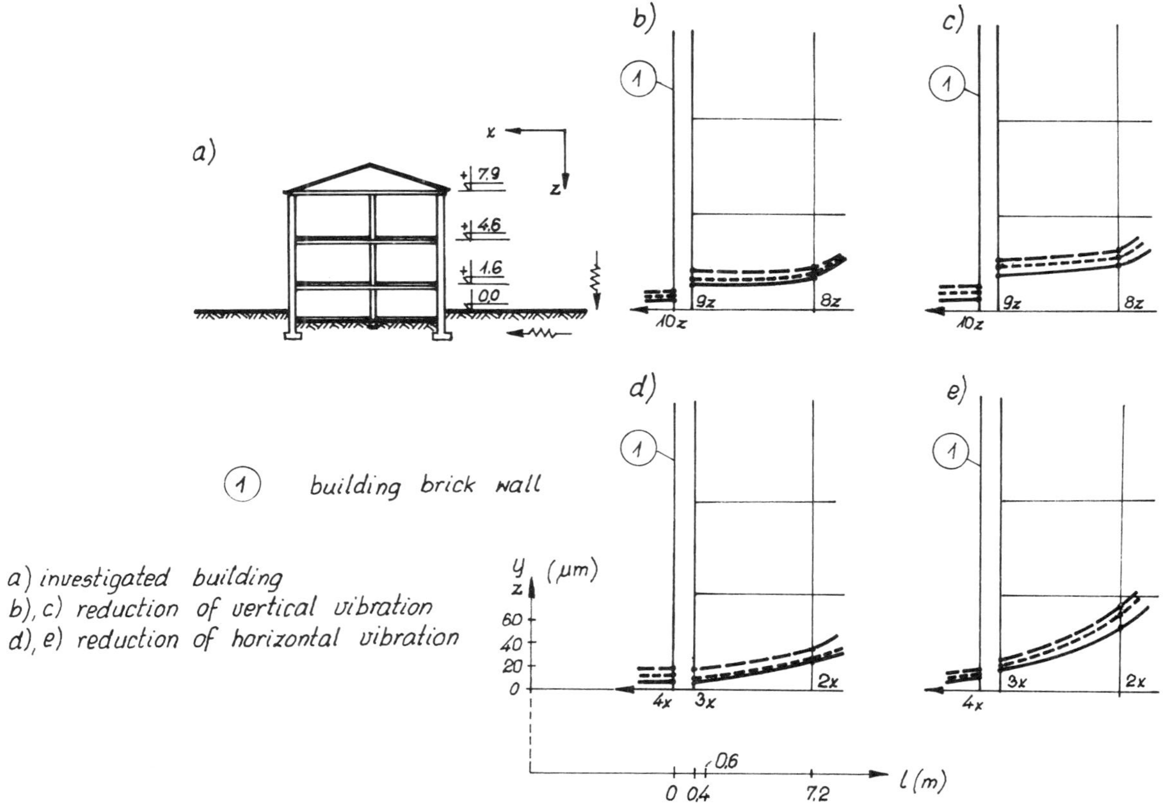

Fig. 12 The reduction of impact vibrations at transition from the ground into the building (6)

Most often for compact buildings we assume discreet models of many degrees of freedom - Figure 13 (4), (12), (21). Such models are mathematically defined by the following relationship

$$[m]\{\ddot{y}\} + [C]\{\dot{y}\} + [K]\{y\} = \{P_{(t)}\} \quad (11)$$

where $[m]$, $[K]$, $[C]$ are the matrices of mass deformation and damping. The solution is simplified by assuming the damping matrix to be

$$[C] = a\,[M] + b\,[K] \quad (12)$$

where a and b are constants.

It is usual in the calculations to apply equation (10), from which the final design loads on buildings can be found. Then for the given loading, internal forces can be calculated taking into account 2 or 3 characteristic frequencies of the structure. Because of the phase displacement of the maxima, the results cannot be added algebraically, but are summed up using various formulae such as

$$N = \sqrt{\sum_{i=1}^{n} N_i^2} \quad (13)$$

or

$$N = \sqrt{N_1^2 + \left(\sum_{i=2}^{n} N_i\right)^2} \quad (14)$$

Equation (14) can be applied when the successive own frequencies are far apart. It is assumed that with the higher frequencies the maxima and their effects arise in the same phase. The selection of appropriate models, and the calculations themselves, introduce a number of problems which cannot be discussed here. In order to render the calculations more accurate, more and more research is devoted to identification of the parameters of the dynamic models of the structures (10). Again, considerable interest is now concentrated on the reduction of the seismic loading allowing deformation beyond the elastic limit, which takes into account the dissipation of the energy of vibrations in plastic deformations. This method however can be recommended for paraseismic effects which are generally of a continuous or long-lasting, repeating character.

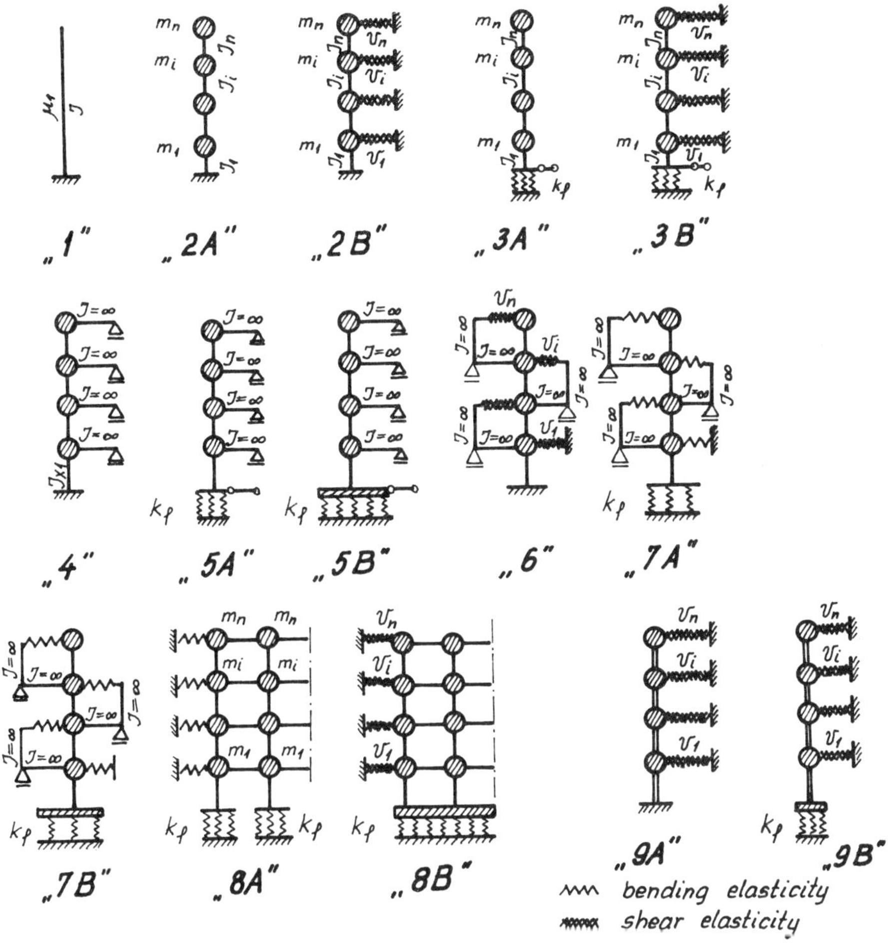

Fig. 13
Examples of physical models for dynamic analysis of multi-storey buildings (4)

THE BEHAVIOUR OF CERTAIN TYPES OF BUILDINGS SUBJECTED TO PARASEISMIC LOADING

The calculations mentioned above may be comparatively tedious and so in order to simplify them, it is usual to introduce scales of seismic influences when defining the dynamic behaviour of buildings. Many such scales have been prepared for earthquakes. The most popular of those being MM (Modified Mercalli) and in Europe M-C-S (Mercalli - Cancani - Sieberg) or M-S-K (Miedwiediew - Sponheuer - Karnik). On the other hand, there are also scales for normal dynamic loading of Zeller, Crandall, Langenfors, and others and vibrars (2), (5), (22) where the measure of the dynamic loading is the index of the force of vibrations χ

$$\chi = \frac{a^2}{n} \quad (cm^2\ s^{-3}) \tag{15}$$

The seismic scales and those for dynamic loads cannot be directly applied to diagnose the buildings, because they give inconsistent results, and what is more important, do not correspond to reality. In practice, however, the use of scales is attractive and because of this new scales have been developed for ordinary masonry and reinforced concrete buildings up to five storeys. These scales, called SWD, (scale of dynamic effects) were first presented at the RILEM Symposium in Budapest in 1963 (3).

BRIEF CHARACTERISTIC OF THE SWD SCALES

In preparing SWD scales it has been assumed that they will not apply to all but only to the most frequently encountered types of building. For this reason also, on the basis of the analysis of different types of building, two buildings have been chosen for reference. For these two, dynamic calculations had been carried out which assisted in formulation of the scales. A brief description of the reference buildings is given in Figure 15 (for scale SWD - I) and in Figure 16 (for scale SWD - II). The calculations took into account various types of building referred to above, as well as various kinds of soil on which the building is founded, the condition of the buildings etc. The criteria of different degrees of damage were based on conditions of stiffness, strength and stability. Reference (3) gives details of definition of the scales corresponding to given degrees of damage and shown in simplified form in Figures 16 and 17. Scales SWD - I and SWD - II have been prepared by assuming

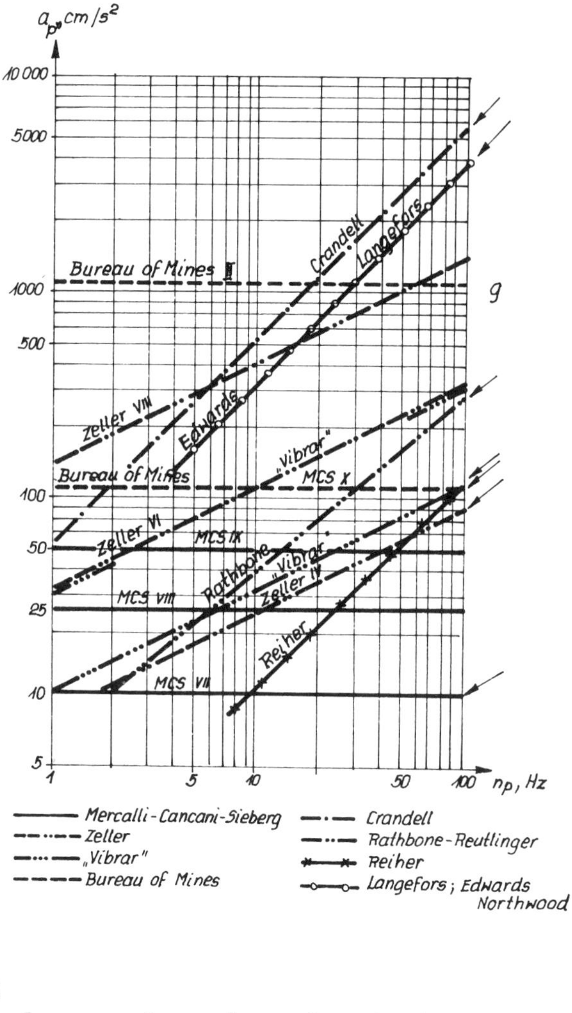

Fig. 14

Graphs of several scales of seismic and para-seismic effects on buildings. ↙ denotes a line showing the first damage to buildings

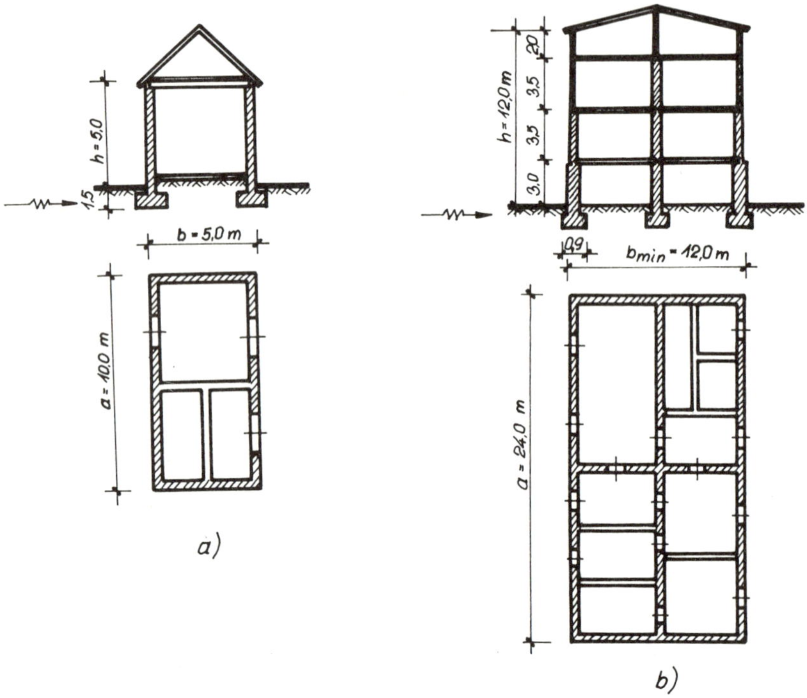

Fig. 15 Details of reference buildings used in the preparation of SWD scales (5)

a) SWD - I
b) SWD - II

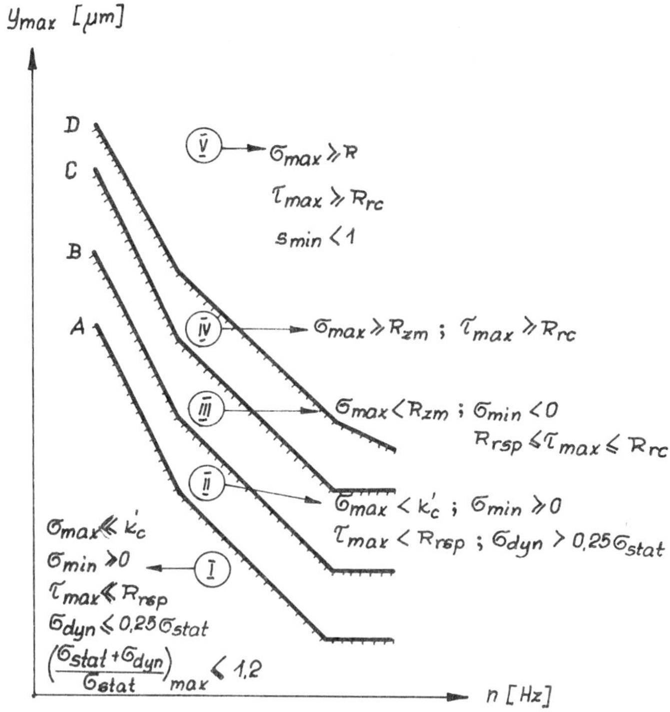

Fig. 16 Basis for the preparation of SWD scales

σ normal stresses

τ shear stresses

R resistance

k_c stresses at first damage

s coefficient of stability

A, B, C, D limits: A limit of detection of vibrations
B Stiffness
C Resistance
D Stability

Substantial damage takes place above line B
Full explanation in reference (3), (5)

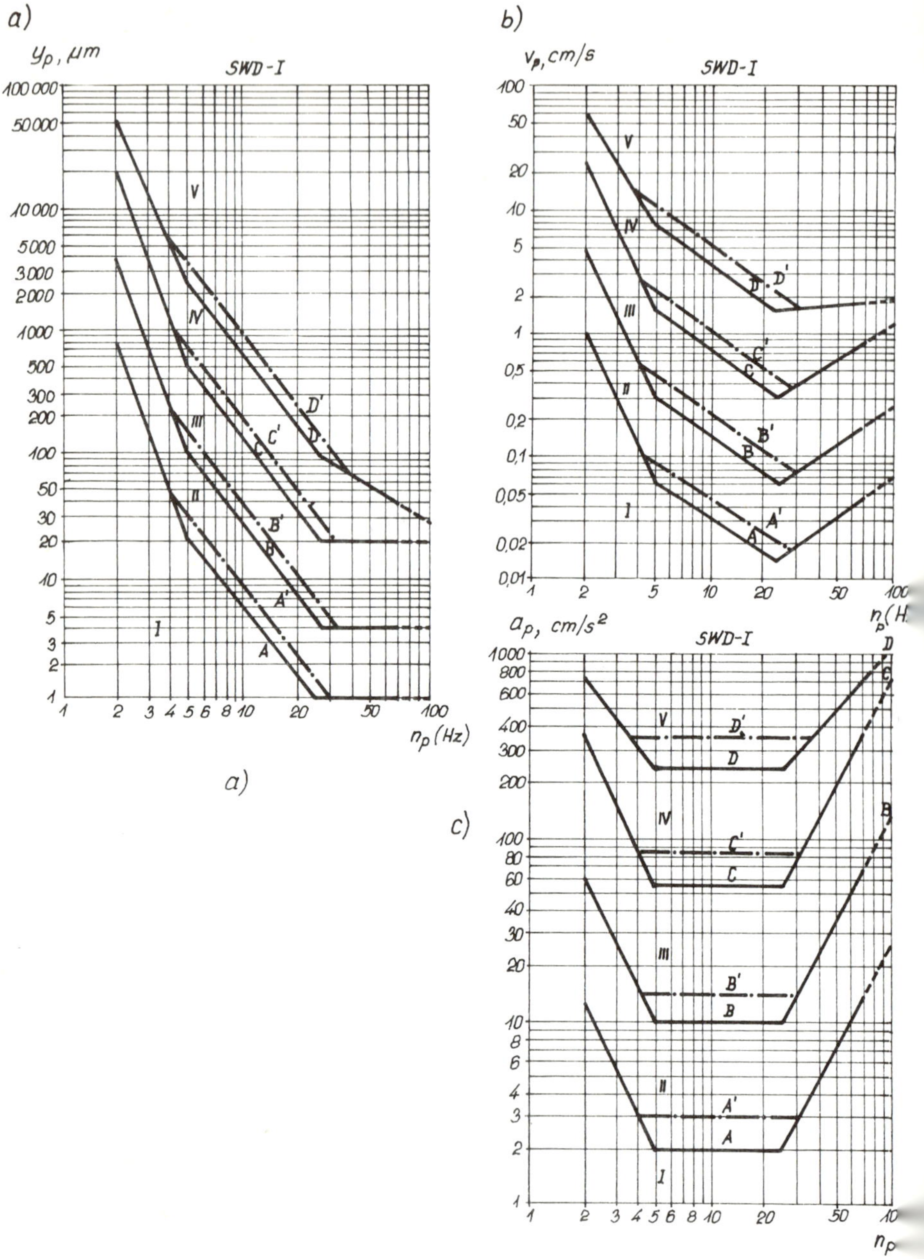

Fig. 17
SWD-I scale for the displacement, velocity and acceleration $y_p - n$, $v_p - n$, $a_p - n$

limiting values of stresses in accordance with relevant constructional standards and the internal material damping defined by Figure D = 0.05 (5%). The use of the curves defining different zones of degree of damage has also been applied to buildings similar to those analysed in the calculations. Thus it has been proposed, (3), that scale SWD - I Figure 17, can be applied to compact buildings of small plan area, 1 or 2 floors, with the overall height not exceeding either of the plan dimensions. Scales SWD - I refer to masonry buildings constructed of bricks, blocks, hollow blocks and no fines concrete.

Scale SWD - II (Fig. 18) is intended for buildings of several storeys (up to 5), of masonry or mixed type where $h > l_{min}$.

Thus, although they do not solve in detail the problem of the assessment of vibration damage for all buildings, the SWD scales take into account a comparatively large variety of structures.

The use of the scales depends on the measurement of vibrations of the building under consideration during ground tremors. The horizontal component is measured at the bottom floor, at ground level. Maximum measured values (y - n, V - n, a - n) plotted on the appropriate SWD scale diagram, fall into one of five zones of the diagram thus enabling the diagnosis to be carried out. In practice, we are mainly concerned with zones I, II and III, and only exceptionally with zone IV.

VERIFICATION OF THE SWD SCALES

Between 1960 and 1979 the Technical University of Cracow carried out almost 200 cases of diagnosis of buildings according to the flow diagram in Figure 19. The sources of vibrations were: traffic, machines installed near buildings, pile driving, drop-hammers and quarry blasting. This gave sufficient material to enable statistical analysis of the agreement between the assessed and the actual values. As shown in table 2, agreement is very good. In over 80% of cases the diagnosis based on SWD scales is accurate in practice. It must be stressed that the verifications were carried out by the same team or by other teams under our control so that there was no difficulty or difference in interpretation of various data. Generally the measured vibrations are of an irregular character and

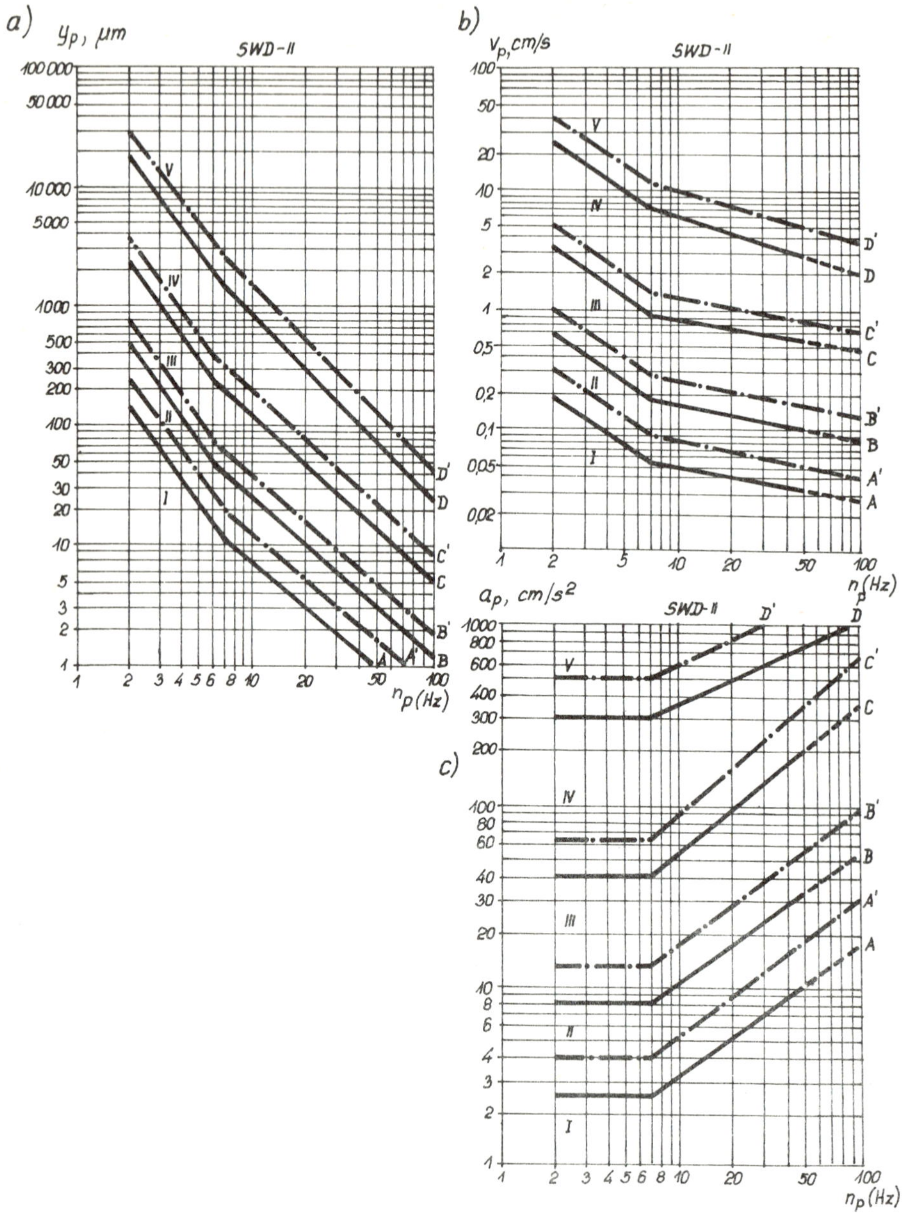

Fig. 18

SWD - II scale for the displacement, velocity and acceleration y_p - n, v_p - n, a_p - n

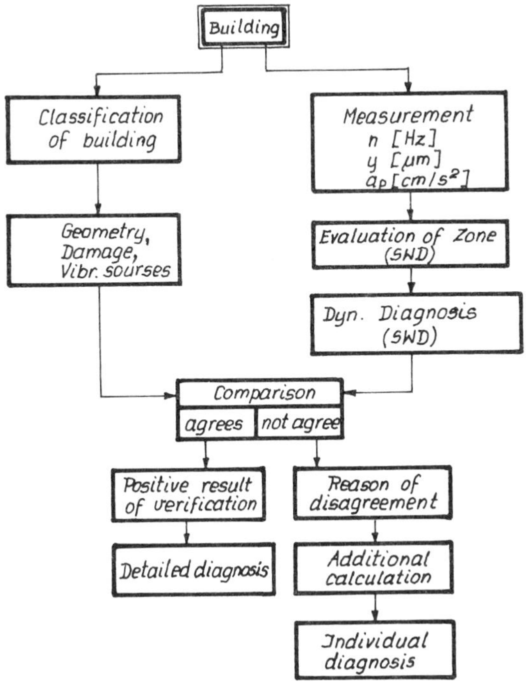

Fig. 19 Scheme for verification of the scales

Tab. 2 Results of Diagnosis Using SWD Scales (1960 - 1970)

No. of investigated buildings / Results of diagnosis according to SWD	Total		Acc. to SWD - I		Acc. to SWD - II		Remarks
	No.	%	No.	%	No.	%	
Total	198*	100	131	100	67	100	* From our own experiments or under our control
Results in agreement with measurement	159	80.3	108	84.4	51	76.1	
Under-estimated results	10	5.0	8**	6.1	2	3.0	** In most cases the history of loading prior to investigation was not known
Over-estimated results	29	14.7	15	11.5	14***	20.9	*** Together with buildings included by analogy

different from harmonic distribution. It is therefore necessary to consider the greatest amplitude of vibrations and the corresponding frequencies In the case of several distinct frequencies arising - all must be checked. The SWD scales have been incorporated in the Polish Building Standards PN-B-02170 and have now been used in many cases. It must be emphasised that the scales are applicable only to a certain class of buildings with, if necessary, corrections for various characteristics of the given structure and the ground. Application of the SWD scales to other buildings e.g. steel structures, will not give correct results. For buildings of up to 5 storeys, the scales may be assumed to have been verified in practice. Hence it has been felt worthwhile to draw your attention to them and to recommend their application in practical diagnosis of buildings subjected to paraseismic effects.

REFERENCES

1 J.A. Blume - The motion and damping of buildings relative to seismic response spectra, Bulletin Seismic Society of America No 1, 60 (1970).

2 G. Budugan - Dynamique des Fondations de Machines, Editions Eyrolles, Paris (1972).

3 R. Ciesielski - Charts for evaluating the effects of externally applied vibrations and impacts on masonry structures, RILEM Symposium, Budapest (1963) 2, 19-30.

4 R. Ciesielski and J. Kawecki - Dynamic models of multistorey buildings in comparison with measurement results, KJ PAN XVII Conference, Krynica (1971) 1, 25-36 (in Polish).

5 R. Ciesielski - Evaluation of Dynamic Influences on Buildings, Arkady Publ, Warszawa (1973) (in Polish).

6 R. Ciesielski - Research on vibrations of buildings and propagation of vibration in the soil caused by driving piles, Mechanika, PAN Kraków 9 (1976) 7-66 (in Polish).

7 R. Ciesielski, E Maciąg and K. Stypuła - Propagation of vibrations in soil caused by pile driving, Conference on Dynamics of Engineering Structures, Smolenice (1977) 1, 249-256.

8 R. Ciesielski - Analysis of seismic and paraseismic vibration sources in Poland, Symposium on Seismic Influences on Structures, Kraków (1978) 1, 7-26 (in Polish).

9 G.W. Housner - Interaction of building and ground during an earthquake, Bulletin Seismic Society of America No 2, 47 (1957).

10 J.L. Korczinski (Ed) - Issledowanija po seismostojkosti zdanij i sooruženij, Gosstrojizdat, Moskwa (1961).

11 D.L. Lycan and N.M. Newmark - Effect of structure and foundation interaction, Journal Engineering Mechanics Division, ASCE, EM No 5 (1961) 87.

12 S.W. Medwiediew - Inženiernaja Seismołogija, Strojizdat, Moskwa (1962).

13 K. Muto (Ed) - Earthquake loading and response, ASCE - ABSE T. Conference, Lehigh Pa. (1972) 6, Ib-6.

14 F.P. Müller - Baudynamik, Beton Kalender V II, W. Ernst u. Sohn Berlin (1978).

15 F.P. Müller and E. Keintzel - Erdbebensicherung von Hochbauten, W. Ernst u. Sohn, Berlin, München, Düsseldorf (1978).

16 N.M. Newmark and E. Rosenblueth - Fundamentals of Earthquake Engineering, Prentice Hall Inc. Publ., Englewood Cliffs, N.J. (1971).

17 N.M. Newmark, J.A. Blume and K.K. Kopur - Seismic design spectra for nuclear power plants, Journal of the Power Division, ASCE, No 11 (1973).

18 R.A. Parmelee and J.H. Wronkiewicz - Seismic design of soil-structure interaction system, Journal of the Structures Division, ASCE, No 10 (1971) 97.

19 S.V. Polyakov and L.S. Klimnik - On the unification of codes for structural design in seismic areas, Bulletin of EAEE 4 (1978) 3-12.

20 R.V. Whitman - The current status of soil dynamics, Applied Mechanics Review 22 (1969) 1-8.

21 R.L. Wiegel (Ed) - Earthquake Engineering, Prentice Hall Inc. Publ., Englewood Cliffs, N.J. (1970).

22 W. Zeller - Stärkebestimmung von mechanischen erschütterungen, Bauingenieur No 12 (1931).

INTERFACE BEHAVIOUR IN A DYNAMIC SOIL/STRUCTURE INTERACTION PROBLEM

Edward G. Prater, Ph D

Research Engineer, Institute of Foundation Engineering and Soil Mechanics, ETH-Zurich, Switzerland

INTRODUCTION

Regulatory agencies for the safety of nuclear power plants, in particular, and earthquake design codes require that important structures should maintain a certain factor of safety with respect to overturning and sliding under conditions of strong seismic excitation. For example, the new ATC Proposal (1978) states that the resultant of the seismic forces and the vertical loads at the foundation-soil interface shall not fall outside the middle one-half of the base. A slight uplifting of the edge of the mat foundation during earthquake vibration is not generally regarded as being harmful with respect to structural distress. On the contrary, such behaviour normally leads to a sizeable reduction of the horizontal acceleration and the seismic forces (shear force and overturning moment), though with an increase of lateral displacements and vertical seismic response. The benefits associated with base mat uplift may also obtain for more than moderate separation of the base from the soil. Theoretically, under extreme seismic loading the point of action of resultant force at the base of the mat may lie outside the edge of the mat without the structure tipping over completely (Wolf and Skrikerud, 1978). This result is due to the oscillating nature of the seismic excitation. As the amount of lift-off increases, however, there will be a corresponding increase in contact pressure between the mat foundation and the soil, until at the point when the foundation stands practically on its edge the bearing pressures would be so high that they would cause ground failure under static conditions. It may be assumed that under dynamic conditions too the shear resistance of the soil is exceeded for this extreme case of loading, with respect to both tilting and sliding. The problem then is to assess the total deformational behaviour of the structure, taking into account a certain amount of plastic deformation in the ground. This is a somewhat intractable problem and up till now mainly relatively simple non-linear structural models have been used to investigate it.

The usual practice in the earthquake resistant design of structures is to assume that the seismic input acceleration may be applied directly at the base of the structure. This is tantamount to assuming that the structure rests on non-deformable soil. Provided the ground is much stiffer than the supported structure the usual assumption is adequate. The problem of base uplift can then be handled quite easily (Meek, 1975). Many cases exist, however, in which the soil foundation cannot be assumed rigid in comparison with the structure and it is no longer permissible to apply the free field motion at the base of the structure, as the presence of the structure modifies the free field motion at this level. This interaction effect between the structure and the soil has received a great deal of attention in the last decade or so. The majority of solutions assume that the base mat is fully contiguous with (i.e. "welded" to) the underlying flexible soil stratum. In terms of structural response the soil-structure interaction usually leads to an increase in the natural periods of the system (although due to half-space or transmitting boundary considerations the composite system does not possess normal modes in the classical sense) and the appearance of rotational components at the level of the base mat. Energy is radiated away from the structure at the interface with the flexible soil foundation giving an increase in the apparent damping.

A complete consideration of interface behaviour would require, therefore, an inclusion of uplift and slip effects in the context of soil-structure interaction.

SOIL-STRUCTURE INTERACTION ANALYSES

Several methods of soil-structure interaction analysis are available, which may be broadly classified in two groups: impedance solutions and the numerical one of a discretized model. In the case of the former simple lumped parameter models of the half-space incorporating constant, frequency-independent parameters are also commonly used, although for a layered elastic half-space frequency effects become more important and may not be neglectable. The models which discretize the soil domain surrounding the structure may be subdivided into two groups: in which (i) the field variable is defined specifically on the boundary of the soil domain (e.g. rigid, reflecting condition) and (ii) compatible wave transmission conditions are introduced at the boundaries. Wave transmission is equivalent to an absorption of energy. The introduction of exact transmitting boundary conditions, however, is only possible for layered elastic system, including the special case of a flexible elastic medium on a rigid base (Kausel et al 1975; Vaish and Chopra 1974). The same results should be obtained in theory using either this approach or the impedance function approach. The raison d'être of the finite element approach, however, is to investigate non-homogeneous, nonlinear problems. Its use for the case of a layered elastic half-space is warranted sometimes because the impedance functions are not readily available for a flexible base mat and for arbitrary values of Poisson's ratio, layer thickness etc.

Some representative papers discussing the application of the various solution methods are (Hall and Kissenpfennig, 1976; Hadjian, 1976; Lysmer, 1978; Gutierrez and Chopra, 1978; Holzlöhner, 1977; Jennings and Bielak, 1973; Whitman and Bielak,1980).

Impedance methods

The impedance method involves obtaining the frequency-dependent force displacement relationships for the soil foundation. To arrive at a solution the following assumptions are usually made: (1) The soil media can be represented by a homogeneous, isotropic linear elastic half space. A system of bonded elastic layers can also be treated and more recently material damping effects have been included (Luco, 1977). (2) The forces are assumed to be those acting on a rigid, weightless disc (or rectangular shape of mat) resting on the surface of the half-space. The problem belongs to the category of mixed boundary elastodynamic problems. To simplify the evaluation of the impedance functions a relaxed boundary condition is assumed, i.e. the contact is frictionless for vertical and rocking vibrations and pressureless for horizontal vibrations. The effect of this relaxed boundary condition on the solution has been discussed by Jennings and Bielak (1973) who showed that for a simple half-space it is not great. Gazetas and Roesset (1976), however, using the finite element method of solution, indicate that in the rocking mode differences may be more significant, especially if a rigid base underlies the elastic foundation layer.

The force relationships for harmonic loading of a circular mat of radius r may be expressed through the impedance functions and displacements as follows (Luco, 1977):

$$H(t) = K_h \{k_{hh}(a_o) + i\ a_o\ c_{hh}(a_o)\}\ u_f \qquad (1a)$$

$$M(t) = K_\theta \{k_{mm}(a_o) + i\ a_o\ c_{mm}(a_o)\}\ \theta_f \qquad (1b)$$

$$V(t) = K_v \{k_{vv}(a_o) + i\ a_o\ c_{vv}(a_o)\}\ v_f \qquad (1c)$$

where a_o is a dimensionless frequency parameter equal to $(\omega r/V_s)$. ω is the frequency of excitation and V_s is the shear wave velocity of the foundation medium. The coefficients K_h, K_θ and K_v represent the static values of the foundation stiffnesses, i.e.

$$K_h = \frac{8\ G\ r}{2 - \nu} \qquad (2a)$$

$$K_\theta = \frac{8\ G\ r^3}{3(1 - \nu)} \qquad (2b)$$

$$K_v = \frac{4\ G\ r}{1 - \nu} \qquad (2c)$$

where G and ν are the shear modulus and Poisson's ratio of the elastic medium respectively. Luco (1977) presents some results for a two layer system (see Figs. 1), where a_o and the constants in eqn. (2) relate to the upper medium (denoted by suffix 1) and the functions k_{hh}, c_{hh} etc. are plotted against a_o

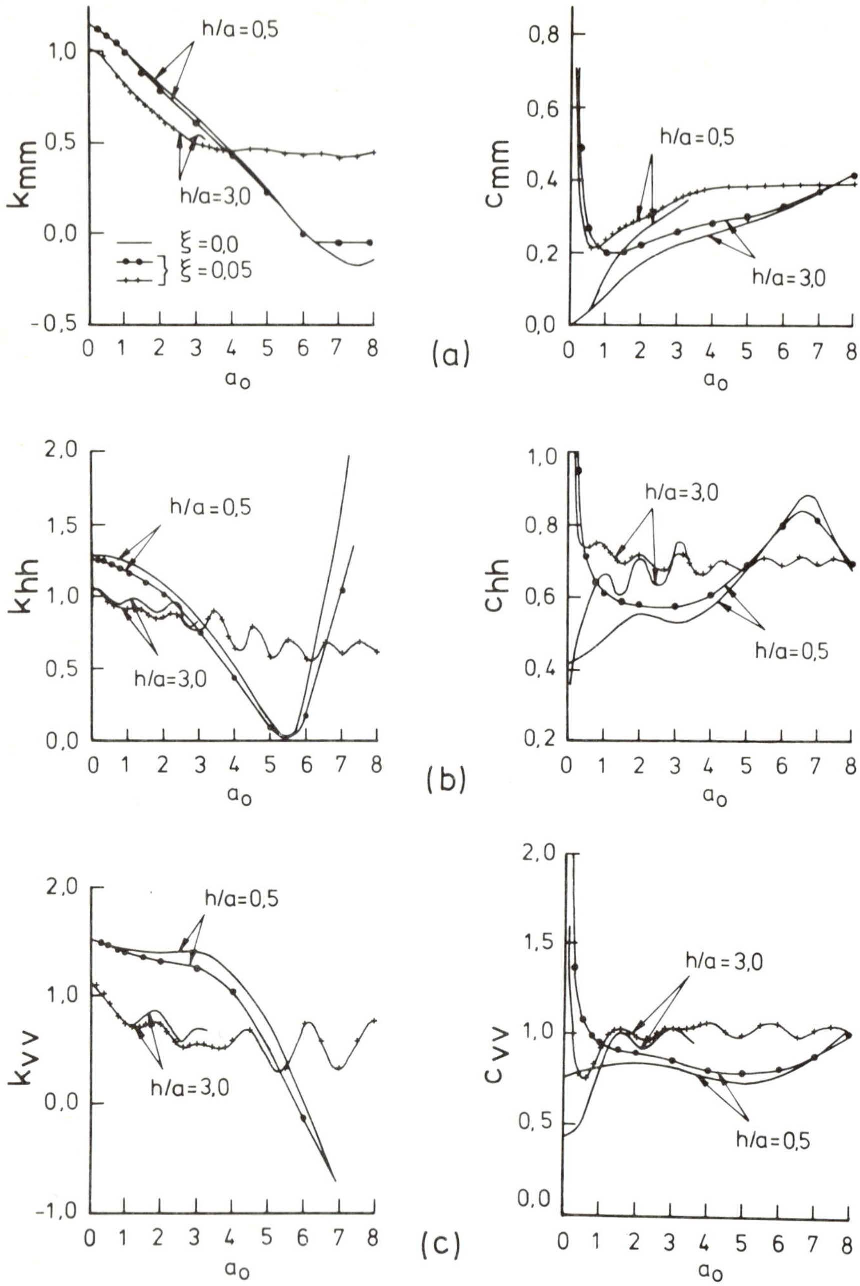

FIGURE 1. Impedance functions for 2-layer hystereticall damped half space. For upper (1) to lower (2) layer: de sity ratio = 0.85,shear wave velocity ratio = 0.8. Pois on's ratio = 0.25. Damping and geometrical relations (d th of layer 1 to base radius) as shown (after Luco,1976

for a given value of V_{s1}/V_{s2}, ξ the material damping constant and the ratio of thickness of upper layer to mat radius h/r. Generally, the frequency dependence of the impedance functions is more marked for a layered system than for a half-space, which would suggest that the simplification of using constant values of stiffness (eqn. 2) together with a certain value of damping is no longer valid. However, Werkle and Rönnberg (1978) have produced results to show that sufficient accuracy may still be achieved, provided the impedance functions are evaluated at the value of the fundamental eigenfrequency of the coupled structure. For structures, whose first mode gives the major contribution the approximation is good.

The application of impedance methods to embedded structures is not very advanced due to the complicated boundary conditions. By simplifying the shape of the embedded part to a semi-ellipse Wong (1975) has obtained an exact solution for a strip foundation, handling at the same time SH-waves with an arbitrary angle of incidence. For rectangular embedded shapes only approximate analytical solutions are available.

A simple lumped mass model, in which the masses and additional horizontal springs are adjusted to simulate the free field motions, has been presented by Hall and Kissenpfennig (1976). Generally, however, embedment has to be handled by finite element models (Kausel et al, 1977). Lumped parameter models can be calibrated using finite element solutions. For example, in the new ATC Proposal (1978) the static spring constants (eqn. 2a and 2b) are modified as follows to account for embedment to depth d in a layer of thickness D_s resting on a rigid base

$$K_h \cong \frac{8\ G\ r}{2 - \nu}\ \{1 + \frac{2}{3}\ \frac{d}{r}\}\{1 + \frac{1}{2}\ \frac{r}{D_s}\}\{1 + \frac{5}{4}\ \frac{d}{D_s}\} \qquad (3a)$$

$$K_\theta \cong \frac{8\ G\ r^3}{3(1 - \nu)}\ \{1 + 2\frac{d}{r}\}\{1 + \frac{1}{6}\ \frac{r}{D_s}\}\{1 + 0.7\ \frac{d}{D_s}\} \qquad (3b)$$

Having obtained the impedance functions the next step is to formulate the equations of motion for the structure. The combined system of equations may then be solved by different methods (Jennings and Bielak, 1973; Meek and Veletsos 1972). One of the principal advantages of this approach is that the surface free field motion is applied directly as the earthquake excitation, and deconvolution processes are not required. The above approach may be regarded as a special case of the sub-structure method (Gutierrez and Chopra, 1978). The main advantage of the latter, in its general form, is the ability to analyse the structure with the aid of complex finite element models, as the rigid plate assumption is no longer necessary. A simpler approach also permitting consideration of plate flexibility has been proposed by Hall et al (1978), using a higher order Winkler model, in which the foundation mat is discretized using interconnected rigid discs.

Finite Element and Finite Difference Models

A wealth of literature exists on the application of the finite element method to soil-structure interaction problems (Lysmer, 1978). If energy-absorbing conditions on the lateral boundaries are not applied the domain must be made large enough to avoid the so-called box effect (Costantino et al, 1976). Increasing the dimensions of elements towards the edge of the domain tends to make them opaque to the propagating waves, thus defeating the purpose of having an extensive soil domain. This is one reason why improper use of the finite element method may lead to poor correspondence with impedance solutions. A close correspondence is, in any case, not easily attainable over the complete time domain, cf. the results of Arockiasamy et al (1979).

A major contribution to investigating the interface behaviour in soil-structure interaction problems is that of Scaletti (1977), whose finite element solution, under the assumption of plane-strain conditions, incorporates non-linear soil deformation properties both within the soil domain using a Ramberg-Osgood material law and at the interface structure-soil. Slipping and debonding due to low tensile strength of soil was allowed on the base of the structure and on the vertical sides of embedded structures. For a typical nuclear reactor building it was found that the effects of separation and sliding were more pronounced for the embedded case. Only this one class of structure, however, was investigated.

The finite difference code STEALTH (1976) also possesses the capability of handling this type of problem. A simpler finite difference code not allowing for uplift was used by Prater and Wieland (1977) to investigate other aspects of the problem, i.e. travelling waves, non-reflecting boundaries and non-linear soil properties.

Simplified Structural Models for Uplift Behaviour

Several investigators have developed simplified models to investigate the problem of separation, sliding and plastic deformation at the soil-structure interface. Meek (1975, 1978) investigated the response for the first mode approximation of a structure supported on a rigid medium, whereby on recontact impact effects were allowed for.

Several investigators have modified the elastic half space theory for conditions of lifting-off. Takemori et al (1976), Kennedy et al (1976) and Stoykovich (1978) assume that the interface pressure varies linearly across that part of the base mat in contact with the soil. This leads to a non-linear base moment - rotation relationship, since the decreasing contact area reduces the base rotational spring constant. The resulting non-linear elastic stiffness relations do not possess hysteretic damping characteristics. A solution on similar lines, but of a somewhat more sophisticated nature, is due to Wolf and Skrikerud (1978), who introduce the concept of a subdisc for the reduced contact area.

For large overturning moments the question of soil yielding must be considered. This is most easily achieved by supporting the base mat by distributed elasto-plastic springs. Debonding effects are obtained by making the springs tensionless. If it is desired to have in the elastic range correctly coupled vertical and rocking motions in relation to half-space theory a single set of equal uniformly distributed springs will not suffice since the following relation, which may be obtained from eqns. 2(b) and 2(c), must hold for a circular slab

$$3\ K_\theta = 2\ r^2\ K_V \qquad (4)$$

If strips of width dx were taken the constant for the spring is $2\ G\ dx/(1-\nu)$ for the vertical mode and $4\ G\ dx/(1-\nu)$ for the rocking mode (Hall et al, 1978). Thus one must compromise by choosing the constant to match the most important mode of deformation to be investigated.

Wolf and Skrikerud (1978), on the other hand, vary the distributed stiffness such that eqn. (4) is satisfied, i.e.

$$k_V(x) = \frac{K_V}{16r}\ \{5 + 99\ (x/r)^8\} \qquad (5)$$

This expression corresponds to very high edge stresses in the case of a rigid base mat. Theoretically, for an elastic medium the edge stresses would be infinite. In practice soils would not be able to sustain very high stresses and the yield limit would be reached at the edge at fairly low applied loads. A similar expression to eqn. (5) with the power of (x/r) reduced to 4 can be obtained, which still satisfies eqn. (4), but at the expense of k_V reducing to almost zero at the centre, i.e. $x = 0$. However, for sands and gravels the actual interface pressure distribution is approximately parabolic with a maximum value at the centre. Thus, for cohesionless soils it is futile to try to obtain exact agreement with elastic theory locally at the interface, and relations such as eqn. (4) will not hold. There may be some merit, therefore, in a simple approach such as carried out by Byrne and Nathan (1978), in which a set of uniform springs were used to carry out parametric studies for rigid buildings of various height/width ratios. The main drawback of the elasto-plastic springs approach is determining a yield stress for the individual springs. For granular materials the yield stress would depend on confinement stresses and would thus vary with radius x.

For a uniform set of elasto-plastic springs the interface pressure distribution under a rigid circular slab is as shown in Fig. 2. For this case the total normal force V at any instant is the weight of the structure with the addition of the vertical inertia force. The moment M of the soil reaction may be found by integration, which then allows the eccentricity e of V to be calculated, i.e. $e = M/V$.

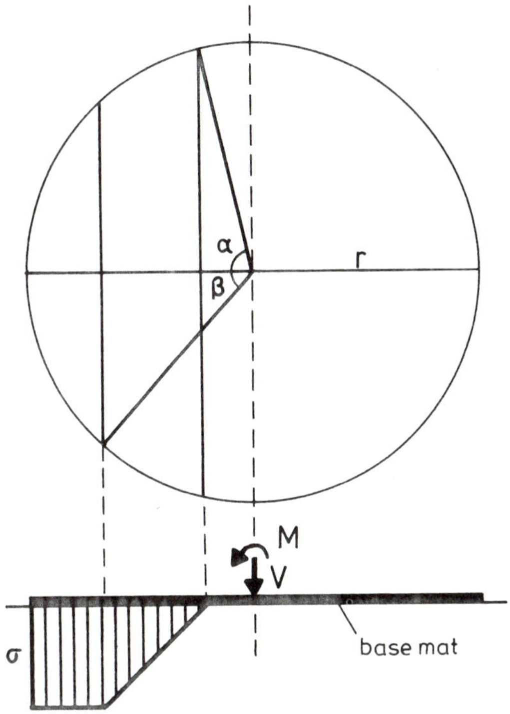

FIGURE 2. Contact stress distribution for a yielding soil medium under a large overturning moment

APPLICATION TO A LIQUID STORAGE TANK

The structure considered here is a refuelling water tank for a nuclear power plant, consisting of a steel cylindrical shell of height 22.86 m and diameter 9.17 m of variable thickness (20 mm over the bottom 6 m, 16 mm over the next 6 m, and 10 mm at the top), a roof of thickness 5 mm, a circular concrete base mat of diameter 10.26 m and thickness 1.0 m embedded 0.9 m into a gravel layer of thickness 14 m. The gravel is underlain by a rock formation (mudstone). In close proximity to the tank is the massive deeply embedded nuclear reactor building. Its influence on the seismic response of the tank is not investigated further here. It could be taken into account, very approximately, by increasing the maximum ground acceleration used as input to the problem. The problem of the interaction of adjacent structures has been studied amongst others by Wong (1975).

The response of liquid storage tanks to seismic loading is a problem deserving special consideration in its own right. The solution for a rigid tank and an incompressible liquid subjected to horizontal earthquake excitation is well-known. The fluid-structure interaction is characterized by a convective force (so-called "sloshing" effect) and an impulsive force. For large tanks the sloshing effect turns out to be relatively unimportant. Simple design equations leading to base bending and overturning moments for tall tanks are given by Epstein (1976). The dynamic pressure distribution on the base of the tank must be included to determine the overturning moment. For a circular tank the equation becomes (neglecting sloshing)

$$OTM = \frac{\gamma\, r^4\, A_O}{g}\,(1.57\,\alpha^2 - 1.37\,\alpha + 1.43) \qquad (6)$$

in which γ = unit weight of liquid, A_O = max. base acceleration, and α is the ratio H/r.

It is well-known that the flexibility of the tank walls can increase substantially the seismic effects (Veletsos, 1974), and that the rigid tank assumption (Housner, 1970) can lead to non-conservative results. Fischer (1979) presents expressions for impulsive mass, overturning moment etc. Assuming various modes of wall deformation he proposes the following procedure for simplified seismic analysis:

(i) determine the added mass (which for an incompressible fluid is frequency-independent) and distribute it uniformaly over the wetted surface (ii) calculate the natural frequencies of the modified shell structure (iii) check that the assumed mode shape is correct, and if necessary iterate (curves are given for three modes shapes) (iv) obtain the maximum acceleration for the calculated natural frequencies from the given design earthquake spectrum (v) calculate maximum seismic forces.

Thus for practical problems it is important to estimate the fundamental frequency of the shell-added impulsive mass system. In this investigation the finite element code ASHSD 2 (1975) was used to analyse the shell structure, both clamped at its base and attached to a discretized portion of the gravel foundation. For a fixed base a comparison was made for various assumption regarding the added mass, whereby the structure was subdivided into 24 and, in case (d), into 12 elements. 10 eigenfrequencies were calculated but only the first three are given below.

a) steel structure alone: 9.2, 16.6 and 16.8 Hertz

b) added mass according to Epstein (1976): 4.2, 8.9 and 12.3 Hertz (here the added mass $\simeq$ total water mass for $\alpha \simeq 5$ but is not evenly distributed)

c) added mass according to Fischer (1979): 5.0, 9.0 and 16.4 Hertz (for $\alpha \simeq 4$, mode 1, no iteration)

d) steel structure with fluid finite elements: 3.3, 5.5 and 6.0 Hertz (Lamé constants $\lambda = 2 \cdot 1 \cdot 10^6$ kN/m^2, $\mu \simeq 0.$, Poisson's ratio $\nu = 0.4995$)

The fundamental frequency is thus reduced by a factor of about 2 to 3 due to the added impulsive mass. The resulting behaviour will then depend on the values of the design response spectra at these frequencies.

Based on eqn. (6) the overturning moment for $\alpha = 4.99$

$$OTM = 33.7 \gamma r^4 (A_o/g) \tag{7}$$

The eccentricity of the resultant force acting on the base mat is for the extreme case, therefore,

$$e = \frac{OTM}{W} \simeq 2.1\ r\ (A_o/g) \tag{8}$$

in which W is the total weight of the steel structure plus fluid. Thus for the rigid tank assumption e lies within the middle radius for a maximum ground acceleration of 0.2 g.

Following Fischer (1979) and assuming the most unfavourable mode shape, the overturning moment for the flexible tank is

$$OTM \simeq 0.33\ \gamma\ \pi\ H\, r^3\ (A_{max}/g) \tag{9}$$

in which A_{max} is the maximum radial acceleration of the tank wall (relative value) corresponding to the first natural frequency on the given (U.S. NRC) response spectrum. Assuming 4 % critical damping of the steel tank structure A_{max} was estimated to be approximately 0.65 g. The maximum eccentricity e of the resultant force is now found to be slightly greater than the radius of the concrete base mat. This result indicates that a closer examination of the problem is required. The approach due to Fischer (1979) does not take into account base mat rocking motions, i.e. the soil structure interaction has been neglected, whereby it is possible to overestimate grossly the overturning moment and base shear.

Meek and Veletsos (1972) indicate that SSI effects are important when the dimensionless shear wave velocity $\sigma = V_s/f_1 \cdot H$ is less than about 20, where f_1 is the fundamental fixed base natural frequency (Hertz). The value of σ for this particular structure with V_s = 370 m/s is approx. 3.25 which shows that interaction effects should be considered. One very significant influence of interaction is to reduce the natural frequencies of the structure. An approximate way of estimating the reduced fundamental frequency is to assume the structure to be modelled by an uniform shear beam on an elastic half space. (Thick beam analysis with bending and shear would of course be better). Based on results presented by Meek and Veletsos (1972) the reduction for $\sigma \simeq 3.25$ and $\alpha \simeq 5$ is about 20 %. Using the finite element model for the system shown in Fig. 3, with the elastic soil parameter, shear modulus G = 300,000 kN/m^2 and $\nu = 0.35$, the first three natural frequencies are 4.0, 6.4 and 6.5 Hertz. The reduced fundamental frequency compares well with the one estimated by the first mode approxi-

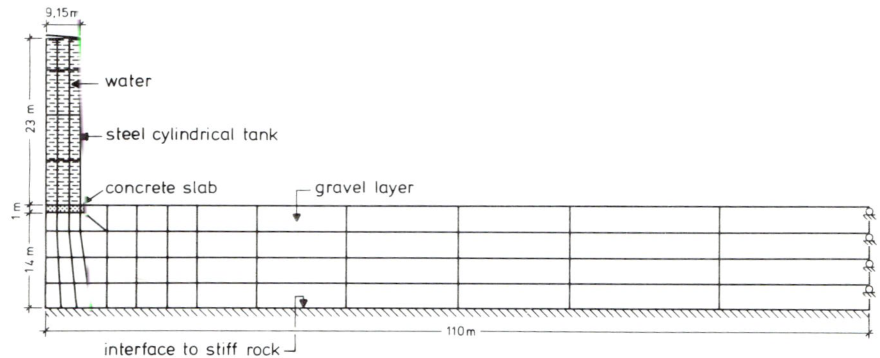

FIGURE 3. Tank structure on gravel layer - due to axial symmetry only half shown - with finite element net for the layer, the water mass being lumped in subsequent calculations to the tank wall

mation for a shear beam. For the latter the mass of the equivalent oscillator is $8m/\pi^2$, in which m = mass of tank plus added water, acting at a height $2H/\pi$.

Uplift and slipping could then be further investigated using the idealized structural model and equations presented by Wolf and Skrikerud (1978). The spring and damping values used to simulate the half-space should, however, take account of the ground layering, i.e. the gravel underlain by bedrock. The actual stiffness properties in the gravel vary with depth, since for granular materials stiffness is influenced by effective pressure σ_m' (both due to overburden pressure and surface loading). Laboratory cyclic triaxial tests on samples 150 mm diameter (max. grain size ≃ 30 mm) showed that the assumption of a square root relationship, i.e. G proportional to $(\sigma_m')^{1/2}$, is applicable (see Fig. 4). The damping ratio appears to be practically independent of σ_m'. Both stiffness and damping are strongly influenced, however, by shear strain (Fig. 4). To a lesser extent the same trends were found for the mudstone. For the purpose of estimating impedance values corresponding to the reduced fundamental frequency of the system (≃ 4 Hz) average soil properties for the two layers must be chosen. For the given system the dimensionless frequency $a_o \simeq 0.35$ and the parameter h/a (thickness of layer/radius of foundation) is about 2.7. The amount of material damping (ξ) is in excess of 8 % for shear strains $\gamma = \varepsilon(1+\nu) >$ approx. 10^{-2} %. The ratio $V_{s1}/V_{s2} \simeq 0.7$, which is less than the chosen value in Fig. 1 by 13 %. The Poisson's ratio values, which are also dependent upon shear strain level, tending towards 0.5 at higher strains, are not known for the materials. Due to the high water table cross-hole seismic tests could only give information about S-wave velocities, and laboratory determinations on the gravel samples were not undertaken. Using Fig. 1 to estimate the impedance functions one finds that the stiffness functions k_{hh}, k_{mm}, k_{vv} are near enough to unity to justify use of the static spring constants, which govern the response in any case at low frequencies. With regard to the damping constants it may be seen from Fig. 1 that for a value of $a_o \simeq 0.35$ the influence of the internal damping of the ground media is appreciable.

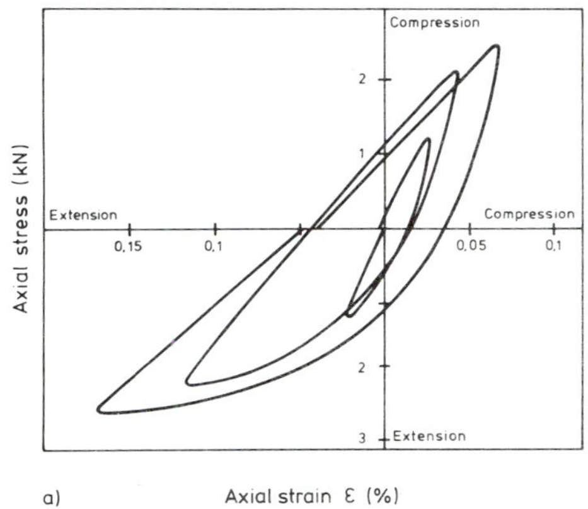

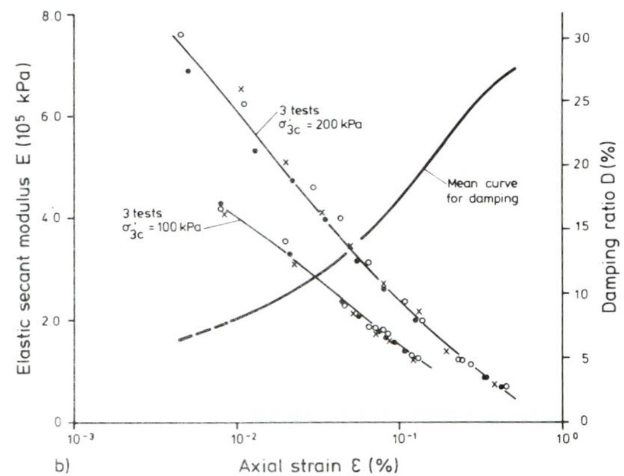

FIGURE 4. Dependence of deformation modulus and damping ratio of the gravel on deviator strain under cyclic triaxial conditions.Cell pressure cycled to maintain mean principal applied stress

Bearing Capacity of Soil

As the tank begins to lift off the ground and the eccentricity of the instantaneous loading on the supporting soil increases, the question of exceeding the allowable bearing pressure arises, and if so what are the consequences in terms of permanent settlements. The most exhaustive way of investigating this problem would be to carry out a full elastoplastic analysis, in which the inelastic behaviour of the ground is considered together with debonding effects at the interface soil-structure. With an appropriate material law the permanent settlements due to cycling at stress levels below failure (i.e. ultimate collapse load) could be determined in a shakedown mode. But the choice of material law to describe the inelastic behaviour might present some difficulties. Apart from this, the expensiveness of the analysis would tend to be prohibitive. An alternative approach would be to employ limit equilibrium methods of analysis. For static problems solutions exist for foundations not subjected to overturning moments, but little progress has been made towards finding solutions when this aspect is present, except in the case of a material obeying Tresca's law (e.g. in connection with offshore gravity platforms founded on clay, Chan 1978). To include the overturning moments therefore there is little choice other than to resort to approximate solutions, based on the concept of a reduced equivalent foundation area. The problem is first considered from a pseudo-static standpoint, using the general bearing capacity equation of Meyerhof (1963). Neglecting the cohesion the bearing capacity is

$$q_D = \gamma\, D\, N_q\, \lambda_q\, d_q\, i_q \;+\; \frac{\gamma}{2}\, B'N_\gamma\, \lambda_\gamma\, d_\gamma\, i_\gamma \qquad (10)$$

in which

- B': effective width $(B - 2e)$
- B : actual width of foundation $(B = 2R = 10.3\text{m})$
- e : eccentricity of loading
- D : depth of foundation (0.9m)
- γ : unit weight of soil $(2.25\ \text{t/m}^3)$;slip surface assume to lie wholly above ground water
- d, i, λ : correction factors; here $i = 1$ and
  $\lambda_q = \lambda_\gamma = 1 + 0.1(B'/B)\tan^2(\pi/4 + \phi/2)$
  $d_q = d_\gamma = 1 + 0.1(D/B')\tan(\pi/4 + \phi/2)$
- N_q, N_γ : bearing capacity coefficients

For the gravel foundation material in question the friction angle ϕ varies between about 51° and 43°, depending upon the confining pressure. Here the higher values would apply but for the sake of conservativeness we take $\phi = 43^{o}$. The corresponding values of N_q and N_γ are 100 and 200 (approx.) respectively. Strictly speaking, higher values could be taken for a circular foundation. However, at greater eccentricity the contact area becomes more strip-like in shape, so that no increase is assumed. Substituting for the given values

$$q_D = 225(0.9 + B')(1 + 0.05 B')(1 + 0.21/B') \quad (11)$$

Assuming a seismic coefficient k = 0.4 acting at the centre of gravity of the tank structure the eccentricity e = 4.4m and B'= 1.46m. The bearing capacity for a foundation of this reduced width then amounts to 652 t/m^2. This value has to be compared with the actual applied pressure (assumed uniform over the reduced area) of 237 t/m^2. Thus for this value of e there is still a substantial safety factor against ground failure. As e tends towards the value R the applied pressure increases rapidly. However, the analysis is too approximate to justify further calculated comparisons.

In choosing a seismic coefficient one has to take account of the fluctuating nature and average effect of seismic motions. Thus k = 0.4 presupposes maximum accelerations of much greater magnitude.

The value of the average maximum acceleration for the tank structure was estimated using the linear elastic FE program ASHSD 2, for the system shown in Fig. 3 with the water mass lumped to the shell structure. Modal damping values of 5 % were assumed. The following seismograms scaled to 1.5 m/s^2 were input at the rock base: El Centro N-S component, first 16 secs; Taft S69E, first 16 sec; San Fernando 1971 (Lake Hughes station), first 16 sec; sine sweep function, 4 sec duration (Fig. 5(a)). The condition of a welded interface implicit in linear analysis overestimates response for horizontal accelerations, so that the values obtained are conservative. The different seismograms excite the natural periods of the gravel layer in shear, i.e. $T_n = \frac{4 D_s}{(2n-1)} (\rho/G)^{1/2}$

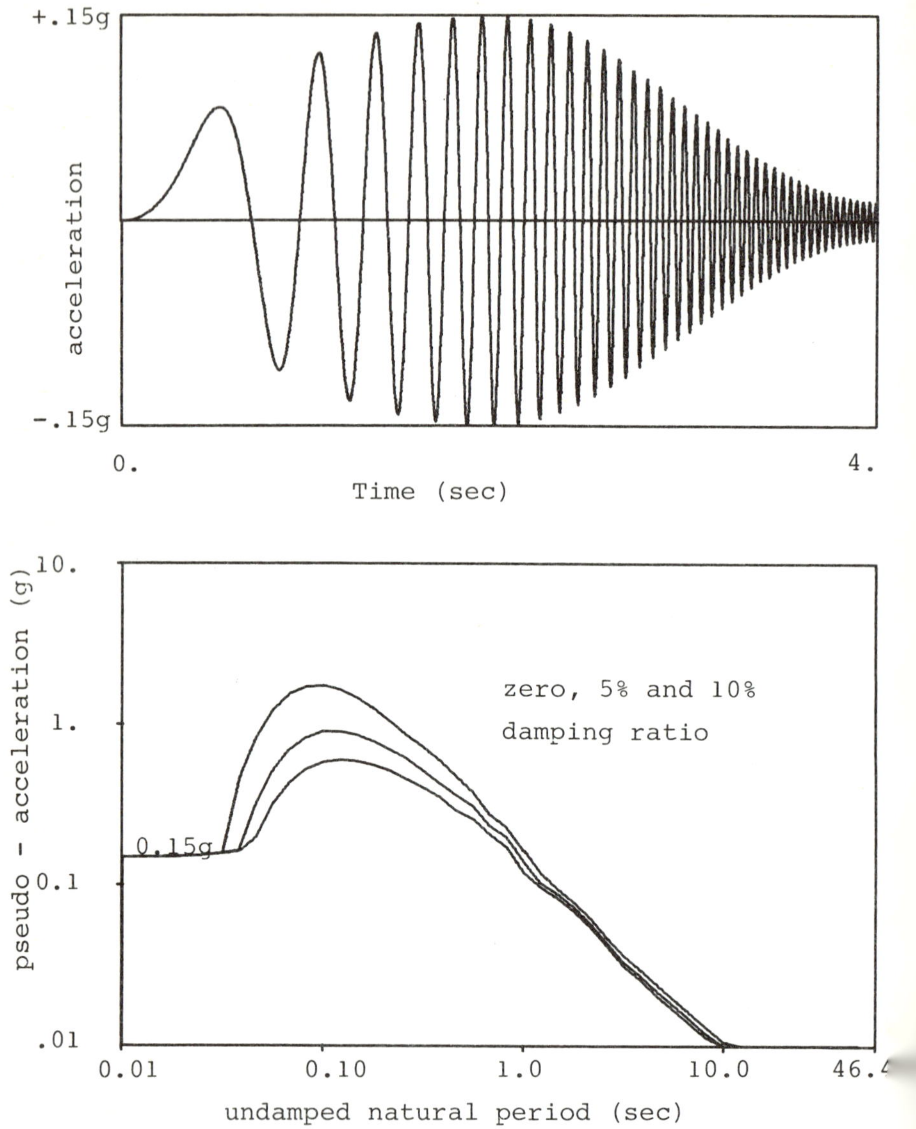

FIGURE 5. Sine sweep function $A = f(t)\sin(\Omega(t).t)$ whe
$\Omega(t) = 0.314 + 5t + 1.5t^2 + 0.3t^3$ and
$f(t) = 0.2 + 0.8t - 0.2t^2$ if t less than 2.
$f(t) = -3.5 + 5.4t - 2.025t^2+0.225t^3$ if t greater than 2.0
with response spectrum for pseudo absolute accelerati

where D_s is the depth of the layer by different amounts. For a density, $\rho = 2.25\ t/m^3$, and the previously given value of shear modulus G the fundamental frequency of the layer is 5.7 Hz. The absolute maximum accelerations, total and relative, are given in Table I for various levels. The sine sweep function (Fig. 5), which covers uniformly a wide range of frequencies, excites, as might be expected, the highest response with amplifications in the layer alone of about 650 %, the non-synthetic records amplifying by a factor of about 2. On the basis of expected earthquake intensity the value of maximum acceleration is specified corresponding to a depth of 3 m. Thus the bedrock accelerations would have to be scaled accordingly, which would involve extra computational work if non-linear soil response were to be considered. For the purpose of comparison the bedrock accelerations were made equal. The accelerations obtained for the real seismograms indicate that foundation overturning and excessive bearing pressures will not be a problem, but that the artificial seismograms (the sine function or a record made to match the NRC design response spectrum) may provide borderline cases.

The limit equilibrium approach, on which the above considerations depend, assumes ideal rigid plastic soil behaviour with the neglect of movements below the failure load. Attempts have been made to calculate the plastic deformations under footing subjected to impact loads by considering the rigid body movement of a block resisted by frictional and cohesive forces (e.g. Triandafilidis, 1965). In comparison to the static bearing capacity calculation the dynamic analysis is based upon the equation of motion of a cylindrical block, viz.

$$M_q = M_c + M_\phi + M_g + J\ddot{\alpha}$$

in which

M_q : moment due to applied load Q(t)

M_c : moment due to cohesive resistance

M_ϕ : moment due to reactive force

M_g : moment due to self-weight

TABLE I Absolute maximum accelerations for soil structure system shown in Fig. 3 under horizontal seismic excitation.

Height above bedrock (m)		Acceleration maxima (m/s^2) Total				Relative			
		a)	b)	c)	d)	a)	b)	c)	d)
Tank shell	36.9	+12.5	-8.8	+6.3	+22.4	+12.0	+9.2	+6.2	+21.7
	30.0	+7.0	+5.0	+3.8	-9.9	+6.6	+5.1	+3.7	-11.3
	26.0	+3.8	+3.5	-2.5	-6.1	-3.7	+3.3	-2.9	-7.4
	20.0	-3.0	-3.2	-2.3	-8.9	-2.8	-2.5	-2.6	-10.1
	14.0	+3.5	-2.9	-2.3	+10.4	-3.1	-2.6	-2.5	+11.9
Base mat sole	13.0	+3.6	-3.0	-2.3	+10.6	-3.1	-2.7	-2.5	+12.1
Gravel Layer	10.5	+3.5	-2.9	-2.2	+10.0	-2.9	-2.6	-2.3	+11.4
	7.0	+2.9	-2.3	-1.7	+7.2	-2.2	-2.0	-1.7	+8.6
	3.5	+2.1	-1.7	-1.0	-3.4	-1.2	-1.1	-0.9	+4.7
Bedrock surface	0.	-1.5	-1.5	-1.5	-1.5	0.	0.	0.	0.

a) El Centro N-S component

b) Taft S69E component

c) San Fernando 1971 (Lake Hughes Station)

d) sine sweep function

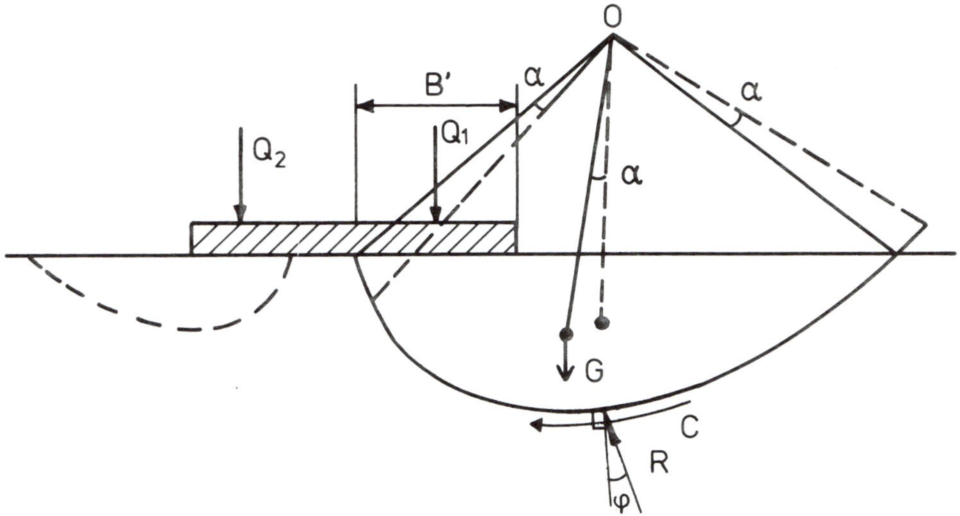

FIGURE 6. Sketch illustrating permanent displacement analysis for excessive contact pressure, showing also the impossibility of exact analysis under rocking conditions due to non-specificness of rupture surface

J : mass polar moment of inertia of block

α : rotation of block

The inertia term provides an additional resistance of the ground and it would also play an important part under seismic loading. Under the action of overturning forces, however, it is virtually impossible to calculate the actual movements as one is not able to define the actual sliding block, since the eccentricity (which gives B') is continually changing both magnitude and sign (cf. Fig. 6).

This kind of analysis is similar to one currently employed in the analysis of earth dams, i.e. the Newmark procedure (Seed 1979). With regard to nuclear power plants it has also recently been employed to estimate horizontal sliding deformations (Kausel et al, 1979).

CLOSING REMARKS

The general problem discussed in this paper, namely that of foundation overturning, in which the possibility of large permanent displacements or ground failure under severe earthquake loading may be a matter of concern, is one which has received relatively little attention to date. Full nonlinear analysis may be the only method to realistically tackle the problem, as the simplified model , in which the ground is replaced by elastoplastic springs, is not adequate to investigate excessive ground displacements, except in the case of horizontal sliding. The dynamic bearing capacity is not amenable to analysis in this case.

The particular problem given some attention here involved a reserve refuelling water storage tank. The liquid storage introduces added complexity to the problem, although the seismic safety of tanks, it would seem, is rarely endangered (Clough et al, 1979).

There is a definite need for more research into nonlinear soil-structure interaction (Whitman and Bielak, 1980) as design provisions may be too conservative in many cases.

REFERENCES

M. Arockiasamy, D.V. Reddy, W. Bobby and A.K. Haldar - Comparison of finite element and lumped parameter modelling for seismic response of reactor building-foundation systems, Third International Conference of Numerical Methods in Geomechanics, Aachen (1979), 817-829.

ASHSD2 (S. Ghosh and E. Wilson) - Dynamic stress analysis of axisymmetric structures under arbitrary loading, Report No. EERC 69-10, University of California, Berkeley, Revised program: C.J. Lin (1975).

ATC 3 - Tentative provisions for the development of seismic regulations for buildings, Applied Technology Council Publication ATC 3-06, Palo Alto (1978).

P.M. Byrne and N.D. Nathan - Earthquake induced overturning moments and their effects, Sixth European Conference on Earthquake Engineering, Dubrovnik (1978) 2, 85-92.

A.K.C. Chan - Discussion, Foundation stability on nonhomogeneous clays (Proc. Paper 13302: J.D. Murff and T.W. Miller), Journal of the Geotechnical Engineering Division, ASCE, 104 (1978) GT12, 1530-1532.

R.W. Clough, A. Niwa and D.P. Clough - Experimental seismic study of cylindrical tanks, Journal of the Structural Division, ASCE, 105 (1979) ST12, 2565-2590.

C.J. Costantino, C.A. Miller and L.A. Lufrano - Soil-structure interaction parameters from finite element analysis, Nuclear Engineering and Design, 38 (1976) 289-302.

H.I. Epstein - Seismic design of liquid-storage tanks, Journal of the Structural Division, ASCE, 102 (1976) ST9, 1659-1673.

D. Fischer - Dynamic fluid effects in liquid-filled flexible cylindrical tanks, Earthquake Engineering and Structural Dynamics, 7 (1979) 587-601.

G.C. Gazetas and J.M. Roesset - Forced vibrations of strip footings on layered soils, Methods of Structural Analysis, ASCE Publication, 1 (1976) 115-131.

J.A. Gutierrez and A.K. Chopra - A substructure method for earthquake analysis of structures including soil-structure interaction, Earthquake Engineering and Structural Dynamics, 6 (1978) 51-69.

A.H. Hadjian - Soil-structure interaction - an engineering evaluation, Nuclear Engineering and Design, 38 (1976) 267-272.

J.R. Hall, Jr. and J.F. Kissenpfennig - Special topics in soil-structure interaction, Nuclear Engineering and Design, 38 (1976) 273-287.

J.R. Hall, Jr., I.V. Constantopoulos and A.P. Michalopoulos - Higher order Winkler model for soil-structure interaction, Third International Conference on Numerical Methods in Geomechanics, Aachen (1979) 933-938.

U. Holzlöhner - Dynamic soil-structure interaction, Proceedings of Dynamic Methods in Soil and Rock Mechanics, Karlsruhe, 1 (1977) 103-126.

G.W. Housner - Design spectrum (in Earthquake Engineering), Ed. R.L. Wiegel, Prentice-Hall Inc., (1970), 103-104.

P.C. Jennings and J. Bielak - Dynamics of building-soil interaction, Bulletin Seismological Society of America, 63 (1973) 1, 9-48.

E.A. Kausel, J.M. Roesset and J.T. Christian - Nonlinear behaviour in soil-structure interaction, Journal of the Geotechnical Engineering Division, ASCE, 102 (1976) GT11, 1159-1170.

E.A. Kausel, R.V. Whitman, F. Elsabee and J.P. Morray - Dynamic analysis of embedded structures, Fourth Conference on Structural Mechanics in Reactor Technology, San Francisco, (1977), Paper K 2/6.

E.A. Kausel, A.S. Lucks, L. Edgers, W.F. Swiger, and J.T. Christian - Seismically induced sliding of massive structures, Journal of the Geotechnical Engineering Division, ASCE, 105 (1979) GT12, 1471-1488.

R.P. Kennedy, S.A. Short, D.A. Wesley and T.H. Lee - Effect of nonlinear soil-structure interaction due to base slab uplift on the seismic response of a high temperature gas-cooled reactor (HTGR), Nuclear Engineering and Design, 38 (1976) 323-355.

J.E. Luco - Vibrations of a rigid disc on a layered visco-elastic medium, Nuclear Engineering and Design, 36 (1976) 325-340.

J. Lysmer - Analytical procedures in soil dynamics, ASCE Speciality Conference on Earthquake Engineering and Soil Dynamics, Pasadena (1978) III, 1267-1316.

J.W. Meek - Effects of foundation tipping on dynamic response, Journal of the Structural Division, ASCE, 101 (1975) ST7, 1297-1310.

J.W. Meek - Dynamic response of tipping core buildings, Earthquake Engineering and Structural Dynamics, 6 (1978) 437-454.

J.W. Meek and A.S. Veletsos - Dynamic analysis and behaviour of structure foundation systems, Report No. 13, Department of Civil Engineering, Rice University, Texas (1972).

G.G. Meyerhof - some recent research on the bearing capacity of foundations, Canadian Geotechnical Journal, 1 (1963) 16-26.

E.G. Prater and M. Wieland - Some factors affecting the response to seismic loading in a soil-structure interaction problem, Conference on Dynamical Methods in Soil and Rock Mechanics, Karlsruhe, (1977), A.A. Balkema, 3, 237-254.

H.B. Seed - Considerations in the earthquake-resistant design of earth and rockfill dams, Géotechnique, 29 (1979) 3, 215-263.

H. Scaletti - Nonlinear effects in soil-structure interaction, Ph.D. Thesis, Department of Civil Engineering, Massachusetts Institute of Technology (1977).

STEALTH - A Lagrange explicit finite-difference code for solids, structural and thermohydraulic analysis, R. Hofmann, Science Applications, Inc., San Leandro, California (prepared for Electric Power Research Institute, Palo Alto) 1976.

M. Stoykovich - Nonlinear effects in dynamic analysis and design of nuclear power plant components: research status and needs, Nuclear Engineering and Design, 50 (1978) 93-114.

T. Takemori, K. Sotomura and M. Yamada - Nonlinear dynamic response of reactor containment, Nuclear Engineering and Design, 38 (1976) 463-474.

G.E. Triandifilidis - The dynamic response of continuous footings supported on cohesive soils, Sixth Internation Conference on Soil Mechanics and Foundation Engineering, (1965) 2, 205-208.

A.K. Vaish and A.K. Chopra - Earthquake finite element analysis of structure-foundation systems, Journal of the Engineering Mechanics Division, ASCE, 100 (1974) 1101-1116.

A.S. Veletsos - Seismic effects in flexible liquid storage tanks, Fifth World Conference on Earthquake Engineering, Rome, (1974) 1, 630-639.

H. Werkle and K. Rönnberg - Erdbebenbeanspruchung steifer Bauwerke auf geschichtetem Baugrund, 10. Forschungskolloquium des deutschen Ausschusses für Stahlbeton, Karlsruhe (1979), 79-84.

R.V. Whitman and J. Bielak - Foundations (in Design of Earthquake Resistant Structures, ed. E. Rosenblueth, Pentech Press, London, (1980) 223-260.

J.P. Wolf and P.E. Skrikerud - Seismic excitation with large overturning moments: tensile capacity, projecting base mat or lifting-off?, Nuclear Engineering and Design, 50 (1978) 305-321.

H.L. Wong - Dynamic soil-structure interaction, Report No. EERL-75-01, California Institute of Technology, Pasadena (1975).

Additional relevant reference

J.S. Dalal and P.R. Perumalswami - Overturning behaviour of nuclear power plant structure during earthquakes, Sixth World Conference on Earthquake Engineering, New Delhi, (1977) 8, 39-44.

SLOPE STABILITY UNDER EARTHQUAKE LOADING

E.G. Prater, Ph.D.

Research Engineer, Institute of Foundation Engineering and Soil Mechanics, ETH-Zurich, Switzerland

INTRODUCTION

The seismic stability of soil slopes is a problem area with regard to which much uncertainty still exists. The problem, however, is particularly important for engineers concerned with the design of water-retaining structures, which have a large hazard potential. Traditionally, simple pseudo-static (or static equivalent load) methods of analysis have been incorporated into design codes. It is now well-known, however, that peak accelerations approaching that due to gravity (g) are possible in the near field of even moderately intense earthquakes (Ambraseys, 1974). Using traditional methods, however, it is uneconomical to design for seismic coefficients greater than about 0.20.

In trying to allow for the effects of transient inertia loading the problem is to try to relate the design seismic coefficient to the expected peak acceleration together with the other significant features of the time history of ground acceleration (e.g. duration, maximum velocity and frequency characteristics). It is clear that due to the oscillatory nature of earthquake loading on a flexible structure like a rockfill dam unrealistically conservative designs would be obtained if seismic coefficients corresponding to maximum credible earthquake accelerations were used. Thus values appearing in various codes even in areas of very high seismicity are usually not in excess of 0.10-0.25. Historically, these values are no doubt to be accredied to an insufficient knowledge of earthquake motion characteristics and economic pressures. But retrospectively, it may be observed that the real effect of an earthquake was represented perhaps fortuitously by a reduced value of acceleration, since in general damage has been light.

Today, it is recognized that the problem must be approached more from the point of view of the material-technological aspects of structural behaviour (Seed, 1979). Certain materials are able to withstand high stresses for short durations, whereas other more brittle or liquefiable materials are not. However, in terms of the aseismic design of dams and embankments, if a complete loss of strength of the constituent materials can be ruled out (examples where this may not be the case are loose hydraulic fills and tailings dams), then the problem is reduced to one of finding the plastic irrecoverable deformations due to inertia effects. The impetus to work in this field was given mainly by the contribution of Newmark (1965). By using this method of approach the vagueness and uncertainty attached to the pseudo-static method of analysis is avoided, and a greater measure of rationality is given to the aseismic design of slopes.

Ideally, it would be better to carry out a full non-linear dynamic analysis as already accomplished for example by Dibaj and Penzien (1969), Finn and Miller (1973), who used an ideal elastoplastic soil model. This approach, however, is generally very costly in terms of computer time, and has not yet found favour in practice. Newmark's approach, on the other hand, is based on simple limit equilibrium principles, and does not involve extensive computational work. The basic idea is that if the factor of safety of a potential sliding mass is reduced temporarily below unity a certain amount of movement will take place. The contribution of the various pulses of earthquake motions are added together to determine the total permanent deformation, which may then be adjudged acceptable or unacceptable depending upon the design criterion. A unit factor of safety is defined in terms of the so-called "yield acceleration". If the acceleration of the soil mass exceeds the yield acceleration motion is induced. The yield acceleration could, in fact, be obtained by any conventional method of slope stability analysis, and for general situations (inhomogeneity, seepage, action of external forces, etc.) one of the various methods of slices would normally be employed. For a homogeneous slope it has been found that the log spiral method of analysis originally suggested by Rendulic (1936) yields a simple mathematical solution (Prater, 1979). In the following a comparison is made between the assumptions of a log spiral and a plane sliding surface, the main results being presented in graphical form.

YIELD ACCELERATION

Plane slip surface

Newmark and Rosenblueth (1971) have deduced theoretically that for cohesionless soils the most critical conditions of yielding are associated with a plane sliding surface. Cohesion will produce a small curvature of the sliding surface. Nevertheless, for comparison purposes, we also consider the case of a plane surface for cohesive soils. Further, the computation of displacements is simple for this model. It may be easily shown (Prater, 1973) that an arbitrarily directed seismic coefficient k required to cause yield is given by

$$k.f(\alpha) = \frac{2\,N_s \cos\phi \sin i}{\cos(\theta-\phi)\,\sin(i-\theta)} - \tan(\theta-\phi) \qquad (1)$$

where N_s is the stability number ($c/\gamma H$) and c is the cohesion, γ the unit weight, H the height of slope and ϕ the angle of internal friction. θ and i are the inclinations of the sliding surface and embankment slope respectively, α is the angle subtended by the seismic coefficient with the horizontal, and

$$f(\alpha) = \sin\alpha\,\tan(\theta-\phi) + \cos\alpha \qquad (2)$$

For horizontal excitation only $f(\alpha) = 1$.

The inclination of the most critical wedge may be obtained from the following analytic expression

$$2\,N_s \cos\phi\,\sin i \quad \cos(i+\phi-2\theta) - \sin^2(i-\theta) = 0 \qquad (3)$$

Logarithmic spiral slip surface

For a spiral defined as below the resultant of the normal and frictional forces on the sliding surface passes through the pole. This useful property means that in the moment equation only the weight, cohesive and inertia terms appear. The radius r of the spiral is

$$r = r_o \exp(\theta \tan\phi) \qquad (4)$$

and equilibrium of moments is given by

$$(1 \underset{-}{+} k_v) M_{weight} + k_h M_{inertia} - M_{cohesion} = 0 \qquad (5)$$

where k_h and k_v are the component seismic coefficients.

This equation differs from the one derived by Taylor (1937) by the inclusion of k_v and k_h. The trigonometric expression for the term $M_{inertia}$ is given by

$$M_{inertia} = M_4 - M_5 - M_6 \tag{6}$$

where

$$M_4 = \frac{B\ d^3}{9\tan^2\phi+1}\{(m^3\cos j+\cos q) + 3\tan\phi(m^3\sin j-\sin q)\} \tag{7}$$

$$M_5 = B\ d^3\ \sin^3 q\ (\cot q + \cot j) \tag{8}$$

$$M_6 = \frac{B}{2}\ (3\ d\ \sin q + 1)\ (\cot i - \cot j) \tag{9}$$

in which $B = \gamma\ H^3/3$ and following Taylor (1937)

$$m = \exp(z\ \tan\phi)$$

$$d = (1 + m^2 - 2\ m\ \cos\ z)^{-1/2}\ /\ \sin t$$

$$j = t + \sin^{-1}\ (d\ \sin z\ \sin t)$$

and $q = \pi - z - j$

The angles z and t are shown in Fig. 1. i is the slope of the embankment. The expressions for M_{weight} and $M_{cohesion}$ may be found in Taylor (1937) or Prater (1979). The simplest procedure for finding the most critical surface is to assume that the time variation of k_v matches that of k_h (or is constant). The minimum value of k_h found from equ. (5) gives the critical case. Practically, this is achieved by varying the location of the pole using small increments of the angle z, keeping the angle t constant, and then repeating for small increments of t. The initial value of t is just less than i. It should be noted that for seismic loading there is usually an amplification of acceleration above the toe of the slope. There are two consequences of this: (i) deep-seated failure surfaces need not be considered, (ii) the seismic coefficient should be averaged. To accomplish the latter, vibration analysis using finite element methods or shear beam theory has to be carried out (Makdisi and Seed, 1979).

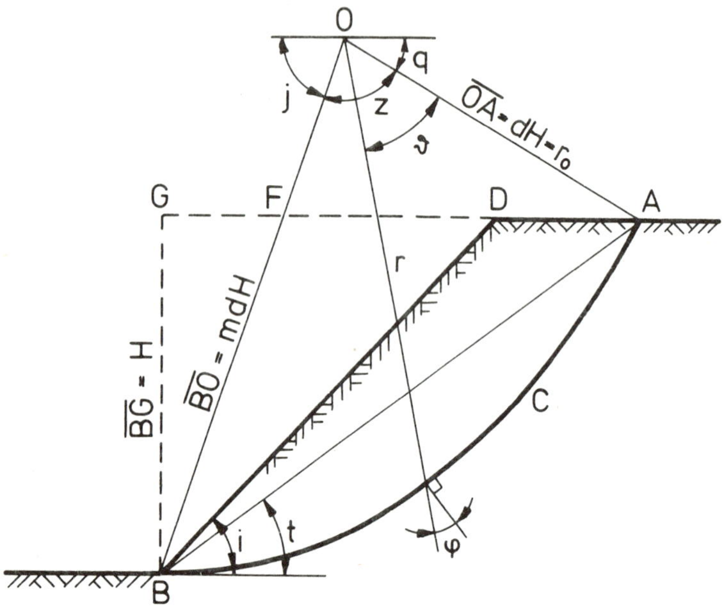

Fig.1 Log spiral slip surface for displacement of rigid mass of soil

Some results have been calculated for the assumption $k_v = 0$. These are presented for various slope angles and values of stability factor N_s in Figs. 2, in which a comparison is made with those obtained for the assumption of a plane sliding surface. An examination of these results shows that for shallow slopes the yield acceleration can be considerably overestimated, especially at smaller values of k_h. For a slope in which under static conditions the factor of safety against sliding failure is 1.5 the yield acceleration calculated on the basis of the log spiral generally lies in the range 0.08 g to 0.15 g. For these values of k_h the planar slip surface assumption leads to much higher values of yield acceleration, for $i < 30^o$ by 50 % - 100 %.

CALCULATION OF PERMANENT DISPLACEMENTS

Displacement is initiated whenever the ground acceleration, A, (given by the slope of the ground velocity curve) exceeds the yield acceleration, N = k.g. Franklin and Chang (1977), working directly with the ground velocity curve, have used numerical integration techniques to compute the permanent displacement of a simple block for 179 seismograms from the Western United States together with a few artificial seismograms. The results were compared with the upper bounds established by Newmark (1965), which were found to be too low.

In this study explicit integration of the ground accelerogram has been used, assuming that it is piecewise linear between digitized values. Only the horizontal component of earthquake motions is considered. The accelerogram consists of a number of supercritical pulses of relatively short duration. Thus the activating force lasts for only a short time, whereas the decelerating force due to the earth resistance acts until the relative velocity between the sliding wedge and the ground becomes zero. The motion is more quickly retarded if the effect of negative pulses is also considered. Resistance to uphill motion is assumed to be infinite, so that relative motion is completely unsymmetrical.

The equation of motion of a wedge is given by

$$\ddot{u} = \beta\ (A-N) \tag{10}$$

in which $\ddot{u}$ is the relative motion and β is a parameter depending upon the assumed model. For instance, for a block sliding on a horizontal surface $\beta = 1$, whereas on a plane sloping surface of inclination θ_c (cf. Sarma, 1975)

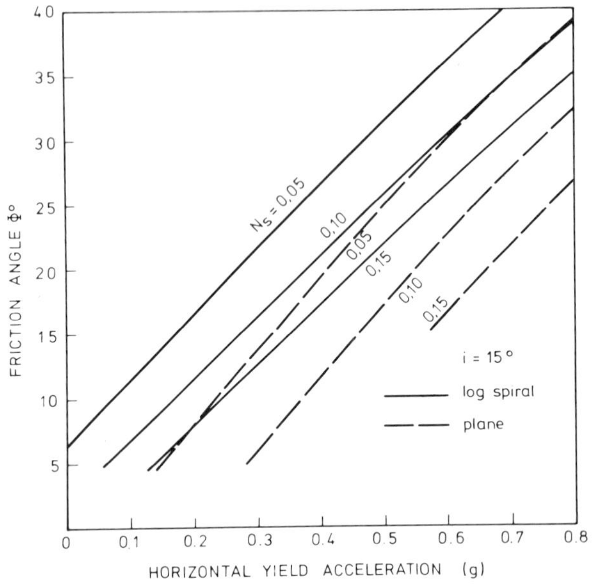

(a) i = 15°

Fig.2 Yield acceleration for horizontal seismic excitation for plane and log spiral sliding surfaces as a function of slope angle i and strength parameters

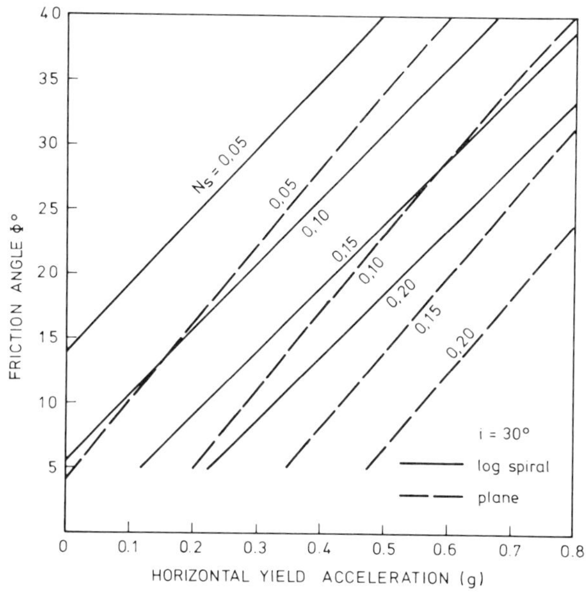
FRICTION ANGLE Φ°
HORIZONTAL YIELD ACCELERATION (g)
Ns = 0,05
0,05
0,10
0,15
0,10
0,20
0,15
0,20
i = 30°
log spiral
plane

(b) i = 30°

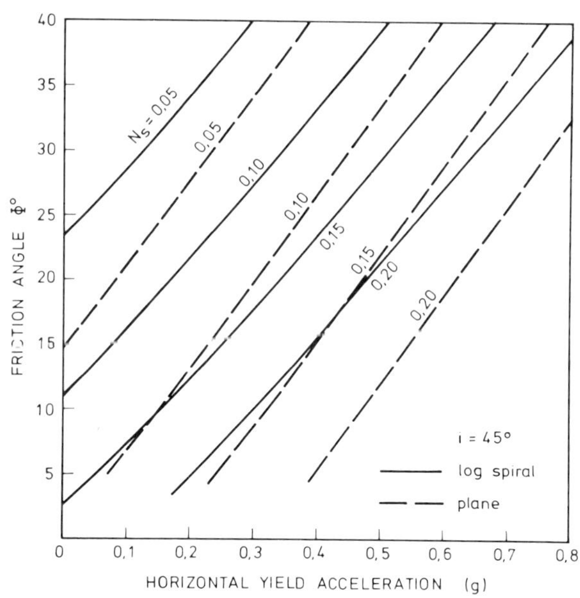
FRICTION ANGLE Φ°
HORIZONTAL YIELD ACCELERATION (g)
Ns = 0,05
0,05
0,10
0,10
0,15
0,15
0,20
0,20
i = 45°
log spiral
plane

(c) i = 45°

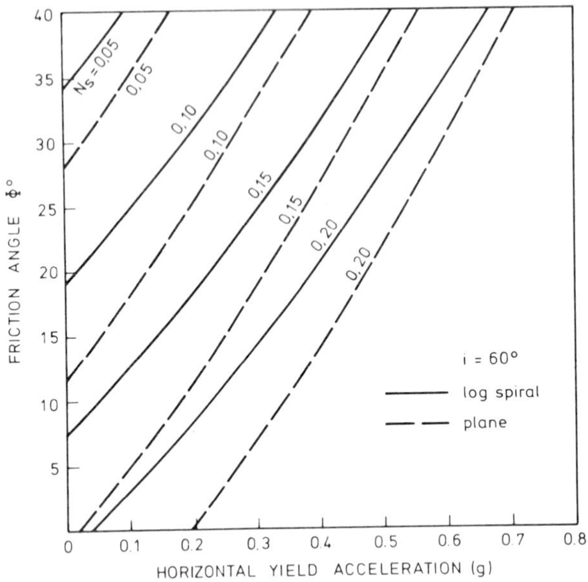

(d) i = 60°

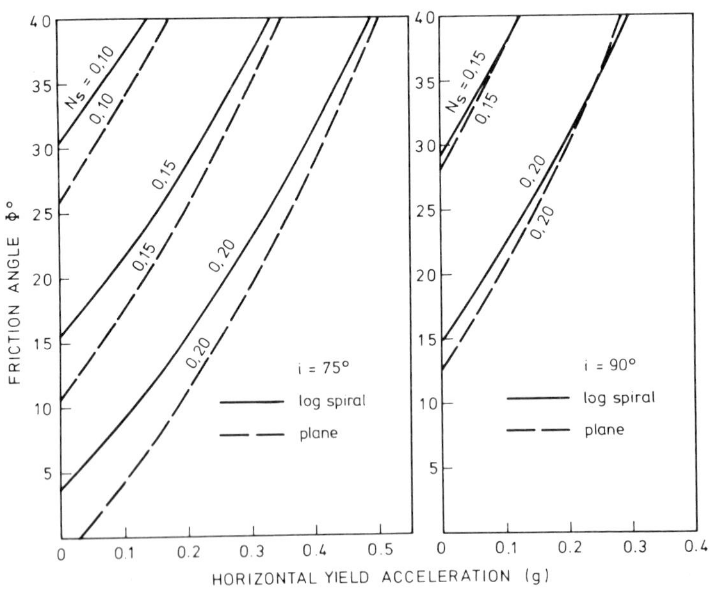

(e) i = 75° and 90°

$$\beta = \frac{\cos(\theta_c - \alpha - \phi)}{\cos\phi} \tag{11}$$

wherein, for horizontal excitation only $\alpha = 0$. For most cases of θ_c (as calculated from equ. 3) β is near to unity ($1 < \beta < 1.4$ approx.).

In the case of a non-planar slip surface (log spiral), if Θ represents the rotation of the wedge the inclusion of the inertia effect leads to the following equation (Prater, 1979)

$$J\,\ddot{\Theta} = M_{inertia}\,(A-N) \tag{12}$$

in which J is the mass polar moment of inertia of the sliding body. This equation is of the same form as equ. (10), whereby now $\beta = M_{inertia}/J$. It is clear that it is possible to evaluate for $\beta = 1$ any accelerogram with chosen values of N and then final adjustment of the displacement (angular rotation) is made by multiplying by the given value of β.

The procedure adopted for handling a piecewise linear accelerogram is set out in the following steps:

1. Consider the first interval in which movement is initiated. A varies from A_i to A_{i+1} in Δt such that $A_i < N$, $A_{i+1} > N$. Initially $\dot{u} = u = 0$. Relative motion is produced over the reduced time interval

$$\Delta t_r = \frac{A_{i+1} - N}{A_{i+1} - A_i} \cdot \Delta t \tag{13}$$

At time t the relative velocity is given by

$$\dot{u}_t = \dot{u}_o + (A_i - N)(t - t_o) + \frac{(A_{i+1} - A_i)}{2\,\Delta t}\{(t-t_i)^2 - (t_o - t_i)^2\} \tag{14}$$

in which t_o is the time at which A = N.

At the end of the interval the velocity is

$$\dot{u}_{i+1} = \frac{\Delta t_r}{2}\,(A_{i+1} - N) \tag{15}$$

2. Equ. (14) is integrated to find the displacement, viz. at $t = t_{i+1}$

$$u_{i+1} = \frac{(A_i - N)}{2} \Delta t_r^2 + \frac{(A_{i+1} - A_i)}{2\,\Delta t} \left\{ \frac{\Delta t^3}{3} - \frac{(\Delta t - \Delta t_r)^3}{3} - (\Delta t - \Delta t_r)^2 \Delta t_r \right\} \qquad (16)$$

3. In the next interval ($i+1 \rightarrow i+2$) equ. (14) remains valid except that the value of $\dot{u}_t$ is non-zero at the beginning of the interval, i.e. $\dot{u}_o = \dot{u}_{i+1}$ and $t_o = t_{i+1}$, so that at the end of the inverval

$$\dot{u}_{i+2} = \dot{u}_{i+1} + \Delta t \left(\frac{A_{i+2} + A_{i+1}}{2} - N \right) \qquad (17)$$

Note, if the average acceleration is less than N the relative motion begins to slow down.

4. If $A_{i+2} < N$ a check is made on $\dot{u}_{i+2}$ for a negative value. If the velocity at the end of the interval has become negative the time t_m at which $\dot{u} = 0$ is calculated assuming a linear variation of $\dot{u}$. This is sufficiently accurate and resort does not have to be made to equ. (14), which is quadratic in t.

5. The displacement is given by

$$u_{i+2} = \dot{u}_{i+1} \Delta t + \frac{(A_{i+1} - N)}{2} \Delta t^2 + \frac{(A_{i+2} - A_{i+1})}{6} \Delta t^2 + u_{i+1} \qquad (18)$$

Note, if $\dot{u}$ becomes zero in this interval Δt is reduced and the value of A_{i+2} equals A(t) at $t = t_m$.

6. The whole procedure is then repeated for the various calculated velocity pulses.

Some typical results are shown in Figs. 3-5 for a triangular pulse, with and without the influence of negative pulses, a sine sweep function and an artifical seismogram. Some results are also tabulated in Table 1, whereby input accelerations have been scaled to 0.5 g, the frequency content being left unchanged (i.e. there is no scaling of peak ground velocity). The presented values must be multiplied by β g to obtain the displacement in length units corresponding to the units of g. The most intense

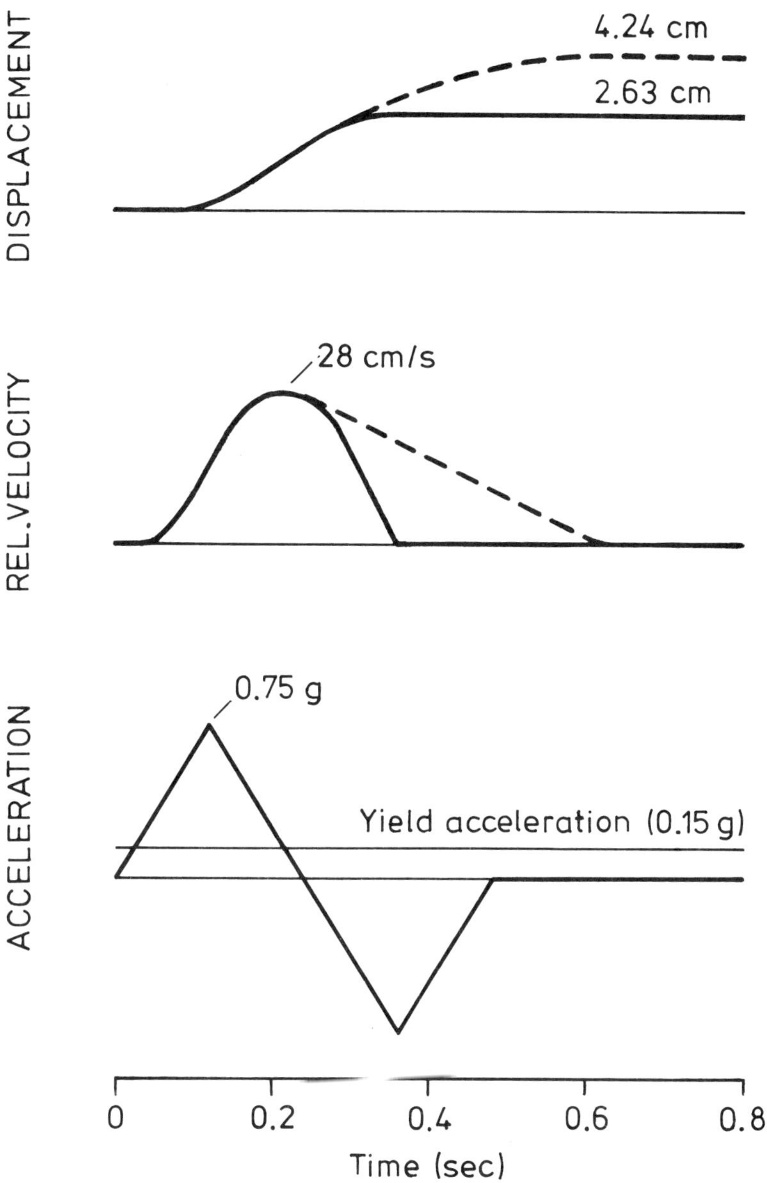

Fig.3 Response of sliding block to triangular pulse for N/A_{max} = 0.2 with (full line) and without (dashed line) the effect of negative pulses in seismic loading function

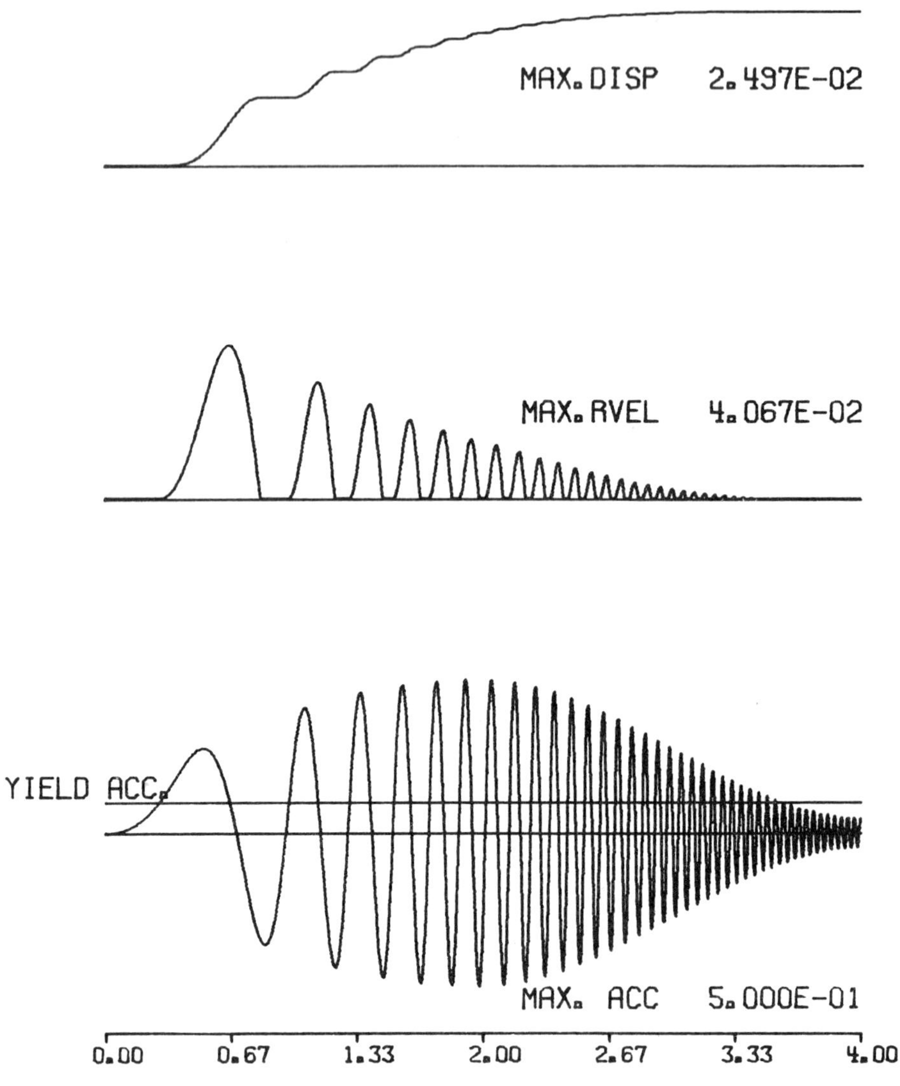

Fig.4 Response of sliding block to sine sweep function with $N/A_{max} = 0.2$ including the effect of negative acceleration pulses

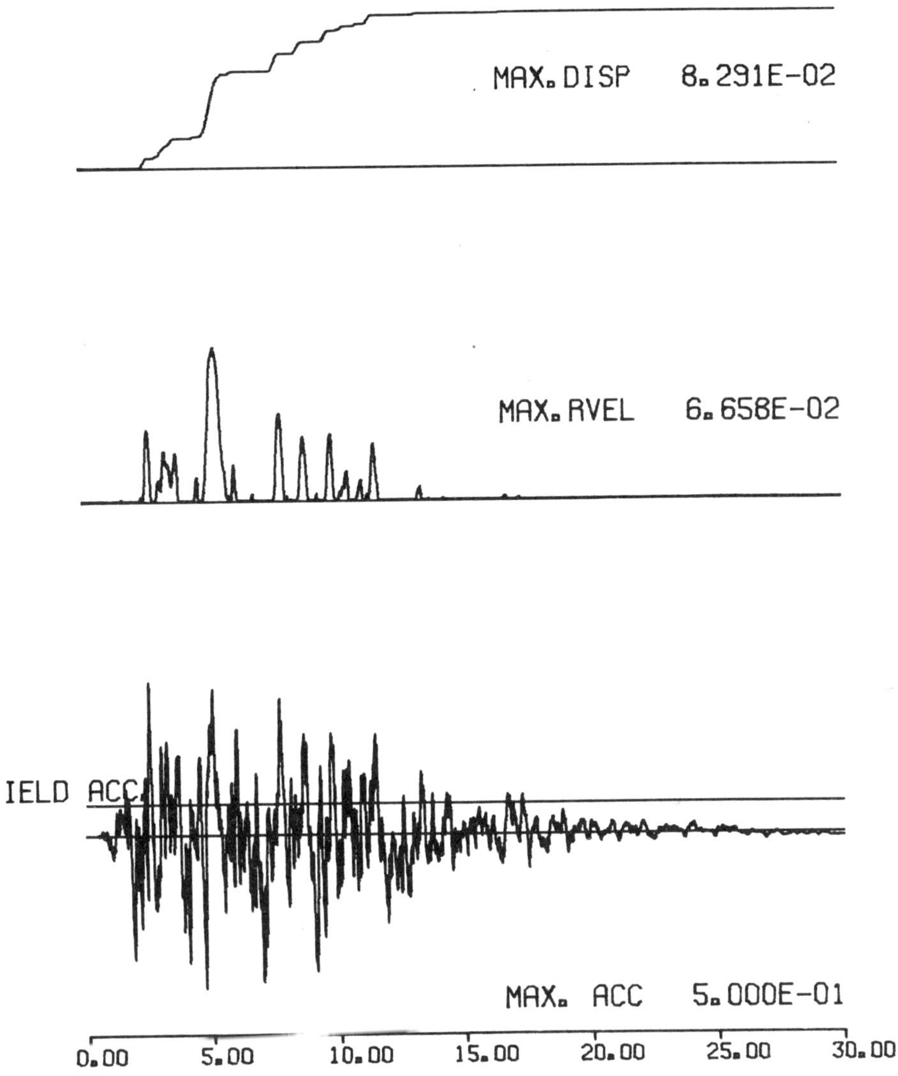

Fig.5 Response of sliding block to synthetic seismogram (Program of Ruiz and Penzien, 1969) with N/A_{max} = 0.2 including the effect of negative acceleration pulses

TABLE I Permanent displacements for some typical records.
$A_{max} = 0.5\ g$, $\beta = 1$

DISPLACEMENTS (x β g) units of length as in g					
negative pulses:					
considered				not considered	
N/A_{max}	0.3	0.2	0.1	0.2	0.1
E 1	.040	.126	.304	.239	9.20
	.046	.099	.324	.147	8.61
E 2	.010	.024	.070	.029	.233
	.007	.025	.073	.030	.226
E 3	.015	.029	.103	.039	.668
	.013	.036	.076	.048	.719
E 4	.019	.051	.086	.130	1.21
	.020	.037	.133	.062	0.90
E 5	.033	.083	.250	.119	2.52
	.039	.091	.233	.122	2.42
E 6	.004	.010	.036	.011	.101
	.002	.008	.032	.009	.062

E 1 : Seed-Idriss synthetic record M = 8 1/4, duration 75 sec.

E 2 : Seed-Idriss synthetic record M = 7 1/4 (modified Pacoima and Taft records) duration 37 sec.

E 3 : Taft S69E record, duration 30 sec.

E 4 : El Centro record, first 16 sec.

E 5 : an artifically generated record using computer program of Ruiz and Penzien (1969), duration 30 sec.

E 6 : San Fernando, 1971. Pacoima dam S16E.

of these records (the scaled Seed-Idriss magnitude M = 8 1/4 synthetic record) gave a maximum displacement for $\beta = 1.$ and $N/A_{max} = 0.1$ of 318 cm. This is about 30 times smaller than when negative pulses are neglected. At $N/A_{max} = 0.3$ this effect is much smaller, e.g. for the El Centro record the difference is only 35%.

Ambraseys (1974) gives an upper bound relation, viz.

$$\log_{10}(u) = 2.3 - 3.3\ (N/A_{max}) \tag{18}$$

for displacement u in cm units. At $N/A_{max} = 0.1$ this gives about 100 cm. Upper bound curves for various values of M may be found in Seed (1979) and Makdisi and Seed (1978), wherein the displacement is normalized with respect to A_{max} and the first natural period of the embankment to reduce the wide scatter that is otherwise obtained.

DISCUSSION AND LIMITATIONS OF APPROACH

Deformation Characteristics and Failure Mechanism

The method due to Newmark assumes perfectly plastic soil behaviour, which is only approximately satisfied for most soils. Seed (1966) proposes a method by which the non-linear stress-strain characteristics of soils are better accounted for. In its more advanced form (Seed et al, 1973, Seed, 1979), the determination of the stress history in soil elements in the dam may involve iteration on soil properties (both stiffness and strength degradation) and the advantage of simplicity is lost, especially if at the end a strain-harmonizing FE-analysis is required. A full elastoplastic analysis may then be preferable, whereby nowadays a loss of shear strength upon cyclic loading may also be incorporated in the analysis (Zienkiewicz et al, 1978). A further developement of Seed's (1966) approach is found in Reséndiz (1975).

The simplified and more elaborate elasto-plastic and non-linear analyses lead to deformations within the whole of the dam, on the basis of which the loss of freeboard can be estimated. Newmark's method, on the other hand, utilizes the limit equilibrium principle and considers a rigid block sliding on an inclined plane or along a curved surface. In finding the yield acceleration for cohesive soils a log spiral surface has been assumed. This has been done in preference to a circular surface because the resulting analytical expressions are more easily obtained. In many cases a log spiral can be closely approximated by a circular arc, the centre of the circle lying, of course, in a different position to the pole of the spiral. The log spiral gives better upper bounds and it is also the only curved surface adopted in the limit analysis (Chen, 1975), since it satisfies the normality condition of the theory of plasticity for the thin slip band (i.e. zone of velocity discontinuity) separating the rigid blocks. The kinematic slip condition allows for plastic volumetric expansion ($\phi \neq 0$) during shearing. The amount of slip still has to be fairly small, however, in comparison to the overall dimensions of the sliding block. This is obviously true for another reason, since the contact area between the rigid blocks decreases with increasing slip movement. In nature the material forms a heap at the toe of the slope. Thus larger deformations can only be roughly estimated using Newmark's method.

Comparison with observed Behaviour and Model Tests

The application of a Newmark-type method of analysis involving sliding on distinct failure surfaces must be judged against the background of the actual observation of the seismic behaviour of embankments and of the results of model studies.

Most dam failures that have been observed have involved poorly compacted dam materials and the failure has resulted from the liquefaction and collapse of these materials (Seed et al, 1978). Otherwise, there are practically no reports of slides occurring in embankment dams, although in natural slopes with low safety factors slides are not uncommon.

With regard to model tests, relatively few studies have been undertaken. The main emphasis has been placed on cohesionless materials. Bustamante (1965) and Goodman and Seed (1966), using shaking tables, have investigated the types of failure of sand embankment models. Two failure modes were observed, a progressive type of failure affecting surface grains and the triggering of shallow surface wedge-like slides. In connection with the Nurek dam in Russia Nichiporovitch and Melnik (1968) tested gravel models under impulsive load with maximum acceleration of 0.7 g. Induced deformations resulted mainly from the avalanching of grains near the surface of the slope. If pore pressures can build up the failure zone may extend more deeply into the body of the embankment. Model studies of zoned rockfill dams have been reported by Clough and Pirtz (1958) and Okamoto (1973). The first authors emphasize the flexibility of rockfill structures. In their tests shaking with accelerations up to about 0.4 g were withstood quite satisfactorily. Okamoto observed that the surface grains in the slope tend to roll and slide until a more stable cross-section is obtained. If the geometry changed greatly secondary failure in the form of cracking of the central core could be observed. Okamoto, ibid., also investigated the influence of grain size on the deformational behaviour. However, one probably cannot, to this extent, extrapolate the results of laboratory simulations to field behaviour. The main problem is to simulate the shear strength, which for granular materials depends upon the effective pressures. The slope of the Mohr strength envelope is known to decrease considerably at high confining pressures for rockfill materials, ϕ reducing by as much as 15° in pressure ranges encountered in very high rockfill dams.

Shaking table tests on slopes of cohesive material have been reported by Arango and Seed (1974). For the short duration of earthquake loading the undrained shear strength of clay (s_u) may be used in the analysis, and the difficulty of simulating the intergranular pressure is not such a drawback. It was shown that if the yield acceleration for the given slope configuration is exceeded deformations occur which may be significant. Generally, however, it is known that clayfill dams are much less vulnerable to failure under seismic loading (Seed et al, 1978), though failure of the foundation materials may require clarification if the sediments are of relatively low strength.

FINAL REMARKS

Embankments could be designed so that the yield acceleration is greater than the possible earthquake acceleration, i.e. simply by the limit equilibrium approach with the factor of safety always exceeding unity. Ambraseys (1965) suggests that this should be done for full reservoir conditions, applying the displacement approach (Newmark's method) only when there is no danger of large amounts of water escaping. This design method, however, may not be economically feasible, especially if the materials of the embankment have a "self-healing" capability and can tolerate a certain amount of deformation. Newmark's method shows the way to a less conservative approach. It is based, however, on an effective stress analysis and for cohesionless soils, unless the pore pressures developed during the earthquake can be estimated, its use is limited to dry materials (Seed, 1966), for example the downstream side of a rockfill dam.

In principle strength degradation could be incorporated into Newmark's method if the phenomenon were only better understood. With the displacement method the current practice is to work with upper bound curves (Newmark, 1965, Ambraseys, 1974, Franklin and Chang, 1977) obtained using a number of seismograms normalized both with respect to peak ground acceleration and peak ground velocity. Conservatism also results from the assumption of unsymmetrical movements and neglecting the negative pulses in the accelerogram. For ratios N/A_{max} greater than about 0.3 displacements are generally small (less than 1 m). As the ground acceleration increases in relation to the yield acceleration (N/A_{max} <0.2) displacements begin to increase rapidly. In this case it may be appropriate to consider the effect of negative pulses. The final displacement will often be found to be small enough to prevent overtopping. Other problems, such as rupture of the core and damage to appurtenant structures, however, must not be overlooked.

REFERENCES

N.N. Ambraseys - A review of the seismic stability methods of earth dams, Proceedings of the 3rd World Conference on Earthquake Engineering, New Zealand, (1965) 3, IV: 391-399.

N.N. Ambraseys - Dynamics and response of foundation materials in epicentral regions of strong earthquakes, Proceedings of the 5th World Conference on Earthquake Engineering, Rome, (1974) 1, CXXVI-CXLVIII.

I. Arango and H.B.Seed - Seismic stability and deformation of clay slopes, Journal of the Geotechnical Engineering Division, ASCE, 100, (1974) GT2, 139-156.

J.I. Bustamante - Dynamic behaviour of noncohesive embankment models, Proceedings of the 3rd World Conference on Earthquake Engineering, New Zealand, (1965) 3, IV: 596-612.

W.F. Chen - Limit Analysis and Soil Plasticity, Elsevier Scientific Publishing Company, Amsterdam, (1975), pp. XIV+ 638.

R.W. Clough and D. Pirtz - Earthquake resistance of rockfill dams, Transactions of the American Society of Civil Engineers, 123, (1958), 792-810.

A.G. Franklin and F,K. Chang - Permanent displacements of earth embankments by Newmark sliding block analysis, Miscellaneous Paper 5-71-17, Report 5, U.S. Army Engineer Waterways Experiment Station, Vicksburg, (1977).

R.E. Goodman and H.B. Seed - Earthquake induced displacements in sand embankments, Journal of the Soil Mechanics and Foundation Engineering Division, ASCE, 92, (1966) SM2, 125-146.

F.I. Makdisi and H.B. Seed - Simplified procedure for estimating dam and embankment earthquake-induced displacements, Journal of the Geotechnical Engineering Division, ASCE, 104, (1978) GT7, 849-867.

F.I. Makdisi and H.B. Seed - Simplified procedure for evaluating embankment response, Journal of the Geotechnical Engineering Division, ASCE, 105, (1979) GT12, 1427-1434.

N.M. Newmark - Effects of earthquakes on dams and embankments, Géotechnique, London, 15, (1965) 2, 141-156.

N.M. Newmark and E. Rosenblueth - Fundamentals of Earthquake Engineering, Prentice Hall, Inc., Englewood Cliffs, N.J., (1971), pp. XVI + 640.

A.A. Nichiporovitch and V.G. Melnik - Stability of dam erected of local materials under dynamic loads, Proceedings of the International Symposium on Wave Propagation and Dynamic Properties of Earth Materials, University of New Mexico Press, (1968), 839-845.

S. Okamoto - Introduction to Earthquake Engineering, University of Tokyo Press, (1973), pp. XVI+571.

E.G. Prater - Die Stabilität von erdbebenbeanspruchten Boeschungen, Schweizerische Bauzeitung, 91, (1973) 47, 1157-1162.

E.G. Prater - Yield acceleration for seismic stability of slopes, Journal of the Geotechnical Engineering Division, ASCE, 105, (1979) GT5, 682-687.

L. Rendulic - Ein Beitrag zur Bestimmung der Gleitsicherheit, Der Bauingenieur, 16, (1935) 19/20, 230-233.

D. Reséndiz - Optimum seismic design of embankment dams, 5th Panamerican Conference on Soil Mechanics and Foundation Engineering, Buenos Aires, (1975) IV, 437-476.

P. Ruiz and J. Penzien - PSEQGN, Artificial generation of earthquake accelerograms, Report: Earthquake Engineering Research Center, University of California, Berkeley, (1969) March.

H.B. Seed - A method of earthquake resistant design of earth dams, Journal of Soil Mechanics and Foundation Engineering Division, ASCE, 92, (1966) SM1, 13-41.

H.B. Seed, K.L. Lee, I.M. Idriss and F. Makdisi - Analysis of the slides in the San Fernando Dams during the earthquake of February 9th, 1971, Report No. EERC 73-2, Earthquake Engineering Research Center, University of California, Berkeley, (1973).

H.B. Seed, F.I. Makdisi and P. DeAlba - Performance of earth dams during earthquakes, Journal of the Geotechnical Engineering Division, ASCE, 104, (1978) GT7, 967-994.

H.B. Seed - Considerations in the earthquake-resistant design of earth and rockfill dams, Géotechnique, 29, (1979)3, 215-263.

D.W. Taylor - Stability of earth slopes, Contributions to Soil Mechanics 1925-1940, Boston Society of Civil Engineers, (1940), 337-386.

O.C. Zienkiewicz, C.T. Chang and E. Hinton - Nonlinear seismic response and liquefaction, International Journal of Numerical and Analytical Methods in Geomechanics, 2, (1978)4, 381-404.

SEISMIC RESPONSE OF TUNNELS

V.H. Joshi, M E

Commonwealth Fellow, Department of Civil Engineering and Engineering Mechanics, McMaster University

J.J. Emery, Ph D , P Eng

Associate Professor of Civil Engineering and Engineering Mechanics, McMaster University, and Principal, The Trow Group Limited Consulting Engineers

INTRODUCTION

The analysis of underground lifeline systems such as utility pipelines and conduits subjected to spatially distributed strong ground shaking has evolved to the stage where idealized, but practical, designs can be completed (ASCE, 1976; Ariman, et al, 1979 for instance). However, similar aseismic designs may not be appropriate for large underground openings like transportation and water tunnels, and improved methods are required to determine multicomponent, spatially distributed, input seismic motions (incoherent motions) for realistic combinations of inclined shear and surface waves. While seismic response is often emphasized, engineering experience and judgement are still critical to incorporate aspects such as seismic exposure at the site, soil-structure interaction and geotechnical properties, particularly for components which must remain functional during strong shaking (nuclear power plant cooling water tunnels and shafts that initiated this study for instance).

Starting from an improved, closed form, method for evaluating the incoherent motions for a linear, horizontally stratified, rock/soil layer system based on earlier work by Nair and Emery (1976), the dynamic response of typical tunnel sections is completed using the incoherent motions as nonsymmetric, time varying, free field input to the finite element method axisymmetric analysis. While the general case of combined inclined shear and surface wave propagation (local effects of large earthquakes) is readily incorporated, for simplicity only the idealized case of vertical propagation of sinusoidal shear waves is outlined herein. Parametric studies indicate that for typical tunnels there is little opening/liner influence on the free field incoherent motions, assuming no rupture due to discontinuities. This is an important observation since it permits simplified aseismic designs for complicated tunnel/shaft systems using the incoherent motions as displacement inputs in readily available piping systems dynamic response packages (PIPEFLEX for instance). Before describing the approach developed, background aseismic design concepts for tunnels and shafts will be outlined.

ASEISMIC DESIGN OF TUNNELS AND SHAFTS

Types of Potential Damage by Earthquakes

There are two main types of potential damage to tunnels associated with strong ground shaking (Newmark and Rosenblueth, 1971; Okamoto, 1973; ASCE, 1976):

1) rupture or other damage due to differential movements associated with discontinuities (faulting); and

2) damage from stress increases due to intensive shaking. These are:

   a) transverse stresses;
   b) axial tension and compression;
   c) longitudinal bending and shear; and
   d) localized stresses associated with connection of the tunnel with other tunnels, inlets, appurtenances and structures (this includes changes in tunnel direction, geometry and lining thickness).

It appears that the effects of discontinuities on underground openings are normally limited to narrow zones of active faults. Since extensive effort is made during major plant and transportation facility siting to identify and avoid areas likely to be subjected to actual ground breaks, faulting will rarely be a design consideration.

While the seismic response of small-diameter pipes and conduits is not directly applicable to tunnels, the qualitative observations are important (ASCE, 1976):

a) The free-field response of buried pipes and conduits indicates that they move with the ground;

b) The maximum ground deformation in the direction of the conduit axis is the same as that transverse to the conduit axis;

c) In the free field, axial strain is predominant (large

axial stresses indicated by axial mode failures);

d) The maximum strain of the conduit is not observed at the time of maximum ground acceleration, but at a later time (the after-phase of seismic waves); and

e) The bending strain is predominant at connections to structures, but a short distance away the axial strain becomes predominant.

It would appear that there have been few cases of the interior of a reasonably designed tunnel being irreparably damaged by earthquake effects (Duke and Leeds, 1959; Okamoto, 1973; ASCE, 1976). Damage is usually consistent with poor natural ground and/or construction and is greatest at portals and stress raisers. Based on his surveys of earthquake damage to tunnels, Okamoto (1973) concluded that the rate of damage is higher the poorer the geology of the ground and also the thicker the lining. This is in accordance with the major report by Duke and Leeds (1959). The key preventative measures for the aseismic design of concrete lined tunnels and shafts appear to be: locate where possible to avoid eccentric loadings; deteriorated rock zones should be rock bolted or subjected to appropriate reinforcement; the lining should be in good contact with the rock to avoid eccentric loadings due to voids, particularly at the crown; and the lining should be provided with ductile reinforcement at key points (connections, changes in direction, geometry and lining thickness) where localized straining is possible.

Dynamic Interaction

During shaking, the ground is deformed and any structure built on or in the ground will be forced into a distorted shape. This ground deformation becomes particularly important with structures of large lateral extent such as tunnels. If the tunnel is sufficiently flexible relative to the surrounding medium, it will follow essentially the displacements that the medium would have if the structure were absent. Under these circumstances, simple approximate solutions for the stresses set up in the tunnel are available in terms of the ground displacements (Shah and Chu, 1974 for instance). However, the question of dynamic interaction with tunnels rigid relative to the surrounding medium has not received much attention. For the case of lined tunnels and shafts in rock, it would appear reasonable to assume that the seismically induced deformation of the rock is imposed on the lining, i.e., the problem is not complicated by the effects of medium-structure interaction. However, with soils the interaction could become more important. The prime purpose of this study is to consider the degree of interaction and its significance to aseismic design practices, particularly with regard to simplified approaches.

METHODS FOR THE SEISMIC ANALYSIS OF TUNNELS AND PIPELINES

The available methods of seismic analysis may be grouped into the following categories:

1) Methods considering the propagation of either compression waves in the axial direction or shear waves in the transverse direction;

2) Methods assuming plane strain conditions for transverse sections of the structure; and

3) Methods employing the spring-mass-dashpot analogy to represent the system.

Many of these methods assume that the structure has no influence on the response of the medium in its immediate vicinity, which has yet to be shown convincingly through rigorous analyses. The methods of analysis in the first category cannot account for the seismic ground motions due to inclined propagation of shear waves. In the plane strain analyses, the important axial stresses are not predicted. For the methods of analysis in the third category, it is often a difficult, costly and time consuming task to evaluate various spring constants through laboratory and/or field investigations.

SEISMIC ANALYSIS OF TUNNELS USING THE AXISYMMETRIC FINITE ELEMENT METHOD

Although 3-dimensional finite element method seismic analyses of ground-tunnel systems are available and of general application, they are often impractical due to large computer core requirements and costs, even for small problems. However, it is often possible to use the more computer- and time-efficient axisymmetric finite element method for the seismic analysis of ground-tunnel systems by representing the free field acceleration input (or velocities and displacements) through a full Fourier series representation of the general nonsymmetric boundary conditions (companion paper by Hanafy and Emery; Wilson, 1965; Nair and Emery, 1976). The approach adopted consists of two major steps:

1. Evaluating the spatial variations in strong seismic motions for a linear, homogeneous and horizontally stratified ground system (Nair and Emery, 1976). The inclined propagation of shear waves is incorporated using multiple reflection-refraction theory, and the Rayleigh wave amplification factors are computed using an averaging procedure. The method accounts for focal depth and epicentral distance, the corresponding angle of incidence, and the relative contributions of both shear and Rayleigh waves. Locations at the assumed free field boundary of the ground-tunnel system are specified and

the required components of earthquake motion are provided for these points as output to the axisymmetric analysis.

2. Using the free field accelerations as input to the axisymmetric finite element method dynamic analysis based on a step-by-step procedure incorporating the linear acceleration method (Wilson and Clough, 1962). Essentially, the nonsymmetric excitation of the system from Step 1 is separated into a series of analyses in different harmonics that form the time-dependent acceleration boundary conditions (or velocities and displacements) used in the Step 2 response computations (Emery and Nair, 1976).

Considerable improvements and additions to the starting point Nair and Emery (1976) approach have been made that are beyond the scope of the general descriptions given here.

SIMPLE EXAMPLE

For simplicity, the 6 m diameter tunnel in a single soil layer overlying rock as shown in Figure 1 will be considered to illustrate the axisymmetric method of analysis. The finite element mesh (actually a slice across the mesh which is rotated about the tunnel axis) for the ground-tunnel system is shown in Figure 2. The base excitation used in Step 1 to determine free field accelerations to be applied at the free field boundary in Figure 1 during Step 2 was a sine wave with amplitude 0.305 m/s^2 (1.0 ft/s^2) and frequency of 2Hz. The spatial distribution of ground motions (Step 1) was completed for the vertical propagation of shear waves (sine waves) for a soil shear wave velocity of 152 m/s and rock shear wave velocity of 1829 m/s. It should be noted that only the vertical propagation of shear waves with no Rayleigh waves was considered for this simple example.

Response of the System at Various Radial Distances from the Tunnel

The free field motions from Step 1 are assumed to act at a radial distance of 4 times the diameter of the tunnel. Figure 3 shows the axial and radial displacement response along the ground-tunnel cross-section at the mid-length of the tunnel. The response at a radial distance of 2 diameters is nearly the same as the input free field motion. This indicates that it is reasonable to assume a free field motion to occur at a distance of 4 times the diameter of the tunnel.

Influence of the Assumed End Response

For a long tunnel, only a limited portion can be considered in the seismic analysis. To proceed with such an

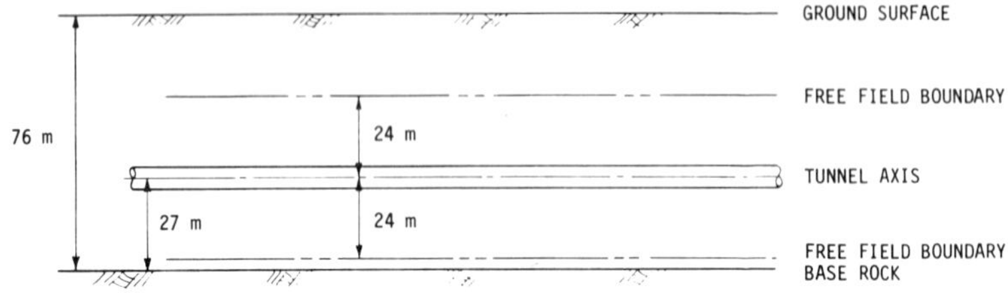

FIGURE 1 TUNNEL IN SOIL LAYER OVERLYING ROCK

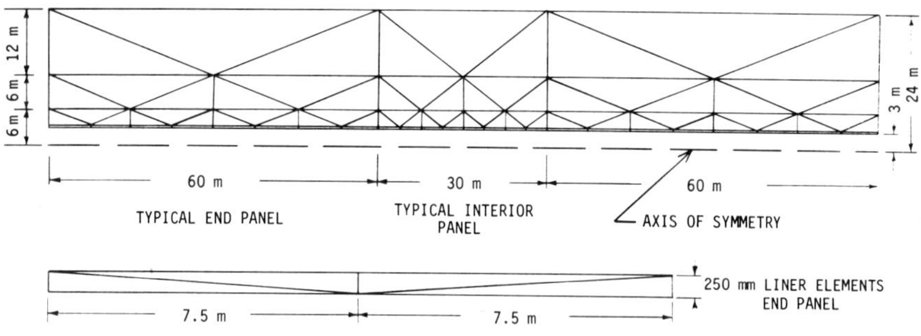

FIGURE 2 TYPICAL FINITE ELEMENT MESH FOR SOIL AND LINER

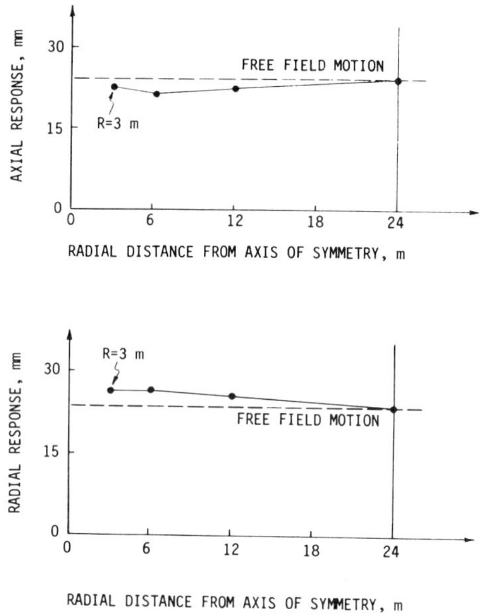

FIGURE 3 RESPONSE AT VARIOUS RADIAL DISTANCES ALONG THE CENTRAL TRANSVERSE SECTION OF THE SOIL-TUNNEL SYSTEM, NO LINER, TIME = 1.0 s

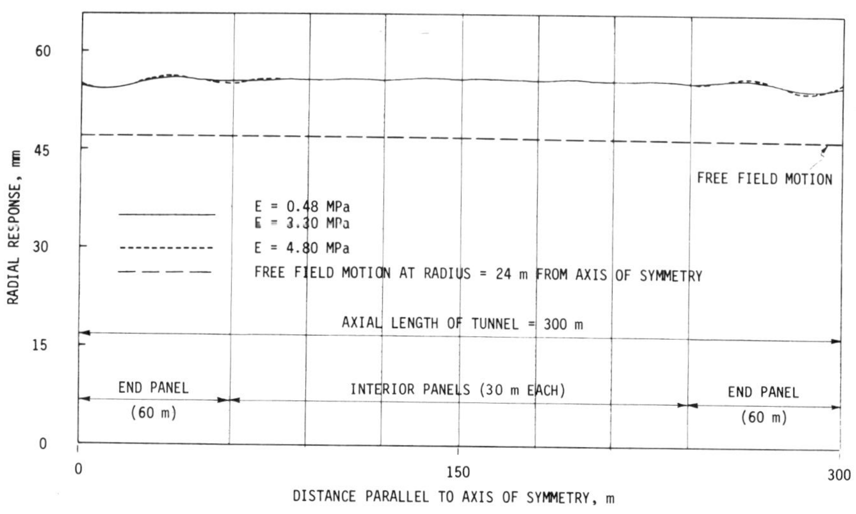

FIGURE 4 RADIAL RESPONSE AT TUNNEL WALL (R = 3m) FOR VARIOUS SOIL MODULI, NO LINER, TIME = 2.0 s

analysis, the actual responses of the system at the transverse end sections, which are not available to begin with, are required. To overcome this difficulty, two methods were tried initially:

1) Assuming the responses along the end sections to be the free field responses; and

2) Allowing the nodes along these sections (except those at the free field boundary) to move freely.

For both methods, the influence of assumed end conditions die out at a distance of 61 m from either end section (the length of the tunnel is 305 m). The results obtained for the second method are significantly better than those obtained for the first method.

It can be noted from Figure 3 that the axial response at the tunnel periphery (mid-length) is very close to the free field motion at that radial distance. However, the radial response is not quite as close to the corresponding free field response. This is logical since the presence of the tunnel cavity makes the system relatively "weak" in the radial direction. It is also for this reason that the response of the system is distorted near the end sections when free field motions are imposed along the end sections.

Influence of Soil Properties on Tunnel Response

The different material properties for the soil considered are given in Table I. The free field motions obtained for a shear wave velocity of 152 m/s was used for all of the trials. The radial response of the tunnel obtained is shown in Figure 4 where the response in these trials is very nearly the same. In all cases, the axial response was close to the free field axial response at the same radial distance. This is in conformity with findings reported by other researchers.

Influence of Tunnel Liner Stiffness and Thickness

The data used to consider liner stiffness influences are given in Table II. It was found that the liner stiffness does not have an appreciable influence on the computed response of the tunnel. This indicates that the response of the tunnel may be obtained for a representation section, and then this response used to study the performance of different tunnel sections without repeating the costly dynamic analyses.

CONCLUSIONS

It has only been possible to give a general outline of the method for evaluating spatial variations in strong seismic motions and their use as input free field boundary conditions in an axisymmetric finite element method dynamic analysis.

TABLE I

PROPERTIES OF SOIL

YOUNG'S MODULUS (E_s) MPa	POISSON'S RATIO	DENSITY kg/m^3
0.48	0.33	1866.8
3.30	0.33	1866.8
4.80	0.33	2029.2

TABLE II

PROPERTIES OF THE CONCRETE LINER

YOUNG'S MODULUS (E_c) MPa	DENSITY kg/m^3	POISSON'S RATIO	THICKNESS OF LINER mm
20.7×10^3	2435	0.2	254
27.6×10^3	2435	0.2	305
34.5×10^3	2435	0.2	305

Additional refinements to the analysis procedures are being made (improved consideration of the infinite extent of the system for instance), and it is anticipated that further design guidance will be available at the time of the Conference. While the results to date indicate little tunnel/liner influence on the free field dynamic movements, this must be confirmed for a broad range of ground-tunnel systems and input excitations including inclined shear waves and Rayleigh waves. If there is little interaction, the goal remains to develop a much simplified design approach using the readily-determined free field acceleration inputs directly with simplified piping systems response analysis packages.

The financial support for this work provided by the Commonwealth Scholarship Programme and by the Natural Sciences and Engineering Research Council, Canada, is gratefully acknowledged.

REFERENCES

T. Ariman, S.C. Liu and R.E. Nickell (Editors) - Lifeline Earthquake Engineering - Buried Pipelines, Seismic Risk and Instrumentation, ASME, New York (1979) pp. 292.

ASCE - Structural Analysis and Design of Nuclear Plant Facilities, Editing Board and Task Groups of the Committee on Nuclear Structures and Materials of the Structural Division of the ASCE, New York (1976), Draft.

C.M. Duke and D.J. Leeds - Effects of earthquakes on tunnels, Proceedings, RAND Second Protective Construction Symposium, Santa Monica (1959) 303-328.

G.P. Nair and J.J. Emery - Spatial variations in seismic motions, Canadian Journal of Civil Engineering 3 (1976) 32-46.

N.M. Newmark and E. Rosenblueth - Fundamentals of Earthquake Engineering, Prentice-Hall, Englewood Cliffs (1971) pp. 640.

S. Okamoto - Introduction to Earthquake Engineering, John Wiley, New York (1973) pp. 571.

H.H. Shah and S.L. Chu - Seismic analysis of underground structural elements, Journal of the Power Division, ASCE 100 (1974) 53-62.

E.L. Wilson - Structural analysis of axisymmetric solids, AIAA Journal 3 (1965) 2269-2274.

E.L. Wilson and R.W. Clough - Dynamic response by step-by-step analysis, Proceedings, Symposium on Use of Computers in Civil Engineering, Lisbon (1962) 45.1-45.14.

SESSION VI

Chairman: Mr E M O'Leary, BE, CEng, FICE, MIHE, MConsE
Veryard and Partners

Introduction to Session: Dr W J Larnach, MSc, PhD, CEng, FIStructE, FICE
Department of Civil Engineering,
University of Bristol

SESSIONAL PAPERS

PROTECTION AGAINST SETTLEMENT AND THE RECONSTRUCTION OF THE FOUNDATIONS OF THE HISTORIC BUILDINGS IN THE OLD TOWN OF THE CITY OF CRACOW

Roman Ciesielski

Technical University of Cracow, Poland

Michael Fuksa

Mining Academy, Cracow, Poland

INTRODUCTION

Cracow, formerly the capital of Poland, one of the oldest and finest cities in Europe, is included by UNESCO in a list of the world's most valuable historical places for world cultural tradition. Old buildings, particularly in the centre of the city's Old Town, have to be renovated. The necessary renovation works have been started. This paper aims to make representatives of the engineering-building environment acquainted with the technical problems associated with the protection of the buildings in the Old Town against the effects of settlement of the subsoil; it also discusses the principles being adopted to find solutions.

ORGANISATION OF THE TECHNICAL WORKS CONNECTED WITH RENOVATION

The renovation of historical buildings in Cracow is being treated comprehensively. The damaged buildings are to be repaired and protected; the interiors are to be redecorated according to the former or future uses. These measures include the introduction of new, modern indoor water/sewerage and heating installations to protect the historical and architectural features. Within the boundaries of the Old Town of Cracow, 638 historical objects from the Gothic and Renaissance periods and about 1200 newer items of the XIXth century and some even from the beginning of the XXth century have been included in the renovation plan. All these buildings are in permanent public use. The future plans for the development and transformation of the physical structure of Cracow are treated comprehensively with the works of renovation.

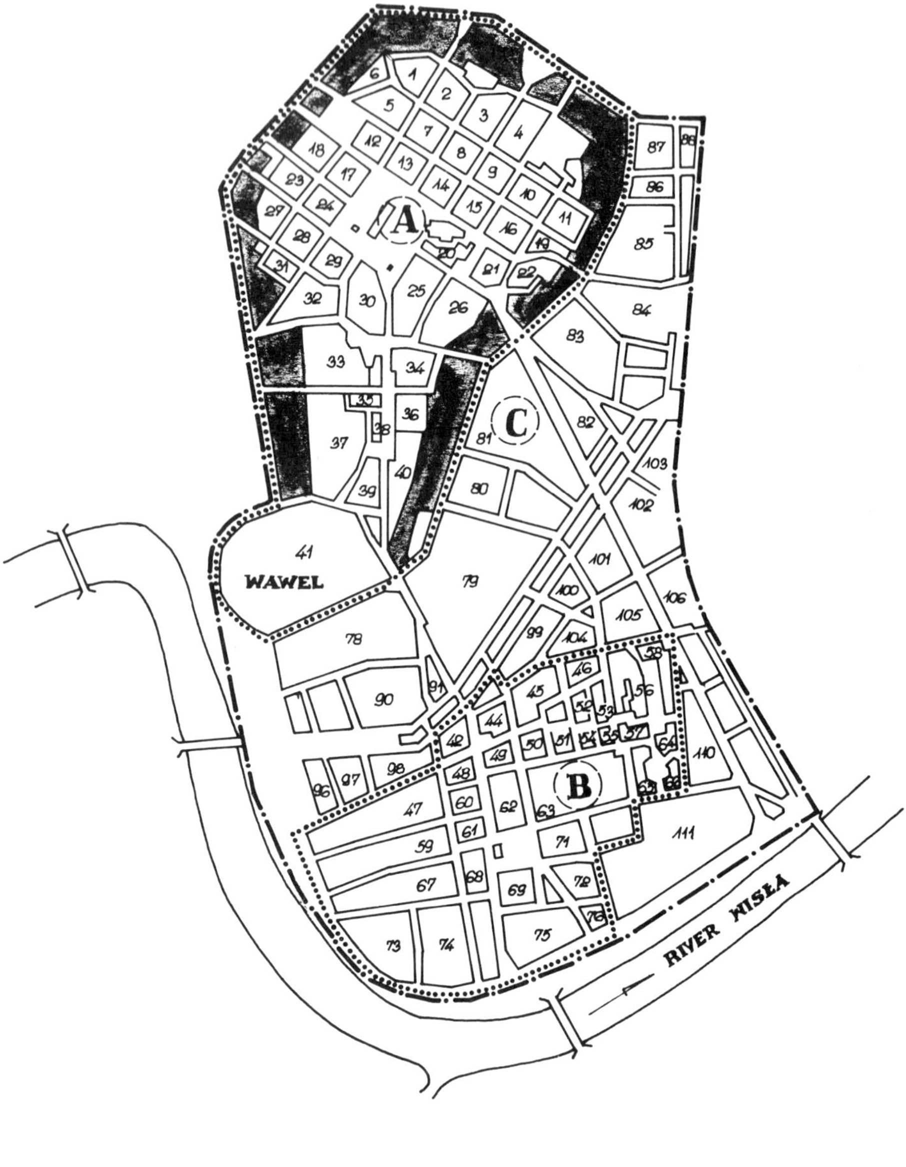

Fig.1 Cracow - renovation area
A - old historical centre, within the green belt of Planty
B - old quarter of Kazimierz
C - other old parts

For planning purposes, the Old City has been divided into sectors A, B, C (Figure 1) in which blocks (to a total of 111) have been delineated in conformity with the divisions made by the streets.

Sector A is the historical centre of Cracow, within the green belt of the "Planty", in which 40 blocks have been identified, the Castle Wawel being an additional block. Sector B is the Kazimierz quarter. Sector C is shown in Figure 1. More recent relicts are also found in other parts of the city; in Podgorze, Kleparz, and Krowodrza. The most urgent tasks concern the Centre within the Planty, and in this part of Cracow the building works are currently concentrated. These works have already been in progress for a long time but within a small area todate.

CAUSES OF DAMAGE

The causes of existing damage to the buildings can be grouped as follows:

- Uneven settlement of buildings founded on built up layers of high compressibility and varying water contents and plasticity; inflow of water from leaking sewer systems with stonework sewers; loosening of the subsoil in the neighbourhood of leaching cesspits. In some cases damage is caused by the deformation or collapse of stonework sewers passing through the walls of buildings.
- Unrepaired and incomplete disposal of surface water which, finding its way into the compressible subsoil, also causes additional changes in the conditions of the foundation and settlement. It should be stressed that the above account does not concern buildings properly founded on noncompressible layers.
- Lack of vertical and horizontal damp-proof courses in foundation walls which causes dampness of the walls and consequential deterioration of bricks and mortars.
- Faults in the building structure itself, such as lack of provision for movements, lack of gable walls, etcetera, cause damage, especially when settlement takes place.
- Reconstruction, breaking out new openings, undercutting walls, breaking through for services, erecting new walls, deterioration of old walls, etcetera, are also a cause of cracks and damage. Moreover, there is also the influence of normal wear and ageing.
- Additional and excessive imposed floor loadings or their extreme variation and general misuse. Superstructures over parts of the buildings, rebuilding of outbuildings etcetera on weak foundations and high and differential pressures on the subsoil are responsible for uneven settlement and damage. Equipment, installed sometimes in flats, causing dynamic loads (engines, presses, hammers, machine tools) have adversely affected the condition of supporting structures not adapted for such loads.

- Vibrations from transport systems, which in postwar years have increased dramatically, cause additional, variable stresses in walls, which are very harmful when tensile in character. In addition, in being transmitted by the subsoil they cause changes in it which can lead to excessive building settlements.
- Considerable mechanical pollution and activation of chemical aggression of the air by coal furnaces, and most importantly, air pollution by sulphur and fluoride compounds by the steel works in Nowa Huta and the aluminium works at Skawina cause penetration by rain of these chemical compounds into the plasters and walls and their subsequent corrosion and decomposition. This is also visible in all external stone elements.

Engineering-building works which are required comprise: rebuilding of foundations, repairs to damaged walls, repair or replacement of ceilings and roofs; all these to be executed with the functioning of the building taken into consideration and with careful preservation of the relictic features. All these works are only apparently simple, since each situation must be dealt with on an individual basis.

SUBSOIL UNDER THE CENTRE OF THE CITY OF CRACOW

The subsoil within the green belt of the Planty consists of, from the surface downwards: a 2 to 8 m thick layer of clayey and dusty fill, which is compressible and not suitable to carry foundations; below it a layer of sand about 8 m thick, of moderate to high density; this is underlain by a 10 m layer of gravels; finally Miocene or Jurassic limestone occurs, which in places emerges at ground surface (the Wawel Skalka). This is the so-called medium terrace (Figure 2). Other parts of the town already mentioned, lie lower on the flood plain of a more differentiated subsoil as far as the geotechnics are concerned. Some of the buildings within the green belt of Planty are founded, in accordance with building principles, on a bearing layer of sands or gravels and behave well as a rule. They include almost all the churches, former palaces, buildings of special use e.g. the Cloth Hall, the Town Hall Tower and the Barbican. On the other hand, a considerable number of the dwellings are founded at shallow depth on compressible ground, and these more than others have been exposed to damage by uneven settlement. The compressible layers of the foundation soil have been adversely affected by human activities and by the improper drainage of rainwater and by leaking sewers. In this part of the town the groundwater table lies below the level of the upper surface of the bearing layer, this being favourable for foundations.

The currently operating sewer system consists of ducts built in the XIXth century. Earlier, in the Middle Ages, the sewerage system consisted of open ditches carrying rainwater and sewage into fortress ditches or branches of the Rivers Vistula, Rudawa

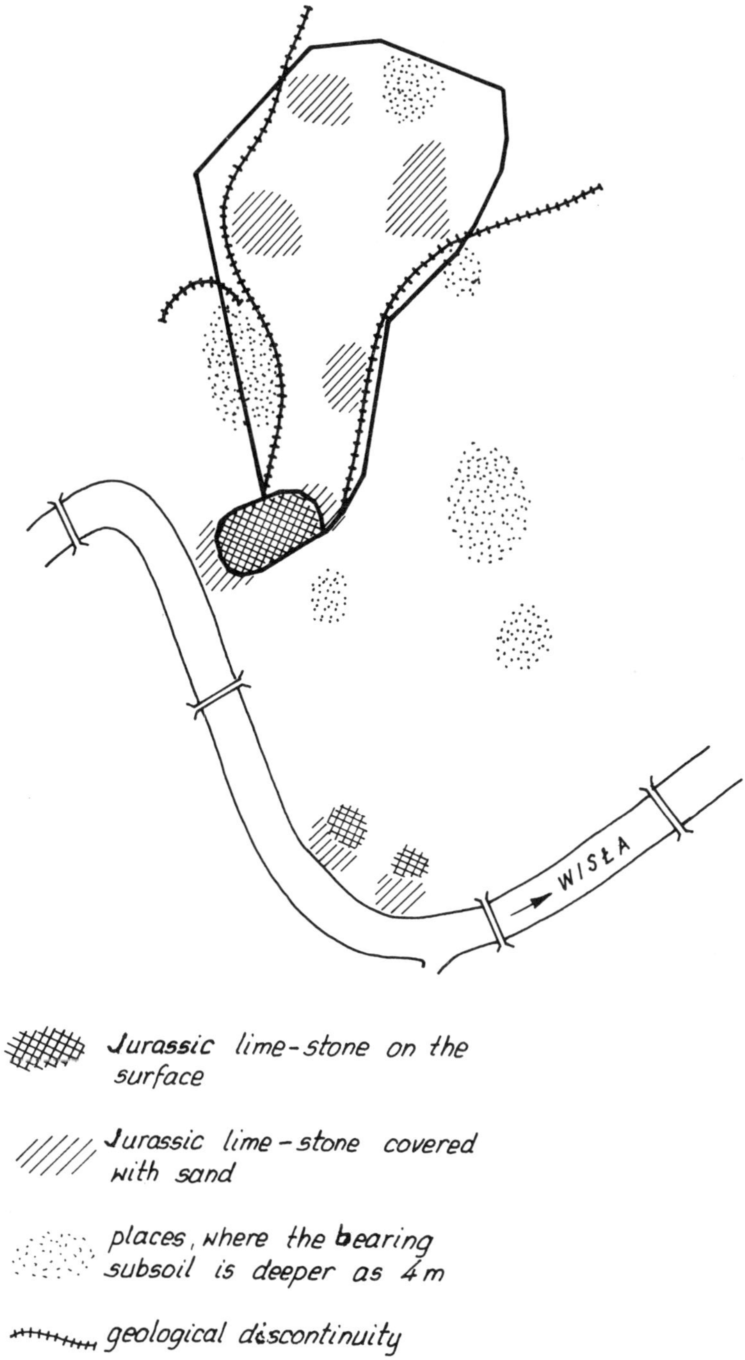

Fig.2 Subsoil conditions in the Old City

and Bialucha. Later, in order to carry away the sewage from particular housing blocks, shallow sewers, the so-called stonework ones, were built. This sewer network is over 11 km long (2). Stonework sewers are, with few exceptions, at present unexplored. It has been established that when they were built, the foundation walls of buildings were damaged and that because of leakage from them a softening of the subsoil (in the fill layer) has occurred under the foundations, thus creating settlements, except where the foundations have rested on noncompressible subsoils. Apparently, before the introduction of the stonework sewers, sewage was carried to local cesspits, dug under or near to buildings. In this part of the town there are more than 2000 of them. These cesspits, 3 x 4 m in plan and up to 12 m deep were backfilled with soil after use and new ones were dug. The contents of these cesspits fermented through the ages and at present constitute a soft, plastic layer. The cesspits, draining water, also weakened the subsoil in their vicinity and this has caused uneven settlement of buildings. Several examples of cross-sections of stonework sewers and a diagrammatic sketch of the influence of a cesspit on the subsoil are shown in Figures 3 and 4.

TECHNICAL PRINCIPLES ADOPTED IN RECONSTRUCTING FOUNDATIONS OF DAMAGED BUILDINGS

The best building foundation is one supported on a non-compressible subsoil, i.e. on a sand or gravel layer. Taking appropriately prepared foundations down to a suitable foundation soil requires complete penetration of the superimposed compressible layer (in the majority of cases 2 to 4 m) by means of established building techniques such as underpinning with a brick wall or concrete, or reinforced concrete elements; or an application of the so-called concrete plugs or wells or short piles (e.g. jacked-in Mega piles). The choice of method varies, being dependent on the subsoil situation and the type of building (3). This procedure makes the building completely independent of all harmful influences brought about by water content and compressibility of the upper layers. With these protective measures, no additional settlements of the buildings are likely to occur, since they are based on a non-compressible layer.

It might be argued that the alternatives of strengthening the existing subsoil , for example by electro-osmosis,or in locations where the upper layers of the foundation soils are not soft-plastic, by increasing the size of the foundations. These two approaches are not adequate in the situation under consideration since the first is too costly and not always applicable, and the other does not protect against possible further changes in the upper layers of the soils, causing settlement and additional damage.

Works are therefore being carried out at present on the rebuilding of the foundations by carrying them down to more suitabl

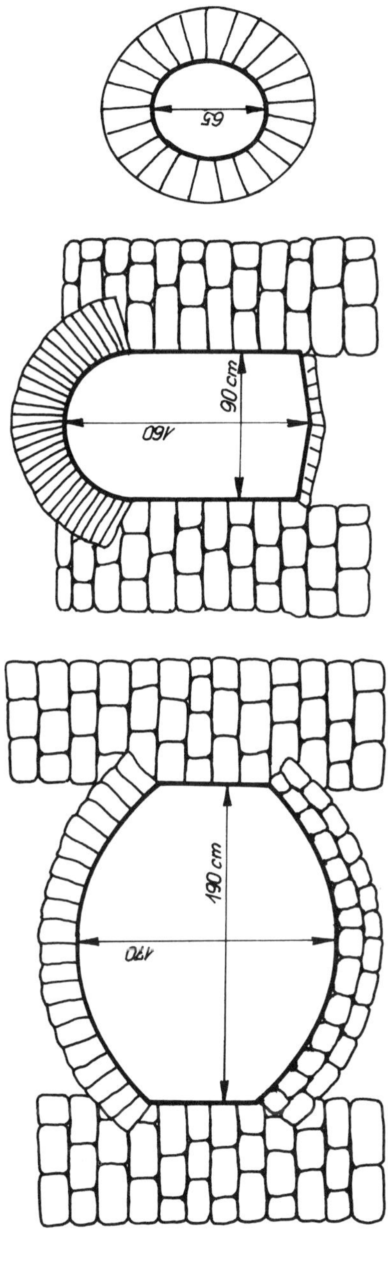

Fig.3 Typical cross-sections of the stonework sewers

Fig.4 Leaching cesspit and its influence on the subsoil and brickwork walls

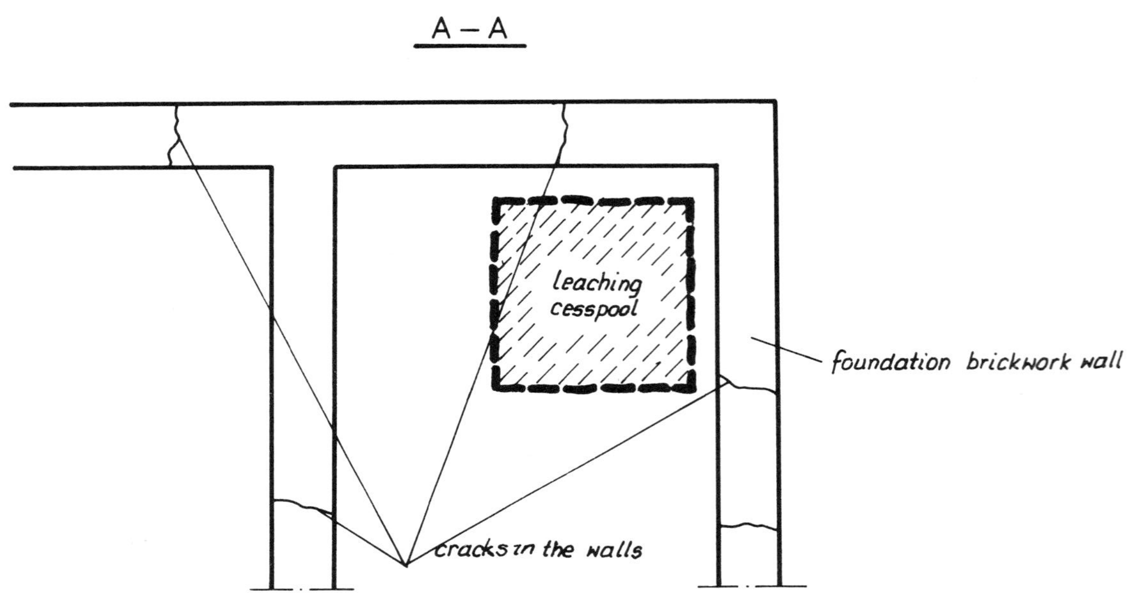

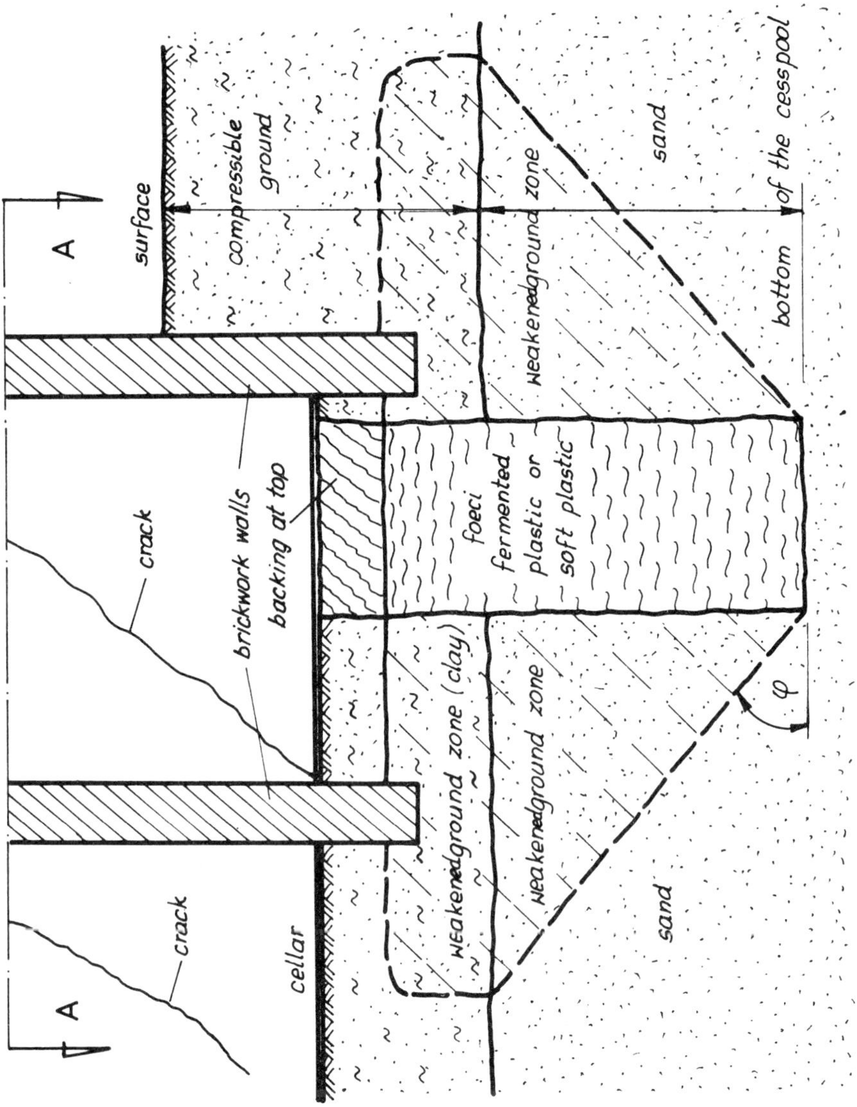
A
surface
compressible ground
weakened ground zone
sand
bottom of the cesspool
foeci fermented plastic or soft plastic
brickwork walls
backing at top
crack
weakened ground zone (clay)
weakened ground zone
φ
sand
crack
cellar
A

founding layers. Examples of the methods in use are presented in Figures 5 to 12. Some of these are very simple, but in such difficult conditions, use of mechanical equipment is not possible. The upper parts of the buildings and other relicts must not be disturbed.

COMPLEX MINING PROTECTION OF THE SUBSOIL

Consideration was also given to the possibility of using mining techniques based upon the principle that the subsoil conditions should be improved first of all, followed by works at the surface. With this principle in mind the plan foresaw preservation being effected by the elimination of the major part of the existing network of stonework sewers. These operations were to be carried out by groups of miners. Investigations of the conditions of the sewers and their subsequent cleaning were made. Some repairs were also planned as preparation for their filling by a fluid material. The filler was to be a cement mortar (waterproof and noncompressive). An extensive research programme was conducted to determine the composition of a suitable mortar. It was visualised that this mortar would be placed under pressure in the stonework sewers. The objective was to prevent the sewers from acting as ducts draining water, with the detrimental effects described earlier. It was finally decided that this was not a feasible proposition, not only because the mortar filling cannot guarantee a complete filling of the free space (also known from the grouting of pre-stressed concrete cable ducts) and this would not eliminate the sewers as channels draining water, but also because it would have required the pre-sealing of the damaged sewers to prevent penetration of the mortar under pressure into the ground. Primarily however it would not have changed the situation in relation to building settlements because it would not have eliminated the other causes of settlement and damage to buildings even if successfully accomplished. It should be added that the stonework sewers do not form a continuous network (they occur as a series of separate sections) and their present state is very bad. It was concluded that rebuilding of the building foundations independent of any filling operations was necessary. The cost of such works, being related to the length of the sewers would also have been very high.

Another aspect of the work for which miners were used as professionals was the so-called shaft digging. It is well-established that in works such as these, examination of subsoil conditions is necssary, for which borings and even shafts may be used. The depth of such shafts may be several metres, exceptionally ten or more. An additional problem in Cracow was created by the fact that for some unexplained reasons, a great number of shafts (over 2000) had been sunk even where the subsoil had already been examined and after being constructed outside the buildings, they were, in the majority of cases, left open since further examinations were to have been carried out in them. They were left open for the whole heavy winter of last year and provided a ready route for water from the melting snow to flow into the foundation soil, soaking it with water and aggravating the damage which was to be eliminated. Fortunately the idea of filling the stonework sewers was abandoned and this form of reconstruction was restricted to places where the sewers passed through or under walls.

CONCLUDING REMARKS

In the tasks connected with the rebuilding of the foundations and the complete renovation of the buildings, some difficult and complex technical and economic problems must be solved. The renovation itself requires the carrying out of various researches concerning, amongst other items, economical methods for the proper renovation of foundations, the properties of old materials, the strength and stability of old structures and the technology of work's organisation. These tasks are being undertaken and it is expected that results from them will lead to advances which will be presented in the technical press or at future technical conferences.

REFERENCES

1 R. Ciesielski and M. Fuksa - Engineering building problems of the renovation of old historic buildings in the centre of Cracow, Building Review No 11 (1979) 595-607.

2 A. Gralak (Ed) - Sewers of Historic Cracow, KAW, Cracow (1976).

3 Miscell. Auth. - Special number devoted to the problems of building in Cracow, Technical Periodical No 5 (1963) 6-64.

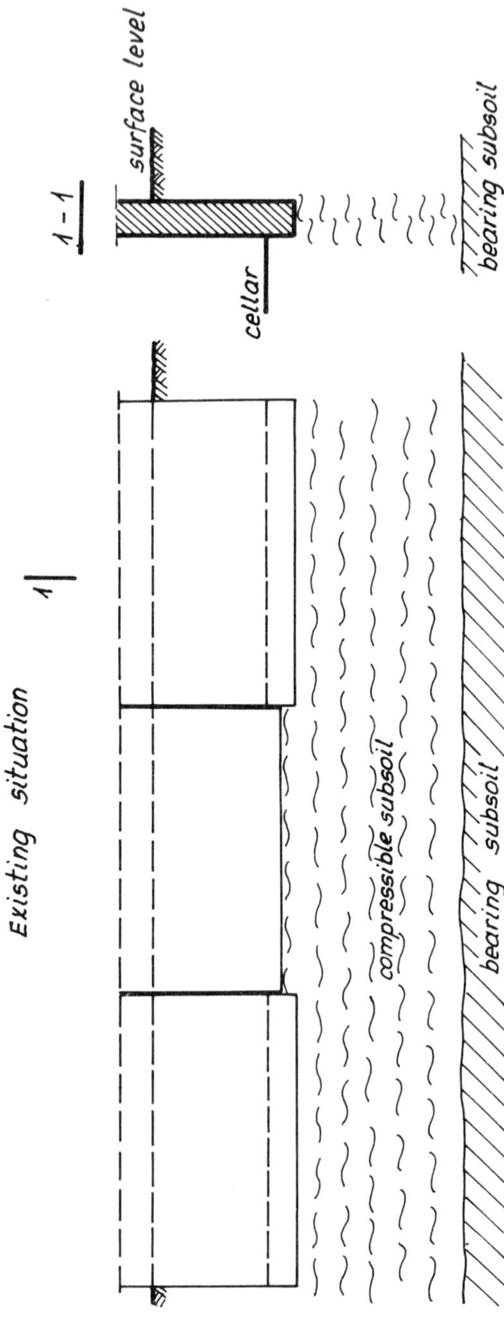

Fig.5 Methods of underpinning existing foundations

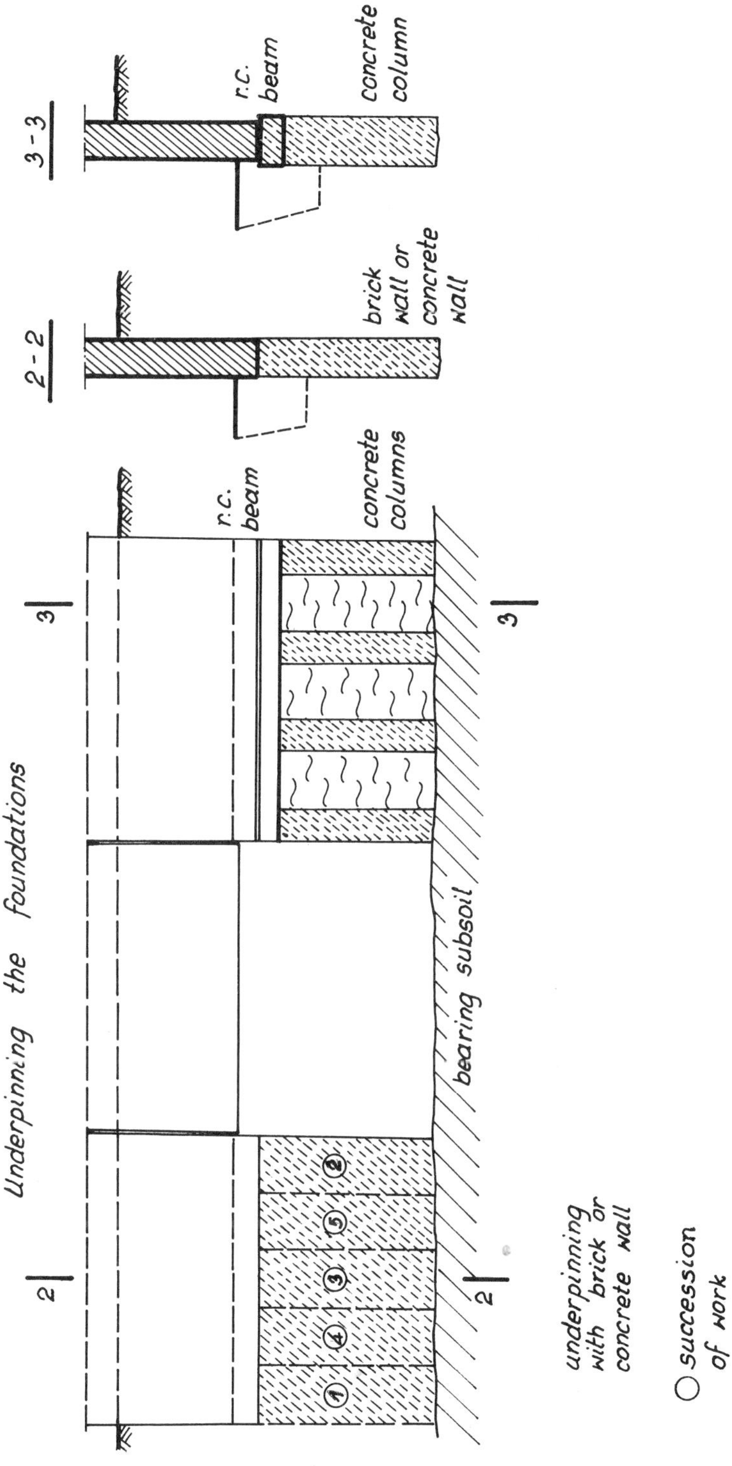

Underpinning the foundations
2-2
3-3
r.c. beam
concrete column
brick wall or concrete wall
r.c. beam
concrete columns
bearing subsoil
3
3
2
2
1
4
3
5
2
underpinning with brick or concrete wall
succession of work

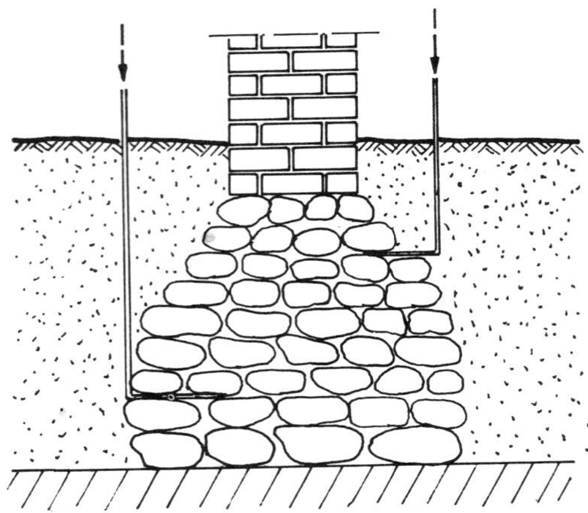

Fig.6 Filling a stone foundation wall with mortar

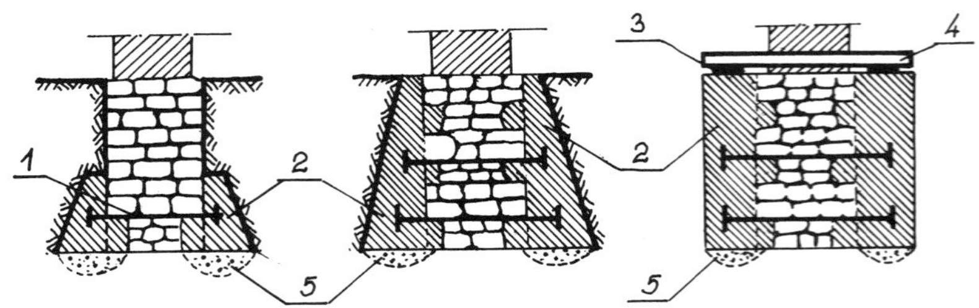

1 anchor element
2 new concrete wall
3 wedge blocks
4 steel beam
5 compressed ground

Fig.8 Strengthening stone foundation. Phases of the work

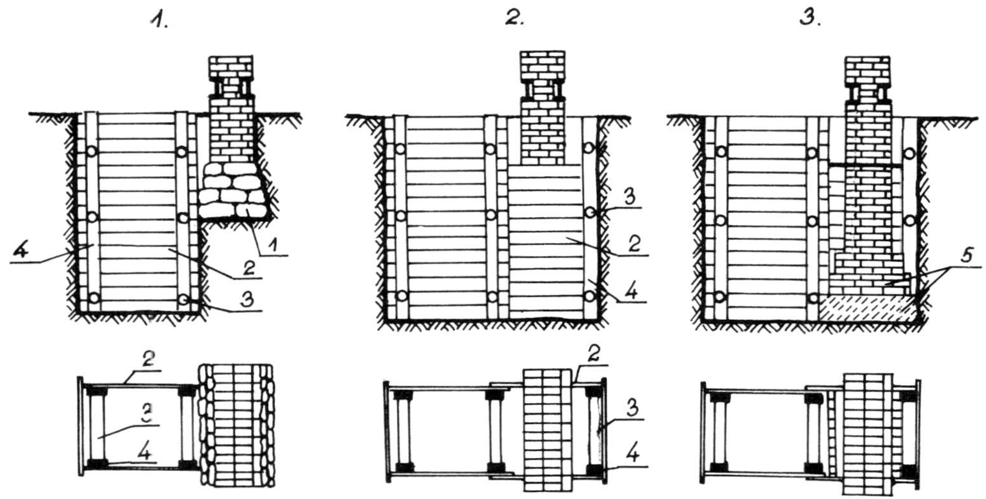

1 old stone foundation
2,3,4 trench timbering
5 new foundation

Fig.7 Converting an old stone foundation to a new brick/concrete one. Phases 1,2,3

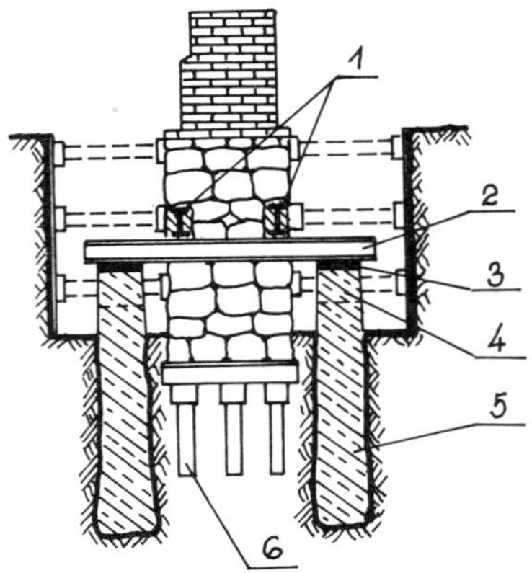

1,2 steel beams
3 new concrete layer
4,5 concrete piles
6 old timber piles

Fig.9 Strengthening with new piles

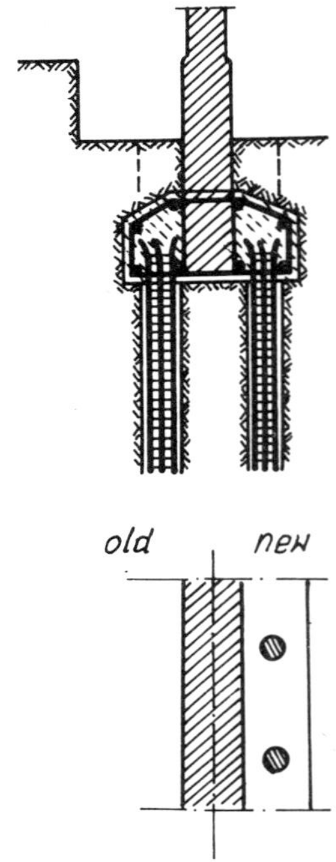

Fig.10 Transfer of wall footing to new bored pile

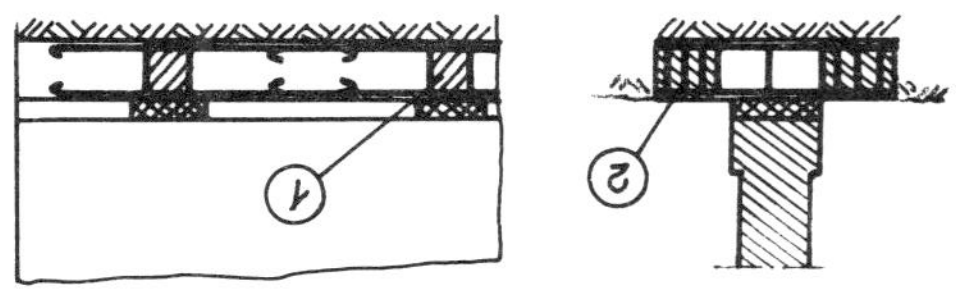

1 cross beams
2 longitudinal footing beams

Fig.11 Strengthening of foundation with RC girder system

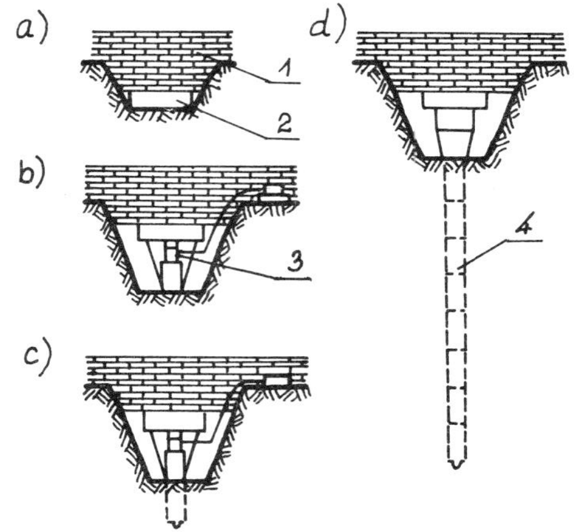

1 existing footing
2 concrete cushion
3 hydraulic jack
4 sections of pile

Fig.12 Strengthening brick wall footing with Mega sectional pile

CONSTRUCTION AND LONGTERM SETTLEMENTS OF FOOTINGS FOR A 4-STOREY BUILDING

Gunther E. Bauer, Ph D , Associate Professor

Department of Civil Engineering, Carleton University, Ottawa, Canada K1S 5B6

INTRODUCTION

In this paper the settlement behaviour, in particular the differential settlements of spread footings, of a 4-storey concrete frame structure is studied over a 10-year period. Initially a piled foundation was considered at an additional cost of several ten thousands of dollars. The selection of spread footings was based mainly on the results obtained from a full-scale footing test (Bauer et al, 1973). Special construction procedures had to be employed in order to minimize the differential settlements. The fissured nature of the clay makes it difficult to predict the construction and consolidation settlements by conventional analyses. The 10-year settlement record has proved the assumption that most of the settlements occurs during the construction period of the building.

SOIL CONDITIONS

The building is located in the western part of Ottawa, Canada on a flat clay plain with a surface elevation of approximately 90 metres. The clay, referred to as Leda clay or Champlain Sea clay in the literature, is part of the lacustrine and marine clay deposit whose geotechnical properties have been described by several researchers (for example Eden and Crawford, 1957, Eden and Bozozuk, 1969 and Hanna, 1976).

A soil profile and soil properties are given in Figure 1. This figure shows that the desiccated crust which is overconsolidated by more than 300 kPa extends to a depth of 3 metres. The softer clay below the crust extends to a depth of 8.5 metres, is overconsolidated by about 150 kPa and has an average undrained shear strength of 40 kPa as determined by

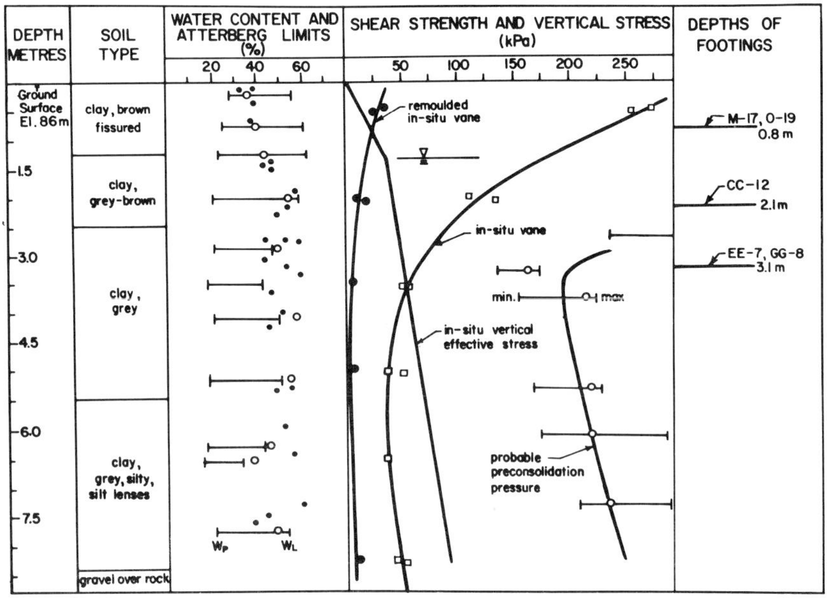

Figure 1 Soil Profile

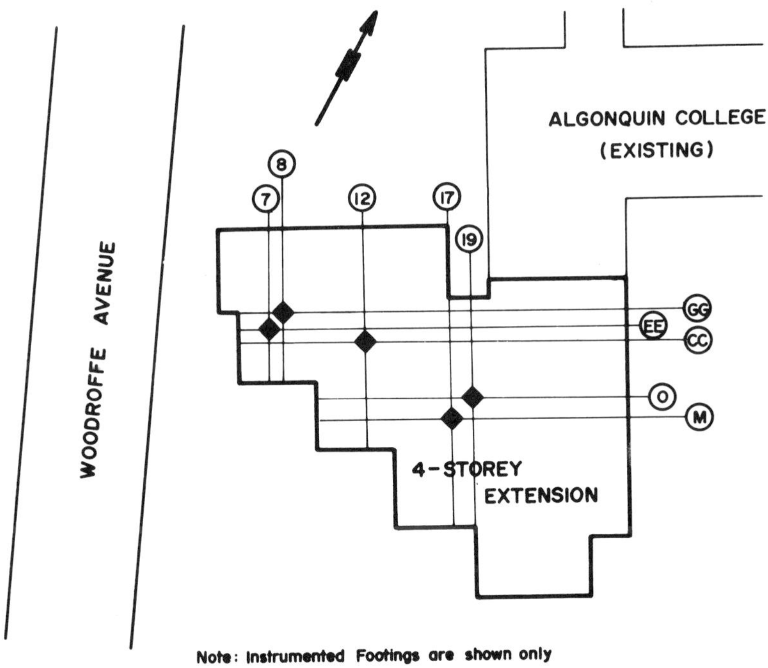

Figure 2 Layout of Instrumented Footings

field vane testing. Also indicated in Figure 1 is the probable preconsolidation pressure with depth as determined by conventional oedometer tests. The last column shows the relative elevation of the five instrumented footings. The footing loads were such that the total vertical stress in all cases was below the preconsolidation pressure of the clay. Bedrock was encountered at a depth of 18 metres below ground surface. The water table fluctuated according to the season and generally varies between 1.5 and 2.5 metres depth but must have been at 3 metres depth in the past as indicated by the oxidized clay crust. Even though the crust shows a very high shear strength compared to the underlying softer material, it is more susceptible to compression under light loads due to closing of the fissures.

LAYOUT OF FOOTINGS

The plan location of the five instrumented footings is shown in Figure 2. For the sake of clarity only the test footings are shown. Similar square footings were used throughout the new extension. The distance between adjacent column lines was 4.6 metres (15 feet). The interior footings were located at different depths in order to achieve compatibility of settlement. The elevations of the five instrumented footings were given in Figure 1. These footings were chosen for settlement observation mainly because their loads could be easily calculated and their variation in size and depth. All footings were made of reinforced concrete with a thickness varying between 750 and 915 mm (2.5 to 3 feet). They were considered to be rigid and rough. As soon as the soil was excavated to footing level, a skin of lean concrete was poured onto the soil in order to prevent any disturbance and remoulding of the clay.

SPECIAL CONSTRUCTION PROCEDURES

Even footings located in so called homogeneous soils or uniform deposits will undergo differential settlements. Unfortunately there is very little in the present literature which deals specifically with differential settlements, that is, the relative downward displacement of footings (or points on the foundation) with respect to one another. For framed structures with masonry walls or other fragile material a commonly accepted criterion to prevent cracking is that the differential settlement δ should not exceed L/300, where L is the span length between columns (Bjerrum, 1970; Polshin and Tokar, 1957). The column loads varied between 220 and 2620 kN and the soil bearing pressures were adjusted from 107 kPa to 375 kPa in order to obtain compatible differential settlements for the footings. The footing elevations varied by 3 metres over short distances as indicated in the last column of Figure 1.

Four special measures had, however, to be included in the design where the predicted differential settlements were unacceptable. The first occurred at lightly loaded exterior walls which supported only the first floor. A detail was developed providing a concentration of loading (536 kPa) at a 30 cm square 'high pressure point' near the centre of these wall spans. A yielding layer of foam plastic was placed beneath the remaindar of the wall to prevent load transfer to the soil except at these points. It was necessary to 'sharpen' the underside of the wall to create pressures equal to the yield point of the foam plastic.

A second special measure was to specify the inclusion of a 5 cm lightly compacted sand fill layer below a group of intermediate width strip footings. This technique was intended to produce additional settlements of 0.8 to 1.3 cm on the first load cycle.

The third special measure was less economical and consisted of specifying a change in the normal construction schedule to delay the construction of section of the lowest floor system. The lightly loaded, small sized footings would not settle sufficiently to be compatible with the perimeter walls and main frame. Hence their construction was deliberately delayed until the upper floors and frame had been built and hence until most of the settlements of the main structure had taken place. This procedure was only valid if, as later confirmed, the major part of the total settlement took place as immediate deformation during construction with little delayed settlement occurring after completing of the building.

The fourth special measure was again a regulation of construction procedures. The location of reshoring struts used to support the uncured concrete floor slabs above the first floor were controlled so that weight would be applied by them to the upper side of the largest footings. Without this provision, most of the structure load could be carried by the reshores on the ground between footings for several months. This would put the settlements of the large footings out of time phase with the other footings and cause temporary but unacceptable differential movement.

The results of all the special measures were unfortunately not measured. The only evidence of their effectiveness is that no visible distress has appeared in the building frame or finishes over ten years of the building's life.

INSTRUMENTATION OF FOOTINGS

The five square building footings were instrumented with surface gauges and deep-seated settlement anchors similar to the test footing which had been tested prior to the building construction. The test footing, 3 by 3 metres in size had been tested to aid the design of the building footings as reported by Bauer et al (1973).

The deep-seated settlement anchors were at depths of 0.25B, 0.50B, 1B, 1.5B, 2B and 2.5B below the base of the building footings; where B is the footing width. Two sets of settlement gauges were installed at each footing, one set as close to the centre of the footing as possible and the other set at the edge of the footing. Both sets showed the same settlement behaviour as the footings are quite rigid. For simplicity sake, the centre settlements only are reported in this paper.

OBSERVED SETTLEMENTS

Construction Settlements

The total settlement of a footing located on a saturated clay mass is composed of three parts, the immediate settlement, consolidation settlement, and creep. The immediate settlement or initial settlement occurs, by definition, at constant volume, that is at no change of pore voids and therefore no dissipation of porewater pressure. In real-life-conditions both the immediate and consolidation settlements will overlap, in particular, if the building is constructed over several months or even years. It therefore would be more practical to speak about construction settlement.

Figure 3 shows the settlement behaviour of footing M-17 over the last ten years. This footing was located high in the crust at a depth 0.8 metres below the ground surface. Approximately 61 percent of the total settlement occurred during construction over a period of 15 months. Similarly for footing O-19 (see Figure 2) which was located at the same depth also 61 percent of the settlement occurred during the construction period. For the same period footing CC-12 located at a depth of 2.1 metres experienced 56 percent of the total settlement and footings EE-7 and GG-8 at a depth of 3.1 metres had settlements of 46 and 48 percent of total settlements respectively. The settlements are summarised in Table I. As pointed out earlier, the upper crust of the clay was heavily fissured which accounts for the higher percentage of construction settlement of the footings at shallow depth.

Long-Term Settlements

The settlement behaviour over the last ten years for footing M-17 located in the crust and a deeper footing GG-8 situated below the crust are shown in Figures 3 and 4

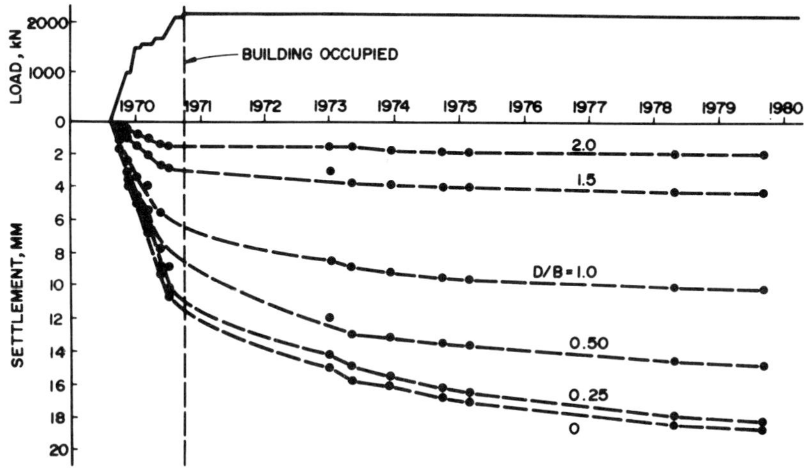

Figure 3 Settlement of Footing M-17

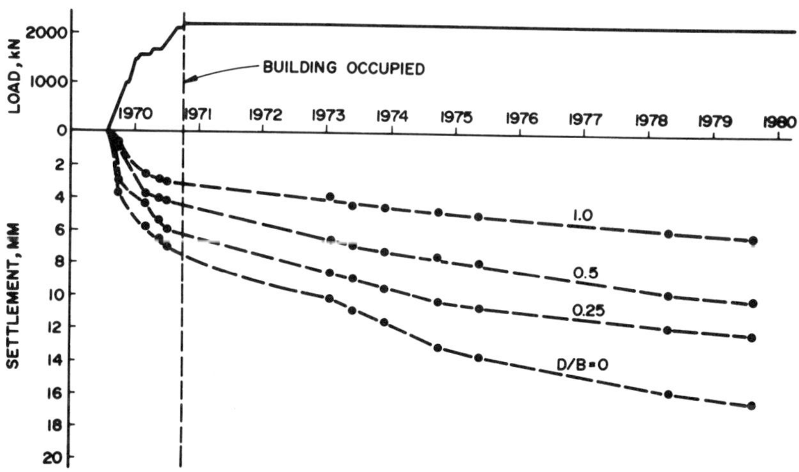

Figure 4 Settlement of Footing GG-8

respectively. It should be pointed out that the main objective of locating the footing at different depths was to achieve compatible settlement or to minimize the differential settlement. The maximum total settlement recorded was about 19 mm for footing M-17 and the smallest value was 11 mm for footing CC-12. From the full scale footing test, reported by Bauer et al (1973) and the special design measures discussed earlier, the total settlement was estimated to be 1 inch (25 mm) with the differential settlement not expected to exceed one half of an inch (12.7 mm). Even if the present rate of settlement of 0.7 mm per year continues, the estimated values should be reached in about another 10 years.

Settlement Distribution with Depth

The distribution of settlement below the centres of the five instrumented footings is shown in Figure 5. The distributions are plotted as percent settlement versus the ratio of depth to the foundation width of the square footings. The depth beneath the foundation in which 90 percent of the soil compression takes place is called the "critical depth of settlement" (Scott et al, 1979). This concept allows the comparison of "critical depth" at different foundation loads and for different foundation sizes and it is also more easily visualised than the stress dissipation concept. Figure 6 shows the theoretical distribution for both rigid and flexible circular footings located at the surface of an elastic isotropic half-space. From Figure 5 it is seen that the average critical depth occurs at $Z/B = 1.6$, where the theoretical critical depth occurs at $Z/D = 5$. It is obvious that the elastic theory with an isotropic medium will overestimate the critical depth by more than twice.

The critical depth of footing M-17 for various load levels is shown in Figure 7. It changed little with load application and time. The critical depth, Z_c at end of construction and at long-term conditions are also given in Table I for the five footings.

CONCLUSIONS

The settlement behaviour of five footings of a four storey building have been measured for 10 years and the maximum total settlement is less than 20 mm and the differential settlement less than 10 mm. Special design measures were used to keep the differential settlement in acceptable limits.

The "critical depth" for the five footings varied between 1.5 and 1.8B for long-term conditions, compared to 5B for the elastic theory.

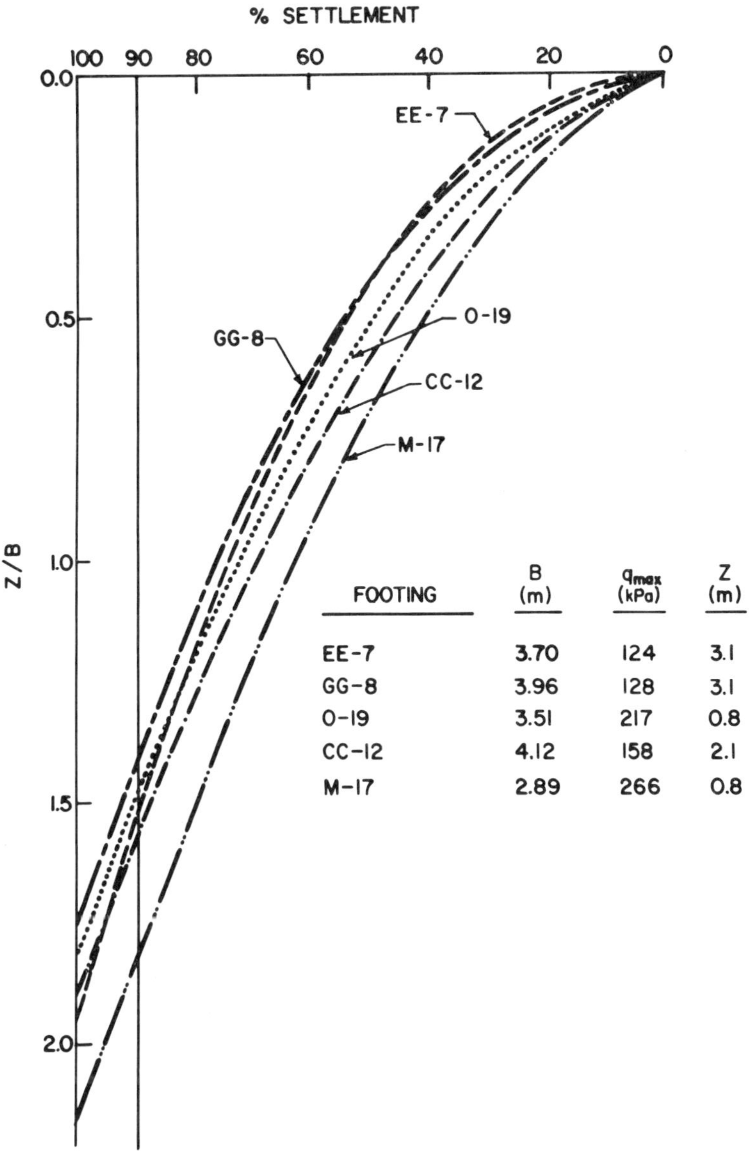

FOOTING	B (m)	q_{max} (kPa)	Z (m)
EE-7	3.70	124	3.1
GG-8	3.96	128	3.1
0-19	3.51	217	0.8
CC-12	4.12	158	2.1
M-17	2.89	266	0.8

Figure 5 Settlement Distribution of Five Footings

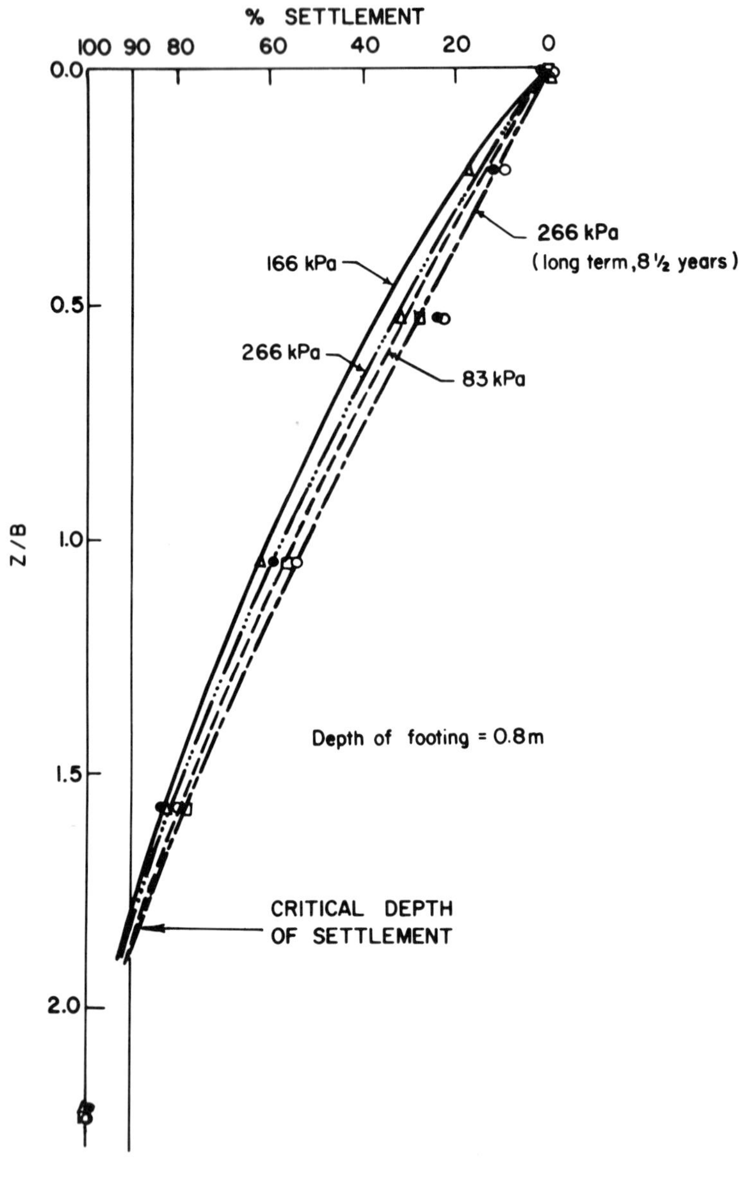

Figure 6 Settlement Distribution for Various Loads

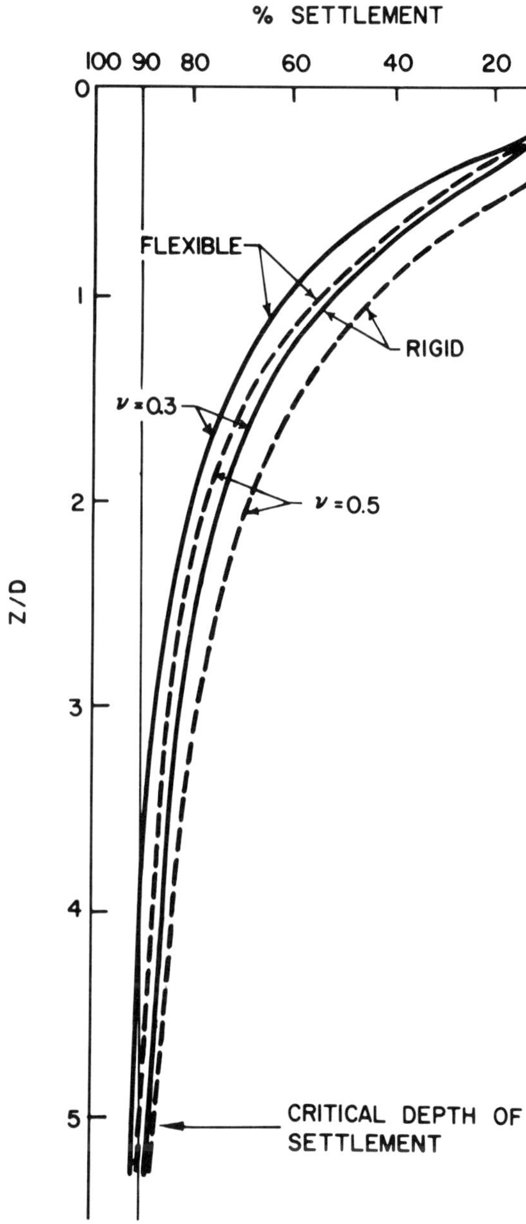

Figure 7 Theoretical Settlement Distribution

TABLE I - Location and Settlement of Footings

Footing	Width B(m)	Max. Berg. Press kPa	Depth Z(m)	Settlement mm		$\rho_{con}/\rho_{10\ yr}$ %	Critical Depth (Z_c/B)	
				Construct. ρ_{con}	10 years $\rho_{10\ yr}$		Const.	10 years
0-19	3.5	217	0.8	10.2	16.8	61	1.3	1.6
M-17	2.9	266	0.8	11.4	18.7	61	1.4	1.8
CC-12	4.1	158	2.1	6.1	11.0	56	1.6	1.7
EE-7	3.7	124	3.1	6.4	13.9	46	1.2	1.3
GG-8	4.0	128	3.1	7.8	16.2	48	1.3	1.5

The building has performed satisfactorily over the last ten years. No distress signals, such as cracks in the brick and plaster work could be detected. Further settlements should be insignificant and should not hinder the normal function of the structure.

ACKNOWLEDGEMENT

The financial assistance of the Natural Sciences and Engineering Research Council Canada (formerly National Research Council) is acknowledged and was greatly appreciated.

REFERENCES

Bauer, G.E., Scott, J.D., and Shields, D.H. - The Deformation Properties of a Clay Crust, Proceedings of the 8th International Conference of Soil Mechanics and Foundation Engineering, Moscow, Volume 1 (1973) 31-38.

Bjerrum, L. - Allowable Settlements of Structures, Proceedings of the 5th European Conference of Soil Mechanics and Foundation Engineering, Wiesbaden, Volume 2 (1970) 135-137.

Eden, W.J. and Bozozuk, M. - Earth Pressures on Ottawa Outfall Sewer Tunnel, Canadian Geotechnical Journal, 6, Ottawa (1969) 17-32.

Eden, W.J. and Crawford, C.B. - Geotechnical Properties of Leda Clay in the Ottawa Area, Proceedings of the 4th International Conference of Soil Mechanics and Foundation Engineering, London, Volume 1 (1957) 22-27.

Hanna, A.J. - Deformation Characteristics of Soil, M.A.Sc. Thesis (1976), University of Ottawa, Ottawa, Canada.

Polshin, D.E. and Tokar, R.A. - Maximum allowable non-uniform settlement of structures, Proceedings of the 4th International Conference of Soil Mechanics and Foundation Engineering, London, Volume 1 (1957) 402-405.

Scott, J.D., Haile, G., Bauer, G.E., and Shields, D.H. - The Critical Depth of Settlement, (to be published by the American Society of Civil Engineers, Foundation Engineering Journal).

Skempton, A.W. and MacDonald, D.H. - Allowable settlement of buildings, Proceedings, Institution of Civil Engineers, Part 3, Volume 5 (1956) 727-768.

A COMPARISON OF THE MEASURED AND PREDICTED PERFORMANCE OF A HEAVILY LOADED WAREHOUSE FLOOR

L.A. Wood, B Sc , Ph D , M I C E

Lecturer, Department of Civil Engineering, Queen Mary College, London

W.J. Larnach, M Sc , Ph D , F I C E , F I Struct E

Senior Lecturer, Department of Civil Engineering, University of Bristol

SUMMARY

A fertilizer and animal feedstuffs warehouse was constructed in 1976 in an area underlain by highly compressible soils. Piled foundations were adopted for the superstructure, the floor slab being ground-supported. Monitoring of the slab settlements has taken place since completion, using precise levelling related to a deep datum, and settlements of the order of 150 mm have been observed. The paper provides a full documentation of this case history. A theoretical analysis of the slab/soil interaction has been made (taking account of the time-varying loading and the effects of heavy transport vehicles) which allows comparisons to be made of predicted and measured settlements.

INTRODUCTION

It is generally accepted in geotechnical engineering that progress in the validation of hypotheses or the accumulation of much-needed design data will be closely related to the study of fully documented case histories describing actual behaviour. This attitude was much encouraged by the Conference on Settlement of Structures in Cambridge (COSOS, 1974), and the present paper attempts first to provide full documentation and second, to attempt comparison between predicted and measured settlements.

THE SITE AND STRUCTURAL CONCEPT

The portal frame warehouse is located within a light industrial estate on a flat site near Plympton, Devon. The site is the abandoned railway downyard at Tavistock Junction and lies immediately to the south side of the main line from Plympton to Plymouth, approximately 1.5 kilometres from Plympton Station. The Institute of Geological Sciences maps show the sand formation to consist of slate beds of the Upper Devonian, and this site lies over a buried channel consisting of alluvial silt with zones of granular material. Because of the heavy proposed loading it was clear that settlements of shallow foundations would be large. Piling of the floor was ruled out on economic grounds, and with the agreement of the client (who intended operating the store using fork-lift trucks which could work successfully on a settled floor slab) the Consulting Engineer decided to adopt the compromise solution of piling the column footings and providing pile-supported peripheral ground beams, whilst leaving the floor slab ground-supported. Careful detailing at the slab/pile cap or ground beam interfaces was necessary to ensure that the slab could settle freely relatively, without load transfer.

THE SITE INVESTIGATION

A simple layout plan of the 40 m x 45 m warehouse is shown in Fig. 1 and the locations of the two 150 mm diameter boreholes made using conventional shell and auger techniques are indicated approximately. (The deep datum point was afterwards installed in Borehole 1). The simplified borehole logs are shown in Fig. 2, and the results of in situ standard penetration tests are included. Undisturbed U100 samples were extracted at invervals and the results of undrained triaxial compression tests and consolidation tests are summarized in Table I.

THE MONITORING OF SETTLEMENTS AND LOADING

The levels were referred to a deep datum installed in Borehole 1 (see Fig. 1), the details being generally in accordance with the suggestions of Cheney (1973). Because all of the levelling points were to be installed in the ground floor slab it was considered more appropriate to use as levelling stations Hilti pins fired explosively into the slab, rather than the removable Building Research Establishment plugs (Cheney, 1973). The location of the measuring points, 39 in all, are shown in Fig. 3, in which the floor bay layout corresponds with that in Fig. 1. Settlements related to the deep datum have been made at regular intervals since construction was completed in July 1976 until the time of writing (September 1979), using the

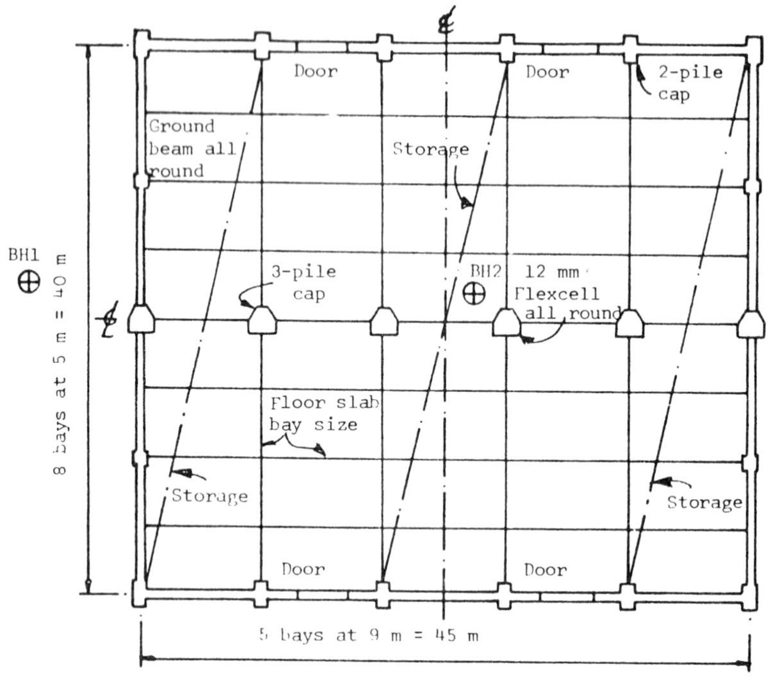

FIGURE 1 Warehouse Layout

TABLE I Laboratory Test Results

Hole	Depth m	Pressure Range - kN/m²								Strength c_u kN/m²
		27 - 54		54 - 111		111 - 222		222 - 444		
		m_v	c_v	m_v	c_v	m_v	c_v	m_v	c_v	
1	3.0	5.05	8.5	3.18	8.3	2.21	53.5	1.1	30.1	25.0
	5.0	7.59	9.9	5.48	11.6	3.30	18.0	1.7	11.1	17.4
	7.0	8.52	9.4	6.18	8.8	3.55	8.1	1.9	6.7	12.8
	9.0	6.22	21.2	3.93	39.6	1.83	31.1	1.1	29.4	29.0
	11.0	4.83	30.9	3.25	24.9	2.61	20.2	1.7	14.1	27.1
2	2.5	5.77	32.9	3.77	24.1	2.12	33.1	1.1	31.1	12.1
	4.0	11.22	2.8	8.88	3.93	5.41	3.5	2.8	3.1	15.4
	7.0	4.85	13.8	7.56	12.31	6.72	3.9	4.1	2.1	20.8
	8.5									20.6
	10.5									28.3

Value of m_v $m^2/kN \times 10^{-4}$

Value of c_v in m^2/year

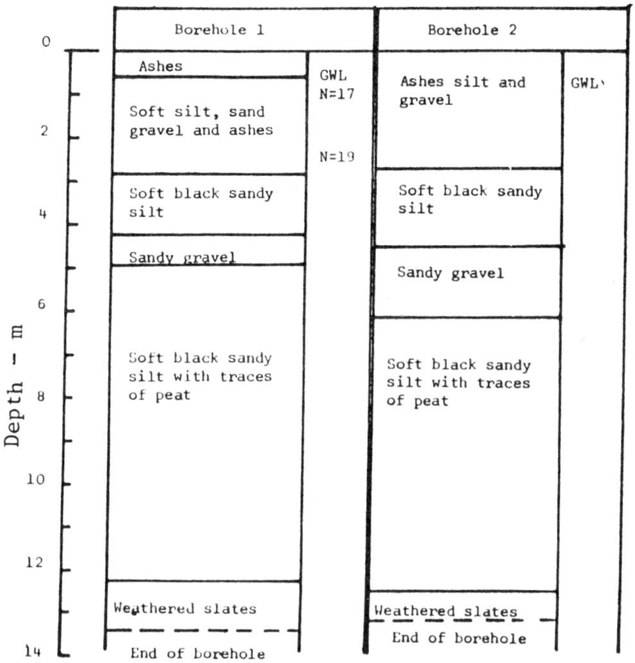

FIGURE 2 Borehole Logs

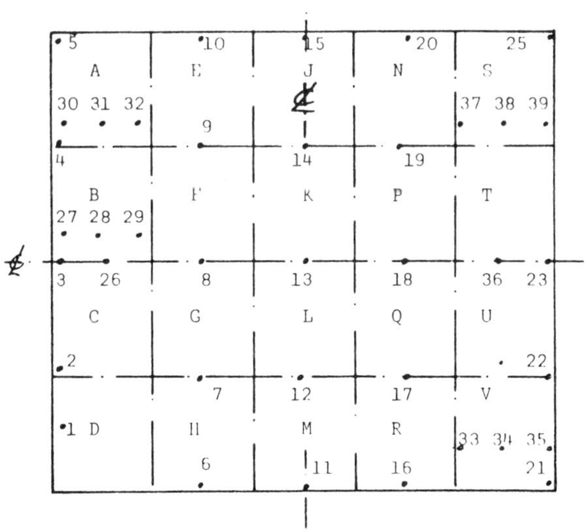

FIGURE 3 Levelling Stations and Bay References

Wild N3 precise level with invar staff. The results are presented in their entirety in Table II.

A particular difficulty in studying warehouse behaviour is the evaluation of the loading history. Due to the continuous arrival and despatch of materials the loading varies considerably with time, and although at each visit the loading per bay was accurately assessed, the loading in the interval is likely to have been different. It is considered, however, that because of the irregular intervals between visits and the fact that measurements have been made over 3 years (which would account for variations in loadings due to seasonal variation in demand for the stored products) the recorded loadings (Table III) represent a reasonable loading history.

A totally unaccountable effect is that of loadings due to fork-lift truck traffic, but some attempt has been made in the subsequent analysis to account for the effects of the heavy (3400 kN all-up weight) delivery lorries occupying and traversing the access aisles of the warehouse.

SETTLEMENT RESULTS

The measured settlements for a number of locations are plotted against time in Fig. 4, the points being confined, as shown, to one quarter of the warehouse plan area. Not every point was accessible on every occasion because of the loading, so that some parts of each curve are not so well defined as others. However, the characteristic shape of a settlement process dominated by consolidation effects is clearly demonstrated. The estimated final settlements indicated in Fig. 4 are the result of simple "hand" calculations, based on grossly simplifying interpretations of the soil succession and properties (see Table I), viz. 0-4 m, $m_v = 6.32 \times 10^{-4}$ m^2/kN; 4-8 m, $m_v = 6.94 \times 10^{-4}$ m^2/kN and 8-12 m, $m_v = 3.6 \times 10^{-4}$ m^2/kN; with ground water level at ground surface. A time-averaged mean loading was used. It can be seen that whereas the order of the maximum settlements is well estimated, many gross anomalies exist, particularly for points 16, 17 and 18 which are located in nominally unloaded access aisles. It would seem that some account should be taken of the effect on settlements of the heavy delivery traffic which uses these areas.

COMPUTATION OF SETTLEMENTS: INTERACTIVE ANALYSES

In a number of papers on soil/structure interaction the authors have developed methods (e.g. Wood and Larnach, 1974, 1975) which involve soil settlement calculations based on treating the soil as a linear or non-linear elastic continuum considered either as homogeneous or as a layered

Posn	1976				1977						1978		1979		
	30 July	14 Sept	26 Oct.	29 Nov	11 Jan	22 Feb.	21 Apl.	21 June	13 Sept	22 Dec.	12 Apl.	7 Sept.	9 Jan.	21 Apl.	6 Sept.
	Weeks 0	6.5	17.5	23.5	29.5	38	48	60	74	90	111	129	144	164	
1	0	3.200		5.76											
2	0	3.49											53.13	52.39	
3	0	10.51	26.04	32.27	37.62	44.80	41.92	49.19	53.04	52.87	56.14			61.03	52.36
5	0									23.57	28.17				
6	0	3.06	5.49	10.88	11.60	12.49	15.86	20.80	23.85	25.91	29.87	32.03	34.89	32.81	33.44
7	0	6.19	17.34	31.34	39.42	42.17	48.30	62.48	72.29	76.19	84.26	93.71	96.31	94.86	94.59
8	0	4.48	11.83	17.85	21.42	25.46	28.83	41.32	45.59	46.99	54.48	60.64	63.03	62.51	62.44
9	0	6.46	20.18	27.00	40.55	42.74	37.98	47.15	51.57	51.80				61.17	61.51
10	0	2.82	6.30	9.38	10.21	10.38		15.78	18.39	19.06	23.66	25.83	26.55	24.35	24.69
11	0				41.59	40.17		56.73	57.23	63.63	66.48				
12	0	6.33				39.35		111.73							157.29
13	0	4.20	23.97	40.61	49.08	58.85	63.72	81.51					124.80		123.27
14	0	3.91											105.13		105.32
15	0	2.37	12.27	31.22	25.24	23.20	20.35	22.68	27.24	28.08	34.88	31.73	32.33		33.13
16	0	2.81	5.60	16.38	32.38	38.69	31.17	38.25	38.59	42.85	47.77	49.57	54.02		55.07
17	0	7.07	29.30	52.03	71.73	74.89	34.38	39.34	96.66	107.90		133.52	135.82		136.72
18	0	5.31	23.64	28.53	-3.20	-6.04	50.93	57.13	66.84	68.69	39.03	93.96	94.36		95.85
19	0	3.81	20.81		-7.35		58.31	60.37	71.04	71.06		91.77	90.98		90.58
20	0	2.27	5.16	15.84		15.52	16.73	18.17	20.48	23.19	24.44	28.31	28.12		27.72
21	0		1.22						2.33	5.16	4.28				
22	0	2.18	27.50	34.82	43.46	41.49	41.06	49.46			50.52	75.63			74.34
23	0	2.28	16.62	20.26	22.92	25.35		31.38		46.87			59.03		66.85
26	+	0	0.15	25.44	32.25	36.59	40.26	48.68	53.52	52.84	57.65		64.72	62.51	63.37
27	+	0	9.50	13.41	15.78	17.31	22.05	26.47	27.36	28.87	30.43	33.33	32.88	35.17	35.39
28	+	0	17.36	24.54	27.70	31.09	36.94	43.69	45.63	46.89	50.26	55.10	53.96	56.49	56.60
29	+	0		16.97	24.59	27.30	34.24	39.88	43.67	44.99	49.28	53.28	52.63	55.62	54.73
30	+	0										23.77	24.90	24.67	24.57
31	+	0										35.29	38.08	41.28	37.50
32	+	0		6.10	16.56			28.91		31.99	32.69	37.32	38.81	40.83	39.06
36	+	0	31.30	42.56	52.02	56.88		65.96		77.91	82.12		87.10		35.16
37	+	0	8.80	16.79		24.99	27.16	26.86	30.17			40.95		39.27	
38	+	0	9.52	13.37	21.67	20.46	23.07	23.11	27.58		31.45	33.42		34.18	
39	+	0	1.89	3.87	9.42	6.43	8.00	9.39		15.04	12.97	16.79		22.75	29.41

Note (1) A blank entry indicates measuring point inaccessible due to loading.
(2) + indicates measuring point not yet installed.
(3) Points 4, 24, 25, 33, 34 and 35 remained inaccessible throughout and are omitted.

TABLE II Record of Measured Settlements (mm)

	1976				1977						1978		1979		
Posn.	30 July	14 Sept.	26 Oct.	29 Nov.	11 Jan.	22 Feb.	21 Apl.	21 June	13 Sept.	22 Dec.	12 Apl.	7 Sept.	9 Jan.	27 Apl.	6 Sept.
A	0	26	26.6	26.6	24.8	29.9	38.7	38.7	27.1	12.2	10.8	6.3	15.1	3.5	1.7
B	0	26	22.1	17.7	17.7	26.0	22.1	16.6	11.1	16.6	17.7	22.0	3.3	18.2	4.4
C	0	0	29.9	18.3	25.5	28.8	22.1	10.6	15.5	6.7	15.6	0	19.9	0	0
D	0	0	0	5.0	5.5	7.2	13.3	8.9	19.4	6.7	9.3	0	13.3	0	0
E	0	0	0	0	0	0	0	0	0	0	0	0	0	0	0
F	0	0	0	0	0	0	0	0	0	0	0	0	0	0	0
G	0	0	0	0	0	0	0	0	0	0	0	0	0	0	0
H	0	0	0	0	0	0	0	0	0	0	0	0	0	0	0
J	0	0	10.6	29.2	37.2	21.0	14.4	14.4	38.6	11.7	33.2	6.6	0	12.6	4.3
K	0	0	42.5	48.7	29.2	33.2	36.5	4.6	27.1	11.1	29.2	22.5	17.5	23.6	3.3
L	0	0	29.3	28.5	27.9	19.9	0	20.6	35.2	4.2	37.6	29.3	9.3	16.6	8.3
M	0	0	29.3	38.0	25.5	8.0	49.8	0	19.9	28.2	29.9	29.9	17.3	26.0	10.0
N	0	0	0	0	5.8	13.7	0	0	0	0	8.6	0	0	0	0
P	0	0	0	0	0	0	0	0	0	0	8.6	0	0	0	0
Q	0	0	0	0	0	0	0	0	0	0	8.6	0	0	0	0
R	0	0	0	0	5.6	0	0	0	0	0	8.6	0	0	0	0
S	0	17	26.6	11.1	29.2	26.6	25.4	1.3	24.3	7.7	14.9	0	15.5	0	17.7
T	0	0	17.7	44.3	29.2	31.0	29.3	1.7	31.9	0	14.9	0	0	13.3	4.7
U	0	0	28.8	29.9	23.9	15.5	23.2	12.1	21.7	5.5	14.9	18.8	6.9	26.9	0
V	0	0	26.6	29.9	23.9	36.1	23.2	13.8	0	14.4	14.9	0	14.6	16.0	0

TABLE III Mean Loadings/Bay (kN/m^2) on Dates Stated

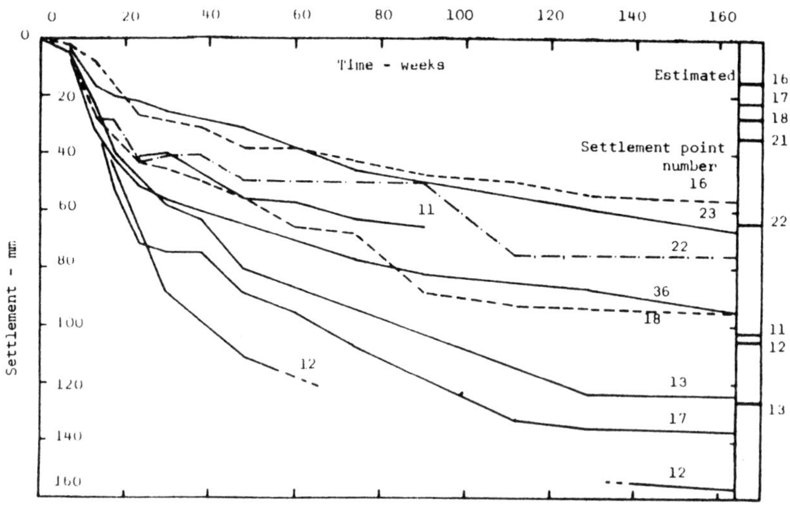

FIGURE 4 Settlements of Selected Points

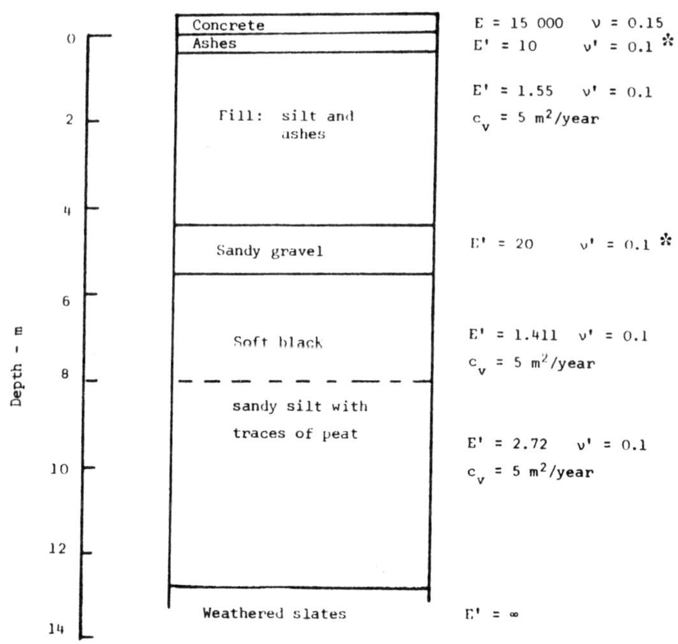

FIGURE 5 Assumed Soil Profile and Properties
E values in $kN/m^2 \times 10^3$
* Permanently drained

system. The second approach is clearly desirable for the site considered here, and the idealization shown in Fig. 5 has been adopted. It can be shown that the elastic moduli for the soil E' (drained) and E_u (undrained) are related to m_v:

$$E' = \frac{1}{m_v}\left[\frac{(1 - 2\nu')(1 + \nu')}{(1 - \nu')}\right]$$

$$E_u = \frac{3}{2(1 + \nu')} E'$$

and the values shown in Fig. 5 correspond to the three values of m_v already mentioned and taking the drained Poisson's ratio as $\nu' = 0.1$. The surface layer of ashes and the lower sandy gravel layer were assumed to be relatively very stiff and permanently drained. The underlying weathered slates were treated as incompressible, and also as providing a further drainage boundary. The ground floor slab was 250 mm thick, and the concrete was presumed to have an E value of 15×10^6 kN/m^2.

The complete three-dimensional system was analysed allowing for structural interaction between slabs (with and without joints) and the underlying soils: in effect the problem is one of an elastic plate on a non-homogeneous elastic foundation. The slab was modelled using plate bending finite elements, with the interface taken as smooth; the soil was modelled, as discussed above, as an isotropic layered elastic continuum of finite depth.

DISCUSSION OF RESULTS

Essentially two basic computations have been made:

(i) assuming undrained conditions on 26th October, 1976, with the then current load assumed to have been in place from the start of observations;

(ii) assuming drained (ultimate) conditions, and with either the maximum recorded load per bay or with an "average" load per bay (taken, somewhat arbitrarily, as the load per bay operative on 22nd February, 1977, which, by inspection, seemed to represent a typical condition).

The results of these calculations appear in Table IV. As can be seen, the agreement between computed and measured values is very variable although generally better than for the hand calculated results, with generally less agreement for the undrained case, which is comparatively a less important condition. The comparison for the drained (final)

Posn.	Final settlements			Undrained settlements	
	Measured (1)	Computed		Measured (2)	Computed
		Average load	Max. load		
3		70.0	70.1	26.0	22.5
6	33.4	6.6+	28.5+	5.5	- 3.5+
7	94.6	29.4+	61.0+	17.3	3.1+
8	62.4	48.7+	65.5+	11.8	7.0+
9	61.5	49.0+	67.1+	20.2	4.8+
10	24.6	21.1	31.3	6.3	- 3.1
13	123.3	86.1	142.6	24.0	46.1
14	105.3	87.4	138.3		32.2
15	33.2	39.4	69.3	12.3	1.5
16	56.1	18.0+	35.2+	8.6	- 1.2+
17	136.7	36.2+	68.7+	28.3	6.1+
18	95.9	45.9+	68.4+	28.6	5.6+
19	90.6	49.4+	65.2+	20.8	4.1+
20	27.7	20.0+	27.6+	5.2	- 3.0+
23	66.9	61.0	78.7	16.6	18.4
26	63.4	89.4	91.1		33.3
27	36.0	70.0*	70.1*	9.5	22.5*
28	56.6	89.4*	91.1*	17.4	33.3*
29	54.7	69.2*	74.2*		18.5*
30	24.6	71.0*	81.9*		21.0*
31	37.5	91.0*	104.8*		31.3*
32	39.0	70.0*	83.1*		17.5*
36	95.2	78.0	101.2	31.3	28.0
37		71.0*	78.1*	8.8	14.7*
38		92.3*	97.8*	9.5	27.5*
39	28.4	72.7*	76.4*	1.9	18.3*

(1) On 6th Sept. 1979.
(2) On 26th Oct. 1976.

+ In access aisle: effect of lorry to be added

* In fire access: actual points unloaded.

TABLE IV Calculated and Measured Settlements

condition is better but it should be noticed that for points located in the access aisles the effect of the lorry traffic is not accounted for.

In Fig. 6 are presented the results of the calculated effect at the centreline of the building of the passage of a loaded lorry along an aisle, for both an undrained (transient) loading and a drained loading, which is tantamount to assuming that the aisle is trafficked continuously by the vehicle. It will be seen that (remembering that the lorry can occupy similar positions beyond the centreline) the centreline settlements can amount to some 15 mm and 36 mm at the centre of the loaded aisle in the undrained and drained conditions respectively.

Figures 7(a) and (b) display computed and measured settlements along the 45 m and 40 m centreline axes respectively. These diagrams indicate that the computed values (amended to allow for the lorry loadings and any additional loading in the aisles, see Table III) generally "bracket" the observed values. The bracket width is not as narrow as could be wished, but it seems clear that by carefully accounting for actual loading variations throughout the observation period, a closer agreement might well be achieved. It is intended to pursue this approach in a subsequent paper.

In a fully interactive computation such as that employed here the stress resultants in the structure (in this case the floor slab) are calculated as part of the solution. Figure 8 shows the slab bending moments along the centreline of the central loaded strip, corresponding to drained conditions and average load conditions, defined in (ii) above. The absolute values of the moments are not high, but the effects of the presence of joints between bays can be seen. It should be noted that due to the rather coarse discretization of the slabs into elements small residual moments are calculated at the joints. Clearly such information could be used in the rational design of ground-supported floor slabs, and this matter has been explored elsewhere (Wood et al, 1979). Finally, Fig. 9 shows the results of attempting to predict the time-settlement curve for a particular point. The central point 13 (see Fig. 3) was chosen for this exercise, in which account was taken of variable bay loading and pore pressure dissipation due to one-dimensional (vertical) drainage in the multi-layer system of Fig. 5.

A central problem in time-settlement calculations is the choice of a proper value for the coefficient of consolidation, c_v. There is evidence to suggest that in a three-dimensional strain field, such as exists under the

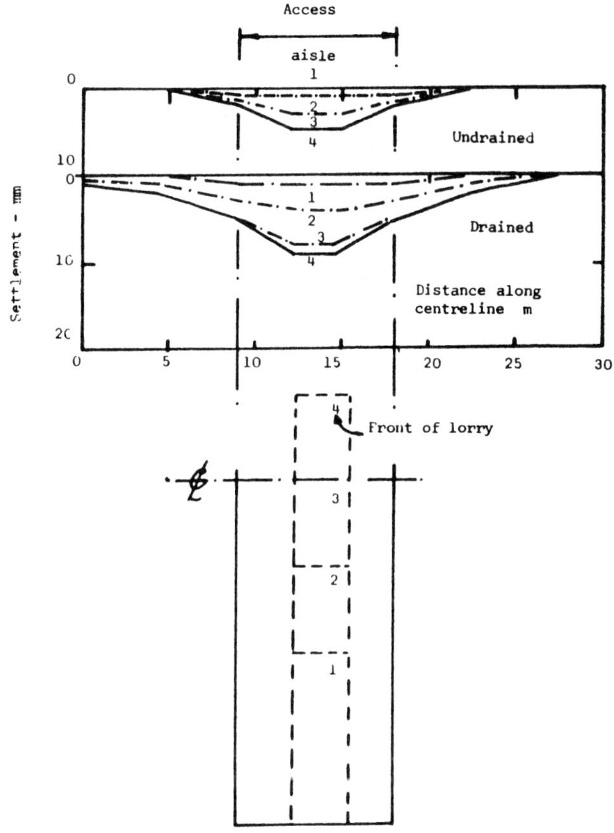

FIGURE 6 Effect of Lorry Loading in Access Aisle

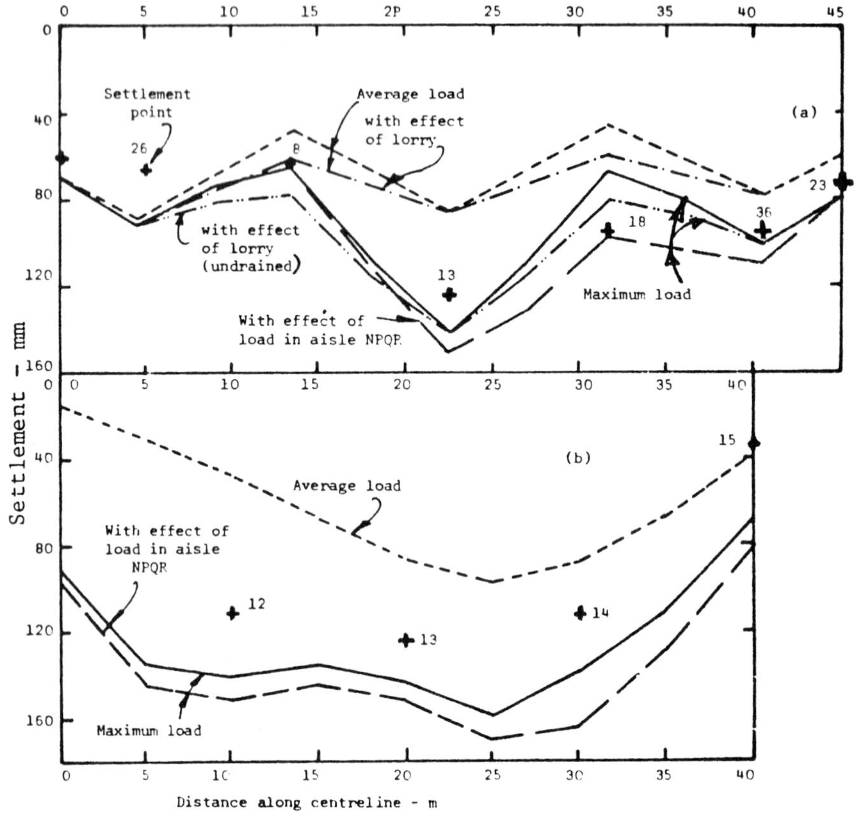

FIGURE 7 Calculated and Measured Settlements

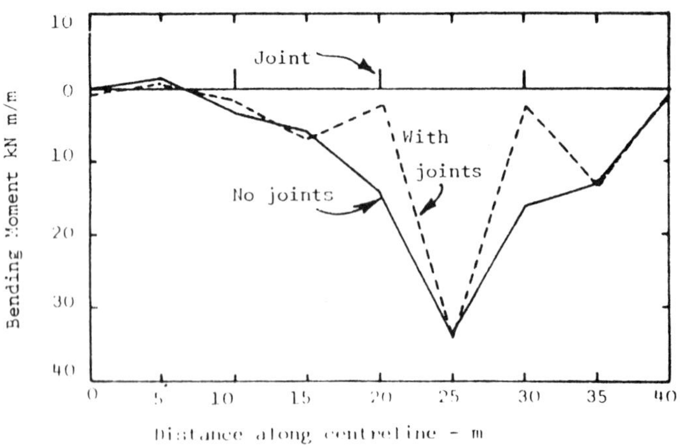

FIGURE 8 Calculated Slab Bending Moments

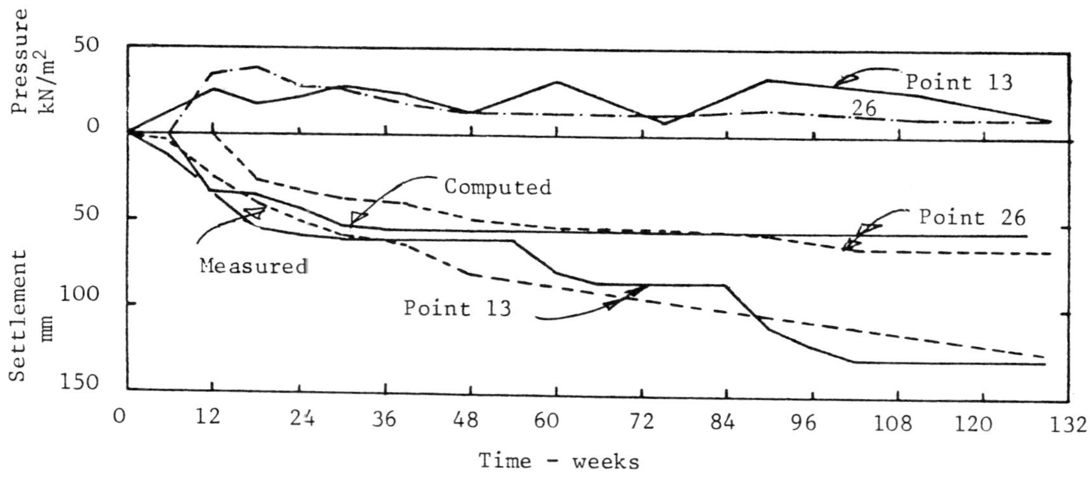

FIGURE 9 Comparison of computed and measured settlement - time curves for two points

(Note: the illustrated loadings refer to the mean loading in adjacent bays. The calculations were performed using actual loadings on all bays.)

warehouse, the corresponding c_v is significantly lower than that measured in conventional one-dimensional tests, Davis and Poulos (1972). On the other hand, c_v values measured on small laboratory specimens do not necessarily reflect the more rapid drainage actually occurring in the field due to structural features of the soil fabric. It is considered that $c_v = 5\ m^2/year$ is a reasonable compromise between these effects for the soils encountered on this site. The resulting computation appears to produce reasonable agreement between measured and predicted behaviour.

CONCLUSION

A case history has been described for which, it is considered, all relevant data have been provided: borehole records; conventional soil properties; loading details; and settlement records. The paper has presented the results of a fully interactive three-dimensional analysis of the ground floor slab/soil system, which has indicated that at the design stage settlement predictions can be made which are adequate for the making of design concept decisions; and furthermore the analysis has shown that the effect of vehicular traffic in otherwise unloaded areas can be significant and can be accounted for.

It is to be hoped that others interested in the prediction of settlements may be encouraged to make their own interpretations of the available data.

The structural concept has been fully justified by the performance of the building. Although floor settlements as high as 150 mm have occurred no adverse effects on piles or ground beams have been noted, and the working of the warehouse has not been interrupted or inconvenienced in any way.

ACKNOWLEDGEMENTS

The senior author expresses his thanks to Mr. Cooper, Director of South Western Storage Ltd., for his ready co-operation in allowing the settlement measurements to be taken; to Mr. Paul Blundell of Blundell Saunders Partnership, Consulting Engineers for the project; and to the Building Research Establishment for its financial support for the monitoring project.

REFERENCES

Cheney, J.E., Techniques and equipment using the surveyor's level for accurate measurement of building movement. Proc. Symp. on Field Instrumentation in Geotechnical Engineering, Butterworth, London (1974) 85-99.

COSOS, Proc. Conf. on Settlement of Structures. Pentech Press, London (1974) xii + 811.

Davis, E.H. and Poulos, H.G., Rate of settlement under two- and three-dimensional conditions. Geotechnique, Vol XXII (1972) 95-114.

Wood, L.A. and Larnach, W.J., The effects of soil/structure interaction on raft foundations. Proc. Conf. Settlement of Structures, Cambridge. Pentech Press, London (1975) 460-470.

Wood, L.A. and Larnach, W.J., The interactive behaviour of a soil-structure system and its effect on settlement. Symp. Analysis Soil Behaviour and its Application to Geotechnical Structures. Univ. N.S.W. (1975).

Wood, L.A., Burley, E. and Harvey, R.C., The design of ground-bearing slabs in warehouse construction. Presented at Int. Conf. on Concrete Slabs. Univ. of Dundee, Proceedings to be published (1979).

SETTLEMENT BEHAVIOUR OF A TEST FOUNDATION SLAB FOUNDED ON SOFT CLAY

S.R. Cochrane, M Sc , C Eng , MICE, MIWES, MIPHE

Senior Lecturer, Department of Civil Engineering, Queen's University of Belfast

A.L. Bell, B Sc , Ph D , FGS

Lecturer, Department of Civil Engineering, Queen's University of Belfast

SUMMARY

The paper deals with a laboratory and field investigation of the settlement of structures founded on the Belfast estuarine deposits.

The Queen's University of Belfast in the period 1971-1976 conducted extensive laboratory investigations of the Belfast soft materials. The results indicated that the deposits are lightly pre-consolidated and are stiffer and stronger than is commonly assumed. However at stress levels above the preconsolidation pressure the soils exhibit high compressibility and time-dependency.

In 1977, an investigation of the field settlement behaviour of full-scale structures was begun. A test load slab was constructed at Kinnegar, Holywood, Co. Down which is underlain by the Belfast estuarine deposits. Measurements of settlement, heave, porewater pressures and in-situ permeability have been made during a series of loadings. Comparison has been made between oedometer predictions of settlement and observed settlement. An elastic behaviour for stress levels below the preconsolidation pressure has been verified and evidence of continuing creep deformations has been obtained.

INTRODUCTION

In the past the low strength and high compressibility of soft soil deposits has ensured that investigation of their fundamental geotechnical properties has been largely ignored since major construction upon them was not normally considered to be feasible. More recently however, land shortages have resulted in the need to develop previously unsuitable sites underlain by soft soils. This has certainly occurred in Belfast and its environs, much of which are underlain by a soft estuarine clay locally known as 'sleech'.

As part of a continuing program of research into the more important geotechnical properties of soft clays in Northern Ireland a detailed investigation of the estuarine clays underlying a site at Kinnegar, on the southern shore of Belfast Lough, was commenced in 1971. Extensive laboratory investigations were carried out in the period 1974 to 1976 and a field test slab study which has been running since 1976, is now producing interesting information.

LABORATORY STUDIES OF THE BELFAST ESTUARINE CLAYS AT KINNEGAR

During the construction of a large sewage works at the site several excavations were opened and these provided an opportunity to recover several block samples of the clay. Each sample was very carefully hand-cut from undisturbed clay at the base of an excavation and wax-sealed on site. Borehole sampling was also used at the site. Over forty samples 457 mm long and fourteen 1000 mm long samples were recovered from a total of nine boreholes using a 100 mm dia. thin-walled stationary piston sampler of Norwegian design. Each boring was advanced, where possible, using a hand auger to minimize disturbance.

All of the samples were carefully transported from the site to the laboratory 15 kms distant in a purpose-made shock-absorbing container. They were stored at a constant temperature of 14°C and humidity of 100 per cent for periods varying from several days up to, in a small number of cases, several months. Control tests have indicated that the influence of differing storage times on laboratory test results is not likely to be high (Bell, 1977).

Visual inspection of the samples revealed that with the exception of those from close to the ground surface they were fairly homogeneous. However many samples contained shells or shelly fragments most of which were small, about 5 mm dia., although some were as large as 00 mm dia. Organic matter could also be seen in some samples and a strong odour was often present. The organic matter appeared as near-vertical decayed rootlets or as networks of strands, channels, fibres and fragments each a few mm long. Small fissures were observed in a few samples but these did not appear consistently at any particular depth. Sand or silt lenses and laminations seemed to be completely absent.

Particle size analyses (fig.2) reveal that the percentages of silt and clay sizes in the ground profile at Kinnegar increase steadily from very low values at 0.9m to over 40 per cent at 2m depth. Apart from minor fluctuations the percentages vary little from 2.0m to 6.0m with clay the most predominant size ranging from 40 to 50 per cent; silt ranging from 38 to 48 per cent; and

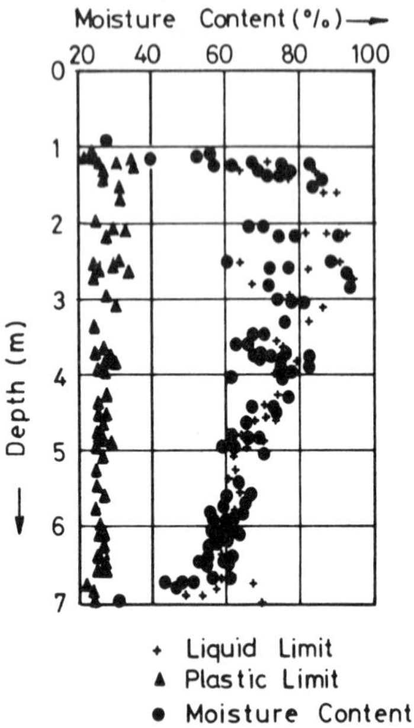

L.L, P.L, M.C. v DEPTH.

Fig. 1

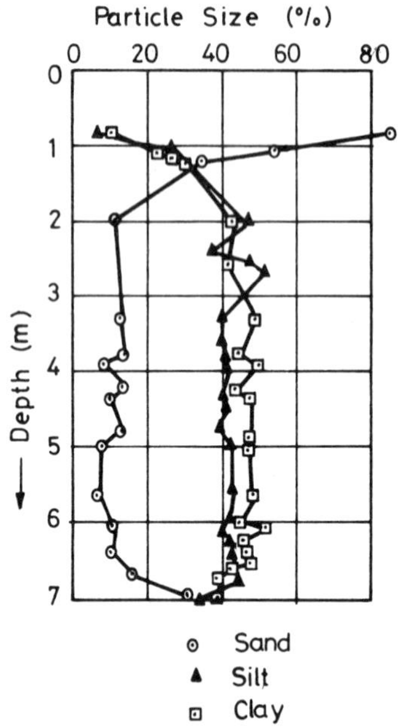

PARTICLE SIZE ANALYSIS.

Fig. 2

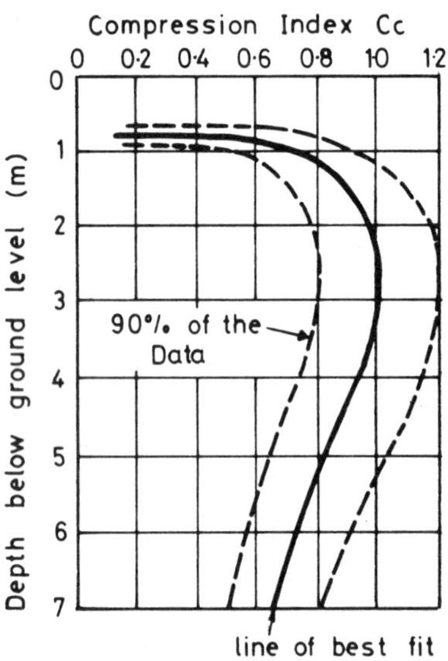

COMPRESSION INDEX v DEPTH

Fig. 3

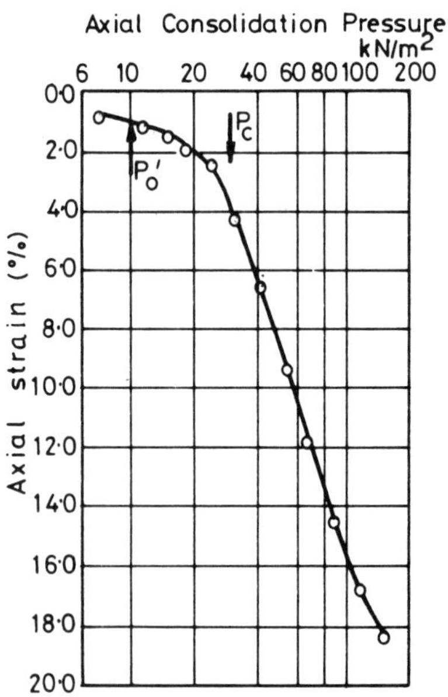

OEDOMETER CONSOLIDATION.

Fig. 4

sand from 8 to 15 per cent. Below 6.5 m the sand content increases and at 7.3m the base of the deposit is, like the top, almost entirely composed of sand. The variations in grain-size with depth are consistent with the current belief that the estuarine deposits were formed during the early part of a marine transgression, its eventual maximum, and final return to more littoral conditions.

Mineralogy was investigated by Crooks (1973) who studied the deposits at a number of sites, including Kinnegar. It was found that illite was by far the most predominant mineral although chlorite and kaolinite were also detected. It seems likely that the clay minerals present in the estuarine clays are derived primarily from local glacial tills (Bell, 1977).

Consistency limits and natural moisture contents are shown plotted on fig. 1. It can be seen that the natural moisture contents are quite high and are generally very close to the liquid limits. These increase from 1m depth to about 1.5m. From this depth to 3.0m the results are scattered although maximum moisture content and liquid limit of about 93 per cent occur at 2.5m. Through the depth interval 3.0m to 6.5m the values decrease steadily to about 60 per cent. Below this considerable variation is again exhibited. Plastic limits range from 22 per cent at 1.0m to 25-30 per cent at 1.5m and mainly vary between these values throughout the remainder of the deposits.

Below 2m the index of plasticity becomes reduced with depth even though the percentage of clay and silt sizes remain essentially constant. Such changes are associated with variations in cation content (Crooks, 1973) and with organic content (Bell, 1977). The high moisture contents although clearly indicative of high voids ratios are also partly caused by organic matter. Organic content, expressed as a percentage of dry unit weight of soil varies from about 1 per cent in the sandy portions of the deposit to a maximum of about 5 per cent just below 2m depth. The mean value is about 3.0 per cent.

Laboratory fall-cone tests were used to determine values of undrained shear strength, S_u, and index of sensitivity. The S_u values show quite wide variation which has been attributed to unavoidable disturbance. Allowing for this the strengths appear to increase with depth from 1.0m to 3.0m with a mean value of about 12 kN/m^2. Below 4m the strengths increase from about 13 kN/m^2 to 23 kN/m^2 at about 7m. The index of sensitivity, defined as the ratio of the undisturbed to the remoulded undrained shear strength varies from 5 to 15 with a mean value of about 8. The sensitivity is relatively high for a British Isles clay deposit and has been attributed to a reduction of magnesium cation concentration caused by leaching (Bell, 1977).

Specimens for 76mm dia. x 19mm oedometer testing and 50mm dia. x 100mm triaxial testing were trimmed from both block and piston samples using a purpose-made apparatus designed to minimize trimming and handling disturbance. Each triaxial specimen, before further testing, was consolidated under a stress system similar to that to which it was subject in the field, i.e. the axial consolidation pressure was chosen to be equal to an estimate of the vertical effective overburden pressure P'_o and the lateral consolidation pressure (cell pressure) was chosen to equal $K_oP'_o$ where K_o is the coefficient of earth pressure at rest, measured in triaxial consolidation tests ranged from 0.45 to 0.52 (Crooks 1973, Bell 1977).

A large number of triaxial and oedometer consolidation tests were carried out. A 'yielding' of each specimen occurred at a preconsolidation pressure, P_c, in excess of the existing in-situ vertical effective pressure P'_o (e.g. fig.4). For pressures below P_c the clay exhibits low compressibility in contrast to that beyond P_c. The clear indication of these tests is therefore that, although previously viewed as normally-consolidated, the Kinnegar deposit is in fact lightly over-consolidated. Such over-consolidation is believed to have been caused in the main part of the profile by a combination of past changes in the groundwater table and secondary compression acting since formation of the deposits. The overconsolidation ratio (P_c/P'_o) reduces from 2.5 to 3.0 near the top of the deposit, to a minimum of about 1.2 at 4.0m below which it increases steadily to about 1.5 at 7.0m.

The compressibility of the soil at stress levels beyond P_c can be expressed in terms of the index of compressibility, C_c. The variation of C_c with depth is illustrated on fig.3. It is worth noting that, in the main portion of the deposit, these values are large even for a soft clay deposit.

Rates of consolidation were also influenced by the over-consolidation effect. At stress levels below P_c the soil consolidated very much more rapidly than for stress levels above P_c. Coefficients of consolidation C_v were measured at 3 to $8m^2$/year below P_c and 0.05 to $1m^2$/year above P_c. Values of C_v at stress levels close to P_c were found to be quickly changing from the higher to the lower values. A further relevant finding is that rates of secondary consolidation or 'creep' were very much higher beyond P_c than below it. Indeed maximum creep rates appeared to occur at or near P_c. Time effects at the P_c stress level were additionally investigated using strain controlled oedometer tests in which the strain rate was step changed by a factor of ten from time to time. All of the rates of strain used were carefully chosen so that excess porewater pressures were essentially zero during test. The results strongly implied that P_c is dependent on the rate of strain - the higher the rate, the higher P_c.

Triaxial tests can only subject soil specimens to axially-symmetric states of stress. Such stress states are confined to a single plane in principal stress space, herein referred to as the 'triaxial plane'. Drained triaxial tests were conducted using a variety of stress path directions in the triaxial plane. It was found that regardless of stress path direction there is an initially 'stiff' or 'small-strain' response to loading beyond the in-situ stresses until a distinct 'yield' occurs beyond which a large strain behaviour is exhibited. Undrained triaxial tests also revealed yield behaviour since the undrained shear strength, S_u, for each sample tested, was attained at very small axial strains.

The combination of effective stresses at which yield for each specimen takes place may be represented as a single point on the triaxial plane. The points from specimens taken from the same depth but subjected to differing stress paths may be connected by a line which forms a yield envelope i.e., the line is a 'P_c' generalized in the triaxial plane. Stress changes within the yield envelope cause only small strains, and those outside cause large strains. Similar yield envelopes have been used to help describe the stress-strain behaviour of several Scandinavian and North American clay deposits (eg Mitchell 1970; Bjerrum 1973).

Many investigators link yield in clays with a change in the microstructure. For the Kinnegar deposit, this idea seems to have many attractions. Before yielding the intact microstructure behaves in an approximate 'elastic' manner, with low compressibility and a low rate of secondary consolidation. Furthermore the intact microstructure transfers little load to the porewater, hence high rates of consolidation would be expected and as previously mentioned this has been observed. At yield, the microstructure breaks down and no longer operates as an elastic framework. High porewater pressures and low rates of consolidation coupled with high rates of secondary consolidation and high compressibiility would be expected, and again this seems to fit the observed laboratory behaviour. At high pressures the microstructure would have become largely re-adjusted to the prevailing stress system and there is some laboratory evidence that this is so. Rates of primary consolidation have been observed to improve and time effects appear to lessen at higher pressures.

CONSTRUCTION AND INSTRUMENTATION OF THE TEST SLAB

Although the laboratory studies of Crooks (1973) and Bell (1977) had indicated that the settlements of lightly loaded foundations on the Belfast estuarine deposits would be appreciably influenced by the preconsolidation effect there was little reliable field data available, particularly for realistic design pressures of perhaps 50kN/m^2 or more. Furthermore, conflicting local opinions were available concerning both magnitudes and rates of settlement which might be expected.

Some of these opinions were formed as a result of research by the Transport and Road Research Laboratory which had suggested that embankments constructed on soft alluvial soils were likely to settle at a very much higher rate than that predicted by conventional laboratory tests although the magnitudes of settlement were quite well predicted (Lewis et al 1975). The higher field rates of settlement were attributed to a natural drainage structure in the ground which small oedometer samples fail to represent. Rowe (1972) had earlier drawn attention to this problem.

In late 1975, in connection with the construction of a new sewage works at Kinnegar, the Department of Environment (Northern Ireland) completed several embankments which were founded on the estuarine clays. As settlements were expected to be considerable for both embankments and unpiled structures, a continuous record of levels was kept from October 1975. In 1976 the DoE (NI) agreed to make a site at Kinnegar, undisturbed by any engineering works, available to a research team from the Queen's University, Belfast, so that a test slab could be constructed, loaded and monitored. The slab was intended to provide worthwhile field information concerning the settlement of structures and to support the information being obtained from the sewage works embankments.

It was decided to construct a 3m square test slab, which was considered to be large enough to represent an individual column or stanchion foundation. Initially light loadings well below P_c stresses in the ground were to be applied so that low-stress behaviour could be investigated. The second loading was to bring the ground stresses close to P_c, at least in the upper part of the deposit. Two final load increments were then to be applied at suitable intervals. As the top of the estuarine clay lies just below ground level it was decided to minimize foundation disturbance by loading the slab with precast concrete blocks placed by hand.

Levelling points were to be provided at each corner of the slab and ground heave or depression around the slab area was also to be recorded. A series of piezometers were to be located at various depths in the clay beneath and to the sides of the loaded area so that porewater pressure changes particularly during loading periods could be monitored. Additionally groundwater levels were to be regularly recorded to allow for the influence of porewater pressure variations.

The initial work at the test site consisted of sinking boreholes for the installation of the piezometers. Each piezometer was placed at its appropriate depth and surrounded by sand, at which stage the borehole was plugged with a bentonitic slurry. The piezometers, arranged in the ground as shown in Fig.5, were connected to mercury manometer measuring systems which are situated in a control hut close to the slab. The hut also houses a de-airing unit and equipment for measuring in-situ permeability via the piezometers. Suitable precautions were taken to ensure protection against frost during severe winter weather. The boreholes were carried out by the Department of Finance (NI) under the supervision of Queen's University staff who installed the piezometers.

On completion of the piezometer installation in the Autumn of 1976 the topsoil was stripped from the slab area which was then covered with a layer of permeable construction fabric. To provide a suitable working surface above possible flood levels for the later construction and loading of the slab 600mm of hardcore filling was placed on the construction fabric. The filling represented a loading of some 8.6kN/m^2 on the clay and from this

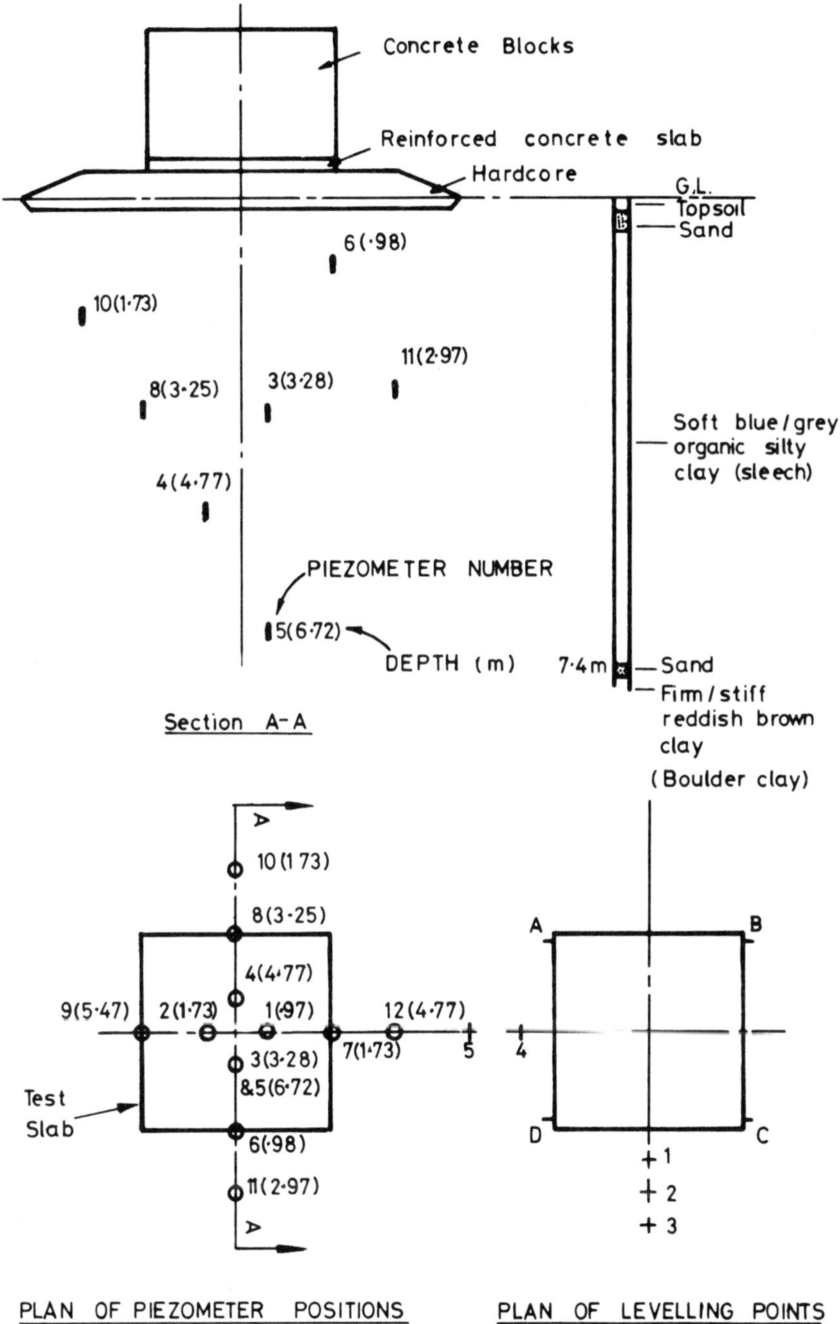
Concrete Blocks
Reinforced concrete slab
Hardcore
G.L.
Topsoil
Sand
6(·98)
10(1·73)
11(2·97)
8(3·25)
3(3·28)
Soft blue/grey organic silty clay (sleech)
4(4·77)
PIEZOMETER NUMBER
5(6·72)
DEPTH (m)
7·4m
Sand
Firm/stiff reddish brown clay
(Boulder clay)
Section A-A
A
10(1 73)
8(3·25)
4(4·77)
9(5·47)
2(1·73)
1(·97)
12(4·77)
7(1·73)
5
3(3·28)
&5(6·72)
Test Slab
6(·98)
11(2·97)
A
B
4
D
C
1
2
3
PLAN OF PIEZOMETER POSITIONS
PLAN OF LEVELLING POINTS

Fig. 5

stage onwards records of settlement, excess porewater pressures and ground water levels were kept. By midsummer 1977 a settlement of 26mm due to the hardcore layer had been recorded.

At this stage the 3m reinforced concrete slab was constructed, complete with levelling points (fig.5) and loaded with concrete blocks until a total pressure of 25kN/m^2 was applied to the soil. This load was maintained for about 18 months until December 1978 when a total settlement of about 90mms had been recorded. The blocks were then removed and the recovery measured for a few weeks. In January 1979 further loading up to 35kN/m^2 was carried out. The loading phases and the mean settlement of the slab are shown versus time on figs 6&7. It should be noted that the settlement recorded at each levelling point was very similar to that at others, and differences indicated only that a very small tilting of the slab is taking place.

DISCUSSION AND ANALYSIS OF SLAB SETTLEMENTS

In the previous section it was noted that the test slab was constructed largely to provide reliable field data so that, linked with local experience and the extensive laboratory studies, a greater understanding of the settlement of structures founded on the Belfast estuarine clays could be achieved. Although the test slab experiment is not yet complete, several interesting points have already emerged and these are dealt with in the following discussion.

Magnitude of Settlements

For the hardcore loading alone it was found that initial excess porewater pressures were quite small, less than 0.4 x the estimated vertical stress increase at any depth, and were essentially dissipated after 100 days. The low excess porewater pressures are thought to be mainly due to an initial small strain 'elastic' response observed in the laboratory undrained and drained tests. They may also, in part, be due to the small amount of unloading caused by excavation prior to placing the hardcore.

Fig.8 illustrates the variation of P_o' and P_c with depth. Also shown is the variation of $(P_o'+\Delta p)$ with depth for a range of foundation pressures. The $(P_o'+\Delta p)$ values indicate the magnitude of the effective vertical stresses at a given depth, assuming full dissipation of excess porewater pressures, and have been calculated using Boussinesq's elastic theory. The real variation of stresses may, of course, be somewhat different from that shown because of the simplifying assumptions, which include the high stiffness of the reinforced concrete slab relative to the estuarine soil. However for the hardcore loading there seems little doubt that P_c is nowhere exceeded. Settlement calculations were therefore based on soil behaviour at stresses beneath P_c.

The existence of an immediate settlement caused by the undrained elastic compression of a clay stratum under an imposed loading has been recognised for several years (eg Skempton & Bjerrum, 1957). In view of the uncertainty concerning the dis-

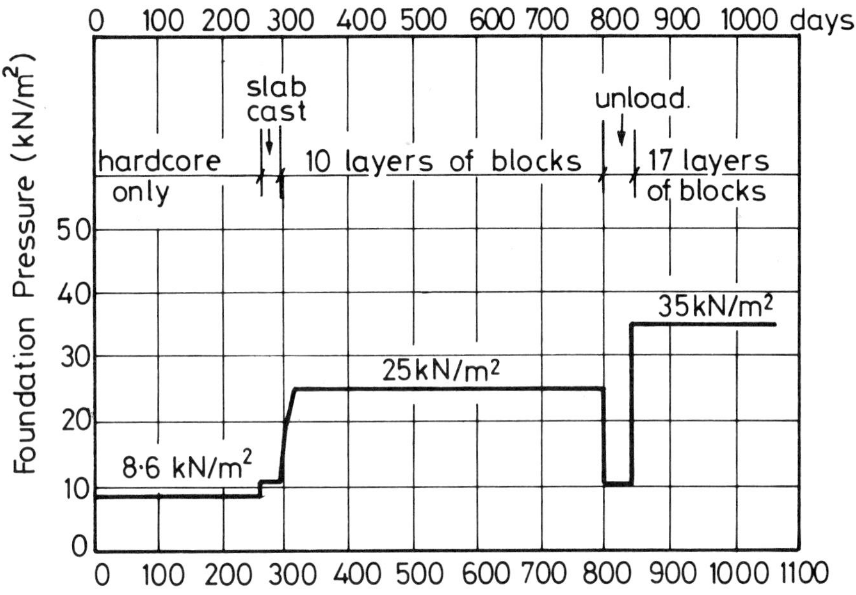

Fig. 6 LOAD / UNLOAD SEQUENCE.

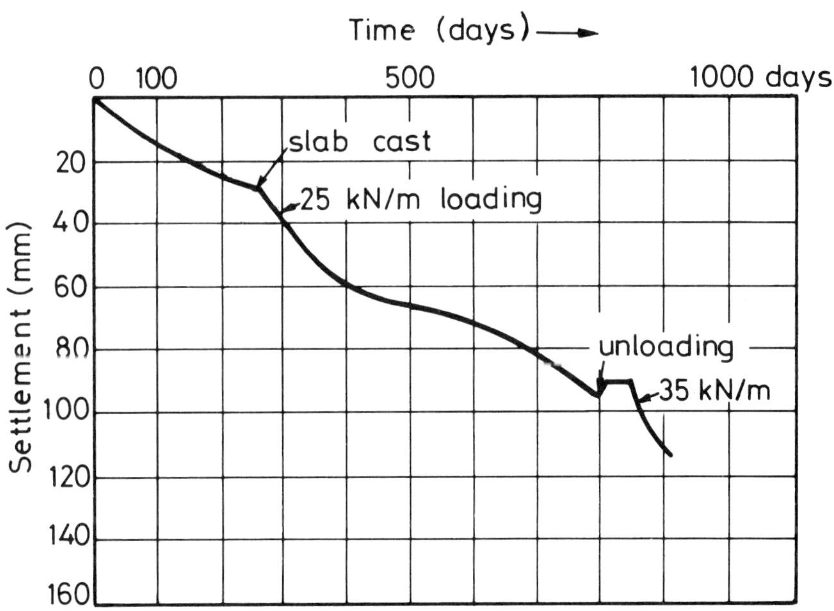

Fig. 7 RECORDED SETTLEMENT.

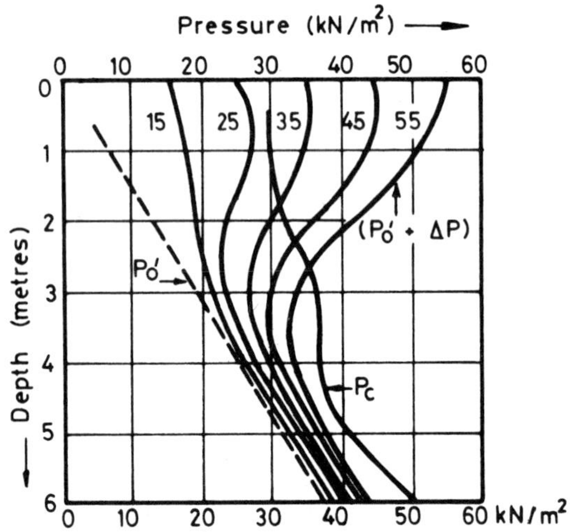

Fig. 8 VARIATION OF P_o', P_c AND $\Delta P + P_o$ WITH DEPTH.

tribution of vertical stress increases two methods were used to predict immediate settlements. Janbu et al (1956) proposed that the average immediate settlement ρ_i beneath a flexible rectangular footing is given by:

$$\rho_i = \frac{q\ B\ \mu_o\ \mu_1}{E_u} \qquad (1)$$

where B - footing width; q - applied pressure on footing; $\mu_o\mu_1$ - dependent on geometry of the problem. Whitman and Richart (1967) provide an approximate solution for the vertical displacement beneath a rigid uniformly loaded rectangular footing:

$$\rho_i = \frac{P(1 - \nu^2)}{B_z\ E_u\ \sqrt{BL}} \qquad (2)$$

where P - total vertical load on footing; B,L - rectangle dimensions; B_z - factor dependent on $^L/_B$; ν - poissons ratio (= 0.5 for undrained conditions).

For both expressions E_u is the undrained stress-strain modulus. From a series of strain-controlled consolidated undrained (cau) triaxial tests Bell (1977) found that E_u for the Kinnegar deposit is primarily influenced by shear stress level, consolidation pressure, and overconsolidation ratio. Using mean values of P_o' and overconsolidation ratio for the upper 4m in which stress increases are significant, the undrained secant modulus of the soil for a factor of safety of 2.0 ranges from 2700 to 3500kN/m^2. Predictions made using these values in eqns. (1) and (2) produce similar answers for ρ_i, which for each of the slab loads, appear to provide reasonable estimates of observed immediate settlements. Table 1 lists the various immediate settlement predictions along with consolidation settlement predictions.

Footing Pressure (kN/m^2)	Predicted Settlement (mm)			Observed Settlement (mm)	
	Immediate		Consolidation	Immediate	Total
	Janbu et al (1956)	Whitman & Richart (1967)	$\sum \frac{H.C_s}{1+e_o} \log_{10} \frac{P_o'+\Delta p}{P_o'}$	(per load period)	(per load period)
0 to 8.6	c.5	c.6	26	c.4	26
8.6 to 25	8-10	9-12	33	c.10	67
8.6 to 35	12-16	15-19	-	c.14	-

Table 1

The soil stratum was divided into 500mm thick sub-layers and the consolidation settlement Δc for each sublayer calculated from:

$$\Delta c = H.\frac{c_s}{1+e_o} \cdot \log_{10}\left(\frac{P_o' + \Delta P}{P_o'}\right) \qquad (3)$$

where H is the sublayer depth; e_o the initial voids ratio; $P'_o + \Delta p$, P'_o as described previously; C_s is the index of compressibility describing the swelling/recompression part of the log pressure voids ratio curve. From oedometer tests it was found that $e_s \approx Cc/6$ and this has been used in the calculations. The total consolidation settlement ρ_{oed} for the hardcore loading, obtained by summing the Δc values, was found to equal 26mms.

Skempton and Bjerrum (1957) suggested that the total settlement under a foundation is equal to $\rho_i + \mu.\rho_{oed}$ where μ is a factor which attempts to account for the difference between the porewater pressure generated in an oedometer sample and that generated in real soil beneath a foundation. For stresses below p_c both field and laboratory porewater pressure measurements lead to a value of about 0.7. Therefore $\rho_t = 6 + 0.7 \times 26 =$ 24mm which compares well with an actual settlement of 26mm. Burland et al (1977) report that there is now much evidence to indicate that for soils which are approximately elastic in their response to vertical loads the total settlement can be estimated with sufficient accuracy using one-dimensional methods such as those based on the oedometer i.e., $\rho_t = \rho_{oed}$. This approach furnishes ρ_t = 26mm which again agrees well with the observed settlement. It should be noted that although initial excess porewater pressures dissipated rapidly the settlement-time curve (fig.7) reveals that settlements were continuing, albeit at a small rate, when the slab was cast. No account was taken of this when applying the above methods.

For the second load increment, from 8.6 to 25kN/m^2 maximum excess porewater pressures were again quite small. Laboratory undrained tests had shown that prior to yield (or undrained shear strength) the excess porewater pressure Δu is equal to $m \times \Delta\sigma_{oct}$ where $\Delta\sigma_{oct}$ is the increase in the mean total normal stress (or octahedral stress) and 'm' is a constant primarily dependent on overconsolidation ratio (Bell 1977). Similar 'elastic' porewater pressure generation seems to have occurred in the field. For example a maximum excess porewater pressure of 3.9kN/m^2 was recorded from piezometer no.2 at a depth of 1.73m (fig.5). The vertical and horizontal stresses, calculated from simple elastic theory, were 8.9kN/m^2 and 2.7kN/m^2 respectively for the load increment under consideration. Assuming axially-symmetric ground stress conditions $\Delta\sigma_{oct} = \frac{1}{3}(8.9 + 2.7 + 2.7)$kN/m^2 allowing the estimation of an 'm' value of 0.81. Laboratory estimates for 'm' at a depth of 1.73m range from 0.8 to 1.0, showing good agreement with the field value, even allowing for inaccuracies in stress increases. By the 400th day most of the initial excess porewater pressures had dissipated.

In view of the relatively small initial porewater-pressures and the estimates of ground stress which do not exceed P_c for the 25kN/m^2 loading (fig.8) settlements for this loading were calculated, as previously, using equations (1), (2) and (3). The estimates are listed on Table 1. The Skempton-Bjerrum method supplies ρ_t = 31 to 35mm and that of Burland et al supplies ρ_t = 33mm, and these considerably underpredict the observed total settlement of 67mm for this loading achieved by the 800th day. However they only slightly underpredict the settlement (≈ 36mms) at 400 days by which stage the excess porewater pressures had dissipated.

This has led to the belief that the discrepancy between the predicted and observed settlement is mainly due to yielding of the clay microstructure. After the dissipation of the excess porewater pressures fig.8 suggests that the effective stresses in the ground come close to the P_c values in the upper part of the deposit. However it is possible that the effective stresses were sufficient to cause slow yielding of the soil microstructure i.e., to exceed the P_c values. This is because the laboratory studies had revealed the time dependent nature of P_c, and because the estimates of stress increase are known to be somewhat inaccurate, and the P_c levels shown on fig.8 are mid-values of a 'band' of results. Therefore the later settlements for the $25kN/m^2$ loading phase are considered to be primarily the result of secondary compression or 'creep' of the microstructure since the effective stresses remained essentially constant during this period. Attempts to predict the total settlement at the 800th day by including a secondary compression component calculated from laboratory tests have so far proved unsatisfactory and have resulted in overestimates of final settlement.

Rate of Settlement

Fig.7 indicates an unusual shape for the time settlement graph for the $25kN/m^2$ loading. Creep is considered to have been important in the latter stages of this loading phase and undoubtedly accounts in some measure for the shape. However seasonal groundwater table variations are also thought to have played an important part in the rate at which settlement took place. Groundwater levels at Kinnegar during winter rise to within 0.5m or less of the ground surface and remain quite high throughout spring and early summer, almost certainly due to the high tides which are known to influence water tables in the Kinnegar area. Such high groundwater levels cause a small increase in porewater pressures and thereby decrease the effective stresses. This reduces the total load carried by the soil microstructure and partially 'arrests' creep movements. The reduction in settlement rate from 400-550 days (fig.7) is believed to have been caused by this process. As the groundwater tables begin to fall in the summer effective stresses once again increase causing an acceleration of creep, and indeed, of primary settlement. This is seen between 550 and 650 days.

The settlement of the DoE embankments at Kinnegar (fig.9) provide useful additional information. Differing settlement rates with accelerations often occurring in summer or late summer are seen. Although other factors are clearly involved these also can largely be explained by fluctuations in groundwater level. It is worth noting that settlements for the embankments, which apply pressures ranging from 20 to over $60kN/m^2$

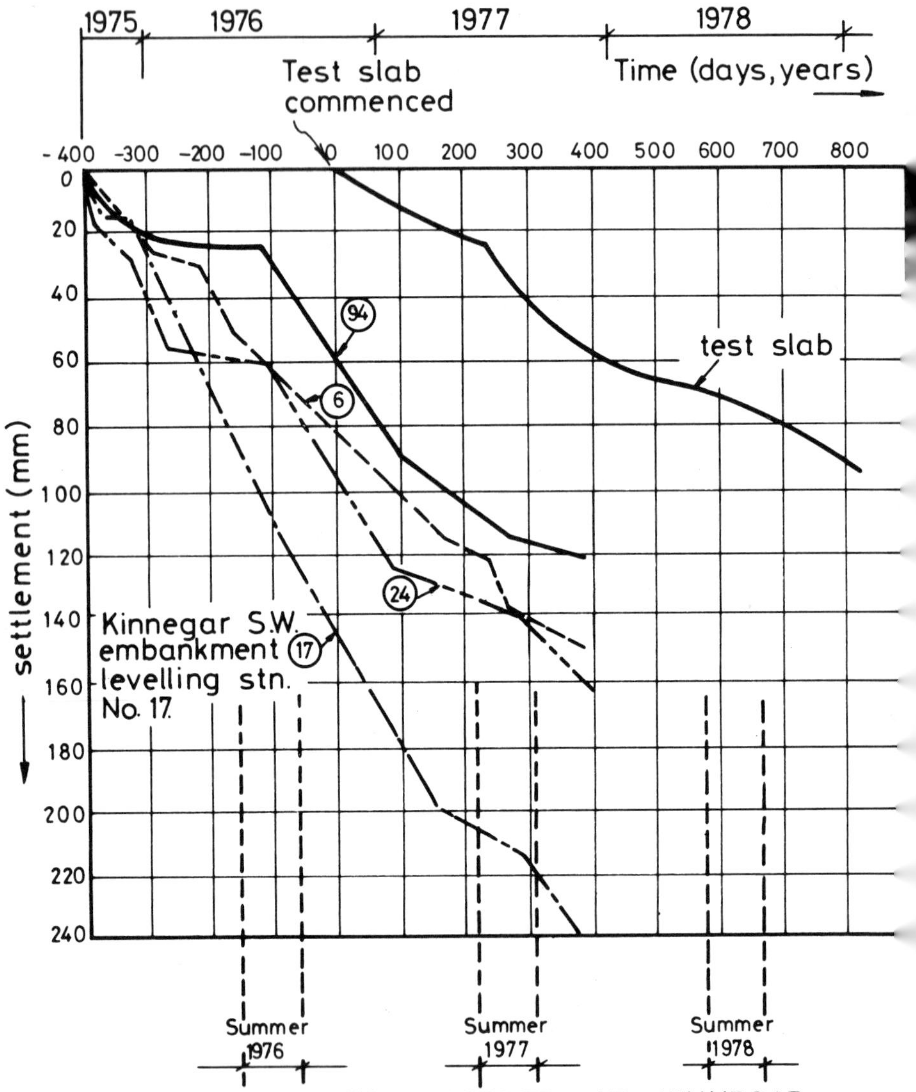

Fig. 9 SETTLEMENTS RECORDED AT KINNEGAR EMBANKMENTS (INDICATING SEASONAL TRENDS IN SETTLEMENT RATES).

have been large and are presently continuing some 3½ years after construction.

Although attempts to predict the rate of creep settlement have to date, like the predictions of magnitude, proven unsatisfactory, light has been shed on some of the problems associated with rates of primary consolidation settlement. Following the work of Lewis et al (1975) it was decided to measure field coefficients of permeability (K) and to use these along with laboratory values of coefficient of compressibility (M_v) to estimate 'field' coefficients of consolidation (C_v). It was found that with two exceptions the laboratory and 'field' C_v values were very similar, ranging from 1.7 to 4.5 m^2/year in the main sedimentary unit, providing that the laboratory M_v values used to calculate 'field' C_v were chosen at the same over-consolidation ratio which applied for the in-situ soil. Piezometer nos.4 and 12 both at 4.77m depth each provided exceptionally low 'K' values. As these correspond very well with backfigured estimates of permeability K from laboratory oedometer tests on disturbed Belfast soft clay, it is thought that installation of piezometers nos.4 and 12 may have caused disturbance of the in-situ soil surrounding the piezometer tips. It therefore appears that at the Kinnegar site good quality laboratory oedometer tests predict C_v values which are representative of the field soil. This is not surprising since an examination of the piston and block samples suggested no evidence of a field drainage structure which would fail to be revealed by oedometer testing.

Encouraging results have so far been obtained using simple 3-D consilidation theory to predict rates of primary consolidation settlement. Work is continuing to try to account for secondary compression and for the variation in C_v with depth.

CONCLUSIONS

Laboratory testing using high quality piston and block samples of the estuarine clays at Kinnegar strongly pointed to an initial small strain 'elastic' response to drained and undrained loading. Upon further loading 'yield' of the soil takes place and large deforamtions occur, some of which are due to creep. The test slab settlement tends to confirm this view of the geotechnical behaviour of the estuarine clays.

For loads causing ground stresses to remain below preconsolidation pressure levels initial excess porewater pressures were small and dissipated rapidly. Immediate or undrained settlements were predicted quite well using values of undrained modulus obtained from laboratory consolidated undrained triaxial tests. Prior to yielding, total settlements could also be predicted using the Skempton-Bjerrum method or that of Burland et al. Slow yielding is thought to have occurred when the ground stresses became close to the preconsolidation pressures, and this was associated with secondary compression at essentially constant effective stress. The magnitude of the secondary settlements were of considerable practical importance and, indeed, evidence of similar behaviour was provided by embankment settlement at

Kinnegar. Research is presently continuing with the aim of formulating accurate methods for predicting the rate and magnitude of such creep settlements. Attempts to predict primary consolidation rates using laboratory oedometer C_v values have been very encouraging. The laboratory C_v values are very similar to 'field' C_v values backfigured from field measurements of coefficient of permeability and laboratory measurements of coefficient of compressibility. Seasonal variations in groundwater level are also believed to influence settlement rates.

REFERENCES

Bell, A.L.(1977), 'A Geotechnical Investigation of Post-Glacial Estuarine Deposits at Kinnegar, Belfast Lough', Ph.D. Thesis, Queen's University of Belfast, January (1977).

Bjerrum, L.(1973), 'Problems of Soil Mechanics and Construction on Soft Clays', State of the Art Report, Session IV Proc. 8th Int.Conf on Soil Mechanics and Foundation Engineering, Moscow, (1973).

Burland, J.B., Broms, B.B., and DeMello, V.F.B.(1977), 'Behaviour of Foundations and Structures' State of the Art Report, Proc. 9th Int. Conf. on Soil Mechanics and Foundation Engng, Tokyo, (1977) 495-546.

Crooks, (1973), 'Laboratory Studies of the Belfast Estuarine Deposits', Ph.D. thesis, Queen's University of Belfast, May (1973).

Janbu, N., Bjerrum, L., & Kjaernsli, B.(1956), 'Norwegian Geotechnical Institute, Publication No.16, 93 pp.

Lewis, W.A., Murray, R.T., & Symons, I.F.(1975), 'Settlement and Stability of Embankments Constructed on Soft Alluvial Soils', Proc. Institution of Civil Engineers, Part 2, Vol.59 Dec. (1975) 571-595.

Mitchell, R.J.(1970), 'On the Yielding and Mechanical Strength of Leda Clays', Canadian Geotechnical Journal, 7 197-213.

Rowe, P.W.(1972), 'The Behaviour of Soil Fabric to Site Investigation Practice', 12th Rankine Lecture, Geotechnique, 22, 195-300.

Skempton, A.W. & Bjerrum, L.(1957), 'A Contribution to the Settlement Analysis of Foundations on Clay', Geotechnique 22 513-520.

Whitman, R.V. & Richart, F.E.(1967), 'Design Procedures for Dynamically Loaded Foundations' Journal of Soil Mechs. and Found. Engng. ASCE 93 SM6 169-193.

THE COLLAPSE OF SLABS ON A COMPACTED HOGGIN FILL

C.R.I. Clayton, B Sc (Hons), M Sc , D I C , Ph D , C Eng , M I C E , M I H E , F G S
Lecturer in Geotechnical Engineering, University of Surrey

N.E. Simons, M Sc (Eng), M A , Ph D , C Eng , F I C E , F G S
Professor of Geotechnical Engineering, University of Surrey

INTRODUCTION

Most engineers will recognise the importance of compaction when placing fill materials, but perhaps rather fewer will readily appreciate the very high standards required of fill to be placed either adjacent to, or beneath, structural members. Typically, vertical movements of the order of 150mm might be acceptable at the surface of a mass fill such as a motorway embankment (Snedker, 1973; Farrar, 1971, 1978; Cox, 1970, 1978), while beneath or adjacent to a structure the maximum tolerable vertical displacement may only be of the order of 5-10mm (Clayton, 1979).

This paper presents a case record of a compacted hoggin fill placed beneath a concrete slab. The fill was placed during an unusually dry Autumn period, and underwent a relatively small collapse upon becoming wetted up in the winter rains.

THE COLLAPSE MECHANISM

The collapse of soils when they become inundated has been recognised for many years, and evidence relating to different soil types can be found in Holtz (1948), Jennings and Knight (1957), Bjerrum (1960), Burland and Jennings (1962), Knight and Dehlen (1963), Baumann (1964), Cox (1970, 1978), Barden (1972) and Clayton (1978).

The types of material suffering the collapse phenomenon are numerous. They include insitu weathered dune sand deposits, and a wide variety of compacted materials including clayey sand, cohesive glacial till, Keuper Marl, Chalk, and rockfill.

All of these materials have one common feature, namely they become significantly weaker if exposed to water.

In fill, the collapse mechanism involves the compaction of a material containing particles or aggregations of relatively high strength. The material must be dry of optimum moisture content, and either silty or with a cohesive content. When loose tipped the contact points between aggregations are few, and the inter-aggregation forces are small.

Compaction improves the fill by increasing the number of contact points, and also by enlarging them by causing local failure. However, this improvement occurs only below the stress level imposed by compaction, and provided that the strength of the aggregations is not subsequently reduced. Fill which will undergo collapse typically appears sound after compaction, because it has high undrained shear strength: ironically, it is this high strength which prevents the fill from being brought to a low air void content by compaction plant.

Because of the high undrained shear strength of aggregations, a compacted fill of this type can possess a high air content and also a relatively low compressibility. In addition a high air content means that compression occurs rapidly, as loading is applied. In the field this type of settlement is rarely noticeable because pavements, surfacing or structures will be placed after compression is complete. However, if water subsequently gains access to the fill relatively larger settlements may occur without any increase in vertical stresses.

Water allowed to come into contact with collapsible fill penetrates rapidly into its midst as a result of its high permeability. Soil or rock fills which will collapse consist of material which will weaken upon contact with water. In soils containing a significant clay content this weakening is associated with swelling of the cohesive material into the air voids, and the volumetric increase and negative pore pressure dissipation can be measured. Figure 1 gives results of such measurements made by Cox (1978) on Keuper Marl.

The swelling of the soil aggregations as negative pore water pressures are dissipated leads to loss of contact strength, and local failure at the contact points leads to rapid collapse and net volumetric decrease as the air content is compressed. As Burland and Jennings (1962) note, this behaviour .

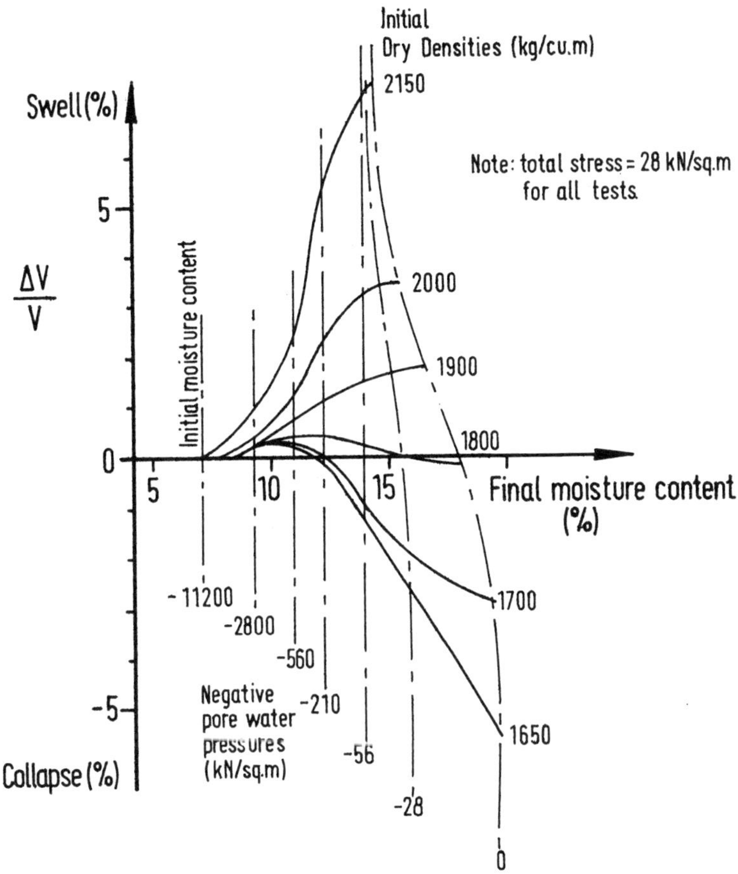

FIGURE 1 Volumetric change and negative pore pressure dissipation in specimens of Keuper Marl during progressive wetting (from Cox(1978))

is not obviously compatible with the principle of effective stress, as effective stress decrease in a collapse mechanism leads to volume reduction.

It is clear from the nature of the collapse mechanism, that this type of problem is avoidable. Firstly, fill types can be selected that do not exhibit loss of strength upon wetting up. Secondly, the fill material can be compacted at such a moisture content as is compatible with the compaction plant and will therefore easily allow the expulsion of air. When this approach is taken, the range of permissible moisture contents will be very small during compaction. As Holtz (1948) demonstrated, the use of a material in too wet a condition results in a high ratio of pore pressure to applied total stress (Skempton's 'B' value - Skempton, 1954) which gives the appearance of sponginess beneath the roller during compaction. Figure 2 shows a simplified version of Holtz's results obtained from a laboratory test programme on a sand with an excess of clay fines, which indicates that based on no collapse upon flooding (shown as $\Delta H/H_o = 0\%$) and a maximum B value of 0.30, the placement moisture content for this material can lie only in the range 12% - 15%.

These close tolerances are not normally required. Many heavy clay materials will undergo both significant overall swelling (i.e. net volumetric increase) and collapse at the same moisture content and dry density, depending on the imposed total stress level. This means that in a fill of any significant height the upper layers may be swelling, while the lower layers collapse. Perhaps for this reason Cox (1978) comments:

"Volume changes are not normally a problem within dry clay road embankments. However, differential movements just below the subbase are a common cause of pavement failure".

The tolerance of the structure to both total and differential settlement is crucial in setting performance limits for the fill, but it is also necessary to recognise that where total settlements are important, some collapsing materials, which contain sufficient "cohesive" fines to act as a binder but insufficient to give net volumetric increases even at low imposed total stress levels, will give greater problems than others. Thus during the flooding of materials such as Holtz's clayey sand, or of rock or chalkfill, all movement is compressive.

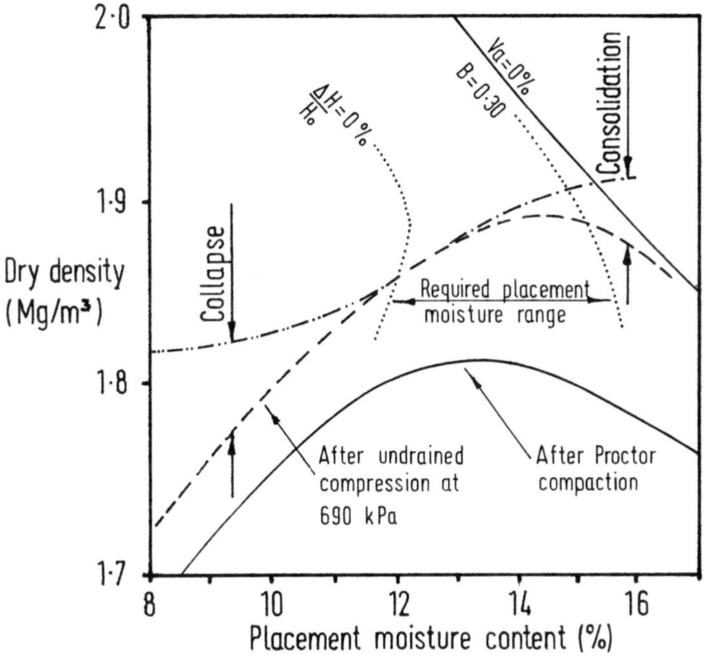

FIGURE 2 Placement Moisture Control (after Holtz [1948])

CASE STUDY

The case record relates to a series of concrete apron slabs constructed abutting a major structure in West London. The structure, rising some 24m above ground level, is supported partially on piles and partially on spread footings between 5m and 6m below ground level, and the concrete apron slabs therefore overly some 5-6m of compacted fill.

Ground Conditions

Finished ground level and original ground level in the area are approximately the same, and the results of shell and auger borings on the site give the generalised succession as follows:-

0-13m	Brown or grey-brown fissured silty CLAY (Weathered London Clay)
13-18m	Mottled multicoloured silty sandy CLAYS (Woolwich and Reading Beds)
18-? m	Silty SAND or weakly cemented SANDSTONE (Woolwich and Reading Beds)

Groundwater was encountered during boring at 12-15m below ground level.

Laboratory tests carried out on samples of the subsoil indicate a 38mm dia. undrained triaxial shear strength of about $35kN/m^2$ at ground level, increasing to more than $200kN/m^2$ at 10m below ground level. At depths of more than 3m below ground level all the measured shear strengths exceeded $100kN/m^2$. Oedometer tests indicated the London Clay to be of low to medium compressibility.

Backfill

The material excavated from the site was not reused as backfill adjacent to the structure, beneath the apron slabs. Instead, the areas were refilled with a selected backfill, which consisted of a naturally occurring and locally available hoggin. Sieve analyses showed this material to be variable, as might be expected, but in general it consists of a well-graded sand and gravel, with 6-12% clay content (Figure 3).

The material was compacted to a method specification based on the results of work carried out at the Road Research Laboratory (Lewis, 1961; Tombs, 1966, (a) to (f)). In areas of fill adjacent to walls and services, compaction was achieved with the use of a BS50k 55kgf vibrotamper, by

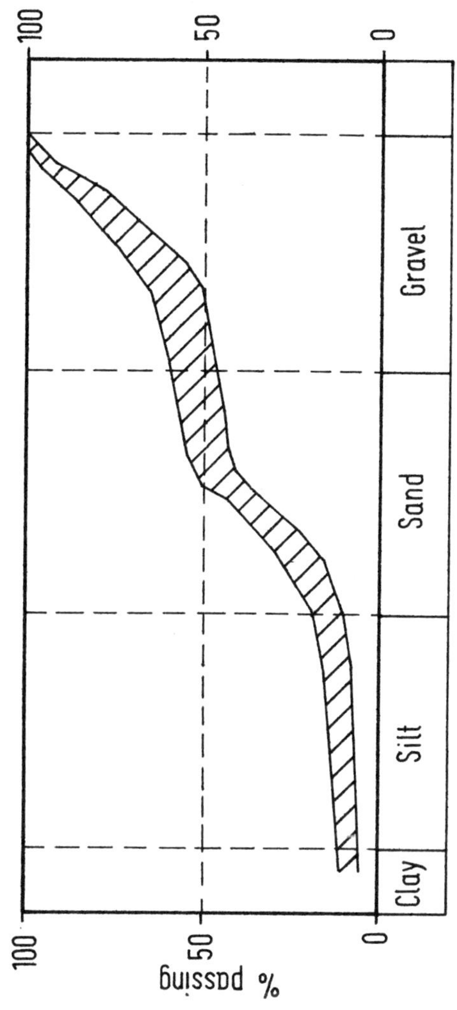

FIGURE 3 Envelope of particle size distributions for hoggin

applying 3 passes to each 100mm layer of fill. In areas of less limited space Bomag BW35S, BW60 and BW90 vibrating rollers were used. The most effective compaction was visually assessed as being provided by vibrating rollers, but these could not be used adjacent to structures. After compaction the hoggin appeared to have produced an excellent fill, with a high bearing capacity.

Backfilling in the area subsequently to be covered by, and to provide support for, the apron slabs was carried out in unusual weather conditions. Site records indicate

Commencement of filling	22nd August, 1978
Completion of filling	23rd November, 1978

with the majority of backfill being placed in October and November. Rainfall measured on site was 2.0 and 9.9mm during these two months, while Meteorological Office Records indicate an average for October of 60mm and November of 65mm. Rainfall between the 3rd and 10th of December was 58.5mm, with 36.5mm of this falling during the weekend of the 9th/10th.

Because the hoggin was compacted to a method specification there are very few records of as-placed moisture content and dry density. A few insitu density measurements made in the area give average values of

moisture content	7%
dry density	1.99 Mg/m^3
air void content	12%

but these figures are not thought typical of the mass of the fill, which often arrived on site during October with a moisture content of about 5%. Assuming that this material was subjected to the same compactive effort, lower densities and higher air void contents would result.

Following the heavy rain during the 9th/10th December, the settlements of apron slabs and fill were first noticed on 11th December. At this stage the general settlements of the apron slabs were of the order of 40mm, while adjacent to walls and services, where compaction had been restricted, settlements were more severe and up to 200mm. By the 21st December settlement appeared substantially complete, although some slight further settlement was noted on January 3rd, after the Christmas site holiday.

Laboratory Study

Laboratory testing of this hoggin requires rather large samples, since 20-30% of the material exceeds 20mm size, and 6-12% exceeds 37mm. In particular, moisture content samples which might have a dry weight of only 50g in a fine grained soil, must have a minimum weight of about 4kg. If smaller samples are used, then measured moisture contents will be much higher than the average moisture content, since the bulk of the moisture is held in the fine fraction. This is a further indication that if site moisture contents are in error they are likely to be an overestimate of the actual placement moisture content.

A series of tests was carried out to classify the fill. The following values were determined:-

Particle Size Distribution	
Gravel	40-55%
Sand	35-45%
Silt	3- 6%
Clay content	6-12%
Coefficient of uniformity	60-200
Specific gravity (av.)	2.67
Compaction to BS1377:1975 test 14 (vibrating hammer method)	
Max. dry density	2.10-2.16Mg/m^3 (Av.2.13Mg/m^3)
Opt. moisture content	8%

Compaction tests to BS1377:1975 test 14 require the exclusion of material in excess of 37mm diameter, which in this case represents about 5-6% by weight, on average. "Soil Mechanics for Road Engineers" (R.R.L., 1952) indicates that this is not likely to give unrepresentative results. For this reason, it was decided to base further tests on the same samples, with material exceeding 37mm excluded.

A series of compression tests was carried out on the hoggin, using 250mm diameter hydraulic compression cells. The maximum sample height obtainable is only 125mm, which is rather too small, considering that the maximum particle size of the fill was found to be in the range 50-60mm, but some compromise is necessary to allow complete flooding of the less permeable specimens. Soil was laid out to air dry to the required moisture content, and was then compacted to give a specified dry density. Each sample was compacted directly into the hydraulic compression cell, with sintered bronze discs being placed top and bottom of the specimen: the specimen was then subjected to a total stress of 75kN/m^2, and compression as a function of time was noted. This stage is equivalent to the imposition of self weight of the fill, in the lower half of the material. Typically a height decrease of about 0.5% was recorded, although this was lower at lower moisture contents.

Collapse was initiated by flooding the sample from the base of the cell, using pressures of up to 40kN/m^2. In the field it is unlikely that the maximum water pressure would exceed one half of the vertical total stress. The observed collapse was very variable, with height changes of between zero and 5% being recorded.

Figures 4 and 5 show the results plotted in the form of collapse contours on a moisture content dry density plot, and of collapse as a function of air void content. Because the clay content of the hoggin is not large enough to give net volumetric increase during swelling, even at low total stress levels, the curves are much simpler than those found by Cox (1970) for compacted Keuper Marl. The collapse contours on the moisture content v. density plot appear parallel to the zero air voids line (Figure 4), and the scatter of results of collapse as a function of percentage air voids content is therefore small (Figure 5). This is very similar to the result obtained by Clayton (1978) for compacted Chalk, except that in the case of compacted Hoggin, there does seem to be reason to believe that provided compaction can reduce the air void content to less than about 10%, collapse movements will be relatively small. The relationship between collapse and air voids for compacted Chalk is superimposed on Figure 5, and it can be seen that quite significant volume changes can occur at very low air void contents.

Post Collapse Behaviour

Apart from the obvious problems associated with the vertical movements occurring at the surface of a collapsing fill, it is important to note the other effects of flooding a collapsing fill. The mechanism involves swelling, and therefore it is associated with an increase in compressibility and a decrease of shear strength. With moisture contents of 5-7% and air voids contents of 10%-15% the CBR value of the hoggin is about 50-80%, but after flooding this value drops to about 15%. Normally, this decrease in strength, although very large, will not give rise to problems provided the material is not placed with steep side slopes. However, most fills will be constructed to take live loads. Figure 6 shows the relative size of initial and post collapse compression, which for this case are seen to be in the ratio of about 1:10. Pre-collapse compression is related to both moisture content and density, with high moisture content and low density fills giving the greatest compressibility: post collapse compression is related to the amount of collapse, since this dictates the change in moisture content that can occur, as the following figures indicate.

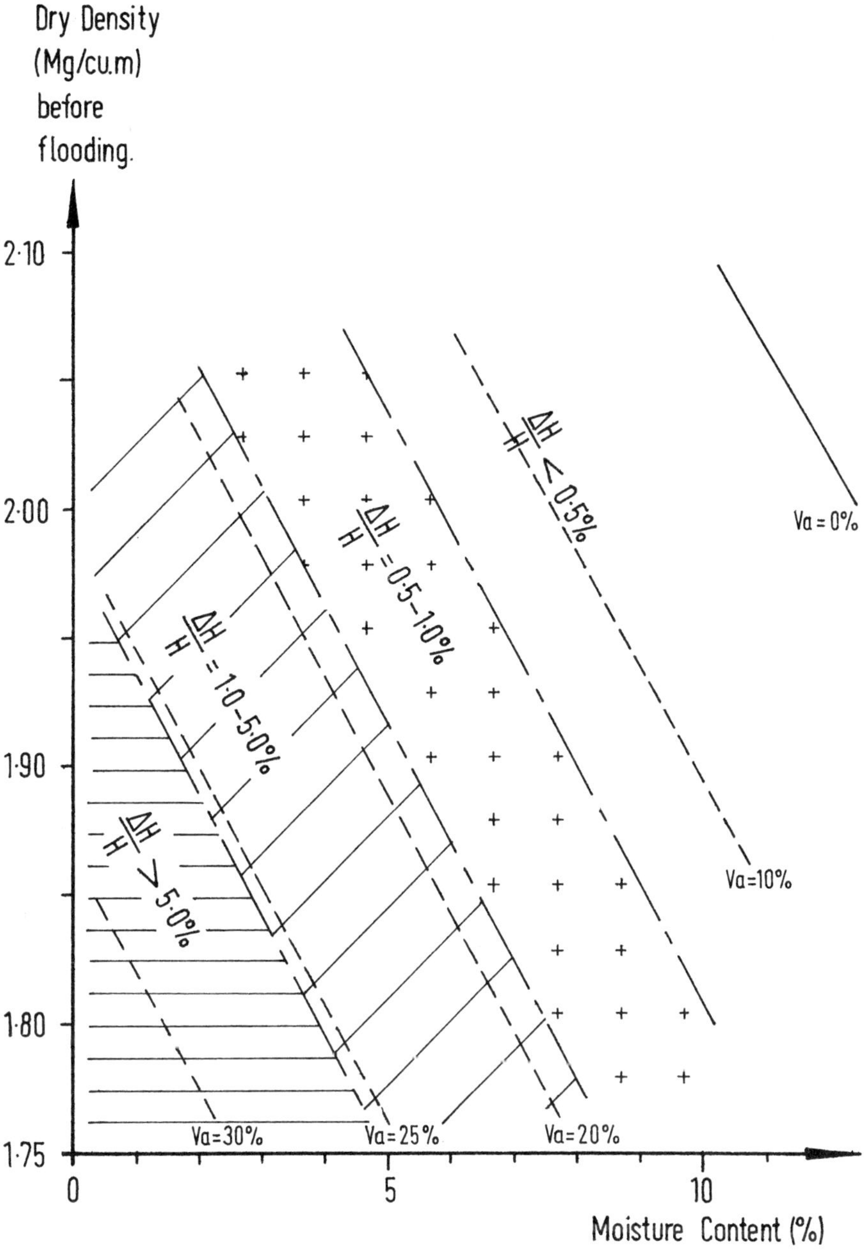

FIGURE 4 Collapse upon flooding of compacted hoggin fill

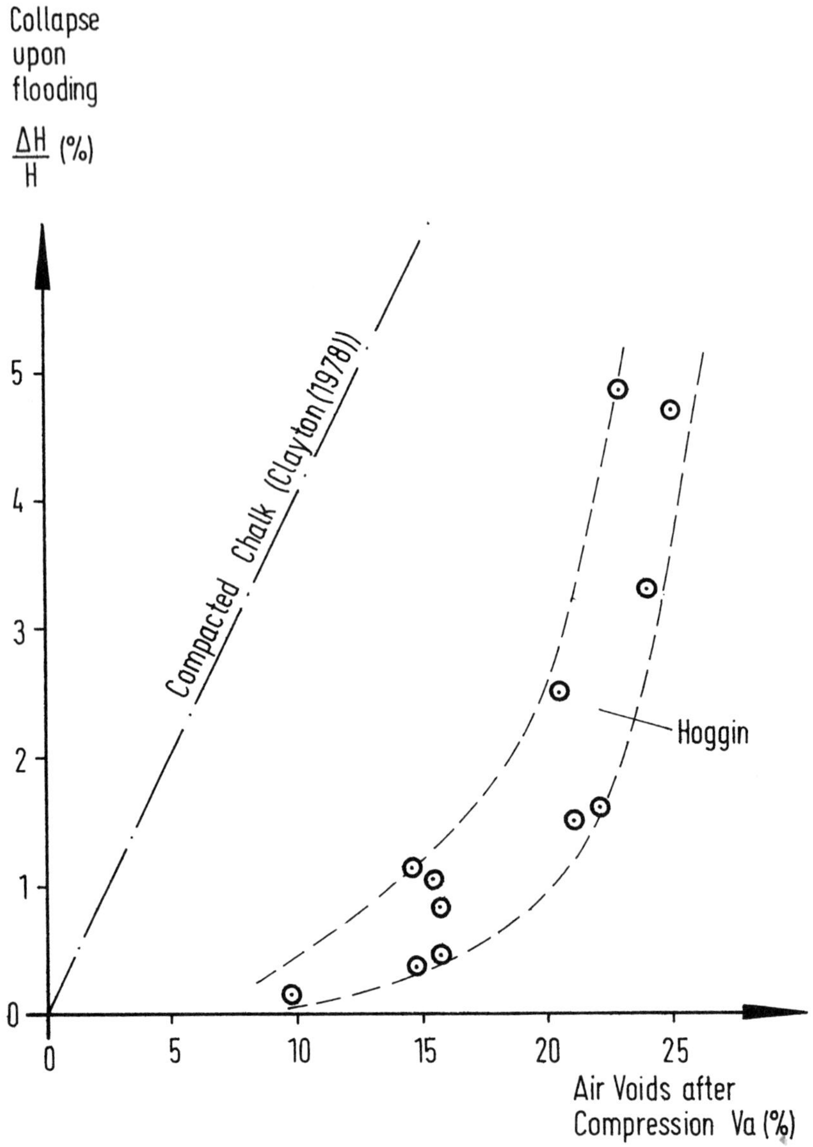

FIGURE 5 Collapse of hoggin as a function of air void content

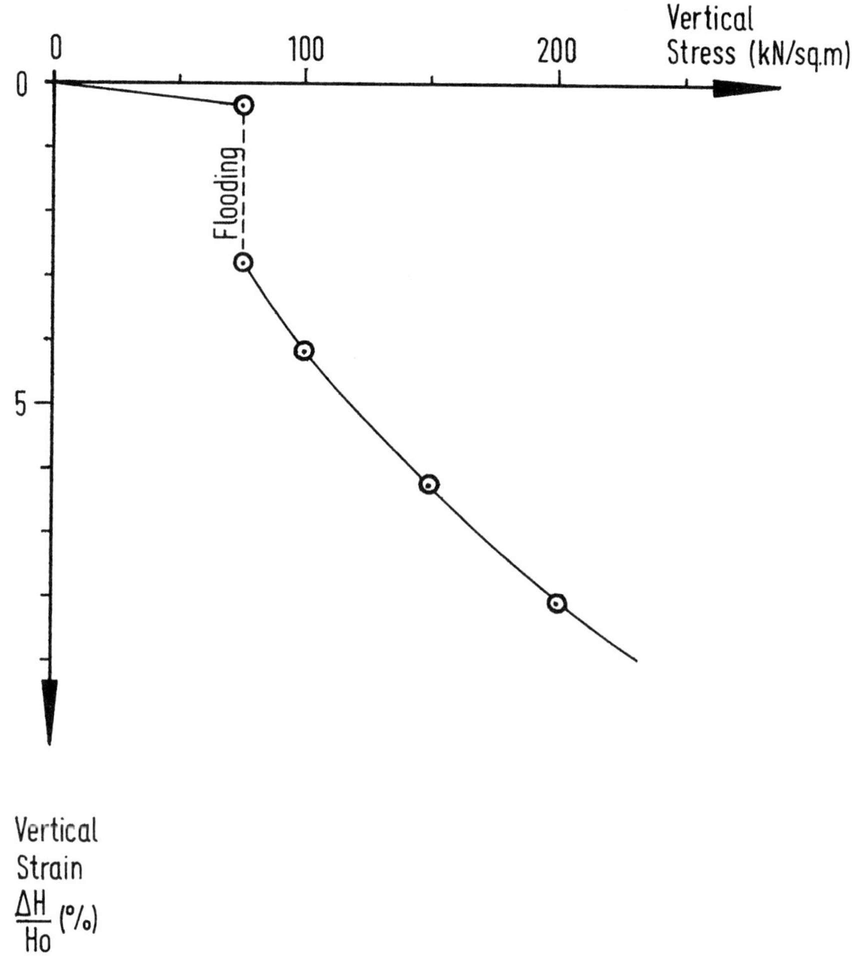

FIGURE 6 Initial and post-collapse compression of hoggin

Initial m/c %	Va %	Collapse upon Flooding (σ_v=75kN/m^2) ΔH/H (%)	Post Collapse Compression 75 - 100kN/m^2 ΔH/H (%)
5	18	0.57	0.16
7	20	2.47	1.38
6	25	4.85	1.65

CONCLUSIONS

Although a wide variety of compacted soils may collapse under constant vertical stress when wetted up, granular materials with a small clay content represent a particularly troublesome group. When these types of materials are to be used adjacent to structures, where self-settlements must necessarily be restricted to very low levels, it is essential that the effects of compaction are monitored closely to ensure that the air content of the fill is being reduced to an acceptably low level.

The use of method specifications for compaction will not guarantee that a satisfactory fill will be achieved, unless extensive field trials and laboratory tests are first carried out on the soil to be used for construction.

REFERENCES

P.F. Baumann - Limit to height criteria for loose-dumped rock-fill dams, Proceedings of the 8th International Congress on Large Dams, Edinburgh (1964), Vol. III, p. 599-613.

L. Barden - Examples of clay structure and its influence on engineering behaviour, stress strain behaviour of soils, Proceedings of the Roscoe Memorial Symposium, Cambridge, (1972), 195-205.

L. Bjerrum - Geotekniske Egenskaper Ved Komprimert jord, Norwegian Geotechnical Institute (1960), reproduced from Den Norske Ingeniorforenings Kurs : Bygging av Fyllings-dammer, 23-26/3/1960.

British Standards Institution - Methods of test soil for Civil Engineering purposes, BS1377, (1975).

J.B. Burland and J.E.B. Jennings - Limitations on the use of effective stresses in partly saturated soils, Geotechnique 12 (1962) 125-144.

C.R.I. Clayton - Chalk as fill, Ph.D. thesis, University of Surrey (1978).

C.R.I. Clayton - Soils, chapter 3 of the House Builders Reference Book, ed. M.J.V. Powell, Newnes-Butterworths, London (1979).

D.W. Cox - Settlement within embankments, Ph.D. thesis, University of Birmingham (1970).

D.W. Cox - Volume change of compacted clay fill, I.C.E. Conference on Clay Fills (1978), 79-86.

D.M. Farrar - A laboratory study of the use of wet fill in embankments, Road Research Laboratory Report 406, (1971).

D.M. Farrar - Settlement and pore-water pressure within an embankment built of London Clay, I.C.E. Conference on Clay Fills (1978), 101-106.

W.G. Holtz - The determination of limits for the control of placement moisture in high rolled earth dams, Proceedings A.S.T.M. (1948) 48, 1240-1248.

J.E.B. Jennings and K. Knight - The additional settlement of foundations due to a collapse of structure of sandy sub-soils on wetting, Proceedings of 4th International Conference on Soil Mechanics and Foundation Engineering (1957) 1, 316.

K. Knight and G.L. Dehlen - The failure of a road constructed on a collapsing soil, Proceedings of 3rd Regional Conference for Africa on Soil Mechanics and Foundation Engineering, Salisbury, (1963) 1, 31-34.

W.A. Lewis - Recent research into the compaction of soil by vibratory compaction equipment, Proceedings of the 5th International Conference on Soil Mechanics and Foundation Engineering, (1961), Vol. 2, p. 216-8.

A.W. Parsons - An investigation into the performance of a Winget Dynapac CG.11 $2\frac{1}{4}$ ton tandem vibratory roller in the compaction of soil, Road Research Laboratory Report 15, (1966).

Road Research Laboratory - Soil mechanics for road engineers, H.M.S.O., London, (1952).

A.W. Skempton - The pore-pressure coefficients A and B, Geotechnique (1954) 4, 143-147.

E.A. Snedker - The choice of an upper limit of moisture content for highway earthworks, Highways Design and Construction, January (1973).

A.F. Toombs - An investigation into the performance of a Winget Dynapac CO.10 1¼ cwt. vibro-tamper in the compaction of soil, Road Research Laboratory Report 10, (1966)(a).

A.F. Toombs - An investigation into the performance of a Winget Dynapac CH.44 4¾ ton towed vibratory roller in the compaction of soil, Road Research Laboratory Report 13, (1966)(b).

A.F. Toombs - The performance of a Whacker DVPN 75 3½ cwt. vibrating-plate compactor in the compaction of soil, Road Research Laboratory Report 26, (1966)(c).

A.F. Toombs - The performance of a Whacker BS50K 1 cwt. vibro-tamper in the compaction of soil, Road Research Laboratory Report 27, (1966)(d).

A.F. Toombs - The performance of a Whacker BS100 2 cwt. vibro-tamper in the compaction of soil, Road Research Laboratory Report 28, (1966)(e).

A.F. Toombs - The performance of a Delmag SV2 8¾ cwt. vibrating-plate compactor in the compaction of soil, Road Research Laboratory Report 53, (1966)(f).

THE DESIGN OF STIFFENED INTEGRAL SLABS

N.J. Withers, M A, M I E Aust, A M I C E

Engineer - Coffey & Partners Pty.Ltd., Brisbane, Australia *

SUMMARY

This paper describes a rational method for designing stiffened integral slabs for expansive soils, based on an analysis of the soil-structure interaction. Models of soil and slab behaviour are described before outlining the assumptions and steps in the design process. A worked example is used to demonstrate the procedure and the results of a successful back analysis of the reported behaviour of two test slabs are presented.

The method increases the understanding of slab behaviour and by allowing a more discerning use of engineering judgment represents an improvement on the semi-rational and empirical methods currently in common use.

Another advantage is that the degree of refinement and sophistication which can be applied to improve the analysis of the soil-structure interaction is limited only by the funds available for the site investigation, laboratory testing and design processes.

The method lends itself to the utilisation of suites of standard designs while still permitting the economic design of "one-off" foundation slabs.

INTRODUCTION

In countries whose regions experience a wide range of climatic conditions, large seasonal soil deformations can occur. As differing soils react to widely varying degrees, a problem exists for the foundation engineer in designing economical but

---

* Assisted by P.T. Stocker, B Sc App , PhD

Senior Engineer - Coffey & Partners Pty.Ltd., Brisbane, Australia adequate foundations. The problem is most acute for the most expansive soils and for houses or similar scale structures for which the funds available are strictly limited.

A common approach has been to develop suites of standard designs each member of which is intended to cater for a specified range of soil conditions. This approach has certain drawbacks. Common design practice for selecting a particular slab design characterises the soil using the insitu moisture content, Atterberg indices and the depths and thicknesses of the layers of expansive soil. The use of index tests at best produces only a qualitative measurement of soil properties, and requires the application of empirical judgment to categorise the soil for design purposes. This leads to widely varying boundaries between categories of the standard design suites and to quite different designs for similar ranges of soil conditions. It is now considered more realistic to characterise soils by directly measuring responses to changes in applied load and suction.

Several semi-rational design methods have been published (9), the most realistic of which is reputed to be by Walsh. A general feature of such methods is that, based on the soil characteristics appropriately expressed, a deformed shape is adopted for the soil surface beneath the covered area. Foundation loads are assumed to modify this shape. The deformed shape and an assigned modulus of subgrade reaction are used to obtain a "support index" for the slab to calculate the slab deflections, bending moments and shear stresses. Although the correlation between deformed shapes and suport indices have been established using computer modelling of the soil-structure interaction, the accuracy of the design process in assessing the actual interaction between the real slab and real soil, responding to the expected variations in environmental conditions, depends critically on the accuracy of an initial assumed edge distance 'e' which defines the slab support conditions. A realistic assessment of this interaction is necessary to avoid overly conservative designs. Guidelines for selecting 'e' are only of a general nature and the computation process provides no feedback to confirm the correctness of the value of 'e' selected.

In the following sections models of soil and slab behaviour are described which are combined to form a method of design explicitly incorporating their interaction.

The straight forward nature of the design process is illustrated by a worked example and the increased understanding resulting from its use is demonstrated by a back analysis of the performance of two test slabs. In appendices to the paper details are provided of: a laboratory testing procedure; the calculations used to derive soil stress-deformation characteristics; presumptive soil suction profiles; the approach used to analyse the slab and to derive ultimate moment capacities and effective moments of inertia; a suite of standard designs; and finally indicative allowable deflections and house loads.

MODEL OF SOIL BEHAVIOUR

The design of stiffened integral slabs involves an assessment of the range of deflections under the applied structural loads and the various environmental conditions likely to arise during the life of the building. Thus the long term deformation response of the foundation soils must be determined.

The three main causes of volumetric change in an element of clayey soil are the applied stresses and the matrix and solute suction stresses. The applied stresses consist of the weight of overlying soil and any externally applied loads. Matrix suction is the pressure applied to the soil skeleton arising from negative pore water pressure. It is sensable by pressure gauges. Solute suction is caused by the chemical solutes in the water and is equal to the pressure difference that would exist between a fluid of the same composition as the pore water and pure water separated by a membrane permeable only to water molecules. The matrix and solute suction stresses in combination form the total suction stresses acting on the soil.[1]

A detailed consideration of the nature and effects of soil suction is beyond the scope of this paper. The South Australian Division of The Institution of Engineers Australia has published an excellent introduction to the subject, also including a recent review of aspects of foundation design for expansive soils (9).

Early researchers (1), (5) into the stress-deformation response of expansive soils attempted to use the concept of a unified effective stress based on the equation developed for saturated soils. Specifically, the relationship

$$\sigma' = (\sigma - u_a) + \chi\,(u_a - u_w)$$

was introduced where

σ' is the net effective stress on the soil skeleton

σ is the total stress on the soil mass

u_w and u_a the pore water and pore air pressure

and χ was called the suction parameter which ranged from 0 to 1.

Many experiments were made to measure the parameter χ and it was found that it depends on both the soil suction and soil type. Therefore, without extensive laboratory testing, it is difficult to define χ adequately for design calculations.

---

[1] In this context the suction component due to vertical elevation is not explicitly considered.

Later work (3) showed that the nature of soil responses to changes in applied and suction stresses were fundamentally different. Increases in applied stress are inherently anisotropic and tend to increase the continuity of the soil mass and eventually induce shear in preferential directions. Soil suction acts isotropically and tends to decrease the effective continuity of the soil by producing tension cracks.

Direct measurement with a membrane oedometer of the deformation of soils in response to changes in applied stress and both components of the total suction stresses have shown that the responses to changes in any one of these stresses are dependant on the ambient values of the other two. Figure 1 which has been adapted from Figures 4.10 to 4.12 of reference (9) shows the results of some tests on a pleistocene clay.

Such experimental work has shown that within a wide range of suction values, the deformation responses can be approximated satisfactorily by straight lines on semi-log plots. Outside this range the response becomes non-linear and much reduced. This linearity allows the definition of Instability Indices, I, with respect to suction changes which are complementary to the familiar Compression Index C_c describing response to changes in applied stress. Specifically, for changes in matrix suction, s_m, solute suction, s_s and applied stress p,

$$I_m = \frac{\Delta\varepsilon_m}{\Delta\log s_m} \quad \text{for given } p,\ s_s$$

$$I_s = \frac{\Delta\varepsilon_s}{\Delta\log s_s} \quad \text{for given } p,\ s_m$$

$$C_c = \frac{\Delta\varepsilon'_p}{\Delta\log p} \quad \text{for given } s_m,\ s_s$$

Thus, with sufficient laboratory testing the effective stress-deformation characteristics of a soil can be quantified within the limits of applied and suction stresses likely to obtain at a particular site. Fortunately the proportion of the deformation response which occurs outside the log linear region is generally negligible.

MEASUREMENT OF SOIL CHARACTERISTICS

The laboratory testing techniques are well recognised and are described elsewhere (3), (9). Nevertheless, the cost and time involved with this scale of testing renders its use prohibitive for individual house or small factory developments. A less rigorous approach is needed.

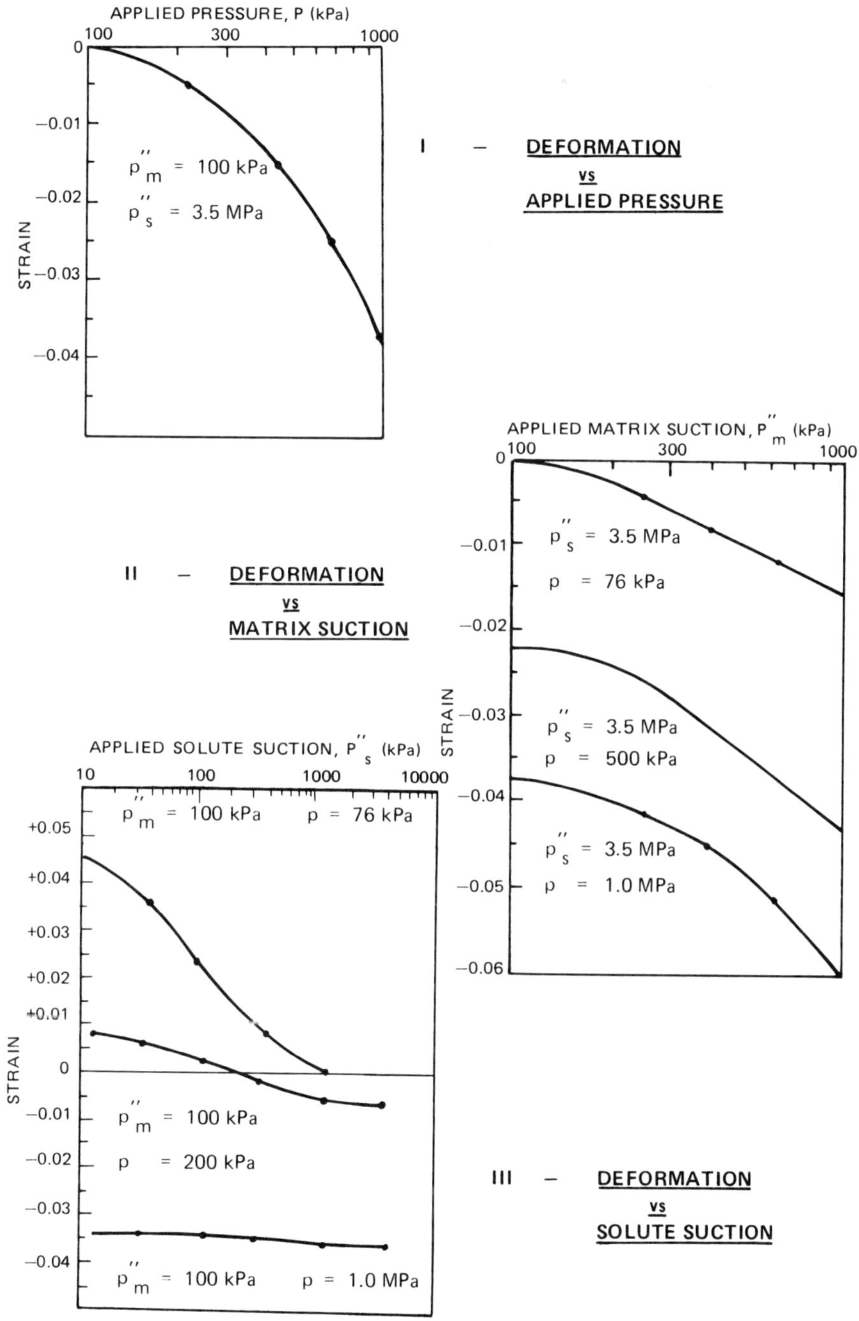

FIGURE 1 **RESULTS OF TESTS WITH A MEMBRANE OEDOMETER ON A PLEISTOCENE CLAY**

One approach is to use typical values of Instability and Compression Indices for soils of varying reactivity. The main deficiency of this approach is that it depends on a reliable categorisation of soils based on more readily determined characteristics. As indicated above, plasticity indices are not a reliable indicator. The C.S.I.R.O. of Australia has devised a Sorption Limit Test which would appear to provide a fairly reliable indication but it is unlikely that this method will find widespread commercial use as it takes three to four weeks to perform.

A more direct approach is to use laboratory measurements of deformations of soil samples in response to changes in suction and applied load. The response of the soil to changes in suction may represent a response to either matrix or solute components or both, for which the individual Instability Indices can differ widely. Laboratory tests can be designed to simulate changes in total suction and/or simulate leaching in the case of saline soils. By assigning suitable specific values of suction to extreme test conditions and assuming that between these the deformation response is log-linear, the suction deformation response can be quantified quite adequately.

Most commonly simple wetting and drying without leaching will be required. This testing directly measures the deformation response of the soil between the conditions of interest, namely the extreme dry and wet conditions. It should be noted that both matrix and solute suctions are altered. The majority of the suction change occurs in the matrix suction component with the solute suction change being small. This mirrors the situation at most sites where leaching does not occur: saturation reduces, and drying increases, solute concentration without changing total quantity. Perusal of the literature suggests that to obtain a good log-linear approximation values of 10 and 10^4 or 10^5 kPa can be assigned, in the absence of specific local data, to the extreme wet and dry conditions.

There will also be sites, a minority, in which volume changes due to leaching are of equal or prime interest. During leaching of saline soils total suction may change very little whilst solute suction is greatly reduced. Changes in solute suction at constant total suction will cause complementary changes in matrix suction to which the volume of the soil skeleton is sensitive. Changes in solute suction can in some soils also change the mechanical properties of the soil skeleton altering its responses to unit changes in matrix suction. It is therefore expected that the simple wetting and drying approach outlined will be satisfactory for the majority of sites, with more sophisticated testing being required for those where leaching is a significant problem.

If volume change characteristics can be assumed to be the

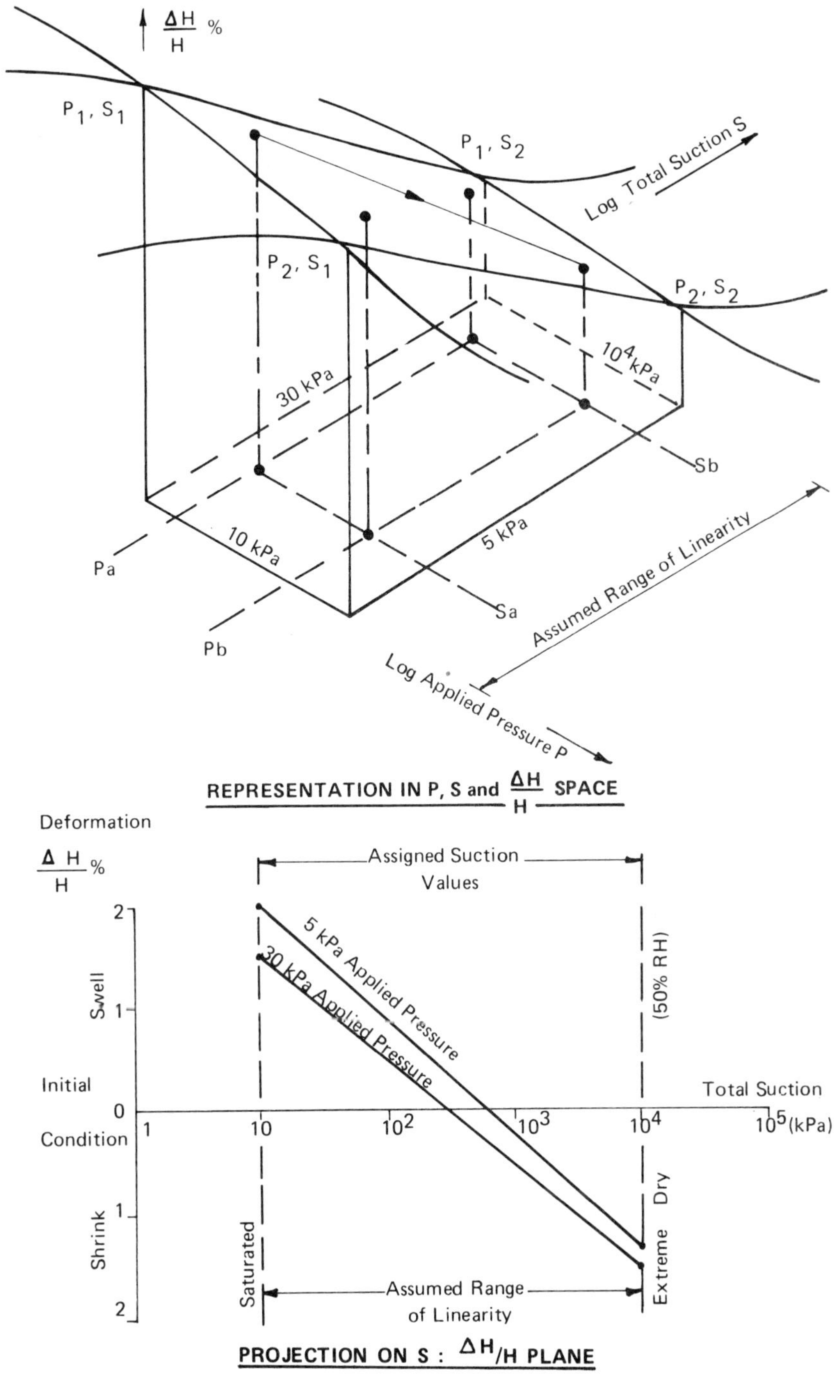

FIGURE 2 REPRESENTATION OF LABORATORY TEST RESULTS

same for both swelling and shrinking within the same limits of stress change, and if the stress-deformation response can be considered to have two interdependant components - applied stress and total suction, then the following stress-deformation characteristics can be calculated from the results of simple tests in which undisturbed soil samples are wetted and dried, from the as-sampled state under chosen external loads

$$I = \frac{\Delta\varepsilon}{\Delta\log s} \quad \text{- for p constant}$$

and Compression Index for applied stress change

$$C_c = \frac{\Delta\varepsilon}{\Delta\log p} \quad \text{- for s constant}$$

Where s is total suction and p is applied vertical pressure and ε is measured vertical strain.

It is possible to sufficiently characterise the soil by measuring the deformation of a sample under two applied stresses as the sample is taken from an intermediate to extreme dry and from intermediate to extreme wet conditions. Figure 2 shows graphically the form the results of such tests might take with applied pressures of 5 and 30 kPa. Of minor interest an indication of the initial suction state of the sample can be obtained directly from such a plot of the laboratory results. Appendix 1 provides details of the laboratory testing procedure. Appendix 2 shows the development of the following expression using such results linking the change in vertical strain of the soil with changes between two stress and soil suction states denoted p_a, s_a and p_b, s_b

$$\Delta\varepsilon = (A + B \log p_a)\ \Delta\log s + (C + B \log s_b)\ \Delta\log p$$

This expression can be written as

$$\Delta\varepsilon = M + N\ (\Delta\log p)$$

where M and N are the appropriate coefficients evaluated for each layer of soil.

It should be emphasised that if the assumptions outlined in the preceding paragraphs are found inappropriate for any reason then the stress-deformation characteristic of the soil can be determined by more comprehensive and sophisticated testing.

SUCTION CHANGES IN FOUNDATION SOILS

It is now necessary to consider the range of field soil suction profiles that are likely to exist. The soil suction at any point within a soil mass represents a state of dynamic equilibrium between the various processes which can cause change. These include the effects of fluctuations of the water table, applied stresses, air humidity, temperature and pressure, soil temperature, soil structure and composition, chemical composition of the pore water, drainage characteristics and seasonal variations in precipitation and evaporation.

If the surface of the soil is extensively covered then in the absence of any other changes the environmental controls become heavily damped and the two components of suction tend towards equilibrium values. In these circumstances the equilibrium values are determined mainly by the depth to the water table or in dry areas by the balance between precipitation and evaporation. Clearly at the edge of a covered area the soil remains more responsive to natural environmental fluctuations, wherein lies the problem to be solved.

In addition, such artificial influences as tree planting, subsurface drains or leaking pipes can cause pronounced changes in the suction profile. Consequently from either natural or man-made influences, the soil may become either extremely wetted or extremely dry on occasions. Figure 3 shows in indealised form some typical extreme and equilibrium suction profiles.

A rigorous analysis of soil movements beneath a foundation slab would involve the measurement of the insitu soil suction profile with additional observations and experiments over a long time span to determine the likely range of soil suction profiles under operational conditions. The techniques required are well established, but generally involve too much time and expense for individual house or similar scale foundation designs.

A less expensive approach would be to cater for changes in soil suction between estimated equilibrium and extreme conditions. This would allow the use of explicit "design" or assumed soil suction profiles and obviate the need for extensive soil testing for the majority of situations.

The choice of design soil suction profile depends on the engineer's assessment of the likely variations in conditions under the centre and edge of the building following construction. For example, if the pre-construction profile is judged to be near either the extreme dry or extreme wet conditions then the change between the two extreme conditions should be used. Otherwise the change from equilibrium to either extreme may be appropriate.

Appendix 3 presents charts of design suction profiles for sites in which the equilibrium suction is controlled primarily by the water table depth. Similar charts can be compiled for sites where the equilibrium suction is not water table controlled; for example where it depends primarily on the balance between local precipitation and potential evaporation.

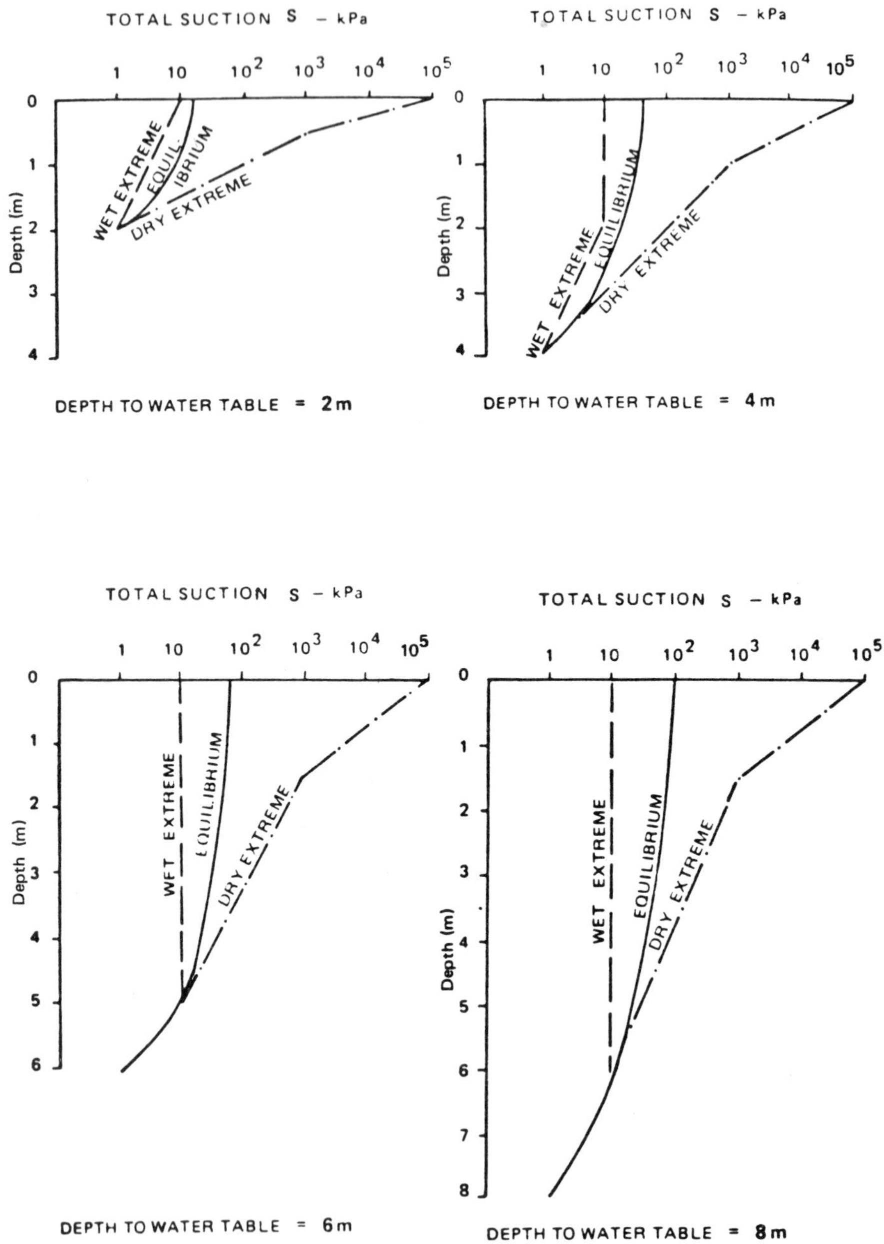

FIGURE 3

SOME TYPICAL EQUILIBRIUM & EXTREME TOTAL SOIL SUCTION PROFILES

STRUCTURAL ANALYSIS OF SLAB

An integral slab consists of a reinforced concrete slab provided with thickened beams around the perimeter and in some grid pattern internally. The beams and slab are continuous one with the other. This slab deforms in response to the building loads and pattern of soil reaction stresses acting on it.

Finite element or grillage analyses which can account for complete compatibility and include shear stresses at the soil-slab interface are usually prohibited by cost. Some simplifications are necessary for design purposes.

The slab is idealised as an assemblage of 'T' beams acting in a unified manner. As a design criterion, the deflection of a unit width central strip is calculated and compared with that tolerable by the superstructure. It is assumed that the strip acts as a beam spanning between the external walls of the house, partially restrained by the effect of the edge beams and acted on by the structural loads and soil reaction stresses. The shear and normal stresses along its sides are ignored. The pattern of soil reaction stresses are simplified into a number of rectangular stress blocks (see Figure 4). Elastic analysis then gives the central deflection and bending moment. The strip is assigned a moment of inertia per unit width equal to the moment of inertia of each 'T' beam divided by the beam spacing. The portion of the slab between the 'T' beams is ignored.

The structural capacity of the slab is assessed using yield line analysis of the complete slab simply supported at the four edges. The moment capacities per unit width in the two principal directions are taken to be those of the component 'T' beams divided by the beam spacing.

Appendix 4 provides details of an analysis suitable for hand or programmeable calculator manipulation.

Appendix 5 provides details of the methods used to establish the moments of inertia and the structural capacity of the slab.

OUTLINE OF DESIGN METHOD

The slab and foundation soils are an interactive system. The slab deforms in response to the building loads and soil reaction stresses. The soil deforms in response to the applied and soil suction stresses.

The design proceeds by searching for a pattern of soil reaction stresses that is in equilibrium with the building loads and which, in combination with the soil suction stresses, produces compatibility of deformation at the interface between

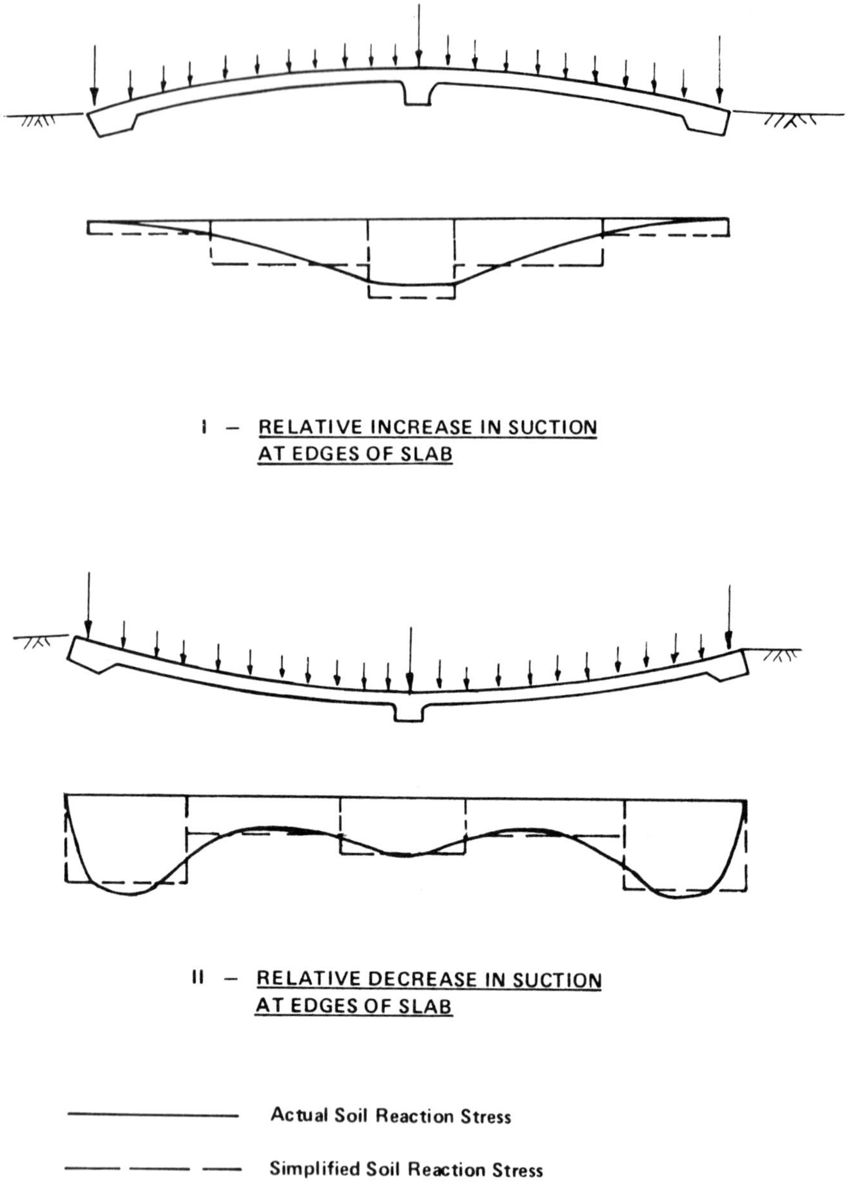

FIGURE 4
SOIL REACTION STRESS PATTERN BENEATH SLAB

the slab and soil. The tolerance of the superstructure to the imposed deformations and the capacity of the slab for the maximum imposed bending moments are then checked.

The previous sections have described certain assumptions made to render the design process amenable to hand calculation viz:-

* The development of log-linear stress-deformation equations from laboratory test results to characterise layers of the foundation soils.

* The use of "design" soil suction profiles.

* The simplification of the soil reaction stresses into a pattern of rectangular blocks for the elastic analysis of a central strip of the slab to obtain the central deflection and bending moment.

* The use of yield line analysis to provide an estimate of the moment capacity of the slab.

Three further assumptions now need to be added to allow completion of the analysis by hand calculation viz:-

* Compatibility of deformation need only be sought for the centre and edge of the slab and not for the entire slab-soil interface.

* Shear stresses between the slab and soil can be ignored.

* The vertical distribution of applied stresses in the soil mass can be approximated by elastic analysis of a homogeneous material. The principle of superposition is used to sum the effects of the rectangular soil reaction stress blocks.

It should be reiterated that each of these assumptions can be refined, by using more sophisticated measurements of soil properties and finite element analyses of the soil-slab interaction to more accurately establish soil and slab deflections and slab moments. Such improvements do not impair the logic of the design process, the steps of which are outlined below.

(1) A slab design (or a number of designs) is chosen for analysis.

(2) The soil beneath the foundation is divided into a number of layers. Each layer is considered to be homogeneous and isotropic.

(3) Any soil pressure distribution is selected that is in equilibrium with the structural loads.

(4) The central deformation of the slab and the central bending moment are computed.

(5) The deformations of the soil surface beneath the centre and edge of the slab are calculated and the difference compared with the slab deformation.

(6) If the deformations are incompatible the soil pressure distribution is modified to reduce the misfit and a second trial is commenced.

(7) Modification is continued until compatibility is satisfied, or until the designer is satisfied that further changes to soil pressures would produce an unrealistic distribution.

(8) If the indicated central bending moment in the slab is excessive the analysis for that slab design is terminated.

(9) If the bending moment is satisfactory the slab deformation is compared with that tolerable for the superstructure.

(10) If the slab proves to be adequate in all respects a lighter slab may be analysed to determine if this too would be adequate: or, more conveniently, if a programmable calculator is used a number of slabs may be analysed in parallel.

WORKED EXAMPLE

This worked example has been included to demonstrate the straight forward nature of the hand calculations. For brevity, only the design condition of a change in soil suction from equilibrium to extreme dry conditions under the edge of the house is studied. The soil beneath the centre is assumed to remain at equilibrium.

Data Single Storey Brick Veneer Home; 12m by 12m in plan; Single Ridge Roof supported on the external walls.
Foundation Soil; CH Clay, high plasticity, red brown, moist, stiff; Water Table not encountered in borings but assume ⩾ 5m; γ_b = 18 kN/m^3.
Results of Laboratory Tests:

	Applied Load (kPa)	Assigned Suction (kPa)	Deformation ΔH/H%
Saturated	5	10	- 2.0
Swell Test	30	10	- 1.5
Shrinkage Test to	5	10^4	+ 1.4
air dry at 50% RH	30	10^4	+ 1.5

Inferred Stress-Deformation Characteristics

The General Equation is (see Appendix 2):

$\Delta\varepsilon = M + N \log p_b/p_a$ for specified suction change

where $M' = A + B \log p_a$, and $M = M' \log s_b/s_a$

$N = C + B \log s_b$

From the laboratory test results

$M' = 0.013 - 0.00171 \log p_a$

$N = 0.0081 - 0.00171 \log s_b$

Strictly speaking, the initial applied stress, p_a, in the foundation soil should comprise the overburden pressure plus the soil reaction stresses induced by the construction of the house and any changes in soil suction subsequent to soil testing. For this calculation, it is assumed that p_a consists of the overburden stress and a uniform increment equal to the average bearing pressure beneath the house. For this house the average bearing pressure is computed to be 8.85 kPa and therefore $p_a = p'_o + 8.85$ kPa.

The soil profile is divided into five layers each 1.0 metre thick and values of M and N are computed for each layer with the required values of s_b, and s_b/s_a taken from the appropriate design charts (Appendix 3). The results are summarised in the following table.

TABLE I

PARAMETERS M AND N BENEATH CENTRE AND EDGE OF SLAB

Depth (m)	0 - 1	1 - 2	2 - 3	3 - 4	4 - 5
p_a (kPa)	17.85	35.85	53.85	71.85	89.85
M'	0.0109	0.0103	0.0100	0.0098	0.0097
			Under Edge of Slab		
$\log_{10} s_b/s_a$	+ 2.00	+ 1.00	+ 0.55	+ 0.23	+ 0.05
M	0.0218	0.0103	0.0055	0.0023	0.0005
$\log_{10} s_b$	+ 4.10	+ 2.30	+ 1.45	+ 0.90	+ 0.40
N	0.0011	0.0042	0.0056	0.0066	0.0074

Under Centre of Slab

Depth (m)	0 - 1	1 - 2	2 - 3	3 - 4	4 - 5
M	0	0	0	0	0
$\log_{10} s_a$	+ 2.1	+ 1.3	+ 0.90	+ 0.67	+ 0.35
N	0.0045	0.0059	0.0066	0.0070	0.0075

House Loads and Soil Reaction Stresses

For the house loadings, the following typical values have been adopted (see Appendix 6).

Side Walls	-	16 kN/m
End Walls	-	9 kN/m
Internal Walls	-	0.6 kN/m
Floor Load	-	4.6 kN/m^2
Average Intensity	-	8.85 kN/m^2

As the calculations are investigating the case of a hogging slab, a unit width strip across the house through the roof-bearing walls is considered.

In view of the small value of the internal wall loads, the intensity of the central stress block R_2 is set equal to that of the adjacent stress blocks r (i.e. $R_2/r = \alpha = 1$). Thus the soil stress pattern consists of two edge blocks of intensity R_1 kN/m^2 and width mL (metres) and one central block of intensity r kN/m^2 and width (1-2m)L (metres) where L is the width of the house slab.

Trial 1

Slab Type E3 from the suite of standard designs (Appendix 5) with a beam spacing of 4m c/c in both directions.

(a) Check Soil Deformations

Try $R_1 = 0$, m = 0.1 whence equilibrium gives r = 9.15kN/m^2.

Using design charts (Appendix 3)

TABLE II

COMPUTATION OF SOIL DEFORMATIONS

Depth (m)	0 - 1	1 - 2	2 - 3	3 - 4	4 - 5
z/L	0.4	0.125	0.208	0.292	0.375
p_o kPa	9	27	45	63	81
p_a (kN/m^2)	17.85	35.85	53.85	71.85	89.85
Under Edge of Slab					
$\frac{\Delta\sigma}{R_1}$	0.9	0.45	0.30	0.21	0.18
$\frac{\Delta\sigma}{R_1}.R_1$	0	0	0	0	0
$\frac{\Delta\sigma}{r}$	0.1	0.35	0.38	0.40	0.41
$\frac{\Delta\sigma}{r}.r$	0.91	3.18	3.45	3.63	3.72
p_b	9.91	30.18	48.45	66.63	84.72
$\log_{10} p_b/p_a$	-0.256	-0.075	-0.046	-0.033	-0.026
$\Delta\varepsilon$	+0.022	+0.010	+0.005	+0.002	+0.000

$$\therefore \delta_e = \Sigma\Delta\varepsilon.h = 0.039m$$

Under Centre of Slab					
$\frac{\Delta\sigma}{R_2}$	0.9	0.45	0.30	0.21	0.18
$\frac{\Delta\sigma}{R_2}.R_2$	8.17	4.09	2.72	1.91	1.63
$\frac{\Delta\sigma}{r}$	0.1	0.35	0.38	0.40	0.41
$2\frac{\Delta\sigma}{r}.r$	1.82	6.36	6.90	7.26	7.44
p_b	18.99	37.45	54.62	72.17	90.07
$\log_{10} p_b/p_a$	+0.0269	+0.0189	+0.0062	+0.0019	+0.0011
$\Delta\varepsilon$	$+1x10^{-4}$	$+1x10^{-4}$	$+4x10^{-5}$	$+1x10^{-5}$	$+1x10^{-5}$

$$\therefore \delta_c = \Sigma\Delta\varepsilon.h = 2.6 \times 10^{-5}m$$

Thus the soil beneath the edge has gone down 39mm while the soil beneath the centre has gone down 0mm (i.e. the differential movement, $\delta_e - \delta_c$, = 39mm.)

(b) Check Slab Behaviour

For the given house loads and assumed soil reaction stresses the data in Appendix 4 permits calculation of the moment and deflection factors μ_c and Δ. Assuming a moment at the edge of the slab equal to 0.1 times the fully fixed value to account for some moment transfer into the edge beam, μ_c is 11.0 kN/m^2 and Δ is 18.5 kN/m^2. For Slab Type E3 with beams at 4m spacing the ultimate hogging moment capacity per unit width is 19.54 kNm and the approach given in Appendix 5 indicates that the moment factor should be less than 3.26 kN/m^2 divided by an appropriate load factor (say 1.5).

Therefore the proposed slab design is not adequate and it is necessary to search for one that is. The results of further calculations are as follows.

Slab Type (Refer Appendix 5)	Allowable Moment Factor		Actual Moment Factor (μ_c)
E3 with beams at 4m centres	2.17	<	11 kPa
F6 with beams at 4m centres	7.38	<	11 kPa
F6 with beams at 3m centres	9.84	<	11 kPa
F6 with beams at 2m centres	14.76	>	11 kPa
G5 with beams at 4m centres	8.54	<	11 kPa
G5 with beams at 3m centres	11.39	>	11 kPa

There is loss of support beneath the edge of the slab as the slab deflection is less than the soil deflections. The judgment is made that with a 1.00m deep edge beam the length affected by suction changes to cause a zero soil reaction stress cannot be greater than 1.2m. Therefore further trials with different soil pressure distributions are not warranted.

Calculations using the methods of Walsh and Lytton have also been made for the same conditions. The results are summarised below.

Design Method	Moment/unit width (kNm)	Suitable Beam Type	& Spacing	Slab Deflection
Withers	66	F6	2m	1.8mm
		or G5	3m	0.9mm
Walsh	80-112	F6	2m	0.8mm
		or G6	3m	0.4mm
Lytton	383	Heavier than G6	-	-

All methods show that for this house and site deflection criteria do not govern the slab design.

Thus there would appear to be reasonable agreement between the Withers' design and that due to Walsh which to date has been generally recognised to be realistic.

BACK ANALYSIS OF TWO EXPERIMENTAL SLABS

Over the last six years six experimental slabs, seven actual housing slabs and eight flexible surface covers have been monitored at a number of sites in Melbourne as part of a major research programme at C.S.I.R.O. (an Australian Government Research Organisation) and Swinburne College of Technology. Two recent papers (10) and (11) report the behaviour of four of the experimental slabs, on a test site at Sunshine, a suburb of Melbourne. Detailed deformation versus time data were given for two.

The site is located in an area of Quaternary basaltic clay, is flat and poorly drained becoming water logged in the wetter winter and spring months. The unrestrained seasonal heave is about 65mm. One of the slabs, S3, was constructed on an initially dry site while another, S1, was built when the site was wet.

Figure 5 shows the results of back analyses of the performance of these two test slabs using the soil suction-deformation data provided in the papers. Assumptions had to be made about soil compressibility, house loadings and the slab reinforcement to resolve deficiencies in the available data.

Of particular interest are the following points:-

(i) The calculated differential deflection of slab S3 of 26mm as the edge conditions change from wet to dry extreme and the centre conditions remain at equilibrium (one of the design criteria which would have been used for this site) appears to agree with the long term range of differential deflection of 20-30mm.

(ii) Whether or not the slabs are cracking is clearly indicated.

(iii) The discrepancy between the calculated and observed deflections at the end of the first wet season for slab S3 is probably explained by the soil under the edge of the slab not achieving the extreme wet condition assumed in the calculations.

(iv) It became apparent in the calculations that the small variations in applied pressure caused significant changes in bending moments which are of crucial interest in determining the appropriate moment of inertia of the slabs.

(v) It would appear that slab S1 was built with the soil near to equilibrium conditions as there is good agreement bet-

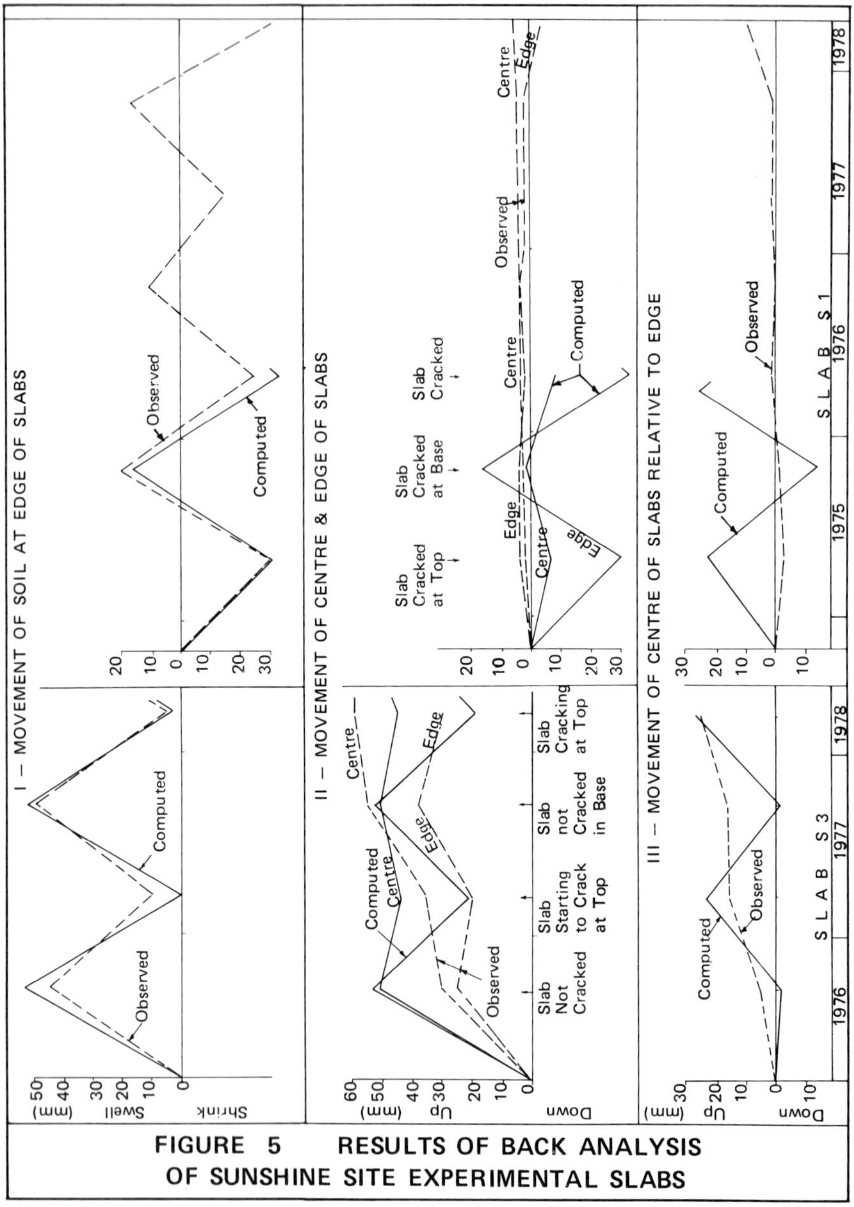

FIGURE 5 RESULTS OF BACK ANALYSIS OF SUNSHINE SITE EXPERIMENTAL SLABS

ween the observed and calculated values of edge soil movements. The calculated slab movements do not agree so well with those observed and the reasons for this are not clear. The field behaviour of the slab is certainly difficult to understand unless for some reason the slab was not cracked and its response to changes in soil suction not yet pronounced. The calculations indicate that the slab should crack and show a greater response to the seasonal variations in suction. If the slab had been assumed to be not cracked the estimates of edge slab deflections would have been about 1/10 of those shown in the figure. The divergence of the slab movements in 1978 suggests that the slab was starting to crack and it would be interesting to see measurements of recent slab movements to know if this trend was permanent or transient.

(vi) Comparison of the curves in Figure 5 shows that the different responses of the two slabs observed in the field, caused by the differences in initial soil conditions, are clearly reproduced by the calculations.

It is expected that both slabs should crack and show the same difference between edge and centre deflections in the long term because of the similarity between the slab details and the fact that they are built on the same soils, subject to the same seasonal changes in suction. While the early response of slabs to environmental change is of great interest, experience with cracked houses indicates it is uncommon for them to show signs of distress in the initial two or more years following construction. It is during this period that the total magnitude of the soil's response reflects the initial environmental conditions and differences between edge and centre slab deflections will be limited while the concrete has not cracked. The seasonal conditions may be such that the concrete may take time to be stressed sufficiently to crack. When it does (if it does) crack, much greater differential movements will occur.

The method proposed explicitly determines whether or not the slab will eventually crack in response to the design soil suction changes. With some knowledge about the different time rates of suction changes under the centre and edge of the slab the method could be used to compute when the slab would crack.

CONCLUSIONS

Following consideration of the behaviour of expansive clays and integral slab foundations, a method of designing stiffened integral slabs has been described. The method uses soil characteristics obtained from direct measurements of deformations under extreme conditions and 'design' soil suction profiles to study the interaction between the slab and the foundation soils, and to compute the deformations of each.

The method has been used to analyse the performance of two test slabs on a site in Melbourne, Australia. The results of this analysis together with other unpublished analyses of cracked buildings in Brisbane, Australia indicate the method produces realistic results.

The greater understanding of the development of deformations and stresses in a slab that the use of this method provides represents an improvement on current approaches and facilitates the use of sound engineering judgment throughout in making design decisions.

As all the assumptions and simplifications are explicit, the design procedure reveals if more accurate procedures are warranted; such as more detailed soil testing to account for variations in soil characteristics, incorporation of measurements of response to leaching in saline soils, evaluation of local soil suction profiles likely to be attained under operational conditions, or more sophisticated modelling of the soil-structure interaction possibly with a finite element analysis.

It is envisaged that the method will be used mostly to check the appropriateness of standard slab designs, for which the moment capacities and flexural stiffnesses are known. However individual designs are readily analysed. The method lends itself to the production of design charts, and to the use of calculator or computer programmes which would enable significant refinements in procedure to be made.

It should also be recognised that as experience with the method is accumulated, empirical correlations between certain terrains, soil types and applicable foundations will develop. Consequently the need to perform specific calculations for each and every foundation will diminish and it will be sufficient merely to carry out the laboratory testing to characterise the soil.

As part of the method a laboratory testing procedure has been outlined which will adequately characterise the soil for most sites. This procedure is straight forward and does not require expensive equipment. It is therefore considered that the unit cost of using the approach proposed in this paper should be no more than that of the less satisfactory methods in current use.

To assist in the evaluations of this method and its use, Appendices to this paper provide details of the analyses and approaches adopted and also include indicative soil suction and applied stress profiles and beam properties.

REFERENCES

1 G.D.Aitchison - Relationships of Moisture Stress and Effective Stress Functions in Unsaturated Soils, Proceedings of a Conference on Pore Pressure and Suction in Soils, Butterworths, London, 1961, 47-52.

2 G.D.Aitchison and B.G.Richards - A Broad Scale Study of moisture conditions in pavement subgrades throughout Australia. Moisture Equilibria and Moisture Changes in Soils Beneath Covered Areas, A Symposium in Print, Butterworths, Australia, 1965, 184-232.

3 G.D.Aitchison and J.A.Woodburn - Soil Suction in Foundation Design, Proceedings of the Seventh International Conference on Soil Mechanics and Foundation Engineering, Mexico City 1969 2, 1-8.

4 G.E.Blight - The time rate of heave of structures on Expansive Clays - Moisture Equilibria and Moisture Changes in Soils Beneath Covered Areas, A Symposium in Print. Butterworths, Australia, 1965, 78-88.

5 G.E.Blight - A Study of Effective Stresses for Volume Change, Moisture Equilibria and Moisture Changes in Soils Beneath Covered Areas, A Symposium in Print, Butterworths, Australia 1965, 259-269.

6 J.B.Burland - Some Aspects of the Mechanical Behaviour of partly saturated soils, Moisture Equilibria and Moisture Changes in Soils Beneath Covered Areas, A Symposium in Print, Butterworths, Australia 1965, 270-278.

7 J.B.Burland and C.P.Wroth - Settlement of Buildings and Associated Damage. Proceedings of the Conference on the Settlement of Structures, Cambridge 1974, 611-653.

8 D.Croney and J.D.Coleman - Pore Pressure and Suction in Soil, Proceedings of a Conference on Pore Pressure and Suction in Soils, Butterworths, London 1961, 31-37.

9 P.J.Fargher, J.A.Woodburn and J.Selby - Footings and Foundations for Small Buildings in Arid Climates, with special reference to South Australia, The South Australian Division of the Institution of Engineers, Australia and the Department of Adult Education, University of Adelaide, 1979.

10 J.E.Holland and P.F.Walsh - Behaviour and Designs of Residential Slabs on the Expansive Clays of Melbourne. Proceedings of the International Conference on Concrete Slabs, University of Dundee, Department of Civil Engineering, 1979.

11 J.E.Holland, W.Pitt, J.Washusen, D.Cameron, J.Jackson - Behaviour of Experimental Housing Slabs on Expansive Clays. Proceedings of the 6th Asian and Regional Conference on Soil Mechanics and Foundation Engineering, 1979.

12 J.E.Jennings - A Revised Effective Stress Law for use in the prediction of the Behaviour of Unsaturated Soils. Proceedings of a Conference on Pore Pressure and Suction in Soils, Butterworths, London 1961, 26-30.

13 F.K.Kong and R.H.Evans - Reinforced and Prestressed Concrete, Thomas Nelson and Sons Ltd., London (1975), pp 54-115 and 170-194.

14 H.G.Poulos and E.H.Davis - Elastic Solution for Soil and Rock Mechanics, John Wiley & Sons Inc., U.S.A. 1974, 36-91.

15 R.K.Schofield - Suction in Swollen Clays, Proceedings of a Conference in Pore Pressure & Suction in Soils, Butterworths, London 1961, 59-74.

16 P.F.Walsh - The Design of Residential Slabs on Ground. C.S.I.R.O. Division of Building Research, 1975.

17 P.F.Walsh, J.E.Holland and T.Kouzmin, The Classification of the Expansive Behaviour of Expansive Soils for Domestic Construction. C.S.I.R.O. Division of Building Research and the Australian & Building Industry Research Association, 1976.

APPENDIX 1 - LABORATORY TESTING PROCEDURE

Two tests are performed on undisturbed samples of the foundation soils - a swell test under saturated conditions with two applied loads and a shrinkage test to air dry conditions.

Saturated Swell Test

Apparatus has been developed for this test. In essence it is a battery of simple oedometers. The sample to be tested is pushed out of the sampling tube, inspected and pushed into a ring 50mm long and 47.5mm diameter. With the use of spacers the sample is trimmed square to 37mm length and located centrally in this ring. The ring is placed on a saturated porous stone mounted on a perforated perspex platen and capped with another. Both are 9.5mm thick and 47mm in diameter. Water is placed in an annular space, provided around the test specimen by a concentric p.v.c. ring, to cover the sample. The sample is loaded to 30 kPa by a 400mm long spring compressed the appropriate amount and the swell monitored by measuring the movement of the top of a central guide rod attached to the spring base plate and passing through the top of the reaction frame. When the swell under 30 kPa load has stopped, the additional swell under a reduced load of 5 kPa is measured.

Shrinkage to Air Dry Test

Shrinkage is measured of undisturbed soil specimens 20mm in height and 47.5mm in diameter dried in a stream of air conditioned to 50% relative humidity i.e. total suction of 100,000 kPa. The specimens are supported on a perforated brass base and are surrounded by perspex covers a little larger than the specimens. Conditioned air circulates below each specimen, passes through a ring of perforations beyond it and through the narrow annulus between the specimen and its cover, then across the top of the specimen to escape through an 8mm diameter hole centrally located in the cover. A dial gauge is used, through this central hole, to measure changes in height of the specimen.

No load is applied to the specimens since at a suction of 100,000 kPa the application of small external loads, of the order of 30 and 5 kPa, is not warranted. By assuming a realistic elastic modulus additional volume changes under load can be calculated, but these are found to be negligible.

The alternative of drying the specimens under vacuum at 50% relative humidity (over saturated calcium nitrate solution) was rejected because a very high vacuum is needed to achieve rapid drying, and suitable high vacuum pumps are seldom available in commercial geotechnical laboratories. As well, measurements of sample height are made much simpler by using air drying.

APPENDIX 2 - DEVELOPMENT OF SOIL STRESS-DEFORMATION CHARACTERISTICS

The laboratory tests produce data in the following form:-

Applied Stress (kPa)	Assumed Total Suction (kPa)	Deformation ΔH/H%
p_1	s_1	$\varepsilon_{p_1 s_1}$
p_1*	s_2*	$\varepsilon_{p_1 s_2}$
p_2	s_1	$\varepsilon_{p_2 s_1}$
p_2*	s_2*	$\varepsilon_{p_2 s_2}$

* Note where $s_2 > p_2$ or p_1, $\varepsilon_{p_1 s_2} \approx \varepsilon_{p_2 s_2}$

Using the definitions of Compression and Instability Indices

$$C_c = \frac{\Delta\varepsilon}{\Delta\log p} \qquad I = \frac{\Delta\varepsilon}{\Delta\log s}$$

Referring to Figure 2:

$$\text{At } p_1 \quad \Delta\varepsilon = I_{p_1}\,\Delta\log s$$

$$\text{At } p_2 \quad \Delta\varepsilon = I_{p_2}\,\Delta\log s$$

$$\text{At } p_a \quad \Delta\varepsilon = I_{pa}\,\Delta\log s$$

$$I_{pa} = I_{p_1} + \frac{\log p_a - \log p_1}{\log p_2 - \log p_1}(I_{p_2} - I_{p_1})$$

$$\text{At } s_1 \quad \Delta\varepsilon = C_c \quad \Delta\log p$$

$$\text{At } s_2 \quad \Delta\varepsilon = Cc_{s_2} \quad \Delta\log p$$

$$\text{At } s_b \quad \Delta\varepsilon = Cc_{sb} \quad \Delta\log p$$

$$Cc_{sb} = C_{c_1} + \frac{\log s_b - \log s_1}{\log s_2 - \log s_1}(Cc_{s_2} - Cc_{s_1})$$

$$\therefore\ \Delta\varepsilon_{ab} = (\log s_b - \log s_a)\left[I_{p_1} + \frac{\log p_a - \log p_1}{\log p_2 - \log p_1}(I_{p_2} - I_{p_1})\right]$$

$$+ (\log p_b - \log p_a)\left[Cc_{s_1} + \frac{\log s_b - \log s_1}{\log s_2 - \log s_1}(Cc_{s_2} - Cc_{s_1})\right]$$

$$= \log s_b/s_a\left[I_{p_1} + \log p_a\,\frac{(I_{p_2} - I_{p_1})}{\log p_2/p_1} - \frac{(I_{p_2} - I_{p_1})}{\log p_2/p_1}\right]$$

$$+ \log p_b/p_a\left[Cc_{s_1} + \log s_b\,\frac{(Cc_{s_2} - Cc_{s_1})}{\log s_2/s_1} - \log s_1\,\frac{(Cc_{s_2} - Cc_{s_1})}{\log s_2/s_1}\right]$$

$$\boxed{\Delta\varepsilon_{ab} = \log s_b/s_a\,(A + B\log p_a) + \log p_b/p_a\,(C + B\log s_b)}$$

$$\text{i.e. } I_{pa}\log s_b/s_a + Cc_{sb}\log p_b/p_a$$

$$\text{where } A = I_{p_1} - B \log p_1 ; \quad C = Cc_{s_1} - B \log s_1$$

$$B = \frac{I_{p_2} - I_{p_1}}{\log p_2/p} = \frac{Cc_{s_2} - Cc_{s_1}}{\log s_2/s_1} = \frac{d_2\varepsilon}{d(\log p).d(\log s)}$$

Note: Alternatively:-

$$\Delta\varepsilon_{ab} = \log s_b/s_a \ (A + B \log p_b) + \log p_b/p_a \ (C + B \log s_a)$$

$$\text{i.e. } I_{pb} \log s_b/s_a \ + \ Cc_{sa} \log p_b/p_a$$

For hand calculation it is convenient to use the follow-equation

$$\Delta\varepsilon_{ab} = M + N \log p_b/p_a$$

where, for each layer,

$$M = (A + B \log p_a) \log s_b/s_a$$

$$N = C + B \log s_b$$

APPENDIX 3 - DESIGN SOIL SUCTION AND APPLIED PRESSURE PROFILES

Two design charts have been composed from equilibrium and extreme soil-suction profiles typical of clay soils, such as those in Figure 3. The first, Figure 6, provides values of $\log s_b/s_a$ at various depths, for different depths to the water table, as the soil suction changes from equilibrium to extreme conditions or between the two extreme conditions. The second, Figure 7, provides values of $\log s_b$.

Figure 8 provides factors for calculating the pressure distributions beneath the centre of the edge stress block and the centre of the slab using the general expression -

$$p_b = p'_o + \Delta_p$$

and the relationships quoted in the figure.

APPENDIX 4 - STRUCTURAL ANALYSIS OF SLABS

Elastic Analysis for Central Deflection and Bending Moment

As stated in the main body of the text, the slabs are considered as sets of one way beams spanning between the external walls of the house for the purpose of evaluating a deflection and bending moment.

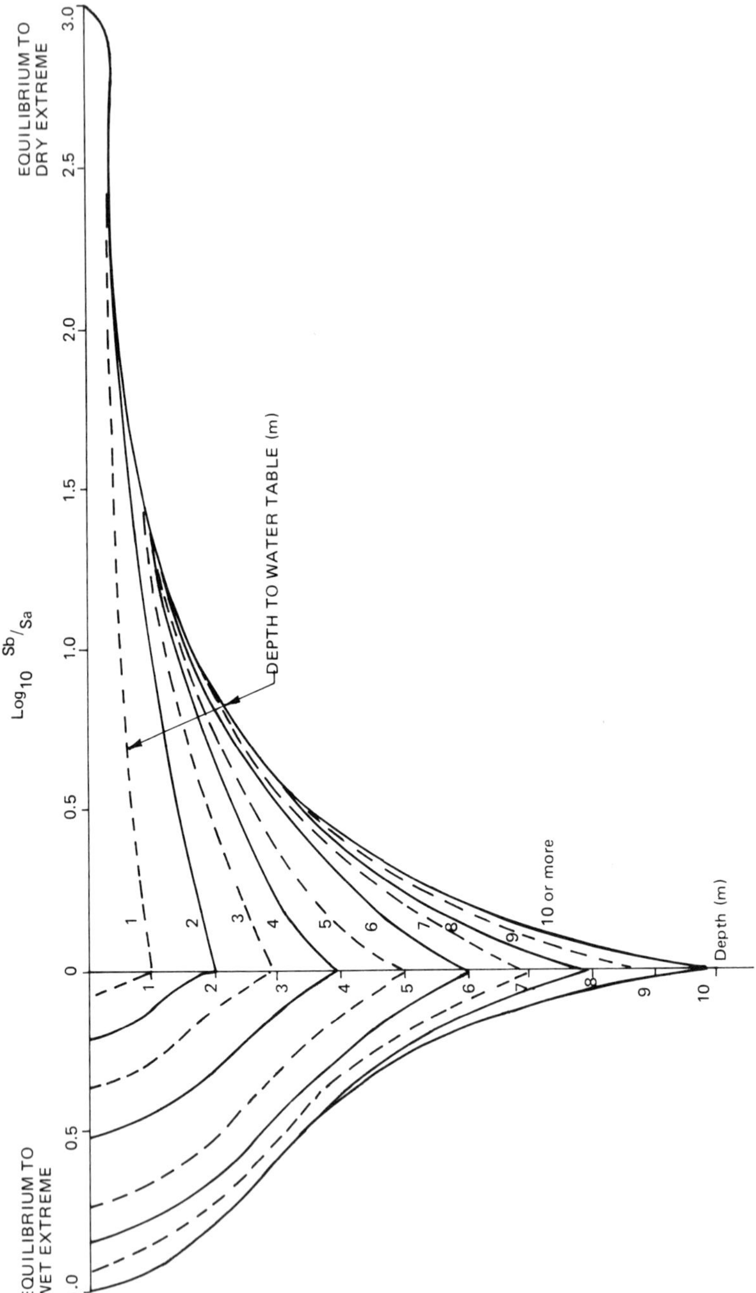

FIGURE 6

DESIGN SOIL SUCTION CHANGES – $\text{Log}_{10}\, ^{Sb}/_{Sa}$

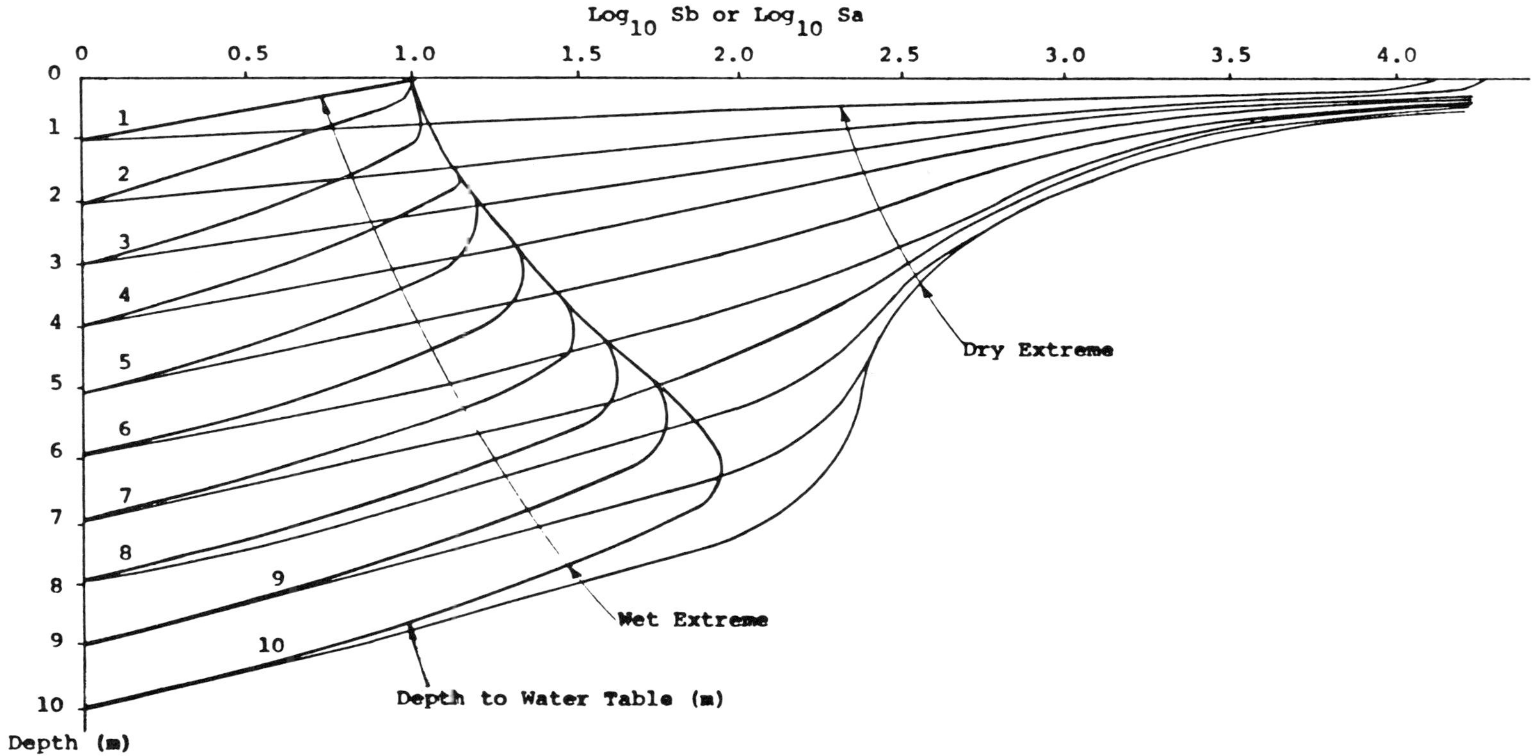

FIGURE 7

DESIGN VALUES OF EXTREME SOIL SUCTION - Log_{10} Sb or Log_{10} Sa

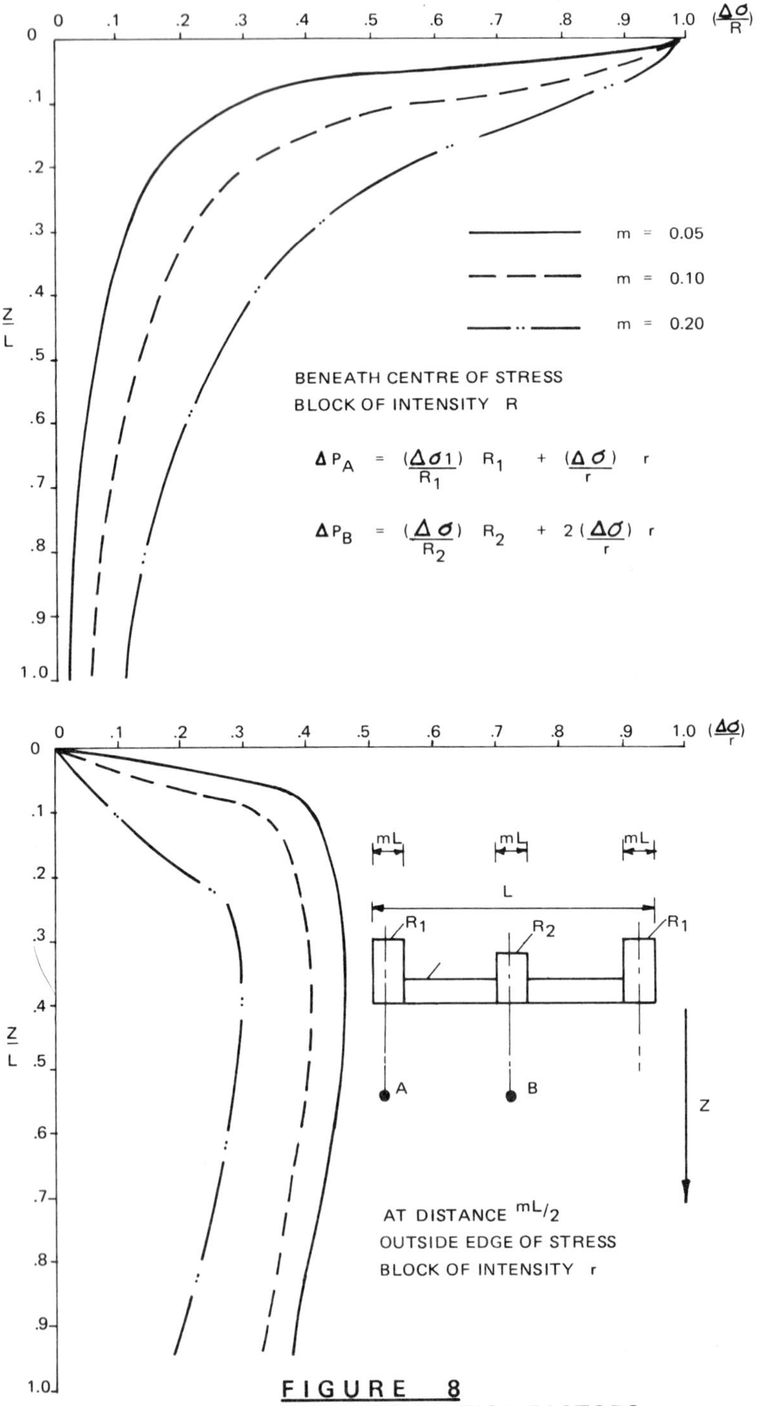

FIGURE 8
APPLIED PRESSURE DISTRIBUTION FACTORS

Treating the real soil reaction stress profile as a set of rectangular stress blocks permits the mid-span deflection and moment to be calculated for each block using standard equations. Superposition then gives the total effect. Equations have been developed for the simple situation of external wall loads, W_1, a central wall load, W_2, a uniformly distributed load, w, and a soil reaction stress profile described by R_1, R_2, r and m = a/L, where a is the width of the edge and central stress blocks. Clearly other equations can be developed for different positions of W_2, but it is considered that this will not be necessary for the majority of cases as W_2 does not play a dominant role. It should also be noted that the central soil reaction stress block R_2 has been expressed as a proportion of value r and its width restricted to the same as that of the edge stress blocks. This is purely for convenience in hand calculations as it restricts the number of variables.

Thus for (1) House Loads W_1, W_2, w

(2) Slab Span L

(3) Soil Stresses R_1, $R_2 = \alpha r$, r

(4) Edge Stress Block Width $a = mL$

Equilibrium is satisfied if

$$2R_1m + R_2m + (1-3m)r = 2W_1/L + W_2/L + w$$

or

$$2mR_1 + (1-m(3-\alpha))r = 2W_1/L + W_2/L + w$$

The following table provides the equations for central deflection, central moment and edge moment if the beam is considered either simply supported or fully fixed at the ends.

Simply Supported Ends	Fully Fixed Ends

Central Deflection, $\delta_c = \Delta . \dfrac{L^4}{384EI}$

Simply Supported Ends	Fully Fixed Ends
$\Delta = 5(r-w)$	$\Delta = (r-w)$
+	+
$(24m^2-16m^4)(R_1-r)$	$16(m^3-m^4)(R_1-r)$
+	+
$(8m-4m^3+m^4)(\alpha-1)r$	$(2m-2m^3+m^4)(\alpha-1)r$
-	-
$\dfrac{8W_2}{L}$	$\dfrac{2W_2}{L}$

Simply Supported Ends

Central Moment, $M_c = \mu_c . \frac{L^2}{24}$

$$\mu_c = 3(r-w) + 12m^2(R_1-r) + (6m-3m^2)(\alpha-1)r - \frac{6W_2}{L}$$

Edge Moment, $M_e = \mu_e . \frac{L^2}{24}$

$$\mu_e = 0$$

Fully Fixed Ends

$$\mu_c = (r-w) + 8m^3(R_1-r) + (3m-3m^2+m^3)(\alpha-1)r - \frac{3W_2}{L}$$

$$\mu_e = 2(r-w) + (12m^2-8m^3)(R_1-r) + (3m-m^3)(\alpha-1)r - \frac{3W_2}{L}$$

For all but the heaviest slabs which possess the deepest edge beams, the slab must be considered as simply supported at the edges or nearly so. With deep edge beams there would be significant torsional restraint at the edges although not to the extent of providing encastre conditions. If it is so desired the design engineer can select values of deflection and bending moment intermediate between the fully fixed and simply supported cases. Charts providing values of μ_c, μ_e and Δ have been drawn up assuming the bending moment at the edge of a slab is equal to 0.1 or 0.25 times the fully fixed value with the central bending moment and deflection computed on a pro rata basis. For brevity these charts have been reduced to parametric equations with the values of the parameters being given in Figures 9 and 10.

APPENDIX 5 - PROPERTIES OF SOME STANDARD BEAM SIZES

Using the procedures outlined below properties of eighteen beam sizes have been calculated for use in standard designs and are summarised in Figure 11.

ρ	α	W_2/L	M	R_1	$\Delta = a + br$		$\mu_c = c + dr$	
					a	b	c	d
0.25	1	0	0.05	0	−18.4	3.96	−11.5	2.47
				30	−17.1	"	−10.8	"
			0.10	0	−18.4	3.82	−11.5	2.41
				30	−13.0	"	−8.7	"
			0.20	0	−18.4	3.30	−11.5	2.13
				30	+2.7	"	−0.2	"
		1	0.05	0	−24.9	3.96	−16.8	2.47
				30	−23.6	"	−16.1	"
			0.10	0	−24.9	3.82	−16.8	2.41
				30	−19.5	"	−14.0	"
			0.20	0	−24.9	3.30	−16.8	2.13
				30	−3.9	"	−5.5	"
		2	0.05	0	−31.4	3.96	−22.0	2.47
				30	−30.1	"	−21.3	"
			0.10	0	−31.4	3.82	−22.0	2.41
				30	−26.0	"	−19.2	"
			0.20	0	−31.4	3.30	−22.0	2.13
				30	−10.4	"	−10.7	"
	2	0	0.05	0	−18.4	4.28	−11.5	2.73
				30	−17.1	"	−10.8	"
			0.10	0	−18.4	4.47	−11.5	2.91
				30	−13.0	"	−8.7	"
			0.20	0	−18.4	4.57	−11.5	3.06
				30	+2.7	"	−0.2	"
		1	0.05	0	−24.9	4.28	−16.8	2.73
				30	−23.6	"	−16.1	"
			0.10	0	−24.9	4.47	−16.8	2.91
				30	−19.4	"	−14.0	"
			0.20	0	−24.9	4.57	−16.8	3.06
				30	−3.9	"	−5.5	"
		2	0.05	0	−31.4	4.28	−22.0	2.73
				30	−30.1	"	−21.3	"
			0.10	0	−31.4	4.47	−22.0	2.91
				30	−26.0	"	−19.2	"
			0.20	0	−31.4	4.57	−22.0	3.06
				30	−10.4	"	−10.7	"

NOTE— $w = 4.6\ kN/m^2$

FIGURE 10

SELECTED VALUES OF PARAMETERS FOR SLAB MOMENTS & DEFLECTION FACTORS

ρ	α	W_2/L	M	R_1	Δ = a + br: a	b	μc = c + dr: c	d
0.1	1	0	0.05	0	−21.2	4.55	−12.9	2.77
				30	−19.5	"	−12.1	"
			0.10	0	−21.2	4.39	−12.9	2.69
				30	−14.7	"	−9.6	"
			0.20	0	−21.2	3.76	−12.9	2.36
				30	+4.1	"	+0.3	"
		1	0.05	0	−28.6	4.55	−18.6	2.77
				30	−26.9	"	−17.8	"
			0.10	0	−28.6	4.39	−18.6	2.69
				30	−22.0	"	−15.3	"
			0.20	0	−28.6	3.76	−18.6	2.36
				30	−3.3	"	−5.4	"
		2	0.05	0	−36.0	4.55	−24.3	2.77
				30	−34.3	"	−23.5	"
			0.10	0	−36.0	4.39	−24.3	2.69
				30	−29.5	"	−21.0	"
			0.20	0	−36.0	3.76	−24.3	2.36
				30	−10.7	"	−11.1	"
	2	0	0.05	0	−21.2	4.92	−12.9	3.05
				30	−19.5	"	−12.1	"
			0.10	0	−21.2	5.12	−12.9	3.23
				30	−14.7	"	−9.6	"
			0.20	0	−21.2	5.21	−12.9	3.38
				30	+4.1	"	+0.3	"
		1	0.05	0	−28.6	4.92	−18.6	3.05
				30	−26.9	"	−17.8	"
			0.10	0	−28.6	5.12	−18.6	3.23
				30	−22.0	"	−15.3	"
			0.20	0	−28.6	5.21	−18.6	3.38
				30	−3.3	"	−5.4	"
		2	0.05	0	−36.0	4.92	−24.3	3.05
				30	−34.3	"	−23.5	"
			0.10	0	−36.0	5.12	−24.3	3.23
				30	−29.5	"	−21.0	"
			0.20	0	−36.0	5.21	−24.3	3.38
				30	−10.7	"	−11.1	"

NOTE— w = 4.6 kN/m^2

FIGURE 9

SELECTED VALUES OF PARAMETERS FOR SLAB MOMENTS & DEFLECTION FACTORS

BEAM DEPTH (mm)	SLAB	TYPE	1				2				3				4				5				6			
		Tension in	ALLOWABLE MOMENT		Effective Moment of Inertia		ALLOWABLE MOMENT		Effective Moment of Inertia		ALLOWABLE MOMENT		Effective Moment of Inertia		ALLOWABLE MOMENT		Effective Moment of Inertia		ALLOWABLE MOMENT		Effective Moment of Inertia		ALLOWABLE MOMENT		Effective Moment of Inertia	
			Mu	Mc	Iuc	Ipc	Mu	Mc	Iuc	Ipc	Mu	Mc	Iuc	Ipc	Mu	Mc	Iuc	Ipc	Mu	Mc	Iuc	Ipc	Mu	Mc	Iuc	Ipc
100	A	Top	4.5	6.3	1.583	–																				
		Bottom	4.5	6.3	1.583	–																				
200	B	Top	12.6	14.2	4.531	–																				
		Bottom	12.2	6.7	4.531	0.3060																				
300	C	Top	22.0	32.9	14.244	–	22.3	33.4	14.608	–																
		Bottom	17.6	13.4	14.244	–	30.4	13.7	14.608	1.538																
400	D	Top	31.8	58.5	33.625	–	31.8	59.3	34.380	–	48.1	59.9	34.863	–	57.0	60.6	35.407	–								
		Bottom	22.9	23.6	33.625	–	41.2	24.2	34.380	3.051	53.1	24.6	34.863	3.888	66.1	25.0	35.407	4.816								
600	E	Top	33.7	122.3	110.566	–	51.2	122.0	112.401	–	78.2	144.7	115.787	–	93.2	124.2	115.07	–	111.7	125.4	116.606	–				
		Bottom	22.3	52.8	110.566	–	62.7	54.1	112.401	–	80.2	52.6	115.787	*(7)	101.1	55.5	115.07	12.187	122.8	56.3	116.606	14.852				
1000	F	Top					89.81	274.9	474.831	–	138	277.1	478.839	–	166.2	278.9	482.945	–	198.6	280.9	487.508	–	265.4	284.3	494.172	–
		Bottom					105.9	145.1	474.831	–	136	146.4	478.839	–	171.0	147.8	482.945	–	209.2	149.3	487.508	–	259.1	151.5	494.172	56.393
1500	G	Top																	307.1	520.0	1480.93	–	411.4	525.2	1496.70	–
		Bottom																	316.7	318.3	1480.93	–	393.7	321.9	1496.70	–
TOP REINFORCEMENT			F62 = 270 mm^2				F62 = 270 mm^2				F62 + 1 x TM8 = 421 mm^2				F62 + 3 x S10 = 506 mm^2				F62 + 3 x S12 = 609 mm^2				F82 + 3 x S12 = 818 mm^2			
BOTTOM REINFORCEMENT			1 x TM8 = 151 mm^2				2 x TM8 = 302 mm^2				1 x TM8 + 3 x S10 = 387 mm^2				1 x TM8 + 3 x S12 = 490 mm^2				3 x S16 = 603 mm^2				3 x S16 + 1 x TM8 = 754 mm^2			

NOTES

1. Effective flange width taken as 1900 mm.
2. Web width = 300 mm.
3. F'_c = 20 MPa, f_y = 410 MPa
4. Maximum Allowable Moment = Mu/1.5, although when Mc < Mu/1.5, it is desirable to use Mc as the limit to restrain deflection.
5. For M < Mc (Cracking Moment) I_ϵ = Iuc (uncracked)
   For Mc < M < Mu/1.5 I_ϵ = Ipc (partially cracked)
6. Units are kNm and $mm^4 \times 10^8$.
7. * Mc ≈ Mu/1.5, beam therefore assumed uncracked.

FIGURE 11 SUMMARY OF ALLOWABLE MOMENTS & EFFECTIVE MOMENTS OF INERTIA OF BEAMS

Moment of Inertia

The moment of inertia of a unit width of the slab is considered to be the moment of inertia of a component 'T' beam divided by beam spacing.

The moment of inertia of the component 'T' beams have been computed using the method proposed by CP110 which uses the concept of a partially cracked section.

The moment, M_c, required to cause the concrete to crack is computed using the expression -

$$M_c = f'_t \times Z_t$$

where f_t is the permissible tensile stress in the concrete, here taken as $0.1\ F'_c$

and Z_t is the section modulus computed on the basis of the uncracked section.

The ratio M_c/M is checked where M is the applied moment. If $M_c/M \geqslant 1$ the section is not cracked and the effective Moment of Inertia of the section is the gross value of the uncracked section allowing for the reinforcement. If $M_c/M < 1$ then the the section has cracked and the moment of inertia is calculated assuming that the concrete carries a certain amount of tension. A maximum level of concrete tensile stress f_{ct} is assigned at the level of the tensile reinforcement, here taken as $0.55 N/mm^2$. It is also assumed that the Neutral Axis position corresponds to that of the fully cracked section. The net moment, M_{net}, resisted by the fully cracked section is then the applied Moment, M, reduced by the contribution of the concrete in tension, M_{ct}

i.e. $M_{net} = M - M_{ct}$

where $M_{ct} = f_{ct} \times Z_{ct}$

and Z_t is the section modulus computed on the basis of the fully cracked section properties.

As the net moment applied to the fully cracked section causes the same curvature as the applied moment to the partially cracked section

$$\frac{M_{net}}{I\ \text{cracked}} = \frac{M}{I\ \text{partially cracked}}$$

Therefore I partially cracked $= \frac{M}{M_{net}} \times$ I cracked

$$= \frac{M}{M-M_{ct}} \times \text{I cracked}$$

In the table of beam properties given in Appendix 6 the value of I partially cracked quoted is the mean value between $M = M_c$ and $M = M_u/1.5$, where M_c is the ultimate moment capacity of the beam.

Ultimate Moment Capacity of Beams & Slab

The Ultimate Moment Capacities of the individual beams have been assessed using conventional analysis

with F'_c = 20 MPa - concrete

f'_y = 410 MPa - steel

and the simplified concrete compressive stress block of CP110.

The ultimate capacity of the slab is assessed with yield line theory for a simply supported slab. For a slab of dimensions L_1 and L_2 where $L_2 \geqslant L_1$, moment capacities per unit width M_1 and M_2 in the L_1 and L_2 directions, and for 45° yield line pattern, the uniformly distributed load, q_{ult}, required to cause yield is given by

$$q_{ult} = \frac{24(M_1(L_2/L_1) + M_2)}{(3\ L_1\ L_2 - L_1^2)}$$

and using a load factor of 1.5, the allowable uniformly distributed load, q_{all}, is given by

$$q_{all} = \frac{q_{ult}}{1.5} - \frac{16\ (M_1(L_2/L_1) + M_2)}{(3\ L_1\ L_2 - L_1^2)}$$

The moment capacities M_1 and M_2 per unit width are given by the moment capacities of the component 'T' beams divided by the beam spacing in the relevant direction. Thus the capacity of the portion of the slab between the 'T' beams is ignored, but calculations indicate the consequent loss of capacity is very small. The calculations also show that the slabs are under reinforced and that at the ultimate state are practically all in tension. Thus membrane action plays a significant role and calculations on the basis of bending theory have a limited relevance. Computations with several yield line patterns to establish a minimum value of q_{all} are therefore not warranted.

For hand calculation the following method of establishing the structural adequacy of a slab has been adopted.

The central bending moment in an elastic simply supported beam acted on by a uniformly distributed load, p, is

$$M_c = p\,\frac{L^2}{8}$$

Therefore for a unit width strip through the centre of the simply supported slab loaded with q_{all}

$$M_c = q_{all}\,\frac{L_2^{\,2}}{8}$$

$$= 2\,L_2^{\,2}\,\frac{(M_1(L_2/L_1) + M_2)}{(3\,L_1L_2 - L_1^{\,2})}$$

Now the moment at the centre of this strip under the real distribution of loads is

$$M_c = \frac{\mu_c\,L_2^{\,2}}{24}$$

The assumption is then made that the slab is adequate provided

$$M_c = \frac{\mu_c\,L_2^{\,2}}{24} \leqslant 2\,L_2^{\,2}\,\frac{(M_1(L_2/L_1) + M_2)}{(3\,L_1\,L_2 - L_1^{\,2})}$$

or $$\mu_c \leqslant 48\,\frac{(M_1(L_2/L_1) + M_2)}{(3\,L_1\,L_1 - L_1^{\,2})}$$

APPENDIX 6 - HOUSE LOADS AND ALLOWABLE DISTORTIONS

House Loads

If specific house loads are not available then reference can be made to the indicative values in Figure 12 which have been computed on the basis of a house 12m square in plan with a single ridge roof spanning across the side walls. Wall loads for houses of different plan dimensions can be estimated on a pro rata basis.

Allowable Distortion

The allowable distortion of various forms of house construction have been studied in detail by a number of authors, principally by Burland & Wroth (7). The following values of angular distortion δ/L are given as an appropriate guide.

HOUSE TYPE	TOTAL WT. (For 12 x 12 m)	AV. INTENSITY
① SINGLE STOREY BRICK VENEER	1275 kN	8.85 kN/m^2

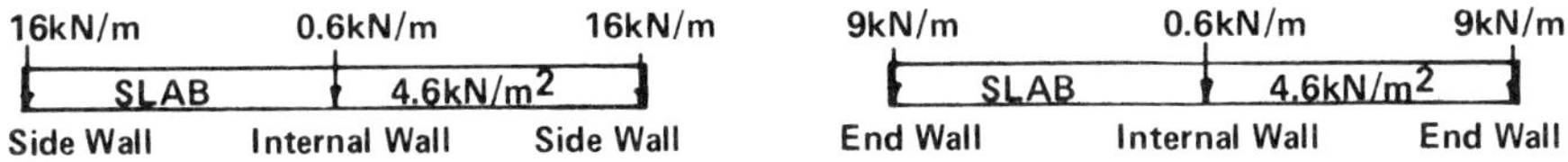

HOUSE TYPE	TOTAL WT.	AV. INTENSITY
② SINGLE STOREY CAVITY BRICK	1563 kN	10.85 kN/m^2

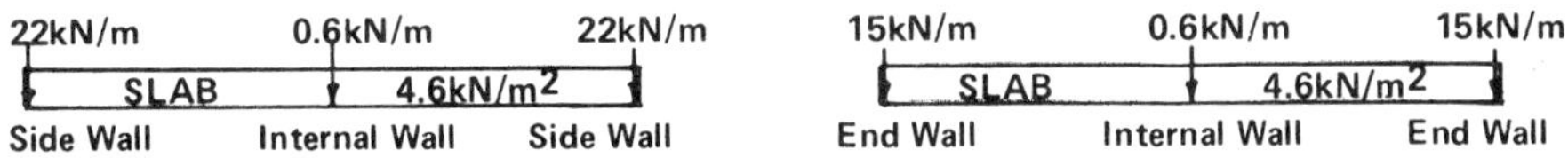

HOUSE TYPE	TOTAL WT.	AV. INTENSITY
③ DOUBLE STOREY BRICK VENEER	1453 kN	10.08 kN/m^2

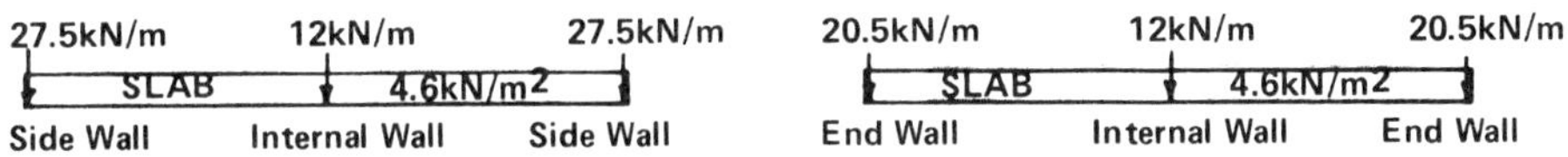

HOUSE TYPE	TOTAL WT.	AV. INTENSITY
④ DOUBLE STOREY CAVITY BRICK	2053 kN	14.25 kN/m^2

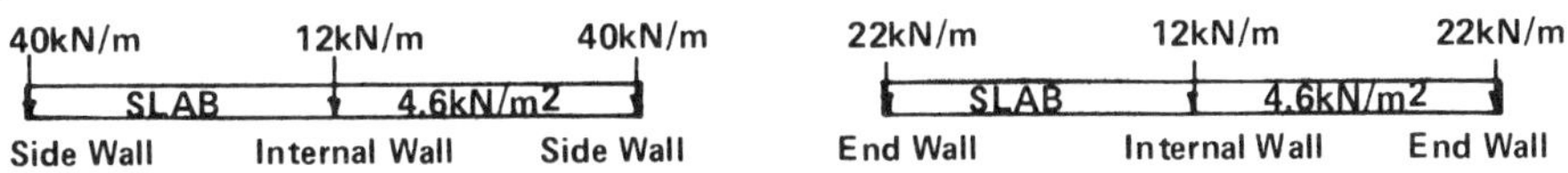

HOUSE TYPE	TOTAL WT.	AV. INTENSITY
⑤ DOUBLE STOREY CAVITY BRICK	2569 kN	17.84 kN/m^2

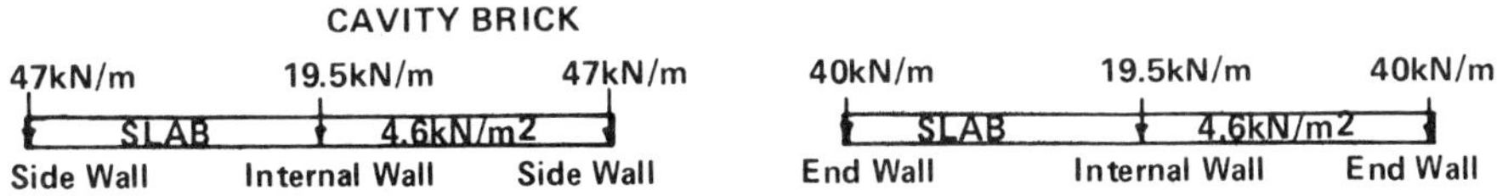

NOTE:
1. The roof load is taken as 1.73 kN/m^2.
2. The first floor loads are taken as – 6.0kN/m^2 – Concrete
   – 2.4kN/m^2 – Timber

   and include the internal wall loads.

FIGURE 12

INDICATIVE HOUSE LOADINGS

Form of Construction	Allowable Angular Distortion δ/L
Timber	1/100
Brick Veneer-Articulated	1/150
Brick Veneer-Solid	1/250
Cavity Brick-Articulated) Masonry-Articulated)	1/400
Cavity Brick-Solid) Masonry-Solid)	1/1000

MODEL STUDIES OF NEGATIVE FRICTION ON THE SINGLE PILE

K.P.R. Chaudhuri, BCE,MSc,PhD,MIE(Ind)
Kulasinghe, Chaudhuri and Partners, Nigeria

K.J. Denman, BEng(Tech),CEng,MIStructE,MICE
Dyfed County Council

R.A. Nicholls, BEng(Tech),PhD,CEng,MICE,FGS
Sir William Halcrow and Partners

M.V. Symons, BSc,PhD,FGS
Department of Civil Engineering and Building Technology, UWIST

SUMMARY

Smooth-shafted model piles, containing an interconnected system of internal strain-gauged load cells, were subjected to negative friction loadings in the laboratory. Long term tests were carried out at different embedment lengths with the resulting settlements, magnitude and distribution of negative friction loads and position of the neutral point measured.

INTRODUCTION

In normal circumstances a pile derives its load carrying capacity by means of end-bearing at its toe and friction on its shaft. For the normal case shown in Figure 1(a), the movement of the pile relative to the soil is downwards for the entire embedded length, and the resulting shear stress or skin friction between soil and shaft is upwards, and conventionally termed positive. Cases can occur, however, in which the movement of the soil relative to the pile is soil downwards for a high percentage of the embedded length and the resulting skin friction between soil and shaft is downwards, and conventionally termed negative. The most commonly occurring instance of this negative friction mechanism is shown in Figure 1(b) in which a pile is installed through a weak, compressible stratum into a firm bearing stratum with the subsequent imposition of fill or other surcharge at the surface of the weak stratum. This leads to large vertical ground movements due to the compression of the weak stratum and the downward settlement of both fill and compressible soil exceeds the downward settlement of the pile, leading to a reversal of the normal frictional mechanism. The

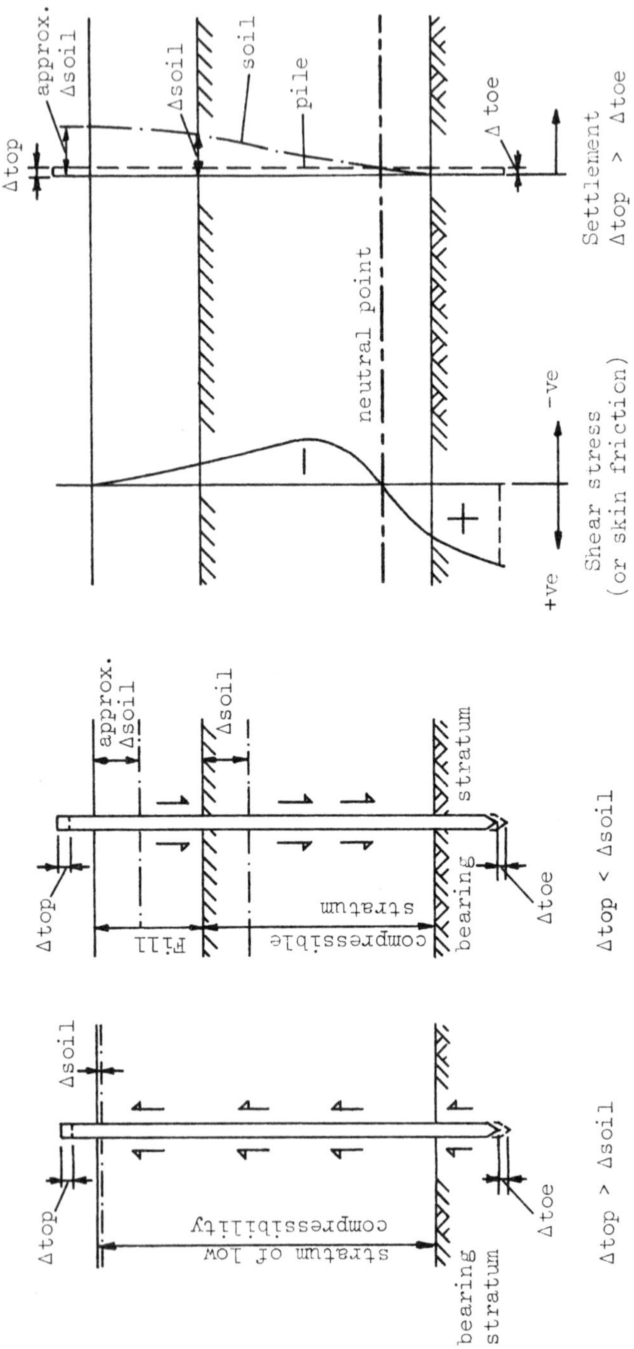

(a) Normal case of positive friction

(b) Common case of negative friction

(c) Shear stress/depth; pile and soil settlement/depth relationships

FIGURE 1

load carrying capacity of the pile is thus reduced and additional applied load is imposed by the negative friction. A further reduction in load carrying capacity also occurs due to the tendency of the downward settling fill and soil to cling to the pile surface. This causes a reduction in geostatic stresses at the level of the pile toe leading to a reduced shearing resistance and, hence, a reduced ultimate bearing capacity in this region (Zeevaert 1973). This increase in loading and complementary decrease in carrying capacity can, in some cases, completely erode the allocated design factor of safety with the result that failure takes place (Florentin and L'Heriteau 1948, Milner 1957, Moretto and Bolognesi 1959, Chellis 1964, Baligh, Vivatrat and Figi 1978).

The actual pile/soil interaction is more complicated than the simple mechanism just described and reference to Figure 1(c) shows that, even when a full negative friction situation has been established, the resulting distribution of skin friction down the pile shaft is partly negative and partly positive. This may be explained as follows:-

the skin friction can be negative only where the relative movement of soil to pile shaft is soil downwards. This condition certainly exists in the middle and upper horizons of the compressible stratum, where the downward settlements are the summation of all the settlements within and below them; and also in the fill which rests on the compressible stratum and follows it down. The lower horizons of the compressible stratum undergo much smaller settlements and the underlying bearing stratum may not compress at all. As a result, in these lower horizons, the relative movement between pile and soil returns to the normal case of pile downwards with a resulting upward or positive skin friction. The transitional zone between negative and positive skin friction is treated as a theoretical horizon of zero relative movement and is termed the neutral point - See Figure 1(c).

A further complexity is introduced by the fact that the settlement of the compressible stratum, due to the imposition of the fill, is time-dependent with the lower horizons ceasing to consolidate before the upper horizons. This produces a change in magnitude of frictional loading and a change in pattern of relative movement between pile and soil with time, causing the neutral point to vary in position until consolidation ceases in the compressible stratum. An understanding of this complex interaction between pile and soil is essential if a rational basis is to be established for the prediction of negative friction loading on the single pile.

This paper presents the results of testing programmes which measured the magnitude and distribution of negative friction loadings and neutral point variations in single piles.

APPARATUS AND TESTING

The results of two different testing programmes are summarised in this paper. The first programme, concerned solely with the measurement of negative friction, will be described in some detail. The second programme, concerned mainly with uplift resistance but which monitored negative friction effects during a period of self-weight consolidation of a clay stratum, will be briefly summarised.

(a) General procedure

(i) First testing programme

The testing arrangement is shown in Figure 2. In order to simulate the most commonly occurring case of negative friction the following test procedure was adopted:-

a bearing stratum, consisting of a 625mm thick bed of saturated sand, was laid and compacted to a dense state inside a circular concrete test bin, 0.76m internal diameter by 1.22m high. A 38mm diameter model pile was installed 250mm into this stratum and held in position by a head clamp. A saturated clay stratum was then formed to a pre-determined thickness by hand-smearing. The inside face of the test bin above the bearing stratum was covered with a layer of silicon grease to minimise friction between the clay and the container. Settlement plates were placed at intervals ($\frac{1}{4}$, $\frac{1}{2}$ and $\frac{3}{4}$ depth) within the clay stratum as laying proceeded. Settlement dial gauges, fastened to fixed datum frames, were connected by fibre-glass threads of low elasticity to the clay settlement plates. The pile head clamp was released as soon as the clay stratum had been laid. The clay was allowed to settle under its own self-weight for at least 7 days in order that an equilibrium state could be reached and all loads and settlements were monitored during this period. A surcharge loading consisting of 250Kg of dry sand, was then poured to a constant density on top of the clay and drainage of water was allowed through drainage holes in the test bin. The resulting negative friction loadings were measured by means of load cells within the pile and the settlements of the pile and consolidating stratum by the dial gauges. Readings were taken at frequent intervals over the first few days of consolidation and then at 2 day intervals until loads and settlements reached constant values or varied by only small amounts, at which time the test was curtailed. This procedure was carried out four times at clay stratum thicknesses of 250, 310, 445 and 625mm with all other details of the test remaining constant (Test 2 - 310mm of clay - was exceptional with 305Kg of dry sand surcharge applied). Because

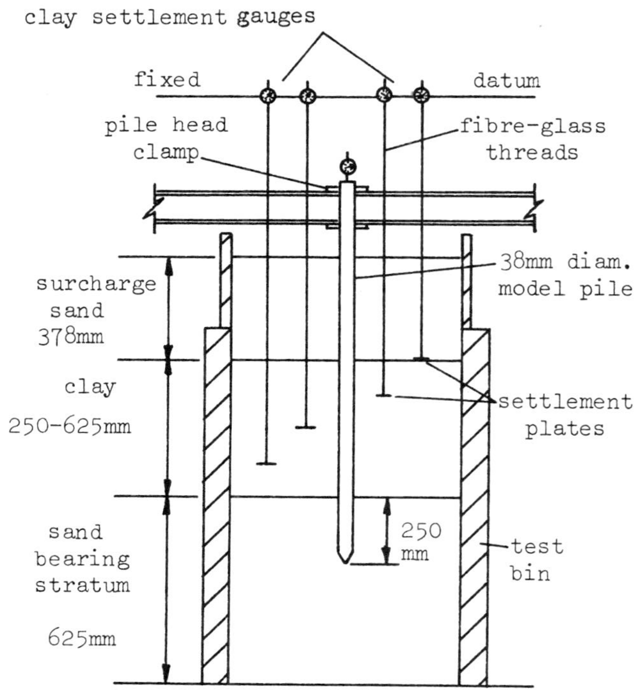

General arrangement of first testing programme

FIGURE 2

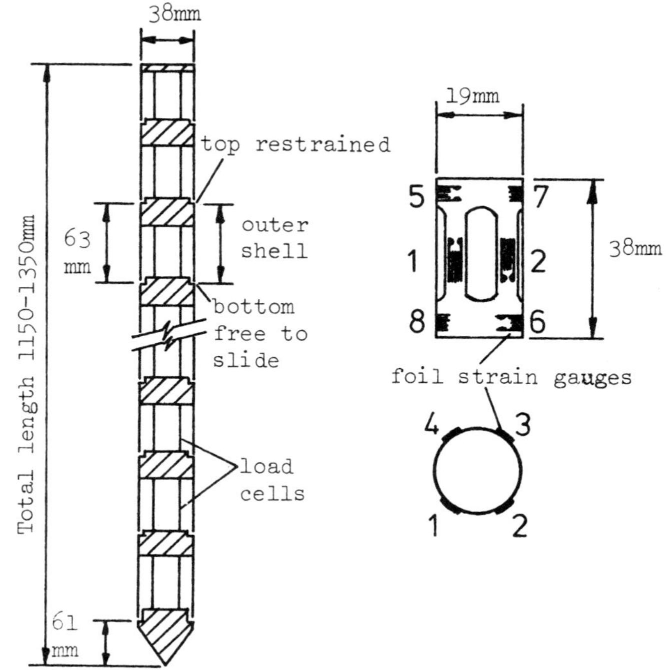

Diagrammatic representation of model pile and load cell

FIGURE 3

the tests could last up to 3 months the entire test rig was surrounded by a double-skin polythene tent and the temperature variation was kept within permissible limits by means of thermostatically controlled radiators.

(ii) Second testing programme

The test bin and settlement plates were identical to those used in the first programme but the model pile was of different diameter (25.4mm) and design. A saturated clay stratum was formed inside the bin, using the same hand-smearing technique, and the pile was completely embedded in this stratum thus simulating the case of a "floating" pile in clay. The clay was allowed to settle under its own self weight for 8 days and the negative friction loads that built up during this period were measured by the load cells in the pile. The pile was then subjected to uplift loading in order to assess its uplift resistance. This procedure was carried out 5 times for pile embedment lengths of 229, 391, 606, 825 and 1151mm.

(b) Model piles and instrumentation

(i) First testing programme

The outer shell of the model pile consisted of a series of hollow brass sections, 63mm long by 38mm external diameter, restrained at their top ends and free to slide at their bottom ends. The total length of the pile varied from 1150mm to 1350mm depending on the thickness of the clay stratum. The inner core of the pile was formed from a series of load cells which measured the friction on the outer shell (Figure 3). The structure of the model pile allowed each load cell to record the total negative friction acting on the pile down to the bottom of its outer shell. This also meant that the difference in recorded load between two successive cells, divided by the surface area of one outer shell, represented the average shear stress or skin friction developed on the lower of the two shells considered.

The load cells consisted of 38mm long by 19mm diameter hollow brass tubes with a wall thickness of 0.51mm. Four slots were cut in the cell walls to increase sensitivity. Because small load variations had to be measured, 4 longitudinal and 4 lateral precision foil strain gauges were cemented to the cell walls and wired in a configuration which minimised the effects of lateral strain and eccentricity. Because tests were being conducted in saturated soils, the cells were waterproofed after completion. The output from the cells was amplified and monitored using a conventional data-logging and print-out system. Instrumentation was developed which activated the load cells only when they were being scanned, to ensure that heat dissipation would not dry out the clay at the critical clay/pile interface. Full details of the piles, load cells and instrumentation are given by Nicholls (1973).

(ii) Second testing programme

The outer shell of the model pile consisted of a series of hollow brass sections, 50.8mm long by 25.4mm external diameter. The internal core of the pile consisted of a series of interconnected load cells which measured the friction on the outer shell in the same manner as the first programme. The load cells consisted of 42.1mm long by 13.9mm diameter hollow brass tubes, provided with screw threads at their ends to make them effective in both tension and compression, and the same configuration of 8 precision strain gauges was employed, (Figure 3). The output from the cells was monitored by a more sophisticated data-logging system which did not require external amplification and which automatically eliminated the possibility of heat dissipation at the clay/pile interface. Full details of the piles, load cells and instrumentation are given by Chaudhuri (1977).

(c) Test media

(i) First testing programme

The sand used for both the bearing stratum and the surcharge was a uniform, medium sea-dredged sand, locally called Portishead Fines, which had been cleaned, dried and sieved before use. Its relevant properties were as follows:-

91% of its particles lay between the Nos. 25 and 100 B.S. sieve sizes; limiting maximum and minimum dry densities = 1.87 and 1.45 g/ml respectively; average unit weight and angle of shearing resistance when compacted to form a saturated bearing stratum = 18.7kN/m^3 and 39° respectively; average unit weight and angle of shearing resistance when poured in a loose, dry state to form the surcharge = 14.4kN/m^3 and 35° respectively; average angle of friction between brass and sand in the saturated bearing stratum state and the loose surcharge state = 14° and 13° respectively.

The clay stratum was formed from a grade of Kaolin clay called Spestone Kaolin and was chosen because it possessed little secondary compression characteristics, thus allowing tests to be completed in a relatively short time (Burland 1967). Distilled, de-aired water was added to the dry clay and mixing was carried out under a vacuum of 103kN/m^2 to ensure complete saturation before the clay was hand-smeared into the bin. The following average properties were obtained:-

liquid limit = 64%; plastic limit = 37%; unit weight = 15.9kN/m^3, moisture content = 55%; coefficient of consolidation c_v = 0.0804mm^2/s; coefficient of volume decrease m_v =

0.00239m^2/kN; apparent cohesion c' = 4.3kN/m^2 (undrained) and 0.35kN/m^2 (drained); angle of shearing resistance ϕ' = 14° (drained); residual cohesion = 4.5kN/m^2 (ring shear) and 5.5kN/m^2 (reversal shear); adhesion between clay and brass = 3.8kN/m^2.

(ii) Second testing programme

The clay stratum in the second testing programme was formed from a grade of Kaolin clay called China Clay Grade E. It retained the same desirable characteristic of little secondary compression, but possessed different engineering properties to the Spestone Kaolin. The mixing and laying procedure was identical to that of the first programme and the following average properties were obtained:-

liquid limit = 43%; plastic limit = 27%; unit weight = 17.3kN/m^3; moisture content = 40.8%; coefficient of consolidation c_v = 0.531mm^2/s; coefficient of volume decrease m_v = 0.00286m^2/kN; apparent cohesion c' = 4.5kN/m^2 (undrained); angle of shearing resistance = zero (undrained); adhesion between clay and brass = 2.5kN/m^2.

RESULTS

The main results of the first and second testing programmes are summarised in Table I.

(i) First testing programme

Negative friction loads - In all four tests the negative friction loads increased with time until settlement of the clay stratum had ceased or was almost complete. The variation in negative friction load with depth for Test 2 (310mm clay) is shown in Figure 4(a) for four different time periods after completion of the laying of the clay stratum. It can be seen that the negative friction load imposed on the pile increased with depth, reaching a maximum value in the lower horizons of the clay stratum; a trend which applied for the period of self-weight consolidation of the clay and after the surcharge loading had been placed. It then decreased in magnitude as positive friction eroded its value. The lowest load cell recorded the resulting negative friction loading acting on the pile. The variations in negative friction load with depth and time for Tests 1-4 are shown at a reduced scale in Figure 5. The same behavioural pattern was obtained in all four tests.

Skin friction - The variation in skin friction with depth for Test 2 is shown in Figure 4(b) for four different time periods after completion of the laying of the clay stratum. It can be seen that the skin friction was negative (downwards), increasing with depth, in the sand surcharge and in the upper horizons of the consolidating clay stratum. It then decreased

TABLE I - SUMMARY OF TEST RESULTS

First testing programme - 38mm diameter end-bearing piles

Test No.	Surcharge sand depth (mm)	Clay depth (H) (mm)	Embedded length/ diameter ratio	Test duration (days)	Final Settlement (mm)		Final position of NP	Final NF loads (N)		Max. negative skin friction (kN/m²)		Notes
					Top of clay	pile		Maximum	Resultant	sand	clay	
1	378	250	23.1	82	11.0	0.40	0.85H	122	50	2.8	3.2	250 Kg surcharge
2	451	310	26.6	105	12.5	0.54	0.85H	187	60	3.8	4.3	305 Kg sur.
3	378	445	28.2	47	23.2	0.30	0.84H	170	43	2.1	2.4	250 Kg sur.
4	378	625	33.1	55	31.0	0.58	0.89H	164	106	1.9	2.7	"

Second testing programme - 25.4mm diameter floating pile

Test No.	Embedded pile length (H) (mm)	Embedded length/ diameter ratio	Test duration (days)	Final Settlement (mm)		Final position of NP	Final NF loads (N)		Max. negative skin friction (kN/m^2)
				Top of clay	pile		Maximum	Resultant	
5	825	32.5	8	7.8	1.6	0.57H	78	17	2.5

25-day curves give negative friction values under self-weight consolidation. Surcharge placed on 25th day

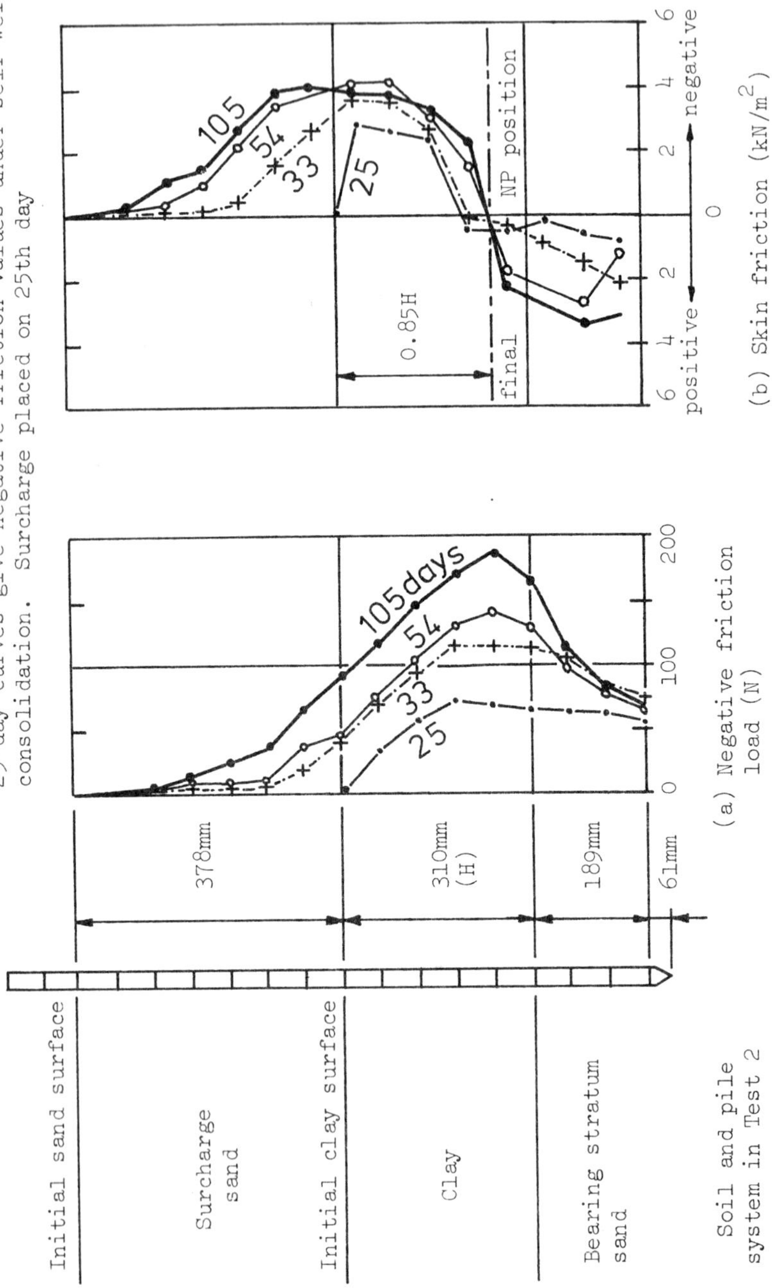

FIGURE 4

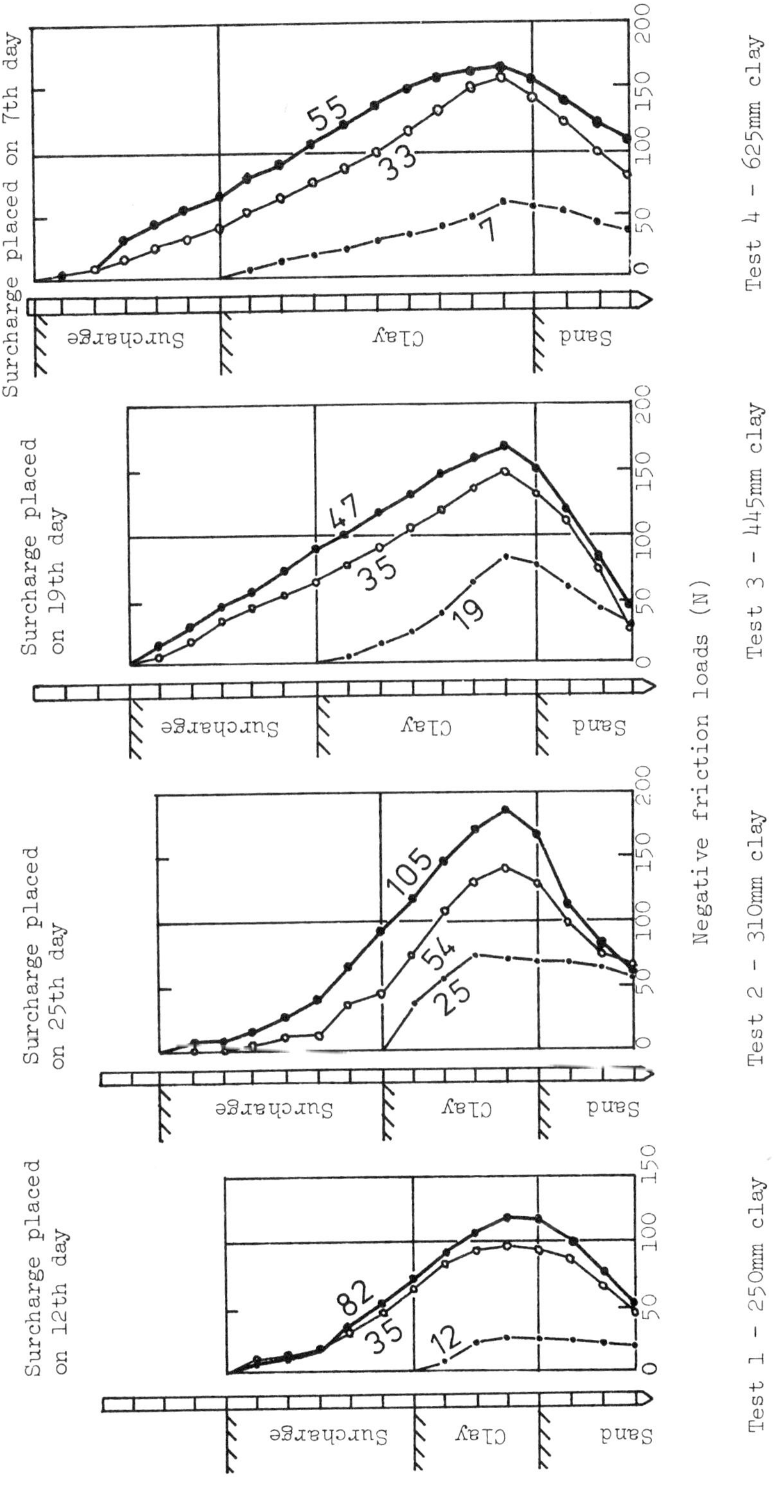
Surcharge placed on 7th day
Surcharge placed on 19th day
Surcharge placed on 25th day
Surcharge placed on 12th day
Surcharge
Clay
Sand
55
33
7
47
35
19
105
54
25
82
35
12
0
50
100
150
200
Negative friction loads (N)
Test 4 - 625mm clay
Test 3 - 445mm clay
Test 2 - 310mm clay
Test 1 - 250mm clay

FIGURE 5

in magnitude and became positive (upwards) in the lower horizons of the clay and in the sand bearing stratum. The variations in skin friction with depth and time for Tests 1-4 are shown at a reduced scale in Figure 6. The same general behavioural pattern was obtained in all four tests.

The neutral point - The point on the skin friction/depth plot at which the shear stress changes from a negative to a positive value is termed the neutral point. Reference to the larger scale plot of Figure 4(b) shows that, in Test 2, the neutral point descended with time, taking up a final position at a level of 0.85 times the initial thickness of the consolidating clay stratum. Reference to Table I and Figure 6 shows that a similar result was obtained in Test 1, but Tests 3 and 4 differed, with the neutral point holding a constant position after the placing of the surcharge. The final position of the neutral·point for all four tests lay in the narrow depth range of 0.84 - 0.89 times the initial thickness of the consolidating stratum.

Settlements - The average settlements recorded by the plates placed at different horizons within the clay stratum greatly exceeded the pile settlements, confirming that the necessary mechanism for negative friction had been established. The average final settlements recorded at the top of the clay stratum ranged from 11 to 31mm with the pile settlements, under the action of the negative friction loadings, lying in the range 0.30 to 0.58mm. The average settlement/time relationships for the plates and pile in Test 2 are shown in Figure 7(a). Similar relationships were obtained in the other three tests.

Figure 7(b) shows the average settlement/clay depth relationships for plates and pile at the end of Test 2. It can be seen that the relative movement between pile and soil changed from soil downwards to pile downwards almost at the bottom of the consolidating clay stratum, whereas the skin friction/depth relationship (Figure 4(b)) showed that this occurred at a depth of 0.85 times the initial thickness of the consolidating stratum; a discrepancy which may be explained as follows:-

the nearest settlement plates to the pile were placed 75mm from the pile face to avoid any interference with the soil/pile interaction, and this was not near enough to measure the degree of "hang-up" that took place between the clay and the pile. This is illustrated in Figure 7(c) which shows the final settlement profiles for the three horizons ($\frac{1}{4}$, $\frac{1}{2}$ and $\frac{3}{4}$ depth of the clay stratum) at which a series of settlement plates were installed. The profiles for the $\frac{1}{4}$ and $\frac{3}{4}$ depth horizons suggest that considerable hang-up of the clay on the pile shaft did occur, although it is not possible to extrapolate the results to give actual values.

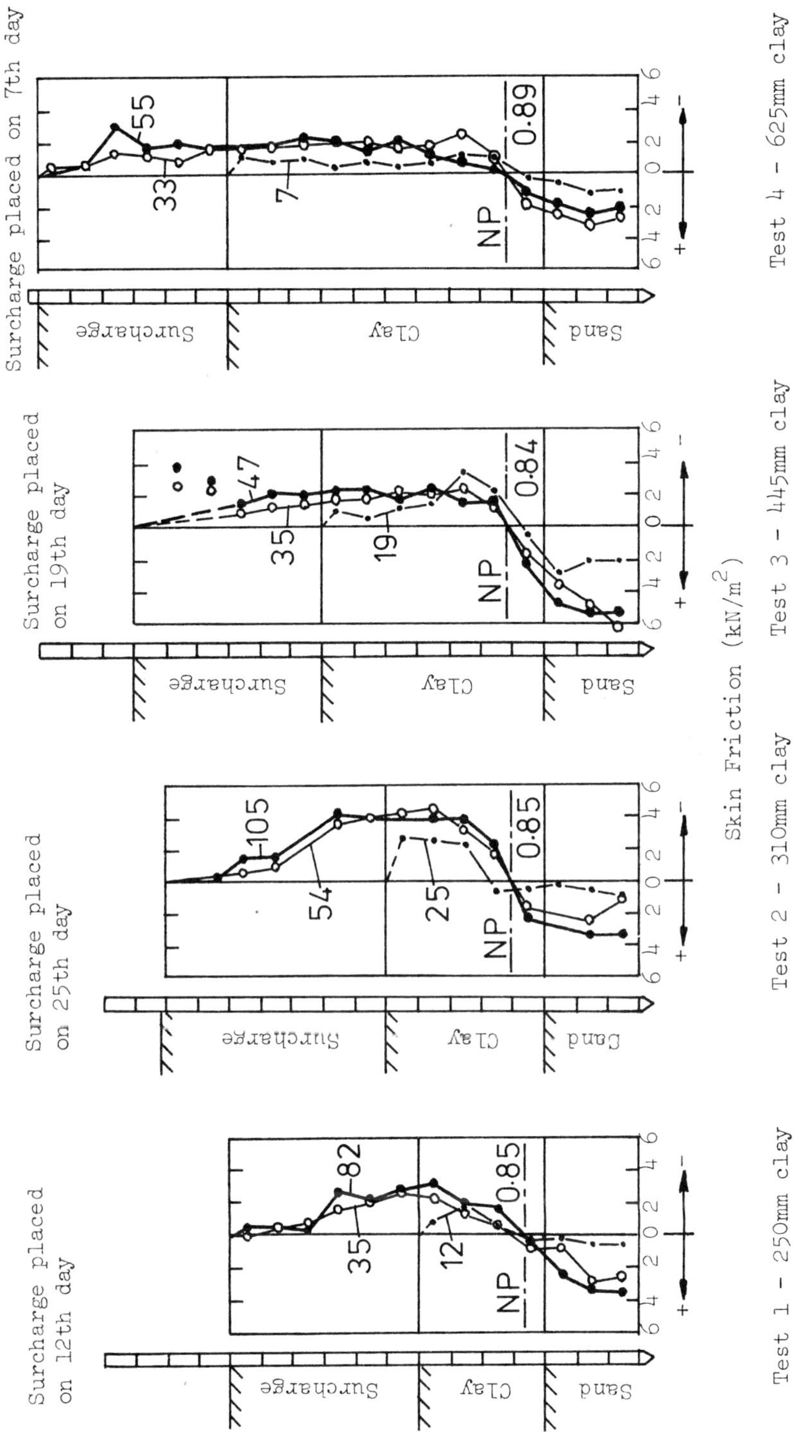
Surcharge placed on 12th day
Surcharge placed on 25th day
Surcharge placed on 19th day
Surcharge placed on 7th day
Surcharge
Clay
Sand
NP
0.85
0.85
0.84
0.89
35
82
12
54
105
25
47
19
33
55
7
Skin Friction (kN/m²)
Test 1 - 250mm clay
Test 2 - 310mm clay
Test 3 - 445mm clay
Test 4 - 625mm clay

FIGURE 6

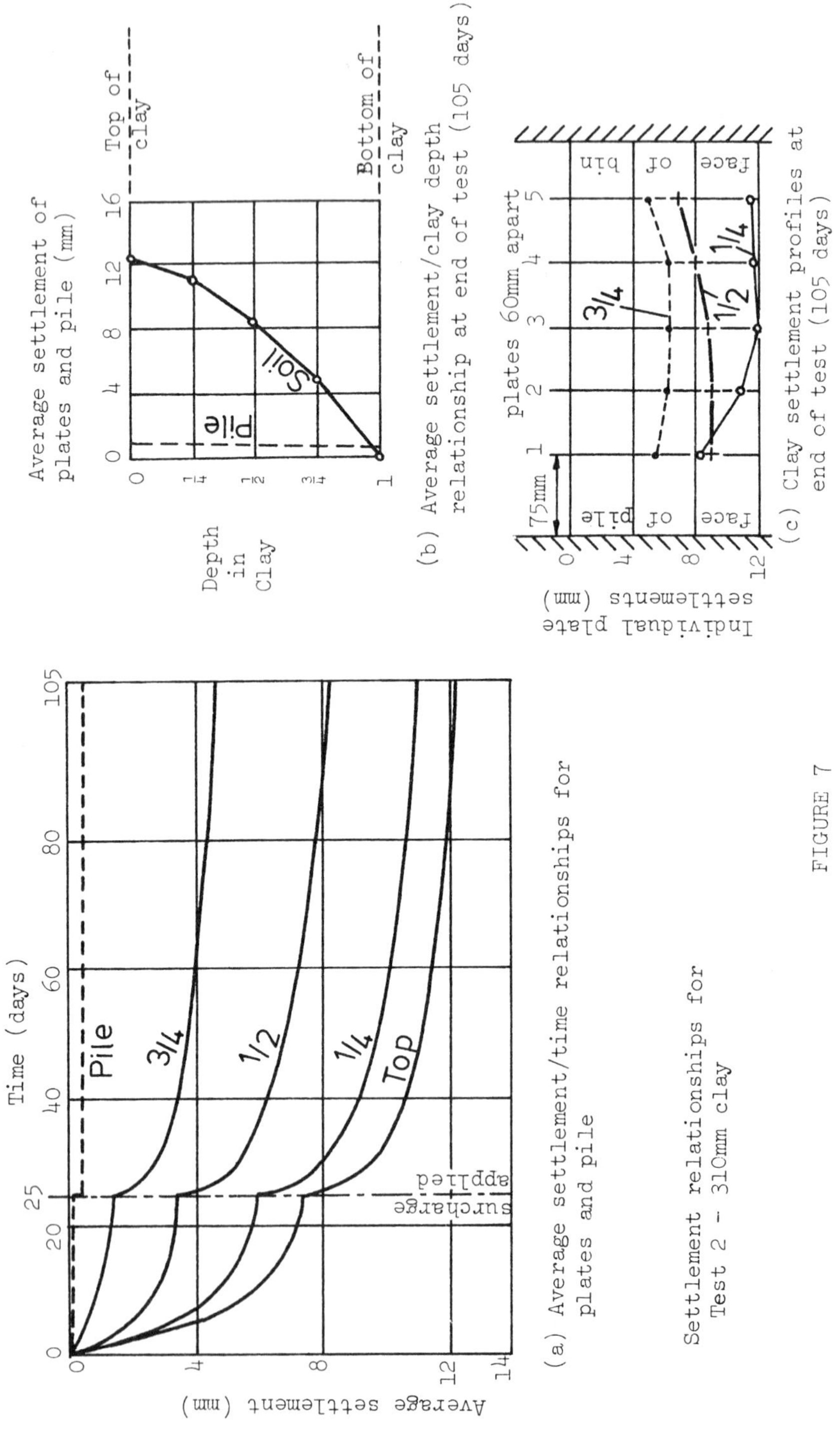

(a) Average settlement/time relationships for plates and pile

(b) Average settlement/clay depth relationship at end of test (105 days)

(c) Clay settlement profiles at end of test (105 days)

Settlement relationships for Test 2 - 310mm clay

FIGURE 7

There seems little doubt that settlements of the clay at the clay/pile interface were less than those recorded by the plates. Visual confirmation of this was provided when the test was dis-assembled and the top surface of the clay was observed to form an elevated cone in the immediate area around the pile shaft despite slippage having occurred between clay and pile.

(ii) Second testing programme

Similar behavioural patterns were obtained in all 5 tests and, because this was an ancillary testing programme, one typical result is reported. Figure 8 and Table I give the negative friction loads, skin frictions, neutral point positions and settlements for the 825mm depth of embedment test.

It can be seen that the higher pile settlements normally associated with "floating" piles in clay occurred. A higher neutral point (0.57 times the depth of the embedment of the pile in clay) was recorded at the end of the period of self-weight consolidation.

ANALYSIS

Comparison of measured and calculated negative friction values

A previous investigation into model pile groups subjected to the identical negative friction mechanism of the first testing programme showed that the most relevant methods for the calculation of negative friction loadings were those based on direct friction or adhesion considerations (Denman, Nicholls and Symons 1977). Many such methods have been proposed (Buisson, Ahu and Habib 1960, Tomlinson 1969, Johanessen and Bjerrum 1965, Endo et al 1969, Bozozuk 1972, Resendiz and Auvinet 1973, Poulos and Davis 1975, Meyerhof 1976), but the basic approach adopted in design practice was summarised in Meyerhof's 1976 state-of-the-art paper. The following values were suggested by Meyerhof for the average ultimate skin friction, f_s, in sand and clay:-

$$\text{Sand:} \quad f_s = K_s \overline{p}_o \tan\delta \qquad (1)$$

in which K_s = average coefficient of earth pressure on pile shaft in sand; $\overline{p}_o$ = average effective overburden pressure in sand; δ = angle of friction between pile shaft and sand.

$$\text{Clay:} \quad f_s = c + K_s \overline{p}_o \tan\emptyset \qquad (2)$$

in which c and $\emptyset$ are effective drained shear strength parameters of clay; K_s = average coefficient of earth pressure on pile shaft in clay; $\overline{p}_o$ = average effective overburden pressure in clay.

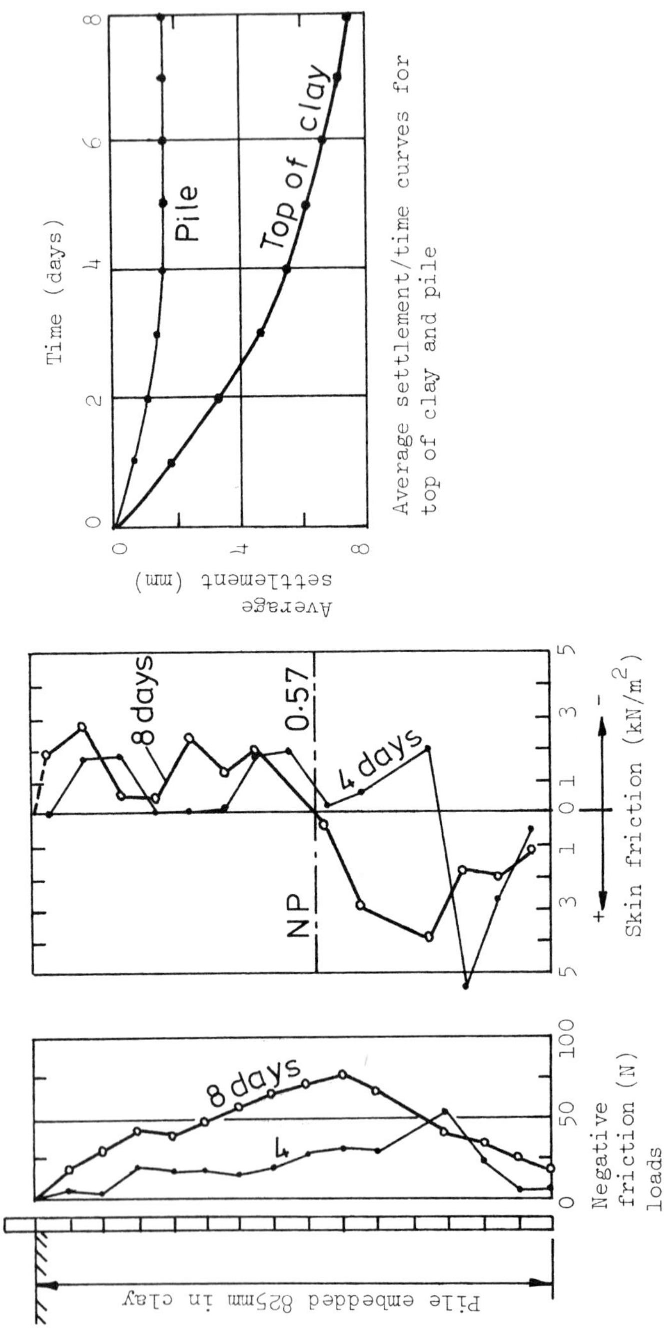

Summarised results for Test 5 - floating pile embedded 825mm in clay

FIGURE 8

c, $\emptyset$, $\overline{p}_o$ and δ can be obtained for any given soil and pile system but a value of K_s must be allocated. Most authors have suggested that, for smooth-shafted piles in loose sands and soft clays, K_s is less than or equal to the coefficient of earth pressure at rest, K_o, normally assumed equal to 1-sin$\emptyset$, a value of less than unity.

Use of equation (1) with an assumed minimum K_s value of 0.5 considerably underestimated the average skin friction measured in the sand surcharge (Table II). K_s values of between 2 and 3 would have to be used in order to obtain correlation between calculated and measured values.

Use of equation (2) with a K_s value of 1-sin$\emptyset$ gave approximate correlation between the calculated and measured average skin frictions in the clay for the shallower Tests 1 and 2, but overestimated the calculated values for the deeper Tests 3 and 4.

The calculated negative friction loads, based on the calculated average skin frictions in sand surcharge and clay, are given in Table II, where they are compared with the measured values. It can be seen that, in all four tests, the calculated resultant negative friction load imposed on the pile exceeded the measured resultant load despite the underestimation of the sand surcharge negative friction. This was because the calculated clay skin friction was assumed negative to the bottom of the clay stratum, without account being taken of positive friction below the neutral point.

Position of the neutral point

Any future development of direct friction or adhesion theories for negative friction estimation must be based on realistic assumptions regarding the position of the neutral point. It is difficult, however, to theoretically determine this position, because the "hang-up" phenomenon at the soil/pile interface invalidates calculations based on the relative compressions of soil and pile, and use will have to be made of published results. A number of investigations, involving full-size piles, theoretical approaches and model testing, have given neutral point positions and their findings are summarised below in terms of Authors' names, year of publication, pile type and position of neutral point.

Full-size tests: Endo et al (1969) and Sawaguchi (1971), end-bearing 0.73 to 0.76, floating 0.78; Bozozuk and Labrecque (1969), end-bearing 0.81; Bozozuk (1972), floating 0.34. Theoretical approaches: Salas and Belzunce (1965), end-bearing 0.85, floating 0.43; Bozozuk (1972), end-bearing 0.71.

TABLE II

COMPARISON OF CALCULATED AND MEASURED NEGATIVE FRICTION VALUES

First testing programme - 38mm diameter end-bearing piles

Test No.	Surcharge sand values				Clay values				Embedded pile values	
	Measured		Calculated		Measured		Calculated		Measured	Calculated
	NF load (N)	average f_s (kN/m^2)	NF load (N)	average f_s (kN/m^2)	NF load (N)	average f_s (kN/m^2)	NF load (N)	average f_s (kN/m^2)	resultant NF load (N)	resultant NF load (N)
1	71	1.6	14	0.3	48	1.6	53	1.8	50	67
2	85	1.6	21	0.4	85	2.3	75	2.0	60	96
3	86	1.9	14	0.3	68	1.3	108	2.0	43	122
4	68	1.5	14	0.3	86	1.2	173	2.3	106	187

Model testing: Silva (1966), endbearing 0.96; Chaudhuri, Denman, Nicholls and Symons (1980), end-bearing 0.85 to 0.89, floating 0.57.

It can be seen that the results of this present model testing programme compare well with the full-size investigation of Bozozuk and Labrecque and confirm the theoretical investigation of Salas and Belzunce.

CONCLUSIONS

The following conclusions relate to model studies involving smooth-shafted single piles subjected to the negative friction mechanisms described:-

1. Negative friction loading is time-dependent, building up gradually to a maximum value as consolidation proceeds.
2. The maximum value of negative friction loading is governed by the final position of the neutral point. This investigation gives final neutral point positions of 0.85 to 0.89 times the initial thickness of the consolidating stratum for end-bearing piles, and 0.57 times the length of the pile in clay for floating piles (under self-weight consolidation); results which confirm full scale and theoretical findings.
3. Existing design practices, based on average shear stress or skin friction values acting down the entire pile length, oversimplify the negative friction mechanism. All estimates of negative friction loading should assume negative skin friction down to the neutral point only, with positive skin friction acting on the pile below this level.
4. The model testing of this investigation closely simulated practical conditions and, in view of the fact that close agreement was obtained with a previously reported full-size test, extrapolation of the results to full-size piles may be valid.

ACKNOWLEDGEMENT

The work reported in this paper was carried out in the Department of Civil Engineering and Building Technology of the University of Wales Institute of Science and Technology and constitutes part of a wider research programme into the model testing of piled foundations.

REFERENCES

M.M. Baligh, V. Vivatrat and H. Figi - Downdrag on bitumen-coated piles, Journal of the Geotechnical Engineering Division, ASCE 104 (1978) 1355-1370.

M. Bozozuk - Downdrag measurements on a 160 ft. floating pipe test pile in marine clay, Canadian Geotechnical Journal 9. (1972) 127-136.

M. Bozozuk and A. Labrecque - Downdrag measurements on a 270 ft. composite pile, Proceedings of Symposium on Performance of Deep Foundations ASTM STP444 (1969) 15-40.

M. Buisson, J. Ahu and P. Habib - Le frottment negatif, Annales de L'Institut Technique du Batiment et des Travaux Publics 145 (1960).

J.B. Burland - Deformation of soft clay, University of Cambridge PhD Thesis (1967).

K.P.R. Chaudhuri - Uplift resistance of foundations, University of Wales PhD Thesis (1977).

R.D. Chellis - Pile Foundations, 2nd Edition, McGraw Hill Book Co. (1964).

K.J. Denman, R.A. Nicholls and M.V. Symons - Model studies of negative friction on pile groups, Conference on Large Ground Movements and Structures, UWIST, Cardiff, Pentech Press (1977) 252-271.

M. Endo, A. Minou, T. Kawasaki and T. Shibata - Negative skin friction acting on steel pipe-piles in clay, Proceedings of the Seventh International Conference on Soil Mechanics and Foundation Engineering 2 (1969) 85-92.

J. Florentin and G. L'Heriteau - About an observed case of negative friction on piles, Proceedings of the Second International Conference on Soil Mechanics and Foundation Engineering 5 (1948) 155.

I.J. Johanessen and L. Bjerrum - Measurement of the compression of a steel pile to rock due to settlement of the surrounding clay, Proceedings of the Sixth International Conference on Soil Mechanics and Foundation Engineering 2 (1965) 261-264.

G.G. Meyerhof - Bearing capacity and settlement of pile foundations, Journal of the Geotechnical Engineering Division, ASCE 102 (1976) 197-228.

R.P. Milner - Discussion on foundations of structures, Proceedings of the Fourth International Conference on Soil Mechanics and Foundation Engineering (1957) 201-202.

O. Moretto and A.J.L. Bolognesi - Pile foundations stressed by negative friction, Proceedings of the First Pan American Conference on Soil Mechanics and Foundation Engineering 3 (1959) 315-325.

R.A. Nicholls - Negative friction on piled foundations, University of Wales PhD Thesis (1973).

H.G. Poulos and E.H. Davis - Prediction of downdrag forces in end-bearing piles, Journal of the Geotechnical Engineering Division, ASCE 101 (1975) 189-204.

D. Resendiz and G. Auvinet - Analysis of pile foundations in consolidating soil, Proceedings of the Eighth International Conference on Soil Mechanics and Foundation Engineering 2 (1973) 211-218.

J.A. Salas and M. Belzunce - Theoretical solution of stress distribution in piles, Proceedings of the Sixth International Conference on Soil Mechanics and Foundation Engineering 2 (1965) 309-313.

M. Sawaguchi - Approximate calculation of negative skin friction of a pile, Soils and Foundations 11 (1971) 31-49.

A.J. Silva - Downdrag on piles, University of Connecticut, U.S.A., PhD Thesis (1965).

M.J. Tomlinson - Foundation Design and Construction, Pitman and Sons Ltd., London (1969).

L. Zeevaert - Foundation Engineering for Difficult Subsoil Conditions, Van Nostrand Reinhold Co., New York (1973).

SESSION VII

Chairman: Mr C J Evans, MA, CEng, FIStructE, FICE, FIHE, FASCE, MConsE
Vice-President The Institution of Structural Engineers
Chairman Wales Branch The Institution of Structural Engineers

Introduction to Session: Dr M V Symons, BSc, PhD, FGS
Department of Civil Engineering and Building Technology
UWIST

SESSIONAL PAPERS

PREDICTED AND MEASURED GROUND MOVEMENTS ADJACENT TO A TRENCH EXCAVATION IN LONDON CLAY

K. Kyrou, BSc(Eng), MSc(Eng)

Research Student, University of Surrey

B.K. Menzies, PhD, MICE, FGS

Lecturer, University of Surrey

A.I. Tarzi, BSc(Eng), MPhil

Lecturer, University of Surrey

THE INFLUENCE OF EXCAVATION SHAPE ON THE PATTERN OF EXCAVATION INDUCED GROUND MOVEMENTS

The ground movements associated with deep wide excavations have achieved prominence in the literature of the past decade since the comment by Peck (1969) about the scarcity of case histories leading to an incomplete understanding of the problem. Particular attention has been focussed on the measurement, prediction and the minimisation of ground movements by appropriate constructional procedures (Cole and Burland, 1972; Burland and Moore, 1973; Huder, 1969; St. John, 1975; Tomlinson, 1975; Simpson et al, 1979). In contrast, the performance and serviceability of structures adjacent to an excavation which interact with the excavation induced soil deformation, has received little attention. The vertical component of movement outside an excavation has often been considered the critical component which could cause damage to nearby structures. As Cole and Burland (1972) point out, however, "there is no reason to believe that structures are less sensitive to horizontal movements than to vertical movements and indeed the reverse may be true". Indeed, St. John (1975) and Creed (1979) show that horizontal movements due to excavations can be considerably higher than vertical ones. Accordingly, horizontal ground movements induced by a deep wide excavation may be of sufficient magnitude to cause damage or loss of serviceability of nearby structures. It may be expected therefore that lateral deformations will be induced in soil adjacent to shallow narrow excavations such as trenches, and that these movements may be sufficient to affect adversely nearby structures such as buried services. In comparison with the ground movements induced by deep excavations, these movements are small although poor control of the works may allow greater magnitudes. Whether the movements are small or large, however, they may be of such a magnitude as to cause excessive straining and subsequent failure of adjacent pipelines and services. Comparatively little has appeared in the literature on the nature of movements associated with trench excavations probably because trench excavations are temporary and are backfilled with soil or filled with concrete as in the case of diaphragm walls. The influence of trench excavation induced ground movements on nearby underground services such as gas and water mains is now receiving attention because failures have occurred some months after reinstatement of the trench (TRRL, 1978). An analytical model has been developed by Crofts et al (1977) which examines the effect on buried pipes of the longitudinal bending stresses induced by adjacent trench excavations. The pipe-soil interaction was modelled by a beam embedded in a bed of springs and was similar to a beam on an elastic foundation made up of a Winkler medium. This model was sensitive to the restraining effect of the end of the trench where the adjacent pipe was subject to greatest bending. The extent of the zone of trench end restraint upon the displacement of the pipe was treated intuitively. To enable this important feature to be studied further, two dimensional and three dimensional finite element continuum models may be employed. These techniques also allow a study to be made of the form of ground movements associated with trenches.

This paper reports on a simple finite element study of deformations associated with the cutting of a slot in a pre-stressed linear elastic medium and considers the validity of this model for predicting the form and magnitude of ground movements associated with trench excavations. The model is calibrated by comparing the predicted movements with those observed during and after excavation, propping and back-filling of a trench cut in London clay (Gumbel and Wilson, 1980).

THE ANALYTICAL MODELS

The soil was modelled by an isotropic linear-elastic solid. The soil behaviour was analysed using the finite element method for a two dimensional model imposing both plane stress and plane strain conditions, and by a three dimensional model.

The assumption of isotropy was found by Creed (1979) to have little effect on the pattern of ground movements where deformations associated with excavations supported by diaphragm walls, was analytically modelled. Non-homogeneity of the modelled soil was incorporated in the analysis since this affects considerably the pattern of movements (Simons, 1974).

Excavation of the trench was simulated by total stress removal along the face of a slot in the elastic continuum.

Using the finite element models described, horizontal displacement patterns of the face of the trench at ground level were obtained and they are shown in Fig. 1. The points to be noted are:

(i) Considerable restraint to horizontal movement is offered by the excavation ends as well as the bottom of the excavation.

(ii) The build up of horizontal movements near the excavation ends is very rapid and most of the movement takes place within a distance equal to the depth of the excavation. Although horizontal movements due to wide and narrow trenches of the same depth may be of different magnitudes, the rate of build up of movement is almost equally rapid.

(iii) The horizontal displacement pattern obtained from a two dimensional plane stress model which essentially represents a deep slot in the ground follows a beam type deflection profile which is typical of deep excavations (Cole and Burland, 1972; St. John, 1975).

Further results showed that:

(iv) Vertical movements were of comparable magnitude and the build up of these movements is almost equally rapid near the excavation ends as in the case of horizontal movements.

(v) Horizontal and vertical movements away from the excavation corners computed from two dimensional plane strain models were of a similar pattern and magnitude as those computed from three dimensional models.

THE TRIAL TRENCH

In order to provide a case history of ground movement data relating specifically to trench excavations, a 4m deep by 1m wide by 15m long trial trench was dug in the basement

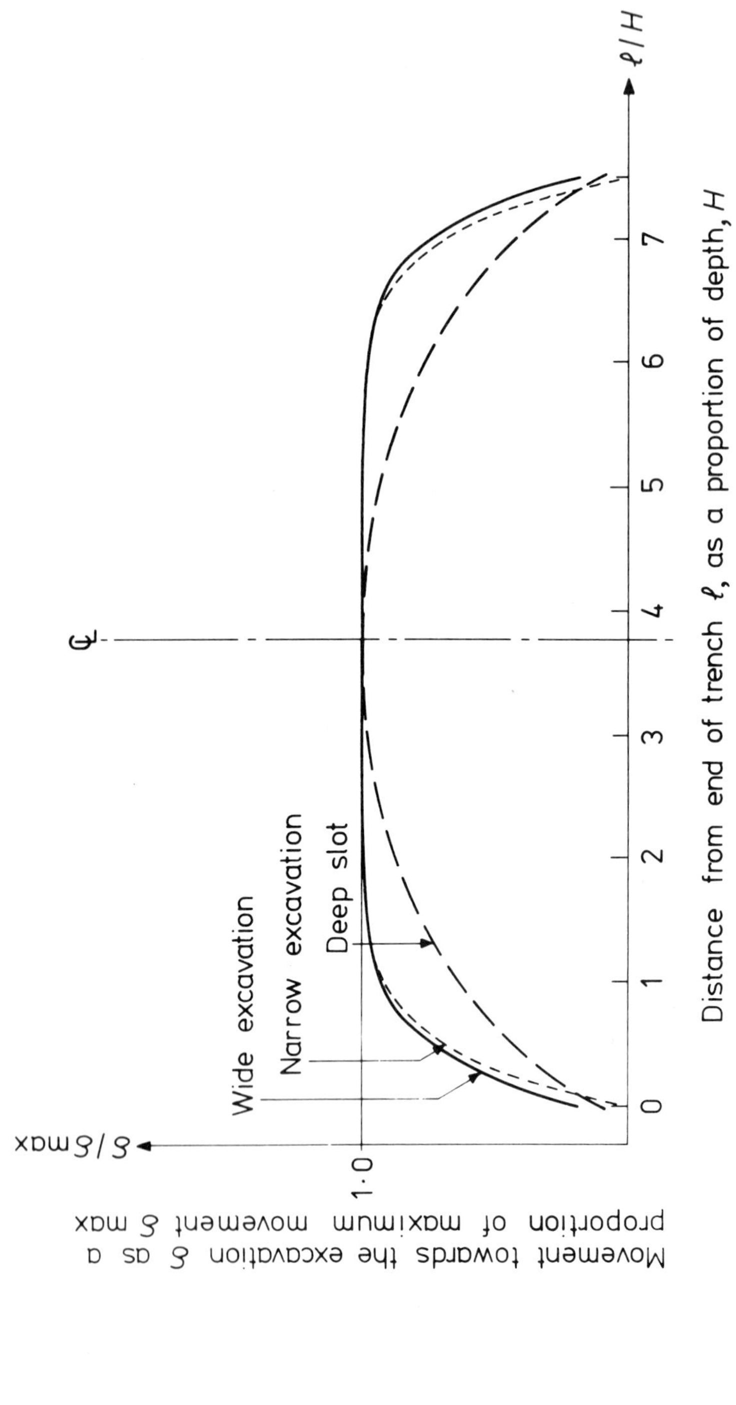

Fig. 1. Horizontal movement of excavation face at ground level

beds of the London clay at the Manor Farm site of the University of Surrey, Guildford. Movements of the ground surface around the trench were monitored using simple surveying methods since the recorded movements were in the order of 10-40mm. The trench was partially shored 4-5 hours after excavation and left open for 28 days, although monitoring of movements continued after back-filling. Some indication of the stiffness of the ground was obtained from insitu plate bearing tests applied to the trench walls about 48 hours after excavation. The stiffnesses were obtained from a loading condition and it is unlikely that they represent the unloading condition arising from the stress release on the excavation faces. In the absence of better data, however, these stiffnesses were used in the analysis.

PREDICTED AND OBSERVED PATTERNS OF MOVEMENT ASSOCIATED WITH THE TRIAL TRENCH

Any correlation between the observed and predicted magnitudes must be regarded as fortuitous owing to the uncertainties relating to the interpretation of plate loading tests and the soil behaviour which may not be that of a continuum. Nevertheless, the predicted form of the ground movements compares well with the observed form of ground movements (Fig 2 and 3). In particular, the predicted and observed horizontal ground movements are in good agreement in showing the rapid build up of trench end restraint. The vertical movements, although smaller in magnitude seem to be of a similar nature.

The following broad conclusions may be drawn from this study:

(i) Sufficient care should be applied in interpreting data from finite element models on a quantitative basis due to the time dependency of the soil stiffness (Burland et al, 1979) and consequently the time dependency of the ground movements.

(ii) An analytical simulation of the problem of trench excavations using two dimensional finite element models is subject to limitations due to the three dimensional nature of the displacement vectors in the field. In particular, the inability of these models to simulate the effect of the bottom of the trench on the ground movements is shown in Fig. 1.

(iii) Good qualitative comparisons can be made between the observed ground movements and the movements

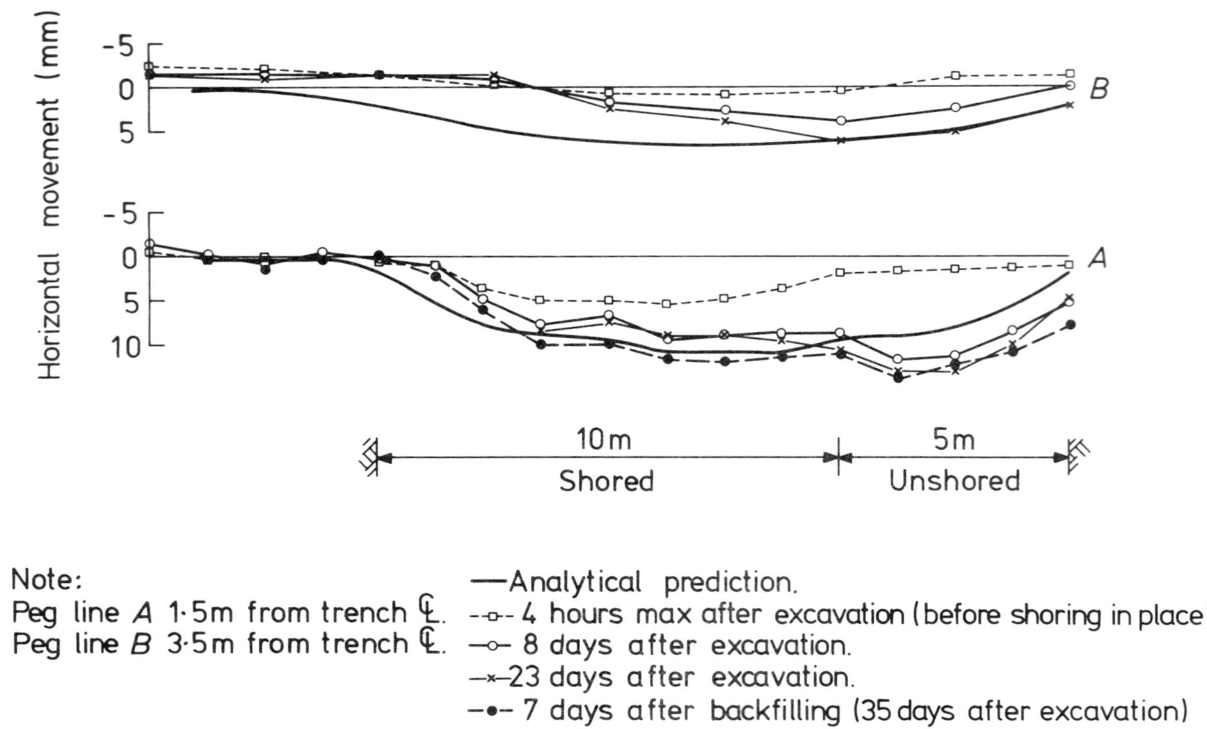

Fig. 2. Manor Farm Trial Trench - Horizontal movements of ground surface pegs

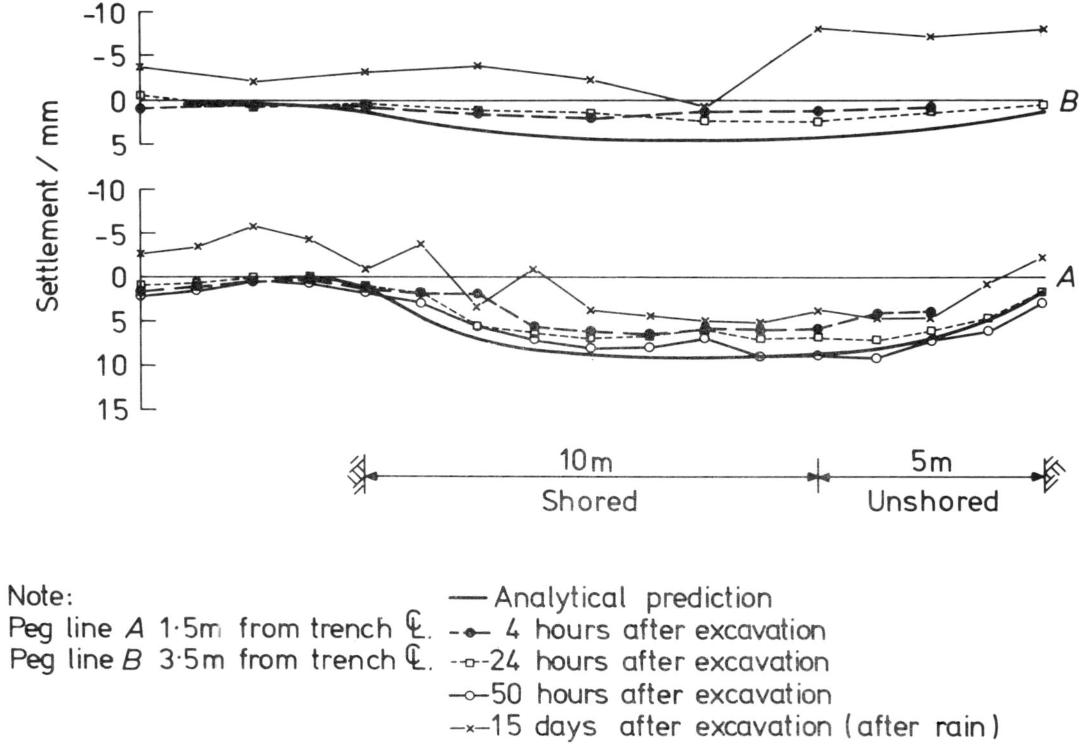

Fig. 3. Manor Farm Trial Trench - Vertical movement of ground surface pegs

obtained using linear-elastic three dimensional finite element models.

(iv) The trench end restraint affects considerably the pattern of horizontal and vertical ground movements.

REFERENCES

J.B. Burland, B. Simpson, H.D. St. John - Movements around excavations in London Clay, Proceedings 7th European Conference on Soil Mechanics and Foundation Engineering, Brighton, (1979), 13-29.

J.B. Burland and J.F.A. Moore - The measurement of ground displacement around deep excavations, Proceedings of Symposium on Field Instrumentation, (1973), Butterworths, 70-84.

M.J. Creed - Analysis of the deformations associated with diaphragm walls, Ph.D. thesis, University of Surrey (1979).

J.E. Crofts, B.K. Menzies and A.I. Tarzi - Lateral displacement of shallow buried pipelines due to adjacent deep trench excavations, Geotechnique 27, No. 2, (1977), 161-179.

K.W. Cole and J.B. Burland - Observation of retaining wall movements associated with a large excavation, Proceedings 5th European Conference on Soil Mechanics and Foundation Engineering, Madrid, (1972), Vol. 1, 445-453.

J.E. Gumbel and J. Wilson - Observations of ground movement around a trench excavation in London Clay, 2nd Conference on Ground Movements and Structures, Cardiff, (1980).

J. Huder - Deep braced excavation with high ground water level, Proceedings 7th International Conference on Soil Mechanics and Foundation Engineering, Mexico, (1969), 443-448.

R.B. Peck - Deep excavations and tunnelling in soft ground, Proceedings 7th Conference on Soil Mechanics and Foundation Engineering, Mexico, State of the Art Report, (1969) 225-290.

N.E. Simons - Normally consolidated and lightly overconsolidated cohesive materials, Proceedings British Geotechnical Society Conference on Settlement of Structures, Session II, General Report, (1974), Pentech Press, 500-530.

B. Simpson, N.J. O'Riordan, and D.D. Croft - A computer model for the analysis of ground movements in London Clay, Geotechnique 29, No. 2, (1979) 149-175.

H.D. St. John - Field and theoretical studies of the behaviour of ground around deep excavations in London Clay, Ph.D. thesis, University of Cambridge, (1975).

M.J. Tomlinson - Foundation design and construction, Third Edition, London : Pitman (1975).

Transport and Road Research Laboratory - Ground movements and their influence on buried pipes, TRRL Leaflet LF681, (1978).

BACK ANALYSIS OF THE BEHAVIOUR OF A DIAPHRAGM WALL SUPPORTED EXCAVATION IN LONDON CLAY

Michael J. Creed, B E , M Eng Sc , Ph D
Lecturer in Civil Engineering, University College, Cork

Noel E. Simons, M Sc (Eng) , M A , Ph D , F I C E , F G S
Professor of Geotechnical Engineering, University of Surrey

Gilliane C. Sills, Ph D , M I C E
Lecturer in Engineering Science, University of Oxford

INTRODUCTION

Observation of the performance of full scale foundations in the field plays an important role in improving understanding of the behaviour of the ground and its interaction with the structure. The field measurements may be used in conjunction with appropriate analytical techniques, to back analyse the in situ deformation properties of the soil. Provided the analysis is qualitatively correct, back analysis can provide valuable quantitative information, which may be of direct use, where similar ground conditions are encountered. Finite element elastic analysis has been used on a number of occasions to estimate the deformation parameters of heavily overconsolidated London Clay, by comparing analytical results with field observations (e.g. Cole and Burland, 1972; Hooper, 1973; St. John, 1975). More recently Simpson, O'Riordan and Croft (1979) have suggested that a non-linear elasto-plastic stress-strain model is more suitable for modelling the behaviour of London Clay around excavations. An excellent review of the movements around excavations in London Clay has been presented by Burland, Simpson and St. John (1979).

Finite element analysis of case records in London has been mainly concentrated on deep excavations, supported by propped or braced diaphragm walls (Cole and Burland, 1972; St. John, 1975; Simpson et al, 1979). The results of a field instrumentation programme on an excavation in stiff London Clay, supported by an anchored diaphragm wall, have been published by Sills, Burland and Czechowski (1977). The end of excavation performance of this excavation is analysed assuming the London Clay to be a cross-anisotropic linear elastic material, which is almost incompressible in the undrained condition.

SITE AND INSTRUMENTATION

The instrumented wall panel, described by Sills et al (1977), forms part of a diaphragm wall, retaining the sides of a cutting, up to 10m deep, which takes Neasden Lane under the North Circular Road, in north west London. Each wall panel is supported by eight prestressed tiebacks, inclined at angles varying between 20° and 40°, and anchored in the London Clay. At the instrumented panel, the excavation base is about 8.6m below the retained ground surface.

The site is covered by 200mm of topsoil and made-up ground, overlying stiff brown fissured silty London Clay, becoming grey-blue at a depth of approximately 8m. Although none of the site investigation boreholes was taken into the Woolwich and Reading Beds, nearby borehole information suggests that they are encountered at a depth of about 30m. This information further suggests that chalk is found at a depth of approximately 50m. The soil properties, published by Sills et al (1977), are believed to be fairly typical of London Clay.

Instrumentation was placed along Elm Way, which lies almost perpendicular to the line of the wall. Displacements of the retained ground were measured with reference to two datum points, 44m and 60m behind the wall. Horizontal and vertical surface movements were measured at five surface survey points, placed between 4m and 38m behind the wall. Subsurface movements were measured by means of magnet extensometers (for vertical movements) and inclinometers (for horizontal movements) at distances of 4m and 19m behind the wall. Wall movements were measured using an inclinometer cast in the instrumented wall panel. Pore water pressures were measured using pneumatic piezometers and Casagrande standpipes, placed just behind the wall, and at distances of 4m and 19m from the excavation. The loads were measured in each of the eight anchors in the test panel, using vibrating wire load cells. A fuller description of the instrumentation may be found in Sills et al (1977).

FIELD MEASUREMENTS

Surface surveys and instrument readings took place at various times before, during and after completion of the excavation. Excavation commenced in mid-February, 1972, and was carried out in a number of steps to completion in early August of that year. In the test panel the four rows of anchors were installed on 1/6/1972, 23/6/1972, 24/7/1972 and 4/10/1972, respectively. A comprehensive record of the field measurements may be found in Sills et al (1977).

During the early stages of excavation, the magnitude of horizontal movement decreased with increasing distance from the excavation. After prestressing of the second and third rows of anchors, the observed movements indicated a zone of compression immediately behind the wall, with extension occurring further from the excavation. The compression behind the wall decreased somewhat, but nevertheless remained in the long term. Significant vertical settlements were concentrated within 20m of the wall, while horizontal movements extended much further back. All movements were observed to be strongly time-dependent, about 50% on average occurring after the end of excavation.

Considerable difficulty was experienced with the operation of the piezometers, due mainly to damage caused during construction. A general drop in pore pressure was observed to continue after the end of excavation, reaching a minimum eight to ten months later.

Anchor loads either remained constant or decreased slightly with time to an eventual constant level.

FINITE ELEMENT ANALYSIS

A finite element analysis is carried out, assuming a condition of plane strain, on a unit width of wall, retained by equivalent anchored tiebacks. The equivalent prestress loads and cross-sectional areas are taken to be the actual values divided by the horizontal spacing of the tiebacks. Such an approach has been shown to be valid for a typical diaphragm wall by Clough and Tsui (1974).

The excavation is assumed to be 8.6m deep, 20m wide and symmetrical about the centre line of the cutting (see Figure 1). Only half the symmetrical excavation is analysed. A rigid horizontal boundary is assumed at the base of the Woolwich and Reading Beds, 50m below the ground surface. A rigid vertical boundary is assumed at a distance of 90m from the excavation. Rodrigues (1975) has investigated the influence of the position of this boundary, and suggested that it has negligible effect on the movements in the vicinity of an excavation, provided it is at least ten times the excavation depth from the excavation.

The finite element continuum is composed of three materials. The reinforced concrete diaphragm wall, which is 0.6m wide and 13m deep, is represented by 14 linear strain triangular elements. This number is sufficient to model accurately the bending of the wall. The concrete is assumed

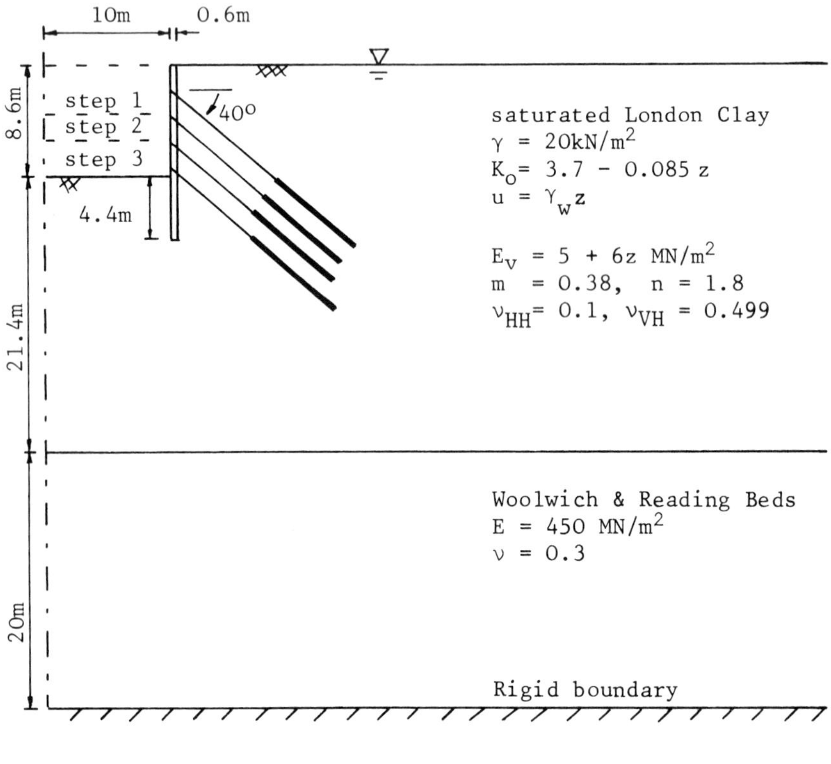

FIGURE 1 ANALYTICAL SIMULATION OF EXCAVATION

to be an isotropic linear elastic material, having Young's modulus, E_c = 20 GN/m^2, and Poisson's ratio, ν_c = 0.2.

There is a good deal of evidence to suggest that London Clay deforms elastically up to about 1% strain (Som, 1968; Atkinson, 1975). Triaxial tests on undisturbed samples of London Clay have shown that the soil exhibits anisotropic behaviour, the horizontal stiffness being significantly greater than the vertical stiffness (Ward et al, 1965; Henkel, 1971; Gibson, 1974; Atkinson, 1975). Interpretation of laboratory tests on London Clay in terms of elasticity suggests that the soil stiffness is strongly stress-dependent, and that the stiffness moduli may be assumed to vary linearly with depth in the field (Wroth, 1971). In the analysis, the heavily overconsolidated London Clay is modelled as a linear elastic cross-anisotropic material, having stiffness moduli increasing with depth below the ground surface. The clay is assumed to be practically undrained, and hence incompressible, at the end of the excavation. It is described by three independent elastic parameters, vertical Young's modulus E_V, horizontal Young's modulus E_H, and shear modulus in the vertical plane, G_{VH}. Gibson (1974) has suggested that these parameters may be approximately related, for London Clay, by $E_H = 1.8E_V$ and $G_{VH} = 0.38E_V$. Taking these values as being appropriate for the analysis, it follows that Poisson's ratio in the horizontal plane, ν_{HH}, is equal to 0.1, and Poisson's ratio in the vertical plane, ν_{VH}, is equal to 0.5, for an incompressible material. To enable the analysis to be carried out it is assumed that ν_{VH} = 0.499.

The Woolwich and Reading Beds which underlie the London Clay are known to be significantly stiffer than the clay. Laboratory tests have indicated vertical and horizontal stiffnesses of the order of 375 MN/m^2 (Ward, 1961) and 580 MN/m^2 (Cole and Burland, 1972), respectively. Cole and Burland (1972), assuming isotropic linear elasticity, obtained a Young's modulus of 590 MN/m^2 from back analysis of a propped excavation. St. John (1975) has used values ranging from 300-450 MN/m^2, in finite element analysis of braced excavations. In the present analysis, the Woolwich and Reading Beds are assumed to be isotropic linear elastic, with E = 450 MN/m^2 and ν = 0.3.

The prestressed anchored tiebacks in the test panel are located at depths of 1.8m, 3.8m, 5.8m, and 7.8m below the top of the wall. A tieback at Neasden consists of four low relaxation Bridon prestressing strands, each of nominal diameter 15.2mm and nominal area 138.7mm^2. The equivalent tieback of unit width is assumed to have a cross-sectional area of 247mm^2 and a Young's modulus of 198 GN/m^2. The prestressing loads are 199 kN/m, 234 kN/m and 215 kN/m for the tiebacks at depths of 1.8m, 3.8m, and 5.8m, respectively.

Although the tiebacks on the instrumented panel are inclined at 20° to the horizontal, all of the adjacent tiebacks are inclined at 40°. In the analysis an inclination of 40° is assumed as it is believed that this figure more accurately reflects the direction of anchor loading on the panel, due to the effect of adjacent panels.

In order to model the actual excavation progress as closely as possible, the excavation is simulated analytically in three steps as follows:-

<u>Step 1</u> Excavate to a depth of 3.8m and prestress the 1st row of tiebacks.
<u>Step 2</u> Lock off the 1st row of tiebacks, excavate to a depth of 5.8m and prestress the 2nd row of tiebacks.
<u>Step 3</u> Lock off the 2nd row of tiebacks, excavate to a depth of 8.6m and prestress the 3rd row of tiebacks.

As the bottom row was installed two months after the completion of excavation, it is excluded from the end of excavation analysis.

Stress release on the newly excavated surfaces is simulated in each step by applying nodal forces statically equivalent to the total stresses to be released. A bulk unit weight of 20 kN/m^2 is assumed for the saturated clay. In situ pore water pressures are taken to be hydrostatic, with the water table at the ground surface, in line with available borehole information. The at rest earth pressure coefficient, K_o, is chosen to lie between the values suggested by Skempton (1961) and Bishop, Webb and Lewin (1965), and is given by

$$K_o = 3.7 - 0.085\ z \qquad ..\ (1)$$

over the depth of interest, where z is the depth below the ground surface.

The excess pore water pressure generated by undrained loading may be computed from the changes in total stresses, provided the elastic deformation parameters, n' , ν'_{VH}, ν'_{HH}, of the soil skeleton are known. In the analysis it is assumed that n' = 2.0, $\nu'_{VH} = 0.2$, and $\nu'_{HH} = -0.1$, for the London Clay. These values are compatible with those assumed for the undrained incompressible clay (Creed, 1979). The change in pore pressure is given by

$$\Delta u = 0.3\ \Delta\sigma_x + 0.7\Delta\sigma_z \qquad ..\ (2)$$

where $\Delta\sigma_x$, $\Delta\sigma_z$ are the total stress changes in the horizontal and vertical directions, respectively.

RESULTS

By a trial and error process it is found that the observed end of excavation displacements at Neasden are most closely approximated by assuming the vertical Young's modulus of the undrained London Clay to be distributed as

$$E_V = 5 + 6\,z\ (MN/m^2) \qquad ..\ (3)$$

i.e. increasing linearly from $5MN/m^2$ at the surface to a value of 185 MN/m^2 at a depth of 30m. The analytical displacements of the wall and retained ground surface are compared in Figure 2 with those observed by Sills et al (1977), at the end of excavation. The similarity between horizontal wall movements into the excavation is achieved by varying E_V for the clay, to obtain a good fit with the observed displacements. The agreement between surface movements is not as good. The horizontal compression of the ground surface observed in the field is not reproduced in the analysis, which suggests a decrease in horizontal movement with increasing distance from the excavation. Apart from the zone in the vicinity of the wall, the analytical and observed horizontal surface movements are qualitatively similar. In the field the top of the wall settled by 10mm, and a maximum surface settlement of 12mm was observed 4m behind the wall. Settlement decreased to zero at 40m from the excavation. The analysis computes a slight heave of the top of the wall, behind which the surface settles to a maximum of 5mm at a distance of 3.5m from the excavation. Thereafter, the computed settlement decreases very slowly with increasing distance from the excavation.

Pore pressures computed at distances of 2m, 5.5m and 16m from the excavation are compared in Figure 3 with pore pressures recorded in the field at distances of 0m, 7m and 16m behind the wall. The field data, chosen for the comparison, are those obtained before and after the end of excavation, the dates of the readings being shown in Figure 3. Both the observed and computed pore pressures are seen to be lower than hydrostatic at all three locations. Contours of computed pore pressure in the vicinity of the excavation are shown in Figure 4.

The soil pressure distribution computed for the back of the wall is shown in Figure 5. A tension zone, 0.5m deep, is found at the top of the wall. A general drop in normal soil pressure from in situ values is computed, except near the bottom of the wall, where there is little change. A down-

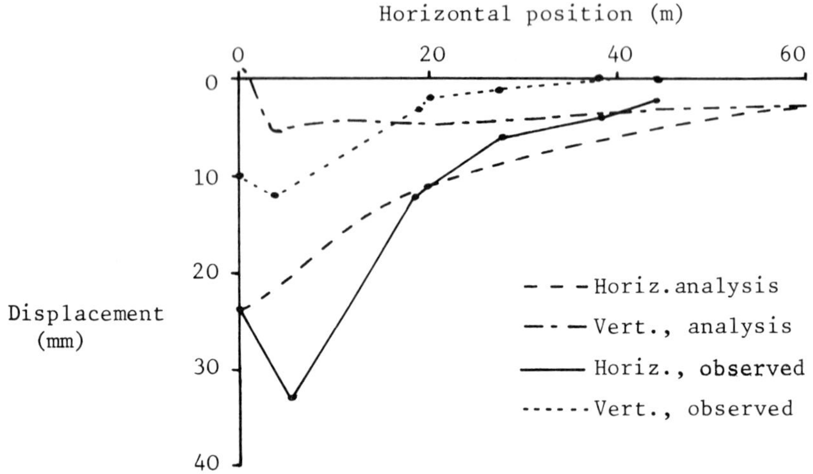

(a) Surface Displacements

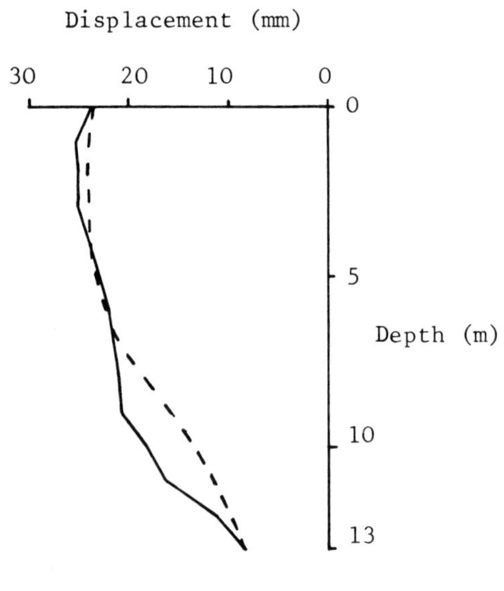

(b) Horizontal Wall Displacements

FIGURE 2 WALL AND SURFACE DISPLACEMENTS

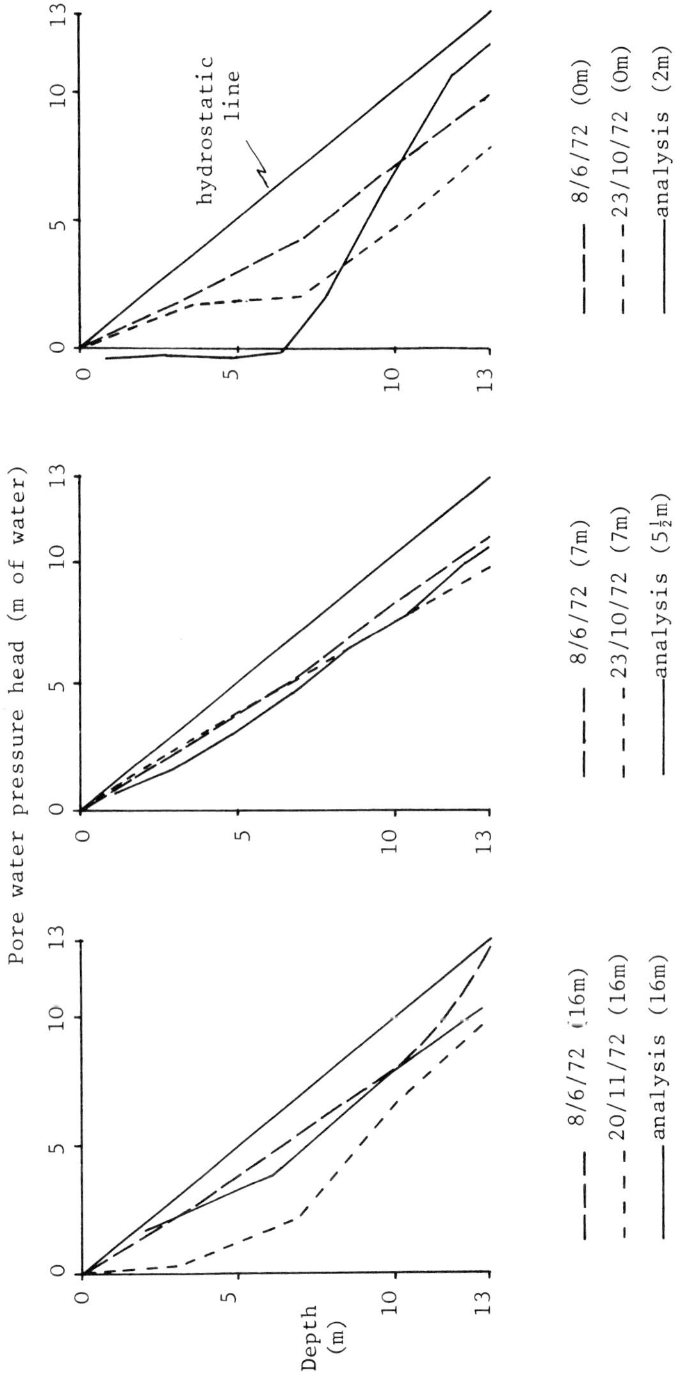

Note: Distance from excavation in brackets

FIGURE 3 PORE WATER PRESSURES

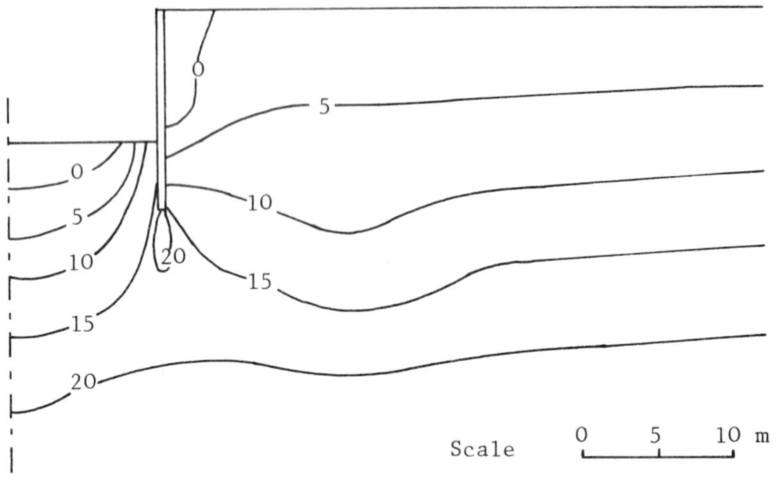

FIGURE 4 COMPUTED PORE WATER PRESSURE CONTOURS

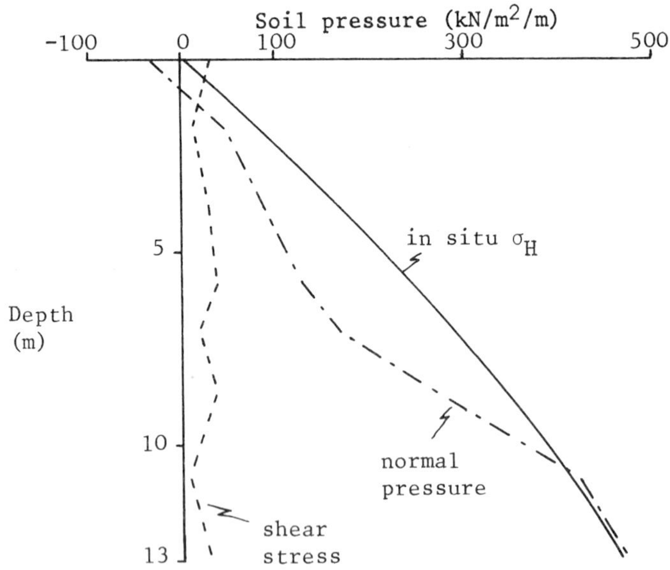

FIGURE 5 SOIL PRESSURES ON BACK OF WALL

ward shear stress, varying from 10-40 kN/m^2, on the back of the wall, is computed. This is caused by soil settlement behind the diaphragm wall.

A slight increase in anchor load for succeeding analytical excavation steps is computed.

DISCUSSION

The distribution of E_V with depth, deduced from the analysis, follows the same general pattern previously obtained from in situ tests (Marsland, 1971) and analyses of full scale foundations (Cole and Burland, 1972; Hooper, 1973; St. John, 1975). These stiffness values are significantly higher than those obtained from laboratory triaxial tests. The reasons for this difference have been discussed by Simpson et al (1979), who suggest that it is mainly due to the threshold effects observed in London Clay.

The horizontal displacement of the top of the wall is sensitive to the value of E_V assumed at the ground surface. If E_V = 0 at the surface and increases linearly to 185 MN/m^2 at a depth of 30m, the same value as in equation (2), a 20% increase in horizontal movement of the top of the wall is computed.

Burland et al (1979) have discussed the distribution of K_o with depth in London Clay, and shown that it is very sensitive to stress history and may vary considerably from site to site. As no measurements of K_o were made at Neasden Lane the assumed distribution is open to question. However it agrees well with the distribution derived by Burland et al (1979) for similar ground conditions.

As the analytical and observed displacements are in reasonable agreement, at least in a qualitative sense, the analysis provides a means of investigating the relative contribution to the displacements of each source of loading on the excavation. The undrained loading on the excavation may be divided into three components:-

(i) the release of overburden pressure on the base of the excavation;
(ii) horizontal stress release on the front of the wall; and
(iii) anchor prestressing.

By means of displacement vector plots, Figures 6(a)-(c) show the contribution to deformation, in the vicinity of the excavation, of vertical stress release, horizontal stress

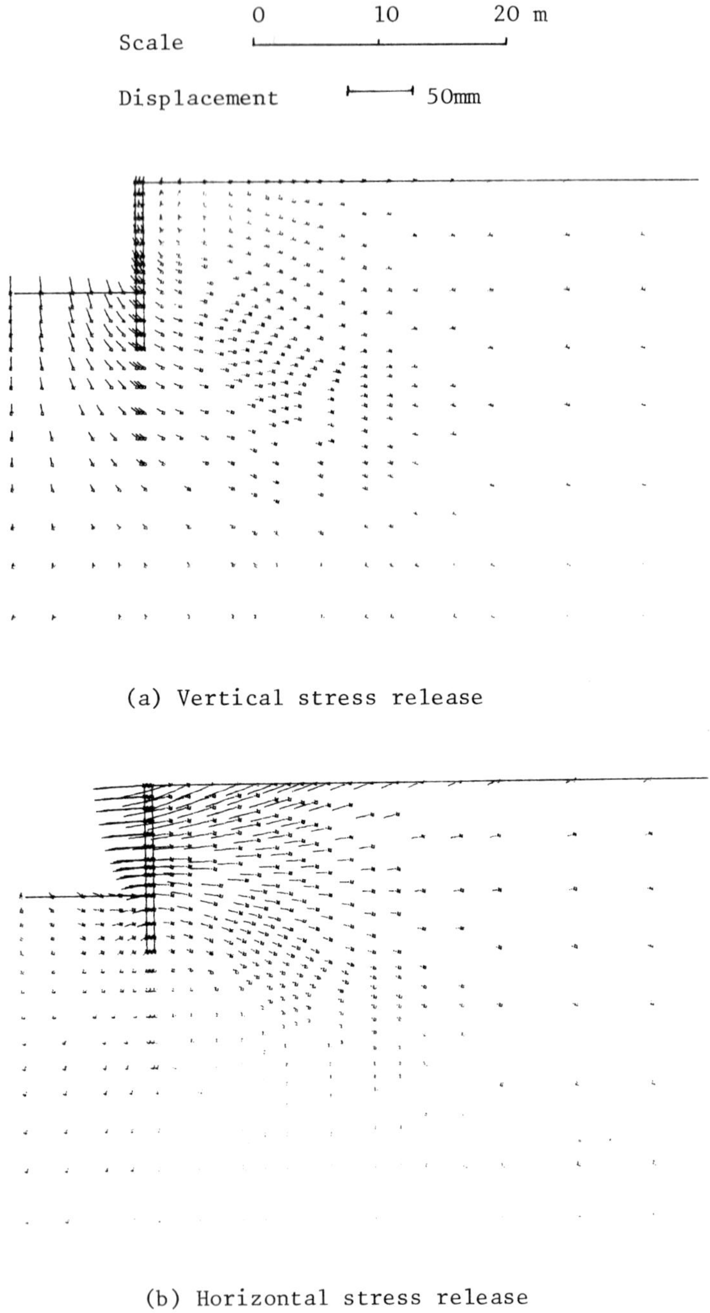

FIGURE 6 DISPLACEMENT VECTORS

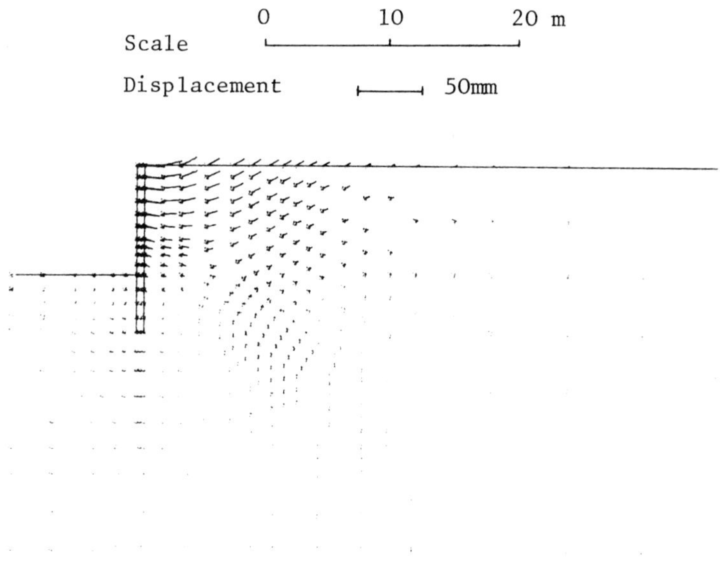

(c) Anchor prestressing

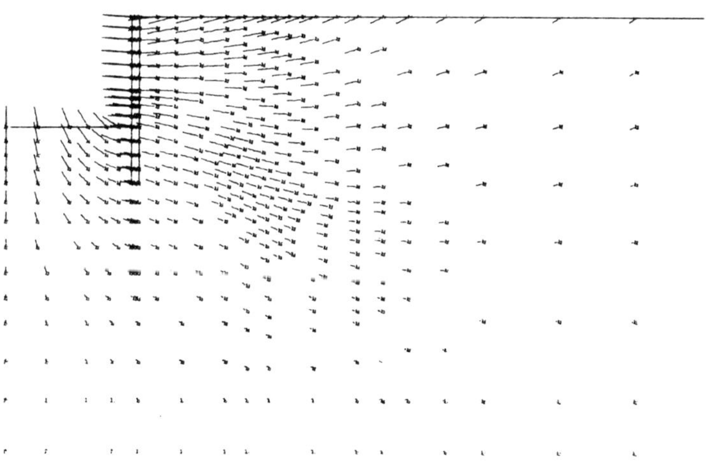

(d) Overall movements

FIGURE 6 (cont'd) DISPLACEMENT VECTORS

release and prestressed anchored tiebacks, respectively. Vertical stress release gives rise to lateral movements below and behind the wall, as the nearly incompressible soil tends to "flow" towards the base of the excavation. It also causes the wall to heave and rotate about its top into the excavation. Horizontal stress release on the vertical face of the wall results in large horizontal movements, for a considerable distance, in the region behind the wall. Some surface settlement occurs, again due to the near incompressibility of the clay. Displacements behind the excavation beneath the level of the toe of the wall are generally smaller than those associated with vertical stress release. Compared with the release of in situ stresses, the region of influence of the prestressed tiebacks is very limited. They basically cause the wall to rotate about its toe away from the excavation, resulting in heave of the ground surface. Displacements in the anchor region are very small. While prestressed tiebacks are effective in reducing wall and surface movements, they do little to prevent the deep seated movements that occur within the soil, as shown in Figure 6(d), which illustrates the overall pattern of movement obtained from the analysis. The reason for large horizontal movements behind the anchor zone appears to be the inability of the prestressed tiebacks to prevent the more remote movements caused by in situ stress release, rather than the prestress loading on the anchor zone.

Pore pressures were measured in the field on a number of occasions, during and after excavation. As no field records are available for early August, 1972, when excavation was completed, it is necessary to compare the computed pore pressures with those field records, obtained before and after the end of excavation. The locations behind the wall of computed and observed pore pressure do not coincide, but are close together, as shown in Figure 3. Reasonably good agreement is obtained between computed and observed pore pressures, at all three locations. Pore pressures behind the wall generally drop below hydrostatic, as a result of excavation. The analysis suggests that pore pressures, 2m behind the excavation, drop to zero for a depth of 6.5m below the retained ground surface, and that little change occurs at the level of the toe of the diaphragm wall. A smaller but more uniform drop in pore pressure is computed at the two locations further behind the wall. The contours of computed pore pressure, shown in Figure 4, present an overall picture of pore pressure changes in the vicinity of the excavation. A general drop in pore pressure is computed except near the toe of the wall, where an increase occurs, probably due to compression caused by axial loading on the wall. Negative pore pressures are computed immediately behind the wall above the level of excavation, and near the base of the excavation. The drop in pore pressure in the anchor zone is also apparent from Figure 4.

CONCLUSIONS

The analysis provides further evidence that the undrained stiffness of London Clay increases with increasing depth below the ground surface.

The overall pattern of movement, at the end of excavation, around an anchored diaphragm wall in London Clay may be computed with reasonable accuracy assuming the clay to be a cross-anisotropic linear elastic material. Due to the near incompressibility of the soil model, significant displacements are computed far away from the excavation, at variance with field observations.

Although the prestressed tiebacks reduce horizontal wall movement considerably, their area of influence is limited to a triangular wedge of soil behind the wall. The large movements occurring behind the anchor zone are more likely due to the inability of the anchored tiebacks to reduce the movements caused by insitu stress release at the excavation, rather than the application of the prestress loads to the anchor zone.

The analytical work described in this paper was financed by a grant from the Science Research Council.

REFERENCES

J.H. Atkinson - Anisotropic elastic deformation in laboratory tests on undisturbed London Clay, Geotechnique 25 (1975) 357-374.

A.W. Bishop, D.L. Webb and P.I. Lewin - Undisturbed samples of London Clay from the Ashford Common Shaft: strength-effective stress relationships, Geotechnique 15 (1965) 1-31.

J.B. Burland, B.Simpson and H.D. St. John - Movements around excavations in London Clay, Design parameters in geotechnical engineering, BGS, London, (1979) 1, 13-29.

G.W. Clough and Y. Tsui - Performance of tied-back walls in clay, Proceedings of the American Society of Civil Engineers, Journal of the Geotechnical Engineering Division 100 (1974) 1259-1273.

K.W. Cole and J.B. Burland - Observation of retaining wall movements associated with a large excavation. Proceedings of the Fifth European Conference on Soil Mechanics and Foundation Engineering, Madrid, (1972) 1, 445-453.

M.J. Creed - Analysis of the deformations associated with anchored diaphragm walls, Ph.D. Thesis, University of Surrey, (1979).

R.E. Gibson - The analytical method in soil mechanics, Geotechnique 24 (1974) 115-140.

D.J. Henkel - The relevance of laboratory-measured parameters in field studies, Proceedings of the Roscoe Memorial Symposium, Cambridge, G.T. Foulis, Henley, (1971) 669-675.

J.A. Hooper - Observations on the behaviour of a piled-raft foundation on London Clay, Proceedings of the Institution of Civil Engineers, Part 2, London, (1973) 55, 855-877.

A. Marsland - Laboratory and in situ measurement of the deformation moduli of London Clay, Proceedings of the Symposium on the Interaction of Structure and Foundation, Midland Soil Mechanics and Foundation Engineering Society, (1971) 7-17.

J.S.N. Rodrigues - The development and application of a finite element computer program for the solution of geotechnical problems, Ph.D. Thesis, University of Surrey, (1975).

H.D. St. John - Field and theoretical studies of the behaviour of ground around deep excavations in London Clay, Ph.D. Thesis, University of Cambridge, (1975).

G.C. Sills, J.B. Burland and M.K. Czechowski - Behaviour of an anchored diaphragm wall in stiff clay, Proceedings of the Ninth International Conference on Soil Mechanics and Foundation Engineering, Tokyo, (1977) 2, 147-154.

B. Simpson, N.J. O'Riordan and D.D. Croft - A computer model for the analysis of ground movements in London Clay, Geotechnique 29 (1979) 149-175.

A.W. Skempton - Horizontal stresses in an overconsolidated Eocene clay, Proceedings of the Fifth International Conference on Soil Mechanics and Foundation Engineering, Paris,(1961) 1, 351-357.

N.N. Som - The effect of stress path on the deformation and consolidation of London Clay, Ph.D. Thesis, University of London, (1968).

W.H. Ward, A. Marsland and S.G. Samuels - Properties of the London Clay at the Ashford Common Shaft, Geotechnique 15 (1965) 321-344.

C.P. Wroth - Some aspects of the elastic behaviour of overconsolidated clay, Proceedings of the Roscoe Memorial Symposium, Cambridge, G.T. Foulis, Henley, (1971) 347-361.

GROUND MOVEMENT ADJACENT TO A DEEP EXCAVATION

W.J.P. Leigh, BSc, CEng, MICE, MIMinE, AMBIM
Senior Operations Engineer, NCB Opencast Executive
M.J. Scoble, ACSM, MSc, CEng, MIMM
Mining Engineering Department, University of Nottingham
G.J. Young, BSc, PhD
Formerly Mining Engineering Department, University of Nottingham

INTRODUCTION

The paper describes ground movements associated with the appearance of a single crack behind the crest of a 60 m deep excavation forming part of an East Midlands Opencast Mine. In a seven-month period, the vertical dimensions of the crack increased from zero to 500 mm and the horizontal from zero to 260 mm. The graphs of both vertical and horizontal movement contain periods of acceleration and quiescence.

The slope of the excavation is made up of some 12 m of red marls with sandstone bands overlying coal measures consisting of mudstones, coal seams, seatearths and sandstones, the inclination of the slope is some 45° over coal measures and 30° over the red marls. The slope runs north-south and forms the western side of the opencast excavation. It is flanked on its crest to the west by a screening mound of topsoil, marking the boundary of the opencast mine, a public road and a farm.

In addition to being made up of a number of materials the slope is structurally complicated. Figs. 1 and 2 show cross sections at right angles to the slope crest compiled from coal exploration boreholes. Fig. 1 is a section approximately on the centre of the zone of movement, Fig. 2 a section close to the southern extremity of the affected area. On both cross sections the apparent locations of the coal seams have been marked by dotted lines and it can be seen that the dip of the seams out of the slope on the southerly cross section is greater than that on the north section. In fact the dip shown on Fig. 2 is an apparent dip and interpolation from other boreholes indicates a local true dip of the order of 1 in 1 in a North Easterly direction. Furthermore, the local area of steep dip culminates in a fault running North West-South West. Additional complication arises from the known presence of old workings in two of the seams in the slope. These old workings from the nearby Old Heather Colliery (circa 1870) had been observed during coal extraction in the form of narrow headings entering the slope. Two boreholes drilled for the installation of monitoring equipment also encountered old workings as did boreholes in the original coal exploration programme. The available information does not however allow the establishment of the complete pattern of old workings.

Explanatory diagrams of the slope circumstances are Figs. 3, 4 and 5. Fig. 3 shows the slope, coal seams, public road, crack pattern, fault and farm buildings (the fault hade estimated from available information). Fig. 4 shows the main features on a plan view and Fig. 5 a longitudinal section through instrumentation boreholes.

The maximum distance behind the crest of the slope at which movement was observed, was 60m. giving a 1 to 1 ratio of distance of measurable movement to depth. Ratios of 2.50 to more than 3 have been observed relating to deep excavations in clay (Ref. 1) presumed to result from elastic recovery following lateral and vertical stress relief. In excavations predominantly in stronger coal measures however, ratios of distance of measurable movement to depth are usually, except in the case of sliding slabs, not greater than some fraction to 1,traceable frequently to shear failure along discontinuities. Consequently, the initial appearance of the crack in the field on the far side of the road was unusual and a matter of concern

In order to monitor the progress of movements to find an explanation for their cause and hence predict the likely ultimate degree of movement, a monitoring programme was initiated comprising precise surveying, ground tilt measurements and borehole instrumentation (extensometer and piezometer). Based upon the results of observations made,

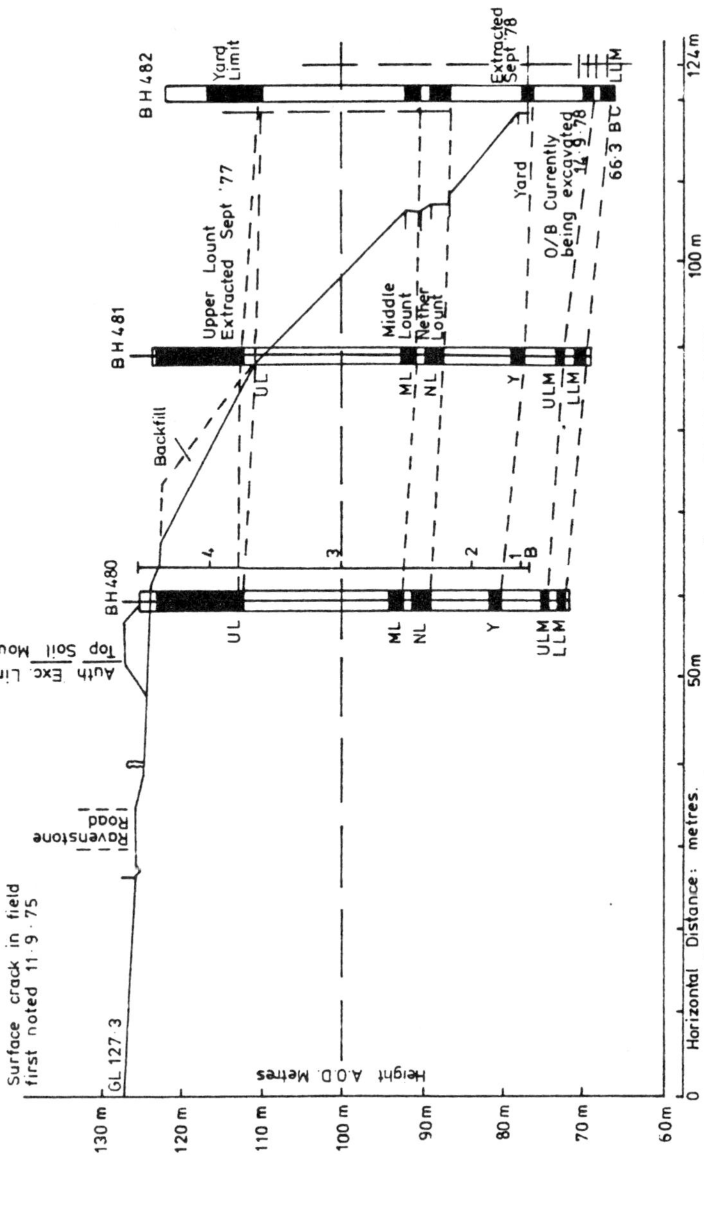

Surface crack in field first noted 11·9·75
GL 127·3
Ravenstone Road
Auth Exc. Limit
Top Soil Mound
BH480
BH481
BH482
Backfill
Upper Lount Extracted Sept '77
Middle Lount
Nether Lount
Yard Limit
Extracted Sept '78
Yard
O/B Currently being excavated 14·9·78
66·3 BC
UL
ML
NL
Y
ULM
LLM
B
130 m
120 m
110 m
100 m
90 m
80 m
70 m
60m
Height A.O.D. Metres
0
50m
100 m
124m
Horizontal Distance: metres.
COALFIELD FARM CROSS SECTION B – B' SCALE: 1:500

Fig.1

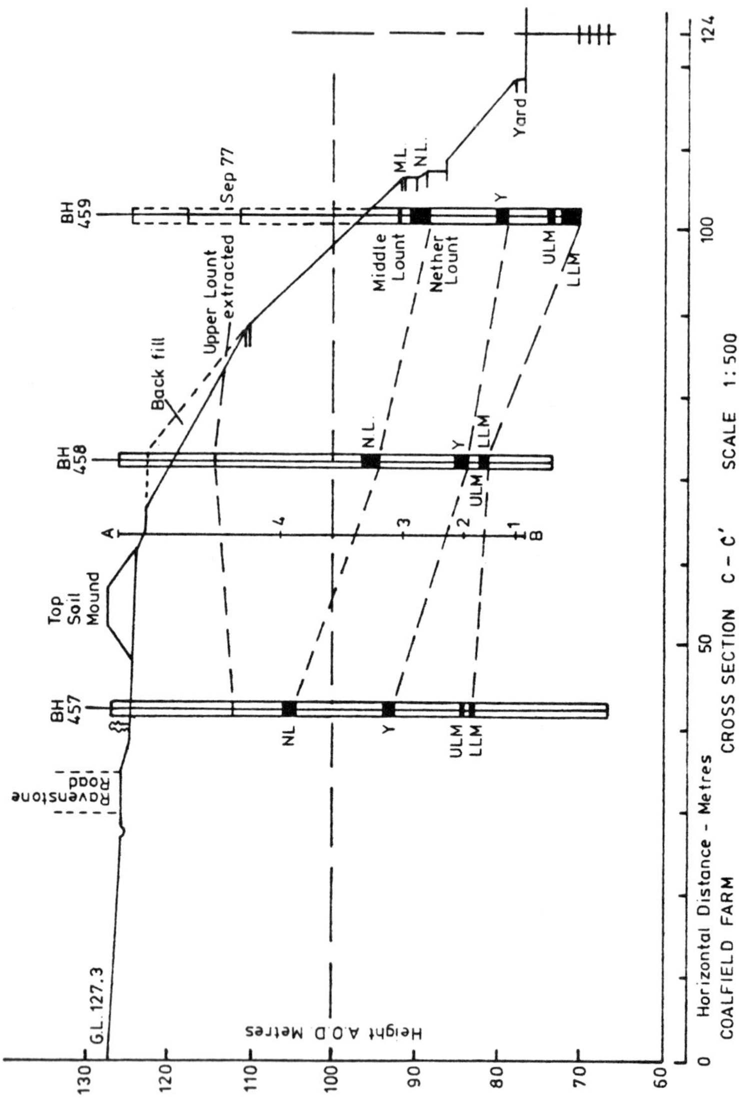

BH 459
Sep 77
Upper Lount extracted
Back fill
BH 458
Top Soil Mound
BH 457
Ravenstone Road
G.L. 127.3
Height A.O.D Metres
Middle Lount
Nether Lount
ML
N.L.
Yard
Y
ULM
LLM
NL
A
B
Horizontal Distance - Metres
COALFIELD FARM
CROSS SECTION C – C′
SCALE 1: 500

Fig. 2

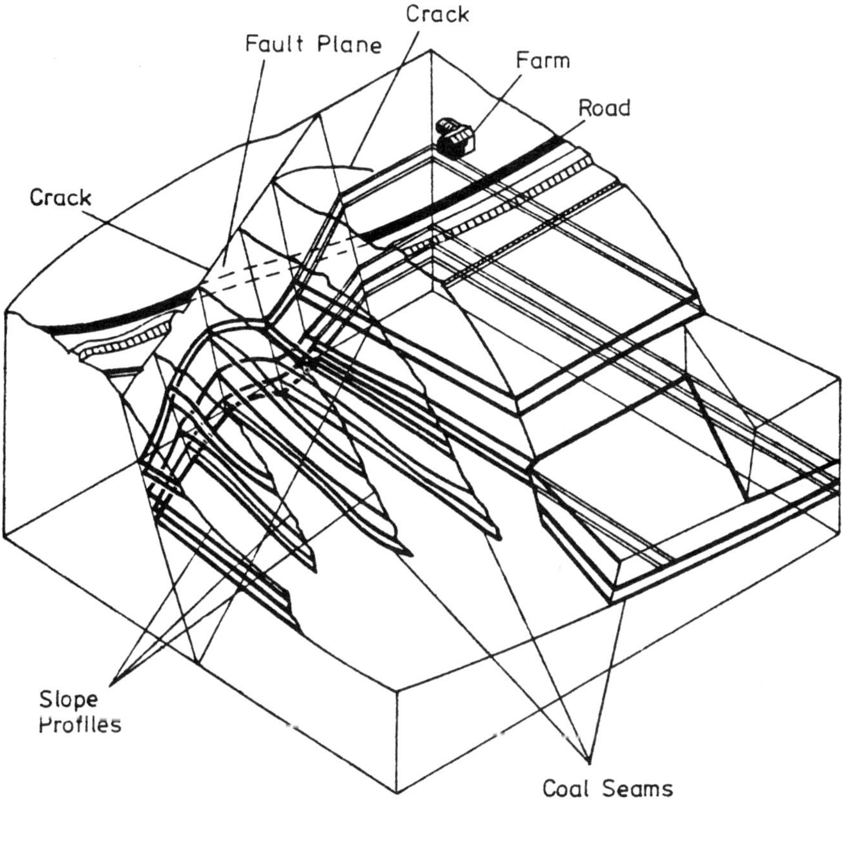
Crack
Fault Plane
Farm
Road
Crack
Slope Profiles
Coal Seams

Fig. 3.

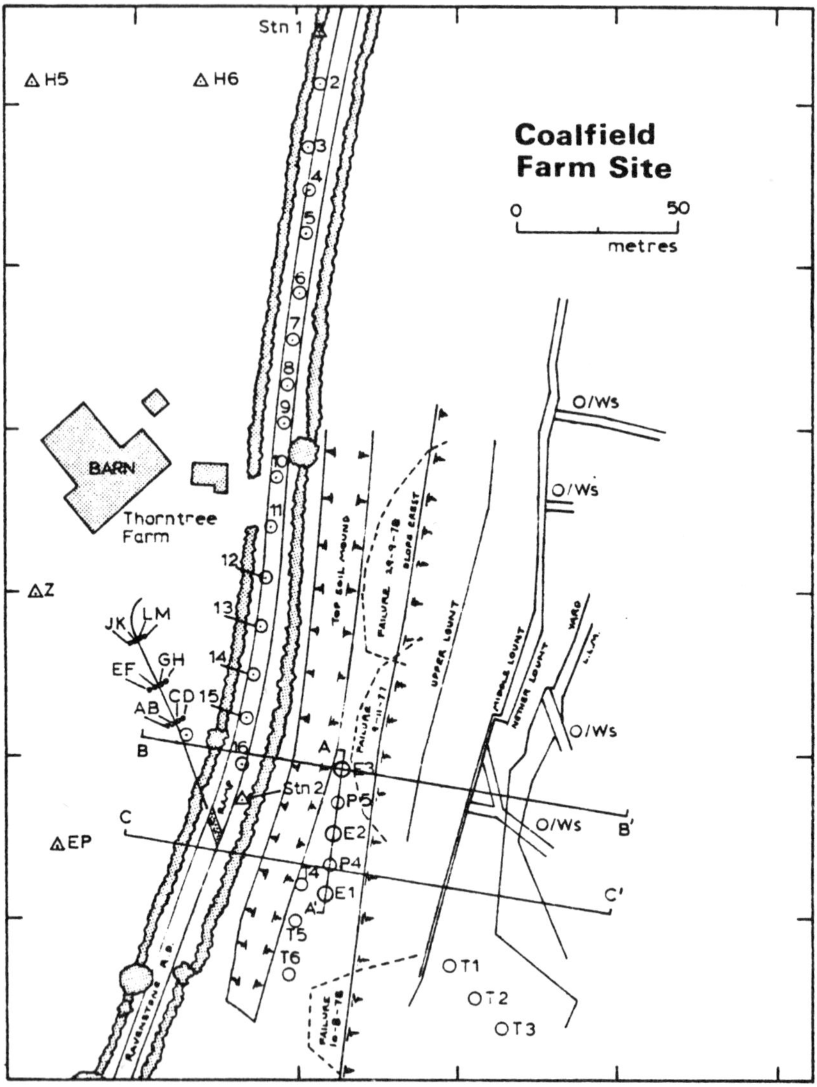

Plan showing positions of survey stations and instruments

Fig. 4

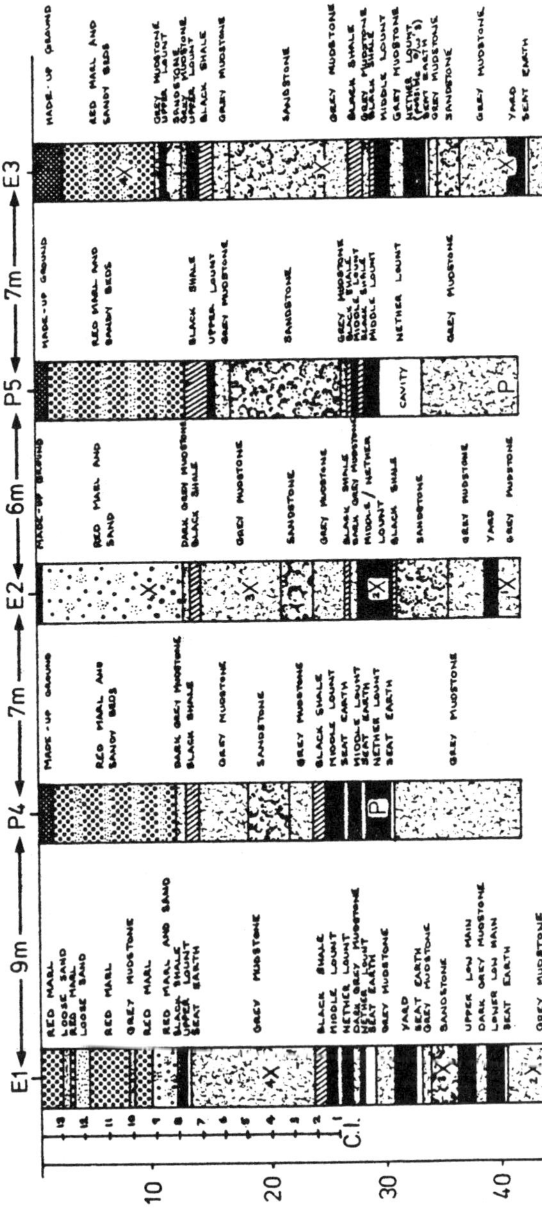

C.I. CONTINUITY INDICATOR
X EXTENSOMETER REF. POINT
P PIEZOMETER TIP

Fig. 5 SECTION THROUGH EXTENSOMETERS AND PIEZOMETERS

various hypotheses which are discussed below, were put forward at various times to account for the movements. The hypotheses were convergence across old coal workings, slope failure by conventional quasi-circular failure, slope failure through sliding on a bedding plane discontinuity and slope failure by bearing capacity failure.

HISTORY OF MOVEMENTS

The slope was initially exposed to the base of the Nether Lount seam in May, 1977, and most of the excavation to the lower seams completed by July, 1978. From time to time, minor slope failures occurred taking the form of circular failures in Marls and bedding plane and wedge failures affecting Marls and coal measures. Three such failures are shown on Fig. 4, the largest consisting of a wedge failure above the Nether Lount Seam. The susceptibility to bedding plane slides of certain horizons was evidenced by test results. A thin band of clay in seatearth below the Nether Lount seam gave shear box results of $C^1 = 25.8$ Kn/M^2 and $\emptyset^1$ of 5.8^o. A similar band below the Upper Lount Rider seam gave $C^1 = 12.8$ Kn/M^2 and $\emptyset^1 = 12.6^o$.

The movements which are the subject of this paper commenced in August, 1978, with the appearance of a crack some 80mm in vertical extent in the field the far side of the site boundary. This initial crack was observed to follow the line of the fault crossing the side wall of the excavation. Since the crack maintained this direction away from the excavation and there was at its western extremity no evidence of it turning back towards the excavation, it was presumed that the cause of the movement was convergence in old workings focussed at the fault which could reasonably be expected to be their limit. At the same time, it was noted that reactivation of subsidence in old workings is rare and no records existed of a similar phenomenon associated with slopes containing old workings. In succeeding weeks however, the dimensions of the crack increased vertically, in width and in length, until the crack extended from the field, across the road and screening embankment to the side of the excavation. Finally in October, 1978, the extremity of the crack in the field extended and changed direction back towards the side of the excavation over a distance of some 5m.

At this stage, the pattern of cracking began to take on the appearance of a tension crack of a large slope failure and the road was placed under 24 hour surveillance.

At the same time, work was started on the construction of a rockfill buttress within available space in the excavation, concentrating at the southern end of the failure zone.

Examination of various types of slope failure was begun. Circular or quasi-circular failure in marls, or in marls and coal measures was largely discounted. Approximate analysis of possible sliding surfaces ending at the crack position using typical strata strengths showed minimum factors of safety in excess of three.

The most likely mechanism postulated was that the seat of the movement was located in the steeply dipping strata adjacent to the fault where a block of strata bounded by the fault was moving out. In view of the low shear strength referred to above the seatearth below the Nether Lount Seam was considered a possible horizon. The Northern limit of the moving block was considered to be indicated by the changed direction of the crack in the field. It was noted also that due to the decreasing eastwards dip to the north the driving forces for a bedding plane failure would reduce and the changed direction of the crack could be the result of a 'tear' fracture resulting from the greater degree of movement to the south. Certain of the surface survey measurements described below supported this hypothesis, but at no time could shear movement be detected at the face of the excavation.

SURFACE SURVEY MEASUREMENTS

Surface survey stations installed to monitor movements are described in the following table (with reference to Fig. 4)

	Station	Purpose
(1)	X,Y,Z, EP.	Control stations outside the presumed bounds of the unstable area.
(2)	Road Station 1, Bolts 2-16 and Road Station 2.	To monitor both vertical and horizontal displacements occurring on the road.
(3)	9,14,4,10	To monitor vertical displacements occurring on the road.

Station		Purpose
(4)	Crack bolts A,B,C,D, E,F,G,H,J,K,L,M.	To monitor vertical and horizontal displacements occurring at the main crack.

Observations of vertical movements were also made on the three extensometer heads, (E1, E2 and E3).

It is not possible in this paper to show the curves of movement versus time for all survey points. Typical of the pattern of vertical and horizontal movements are Figs. 6 and 7 showing vertical and horizontal movements of certain of the road stations (stations (2) above).

With respect to the vertical movement of the road stations (Fig. 6) between July, 1978, and February, 1979, it can be seen that stations to the south and closest to the fault have moved to a much greater extent than those to the north with a maximum value of almost 500mm at road station 2. The point at which the value of vertical movement increases markedly is at bolt 12 which is approximately in line with the change in direction of the crack in the field (Fig. 4).

Vertical movements at the crack in the field also show greater values of movement to the south. The vertical separation between station A (stable side) and station C increasing to over 400mm between September, 1978, and March, 1979. In the same time period the separation between station E (stable side) and station G increased by 320mm.

Vertical movements of the extensometer heads, E1, E2 and E3 between November, 1978, and March, 1979, show greater values of vertical movement to the south E1 settling by 257mm, E2 by 190mm and E3 by 146mm.

A degree of compatibility could be observed in the movements of the various points, thus between November and January inclusive:-

Road station 2 settled by 190mm
Crack bolt C settled by 190mm
Extensometer head E settled by 200mm .

In addition, the curves of movement appeared to be generally parallel with similar increases in the rate of

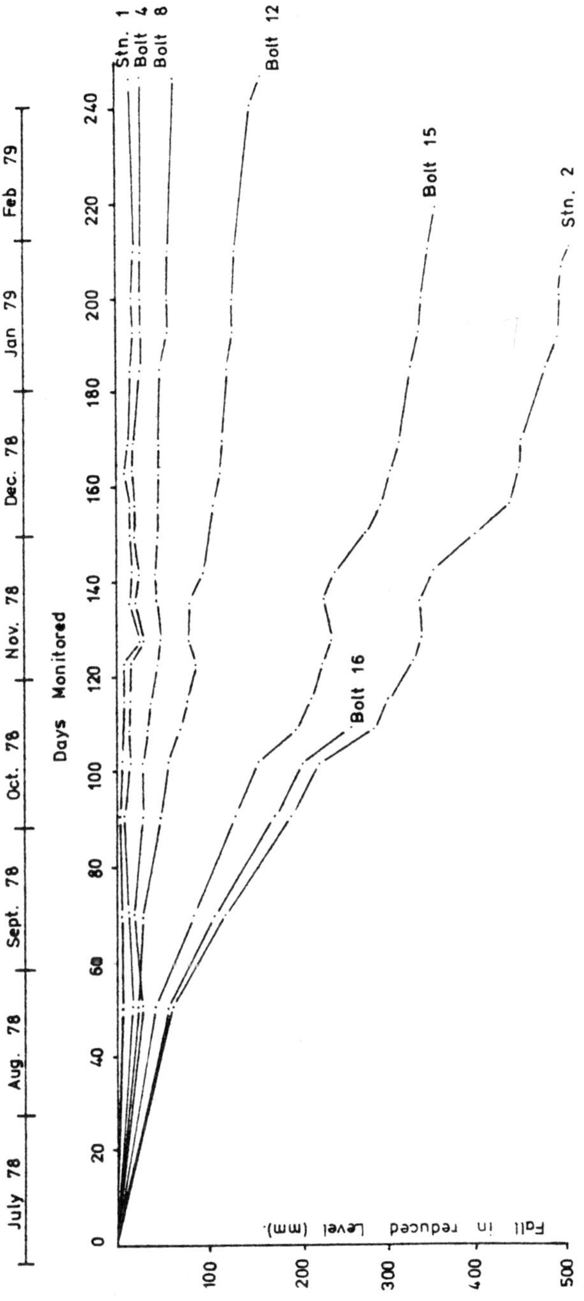

Fig.6 Graphs Showing Vertical Displacements for the Road Stations W. R. T. Time.

Fig. 6

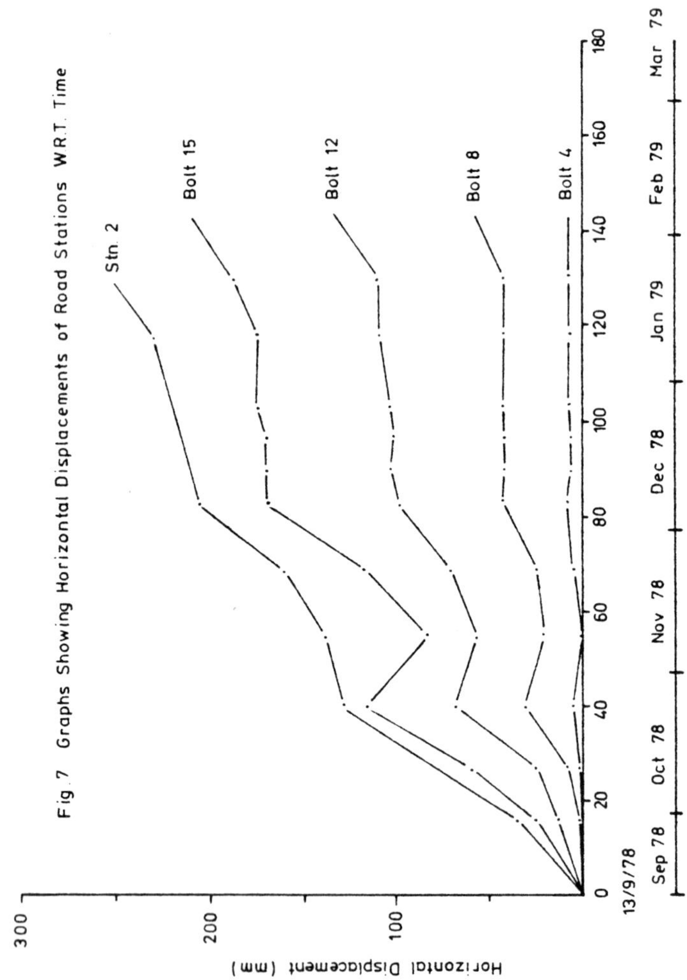
Fig. 7 Graphs Showing Horizontal Displacements of Road Stations W.R.T. Time
Stn. 2
Bolt 15
Bolt 12
Bolt 8
Bolt 4
Horizontal Displacement (mm)
300
200
100
0
20
40
60
80
100
120
140
160
180
13/9/78
Sep 78
Oct 78
Nov 78
Dec 78
Jan 79
Feb 79
Mar 79

Fig. 7

movement and similar quiescent periods.

With respect to horizontal movements of the road stations (Fig. 7) between September, 1978, and March, 1979, it was again observed that the larger values of movement took place to the south. Significant movement again occurring at bolt 12 with a maximum of 260mm at road station 2. Horizontal separation at the crack in the field followed a similar pattern with separation reducing from 260mm at BC to a small value - 20mm at KL.

Horizontal movements at the extensometer heads were difficult to observe due to a difficulty in obtaining sighting positions on stable ground, however such measurements as were taken indicate compatibility with the measurements at road stations and crack bolts thus:- between 13/11/78 and 22/11/78 the horizontal movements of extensometer heads E1, E2 and E3 were 36mm, 83mm and 23mm, and between 13/11/78 and 28/11/78 the movements were 84mm, 67mm and 65mm respectively. The increase between 22/11/78 and 28/11/78 is coincident with a steepening of the graphs of horizontal movement of road station 2 and crack bolt C.

Compatibility of the movements at points to the south end of the zone of movement coupled with a degree of parallelism in all graphs of movement was regarded as being in support of the theory of a sliding block at the south part of the slope.

CYCLIC NATURE OF MOVEMENTS

Another feature of the movements supporting the sliding block theory was the cyclic pattern of the movements observed in surface survey measurements. Broadbent and Ko (Ref. 2) observed in connection with many open pit slope failures that the displacement- time function to ultimate failure consisted of a series of exponential decay curves. In the present case, if envelopes are drawn to the successive stages of movement an approximately similar pattern can be observed on graphs of both vertical and horizontal movement. Fig. 8 of the vertical movement of road station 2 is typical. Broadbent and Ko (Ref. 2) ascribe the cyclic movement to factors such as change in disturbing forces due to rising ground water levels in the slope from rainfall, etc. In this case however, observations of the piezometers showed no marked variations and there was no obvious correlation between rainfall and movement.

EXTENSOMETER OBSERVATIONS

In order to gain more evidence in support of the sliding block theory for the movements, it was decided at the end of

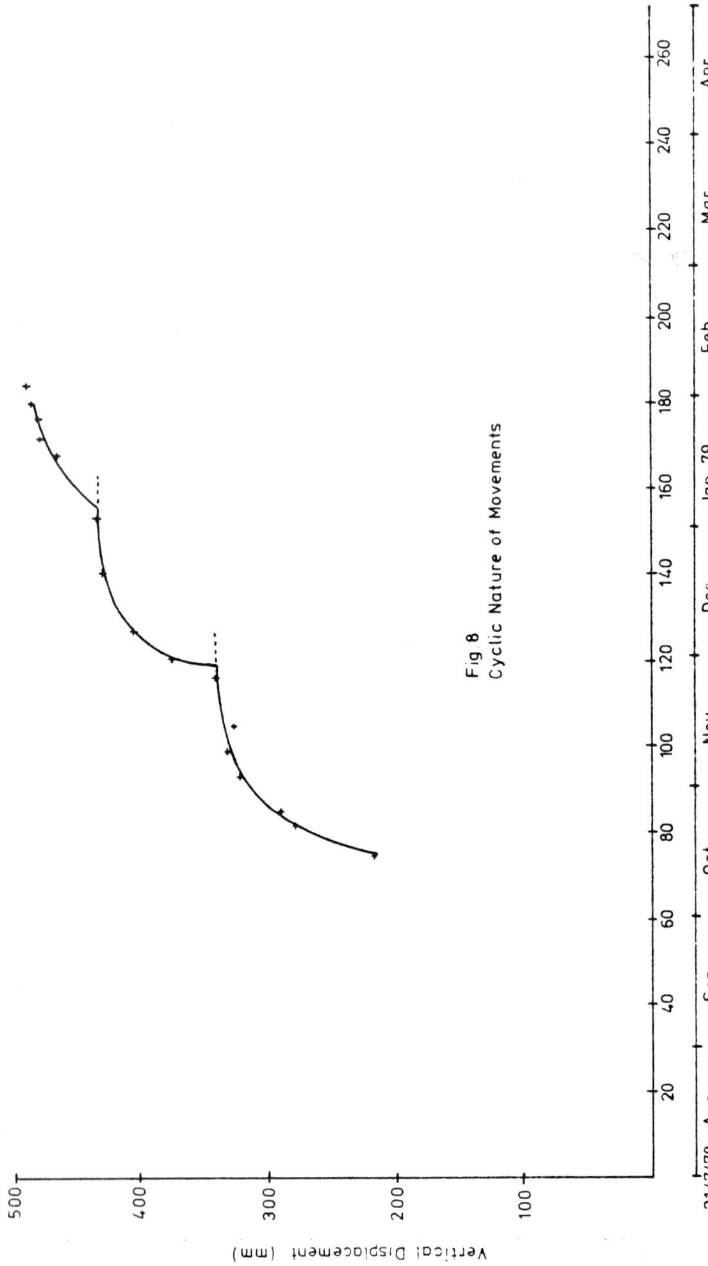

Fig. 8
Cyclic Nature of Movements

Fig. 8

October, 1978, to install extensometers within the slope. Although the evidence of the surface survey measurements supported the sliding block theory there remained two other features not accounted for, namely an apparent settlement of the farm building outside the zone of influence indicated by the crack and the non-appearance of a shear lip in the face of the excavation. Although the observation points on the farm buildings had undergone only small settlements over September and October, the graphs of the movements were steeply sloping (the settlements reached a maximum of 80mm by March, 1979).

Due to difficulty of access for drilling at other locations, the extensometer holes E1, E2 and E3 were drilled immediately in front of the screening embankment. Four anchors were installed in each hole in the positions shown on Fig. 5. A continuity indicator was also installed to a depth of 26 metres in borehole E1, having junctions at 1 metre intervals. Two piezometers were installed in boreholes between the extensometer holes as shown in Fig. 3. The anchor points in the boreholes are connected to the surface by wires passing over pulleys at the extensometer heads and kept taut by suspended weights. Movement of the wires can be obtained accurately by micrometer measurement of the change of distance between suspended weight and a plate at the extensometer head.

The rationale behind the detection of shear movement by the extensometers is that should shear take place across the boreholes containing the wires, the weights at the surface should rise. In examining the results of the extensometer observations however, regard had to be given to the fact that during the period November, 1978, to March, 1979, the extensometer heads settled E1 by 257mm, E2 by 194mm and E3 by 146mm. During this period, the weights attached to all anchor points did not move by comparable amounts indicating that either (a) shear movement occurred simultaneously with surface settlement of amounts such as to maintain the weights in their approximately original relationship to the heads, or, (b) the anchor points settled in a generally similar manner to the surface.

It was observed also that the relative movement of the two upper and two lower points in each borehole indicated a convergence of 25 to 30mm in the vicinity of the Nether Lount seam. This amount of convergence indicated that the whole of the surface settlements could not be ascribed to convergence across old workings which had been one of the initial hypotheses.

An examination of the shear possibility ((a) above) showed that shear movement would have been required above the upper anchor points in all boreholes of amounts in excess of 200mm to account for the behaviour of the weights. It was considered that such amounts would have produced a marked shear lip at the face of the excavation, none was in fact observed. Furthermore, the shear indicator in borehole E1 would have reflected such movement, this remained intact during the observation period. In consequence, it was concluded that the most likely explanation was that at (b) above - settlement of anchor points.

BEARING CAPACITY FAILURE

The absence of positive indication of shear or of convergence across old workings from the extensometer results prompted a search for an alternative explanation. Some consideration was given to the possibility of a "bearing capacity failure" (Ref. 3), the characteristics of this type of failure are large downward and outward movements of the crest accompanied by little apparent movement at the lower slope and toe. This pattern of movement is held to be the result of the penetration of a 'Prandtl wedge' within the slope akin to the model used in foundation failure.

Failures exhibiting the characteristics have been observed in coal measure strata (but with a surface movement to depth ratio of much less than the 1 to 1 in the present case). In the Prandtl mechanism the failure zone in the slope is divided into two parts, an upper active zone and a lower passive zone, these zones separated by the Prandtl wedge. Cyclic movements are also a feature of the Prandtl mechanism. Kuapil and Clews (Ref. 3) state "In many cases especially with large dimension open pit failures the destruction process is slow. The process is often not fluid but intermittent periods of high activity alternate with periods of calm. This effect is also due to Prandtls prism. During the periods of high activity the breaking of the rock mass by secondary shearing consumes energy, after this the rock mass in the slope needs a certain time for the accumulation of energy. This represents the periods of calm in the failure process".

It is also stated that the time taken to reach failure can be long, even a period of years. In the case under consideration, the fitting of a curve to the graph of vertical separation of the crack at road station 2 yields a curve of the type:

$$y = a + b \, \mathrm{Ln}(t)$$

where y = settlement in mm

t = time in days.

The constant a is -1220.15 and b is 319.1 with a coefficient of correlation of 0.9903, the curve fitting carried out by programmable calculator.

The amount of separation at the approach of final failure is difficult to estimate but would probably be in excess of 1 metre. If a 1 metre separation is applied in the above formula a time period of some 3 years results.

While the Prandtl bearing capacity mechanism with large movements at the crest accompanied by little visible movement at lower face and toe and with a cyclic nature of movement over a long time period has some characteristics similar to the present case, there remain some unexplained features. These are:-

(a) On the visible portions of the face it would be expected that some disruption would be visible. In previous cases of slope failure exhibiting the Prandtl mechanism, bulging and vertical cracking was visible at the upper portion of the face although no shear surface appeared to emerge.

(b) Although the crack at the surface curved round towards the slope crest as shown on Fig. 4, repeated and careful observations have not shown it proceeding to the crest of the slope. This however could be a function of time, also another factor to be taken into account is the effect of the rockfill buttress at the southern end of the slope.

(c) Points on the Farm Buildings settled up to 80mm while situated outside the zone of visible movement indicated by the cracking.

(d) Extensometer anchor points at various depths within the slope appeared to settle in a generally similar fashion to the surface.

(e) A phenomena not referred to previously but particularly difficult to explain concerns the appearance from time to time of a continuation of the visible crack from the solid ground at the southern portion of the face on to the surface of backfill filling the excavation at this point. The backfilled portion of the site arises in that the solid ground excavated by large dragline ahead of the coal extraction area (the cut) is placed behind the cut as the cut is advanced leading to a progressive process of excavation and backfilling. Although more backfill was added from time to time and trafficked over, the crack continually reappeared in sympathy with the movement of the crack in the solid slope, and precisely on the same line. At times the crack appeared for some tens of metres across the surface of the backfill. A possible explanation can be obtained by considering the cohesive nature of the backfill (which is developed with time) coupled with the outward movement of the slope. If the face of the slope was in fact moving against the slope the cohesive nature of the backfill might be expected to support a crack, but it is doubtful in these circumstances that the crack would be a single crack extending over the distances and in the direction observed, although the appearance of the cracking reduced with increasing load of backfill against the slope and ceased altogether upon completion of backfilling of the excavation during the summer of 1979.

EXPLANATION OF MOVEMENTS OTHER THAN BY SLOPE FAILURE

If various types of slope failure do not account for all the phenomena associated with the ground movements, the alternative appears to be a mechanism connected with vertical and horizontal stress relief due to unloading by excavation down to the lower coal seams ahead of the cut. It can be visualised that strains resulting from stress relief could well concentrate at the fault which would appear to be in a convenient position for this; a photograph taken of the fault when previously encountered in the workings showed a remarkable degree of opening at the fault plane even at depth. A stress relief explanation with strains concentrating at the fault would account for movements outside the zone of cracking in the slope, i.e. the crack continuing across backfill and movements of Farm buildings. The cyclic nature of the movements is not so readily explained by this however and no apparent correlation exists between excavation operations at the eastern side of the mine adjacent to the slope.

CONCLUSIONS

What started as an emergency situation, where at one stage a major disruption of the slope was considered a real possibility, reached, in the course of time, a situation where it was considered that whatever the cause of movements, backfilling would be complete before disruption was imminent. In fact, backfilling completed in the summer of 1979 removed the possibility of serious disruption. Nevertheless the circumstances have been of geotechnical interest in that no incident of similar proportions has occurred previously and the various measurements taken have not provided a total explanation of events.

With respect to the instrumentation it must be admitted that this was installed in emergency circumstances and directed at obtaining confirmatory evidence of supposed shear failure within the slope. With hindsight other observations could have usefully been made such as those from extensometer anchor points located at greater depths below the slope base, from longer continuity indicators and from inclinometers. Such observations might have shed light on the unexplored possibility of failure below the base of the excavation, this would have accounted for the cyclic movements.

Of interest is the exponential decay pattern of the movements similar to the pattern described by Broadbent and Ko (Ref. 2) which is in opposition to the frequently assumed accelerating rate of movement to failure. Absolute confirmation of exponential decay is however, not provided by the observations in this case since a damping effect may have been provided by the buttressing increasingly provided over the observation period.

Also of interest is the period of months over which movements developed and the difficulty of interpretation of the movements of various points within the moving mass. The case illustrates the difficulties inherent in providing firm geotechnical advice on the outcome of this type of situation.

The authors express their thanks to those who have assisted in the supply of data and to the N.C.B. Opencast Executive for permission to publish this paper.

Opinions expressed are those of the authors and are not necessarily those of the N.C.B.

REFERENCES

(1) Burland, Longworth & Moore - A Study of Ground Movement and Progressive Failure caused by a Deep Excavation in Oxford Clay - Geotechnique Vol. XXVII Number 4 December (1977) 557-591.

(2) Broadbent & Ko - Rheologic Aspects of Slope Failure - Symposium on Rock Mechanics Denver (1971).

(3) Kuapil and Clews - An examination of the Prantdl Mechanism in large slope failures, Transactions/Section A of the Institution of Mining and Metallurgy Vol. 88 (1979).

GROUND SETTLEMENT RELATED TO DEWATERING FOR A SUBWAY EXCAVATION

I H WONG, BSc, MS, ScD, MASCE

Consulting Engineer, Ebasco Services Incorporated, New York, NY

INTRODUCTION

It is commonly known that excavations frequently result in movements of adjacent ground, sometimes to the detriment of structures nearby. Concerns for such ground movements often influence the choice of construction procedure used for the excavation and the design of the support system.

Numerous case histories have been reported in the literature related to ground movements induced by excavations. The vast majority of these case histories deal with situations where settlements can be related to movements of the sides or the bottom of the excavation caused by unloading. Ground settlements related to the lowering of the ground water table, associated with the dewatering of an excavation, have not often been discussed in the literature.

This paper describes a case history involving a subway excavation, using the cut-and-cover method. The subway excavation was along Accolon Way, a part of Boston's Haymarket Square to Charlestown extension of the Massachusetts Bay Transportation Authority (MBTA). The section described herein was at Station 76 + 30, and is known as "Test Section B." It was one of a group of test sections instrumented and studied by the Massachusetts Institute of Technology Department of Civil Engineering under a contract with the MBTA between 1967 and 1972.

Various aspects of the performance of Test Section B have been described by Lambe, Wolfskill and Wong (1970). In the present paper, the effects of dewatering on ground settlement are discussed. It is shown in this paper that the consolidation settlement caused by dewatering constituted a large portion of the total ground settlement.

SUBSOIL PROFILE

Figure 1 shows the profile at Test Section B. This profile was developed from bore hole data and from a study of the soils exposed by the actual excavation. An important feature of the subsoil profile at Test Section B is its variability in both horizontal and vertical directions.

The top layer in the soil profile was a fill, placed in the middle of the last century. It was a loose mixture of gravel, sand, silt and clay. Underlying the fill was a silt, which graded from light gray, slightly organic silt, to black organic silt containing shells. Underlying the silt was a till, which consisted of stratified layers of sandy silt or clay, sand and gravel, and sandy clay. The base rock, which underlay the till, was a dark jointed rock; this was probably a volcanic intrusion into the Cambridge Argillite commonly encountered in the Boston area.

The permeability values shown in Figure 1 were obtained from field sensitivity tests performed on the hydraulic piezometers and on the deep observation wells maintained by the MBTA. The permeability of the till was obtained from values listed in publications relating permeabilities with soil types and particle sizes (Lane & Washburn, 1946, Terzaghi & Peck, 1967). In consideration of the interbedding of sand, silt and clay layers, a ratio of horizontal to vertical permeabilities of 10 was utilized in the analyses.

DETAILS OF EXCAVATION

Figure 1 also shows a view of the excavation at Test Section B. As shown in this figure, the excavation was 11.3m (37 feet) wide and 17.7m (58 feet) deep. It was braced with ZP38 sheeting and five levels of struts. The horizontal spacing between struts was 3.7m (12 feet) on centers. The excavation was started in September 1967 and was completed in November 1968.

The east wall of the excavation was bordered by a vacant lot at the time of the excavation. A one-story building, called the Empire Carpet Building, was located 4.6m (15 feet) to the west of the excavation.

In order to have a dry site for work, the contractor pumped water from sumps within the excavation. The contractor was required to maintain the pore pressure in the till outside the excavation at a total head of +29.0m (+95 feet). The static value of total head prior to construction ranged from +32.9m (+108 feet) to +33.6m (+110 feet). Despite the number of recharge wells placed in the till,

TABLE I CONSOLIDATION PROPERTIES OF COMPRESSIBLE LAYERS

			Slightly Organic Silt	Black Organic Silt
Coefficient of Consolidation	c_v	cm^2/sec	157×10^{-4}	20×10^{-4}
Recompression Index	C_r		—	0.12
Compression Index	C_c		0.15	1.07
Initial Void Ratio	e_o		0.88	1.9
Overconsolidation Ratio	OCR		1	1.2
Maximum Past Pressure	σ_{vm}	kN/m^2		118
Thickness of Compressible Layer	H	m	3.7	1.8

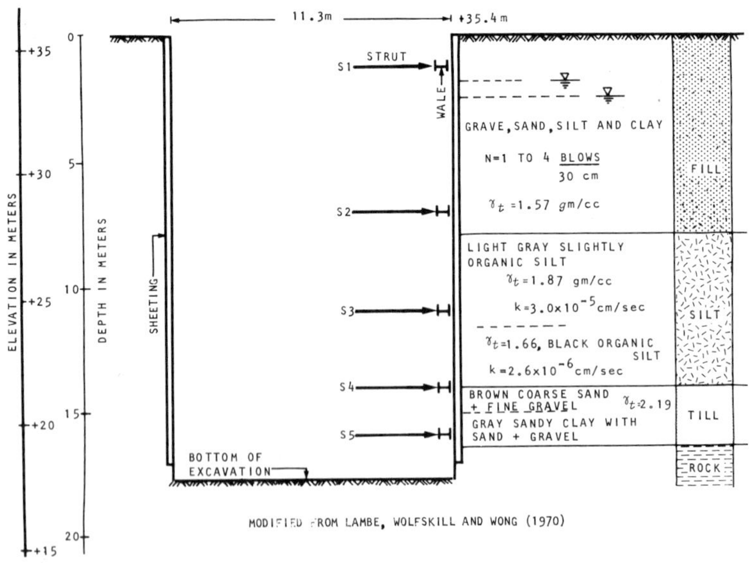

FIGURE 1 CROSS SECTION OF EXCAVATION

substantial drops in total head outside the excavation took place.

The contractor had intended to drive the sheeting approximately 3.1m (10 feet) below the bottom of the excavation to a depth of 21.4m (70 feet). At Station 76 + 30 the sheeting hit refusal at a depth of approximately 17.1m (56 feet). However, approximately 15.3m (50 feet) north of the centerline of Test Station B, the contractor was able to install the sheeting to the design depth.

Numerous instruments were installed at Test Section B. These instruments included piezometers (both hydraulic and electric) and slope indicator wells. The electrical piezometers were the Geonor vibrating wire devices. The hydraulic piezometers were the Casagrande double tube devices. There were 17 slope indicator wells at the test section, 6 in the soil behind the sheeting and 11 attached to the sheeting. Several of the slope indicator wells on the sheeting were perforated to permit them to function as groundwater observation wells. More details of the field instrumentation were described in Lambe, Wolfskill and Wong (1970).

MEASURED PERFORMANCES OF TEST SECTION

The measured performances of Test Section B were described in detail by Lambe, Wolfskill & Wong (1970). Only a brief summary of the pore pressure changes and ground movements will be discussed herein.

Measured Pore Pressures

Significant leakage into the excavation occurred at Test Section B. The groundwater problem was caused by:

A. Highly pervious gravel stratum in the till
B. Fractures and joints in the rocks
C. Ruptured joints between the sheet piles
D. Holes in the sheet piles

Figure 2 shows the contours of the measured total head outside the east wall. It shows that even as far away as 21.4m (70 feet) from the excavation, there was a very marked decrease in total head in the till.

Measured Movements

Figure 3 shows the horizontal movements of the sheeting and the settlements of the ground surface outside the excavation at Test Section B. It also shows the levels

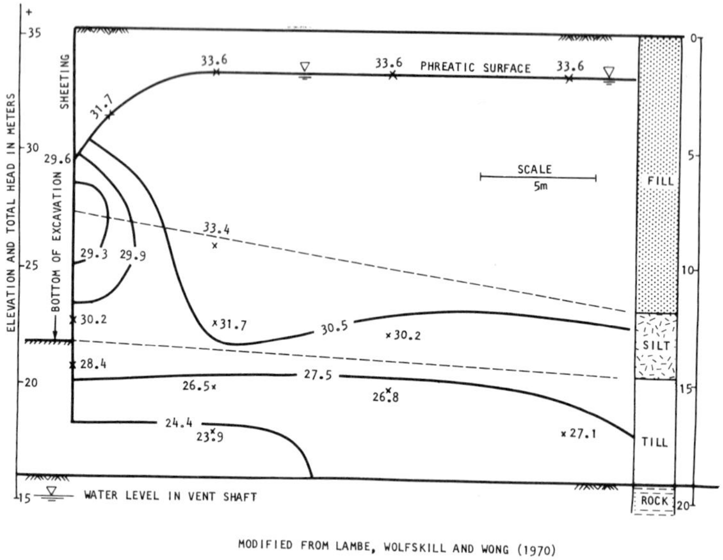

FIGURE 2 MEASURED TOTAL HEAD CONTOURS

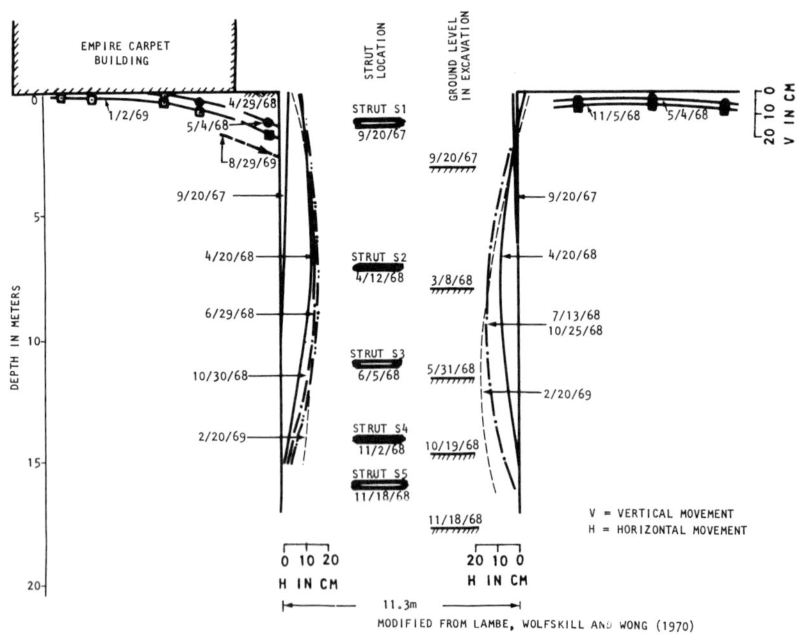

FIGURE 3 MEASURED GROUND MOVEMENTS

of excavation at various dates as well as the dates of installation of the struts.

According to Figure 3, the maximum horizontal movement towards the excavation of the sheeting was about 13 cm (5 inches). This movement was typical of the movements of the sheeting at Test Section B. One slope indicator, however, showed a movement of the west sheeting equalled to 23 cm (9 inches). Figure 3 also shows large settlements of the ground surface outside the excavation. The maximum settlements were 13 cm (5 inches) east of the excavation and 25 cm (10 inches) west of the excavation. Calculations presented later herein show that these settlements were caused by both unloading in the excavation and by consolidation due to dewatering. As discussed in Lambe, Wolfskill and Wong (1970), substantial damage to the Empire Carpet Building occurred because of the large ground movements.

PREDICTED PERFORMANCES OF TEST SECTION

Predicted Pore Pressures

As a basis for predicting consolidation settlement, an analysis was performed to calculate the pore pressure distribution outside of the excavation. The analysis procedure utilized was based on the finite element method, using a computer program developed by Taylor & Brown (1967). The finite element grid used for the analysis is shown in Figure 4. The finite element method was thought to be particularly suitable as it could handle any boundary conditions and soil inhomogeneity and anisotropy. It should be noted, however, that the finite element program used lacks the capability to simulate three-dimensional flow conditions which were believed to prevail in the field because of variable sheeting lengths and rock surface elevations. This was somewhat compensated for by using an average sheeting length and average rock surface elevation in the two-dimensional analyses.

One complicating factor in the analysis was that the sheeting had holes and ruptured joints, and was not watertight. An equivalent permeability of 0.1×10^{-6} cm/sec was chosen for a sheeting of gross thickness of one foot. This was based on the assumption that when the excavation was about 11.6m (38 feet) deep, roughly equal amounts of water flowed through the sheeting as under the sheeting.

Figure 5 shows the predicted contours of the total head when the excavation was 14.6m (48 feet) deep. It is interesting to compare it with Figure 2, which shows the measured contours.

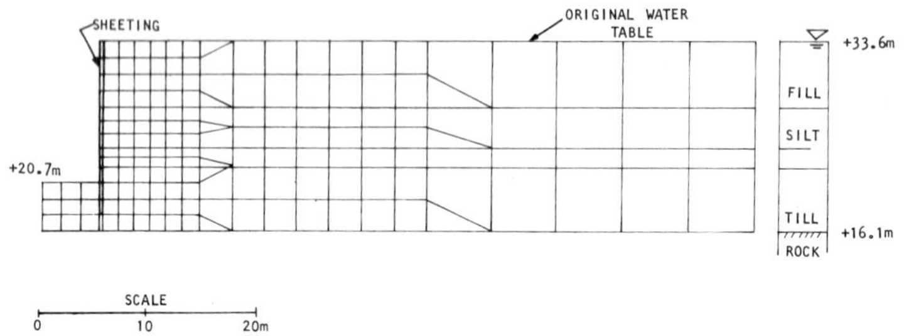

FIGURE 4 FINITE ELEMENT GRID FOR SEEPAGE ANALYSIS

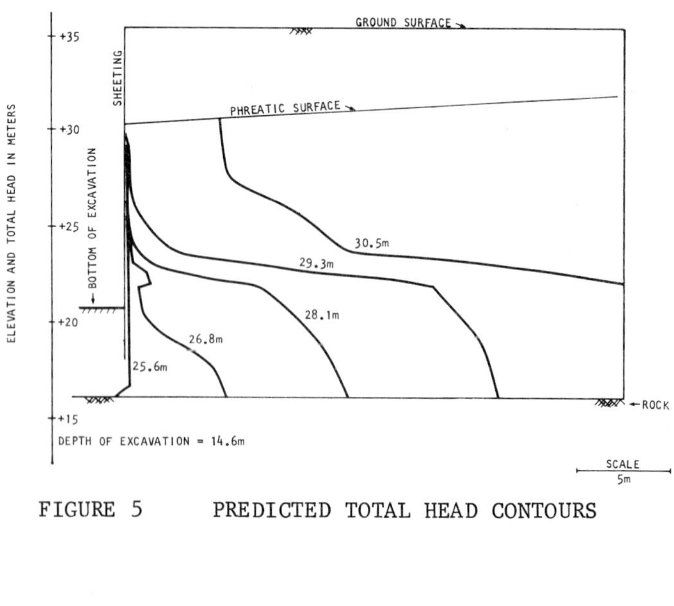

FIGURE 5 PREDICTED TOTAL HEAD CONTOURS

Predicted Ground Settlements

There are two factors which caused settlements of the ground near an excavation. The first factor is the horizontal and vertical unloading resulting from removal of the soil mass from within the excavation. The second factor is the decrease in pore pressures in the ground caused by dewatering operations. Settlement due to the first factor is the initial settlement. The settlement due to the second factor is the consolidation settlement.

Initial Settlements

The initial settlement at Test Section B was predicted using a finite element computer program BRACE. The method of analysis followed in BRACE allowed for soil inhomogeneity and soil nonlinearity, sheeting stiffness, construction sequence, including strut installation and prestressing. The details of this analysis were explained in Wong (1970) and are out of the scope of this paper. The predicted initial ground settlements are presented in Figure 6.

Consolidation Settlements

Based on the predicted head drops shown in Figure 5, and assuming one dimensional consolidation of the silt, the consolidation settlements at Test Section B were predicted.

The soil parameters used in the prediction of the consolidation settlements are listed in Table I. These soil parameters were obtained from laboratory oedometer tests. The predicted consolidation settlements were added to the predicted initial settlements to obtain the predicted total settlements of the ground surface.

Figure 6 shows the predicted total ground settlements, together with the predicted initial settlements of the ground surface.

SUMMARY AND CONCLUSIONS

The piezometer readings indicated that substantial lowering of pore pressures in the ground took place at Test Section B as a result of excavation and dewatering activities. In spite of the recharging, the pore pressures near the sheeting wall were far below the static distribution.

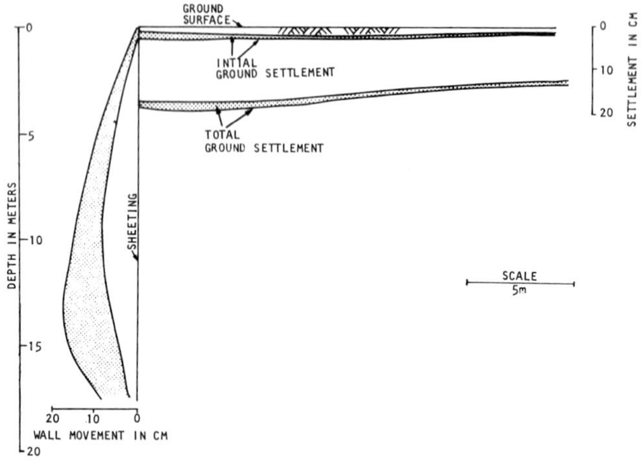

FIGURE 6 PREDICTED GROUND MOVEMENTS

The maximum predicted total settlement shown in Figure 6 was 18 cm. It consisted mostly of consolidation settlement. The maximum measured total settlements were 25 cm and 13 cm, west and east of the excavation, respectively. The overall agreement between the maximum predicted and the maximum measured total ground settlements is good.

It can be seen from Figure 6 that a major part of the total ground settlement consisted of consolidation settlement due to dewatering.

REFERENCES

T W Lambe, L A Wolfskill and I H Wong - Measured Performance of Braced Excavation, Journal of the Soil Mechanics and Foundations Division, ASCE, Vol 96, No. SM3, (1970) 817-836.

K S Lane and D E Washburn - Capillarity Tests by Capillarimeter and by Soil Filled Tubes, Proceedings of the Highway Research Board, (1946).

R L Taylor and C B Brown - Darcy Flow Solution with Free Surface, Journal of the Hydraulic Division, ASCE, Vol 93, No. HY2, (1967).

K Terzaghi and R B Peck - Soil Mechanics in Engineering Practice, John Wiley & Sons, Inc, New York, (1967).

I H Wong - Analysis of Braced Excavation, thesis presented to the Massachusetts Institute of Technology, Cambridge, Massachusetts, in Partial Fulfillment of the Requirements for the degree of Doctor of Science, (1971).

BUILDING SETTLEMENTS AND GROUND MOVEMENTS ASSOCIATED WITH CONSTRUCTION OF TWO STATIONS OF THE MODIFIED INITIAL SYSTEM OF THE MASS TRANSIT RAILWAY, HONG KONG

K. Morton, D I C , M I C E

Senior Engineer (Materials & Geotechnics), Hong Kong Mass Transit Railway Corporation

M.S.M. Leonard, M Sc (Eng), M I C E

Geotechnical Engineer, Hong Kong Mass Transit Railway Corporation

R.W. Cater, M I Struct E

Structural Engineer, Hong Kong Mass Transit Railway Corporation

SUMMARY

The Modified Initial System of the Hong Kong Mass Transit Railway includes thirteen deep excavations in heavily built up areas. This paper describes the settlement of a number of adjacent structures at two stations as a result of construction activities. It also describes the structural behaviour of different types of buildings as a result of ground movements.

INTRODUCTION

The Modified Initial System (M.I.S.) is 16 km long, and is the first stage of a Mass Transit Railway to provide a rapid transport service for the intensely populated colony of Hong Kong. It will also provide a railway link between Hong Kong Island and the Kowloon Peninsula through a 1.7 km long curved twin immersed tube. The route is shown on Fig. 1. A more comprehensive description of the railway is included in a paper by J.T. Edwards (1976).

Construction of the M.I.S. began in 1975 and the first section opened ahead of schedule in early October 1979. The cost of the civil engineering works was 800 million U.S.dollars for construction in very difficult ground conditions, varying from soft silts to sound granites with a groundwater table near the surface. Typical soil properties are shown in Table 1, and

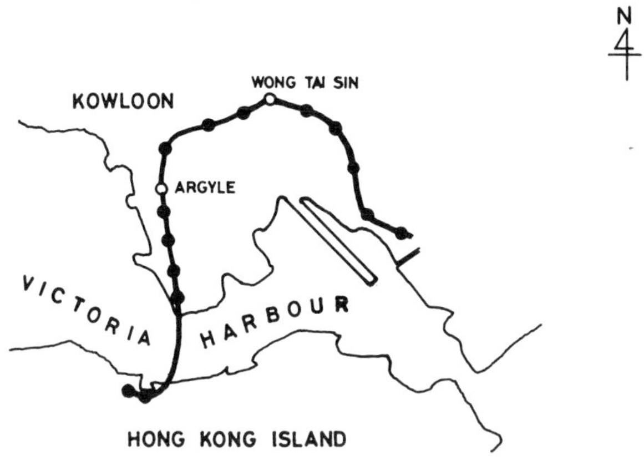

FIG. 1 : MAP OF M.I.S. SYSTEM

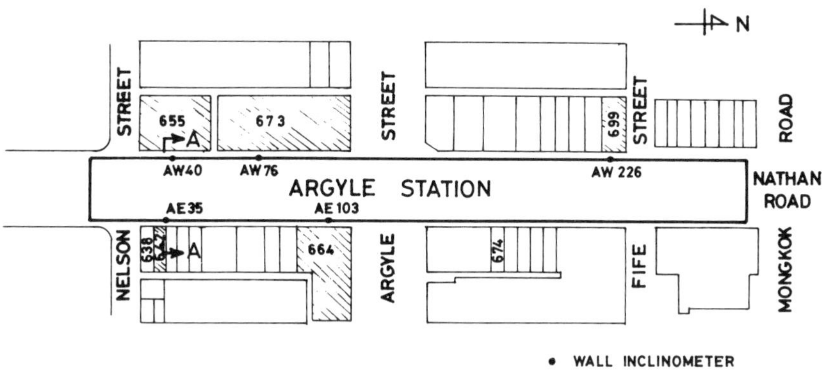

FIG. 2a: PLAN OF ARGYLE STATION

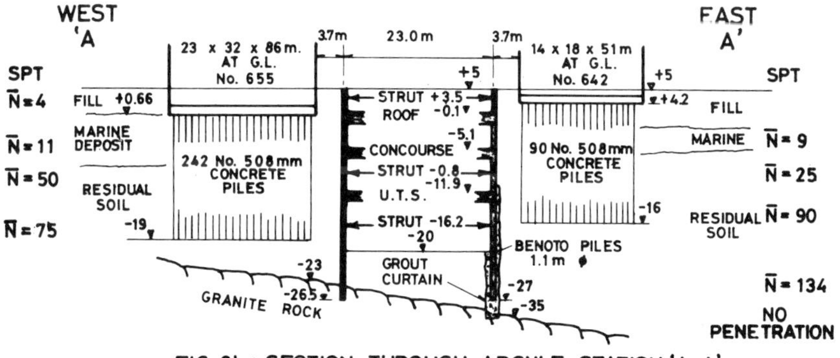

FIG. 2b: SECTION THROUGH ARGYLE STATION 'A–A'

Liquid Limit	Plastic Limit	Permeability m/sec	Compressibility m^2/MN	Strength ϕ deg	Strength C' kPa
5 - 15	30 - 50	2×10^{-5} - 10^{-7}	1×10^{-2} - 6×10^{-2}	33-37	0-5

TABLE I Typical properties of H.K. Residual Soils

Building number	Type of Foundation	Settlement (Front / Rear) To Aug 1977	Settlement (Front / Rear) To Dec 1977	Settlement (Front / Rear) To Feb 1979
638	19m deep precast piles driven 8m into residual soil	36 10	30 13	NA 12
664	18m deep Franki piles driven to top of residual soil	56 7	40 41	44 1
674	No records available. Assumed spread fottings on timber piles	108 38	105 29	113 NA
655	25m deep colcrete piles, 15m into residual soil	8 3	0 2	4 NA
699	Driven piles to 15m depth to 9m into residual soil	46 2	50 0	68 NA
673	Caissons dug to weathered rock	14 4	9 1	9 0
642	Concrete piles to 23m depth driven 10m into residual soil	39 8	30 6	35 10

In August 1977 caisson dewatering stopped
In December 1977 GWL had fully recovered
In February station excavation stopped

TABLE II Argyle Station building settlements

a detailed description of Hong Kong soil conditions has been made by Lumb (1962).

In several instances the M.I.S. passes less than 2 m. from adjacent structures. Building and groundwater levels were measured regularly, and inclinometers were installed in station walls. The relative interaction between wall and building movements at two stations is studied in this paper.

ARGYLE STATION

Argyle Station is situated in the northern half of the Kowloon Peninsula (see Fig. 1) which is probably one of the world's most densely populated areas.

Adjacent Buildings

The station lies under Nathan Road (see Fig. 2a and b) and is surrounded by many types of old and modern structures. Many of the buildings have been affected by deep pumping in private wells as recorded by Lumb (1964). The reported average settlements caused by deep pumping from 1964 to 1970 was 50 mm for shallow founded buildings and 15 mm for piled structures. Therefore many of the structures likely to be affected by the Mass Transit works had already moved before construction commenced.

The buildings adjacent to the station varied from a flexible to a very rigid construction. Although all were concrete beam and column framing, the earlier ones could not be termed rigid due, to the lack of continuity at pile cap level and the use of brickwork non-structural walls, which had deteriorated over the years.

Recent buildings have reinforced concrete shear walls. Non-structural walls are constructed using nominally reinforced concrete.

Station Construction

The station was constructed using a top down method inside 1.1 m diameter secant benoto piles, except in areas where restricted headroom caused by overhanging balconies necessitated the use of hand dug caissons 1.8 or 2.2 m diameter. The piles and caissons were to become the station's permanent walls.

The hand dug caisson is a local technique of manually digging inside cast in-situ rings placed progressively as the excavation descends. The ground water is removed by sump pumps or deep wells.

Building number	Building	Maximum Tilt	Behaviour
638	Built 1959, 12 storeys. Good state of repair. Beam and columns, no shear walls.	1/730	Architectural damage at ground level.
664	Built 1954, 9 storeys. Beam and column frame. Building split by expansion joint.	1/387	A few cracks in plaster, none in structural walls. Expansion joint opened by 100mm due to differential settlement.
674	Built 1955, 5 storeys. Flexible beam and column frame.	1/296	Architectural damage and cracks in minor structural member, not serious. Highest recorded settlement for station.
655	Built 1965, 25 storeys with revolving roof top resturant. Beam column & shear wall frame.	1/4500	No damage reported.
699	Built 1958, 8 storeys. Beam and column frame. Some rigidity, no crosswalls.	1/312	Slight damage to finishes. Some damage to outhouses, not serious.
673	Built 1969, 16 storeys. Rigid beam column & shear walls, 2 level basement.	1/2250	Slight ingress of water in basement. No settlement due to station excavation nor sheetpile station entrance.
642	Built 1962, 15 storeys. Rigid beam and column frame with cross walls.	1/1631	Slight architectural damage.

TABLE III Argyle Station - Details of building and their behaviours

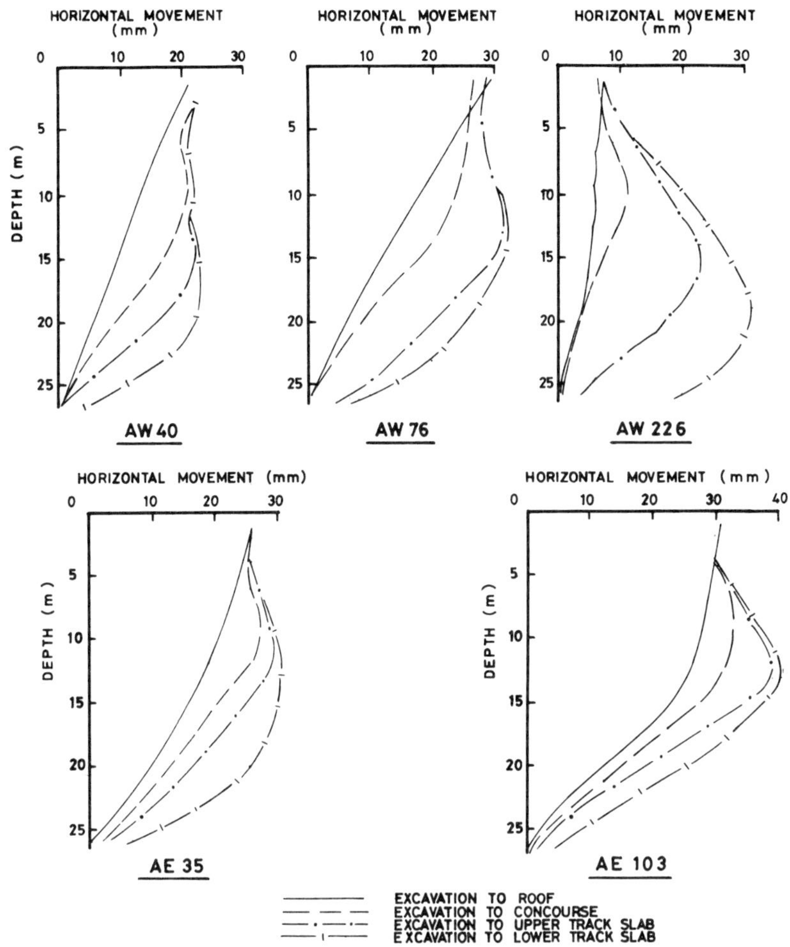

FIG. 3: ARGYLE STATION INCLINOMETER ENVELOPES

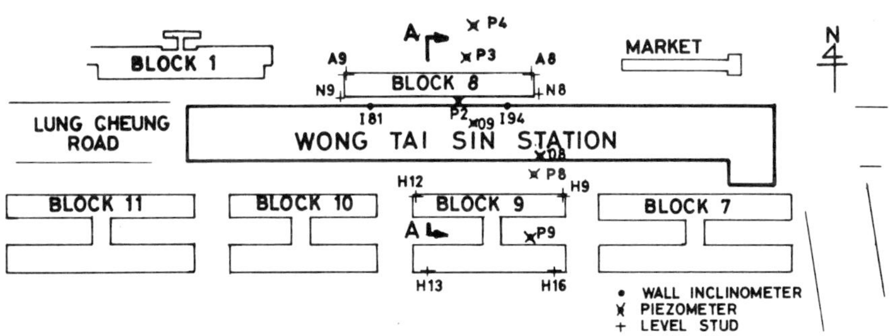

FIG. 4a : PLAN OF WONG TAI SIN STATION

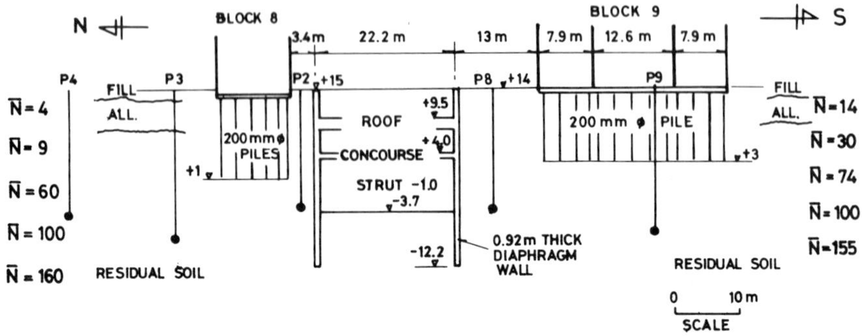

FIG. 4b : SECTION THROUGH WONG TAI SIN STATION A-A

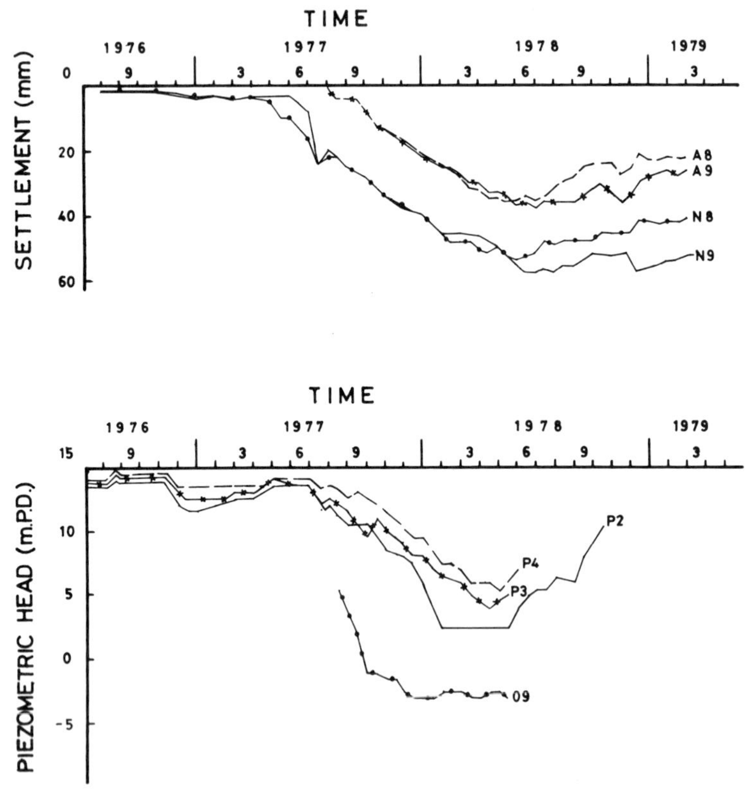

FIG. 5a :
WONG TAI SIN STATION – SETTLEMENT & DRAWDOWN AT BLOCK 8

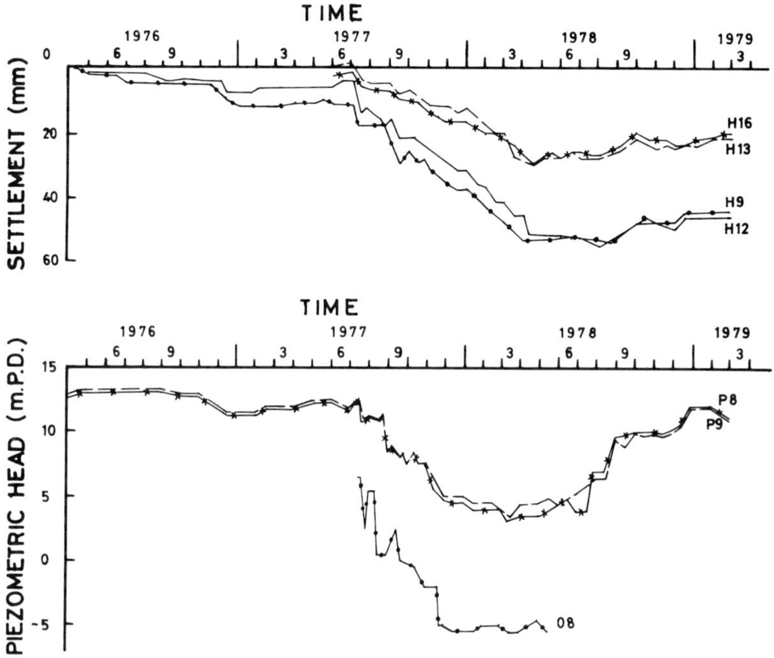

FIG. 5b
WONG TAI SIN STATION - SETTLEMENT & DRAWDOWN AT BLOCK 9

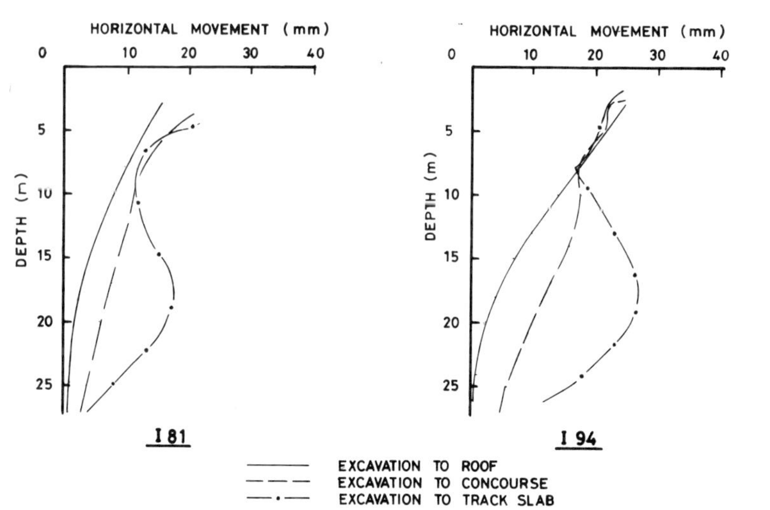

FIG. 6: WONG TAI SIN STATION INCLINOMETER ENVELOPES

Building Number	Inclinometer Number	Settlement during Excavation mm	Lateral Wall Movement mm
664	AE 103	4	43
655	AW 40	4	22
699	AW 226	18	31
673	AW 76	0	33
642	AE 35	5	32

TABLE IV Argyle Station. Building Settlement during Excavation and Wall movement

Building	Settlement				
	during wall constn. mm	observed from dewatering before excavation mm	calculated for dewatering during excavation mm	due to wall movement mm	lateral wall movement mm
Block 8					
N 8	14	6	36	(+4)	28
N 9	14	6	36	2	18
Block 9					
H 9	5	3	36	12	Not measured
H12	5	1	36	10	Not measured

(+4) indicates inaccuracy in assumptions.

TABLE V Wong Tai Sin Station. Building Settlement

Station wall secant pile construction began in July 1976 on the East side. Wall caissons were started in August 1976.

Work on the caissons led to large scale dewatering and building settlements soon became unacceptable, in spite of efforts to reduce pumping rates. It was decided to grout the location of future caisson works, and to grout below the station walls to granite bedrock.

Observed Settlements

The settlements measured on the adjacent buildings varied from 8 mm to a maximum of 113 mm. The largest part of, and in some cases, all of the settlement, was due to the dewatering that occurred during the construction of hand dug caissons. Benoto pile wall construction is not considered to have caused settlements except in isolated incidences, as all the piles were cased.

After the dewatering caused by the construction of the hand dug caissons was stopped, most of the buildings recovered by a significant amount averaging 10 mm. The buildings remained quite stable then for a period of 8 months. Only small settlements were recorded during the main box dewatering and excavation, reflecting the efficiency of the grout curtain.

Table II shows the settlements at the front and rear of seven adjacent buildings and the apparent tilt is given in Table III. Wall inclinometers were installed in front of five of these structures, as shown on Figure 2a.

Lateral Wall Movements

Inclinometer tubes were installed from ground level to the toes of five benoto piles, and the maximum movements are given in Figure 3.

During the period of station box dewatering and excavation, the outside piezometer levels fluctuated by 2 m, but never approached the previous levels of dewatering. It has therefore been assumed that all building settlement during the excavation period was due to wall movement. The building settlement and wall movements during the period of main excavation are given in Table IV. The large settlements at No.699 during excavation are probably due to the influence of an adjacent entrance construction.

Effects of Settlement on Adjacent Structures

The adjacent buildings did not suffer a significant amount of damage, when it is considered that their general condition was often poor, and that they were situated only 3.7 m from the station wall.

There were reports of architectural damage in the ground floor areas, often occupied by shops, where the finishes i.e. windows, marble tiles, etc., are more ornate than usual. In some cases the architectural damage, such as broken windows, rendered the area unserviceable, but this type of damage was localised and easily repairable.

Expansion joints were not generally employed even though buildings were built up tight to each other, and differential movement of the adjacent buildings resulted in the joints opening, causing ingress of water and resultant effects. This was of no serious consequence.

Cracking of plaster, even in the newer (1970) buildings, was attributed to the adjacent MTR work though no conclusive evidence of the cause could be determined.

Buildings appeared to tilt as units, and the differential settlement between front and rear did not cause any angular distortion, the building frames remaining square. This caused some cracking at the joints with low rise outhouses added on to the rear of some buildings. There was also some cracking of basements which caused ingress of water. The settlement of the building frame, causing a hogging mode in the ground slabs, was the reason for the water leakage through the cracked water-proofing screed.

WONG TAI SIN STATION

Wong Tai Sin is a typical two level station located in the north eastern part of the Kowloon Peninsula. A plan and section through the station is shown in Figures 4a and 4b. The station lies under Leung Cheung Road and is surrounded by high density resettlement blocks.

Adjacent Buildings

The resettlement blocks on the south side are seven storeys high, are about 20 years old, and are all founded on piles. Block 9 is 13.7 m from the station.

The buildings on the north side are more modern and block 8 is situated 3.7 m from the station wall.

All the adjacent buildings are of a very rigid (shear wall) construction, and were regulary monitored for settlement. Inclinometers were installed in the station walls infront of block 8 (See Fig. 4a).

Station Construction

The station was constructed using a top down method inside 0.9 m thick diaphragm walls. Wall construction began on the south wall in April 1976 which was completed in October 1976.

Wall construction commenced on the north side in April 1977 and was completed by the next July. Station box dewatering and excavation began almost immediately with the base slab being completed in June 1978.

Station box dewatering caused several metres drawdown beneath adjacent structures. The settlement and piezometer readings for blocks 8 and 9 have been plotted on Figures 5a and 5b.

Observed Settlements

Maximum settlements of between 52 & 58 mm were recorded for Blocks 8 and 9. The buildings also experienced tilt due to differential movement between front and rear. Such was the rigidity of the structures that each unit acted as one and no distortions were detected within the structures.

The majority of settlement was the result of dewatering which occured during station excavation. In order to identify settlements resulting from wall installation, dewatering, and lateral wall deflection, it is necessary to make a theoretical estimation of one of these factors.

This is most easily done by considering settlement resulting from dewatering, and work done by D'Appolonia and Lambe (1970), and Martin (1972), on soil compressibility has been correlated with SPT results, to provide a method of estimating building settlement in Hong Kong.

The empirical relationship

$$1/m_v = 0.8\ N\ m^2/MN$$

has been adopted with

m_v = soil compressibility

N = SPT blow count.

This has been checked at a number of other locations on the project and proved reliable. Using this method, settlements of 1 to 6 mm, and 36 mm, are calculated for dewatering before and during excavation respectively. This then allows a breakdown of the probable settlement resulting from the various causes which is shown on Table V.

It can be seen that at Block 8 relatively large settlements occured during wall installation. It is thought that the installation of diaphragm walls produces ground relaxation in the same way as station excavation. This has been recorded at other stations on the railway.

Lateral Wall Movements

Inclinometer tubes were attached to the reinforcement cages of diaphragm wall panels 81 and 94 in front of Block 8. The results of obervations are given on the horizontal movement envelopes in Figure 6.

The maximum movement at depth is between 18 and 28 mm and it can be seen that the apparent settlements resulting from wall movements are very low.

DISCUSSION

It is apparent that ground water lowering was the predominant cause of building settlement at these stations. At Argyle the dewatering occurred during wall installation activities, and at Wong Tai Sin the dewatering was carried out as part of the main excavation.

It can also be deduced that for Block 8 at Wong Tai Sin, there was a significant proportion of settlement during diaphragm wall excavation. This phenomenum has been noted at other sites on the M.I.S., and confirms that it affects the adjacent buildings in the similar manner to major excavations.

It is noted that the maximum lateral wall movements were significantly higher for the deeper excavation at Argyle than for Wong Tai Sin. The maximum wall movement as a percentage of excavation depth was 0.17% consistent with the case history for a diaphragm wall described by Lambe (1970).

Case histories reported by others normally refer to ground movements adjacent to excavations. It is more useful for designers to examine the effects on adjacent structures. For most of the buildings documented here the ratio of lateral wall movement to building settlement is greater than 5:1. The only exception is one case at Argyle where the building was affected by another excavation.

In the main, observations indicated that even buildings with little stiffness, such as frames on isolated foundations and no crosswalls, were stiff enough to re-distribute significant differential settlements. The exception, confirmed at other M.I.S. sites, was where the buildings crossed the zone of influence indicated by Peck (1969). It was also noted that the effects of wall installation on excavation causes longitudinal differential settlement not normally considered by designers.

CONCLUSIONS

The results of the observations at these two sites show that even well controlled construction of deep excavations causes large settlements and tilts. However, it was also observed that a wide variety of buildings with different stiffnesses were able to accept these movements with only minor damage.

ACKNOWLEDGEMENTS

The authors wish to thank the Hong Kong Mass Transit Railway Corporation for giving their permission to write this paper, but the opinions expressed are those of the authors and not necessarily the Corporation.

REFERENCES

D'Appolonia, D.J. and Lambe, T.W.-Method for Predicting initial Settlement, Journal Soil Mechanics & Foundations Division, American Society of Civil Engineers (1970) Vol. 96, SM 2, 523-544.

Edwards, J.T.-Planning and Design of the Hong Kong Mass Transit Railway, Proceedings Institution of Civil Engineers, (1976), Vol. 60, Pt.1, 9-26.

Lambe, T.W.-Braced excavations, Speciality Conference on Lateral Stresses in the Ground and design of Earth Retaining Structures, American Society of Civil Engineers (1970) 149-215.

Lumb, P.-General nature of the soils of Hong Kong, Symposium on Hong Kong Soils, (May 1962), H.K. Joint Group of the Institutions of Civil, Mechanical and Electrical Engineers.

Lumb, P.-Report on the Settlement of Buildings in the Mongkok District of Kowloon, Hong Kong, H.K. Government Publications, Hong Kong (1964).

Martin, R.E. - Estimating Foundation Settlements in Residual Soils, Proceedings American Society of Civil Engineers (1977) Vol. 103, No. GT3, 197-211.

Peck. R.B.-Deep Excavations and Tunnelling in soft ground, 7th Conference on Soil Mechanics and Foundation Engineering Mexico (1969) 225-291.

Ruxton P. and Berry L. - Weathering of Granite and associated Erosional Features in Hong Kong, Bulletin Geological Society, America (1969) 1263 - 1299.

GROUND MOVEMENTS DUE TO DEEP CUTS AND TUNNELLING IN QUATERNARY SOILS IN WARSAW, POLAND

Henryk Stamatello

Professor of the Warsaw Technical University, Warsaw, Poland

SYNOPSIS

The author's views concerning ground movements observed near deep cuts and over tunnelling construction are analysed in this paper. Relevant observations have been conducted for fifteen years or so from the viewpoint of an interdependence between such movements and the type of soil encountered, the ground-water conditions and the construction method adopted.

Data gathered in this way is found convincing enough to allow the conclusion that it is chiefly the movements arising during the actual earthwork operations and due to engineering errors which determine the extent and scale of the ground movements. The type of soil involved and its properties have, in fact, an indirect effect only, limited by the proper selection of the engineering methods normally chosen to suit the actual field conditions. The author suggests that an empirical formula be adopted for assessing the extent of ground movements, to include certain factors found to be responsible for the occurrence of these movements.

SOIL CONDITIONS IN WARSAW AND ITS SURROUNDINGS

In much the same way as in many other parts of Central Europe, the soils of Warsaw and its surroundings represent Quaternary deposits which result from two successive periods of glaciations (Mindel and Ries), as well as periods following these glaciations. They mostly consist of morainic deposits i.e. sands varying in their grain-size distribution, boulder clays of differing plasticity and varved clays.

From time to time shallow deposits of Pliocene clays occur right under the ground surface. These clays are chiefly hard and durable. Some geologists compare them therefore with the London clays which are considered to be rather advantageous for tunnelling operations.

Geological explorations and earthworks executed more recently in these clays do not confirm such a point of view. The clays lack uniformity and include saturated sand lenses whereby work in them is considered to be far from easy.

GROUND WATER

Ground water of glacial origin is present in the sands and its level remains dependent on the actual position of the permeable soils. It can nevertheless be agreed that the quantity of water available in these soils is rather small. More water can be found in the alluvial deposits, in the neighbourbood of the Vistula river and its tributaries. To drain water from both kinds of deposit offers no special problem today, and the influence which the ground water has on engineering works is by no means such as to exceed the limits of generally accepted hardships.

THE AREA OF GROUND MOVEMENTS

Ground movements occurring in the vicinity of ground structures must basically be considered a problem in soil statics. It has been under investigation for a very long time, its beginnings date back as far as the 18th century i.e. to the time of Coulomb and then Rankine and others. Researches have had one objective in view, to find a method of determining the loading characteristic for civil engineering structures.

It is self-evident that such loadings i.e. ground pressure, must depend on the properties of the soil involved, and this is why the parameters used to describe the mechanical properties of the soil, such as γ (bulk density), ϕ (angle of internal friction) and c (cohesion) play such an important role in the mathematical formulae dealing with ground movements. Most theories concerned with an area of ground in which movements are likely to occur assume that it is the phenomenon of movements produced by external forces i.e. movement of a retaining wall, the sheeting of a cut or tunnel, or sliding, which may bring about stresses in a soil precipitating a state of plastic equilibrium. This, in turn, may readily be understood as theoretical confirmation of an interdependence between the hazard zone in the vicinity of a cut and the technique of excavation, which on the whole is in itself a problem for long well known. There are also several opinions on the question of how to determine the radius of a hazardous zone in the vicinity of a cut. According to a number of sources it is a plane crossing the lower edge of the cut at an angle of 45°, near its bottom, which proves to be a boundary between the hazardous and safe zones (Figure 1). Some other

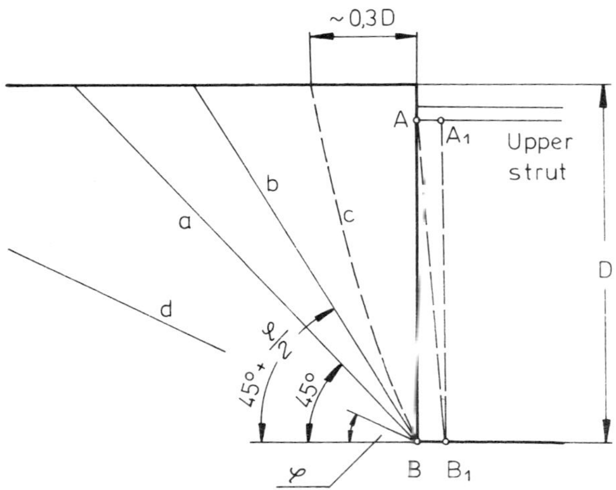

FIGURE 1 Ground movements near an open cut according to different assumptions

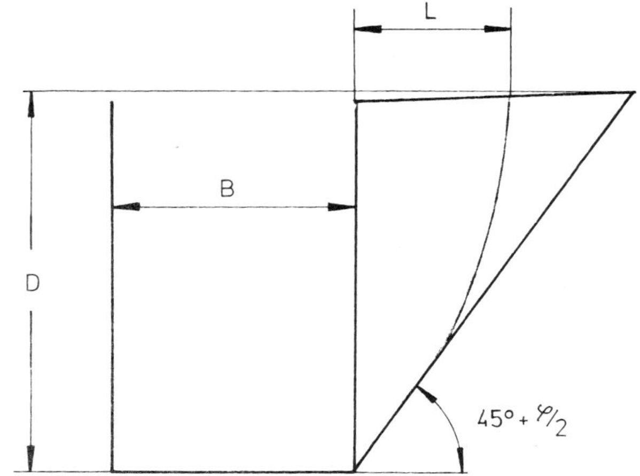

FIGURE 2 Surface settlement along an open cut

sources suggest that the plane in question ought to be inclined at an angle ϕ, equal to that of the internal friction (Figure 1). Both these suggestions seem to be unfounded and not confirmed by practice when the excavation has been performed properly. Particularly glaring and far from customary experience appears to be the latter suggestion, even in the case of unsupported excavations.

Should, following Rankine's theory, the boundary between the area remaining in a state of plastic equilibrium and that in a state of elastic equilibrium be taken as a plane, the angle of its inclination to the horizontal would be $45 + \phi/2$ (Figure 1). Such an approach is cautious too. The conditions specified above are likely to occur exclusively when the wall sheeting of a cut moves towards the cut itself, into position $A_1 - B_1$, thus causing the soil to expand throughout the entire depth of the cut. This might be possible in an unsupported or poorly supported cut.

Originally, the main object of the author's research was to find a relationship between the zone of ground movements and the soil parameters cited above, in order to determine the extent of the hazardous area in the case of deep cuts and tunnels built under specifically urban conditions. Potential dangers due to ground movements relate to buildings located near the earth-cutting operations as well as to underground urban installations such as water or gas pipes.

Observations made covered :

- An area of approximately 10 km alongside cuts 1 to 5 metres in width and 5 to 15 metres in depth, supported in a generally accepted manner;

- Areas over tunnelling, approximately 2.5 km in length executed by classical mining methods to build sewers up to 18 metres in depth (tunnelling bottom);

- Areas over several hundred metres of tunnelling made by the shield method in water-saturated sand, 6 metres in diameter. The shield bottom was approximately 15 metres deep and the air pressure was 2.5 at g.

Measurements taken during tests covered:

- Cracks present at the ground surface, produced as a result of ground movements. Measurements were made of their distance from the edges of the cut.

  These measurements could be taken only outside completely built up areas.

- Accurate surveying of a site on both sides of the cut or tunnel to establish the true width of the trough.

- Surveying of datum points on buildings in the neighbourhood of a cut or tunnelling.
- Deterioration and cracks in buildings bordering the structure. Cracks have been photographed more than once during observations; before the work started, during its progress and after its conclusion.

Surveying of a site has been performed again some time after the operations were concluded, notably after twelve months. Surveying data was of potential value in determining relationships between ground movements and the duration of earthworks.

To provide information that could confirm a relationship between the extent of ground movements and the respective soil properties, laboratory research was launched to establish the actual properties of the soils involved.

Analysis of the observations, measurements and research has led to the following conclusions :

It is beyond any doubt that an interdependence exists between the soil conditions and the ground movements evoked by such structures as deep cuts and tunnels being built on a site. This interdependence is however not direct because the operational techniques adopted tend to be always chosen according to the actual soil conditions and the expected hardships. Therefore, every attempt to describe the nature of the ground movements as a function of the specification of the soils, in the form of their respective parameters, has failed in consequence. Little success has attended attempts to present these movements in the form of a mathematical formula, even though empirical in character. There is nothing strange about this if we consider how diversified are soil properties and how scarce are the available parameters.

As has been proved by the author's research, the ground losses due to civil engineering operations provide the main reason for the occurrence of ground movements and the respective strains in frictional surfaces. Some of these losses seem unavoidable, for example the loss of ground in the shield method due to movements observed inside the shield. Unavoidable will also be the losses caused by the soil loosening over tunnels, this being independent of the adopted operational technique. Such losses are relatively easy to fix.

GROUND MOVEMENTS DUE TO THE DEPRESSSION OF WATER TABLE

The ground strains produced as a result of water losses

which inevitably lead to water level depression in a soil are relatively easy to determine.

As the ground water-table tends to lower, sandy soils may become susceptible to settlement because of their growing bulk density and volume decreasing at the same time. Theoretical calculations as well as numerous observations seem to corroborate the thesis that the size of such movements has little effect on the occurrence of ground movements. They are very small in fact. With the ground water-table lowered by some 10 metres, the actual magnitude of a settlement may range from 10 to 15 mm.

Such observations have, for instance, been made in the sandy soils of Warsaw in places where sands of approximately $k = 10^{-2}$ cm/sec permeability occur over large areas.

This comment does not apply where there are water pumping errors i.e. to cases where particles of the soil are washed away together with water being pumped. In such circumstances, ground movements may possibly occur on an unforeseeable scale.

GROUND MOVEMENTS NEAR OPEN CUT WALLS

In the event of open cuts having sheeted and braced walls, the movement of such walls towards the cut itself is responsible for the ground movements near the walls of such cuts. The soil undergoes a process of expansion, as a consequence, and that is why the state of plastic equilibrium can be initially observed inside the mass of soil. As already remarked, according to Rankine, such an area will stretch to the plane crossing the lower edge of the cut and inclined at an angle of $45^{\circ} + \phi/2$ to the horizontal (Figure 1). The zone of movement may be considerably reduced if rigid bracing is applied to the upper edge of the cut, in the form of a strut or ground anchor. Evidence suggests that the surface defined by Terzaghi, in the form of a logarithmic spiral (Figure 1) represents the boundary of the ground movement zone.

In its lower portion, this spiral will remain tangential to the straight line crossing the horizontal at $45^{\circ} + \phi/2$. In its top portion, the spiral is perpendicular to the ground surface and this will be expected if we assume that the struts prevent the ground from moving in a horizontal direction. As will be seen from Figure 1, the hazardous zone in the vicinity of a cut then becomes a minimum.

The observations carried out in Warsaw during the construction of 15 km of sewers have been helpful in showing that the width of the zone near the side of a cut, in which ground movements are likely to occur, will extend no more than 0.3 H beyond the side of such a cut. This statement refers to sheeted trenches. With extremely careful bracing of the upper edge of a cut made through a cohesive soil, such a width will not exceed 0.20 H; and 0.25 H in the case of non-cohesive soils.

The above-cited methods of determining hazardous zones in the vicinity of supported open cuts apply to operations carried out in dry soil, both in theory and in practical experience. Even wet soil may be assumed to perform in the same way as the dry soil, if the water is pumped from it in a proper manner.

According to Terzaghi's recommendations, the distance between the edge of a cut and the surface point defining the extent of the ground movements should be determined using a method of successive approximations. For this purpose, use must be made of the equations of equilibrium for the moving mass of soil and for the bracing moment, as well as for the resultant action of an immobile mass of the soil.

As can be seen from the suggestions made above, the area of ground movement remains strictly dependent on the technique of the operations i.e. on the sort of struts used for the sheeting. The main feature in this case will include strutting applied to the upper edge of the sheeting. Where such strutting proves to be rigid and is left until the end of the work i.e. until a trench has been backfilled, ground movements will appear to be the smallest. With no such strutting in use, even for a short period of time, the area covered by movements of this kind will increase to the extent stated in Rankine's theory.

It is obvious that the area of ground movements, which undoubtedly carries with it hazards for structures or installations in the near vicinity, may even be much larger when engineering errors are committed. Several examples occur in the literature, some of them having been discussed in the book written by Terzaghi and Peck (1967). On the other hand, it is certain that examples of that kind will be rather sporadic, if a high standard of work is maintained.

As already stated above, the author of this paper initially tried to trace an interdependence between soil parameters and ground movements. He has, however, not been successful in proving such an interdependence by means of a mathematical formula. Numerous observations, on the other hand, authorise him to suggest a simple empirical formula in which the width of a ground movement zone is presented as a function of the wall of a cut made in the earth, its depth, width and the type of sheeting. In the case of sandy, cohesive, medium-hard soils (Figure 2)

$$L = b.\beta.D \qquad (1)$$

wherein:

L - width of affected zone ;

D - depth of cut;

β - coefficient being dependent on the type of strutting applied;

For the cut edge secured by ground anchors

$\beta = 0.3$

For the top edge strut left till the end of work

$\beta = 0.35$

For the cut backfilled after the removal of sheeting

$\beta = 0.7$.

The coefficient b depends on the width B of the actual cut, and

for $B < 5$ m $\quad b = 1$

for $5 < B < 8$ m $\quad b = 1.1$

for $B > 8$ m $\quad b = 1.25$

In stiff cohesive soils, the area of ground movements may be smaller; sometimes much smaller.

THE ZONE OF GROUND MOVEMENTS OVER TUNNELLING

Observations relating to ground movements over tunnels, though not on such a scale, have also been performed in Warsaw. Surface strains over an approximately 1.5 km long tunnel have been monitored in the case of sewers built by the mining method. Another observation, as already stated, was completed over a length of some 500 metres of tunnelling made in saturated sandy soil by the compressed air shield method. In those places where tunnelling ran under completely built up areas, it was difficult to take measurements of the surface strains, so that such measurements had to be repeated more than once.

The results of the observations do not depart from those cited by P. Attewell in his work (1977), nor from those cited by K. Szêchy (1979). They can be said to confirm the former ones, with the proviso that ground surface movements over tunnels made by mining methods have proved to be smaller than those over shield-made tunnels. This happens in all probability because ground loss in the shield is larger than that in the mining method, if the work is carried out properly.

HEAVING AND SWELLING OF CLAYS

Under the Quaternary deposits in Warsaw there occurs a layer of the Tertiary, Pliocene clays. The top surface of this layer appears to be highly uneven and in some places it comes very near to that of the site. Illites are in mineralogical preponderance in these clays and they have been regarded therefore as being only slightly prone to swelling. During tunnelling, it became evident that the repeated presence of glaciers has resulted, in addition to normal surface changes, in the infraction of the original structure of the clays. As with the lapse of time micro-cracks began to appear in their bulk, they became more and more sensitive to the aggression of moisture, even that coming from the atmosphere. Accordingly, intensive swelling of the primarily dry clays has been noticed in the shield-made tunnel, with consequential cracking observed also in the cast-iron tubing enclosure. This movement has also shaken the view that impermeable and cohesive clays are to be regarded as one of the best materials for tunnelling operations.

REFERENCES

P. Attewell - Ground movements caused by tunnelling in soil, Proceedings of the Conference on Large Ground Movements and Structures, Cardiff (1977) 812-948.

K. Szechy - Surface settlement due to the tunnelling method in cohesionless soils, Proceedings of the Conference on Subway Construction, Budapest (1970) 615-625.

K. Terzaghi and R.B. Peck - Soil Mechanics in Engineering Practice, Wiley, New York (1967).

GROUND MOVEMENTS CAUSED BY DEEP EXCAVATIONS AND TUNNELS AND THEIR EFFECT ON ADJACENT MAINS

Malcolm Howe, Ph D , C Eng , M I Mech E , M Inst E , M I Gas E

Assistant Division Manager, Distribution Operations

Peter Hunter, C Eng , M I Mech E , M I Gas E

Principal Engineer, Distribution Operations

Raymond C Owen, B Sc , Ph D , C Eng , M I C E , F G S

Senior Engineer, Distribution Operations

British Gas Corporation
Engineering Research Station
Newcastle upon Tyne

1 INTRODUCTION

Extensive gas distribution systems representing a major national asset are laid beneath the roads, verges and footpaths throughout the towns and cities of Great Britain.

A large proportion of the total system was laid many years ago when environmental conditions were considerably different from those existing today. British Gas has undertaken an extensive research programme to determine how stresses can build up in mains subjected to changing environmental conditions. The knowledge gained is applied to improve the safety of existing gas distribution systems and to support specifications and codes of practice for future main laying operations.

This paper briefly reviews the factors that can lead to high stresses in mains and then concentrates on the effects of excavations and tunnels on the older more vulnerable cast iron distribution system. A set of risk zones is defined from the field and experimental studies on deep excavations. These zones will form the basis of a predictive facility for assessing the risk to mains from proposed works.

2 FACTORS INFLUENCING PIPE FAILURE

The life expectancy and the probability of failure of buried mains is heavily dependent on the stress levels they experience in service. Stress levels arise from an interaction between the pipe and its surrounding soil often acted upon by surface loading. A general review of the causes of mains failure was given by Needham and Howe (1979). Generally speaking a pipe will be subjected to several causes of loading, possibly from excavations, traffic, blasting etc, during its life as its environment changes. Each cause of loading will produce its own increment of stress and the total of all increments amount to the in-service stress at any time. This paper concentrates only on the high stresses induced in mains by ground movements associated with deep excavations and tunnels.

Ground movement can be caused by two factors:-

i) Having an inherently unstable ground as the result of poor compaction of made up ground, poor compaction in the trench, geological deformations, or basically unstable soils associated with over-steep embankments.

ii) The ground being disturbed or forced to deform by applied loads, in particular to movement generated by adjacent excavations/tunnelling or washout and ground loading due to heavy traffic.

A review of the causes of mains fracture indicates that probably one third of the fractures could have been caused by category i) and the other two thirds by category ii). In category ii) there may be a time delay before stresses appear in the main if ground movement/settlement is delayed.

At the present time efforts are being made to improve liaison with all Utilities with the objective of minimising the secondary damage caused by excavating adjacent to buried plant but this will probably take many years to emerge as a useful means of cutting down the rate of pipe fracture. In the interim period deep excavations and tunnels adjacent to existing cast iron systems present a source of danger if adequate precautions are not taken. Even with the newer distribution pipe materials, such as polyethylene and steels, there may be a risk albeit at a much diminished level. The level of risk to mains in practice is very wide because of a large variation in the current quality of trench supporting practices. An obvious aim is to prevent ground movement in the vicinity of the trench so that the risk to adjacent pipework is minimal. Achieving this calls for the side of the excavation to be firmly supported during the excavation process, while the excavation is open and for a controlled withdrawal of support whilst the excavation is being backfilled to a high standard of compaction. Field tests undertaken to-date indicate that small ground movements may be achieved by available support techniques when properly used. The implementation of the draft revision of BS CP2003(1977) should provide a suitable start to specifying desirable standards of workmanship.

Clearly there is a need to eliminate the practices of some contractors such as using gas mains to hold back the trench shuttering as seen in Plate 1, and having deep unsupported excavations adjacent to gas mains as in Plate 2. The average standard of trench support may improve in the future but in the interim period we need to recognise that ground movements occur and we need to assess their effect on the existing system.

A similar argument applies to tunnelling where time-dependent settlements at surface level occur both above and to the side of the tunnel invert and also somewhat in advance of the heading.

The nature of ground movement from trenches and tunnels is complicated and governed by the precise conditions on a given site, but the following examples illustrate the processes involved and their effect on adjacent pipework. The bending effect on mains is expressed in terms of measured strains in microstrain. In bending the strain changes from tensile to compressive across the pipe diameter. The maximum working strain in grey iron mains is considered to be about 300 microstrain (0.03%) tensile, for static load applications. Any changes in strain from deep excavations when added to existing values should not exceed this maximum. An allowance must also be made for additional dynamic strains from traffic, etc.

3 MEASUREMENTS OF GROUND MOVEMENT AND ITS EFFECT ON BURIED MAINS

The field work to-date has embraced situations where excavations or tunnels ran parallel to or crossed existing gas mains. Site instrumentation consisted of inclinometer bores to measure lateral ground movement and magnet extensometers to measure vertical settlement. The gas mains were fitted with vibrating wire strain gauges to measure changes in longitudinal bending strain.

3.1 Effect of Excavations Parallel to Line of Existing Gas Mains

Excavations for other services in an urban street are often parallel to the gas main supplying the street. The excavation creates horizontal and vertical movements in adjacent ground and this can cause bending in the gas main and pulling on the services running at right angles from it.

PLATE 1
Gas main supporting trench sheets

PLATE 2
Deep unsupported excavation

PLATE 3
4 in (100 mm) cast iron main fractured and displaced as a result of intersecting excavation

3.1.1 Site 1

A high standard of workmanship was employed by the contractor at site 1 when installing a rectangular culvert 2 m x 1.5 m adjacent to a 9 in (230 mm) main. A section through the site showing the positions of the instrumentation relative to the main and excavation is shown in Fig. 1. The installation of the culvert approached the measurement site in a series of excavated bays 4.6 m long x 3.5 m wide. The trench was excavated through mainly soft sandy clay to a depth of 5.5 m. It was well-supported using close-fitting trench sheets and 0.3 m x 0.3 m timber walings braced with 0.3 m x 0.3 m timber struts. The support system was applied progressively as the trench increased in depth and where there were voids left behind the trench sheets these were subsequently packed with sandbags.

When the excavation was alongside the inclinometer bore, the Contractor changed to a hydraulic shoring system in place of the timber walings and struts. Two days after installation of this proprietary system, it had to be reinforced by 0.3 m x 0.3 m timber shoring. The hydraulic system had buckled and the ground adjacent to the trench had fissured. The profile of ground movement towards the trench, obtained from the inclinometer bore is shown in Fig. 2.

Immediately after reinforcement of the hydraulic shoring the maximum ground movement was about 15mm towards the excavation. This increased to 76 mm 7 days later and to 120 mm after a further 7 days before becoming stable.

The vertical settlement of the ground given by the magnets fitted around the inclinometer access tube is shown in Fig. 3. It can be seen that the top magnet experienced a vertical settlement of more than 200 mm as a result of the failure of the hydraulic shoring. This movement was far in excess of the horizontal movement of about 20 mm (Fig. 2) recorded by the inclinometer over the same period. Movement after this period was predominantly horizontal. From the time the hydraulic shoring was reinforced by timbering until backfill was complete, lateral movement was 83 mm and settlement 15 mm. This combined pattern of movement is shown in Fig. 4. The adjacent ground fissured and subsided and the road surface moved away from the kerb to leave a 10mm gap. The fissure, shown on Fig.1 was some 50 mm wide and ran for 9 m parallel to the excavation and 2.6 m from it. The depth was difficult to determine because the fissure was repeatedly filled up by the Contractor. Vertical displacement between the edges of the fissure was approximately 150 mm.

This fissuring coupled with the suddenness and degree of settlement indicates that the predominant failure mechanism was a tension crack followed by shearing along a plane between the fissure and the trench. The resulting soil wedge, of estimated

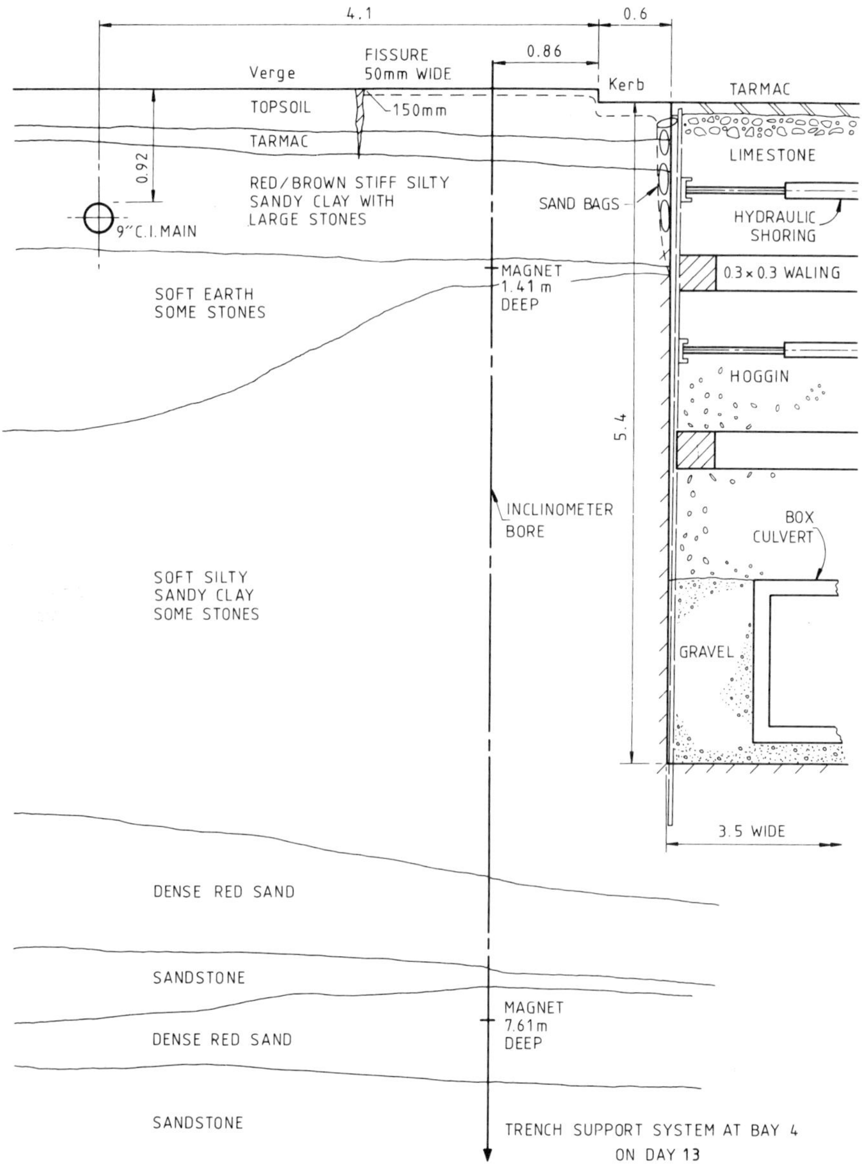

FIG. 1 EXCAVATION PARALLEL TO GAS MAIN
SECTION LOOKING EAST

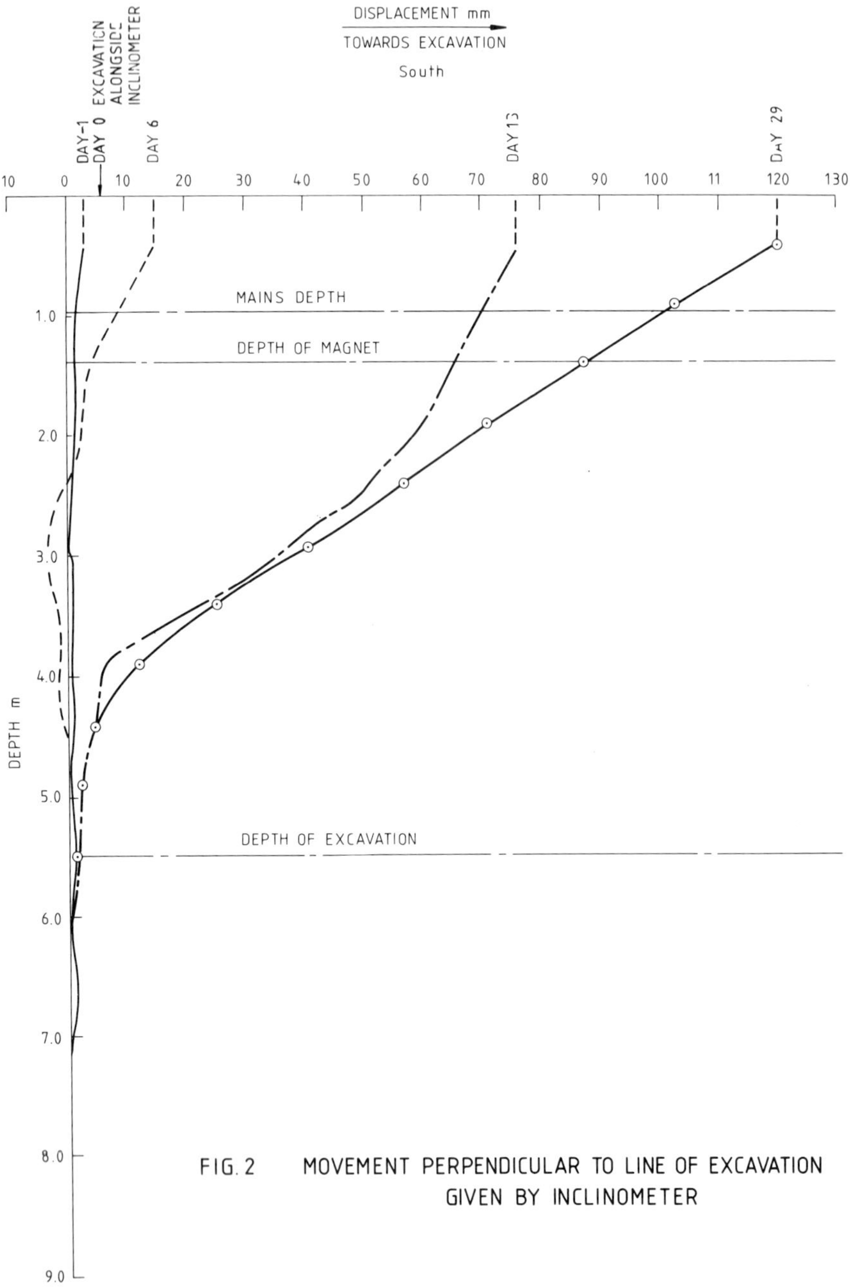

FIG. 2 MOVEMENT PERPENDICULAR TO LINE OF EXCAVATION GIVEN BY INCLINOMETER

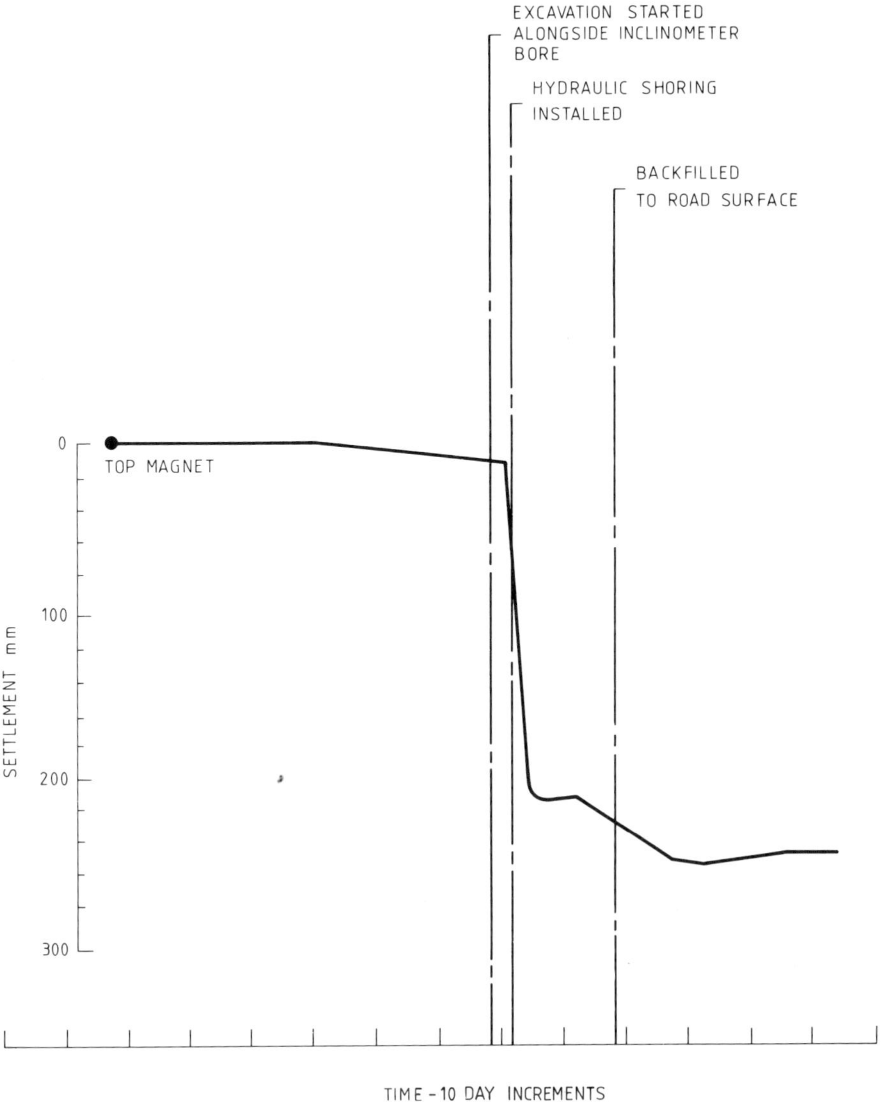

FIG. 3 VERTICAL SETTLEMENT GIVEN BY MAGNET EXTENSOMETERS AT INCLINOMETER BORE

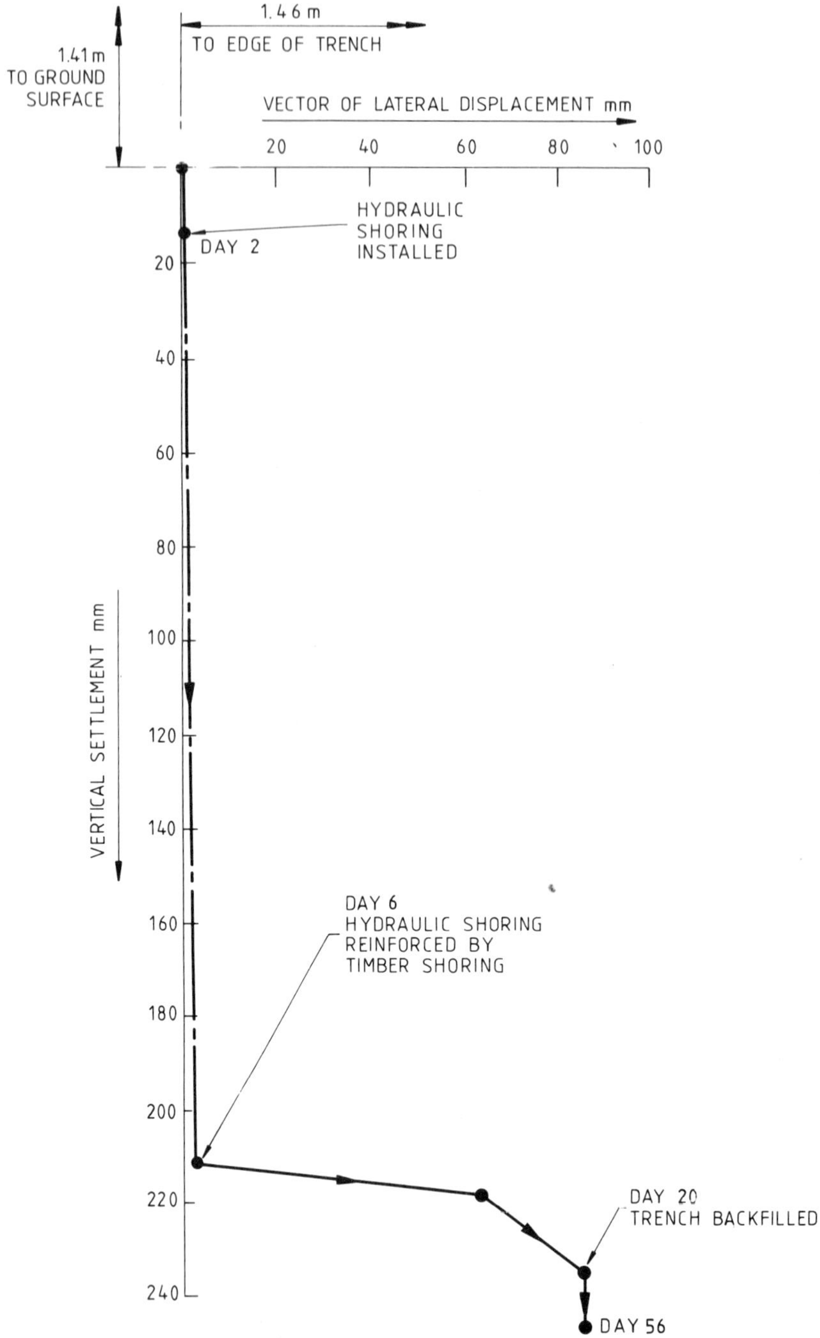

FIG. 4 COMBINED VERTICAL AND LATERAL MOVEMENT OF TOP MAGNET 1.4m DEEP

mass 100 Mg, dropped then pivoted about the bottom trench support frame. Movement continued despite the addition of extra timber frames and only halted when backfill was complete; the trench sheets being left in place.

3.1.2 Site 2

This site was a continuation of the culvert installation described in Site 1 and situated a few hundred metres from it. A section through the site is shown in Fig. 5. The 12" main was 2.7 m from the edge of the excavation, much closer than at site 1. Close-fitted trench sheets, 0.3 m x 0.3 m timber walings and struts were used again and applied progressively as the excavation proceeded.

The culvert sections were laid on a bed of 12-20 mm gravel of a minimum thickness 100 mm. This gravel was also tipped down either side of the culvert sections, completely filling the space between them and the trench sheets. Imported clayey gravel was tipped in on top of the sections and compaction carried out only when the excavation was filled. The road surface was remade by removing the top 0.4 m to 0.5 m of the clayey gravel fill. A sub-base of compacted 50 mm limestone ballast was then applied upon which was laid a layer of 38 mm tarmac. The final surface was made up from 10 mm tarmac compacted using a vibrating roller.

Despite an apparently good standard of trench support, considerable ground movements were recorded at the bore-hole 1.5 m from the trench wall. A surface movement of 70 mm towards the excavation and 15 mm parallel to it were measured. The effect is shown in Fig. 6. Vertical settlement given by the magnet extensometer reached a maximum of 170 mm, before showing a recovery of 70 mm.

Vertical settlement combined with lateral movement towards the excavation is shown in Fig. 7. Comparing this pattern of movement with that produced as a result of the failure of the hydraulic shoring at site 1 (Fig. 4), the vertical component is again the more predominant but the rate of movement is much more gradual. A vectorial movement of 210 mm was generated in 11 days at site 1, whereas a vectorial movement of 180 mm took 28 days to develop at site 2. This gradual movement was made up of almost equal horizontal and vertical movements during a 21 day period while the excavations approached the inclinometer site and remained open alongside. Fissuring and subsidence were observed alongside the excavation whilst it was open. On removing the trench sheets from the adjacent bays during back-filling, the ground movement became almost entirely vertical. A vertical settlement of about 80 mm occurred after 7 days.

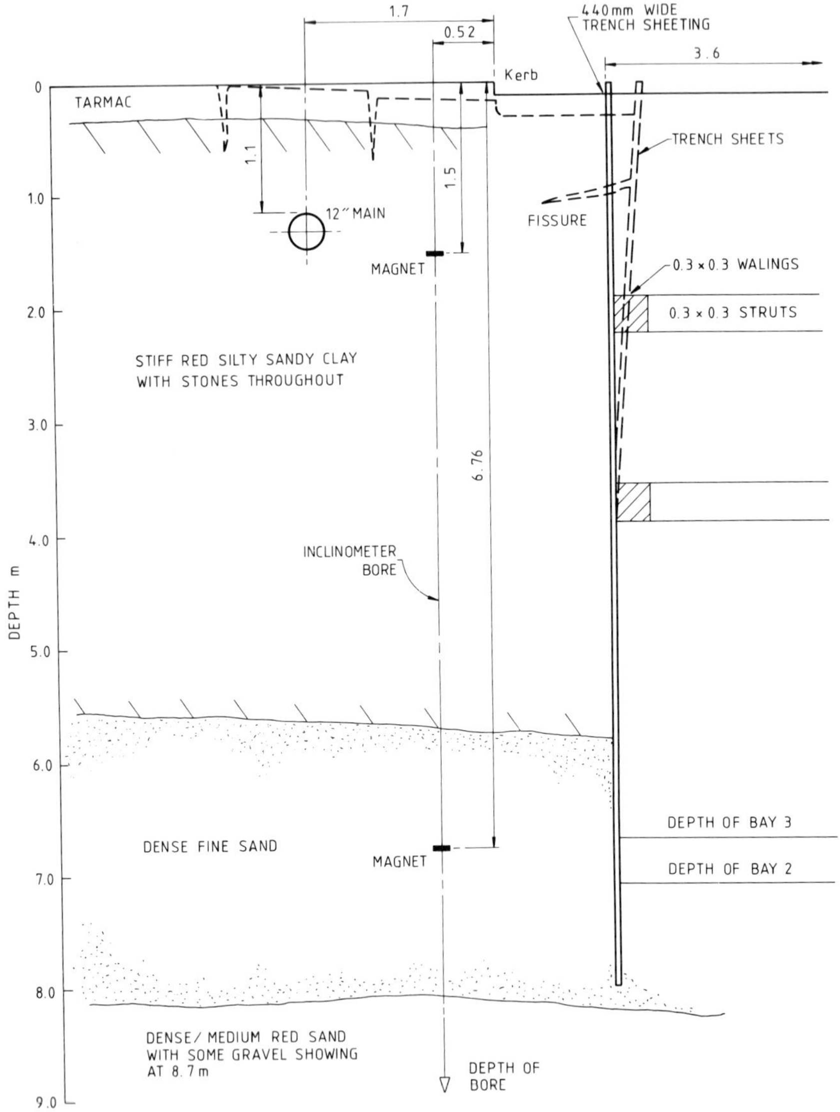

FIG. 5 EXCAVATION PARALLEL TO GAS MAIN
SECTION LOOKING NORTH

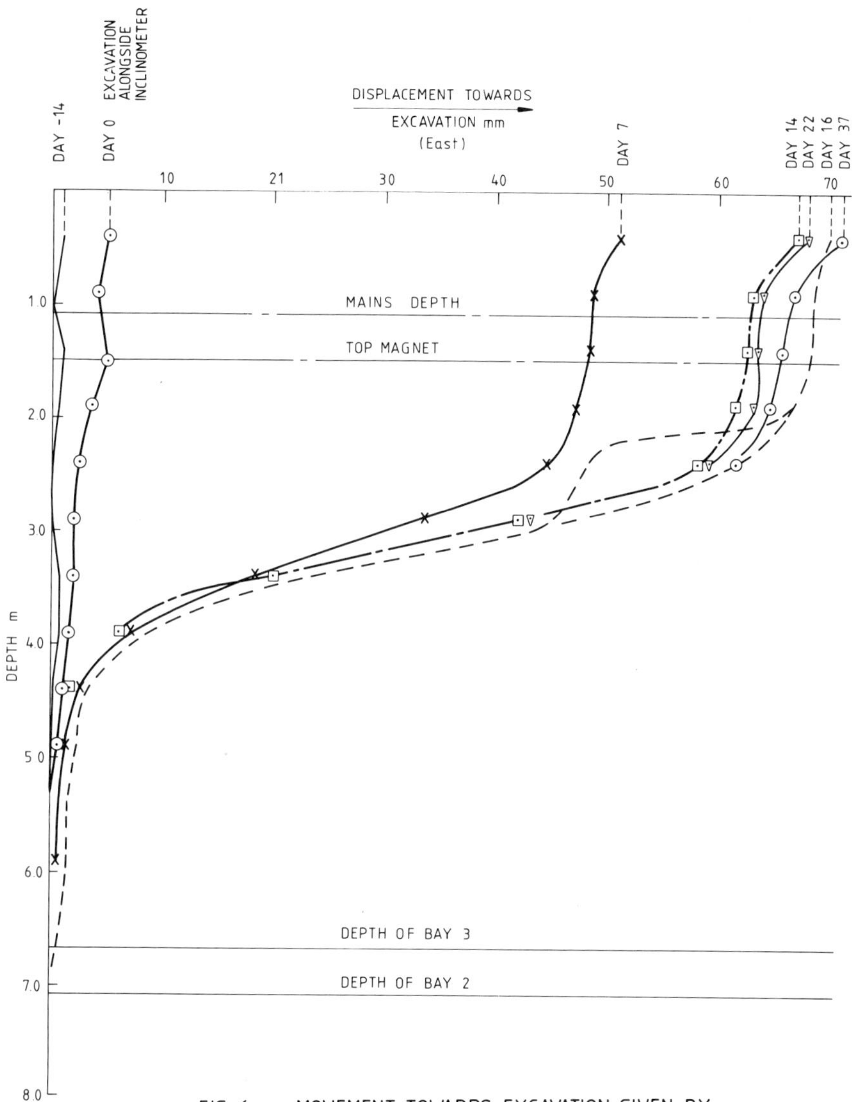

FIG. 6 MOVEMENT TOWARDS EXCAVATION GIVEN BY INCLINOMETER ALONGSIDE BAY 2

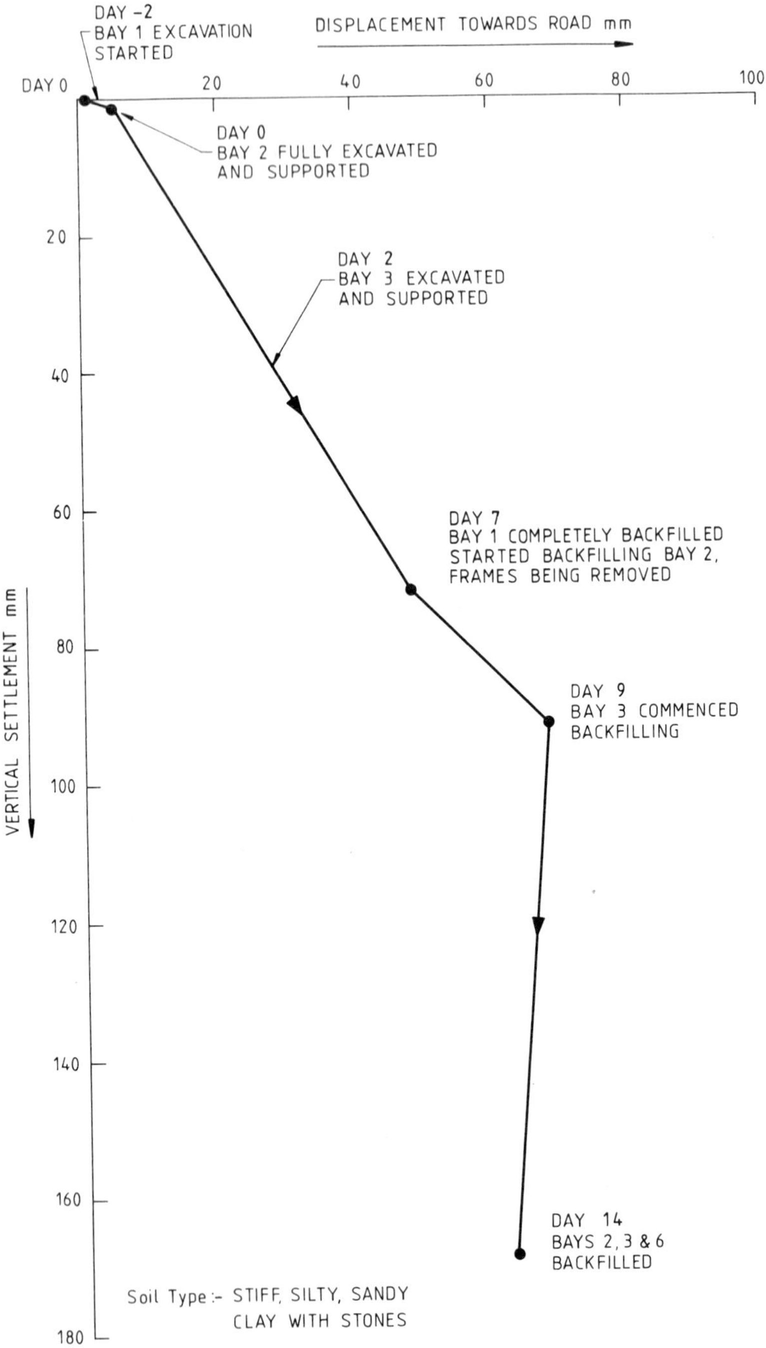

FIG. 7 INCLINOMETER BORE COMBINED VERTICAL AND LATERAL MOVEMENT OF TOP MAGNET 1.5m DEEP

The greatest changes in pipe strain recorded as the excavation proceeded past the 3 strain gauge locations occurred at Position 6 (Fig. 8) where a change of strain of 230 micro strain occurred in the crown of the pipe when Bays 4 and 5 alongside were open. This result indicated that the main was bending with ground movement, although the true maximum strain may not have been recorded because of the relatively few gauges on the long length of main. An increase of 130 microstrain was also induced in the side of the main furthest from the excavation indicating a movement towards the excavation. Resultant movement at this position would therefore appear to be inclined downwards towards the excavation, the strain and ground movement readings indicating that the major component was vertical. Dynamic strains from surface loading by site traffic, etc would be additional to the locked-in values.

3.2 Effect of Excavations Crossing Gas Mains

3.2.1 Site 3

This investigation was carried out at the Research Station to determine the effect on an established buried main, of a deep excavation, intersecting it at right angles. The test was designed to simulate the undermining that could result from a typical deep sewer installation.

The test main, a 5.5 m length of 100 mm ductile iron pipe, had lain buried for two years. Prior to burial the main was fitted with strain gauges along the crown to measure longitudinal bending strain. A trench 5 m long x 0.76 m wide x 2 m deep was excavated across the pipe and subsequently backfilled using the spoil removed during excavation. The soil type was firm clay.

The trench faces were instrumented with displacement transducers and the surrounding ground surface was pinned in a grid pattern. The reference pins were driven into the ground before excavating the trench and were used to monitor ground movement before and after excavation. Compaction during backfill was achieved using the back hoe of the tractor excavator to simulate a not uncommon site practice. Final compaction at the surface was carried out using a reciprocating machine compactor.

The measured ground surface movement along either side of the trench and obtained from the movement of the reference pins, situated 305 mm from the trench face, showed the trench narrowing during excavation and for the day that the trench remained open. Trench narrowing continued during backfilling but ceased 2 days after the backfill was fully replaced. The maximum decrease in width measured across the centre of the trench, was 43 mm which represents 5.6% of the 760 mm original width. After completion of backfill the trench began to widen thus partially reversing some of the previous movement. The overall reduction in trench

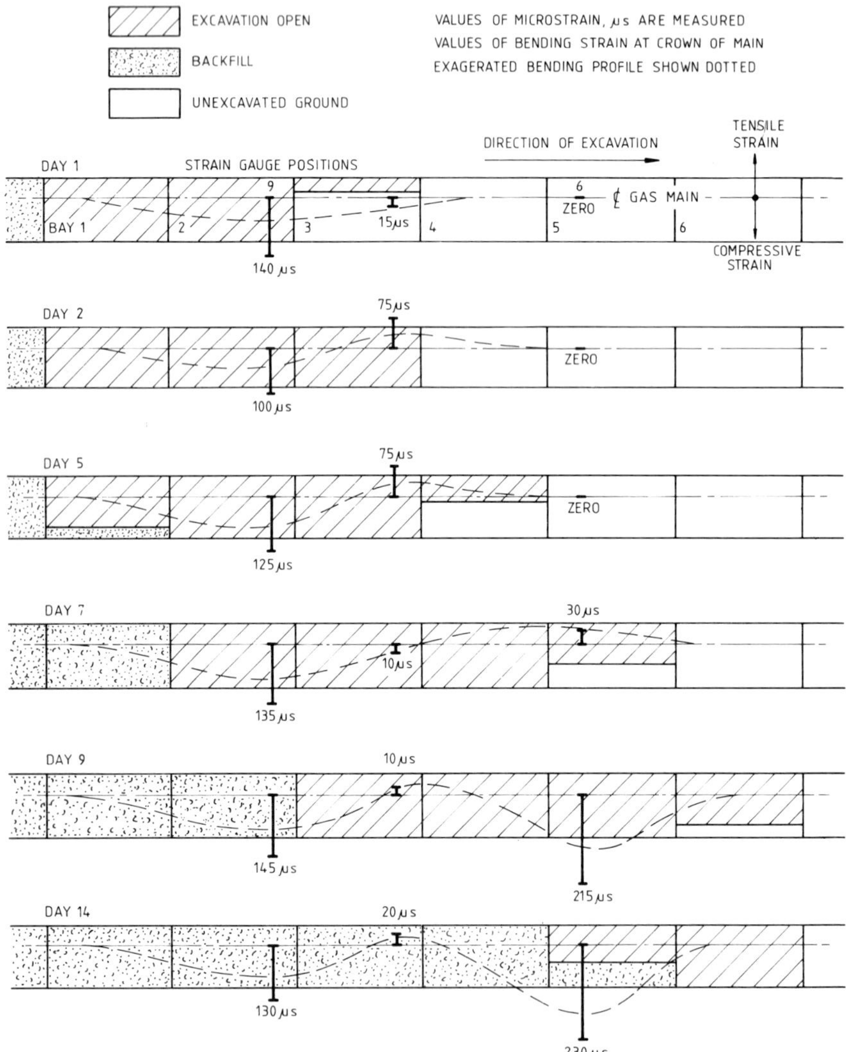

FIG. 8 VERTICAL BENDING INDUCED IN MAIN DURING PROGRESSION OF EXCAVATION

width was then only 4% at the mid-length of the trench. The pattern of movement is shown in Fig. 9.

The displacement transducers began recording 3.5 days after starting the test. Although not operational early enough to detect initial wall movements they did register a narrowing and widening effect similar to that given by the surface reference pins. The reversal of movement was picked up by all transducers even those sited only 150 mm from the trench bottom. The narrowing effect reached a maximum upon completion of backfill 3.6 days after the start of the test. This was confirmed by both the reference pins and the displacement transducers.

The surface movements produced during excavation and back-fill remained stable for approximately 6 weeks, after which the surface began to subside over an area either side of the pipe. After a further 14 days the surface had subsided by about 115 mm. This subsidence occurred after a period of rainfall following a dry period. Over the first 7 days of the 14 day period, the Newcastle Weather Centre recorded a rainfall of 14 mm, whereas during the following week 35 mm was recorded. The subsidence probably occurred as a result of the poor quality of backfilling. The earth removed from the trench and subsequently returned as backfill, consisted mainly of clay lumps about 100-150 mm in size. Compaction of this material, especially in dry weather, would leave considerable void spaces. The water table was allowed to return to its original level after backfilling and the re-wetting softened the clay enabling further compaction under its own weight. This compaction and subsequent consolidation caused the backfill to settle further and the main tended to lose any support restored to it during backfilling.

The measured strain induced in the pipe as a result of the above ground movements are plotted in Fig. 10. Maximum strain, of about 175 microstrain, occurred whilst the trench was open about one day after digging and timbering. This strain was recorded at a point close to the trench wall and showed the pipe to be bending upwards.

Backfilling altered the strain profile considerably. Because of the observed subsidence along the trench either side of the pipe, it appears that the backfill had consolidated vertically and moved down leaving the pipe unsupported over the width of the trench. Of particular interest in this test was the maximum strain of 100 microstrain 'locked-in' to the main at a point about 1.2 m from the trench wall. This result confirmed other field observations that the parts of the main most at risk, as a result of an intersecting excavation, can be in the undisturbed ground away from the excavation itself.

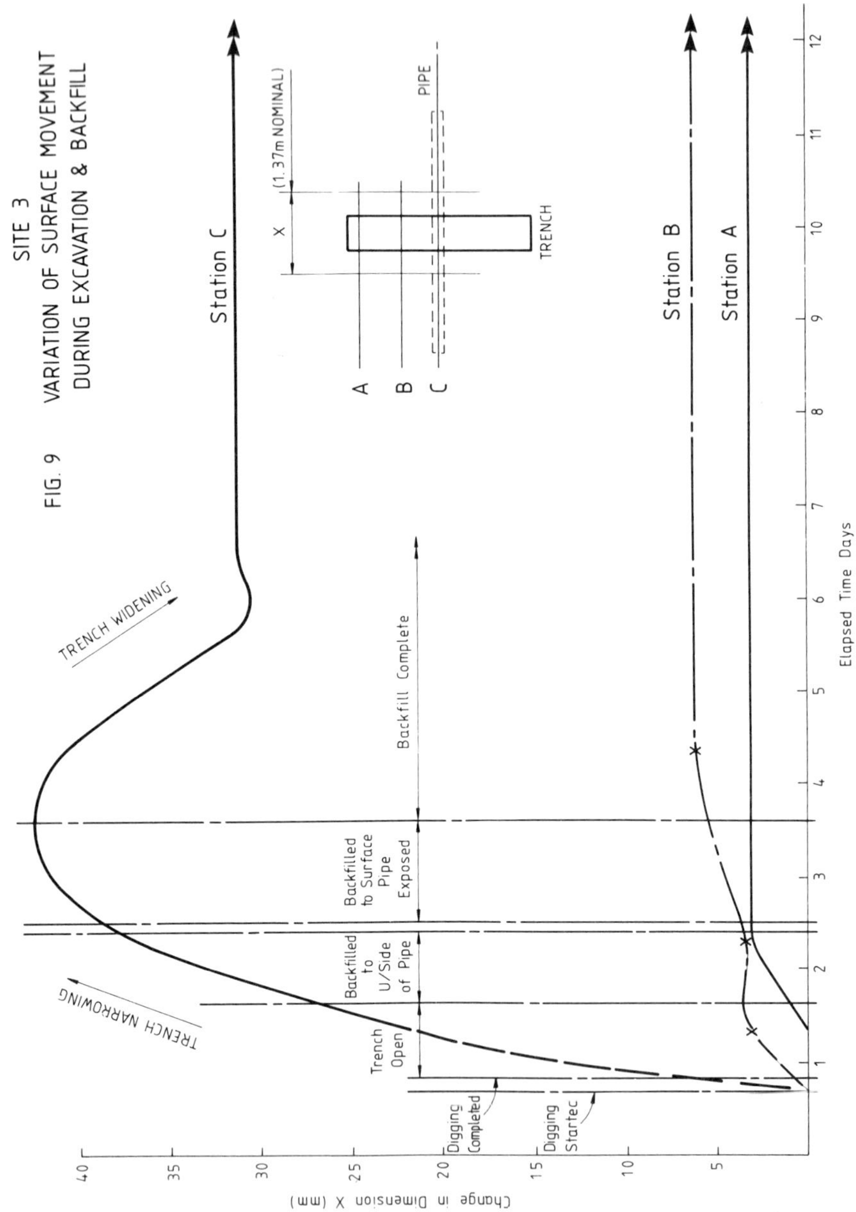

SITE 3

FIG. 9 VARIATION OF SURFACE MOVEMENT DURING EXCAVATION & BACKFILL

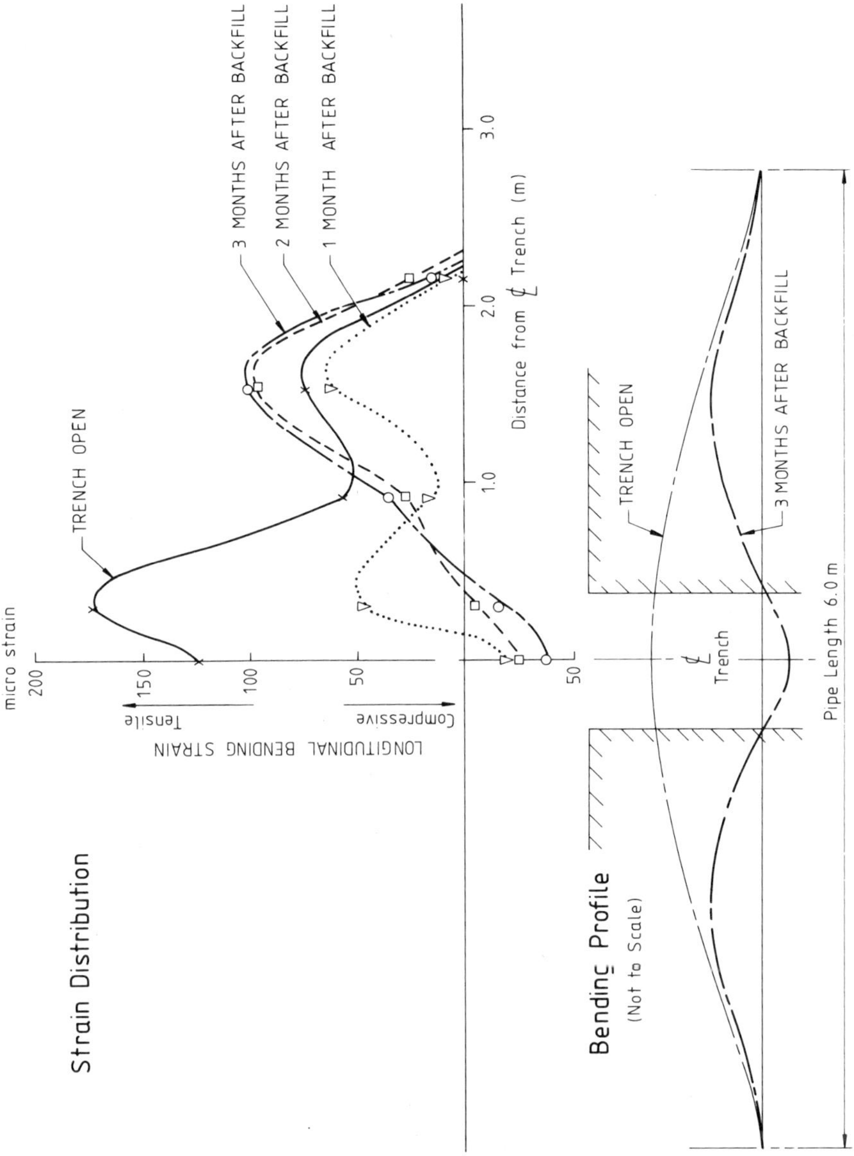

SITE 3

FIG. 10 VARIATION OF BENDING STRAIN ALONG CROWN OF PIPE

3.2.2 Site 4

The excavation for a 1000 mm water main was planned to cross a 12 in (305 mm) cast iron main and a 4 in (100 mm) welded steel main. Both mains were strain-gauged at top and bottom at five stations along a length which lay across the route of the excavation.

The water main excavation crossed the 12 in main first, and the strain levels began to increase 2 days before the excavation intersected the main when the excavation was 3 m from the gas main. The bending strain increased in magnitude towards the trench wall, the highest level being recorded about 0.6 m beyond the wall. Over the 2.23 m length covered by measurement stations, the main 'sagged' but bending strains remained low, the maximum being only 50 microstrain. This low level of induced strain was due to the support given to the trench wall by a concrete 'bridge' which was left in place by the Contractor during the water main excavation. The shoring provided by the Contractor was minimal, two vertical sheets braced with screw-struts were sited either side of the bridge. Movement of the road surface obtained from marker pins showed the surface to the north of intersection to have moved about 16 mm towards the trench.

The 4 in (100 mm) steel main was then crossed by the excavation. As for the previous site the trench supports were minimal. The trench was completely open and without the support of a concrete 'bridge' and substantial strain levels were expected and materialised. The distribution of bending strain along the instrumented section of pipe is shown in Fig. 11. It can be seen that the strain level is highest at station B near the centre of span. The highest levels of strain 1000 micro-strain were recorded when the trench was open. Backfilling reduced this to 600 microstrain.

3.2.3 Site 5

An example of extremely poor workmanship was encountered at a site where a sewer was being installed in a trench 6 m deep x 4 m wide. The trench was completely unsupported.

A 4 in grey cast iron main crossing this excavation at 0.75 m of cover fractured several days after backfilling. An investigation of this failure revealed some of the poorest standards of civil engineering site practice in backfilling the excavation we have come across to date. The infill under the main contained many large voids and large boulders. A boulder 1 m x 0.8m x 0.25 m was found in the backfill over the main and the ground had a surcharge load from a 1.5 m high spoil heap, put there by earth moving plant. The main had fractured in two places near to each wall of the excavation. The backfill had

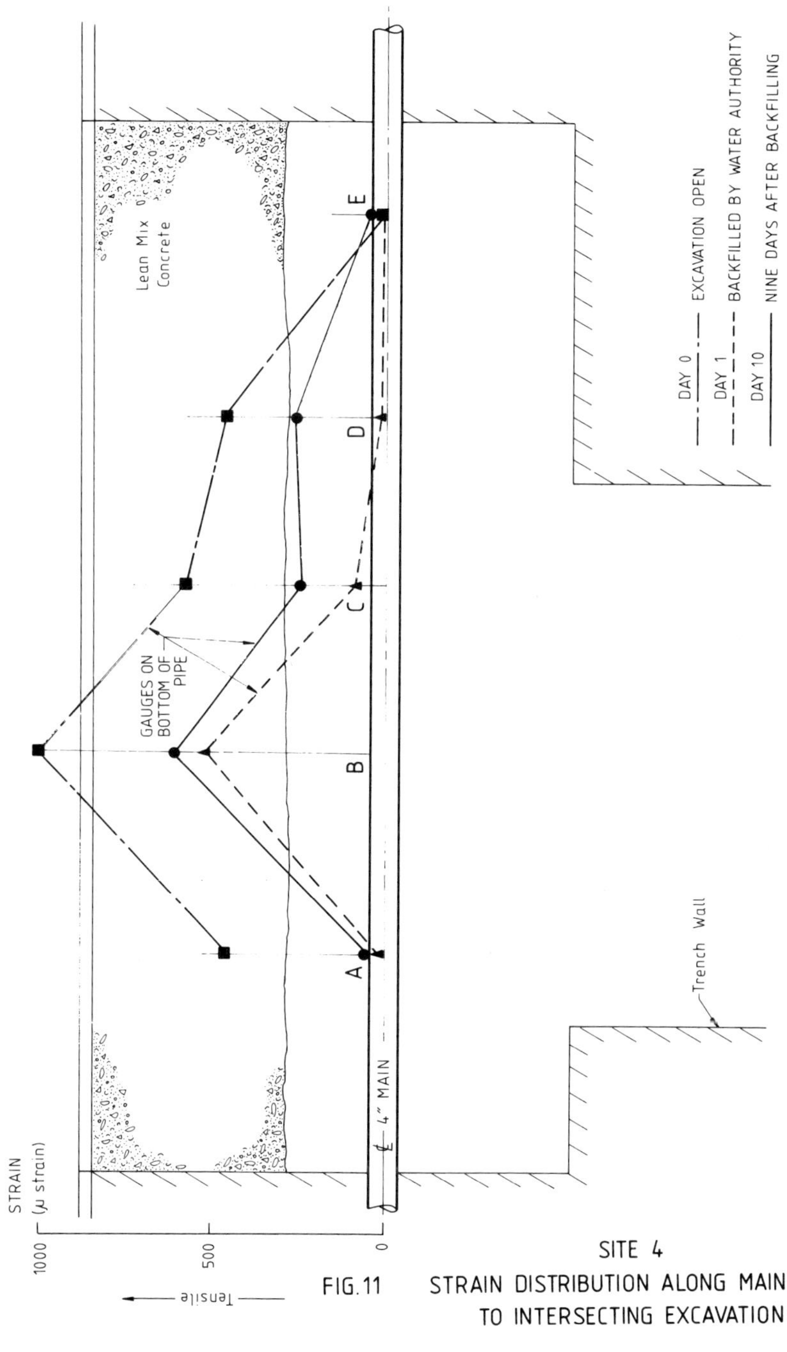

SITE 4

FIG.11 STRAIN DISTRIBUTION ALONG MAIN DUE TO INTERSECTING EXCAVATION settled to such a degree that downward movement equal to about a pipe diameter induced shear failure of the pipe. The fracture, showing the displacement is given in Plate 3.

3.3 Effect of Tunnels

3.3.1 Site A

At this site an excavation was to intersect the line of a 12 in (300 mm) gas main at an angle of about 23°. The gas main was strain-gauged at 5 stations along its length, with gauges at the top, bottom and sides to measure longitudinal strain. The site plan is shown in Fig. 12. Two inclinometer access tubes were installed each fitted with spider magnets to monitor settlement. As the excavation approached the gas main the Contractor decided to tunnel below from a shaft sunk to a depth of 4 m. The tunnel was installed using good standards of construction. Heavy timber frames, 0.3 m x 0.3 m situated at 1 m intervals were used as the main supports and the entire tunnel was boarded behind the frames with close-fitted timber. It was excavated through firm to stiff clay at a depth of 4 m.

Upon completion of the water main installation, Inclinometer bore no. 1 on the north side of the tunnel showed a movement at mains depth towards the tunnel of 4 mm. This movement had increased to about 6 mm some 4 months later. Inclinometer bore no. 2 on the south side of the tunnel moved 7 mm towards the tunnel during and immediately after tunnelling and installation of the water main. A further 7 mm movement in different directions occurred over the next 8 months. The pattern of movement at the bore holes is shown in Fig. 12. Changes in pipe strains associated with these horizontal movements amounted to about 80 microstrain,see Fig 13. In the vertical mode the maximum recorded settlement was about 30 mm which corresponded to maximum changes of vertical strain of 180 microstrain.

3.3.2 Site B

An investigation was carried out after a low pressure cast iron main had fractured adjacent to a tunnel access shaft. Part of the sewer was laid by tunnelling from the base of a shaft. The shaft was 5.5 m deep by 4 m x 3 m and the tunnel was 2 m diameter in clay ground. A 4 in low pressure main fractured in two places next to the vertical shaft. The road surface above the tunnel invert had subsided approximately 100 mm and contributed to ground movement associated with the vertical shaft. This main was relaid in polyethylene to prevent further fractures.

Further progress of the tunnel caused a failure of an 8 in grey iron medium pressure main. However, the risk had been

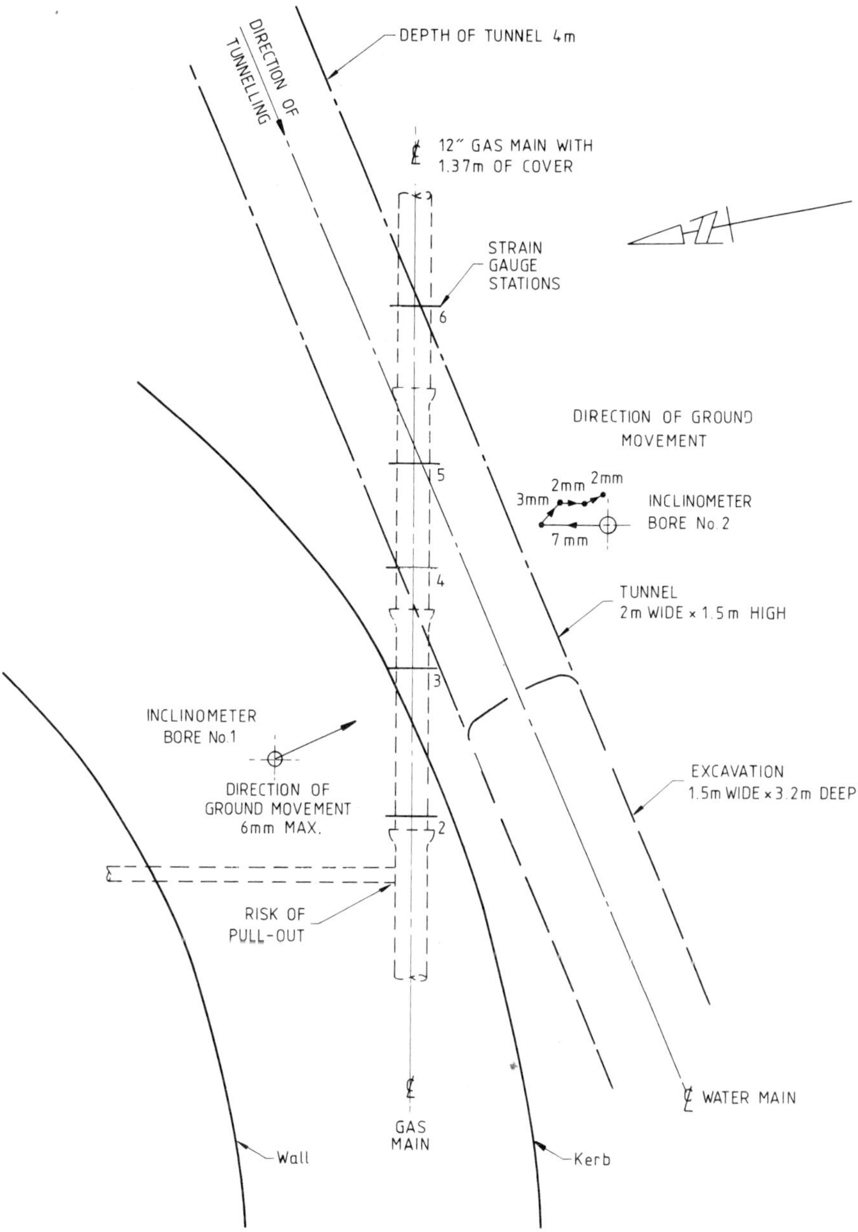

SITE A

FIG.12 TUNNEL INTERSECTING LINE OF GAS MAIN

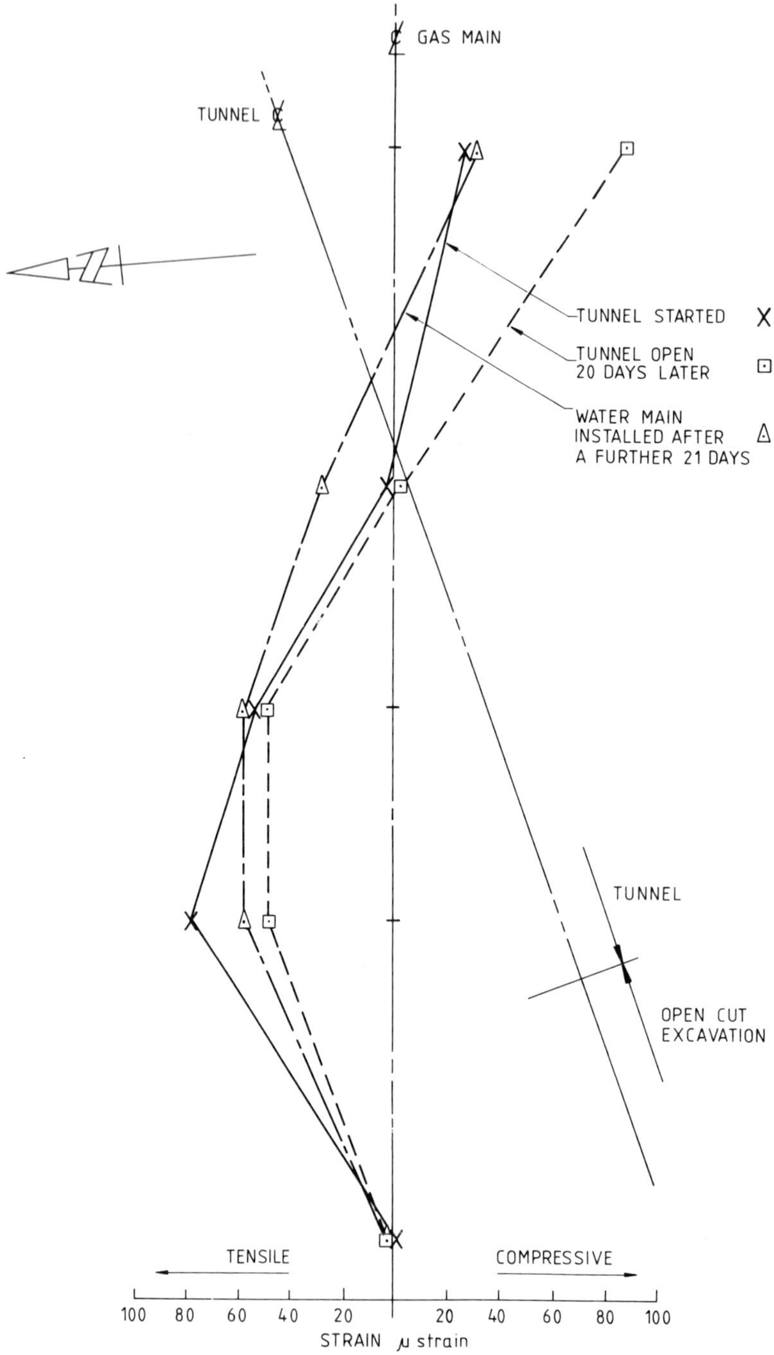

FIG.13 STRAIN CHANGE ALONG SOUTH SIDE OF MAIN DUE TO TUNNELLING AT SITE A

foreseen and arrangements were already in hand to abandon the main when failure occurred. Surface settlement measurements showed that the fracture occurred within the 45° wedge angle ahead of the end of the tunnel as shown in Fig. 14. Where the gas main and tunnel were parallel, ground movement associated with the heading caused services at right angles to the gas main to pull out.

4 DISCUSSION OF FIELD RESULTS

4.1 Excavation parallel to gas mains

Severe ground movements were recorded at these sites at distances of up to 3 m from the main. The strain in the mains attained significant proportions of the maximum working strains without regard to the additional initial strains in the mains or of the subsequent long term strain effects from resumed road traffic, etc. Further strain increments might be expected until settlement ceases. In addition the measured displacements represent a real risk of pull-out from branch mains with associated high localised stress at the branch points on the parallel main. If the mains at site 1 had been situated nearer the kerb close to the inclinometer bore they would have been subjected to movements that would almost certainly have caused failure. Over an 11 day period at site 1 the movement of the magnet located 1.41 m below surface was 223 mm. Thus despite allowances for joint flexibility it is almost certain that a ground movement of this magnitude would have fractured the pipe.

The ground movement at sites 1 and 2 was predominantly vertical and it occurred suddenly, apparently as the result of inadequate trench support. The ground structure did not behave as a continuum but sections sheared and fissured.

Despite the substantial trench support system employed at site 2 and the progressive manner in which it was installed, this site still suffered from quite severe ground movement. The movement occurred in 3 stages; first when the bays were being excavated, secondly when the bays were open during installation of the culvert and thirdly during and immediately after backfilling. The observed subsidence of the ground adjacent to Bays 2 and 3 was greatest during the first two stages. Lateral movement occurred only above the position of the bottom frame which was placed about 3.5 m from the trench bottom.

The majority of ground movement occurred after the trench support system was installed, and it is assumed that robust construction of the support system only allowed very small deflections and that most of the movement was as a result of the ground filling voids left behind the sheets. Fissuring

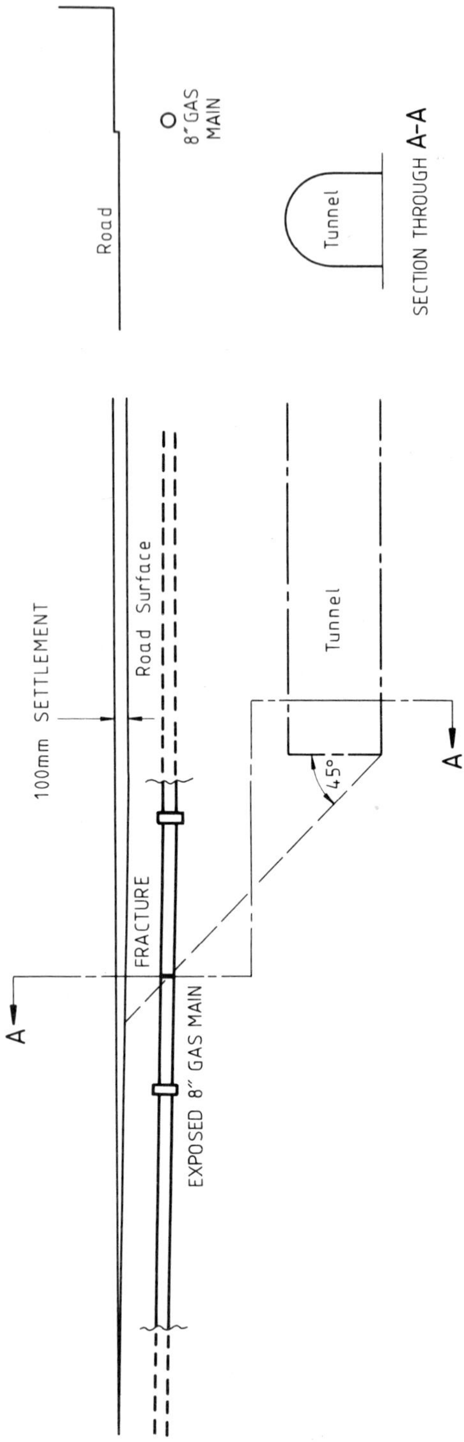

FIG.14 EFFECT OF TUNNELLING AT SITE B

of the trench wall was observed between the trench sheets. These fissures or tension cracks ran horizontally and suggest differential vertical movement due to some of the soil being locally restrained as it moved down and towards the excavation.

Stage three in the movement process occurs during backfill, as the trench sheets are being withdrawn and the fill compacted. In certain conditions the ground movement can be partly reversed possibly by arching caused by excessive compaction of granular fill materials. The resulting lateral stresses appear to be sufficient to reverse the ground strains around the main and in these circumstances the main would be at greatest risk just prior to backfilling. Where the compaction of the backfill is poor and large voids are present, there can be a continuing loss of ground from the sides of the excavation into the backfill. In these circumstances ground strain will continue without reversal after backfilling and the maximum pipe strains will be at some time in the future after completion of the works. Additional settlement of adjacent ground can occur from the action of vibratory compactors and heavy site plant on the adjacent already disturbed ground.

4.2 Excavation crossing gas mains

The effects of inadequate trench support were demonstrated at site 4 where strains as high as 1000 microstrain were induced in the 4 in (100 mm) steel main and at site 5 where trench support was abandoned completely resulting in the fracture of a 4 in grey iron main. Even modern materials, used to replace cast iron mains over trenches, could be at risk under such severe conditions. A length of steel pipe inserted by couplings into a cast iron main prior to the excavation could be pulled out of the couplings under severe deflection, corrosion coatings could be damaged by the large stones and localised buckling could restrict the gas flow at least. This investigation demonstrated the lower extremes of poor quality site practice that have to be avoided in the future.

4.3 Effect of Tunnels, Sites A and B

Despite the apparently good trench support system used in the tunnel construction at Site A, the main experienced a bending strain of almost 200 microstrain. This strain remained 'locked-in' after installation of the water main. The surface settlements recorded at site B were considerable. Water ingress to the access shaft and tunnel indicated the possibility that an adjacent sewer had been fractured and that soil had been washed out. Comparison of the tunnelling procedure as specified by the Local Authority and as carried out on site showed an apparent disparity that would have contributed to the ground settlement. The tunnel lining boards were specified to be 50 mm thick, but in practice apparently thinner timber was used in places and this bulged under the soil loading.

5 THEORETICAL STUDIES AND THE DEVELOPMENT OF RISK ASSESSMENT PROCEDURES

Finite element models are being developed to predict the risk of breakage for buried gas mains due to adjacent tunnelling or excavation work. Results have been obtained for two-dimensional models with linear elastic material properties. Currently a three-dimensional model is being investigated and programmes are available for modelling materials with plasticity.

As an interim procedure for risk assessment an empirical approach is proposed based on field experience and soil mechanics theory. For the design of simple earth retaining structures, calculations of active earth pressure are made using Rankine coefficients. For the limiting case of zero wall adhesion and no surcharge, Fig 15 shows the failure plane to be inclined at an angle of $45^{o} + \phi'/2$ to the horizontal, where ϕ' is the angle of shearing resistance (in terms of effective stress) for the material, and further if we assume a clay soil and undrained conditions then ϕ' is zero and the inclination of the failure plane approaches 45°. Hence we can define a wedge of soil, zone A, which in the fullness of time represents a zone of high risk.

In practice the rupture surface is curved and in order to include all the slipping material we should apply a factor of safety. If there are uncertainties associated with the way the work is to be carried out, such as the quality of the shoring, the length of time the excavation is to be left open, or the probability of a surcharge at the ground surface, then an appropriate factor of safety would be 2. In this case the angle of the failure plane is flattened to 27° and the additional wedge of material zone B, is classified as a zone of intermediate risk. For excavations, where there is the possibility of 'overdig' at the base of the trench, the effective depth (H) should extend to half the width of the trench below the base of the trench. Pipes in zone C would normally not be considered at risk.

6 CONCLUDING REMARKS

Of the factors contributing to damage to buried mains the effect of adjacent excavations is of particular interest to British Gas and will assume increasing importance as the national programme of sewer and drainage renewal gets underway. The extensive field-work to-date has provided first-hand evidence of Civil Engineering site practices backed with detailed measurements of ground movement and resulting pipe strain. The results presented briefly in this paper show that vertical and horizontal ground movements can be significant and induce bending strains in

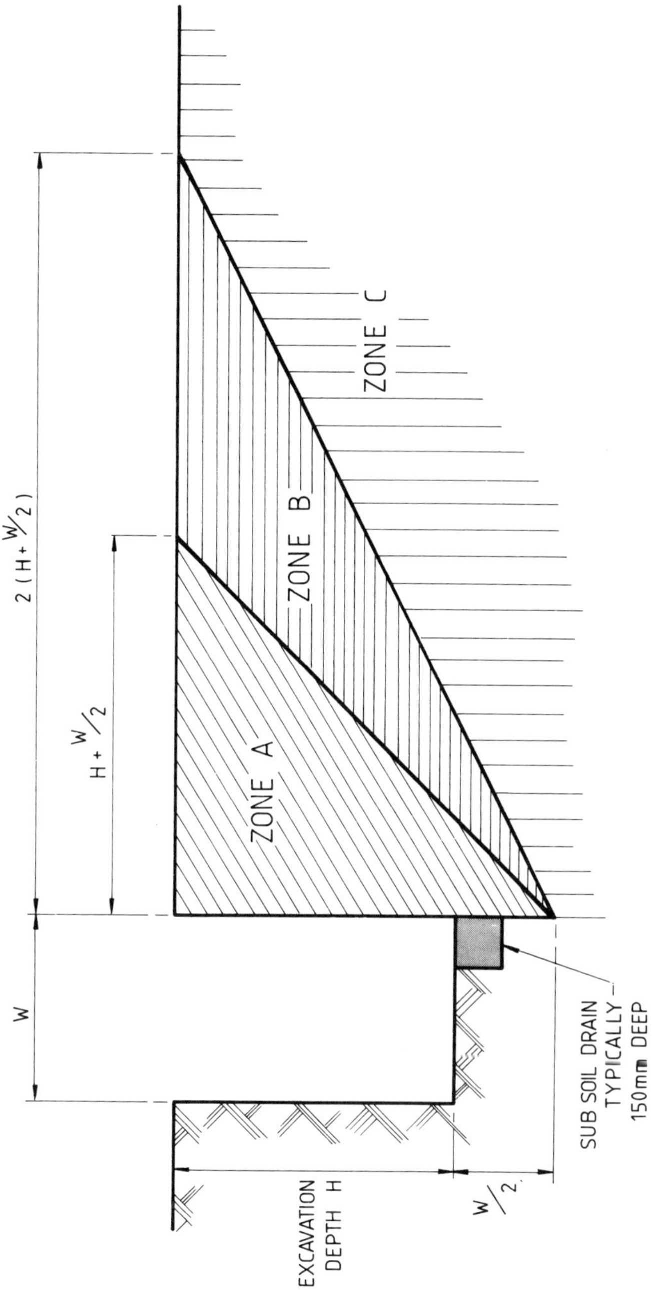

FIG.15 PROPOSED RISK ZONES

adjacent mains. Large proportions of the maximum working strain of grey cast iron mains can be taken up by the ground strains associated with deep excavations and tunnels. Poor standards of support for, and backfilling of excavations and tunnels can lead to mains failure. Although ground deformed by tunnels and excavations can be brought up to original surface levels, the underlying ground remains deformed and induces permanent bending in buried mains. High standards of deep earthworks practice must be encouraged and maintained to minimise the risk to adjacent plant.

Theoretical studies have shown the finite element method of stress analysis to be a useful tool for modelling ground movements. However while programmes are available for both two and three dimensions, data preparation and the interpretation of results for the latter can prove to be counter productive. A range of material properties can be specified from simple elastic to complex elasto-plastic. However at present it is very difficult to measure site properties in detail.

Gas mains tend to be buried in highly variable made ground, and empirical relationships backed up by field observations, would seem to offer the most reliable procedure for risk assessment.

7 ACKNOWLEDGEMENTS

The authors would like to thank their colleagues at the Engineering Research Station and the many Regional staff for their valuable assistance in conducting the experimental work. The authors are also indebted to the British Gas Corporation for permission to publish this paper.

8 REFERENCES

D Needham and M. Howe - Why pipes fail, Institution of Gas Engineers Communication 1103 (1979).

British Standards Institution - Code of Practice for Earthworks CP 2003 (1977).

OBSERVATIONS OF GROUND MOVEMENT AROUND A TRENCH EXCAVATION IN LONDON CLAY

J.E. Gumbel, MA
Geotechnical Engineer, Mott Hay & Anderson
(on secondment to the University of Surrey)

J. Wilson, BSc, DPhil, MICE
Senior Geotechnical Engineer, Mott Hay & Anderson

INTRODUCTION

The identification of ground movement as an important cause of pipeline fractures (King et al., 1977; Roberts and Regan, 1977) has prompted a programme of research into the effects of trench construction on adjacent buried services (TRRL, 1978). The long-term aim of this research is to provide a means of estimating the displacement and consequent risk of failure of an existing pipeline due to trenching works in the vicinity.

Guidance on predicting ground movements caused by trench construction is at present available only from field observations relating to deep excavations (Croft et al., 1977). Simple scaling down of this data for application to shallow excavations may be inappropriate, however, due to differences in construction methods and other factors. The collection of ground movement data relating specifically to trench construction is therefore a prime research objective.

This paper describes a first attempt to generate relevant data by means of a controlled field trial.

AIMS OF PRELIMINARY FIELD STUDY

Due to the lack of previous data on the nature and magnitude of ground movements around trench excavations, the field study described below was planned as a preliminary exercise only, with the following very limited objectives:

(i) to determine the general order of magnitude and extent of ground surface movement around a trench excavation in London Clay, using simple surveying methods,

(ii) to test and develop experimental procedures for use in subsequent more comprehensive field trials.

In the event, the trial excavation yielded much more detailed observational data on ground movements than had been anticipated, and at a cost of less than £800 provides ample evidence of the value to be obtained from a simple field experiment.

DESCRIPTION OF THE SITE

The trial trench was excavated in the London Clay on a site about 2 km west of the centre of Guildford, Surrey. The site lies at the foot of the Hog's Back, where the ground falls at a slope of about 1 in 17 to the north. The trench axis was set along the line of maximum slope to minimise the influence of surface crossfall on lateral movement of the trench walls (Fig.1). The grass-covered surface was slightly uneven, with furrows to a maximum depth of about 250 mm also running parallel to the slope.

A description of the soil profile exposed by the excavation is given in Fig.2. Although the topography of the area suggested possible periglacial transportation of the clay deposit, the finely fissured structure of the trench walls showed no sign of previous mechanical disturbance below 0.75 m.

MEASUREMENT OF GROUND SURFACE MOVEMENT

Prior to excavation, 50 mm x 50 mm x 600 mm long wooden pegs were driven 450 mm into the ground in rows parallel to the trench centre-line to form a grid for the measurement of ground surface movement, as shown on Fig.1. A crosshead screw set into the top of each peg served as a reference point for surveying.

Monitoring of ground movements was carried out during the 28-day period the trench lay open and was continued on peg rows A to D only for a further month after backfilling. The site was revisited just over one year later, by which time most of the pegs were unfortunately missing. It was however possible to measure the lateral movement of the few remaining pegs on rows A and B.

Horizontal movement perpendicular to trench axis

Lateral movement was determined by measuring offsets from a line of sight established along each row of pegs. Theodolite and target stations consisted of a steel plug set in the ground in a 250 mm cube of concrete.

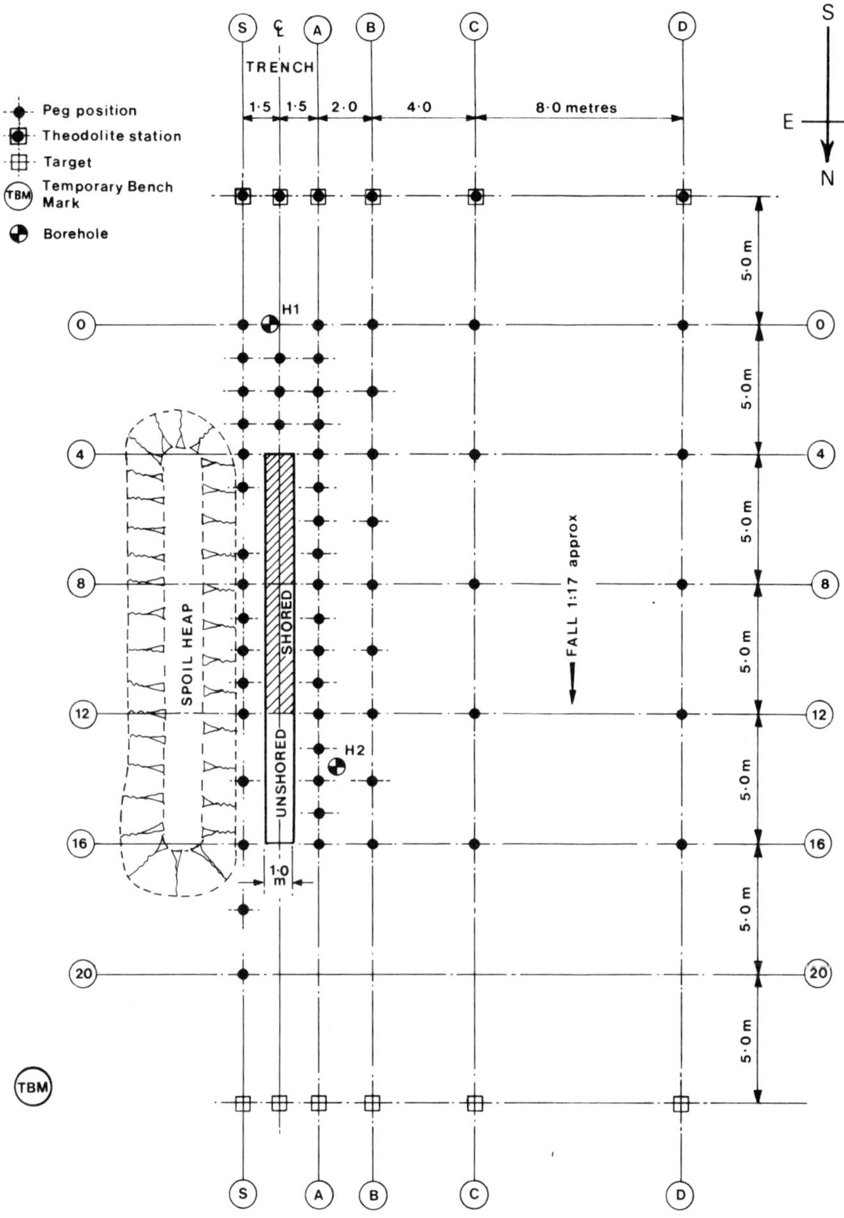

Fig. 1 SITE PLAN

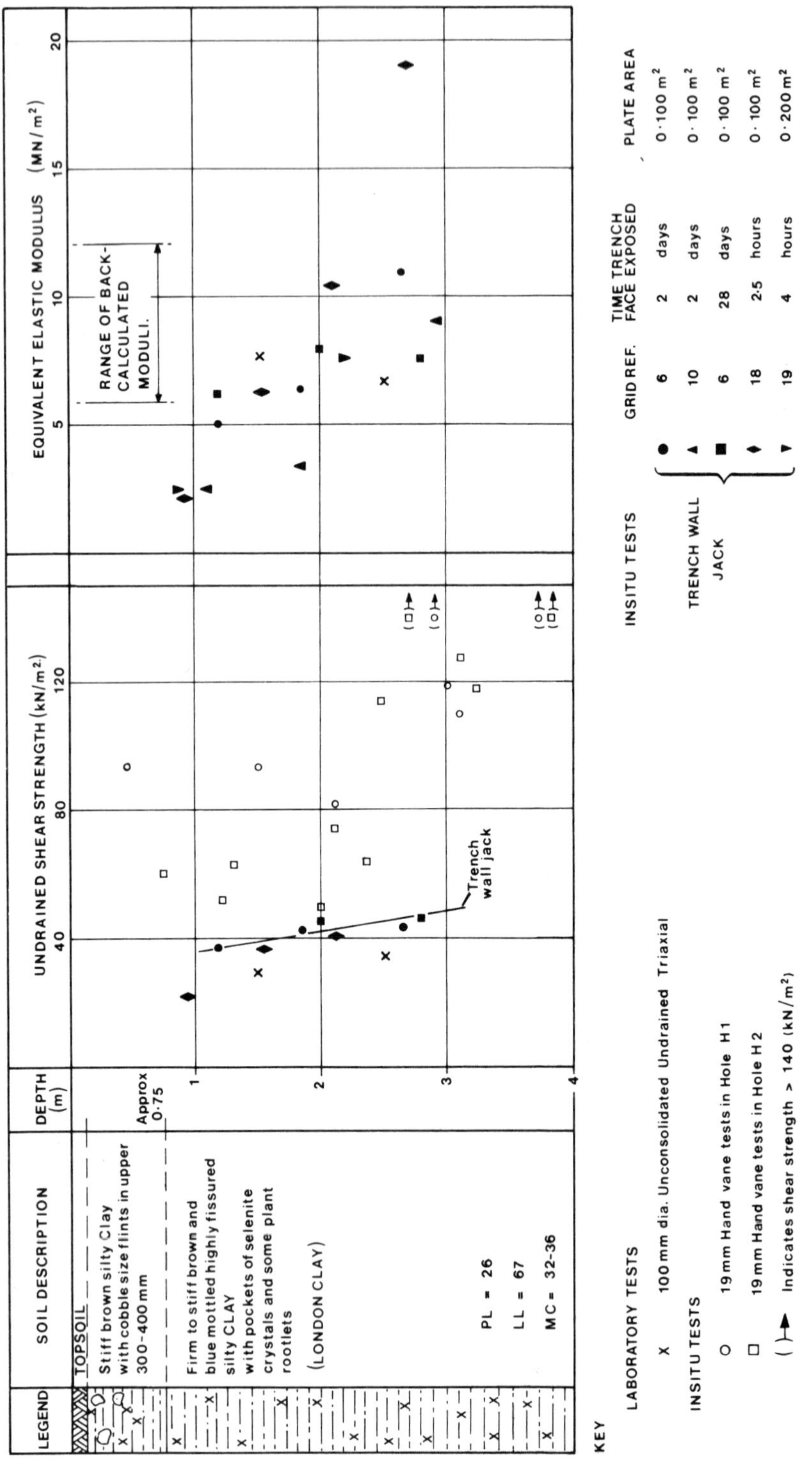

Fig. 2 SOIL DESCRIPTION AND PROFILES OF MEASURED PROPERTIES

Tests for repeatability of the offset readings, using the same theodolite set up by different people on the same day, gave an accuracy better than ± 2 mm. On the day following excavation it was discovered that one theodolite had a misaligned optical plumb. This had affected the initial readings on rows B, C and D and so the measured offsets of these rows include some unknown zero errors.

The accuracy of the readings taken after one year is also uncertain since the reference stations were not designed to remain stable through winter conditions.

Horizontal movement parallel to trench axis

Longitudinal displacements were monitored along each row of pegs with reference to the theodolite station at the end, using a tensioned steel tape. Repeatability of readings was better than ± 2 mm.

Vertical movement

The pegs were levelled with reference to a temporary benchmark located at the north-east corner of the site. The temporary benchmark was constructed by grouting a 3 m long x 25 mm diameter steel bar into the bottom of a 90 mm diameter borehole which was then backfilled with bentonite. The levelling staff used could be read accurately to within ± 1 mm.

TRENCH CONSTRUCTION

Excavation

The trench was excavated on 26th July 1978, hereafter referred to as Day 0. A backacter fitted with a 1 m wide bucket was reversed up between peg rows A and S from the north end of the site and began excavating at grid line 4. From the first excavator position a 2.8 m long by 4 m deep pit was excavated slowly over a period of about 2.5 hours to enable continuous monitoring of movement with increasing depth. The 4 m deep trench was then extended to its full length of 15 m in a continuous operation which was completed about 4.7 hours after excavation commenced.

The spoil was stored alongside the trench in accordance with the usual construction practice, although care was taken not to disturb the row S pegs used to monitor movement of the surcharged shoulder.

Shoring

After 4.2 hours, while excavation was being completed between grid lines 12 and 16, the first 10 m length of trench was spot-braced using 14 units of "Speedshore" aluminium hydraulic shoring. Each unit consisted of a pair of 2.1 m long by 0.2 m wide vertical rails held apart by two hydraulic struts at 1.22 m centres, and could be safely installed by hand from the ground surface. In order to cover the full height of the trench, the units were placed in alternate upper and lower positions along its length (Plate 1). The struts were pumped up to an initial hydraulic pressure of 2.75 MN/m^2, corresponding to a pre-load of about 9 kN in each. The installation was completed within three-quarters of an hour.

Backfilling

The shores were removed and the trench backfilled on Day 28. The excavated clay was dumped back into the trench without compaction by an excavator operating from the east side of the spoil heap, destroying the row S pegs in the process.

SOIL SAMPLING AND TESTING

In-situ tests

Measurements of horizontal stiffness of the trench walls were made using a trench wall jack. The operation of this device, which performs a horizontal plate loading test on opposite walls of a trench, is fully described by Lawrence (1977). Profiles of soil modulus with depth were obtained at several locations in the trench and for different times of exposure of the clay face (Fig.2). A set of readings was first taken on Day 2. Further tests were carried out on Day 28, when the trench was also extended briefly from grid line 16 to grid line 20 to provide a freshly exposed clay face. Setting up the apparatus nevertheless took 2.5 hours; and in view of the very rapid loss of stiffness following exposure of London Clay reported by Marsland (1973), none of the modulus values recorded on Fig.2 should be considered representative of the undisturbed stiffness of the soil.

Where the plate tests reached failure the undrained shear strength c_u of the soil was calculated from the formula given by Meyerhof (1951) for a circular foundation, viz:

$$c_u = 6.2 \times \text{ultimate plate pressure.}$$

These shear strengths, which may once again reflect a degree of disturbance of the clay, are also plotted on Fig.2.

Fourteen months after the trench had been backfilled the site was revisited to make further simple measurements of the strength of the ground. A comparison was sought between areas which were known to have undergone large deformation during trench excavation and areas which had remained relatively undisturbed. Accordingly, two small-diameter auger holes were bored at locations H1 and H2 (Fig.1) and shear strength/depth profiles were obtained (Fig.2) using a 19 mm diameter hand vane.

Comparison of the c_u values recorded in the upper 2.5 m of these two boreholes gives an indication of the effect of disturbance on shear strength. The vane test results for hole H1 (undisturbed) confirm the engineering classification of the clay as stiff.

Beyond these general indications, the shear strength data collected on Fig.2 should be interpreted with care, bearing in mind the differing scales of the two types of insitu test in relation to the fissure spacing, and the many other factors influencing the response of fissured clay (Marsland, 1971).

Laboratory tests

Additional values of shear strength and stiffness (initial tangent modulus) were obtained from unconsolidated undrained triaxial shear tests on two U4 samples taken from the extended trench section on Day 28. The range of natural moisture content was determined from many samples taken along the trench. These results, together with mean Atterberg Limits, appear in Fig.2.

GENERAL OBSERVATIONS

Weather

Fine weather during the first three days after excavation was followed by a period of heavy rain between Days 4 and 7. From Day 8 no further rain fell on the open trench, and the weather remained warm and dry until one month after backfilling.

Groundwater and seepage

Differential drying of the exposed clay surface observed two days after excavation suggested the presence of groundwater at a depth of 2 metres. Following the period of rainfall, over a metre of water collected in the bottom of the trench due mainly to seepage through the upper weathered clay layer. This water was removed by pumping on Day 14. Although there was no further evidence of infiltration through the upper half of the trench wall, water continued to enter the base of the excavation, reaching a level of about 0.6 m before the trench was again pumped dry on Day 27.

Trench wall profile

Within 4 days of excavation, both faces of the trench developed a distinct inward bow, and it was clear from a simple visual inspection that the lateral displacement was greater between 1 to 2 metres below the surface than at ground level.

This was confirmed on Day 8 when horizontal offsets were measured to the trench walls from a plumb-line suspended on the trench centre-line. A bulge at the base of the surcharged wall around grid line 9 was also noted. Due to a lack of datum readings, the precise magnitude of lateral movement at depth could not be ascertained, and attempts to monitor development of this movement with time were hampered by shrinkage and general degradation of the vertical clay surfaces which obliterated reference points. Clearly some form of sub-surface instrumentation installed before the start of the excavation would be a more effective way of monitoring movement of the trench face.

Trench stability

The spoil-surcharged side of the unshored section of trench collapsed overnight within 4 to 18 hours of excavation. The lower three-quarters of the trench wall fell out as a rubble of clay blocks leaving a cleanly broken vertical surface 0.5 m back from the excavated line. The stiffer brown clay crust was left overhanging the void at first, but was later washed down by rain (on Day 5). Spread of the collapse along the trench was arrested at the first unit of shoring (Plate 1).

The remainder of the surcharged side of the trench failed within 10-15 minutes of removal of the shores on Day 28. A vertical crack opened up at the surface along peg row S, and a one metre wide by 3 m deep wedge of clay extending from grid line 12 to grid line 6 slid into the excavation.

The unsurcharged trench wall remained stable throughout the course of the trial.

RESULTS AND DISCUSSION

Selected isochrones of horizontal and vertical movement of the ground surface pegs are plotted on Figs.3 and 4 respectively. Fig.5 shows the development of movement with time for the pegs on grid line 8; the annotated curves are typical of those recorded for other cross-sections through the trench, and illustrate the influence on ground movement of shoring and weather conditions.

Horizontal movement

The only pegs which showed a significant (>3 mm) component of horizontal movement parallel to the trench were those at grid points A15 and A16. These moved southwards by a maximum of 10 mm indicating some inward movement of the end wall of the unshored section of trench on grid line 16. The layout of pegs had been designed to detect the pattern of longitudinal movement around the other end wall of the trench on grid line 4, but here only small (<3 mm) and apparently random movements were recorded. In the following discussion and on Figs.3 and 5,

PLATE 1 VIEW OF TRIAL TRENCH ON DAY 6
SHOWING COLLAPSE OF UNSHORED SECTION

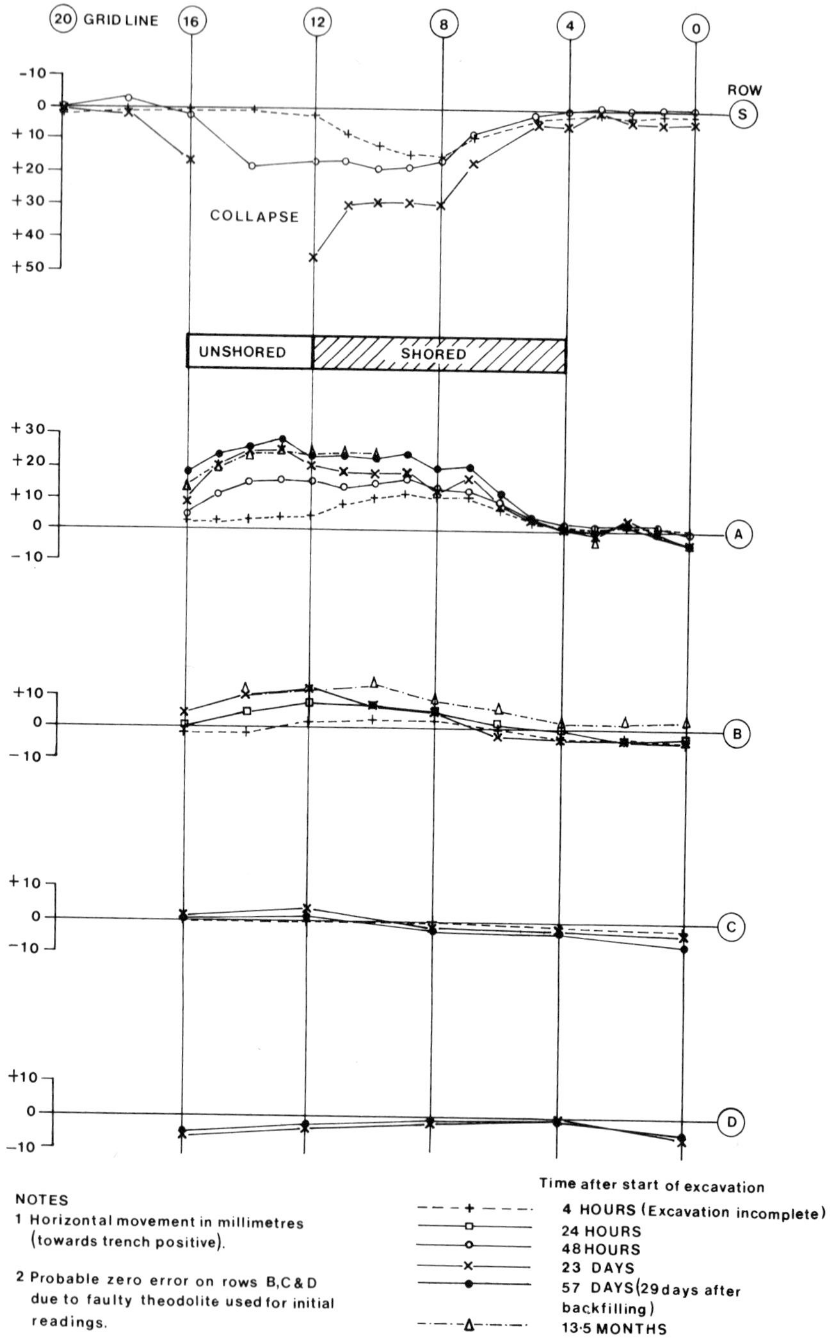

Fig 3 ISOCHRONES OF HORIZONTAL MOVEMENT OF GROUND SURFACE PEGS

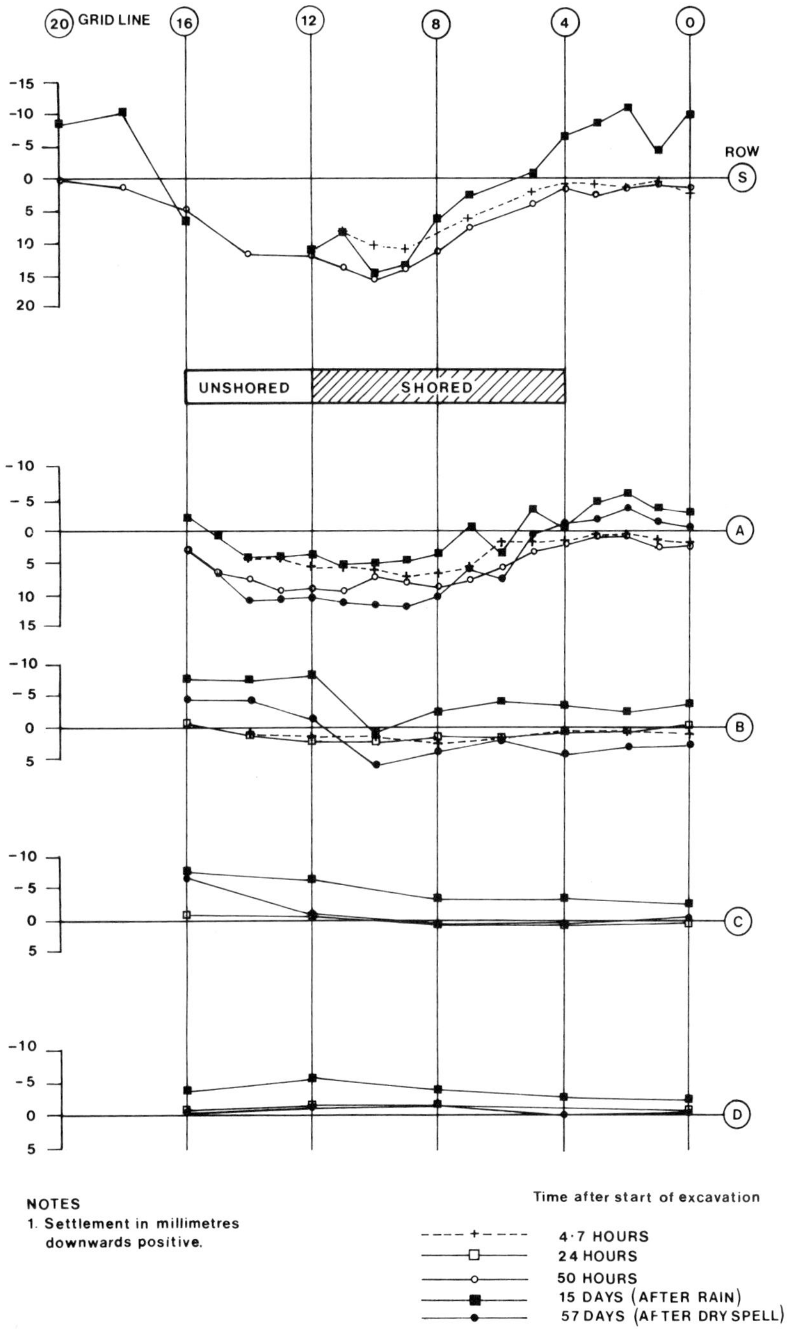

Fig. 4 ISOCHRONES OF VERTICAL MOVEMENT OF GROUND SURFACE PEGS

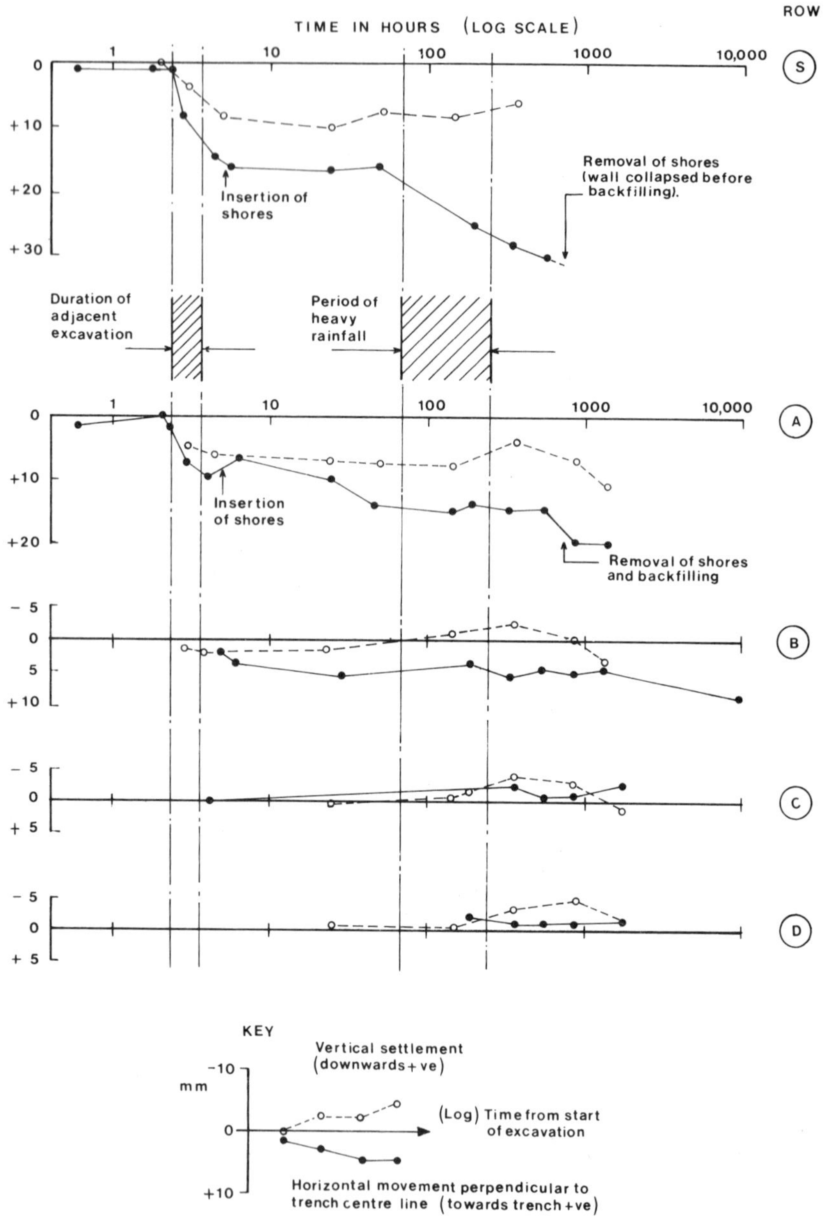

Fig. 5 MOVEMENT-TIME PLOTS FOR GROUND SURFACE PEGS ON GRID LINE 8

the horizontal movement referred to is the component perpendicular to the trench axis i.e. lateral movement only.

Both Figs.3 and 5 illustrate the strongly time-dependent nature of lateral ground movement, and demonstrate the surprising sensitivity of the comparatively crude method of measurement.

Some of the interesting features detected are described on Fig.5. The plot shown for peg A8 typifies the response of the unsurcharged trench shoulder. Excavation to the full 4 m depth caused an immediate horizontal movement towards the trench of some 10 mm. On insertion of the hydraulic shores a small (3 mm) outward movement occurred, but this did not prevent the inward movement continuing with time. On removal of the shores on Day 28, an instantaneous inward movement of 5 mm was recorded. After backfilling, lateral movement of peg row A appeared to stabilise at about 20 mm (see also Fig.3).

Behaviour of the row A pegs adjacent to the unshored section of trench was similar except that there was no outward and inward movement associated with insertion and removal of the shores. From the row A results on Fig.3 it may be seen that the difference in maximum long-term movement adjacent to the shored and unshored sections was only marginal but this may have been because the 5 m unshored section was too short to be free at any point from restraint in the longitudinal direction.

Comparison of the plots for row S and row A on Fig.3 shows that the general effect of the spoil surcharge was to increase lateral movement by about 50 percent. The build-up of movement with time on the surcharged side did however follow a slightly different pattern. This may best be appreciated from the row S plot on Fig.5. Following an immediate lateral displacement of 15 mm on completion of excavation, the insertion of shores produced no outward movement but did arrest inward movement at about 17 mm in the short term. The start of rainfall on Day 4 then triggered further inward movement which was still increasing on Day 28, when removal of the shores led to collapse.

As regards the lateral extent of ground movement due to the excavation, it is clear from Fig.3 that the row B pegs, at 3 m from the trench shoulder, fell well within the range of both short and long-term influence. The horizontal displacements recorded on the row C and D pegs during the first two months after excavation were too small and scattered to be considered significant; but this does not preclude the possibility of ground movement having spread up to 7 m or more from the trench in the long term.

Vertical movement

During the first few hours after excavation, the settlements recorded on peg rows A and S (Fig.4) were of the order of 60 to 80 percent of the lateral movement (Fig.3). Following the insertion of the shores, the ratio of vertical to horizontal movement did not however remain constant (Fig.5).

After the start of heavy rain on Day 4, vertical peg movement became dominated by swelling of the ground surface followed by shrinkage in the subsequent period of warm dry weather. A common pattern of rise and fall was found to affect all the pegs on site between 6 and 57 days (150 and 1370 hours) after excavation, as exemplified by the plots on Fig.5. Some of the scatter in the settlement profiles observed after the rainfall, e.g. on Day 15 (Fig.4) may have been due to pegs creeping upwards relative to the ground as softening of the surface allowed relief of the soil stresses induced by driving. The use of more securely anchored ground surface pegs would be an advantage in future simple trials, although the extent of weather sensitive movements with depth could only be determined by a system of subsurface monitoring.

Shear failure and loss of ground

The fact that the spoil-surcharged side of the trench collapsed whilst the unsurcharged side remained stable could be of practical significance in that the storage of spoil is a relatively straightforward item of construction procedure to control on site. Since the discontinuous movements associated with shear failure of the soil are potentially far more damaging to adjacent structures such as pipelines than are quasi-elastic ground movements, it is instructive to examine the mechanisms of failure in the present case.

The early collapse of the unshored section of trench was a simple fall of material released by the opening of natural fissures in the soil. Such collapses are common in stiff clays (Tomlinson, 1971). The opportunity for the fissures to open is created primarily by the relief of lateral stress, but is enhanced by tensile strains in the clay surface associated with inward movement of the wall of the excavation. The effect of the spoil surcharge was to increase the immediate ground movement and hence also the tensile strains in the clay surface, leading to a critical condition. It is probable that the debris from this initial collapse, which filled the trench to approximately mid-height (Plate 1), effectively stabilised the opposite , unsurcharged wall of the trench which might otherwise have failed by a similar mechanism in time.

The hydraulic shores achieved their stabilising effect by generating surface compression in the clay, but apparently did not prevent stress relief across a vertical plane of fissures about one metre back from the trench face, which opened up and initiated collapse once the shores were removed. This time-dependent stress relief and associated swelling, which led to increasing horizontal ground movement (Fig.5), depended on the availability of water to relieve pore suctions. The role of the spoil surcharge in accelerating this process may simply have been to act as a rainwater reservoir, collecting and storing runoff which on the unsurcharged trench shoulder was removed more rapidly by unobstructed surface drainage, evaporation, and transpiration through the grass.

Back-calculation of equivalent soil modulus

Any elastic analysis of ground movement data raises complex questions relating to the validity of the chosen theoretical model and the values assigned to various soil parameters. Techniques have been developed for the analysis of ground movements around deep excavations in London Clay (as recently reviewed by Burland et al., 1979), but these do not necessarily apply to relatively shallow trench excavations. It has nevertheless been considered of interest to obtain a crude estimate of the mean soil deformation modulus (E) implied by the present data. A back-analysis of the maximum horizontal displacements recorded on the row A pegs was carried out using a simple two-dimensional elastic finite element model of an unsupported trench excavation. Setting $K_o = 1$ and $\nu = 0.49$ yielded equivalent E values ranging from about 12 MN/m^2 for immediate unloading of the soil to about 6 MN/m^2 after one month. This range of values is of a similar order to that of the reloading stiffnesses obtained from the trench wall jack tests (Fig.2), and is a further indication of the degree of bulk softening of the ground caused by the trench excavation.

CONCLUSIONS

Bearing in mind the preliminary nature of the field trial described above, it would be inappropriate to attempt to predict ground movements at other trench sites on the basis of the observations reported so far. The more general conclusions to be drawn at this stage are as follows:

(1) It has been demonstrated that a field trial with very limited objectives and using fairly crude methods of measurement can yield a surprising amount of observational data at a very modest cost. It is hoped that this paper will encourage others to undertake similar experiments in a wide range of soil conditions to establish a broad data base for the prediction of ground movements around trench excavations.

(2) More specifically, it has been found that the ground movements around a trench excavation in London Clay involve both quasi-elastic strains and discontinuous

shear processes which are strongly time-dependent and as much influenced by the weather and the location of the spoil heap as by the provision of supports.

ACKNOWLEDGEMENTS

The work described in this paper was sponsored by the Transport and Road Research Laboratory, who also carried out the trench wall jack tests. The views expressed however, are those of the authors and not necessarily those of the Laboratory.

The assistance freely provided at various stages of the trial by research students and staff of the University of Surrey's Geotechnical Engineering Division is gratefully acknowledged.

REFERENCES

J.B. Burland, B. Simpson and H.D. St. John - Movements around excavations in London Clay, Proceedings of the Seventh European Conference on Soil Mechanics and Foundation Engineering, British Geotechnical Society, London (1979) 1, 13-29.

J.E. Crofts, B.K. Menzies and A.I.Tarzi - Lateral displacement of shallow buried pipelines due to adjacent deep trench excavations, Geotechnique 27 (1977) 161-179.

P.J. King, G.T. Clegg and W.J. Walters - Department of Energy: Report of the Inquiry into Serious Gas Explosions, HMSO, London (1977) 19-21.

G.J.L. Lawrence - Trench wall jack: an apparatus to measure the equivalent elastic modulus of soil, Transport and Road Research Laboratory Supplementary Report 347 (1977).

A. Marsland - The shear strength of stiff fissured clays, Building Research Station Current Paper 21/71 (1971).

A. Marsland - Laboratory and in-situ measurements of the deformation moduli of London Clay, Building Research Establishment Current Paper 24/73 (1973).

G.G. Meyerhof - The ultimate bearing capacity of foundations, Geotechnique 2 (1951) 301-332.

N.P. Roberts and T. Regan - Fractures in Water Mains, City University and Metropolitan Water Division of Thames Water Authority, London (1977).

M.J. Tomlinson - General safety in excavation, Proceedings of the Institution of Civil Engineers Conference on Hazards in Construction, ICE, London (1971) 31-32.

Transport and Road Research Laboratory - Ground movements and their influence on buried pipes, TRRL Leaflet LF 681 (1978).

BENDING OF PIPELINES IN LATERALLY DEFORMING SOIL ADJACENT TO TRENCH EXCAVATIONS

J. E. Crofts, MC, CEng, FICE, FGS

Partner, Nachshen, Crofts & Leggatt

B. K. Menzies, PhD, CEng, MICE, FGS

Lecturer, University of Surrey

A. I. Tarzi, BSc(Eng), MPhil

Lecturer, University of Surrey

INTRODUCTION

Fracture of buried pipelines frequently follows nearby trench excavation, as a result of ground movement. The risk of fracture can of course be avoided by either re-routing the excavation or relaying the pipeline, but where fracture is unlikely such steps are unnecessary. When excavations near services are proposed, decisions have to be taken so as to see that services are suitably protected. This paper sets out a method of assessing the risk of breakage of a pipeline due to ground movement following excavation.

A method has been proposed (Crofts, Menzies and Tarzi, 1977) for estimating the horizontal movement of a long shallow buried continuous pipeline due to the nearby excavation and backfilling of a long deep trench parallel to the pipeline. An elastic model of the problem consisting of a continuous beam embedded in an elastic foundation which is locally displaced laterally was analysed and an approach suggested for estimating the risk of pipe fracture (This approach has been discussed: Symons, 1978; O'Rourke, 1978; Crofts *et al.*, 1978).

Two trench lengths were defined: a 'long trench' where $\lambda L>4.0$ and $A=0.16$; and a 'short trench' where $\lambda L<4.0$ and A lies in the range $0<A<0.32$.

Where L = length of excavation for trench, m

λ = soil-structure interaction parameter $\sqrt[4]{\frac{DK_u}{2EI}}$

A = maximum bending moment parameter, $M_{max}/2EI\lambda^2 x$

The method was extended to articulated or jointed pipelines by Tarzi, Menzies and Crofts (1979) and is described in its entirety in the following sections.

MOVEMENT OF A BURIED PIPE TOWARDS EXCAVATION

The movement x of a buried pipe towards an excavation will consist typically of four components of movement following the construction sequence giving

$$x = x_1 + x_2 + x_3 + x_4 \qquad (1)$$

where x_1 is due to the movement of the unsupported vertical face of the trench
x_2 is due to the movement to bring the face of the trench into contact with the supports
x_3 is due to the deflexion of the supports as a result of the mobilization of the full support of the propping system retaining the trench sides
x_4 is due to the consolidation of backfill used to reinstate the trench.

The assessment of each of these components of pipe movement will now be considered.

The pipe movement x_1 due to the movement of the unsupported vertical face of the trench

This section relates particularly to excavations in cohesive soils cut to their full depth in one operation, but it also applies to the upper part of excavations in cohesive soils made in stages, before supports are put in for the first stage.

Soils having cohesion, such as clays or clayey gravel or clayey sands, will stand vertically for limited periods if the height of the cut face is less than a critical height, H_{cr}. If no cracks occur in the soil above the face, the critical height is $4s_u/\gamma$, but it is recommended that cracks down to half the depth of the excavation be assumed and this will reduce the critical height to $2.6s_u/\gamma$ where s_u is the undrained shear strength and γ is the unit weight of the soil.

It is suggested that a movement of 15 mm be assumed near the top of excavations of critical depth and proportionately less for shallower depths, the amount of bulge of the excavation face x_b depending on both the type of ground being excavated and the depth of excavation.

As a first approximation it is suggested that

$$x_b = \frac{\text{actual depth of trench}}{\text{critical height}}\ 15\text{mm} \qquad (2)$$

Typical values of x_b are given in Table 1.

The movement of a buried pipe parallel to the trench due to bulging of the trench face is x_1 where, owing to the strain gradient in the soil,

$$x_1 < x_b$$

Accordingly

$$x_1 = \alpha x_b \qquad (3)$$

where values of α are measured by using line D of Fig. 1, where the ordinate shows the value of α for any value of the ratio of

$$\frac{\text{distance from excavation}}{\text{depth of excavation}} = \frac{l}{H}$$

This reduction is synthesized from the trends of curves A, B and C which are discussed later.

Table 1 is intended to cover trenches of any length L. It is clear that the ends of a trench will provide restraint against inward movement for some distance along the length of the trench. In the case of short trenches it is possible to apply an intuitive reduction factor to allow for the effect of the end restraint. It is recommended therefore that where L<2H a reduction factor of $\frac{L}{2H}$ be applied to give

$$x_1 = \alpha \frac{L}{2H} x_b \qquad (4)$$

The horizontal movement x_2 due to the movement to bring the face of the trench into contact with the supports.

As a guide, it is recommended that 10mm be assumed, plus the average amount required to fill the troughs of trench sheeting, which might be in the range 2-18mm. These allowances are independent of the depth of the excavation or the strength of the soil, but as the movements take time, they may not be complete in strong soils before the supports are removed.

The earth movement to bring the face of the trench into contact with the supports is thus 10mm +(between 2-18mm) whence the movement of a buried pipe, x_2, due to this earth movement is

$$x_2 = \{10\text{ mm} + (\text{between 2 and 18 mm})\}\ \alpha \qquad (5)$$

TABLE I

Immediate bulging x_b of excavated face of trench

Type of ground*	Depth of excavation	Bulging of the excavation face, x_b
very soft clay very soft sandy clay very soft gravelly clay	0-1.5 m Depths greater than 1.5 m should not be excavated in one stage	0-15 mm
soft clay soft sandy clay soft gravelly clay	0-2 m 2-4 m Depths greater than 4 m should not be excavated in one stage	0-8 mm 8-15 mm
firm clay firm sandy clay firm gravelly clay	0-2.5 m 2.5-5 m 5-7.5 m	0- 5 mm 5-10 mm 10-15 mm
stiff clay stiff sandy clay stiff gravelly clay	0-3 m 3-6 m 6-10 m	0- 5 mm 5- 9 mm 9-15 mm

* For very stiff clays, dense sand or gravel or clean dry sand which will run, bulging does not apply.

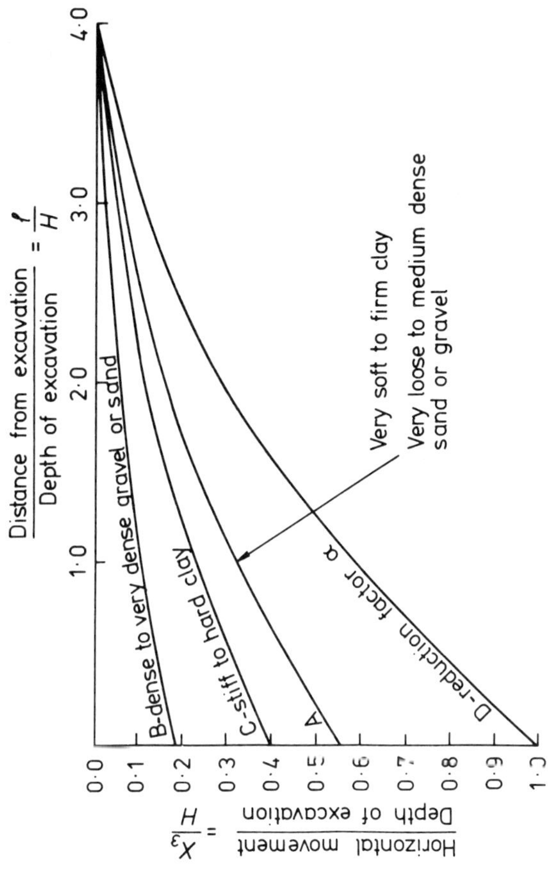

Fig. 1 Horizontal movements related to distance from excavation

where α is given by line D in Fig. 1. Stressing the trench struts and wedging the face supports from the walings may reduce x_2. When sheet piling or concrete diaphragm walls are installed before excavation, $x_2 = 0$. For hard clay and medium to dense sands x_2 should be halved.

The pipe movement x_3 due to the deflection of the supports as a result of mobilization of the full support of the propping system retaining the trench sides.

After allowing for the movements already discussed, an estimate must be made of the movements due to the pressure of the soil on the supports. The movements are due to swelling of the soil, bending of the immediate supports to the sides of the excavation (such as sheet piles, trench sheets, poling boards, etc.), bending of the walings, extension of anchor rods, shortening of the struts and sometimes indentation of struts into timber walings or wedges. Assessment of these movements is made in this Paper entirely on the basis of observations from sites where accurate measurements have been made. These observations are obtained from Peck (1969), Tomlinson (1975) and St. John (1976), and curves A, B and C of Fig. 1 relate to these observations. The curves are envelopes showing maximum movements recorded.

A reduction factor to allow for the effects of trench restraint, similar to that used to obtain x_1, may be applied for trenches with $L<2H$, giving

$$x_3 = \frac{\text{value from Fig. 1}}{100} H \frac{L}{2H} \qquad (6)$$

Where early propping of the trench is assumed and supports are raised and backfill compacted in small stages, a further reduction may be applied as follows:

for trenches up to 3 m deep, reduce x_3 to $\frac{1}{2}$

for trenches 3–5 m deep, reduce x_3 to $\frac{3}{4}$

To make use of this reduction propping should take place within the following times:

firm clay, medium dense sand	1 h
stiff clay, dense sand	2 h
hard clay, very dense sand	4 h

This reduction should not be applied to soft clay or to loose sand or gravel or where the ground water level is not below the level of the trench bottom before excavation begins.

Where the trench is to be formed through dense macadam not less than 250 mm thick or concrete not less than 100 mm thick in firm contact with dry ground a further reduction may be applied as follows:

for trenches up to 3 m deep, reduce x_3 to $\frac{1}{2}$

for trenches 3-5 m deep, reduce x_3 to $\frac{3}{4}$

These reductions to x_3 for the effects of end restraint (equation (6)), time to propping, and surface restraint, respectively, are cumulative. The reduction factors for the effects of time to propping and of surface restraint are intuitive.

The horizontal pipe movement x_4 due to the consolidation of the backfill

The amount of movement can be estimated by assuming that when the backfill is placed, the soil of the walls of the excavation exert active earth pressure against the backfill. With time the backfill will be compressed so offering progressively greater resistance until the earth pressure rises to the at rest value.

Thus

$$x_4 = \alpha \frac{B}{2} \frac{C}{1 + e_o} \log_{10} \frac{K_o}{K_a} \qquad (7)$$

where x_4 is the horizontal movement

B is the width of the trench

$\frac{C}{1 + e_o}$ is the compressibility of the backfill

α is given by line D in Fig. 1

K_a is the coefficient of active earth pressure for the soil in which the excavation is made

K_o is the coefficient of earth pressure at rest for the soil in which the excavation is made

Typical values for these parameters are summarized in Table II. For an excavation in which only one wall is earth, and the other is a material such as concrete, the whole width of backfill B should be used instead of $B/2$.

TABLE II

Presumed soil parameters

Soil description		Coefficient of horizontal subgrade reaction k_u kN/m^3	Coefficient of active pressure K_a	Coefficient of earth pressure at rest K_o		Backfill compressibility $C_c/1 + e_o$	
Clay	Sand			Clay	Sand	Clay	Sand
Very soft to soft	Very loose to loose	4,000	.41	0.5	0.5	0.30	0.15
Firm	Medium dense	8,000	.33	0.8	0.45	0.15	0.10
Stiff	Dense	16,000	.27	1.0	0.4	0.10	0.05
Very stiff or hard	Very dense	32,000	.22	1.5	0.4	0.05	0.03

EFFECT OF THE LOCAL MOVEMENT OF A LONG BURIED PIPE TOWARDS AN EXCAVATION.

In the analysis the local movement of a long buried pipe was simulated by an elastic model. It consists of an elastic beam embedded in an elastic foundation. The elastic foundation may be likened to a bed of springs laid horizontally either side of the beam. One end of each spring is connected to the beam while the other end is connected to a rigid foundation. There are thus two rigid foundations forming a parallel pair, one either side of the beam. The total elastic modulus of the foundation consists of the sum of the moduli of springs on either side of the beam. The elastic modulus of the foundation is thus $2Dk_u$, where D is the depth of the beam normal to the plane of the springs and k_u is the modulus of subgrade reaction for each bed of springs.

It is further assumed that the springs representing the elastic foundation (it is not a continuum) of the elastic model are already pre-compressed corresponding to the in situ at rest horizontal effective stress of the soil prototype. Some of the springs will be unloaded or loaded by the re-adjustments required by the model, and it is further assumed that the local resistance will not fall below the active earth pressure values or rise above the passive earth pressure values, respectively.

For a continuous, unjointed pipeline, Fig. 2 shows the variation with length of trench of the envelope to the maxima of the maximum bending moment parameter. This curve is given by way of a basic design chart.

For an articulated, jointed pipeline the elastic model is made up of an infinite number of pipe sections, each of length p, pinned together end to end to form an infinitely long pipeline. Each pin is rigid to shear distortion and freely flexible to bending distortions up to the maximum allowable angular distortion whereupon the pin behaves as a rigid joint for further increments of loading. As in the case of the continuous pipe, the articulated pipe is first cut at the points corresponding to the extreme ends of the trench excavation and given a uniform lateral displacement x. The problem is thus modelled by a lack-of-fit analysis of an articulated beam on an elastic foundation.

The results of the analysis are presented in Fig. 3, with the maximum bending moment factor A normalized by the corresponding maximum values for a continuous pipe. For a long trench, for an articulated pipeline of section length $\lambda p > 3.0$, the value of A is increased by up to 36.5% For a short trench, for an articulated pipeline with λp within the range $2.6 < p < 2.8$ the value of A is up to 21% higher than the corresponding value for a continuous pipe.

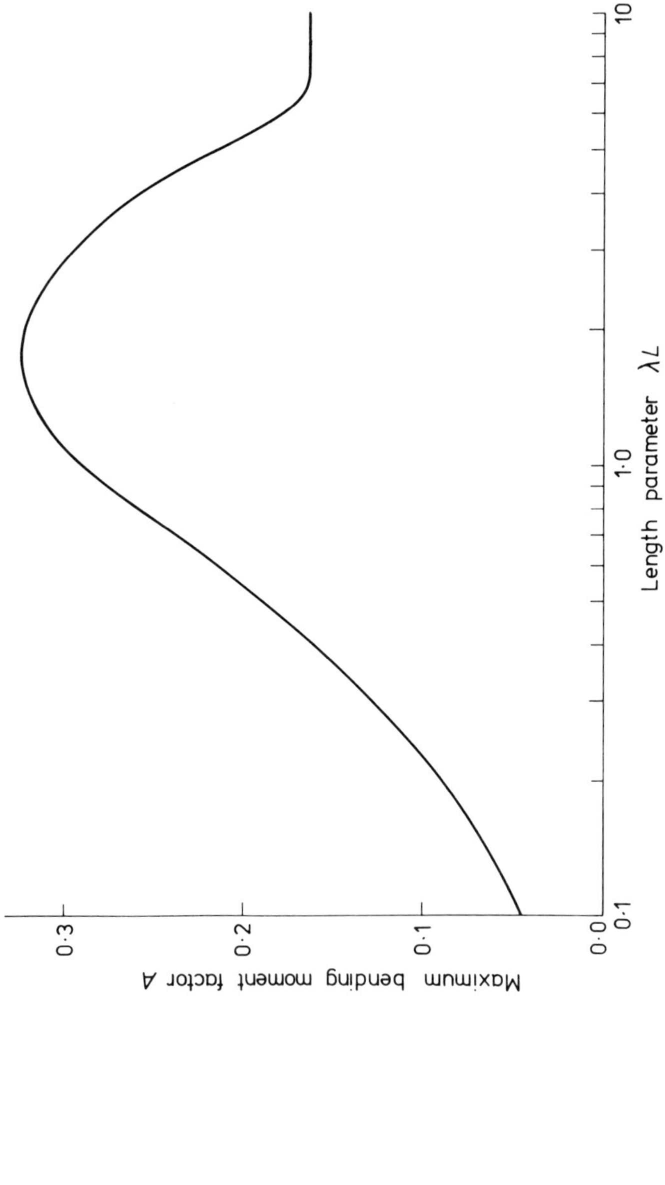

Fig. 2 Envelope to maxima of maximum bending moment parameter A plotted against length of trench parameter λL

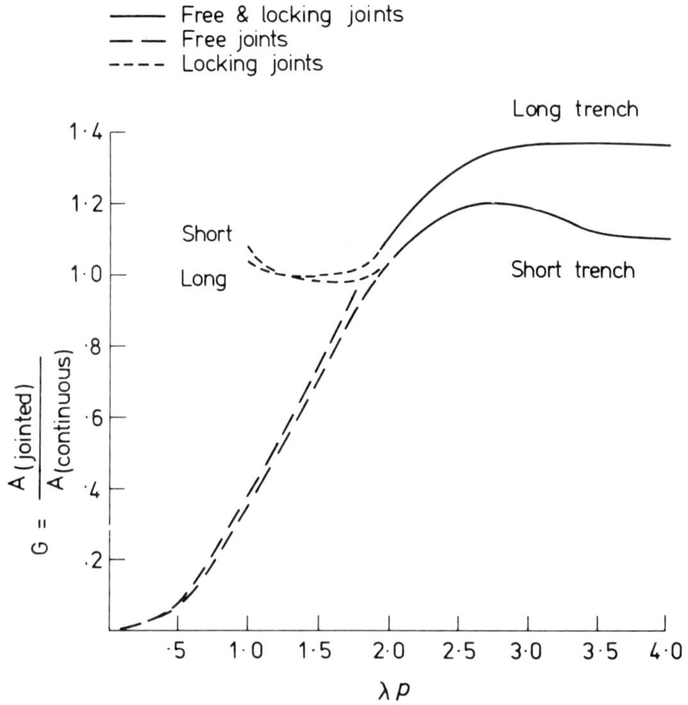

Fig. 3 Relationship between maximum moment factors for continuous and jointed pipelines for varying λp

The elastic model is sensitive to the length of end restraint a, which is the length over which the cut beam is displaced to x from its original position. The extreme condition of a as zero is of course not likely to be consistent with actual performance and can only represent a limiting condition, approaching the case of the pipe immediately adjacent to the face of the trench. The more realistic condition of the pipe near the trench will give rise to a finite length of end restraint which is clearly a function of, among other parameters, the distance l of the pipe from the parallel face of the excavation. As a first approximation, the assumption that $a=l/2$ seems a not unreasonable device for assessing this effect. For finite values of a, A is reduced and this reducttion is given in Fig. 4.

APPROACH TO CALCULATIONS OF THE RISK OF PIPE BREAKAGE OR PIPE BECOMING UNSERVICEABLE.

Unserviceability or breakage of pipes results from ground movement being excessive for the particular pipe concerned. Calculation methods applicable to earth movement and the interaction between earth movement and pipes are not sufficiently accurate to allow clear cut predictions of exactly when a pipe will break or become unserviceable but the methods are adequate to allow a decision to be made of the acceptability of the risk of breakage or unserviceability occurring.

Three categories or risk are proposed:

Category 1 acceptable risk because breakage or unserviceablility of the pipe is unlikely
Category 2 border zone between acceptable and unacceptable risk where breakage or unserviceability of the pipe can probably be avoided if special care is taken with all aspects of the work and its supervision
Category 3 unacceptable risk of breakage or unserviceability.

The strain developed in the pipe as a result of earth movement is now estimated separately for the short term and long term condition, to see which condition develops the greatest pipe strain. The short term condition excludes backfill consolidation; the long term condition includes backfill consolidation.

It may be shown that for the displaced pipe the short term maximum strain is

$$\varepsilon_{\text{short term}} = D\lambda^2 \ (x_1 + x_2 + x_3) \ FGA_{\text{short term}} \qquad (8)$$

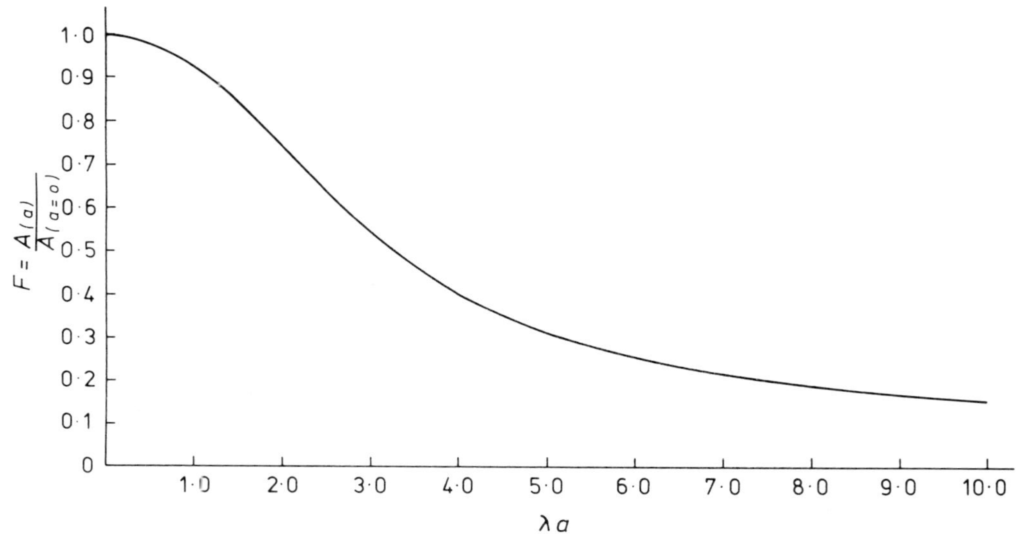

Fig. 4 Bending moment reduction factor F for length of end restraint parameter λa where a may be taken as $\ell/2$

TABLE III

Definition of category as a proportion of breaking strain

Risk category	Assumed proportion of either breaking strain or maximum strain for unserviceability developed by proposed excavation and other causes
Category (1) Acceptable risk because breakage or damage unlikely	0.0 up to 0.25
Category (2) Border zone because breakage or damage can probably be avoided if special care is taken with all aspects of the work and its supervision	Above 0.25 and up to 0.60
Category (3) Unacceptable risk of breakage of unserviceability	Above 0.60

and the long term maximum strain is

$$\varepsilon_{\text{long term}} = D\lambda^2 \ (x_1 + x_2 + x_3 + x_4) \ FGA_{\text{long term}} \qquad (9)$$

where $A_{\text{short term}} = 0.32$ is the peak value of Fig. 2 and where $A_{\text{long term}}$ is read off from Fig. 2 and is not less than the plateau value of *0.16*, *F* is read off from Fig. 4, *G* is read off from Fig. 3.

RISK TABLES

Risk tables have been prepared, in which the risk category is tabulated for trenches up to 6 m deep and various pipes and ground conditions.

ACKNOWLEDGEMENT

The Authors gratefully acknowledge the contribution to the work of the late Mr. M. Nachshen, and the help and encouragement of Professor N. E. Simons.

REFERENCES

J. E. Crofts, B. K. Menzies and A. I. Tarzi, (1977). Lateral displacement of shallow buried pipelines due to adjacent deep trench excavations. Geotechnique 27, No. 2, 161-179.

J. E. Crofts, B. K. Menzies and A. I. Tarzi, (1978). Discussion: Lateral displacement of shallow buried pipelines due to adjacent deep trench excavations. Geotechnique 28 No. 2, 217-220.

T. D. O'Rourke, (1978). Discussion: Lateral displacement of shallow buried pipelines due to adjacent deep trench excavations. Geotechnique 28 No. 2, 214-216.

R. B. Peck, (1969). Deep excavations and tunnelling in soft ground. Proceedings 7th International Conference of Soil Mechanics and Foundation Engineering, Mexico, 259-281.

H. D. St. John, (1976). Recent research into the movement of ground around deep excavations. Proceedings of the Symposium Underground Engineering--The Next Decade, Leamington Spa, England: Foundation Publications Ltd.,

I. F. Symons, (1978). Discussion: Lateral displacement of shallow buried pipelines due to adjacent deep trench excavations. Geotechnique 28 No. 2, 212-214.

Tarzi, A.I., Menzies, B.K. and Crofts, J.E. (1979) Bending of jointed pipelines in laterally deforming soil. Geotechnique 29 No. 2, 203-206.

M. J. Tomlinson, (1975). Foundation design and construction. Third edition. Pitman: London.

APPENDIX

Worked Example

Data:

Soil firm clay. K_u is $8000\ KN/m^3$ (Table II)
Existing pipe 4" cast iron $D = 0.121\ m$ length between joints $p = 5.49\ m$
Depth of existing pipe 600 mm. Pipes assumed articulated at joints. Pipe stiffness EI = 368.44 kNm^2.
Trench in light construction road 3.5 m from pipe = l
Depth of trench H = 4 m
Width of trench B = 1 m
Length of trench L = 20 m
Backfill to be sand compacted to a medium dense state.
Trench to be machine dug to full depth followed closely by placing supports
(From the data the trench is seen to be deeper than the pipe)

a) Movement of unsupported face
By equation (3) $x_1 = 8\alpha$, say
Now $l/H = \frac{3.5}{4} = 0.875$ whence $\alpha = 0.64$ (figure 1) giving
$x_1 = 8 \times 0.64 = 5\ mm$

b) Movement to bring soil into contact with supports allow 3 mm to fill voids. By equation (5)
$x_2 = (10 + 3)\ \alpha = 13 \times 0.64 = 8\ mm$

c) Movement due to deflection of supports
Distance of pipe from excavation = 3.5 m
From Fig. 1 line A and $\frac{l}{H} = 0.875$ equation (6) gives
$x_3 = 0.35\%\ of\ 4\ m = 14$ mm

d) Compression of backfill
From Table II, for firm clay, $\log_{10} K_o / K_a = 0.38$; for medium dense sand $\frac{C_c}{1 + e_o} = 0.10$
then by equation (7)
$x_4 = 0.64 \times \frac{1}{2}\ (0.10)\ 0.38 = 12$ mm
By equation (1) total ultimate movement
$x = x_1 + x_2 + x_3 + x_4 = 39$ mm

e) Bending Moment Factor for articulated jointed pipes.
$\lambda = \sqrt[4]{\frac{DK_u}{2EI}}$ for $K_u = 8000$, $D = 121\ mm$, EI = 368.44 kNm^2
$\lambda = 0.1132 \times 9.46 = 1.071$
see Fig. 4. $\lambda \frac{l}{2} = 1.071 \times \frac{3.5}{2} = 1.87$
$F \doteq 0.78$.

f) Bending Moment factor G for articulated jointed pipes
see Fig. 3. $\lambda p = 1.071 \times 5.49 = 5.88$
since $\lambda L = 1.071 \times 20 = 21.4 > 4$, so trench is a "long

trench" and G = *1.38*.

Calculation of Strains

Short term strain before backfilling

from (8)

$$\begin{aligned}\varepsilon_{\text{short term}} &= D\lambda^2\ (x_1 + x_2 + x_3)A_{\text{short term}} \times F \times G \\ &= 0.121 \times 1.071^2 \times 0.027 \times 0.32 \times 0.78 \times 1.38 \\ &= 0.00129\end{aligned}$$

ultimate strain for 4" C.I. pipe = *0.00187*

proportion of breaking strain mobilized = $\frac{0.00129}{0.00187} = 0.69$

Long term strain mobilized after backfilling

from (9)

$$\begin{aligned}\varepsilon_{\text{long term}} &= D\lambda^2\ (x_1 + x_2 + x_3 + x_4)A_{\text{long term}} \times F \times G \\ &= 0.121 \times 1.071^2 \times 0.039 \times 0.16 \times 0.78 \times 1.38 \\ &= 0.000932\end{aligned}$$

proportion of breaking strain mobilized = $\frac{0.000932}{0.00187} = 0.499$

Short term strain is critical. The proposed work is in category 3 that is there is an unacceptable risk of breakage. NOTE - Had the pipe been assumed to be solidly jointed together, such as by being rusted up solid, then G = 1 and the proposed work would have a risk category 2. This may appear surprising, but the explanation lies in the unknown position of the pipe joints in relation to the trench position. In most possible positions of the joints the bending strain in the iron pipes is reduced below that for a rigid pipe, but there are a small number of critical positions of the joints for which the bending is worse, that is, for which G>1. As the analysis is based on the maximum bending, then when articulation is assumed the maximum strain is increased. However, when the pipes articulate, a value of G less than 1 is statistically probable.

If freely jointed very short pipes were used, then bending strains would tend to zero. This is shown in Fig. 3 where G = 0 for λp below 0.2.

SESSIONS I – VII

DISCUSSION

SESSION I

Chairman : Professor J.D. Geddes, BSc,PhD,CEng,FICE,FASCE,FGS
Department of Civil Engineering and Building Technology, UWIST

DR.D.N. CATHIE - Department of Civil Engineering and Building Technology, UWIST

I wish to supplement the paper by Tsur-Lavie and Denekamp by elaborating a little on boundary element methods in general. In the field of solid mechanics and therefore for mining applications, the BEM is a developing competitor to finite element analysis. Essentially, both methods can solve similar problems but the BEM has certain advantages. The main one in geotechnical applications is that only the boundary of the problem has to be discretised, or divided into elements, and therefore data input is much reduced. Additionally, if the fundamental solution is taken to satisfy the boundary conditions on part of the boundary automatically (e.g. a Mindlin point force solution for a semi-infinite problem) then that boundary does not require discretisation - it is included automatically. This would avoid the "mirroring" required in the paper by Tsur-Lavie and Denekamp. This approach is known as the direct method.

BEM analyses always take place in two stages:

(1) Find the unknown boundary conditions for the problem - in elasticity problems these are tractions and/or displacements.
(2) Compute displacements and stresses elsewhere as required.

The fact that the latter may be computed in zones of interest alone, makes for less unnecessary computer output and hence economy of calculation.

I believe that it is also worth recording, in these proceedings, a little more detail on the BEM, for it is likely that in the next decade practising engineers involved in stress analysis will encounter the method and find it a useful tool.

The BEM has a different mathematical basis from finite elements and herein lies the difference between the two. In the BEM, the governing differential equation of the problem (the Laplace equation for seepage, the Navier equations in elasticity) is transformed into an integral equation, where the integrations for a linear problem need only be evaluated along the boundary. Hence it is sometimes referred to as the boundary integral equation method. The boundary is discretised in order to define the boundary conditions and for the numerical integration procedure.

Two different formulations have been proposed - the direct method (Cruse, 1969; Banerjee, 1976) and the indirect method (e.g. Banerjee and Butterfield, 1977). The paper by Tsur-Lavie and Denekamp, while differing from the indirect method in the reference quoted above is essentially in this category. A direct formulation has the advantage of giving the boundary unknowns directly and presents fewer mathematical difficulties.

For an elasticity problem a system of "constraint" equations is developed using the fundamental solution. These take the form:

$$[A]\{u\} = [B]\{t\} \qquad (1)$$

where the vectors $\{u\}$ and $\{t\}$ are the displacements and tractions on the boundary of the problem. Coefficients of $[A]$ and $[B]$ are evaluated by numerically integrating the fundamental solution (influence function) around the boundary, element by element. The region involved in Equation (1) may be thought of as a "super" or macro finite element.

Since in all boundary value problems half the boundary conditions are known, Equation (1) may be solved for the unknown set. Displacements ($\bar{u}$) and stresses can be evaluated at any other required point using matrix equations of the form:

$$\{\bar{u}\} = [C]\{u\} + [D]\{t\} \qquad (2)$$

and

$$\{\sigma\} = [E]\{u\} + [F]\{t\} \qquad (3)$$

Thus computational effort may be minimised so that Equations (2) and (3) are evaluated only where and if required. This enables stresses, for example, to be evaluated and plotted in a localised region of particular interest.

Inhomogeneity and anisotropy may be dealt with by abutting several macro-elements, and in the case of anisotropy taking a suitable fundamental solution. Programming is the only difficulty. It is rather harder than for finite elements. Once developed the program user will probably prefer the boundary approach. Semi-infinite and infinite boundaries are dealt with automatically and properly by the BEM, in contrast with having to fix boundaries well away from the region of interest in finite element methods. The paper by Wardle and Crotty (1978) gives examples of mining applications using the direct method. Indirect approaches in geotechnical applications have been described by Butterfield (1971), Brady and Bray (1978) and Hocking (1978). The method is now well enough developed for regular application in the fields of potential flow and elasticity. Where non-linear effects such as plasticity or tensile fracture are evident, the method is not yet fully developed but it does show promise (Banerjee, Cathie and Davies, 1979).

Concluding, the boundary element method offers a number of advantages over finite element approaches. It is more efficient computationally because only the boundary unknowns are involved in the simultaneous equations to be solved, thereby reducing matrix sizes. This advantage is particularly evident for three-dimensional applications. In addition, data preparation is reduced very substantially for the 3-D case. The future may see the introduction of 3-D stress analyses using the BEM on a routine basis, where an equivalent finite element solution would be quite impractical.

REFERENCES

P.K. Banerjee - Integral equation methods for the analysis of piece-wise non-homogeneous elastic solids of arbitrary shape, International Journal of Mechanical Sciences 18 (1976) 293-301.

P.K. Banerjee and R. Butterfield - Boundary element methods in geomechanics, Ch. 16 in Finite Element Methods in Geomechanics, Ed. G. Gudehus, Wiley (1977).

P.K. Banerjee, D.N. Cathie and T.G. Davies - Two and three-dimensional problems of elasto-plasticity, Ch. 4 in Developments in Boundary Element Methods I, Eds. P.K. Banerjee and R. Butterfield, Applied Science Publishers (1979).

B.H.G. Brady and J.W. Bray - The boundary element method for determining stresses and displacements around long openings in a triaxial stress field, International Journal of Rock Mechanics and Mining Science 15 (1978) 21-28, and see 29-37.

R. Butterfield - The application of integral equation methods to continuum problems in soil mechanics, Roscoe Memorial Symposium, Cambridge (1971) 573-587.

T.A. Cruse - Numerical solutions in three-dimensional elastostatics, International Journal of Solids and Structures 5 (1969) 1259-1274.

G. Hocking - Stress analysis of underground excavations incorporating slip and separation along discontinuities, Recent Advances in Boundary Element Methods, Ed. C.A. Brebbia, Pentech Press (1978) 195-214.

L.J. Wardle and J.M. Crotty - Two-dimensional boundary integral equation analysis for non-homogeneous mining applications, Recent Advances in Boundary Element Methods, Ed. C.A. Brebbia, Pentech Press (1978) 233-251.

PROFESSOR V. ASHKENAZI - Department of Civil Engineering
University of Nottingham

When comparing the results of measurements made by a particular instrument, or a special procedure, with those of another it is essential to have not just a series of numerical values, but also two sets of estimated accuracies corresponding to these values. Only thus can one interpret the statistical significance of the inevitable differences between the two sets of results. This approach is important when measurements of ground and structural movements made with geotechnical instrumentation are compared with those obtained using surveying techniques. The latter are especially useful when connecting the various geotechnical instruments scattered over a large area, with one another in an "absolute" 3-D coordinate reference system.

The use of photogrammetry is an essential surveying method in some cases e.g. where a structure under study cannot be approached. The major disadvantage is that the accuracy depends on the distance at which the pictures are taken. If this is small the accuracy can be high.

Finally, it is worth pointing out that today, surveying equipment may be highly automated, with direct reading onto magnetic tape, thus eliminating reading errors and speeding up the surveying process.

PROFESSOR R. CIESIELSKI - Technical University of Krakow, Poland

I am interested in the paper by Mr. Burton because of the practical approach to subsidence prediction presented in it. It suggests a method for identifying certain physical parameters that may not be of known value. For example, the physical parameters of the system must be introduced to the model as data parameters such as thickness of layers. However, not all the values are precisely known. Young's modulus is an example.

A physical model is created from input data, the mathematical model then utilises the physical model to produce predictive results. We are also able to measure the actual movements, etcetera, and to compare these measurements with those predicted. The comparison allows a "function of exactness" to be identified, relating to the physical and mathematical models. In practice, measured and predicted results are never the same and such a comparison enables one to answer, at least partly, why this is not so. The new information furnished, permits a return to input with further data and this results in a better solution. Figure D1 shows the procedures involved.

MR.D. BURTON - British Railways Board, Derbyshire

The process of feedback control used in our prediction

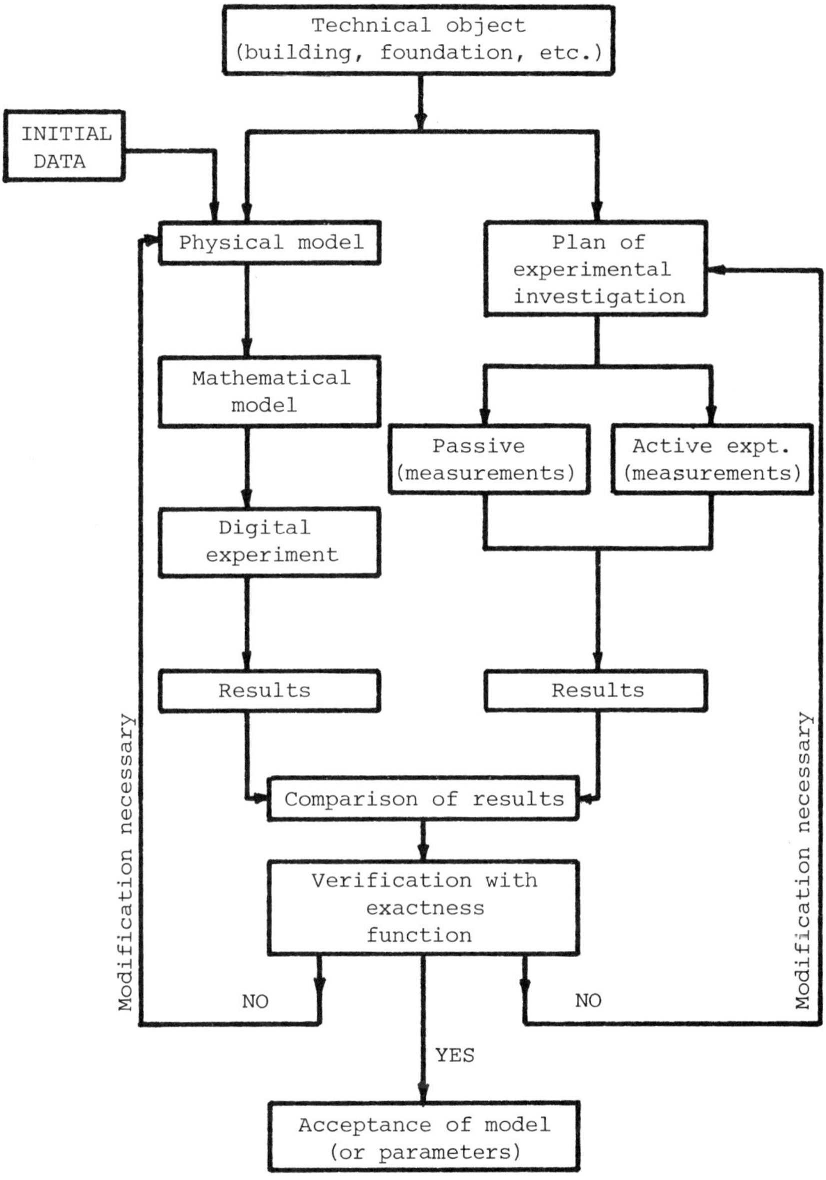
Technical object
(building, foundation, etc.)
INITIAL DATA
Physical model
Plan of experimental investigation
Mathematical model
Passive (measurements)
Active expt. (measurements)
Digital experiment
Results
Results
Comparison of results
Verification with exactness function
Modification necessary
Modification necessary
NO
NO
YES
Acceptance of model (or parameters)

Figure D1

method, makes use of two three-dimensional systems in the computer program, one of which controls the other. Because of certain mathematical properties of subsidence profiles and the differential lateral displacements of surface points, it is possible to construct a very efficient control system by taking advantage of the structure of the mathematical model itself.

MR.J.C. FINNIGAN - Welsh Office, Cardiff

In the paper presented by Dr. Farmer, particular reference is made to the Tupton sandstone but there are at least four other principal sandstones shown in the section illustrated. Is there anything of special significance about the Tupton sandstone?

DR.I.W. FARMER - Department of Mining Engineering
University of Newcastle upon Tyne

The Tupton sandstone is strong and competent and has historical associations with the 1943 Barnborough Colliery rock burst. In Figures 3 and 4 of our paper, one can see the same kind of conditions that could lead to rock bursts. There are low strains in the Tupton sandstone but high strains in the other layers, suggesting that energy is stored in the sandstone. This is an indication that one should be extremely careful when very hard strata are present during longwall mining (see Phillips, 1944).

REFERENCE

D.W. Phillips - Rock bursts or "Bumps" in coal mines, Transactions of the Institution of Mining Engineers 104 (1944) 55-84.

MR.J.C. FINNIGAN

Potential danger must presumably be present in similar situations in other regions. There is too the matter of interbedded sandstones within the coal seams; this must compound the problem somewhat.

DR.I.W. FARMER

Usually one needs a considerable expanse of sandstone before the conditions are created for fracture leading to large-scale release of energy. Persons with local knowledge are often aware of the potential danger.

MR.J.C. FINNIGAN

One must assume then that, in such circumstances, although a sandstone roof may appear to be perfectly safe and in a competent condition, it could conceivably represent a potentially high risk situation because of the stored energy?

DR.I.W. FARMER

That is a good summary of a quite common condition in longwall coal mining, but the risk is usually to production rather than to life.

MR.A.E. WILKINS - Johnson, Poole and Bloomer

In Mr. Burton's mathematical models, it may be possible to evaluate relevant parameters for natural rock strata, but how is it possible to make due allowance for overlying old mine workings which are often highly irregular in area, variable in percentage extraction and which will have collapsed by various amounts over the time since mining ceased?

MR.D. BURTON

Old workings certainly have an effect and often their presence is more important than the local geology. The control of the mathematical model must therefore take into account workings in other seams and in the same seam. Fortunately the combination of control system and model provides some simple rules for anticipating the effect of earlier workings and although the situation becomes very complex, the control system remains comparatively simple and effective.

DR.D. COX - Department of Civil Engineering
Polytechnic of Central London

I wish to address some remarks to Mr. Cook. If two successive photographs are taken of a deforming structure from the same position and then viewed stereoscopically, a 3-D image of the movement is produced. This can be done for a brick wall as discussed in my paper to a later session of the conference. The developing crack pattern can be established by this method because the movements are magnified. Have you tried this technique?

MR.D.A. COOK - School of Architecture and Building Engineering
University of Bath

In answer to Dr. Cox, I must say that I have not tried such a technique which is a new idea to me. However in the context of what I think is implied, it appears to be a very interesting one, particularly for some types of laboratory work, when I should not confine it to brick crack patterns but apply it to a wide range of small movements. Small deformations in brickwork we have always measured directly with a Demec strain gauge; but we could more usefully experiment with Dr. Cox's suggestion for say a soil mechanics experiment behind a glass plate, where the object is less accessible. Dr. Cox's illustration showing the visual effect of a crack movement viewed through a stereoscope is impressive; and I should not anticipate too many difficulties in quantifying the amount of movement.

This of course returns us to the question of accuracy of measurement. Using an unsophisticated method, as described in my paper, one can achieve an accuracy of ± 0.05 mm (measurement on the photograph). We also happen to have a travelling microscope which claims a least count of 0.005 mm. But when you use an instrument of this nature you are in a similar position to holding a one-second theodolite, in that you must take multiple readings if you seriously expect to achieve this sort of accuracy. The final accuracy of the calculation however does not just depend on the refinement of the photo-coordinates, but also on their relative position in the image field. Reference to Equations (i) or (ii) in Appendix 'A' shows this. If a value of angle α, or β, were derived from the photo-coordinates then differentiating the expressions for the object (X_A, Y_A, etc) will show a varying difference for a small (say constant) error $\Delta\alpha$. These aspects are however dealt with in some of the other papers to the conference.

SESSION II

Chairman : Mr.W.S. Allen, CEng,MIMinE,ARICS
National Coal Board, London

MR.J.C. FINNIGAN - Welsh Office, Cardiff

I have a number of comments on several of the papers which have been presented in this session.

I would like to start by congratulating Dr. Symons for his very valuable efforts in the extremely difficult task of correlating coal seam names in the Llanelli area of the South Wales Coalfield as dealt with in the final paper of the session.

I strongly support the conclusion in the paper by Bruhn, Magnuson and Gray that ground subsidence due to old workings cannot be regarded as being complete unless total extraction methods have been employed. The Pittsburgh experience does not appear to be dissimilar to our own here in Wales in that, with partial extraction methods (e.g. pillar and stall), there remains a risk of ground subsidence which can occur long after the workings have been abandoned, depending upon the life-expectancy of the pillars.

Pillar collapse, as is known, can result from a number of factors including, I would suggest, erosion of the pillars by groundwater flowing through the old mine galleries. In the course of time this erosion process can reduce the substance of the pillars to such a degree that they become unable to carry the roof load and thus pillar collapse occurs, progressively, throughout the mine workings. This I think may have been the cause, or at least a contributory factor, in the ground failure at Bagillt described in the paper by Mr. Adams, where the mine workings are over 100 years old.

Pillar (or stoop) collapse also seems to have been the principal source of the very serious mining subsidence problems experienced at Bathgate (Lothian) which was described in the first of this afternoon's papers. Those mine workings are reported as being 200 years old, thus I imagine that we might expect a good deal more subsidence problems from earlier workings in other areas, such as Bagillt for instance, in the future.

Can we have a comment from Mr. Carter on the design-mix of the grout used in the remedial works at Bathgate and can he explain the reason for the change in the mix used?

Would any of the authors care to give their opinion on the traditional method of assessing the 'safe depth' to old mine workings? In the Bathgate paper, the old rule of thumb method (5 to 10 times the seam's thickness) has been referred to but only in respect to 'room' ('stall') collapse. But this can hardly be regarded as satisfactory and also it does not provide for pillar ('stoop') collapse. The Pittsburgh experience quotes a depth figure of 450 feet (137 m) but again this must be regarded as a provisional limit since the authors of that paper are predicting ground subsidence resulting from pillar collapse in workings at depths greater than 450 feet.

My final comment concerns the use, or rather lack of use, of geophysical methods to determine the location, extent and configuration of old mine workings. Were such techniques used at Bathgate? Have the authors of any of the papers any comments they would like to contribute on this topic? As far as I am aware we still have a long way to go in devising a reliable technique(s) for the location of old shafts and workings by geophysical means, or by any other means. But is sufficient being done towards this end? I suspect not.

MR.P. CARTER - Babtie Shaw and Morton, Glasgow

It is our opinion that in the majority of cases, the use of bulking factors to give safe depths of 5 to 10 times the seam thickness is a valid approach, given simple room failure as the mode of collapse. But there are cases, such as Bathgate, where a particularly adverse combination of factors can cause subsidence from workings at substantially greater depths.

With regard to grout mixes, the stiffer grout containing more cement, pea gravel and a little bentonite was used to form a barrier wall in the steeply dipping workings, after which a leaner mix was used to infill the area beneath the flats.

On the question of geophysics, it has been our experience that these methods are of little practical value, particularly in urban areas, but I would be interested to hear of any successful application. I would advise the use of a well-controlled drilling programme using rotary cored holes to verify whether coal if encountered is oxidised, thus indicating the presence of old workings. I am aware of the use of a borehole TV camera in some boreholes which has given limited success in determining the extent of old open workings.

PROFESSOR J.D. GEDDES - Department of Civil Engineering and Building Technology, UWIST

I wish to raise a question with regard to a paragraph in the paper on Bathgate in which the authors state that "the likely structural responses to the predicted maximum subsidence of these buildings have been analysed and found to range between relatively minor cracking for two-storey dwellings well away from the outcrop to major failure of a five-storey building supported on independent column foundations right over the outcrop". Would the authors care to amplify this statement? Prediction of effects is not easy and any further information would be of value.

MR.P. CARTER

Being an engineering geologist rather than a structural engineer, I am not going to be able to give Professor Geddes a full answer to his question. Basically the town was zoned into areas of magnitude of potential mining subsidence and values of ground strain, differential settlement, etcetera, ascribed to these zones using the method described in the paper. The firm's structural engineers then examined the buildings within these zones, particularly those most obviously at risk, and where possible with the use of the as-built drawings. They then analysed these structures using the anticipated subsidence effects, and the make-up of the structure in so far as this could be determined, to ascertain the degree of potential damage. We were primarily interested in structures which presented a threat to public safety if subsidence occurred, such as the Marjoribank Street flats which we subsequently stabilised by grouting. As an aside, it is worth noting that it is sometimes easier to obtain plans of mines over two centuries old, than it is to obtain plans of structures less than two decades old. There should be some way found of ensuring that structural drawings are maintained for the life of a structure, particularly in mining areas.

MR.E.J. WILSON - Consulting Engineering Geologist, Gloucester

Mr. Carter and his colleagues raise an interesting point in relating the competent nature of the roof and floor measures to a preference for pillar failure, in that the normal processes of roof or floor failure, which lead to progressive protection of the remnant pillars, are inhibited. This would seem to be borne out for instance by the pillar failures in deep workings in the Wenlock limestone in the Walsall area.

How far do the authors feel this analogy should be taken and do they feel that we should be generally more wary of pillar failures in those measures with a higher sandstone content?

MR.P. CARTER

Yes, I believe that the high proportion of sandstone, contrasting with the high proportion of shales in Coal Measure areas, did have an important effect in Bathgate, although this cannot be proved. The presence of sandstone would tend to prevent natural mine infilling over the centuries, leaving the small stoops to carry large loads for a long period of time, which could have contributed to failure of the stoops. It could possibly be likened to the Walsall situation, where the relatively strong limestone prevented normal mine infilling in a similar fashion.

MR.J.M. SMITH - Mineral Valuers Department, Inland Revenue Newcastle upon Tyne

In their paper, Mr. Bruhn and his co-authors cite maximum recorded depths of 120 feet (36.6 m) for old workings causing sinkholes and 450 feet (137 m) for those causing troughs as a result of pillar collapse. These appear to be excessive depths in terms of British experiences. It would be appreciated if Mr. Gray could indicate the circumstances giving rise to these extreme cases.

MR.R.E. GRAY - GAI Consultants Inc., Pittsburgh, USA

In most areas around Pittsburgh the colluvial cover of soil is less than 15 feet. However, in the first case mentioned the depth of workings was 150 feet but the rock cover was only 30 feet, the remaining cover of 120 feet being alluvial soil.

In the second situation, the bearing capacity of the floor was found to be very important. Where the floor was of seatearth, pillar punching often took place, giving rise to a surface trough 40 acres in extent.

MR.S. ADAMS - W.G. Curtin and Partners, Mold, Clwyd

At Bagillt the seams were in siltstones and mudstones, therefore there was more cavity migration. The general finding was that seams largely collapsed with soft drilling over. The largest identifiable voids were about 0.5 m thick. At Bagillt, movements were first noticed on council houses twelve years ago, more than 100 years after the seams had been worked out.

In view of this kind of occurrence and evidence presented in other papers before this conference, will the National Coal Board be thinking of reviewing the use of its customary letter in which the phrase "by now all movement can be expected to have ceased" appears?

MR.W.S. ALLEN - National Coal Board, London

The Board's mining reports normally refer to the cessation of movement in respect of active longwall mining situations or where past mining took place at such a depth and in such a manner that it is reasonable to accept that such old workings no longer present a threat to the stability of the surface. The reports do not, however, give a categorical assurance of stability, but comment on the basis of NCB experience.

Where applicable, specific attention is drawn in a report to any shallow workings or old shafts, whether recorded or suspected, and in such cases it is frequently recommended that some form of site investigation would be prudent.

There would therefore appear to be no case for changing the format of the reports for the reason suggested.

PROFESSOR J.D. GEDDES

Does the Chairman have any information on whether the National Coal Board is investigating, or is planning to investigate, the frequency of pillar collapse, and the conditions under which it occurs, in the United Kingdom? The contribution by the Board to our understanding of the movements associated with current or immediate past mining has been extensive and a great service would be rendered if an analysis could be undertaken of incidents which they have presumably had reported to them throughout the country. A study along the lines reported by Mr. Gray and his colleagues would be of value to those concerned with planning and engineering.

MR.W.S. ALLEN

In Great Britain it is my understanding that when it has proved possible to expose old, shallow, room and pillar workings, there has been no evidence of the collapse of pillars, whereas there has, at the same time, been clear evidence of the collapse, into the available rooms, of the strata originally spanning between pillars left.

Much research has taken place, and the results published, as regards the stability of coal pillars and it is generally considered that the relative dimensions of pillars left in old, shallow, pillar and stall workings are much greater than the minimum necessary to prevent pillar failure.

There is little doubt that almost all reported incidents relate to localised failure i.e. room collapse, and as there are so many indeterminate factors in the proposition of an occasional pillar failure, a detailed investigation does not seem to be justified. I am unaware of any plans for an investigation of this nature.

MR.J.A. STUART - Veryard and Partners, Consulting Engineers

I wish to put a question concerning abandoned shallow workings. It appears to be generally assumed that once a seam is at a depth below rockhead such that void migration would not reach rockhead, the surface can be considered to be stable (Piggott and Eynon, 1977).

The papers presented at this conference amplify previously reported instability attributable to pillar failure in worked seams remaining open at depths greater than the conventionally accepted criteria for shallow workings.

In these circumstances is it possible, for example, to say that beyond a certain depth all collapse that is ever likely to give rise to surface movements will have occurred, broadly speaking, contemporaneously with extraction or closely related to it in time? If it is possible to state a value, what is the depth?

REFERENCE

R.J. Piggott and P. Eynon - Ground movements arising from the presence of shallow, abandoned mine workings, Proceedings of the Conference on Large Ground Movements and Structures, University of Wales Institute of Science and Technology, Cardiff, Pentech Press (1977) 749-780.

MR.P. CARTER

I agree that it is generally assumed that once a seam is at a depth below rockhead such that void migration would not reach rockhead, then the surface can be considered stable. This depth is often expressed as a rock cover of ten times the seam thickness and this assumption is satisfactory for nine cases out of ten.

This assumption is not applicable in those few cases where pillar failure, rather than roof failure, is the mechanism of collapse. In such instances, collapses in workings at significantly greater depths may cause subsidence of the ground surface. We consider the Bathgate subsidences may well provide such an instance.

I accept that there is a strong desire on the part of structural and civil engineers to have a similar sort of depth limitation for surface effects due to pillar failure, such as the sometimes quoted figure of 60 metres. I believe however, that such a limit owes more to this desire than to an understanding of old mining methods and the mechanics of collapse.

In my opinion, it is not possible to say that beyond a specific depth all collapse that is ever likely to give rise to surface movement will have occurred, broadly speaking, contemporaneously with extraction, or closely related to it in time. The old miners had no intention of allowing such collapse to occur whilst they were operating the mine and would proportion the pillars to achieve stability, often leaving pillars of width greater than one-tenth of the depth of the workings from ground level to ensure this. Given well-dimensioned pillars and a sufficiently strong roof and floor, old workings may still remain open today at depths of a few hundred metres below ground level. Robbing of pillars on retreat would of course lead to a much more precarious situation if carried out, although even in this case it cannot be necessarily assumed that all movement has now ceased.

I do not believe that we should put all our faith in any such depth rules for subsidence due to old stoop and room (pillar and stall) mine workings. Each case must be judged on its own merits after careful examination.

MR.R.E. GRAY

The following issues have been raised by papers presented to this conference and in discussion:

(1) Should one always assume that the coal has been mined to the outcrop?
(2) Is there a limiting depth of old workings at which it can be assumed that subsidence is complete?
(3) How does one decide on the amount and type of stabilization or the need for special designs?

Our responses to these issues are based on experience with inactive mines in the Appalachian bituminous coal fields, particularly the Pittsburgh Coal where the cover varies from zero at the outcrop to about 1500 ft (460 m). The first mines were developed in seams that outcropped on the hillsides.

The answer to the first question is, yes. Since mines were developed along outcrops, coal was removed under very shallow cover, as little as 5-10 ft (1.5-3.0 m) in some small mines and as little as 25-50 ft (7.6-15.2 m) in some major mines, even as late as the early 20th Century. Our study of subsidence over the Pittsburgh Coal indicates that most subsidence incidents (95 per cent) are sinkholes. These occur in areas of shallow cover where the roof collapses into mine rooms and entries.

The answer to the second question is also yes, in theory. Where the size, distribution and character of the coal pillars and the thickness of the claystone mine floor are known, a limiting cover based on pillar crushing or bearing capacity can be estimated. However, variable pillar size and strength of the mine floor, plus local defects in the rock mass, require conservative use of such calculations. Although nearly 81 per cent of subsidence incidents over the Pittsburgh Coal occurred where the overburden is less than 100 feet (30 m) thick, the maximum reported overburden thickness at a site of damaging subsidence is 450 feet (137 m). Until the late 1960's, the maximum overburden thickness reported for subsidence over an abandoned mine was only 110 feet (33.5 m) (Gray and Meyers, 1970) and some engineers took this as a limiting depth. Figure 5 of our paper summarizes the frequency of occurrence of subsidence in relation to overburden thickness. This sharp increase in the interval at which subsidence damage has occurred over abandoned mines reduces confidence in using a limiting cover concept in evaluating subsidence potential.

Figure 8 of our paper, showing the distribution of periods of mining at sites that have experienced subsidence, reveals that subsidence incidents primarily occur at sites that were undermined before 1940. Prior to 1920 when the block system of mining was introduced and in some mines up to about 1952, long mine rooms were commonly developed with narrow pillars between. Stress concentrations produce spalling, which reduces the pillar's ability to support the overburden and eventually the pillar crushes. The probable strength of a coal pillar has received considerable attention by a number of investigators (Salamon and Munro, 1967; Holland, 1964; Wilson and Ashwin, 1972; Greenwald et al, 1941; Beniawski and Van Heerden, 1975; Hustrulid, 1976). In comparing applied loads on an individual pillar to the computed pillar strength, a factor of safety of 1.6 to 2.0 is usually used for design of long-term support systems. Geotechnical studies have shown that coal pillars could be designed for an allowable bearing pressure of the mine floor in much the same manner as a footing, applying to the calculations a factor of safety of 2 to 3. As far as is known, this concept has only rarely been employed in mine design (Speck, 1979).

In relation to question (3), determining what precautions are required for structures located above abandoned mines is difficult. Decisions are based on the estimated risk of subsidence, probable structural damage, and impact on future use of the facility. In the area underlain by the Pittsburgh Coal, which may be the most extensively mined coal seam in a large urban area of the United States, the number of subsidence incidents documented each year over abandoned mines is relatively small. Documented yearly subsidence incidents in the Greater Pittsburgh area between 1971 and 1975 were one per 10,500 housing units (Bruhn et al., 1978). Unless the risk is judged to be high, little is usually done to stabilize undermined sites or provide special design for residential construction. Schools and hospitals are usually built where it is judged that subsidence will not be a problem or mine stabilization or special structural designs are utilized. Stabilization is often desirable for a specific structure but may not be cost effective for large areal residential developments. Gray et al. (1974)

have summarized coal mine subsidence control procedures in the United States. For single-family dwellings in all but areas of high subsidence risk (areas of very shallow cover) a subsidence insurance program appears more cost effective than extensive stabilization. Subsidence resistant designs have not been used extensively. Occasionally, structures have been supported on raft foundations. Homes, schools, and light industrial buildings have been constructed on continuous footings reinforced at both top and bottom so as to take negative and positive bending over spans of 25 to 30 feet (7.6 to 9.1 m) (Gray and Meyers, 1970).

REFERENCES

Z.T. Beniawski and W.L. Van Heerden - The significance of in-situ tests on large rock specimens, International Journal of Rock Mechanics and Mining Sciences 12 No.4 (1975).

R.W. Bruhn, M.O. Magnuson and R.E. Gray - Subsidence over the mined-out Pittsburgh Coal, American Society of Civil Engineers Spring Convention, Pittsburgh (1978) (ASCE Preprint 3293).

R.E. Gray and J.F. Meyers - Mine subsidence and support methods in Pittsburgh area, Journal of the Soil Mechanics and Foundations Division, ASCE 96, SM4 (1970) 1267-1287.

R.E. Gray et al. - State-of-the-Art on Subsidence Control, Report ARC 73-11-2550 prepared for Appalachian Regional Commission and Pennsylvania Department of Environmental Resources by General Analytics, Inc. (predecessor company to GAI Consultants, Inc.) Monroeville, Pennsylvania (1974). (NTIS No. PB242465/A5).

H.P. Greenwald et al. - Experiments on Strength of Small Pillars of Coal in the Pittsburgh Bed, US Bureau of Mines, Report of Investigations 3575 (1941).

C.T. Holland - The strength of coal in mine pillars, Proceedings 6th Symposium of Rock Mechanics, Rolla, Missouri (1964).

W.A. Hustrulid - A view of coal pillar strength formulas, Rock Mechanics 8 (1976).

M.D.G. Salamon and A. Munro - A study of the strength of coal pillars, Journal of the South African Institute of Mining and Metallurgy 68 (1967).

R.C. Speck - A Comparative Evaluation of Geologic Factors Influencing Floor Stability in Two Illinois Coal Mines, Unpublished PhD dissertation, Department of Mining, Petroleum and Geological Engineering, University of Missouri, Rolla, (1979).

A.H. Wilson and D.P. Ashwin - Research into the determination of pillar sizes, Mining Engineer 141 (1972).

MR.P.N. SPURRIER - South Yorkshire County Council

I would like to support wholeheartedly the three inferences drawn by Bruhn, Magnuson and Gray which are emphasized by italics in their paper. The principles quoted are equally applicable to shallow workings in the United Kingdom and should be remembered when applying the standard "rules of thumb" which are currently becoming more popular, and which can be very dangerous, particularly when applied by a layman or an engineer not experienced in the mining field.

I give one example. To estimate the height above old workings to which voids will migrate before becoming choked, a multiplier can be applied to the thickness of extraction. If the thickness of extraction is close to the seam thickness (which is usually known) a reasonable approximation may be achieved. No account however is taken of the height of the roadways, which may be excavated in strata partly above or below the seam. Even less account is taken of the lengths of roadways or headings in which roof falls have occurred during the working of the mine. Such falls may have been cleared to enable the day to day operations of the mine to continue, and this results in open cavities above the workings which are in height many times the thickness of the mineral extracted. Rules of thumb cannot take such occurrences into account in a realistic way.

MR.W.J.P. LEIGH - National Coal Board, Opencast Executive
Mansfield

Have any of the authors got views on the effect of the dewatering of workings on the subsidence experienced at ground surface level?

MR.R.E. GRAY

Complete details surrounding the incident reported in our paper are not available, but it is speculated that draining the mine may have withdrawn the buoyant support furnished the overburden by the water, causing the weight of the overburden to be transferred back onto the pillars. If the pillars, floor, or roof had weakened since mining, the increased load may have been sufficient to cause failure. Seepage pressures accompanying drainage may also have contributed to failure.

SESSION III

Chairman : Mr.J.A. Derrington, BSc(Eng),DIC,FEng,FIStructE, FICE,FCIArb
President, The Institution of Structural Engineers

MR.B.S. FRANKHAM - New South Wales Department of Mineral Resources, Australia

With reference to the paper by Mr. Davies, what did the mining slip layer consist of?

MR.B.L. DAVIES - South Yorkshire County Council

The slip layer adopted was 75 mm of sand laid on blinding concrete, covered with polythene sheeting and a further blinding concrete layer. The concrete base slab was then reinforced to cater for tensile forces due to the horizontal ground strains, a slip factor of 0.67 being applied to the self-weight. Further details of design principles are given in the ICE and NCB handbooks.

REFERENCES

Institution of Civil Engineers - Ground Subsidence, The Institution, London (1977) pp iv + 99.
National Coal Board - Subsidence Engineers' Handbook, NCB Mining Department, London (1975) pp 111.

MR.L. SLOWIKOWSKI - Victor Slowikowski, James Blackshaw and Partners, Cardiff

Will Mr. Davies please indicate who was responsible for providing information on the likely mining movements shown in Figures 1 and 2 of his paper?

MR.B.L. DAVIES

The procedure adopted in South Yorkshire is for the County Mining Engineer to prepare plans and sections showing potential movements likely to be experienced by a given structure. This information is then made use of by the structural designer. However, the problem sometimes arises that a knowledge of the type of building is not always available in advance and interpretation of these movements and their effects is thus not always easy.

MR.J.E. WHITBREAD - Chief Civil Engineer's Department
British Railways Board

Figures 1 and 2 in Mr. Davies' paper are said to indicate the strain pattern experienced by the ground under the building discussed in the paper. However, for the designer of reinforced concrete ground slabs, a knowledge of strain direction is also essential if an efficient structure is to be built. The figures presented in isolation can also be misleading since there is the implication that between them they represent the worst conditions that will be caused by mining. In my experience, the worst conditions of strain are frequently encountered during the working of a panel, not after its completion.

To illustrate the influence of direction on strain and to show the variations of strain as the workings progress, I have used the model described in Burton (1980) to calculate the ground strains along the edges and the diagonals of the middle bay of the building shown in the Figures.

As stated in Burton (1980), it is possible to vary the parameters to allow for local knowledge. The parameters that I have used are such that the ground movements, where it is possible to compare them, are similar to those predicted using the National Coal Board's Handbook (NCB, 1975).

Figure D2 shows the dimensions of the bay chosen and its position in relation to the workings. The results are shown in Figure D3. As can be seen, both tension and compression are generated in the ground at all times; the sign and magnitude depend on direction. Also in three of the four directions examined, the worst ground strains are experienced soon after the face of X22 panel has passed the building. It is also interesting to note that the strains shown on Mr. Davies' Figures are reasonably similar to those predicted by this model when considering a direction perpendicular to the lines of equal strain, but they are considerably different from strains in other directions.

In conclusion, I would emphasise that the effects of mining can only be adequately allowed for by a system which can determine strains in many directions and also the variations in these as the workings progress.

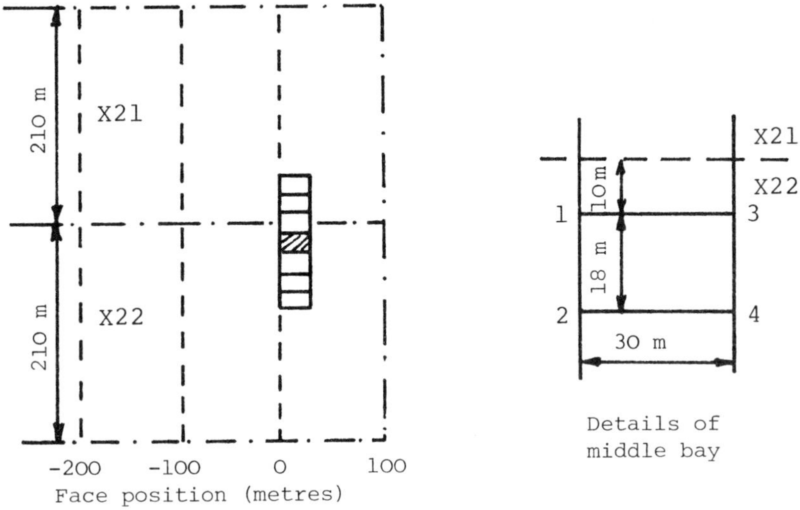

Figure D2

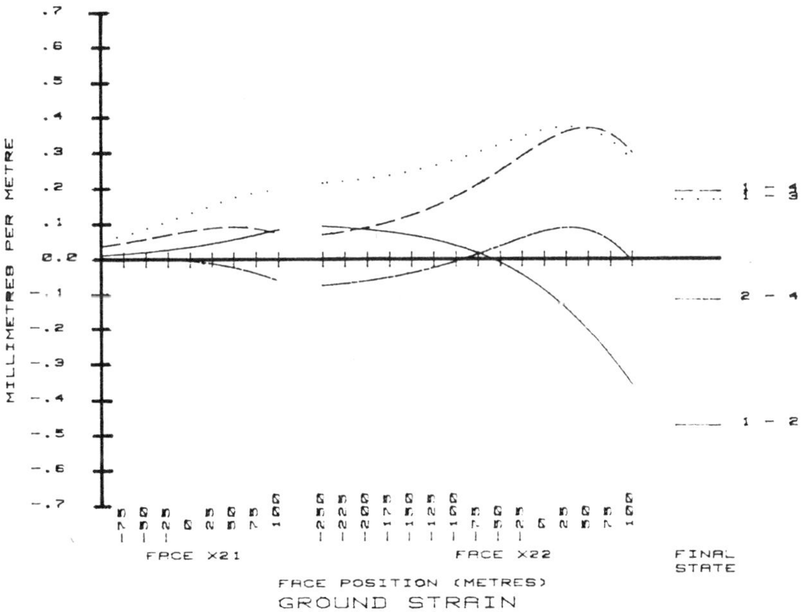

Figure D3

REFERENCES

D. Burton - The introduction of mathematical models for the purpose of predicting surface movements due to mining, Proceedings of the 2nd Conference on Ground Movements and Structures, UWIST, Cardiff, Pentech Press (1981) 50-64.

National Coal Board - Subsidence Engineers' Handbook, NCB Mining Department, London (1975) pp. 111.

MR.D. BURTON - British Railways Board, Derbyshire

I support the comments made by Mr. Whitbread. Engineers should generally bear in mind, with reference to Figures 1 and 2 of Mr. Davies' paper, that horizontal strains affecting structures are governed by the direction in which the observer views the problem, in addition to other factors. Before examining the behaviour of the structure itself it is essential to know the pattern of ground movement in the direction of the surface line or structure that one is interested in.

MR.P.N. SPURRIER - South Yorkshire County Council

In response to the contributions by Mr. Whitbread and Mr. Burton, it is recognised that the ground conditions are a function of orientation and time. The Figures 1 and 2 referred to, show the pattern of the final strain contours. It is agreed that the strain can assume higher values at certain stages before face X22 reaches the final position.

To deal with the intermediate stages of subsidence build-up and the resulting strain variations, it is important that a mining engineer advises the structural engineer and prospective tenderers at all stages of design or negotiation. This is the procedure adopted in our County.

MR.H.K. NG - Public Works Department, Hong Kong

In relation to the use of partial safety factors, would Mr. Davies please clarify whether in the limit state of design, with load factors of 1.7 and 1.4 used, the yield stress of steel is applied without any partial safety factor on material properties?

MR.B.L. DAVIES

In the criterion for Limit State Collapse, current practice has been followed, in that the partial safety factor on material properties has been set at unity. Should the draft recommendations be adopted, it is noted that the partial factor to take account of variability of loads and structural performance are generally lower than the values quoted, but with a partial factor for materials of 1/0.93 .

PROFESSOR J.D. GEDDES - Department of Civil Engineering and Building Technology, UWIST

In his paper Mr. Davies has not referred to the possible effects of ground movement on any cladding or brickwork associated with the steel frame. Would he care to comment?

MR.B.L. DAVIES

The brickwork has not been shown since the paper deals essentially with the possibly damaging effects of horizontal strains on the steel frame. Brickwork and cladding are designed either with movement joints incorporated or with measures to localise damage in such a manner that it does not cause unserviceability of the building and can be easily repairable.

MR.D. BECKETT - Sir Frederick Snow and Partners

Mr. Davies has presented some useful guidelines for the design of steel-framed structures in areas subjected to ground movements. He states that the elastic approach is considered to be very conservative for portal frame design, but this may not necessarily be the case if a non-prismatic design by profile is adopted. An analogy can be made with reinforced concrete design by comparing the reinforcement areas for a yield line analysis with a constant bar spacing and an elastic analysis with variable bar spacing. Fabrication costs are obviously an important consideration, but a tapered frame which in cases of large anticipated movement would be designed as three-pinned, may be a suitable alternative.

MR.B.L. DAVIES

Mr. Beckett is, of course, correct in stating that if one was constrained to adopt an elastic design approach, a satisfactory structure could be obtained by means of articulation and/or reduction of section in areas of low moment. As mentioned in the introductory paragraph, the purpose of developing the criteria was to help in the assessment of submissions made by steelwork contractors which in all cases of continuous portals have been based on prismatic sections and plastic theory.

MR.L. SLOWIKOWSKI

Will Professor Geddes give some indication of the number of individual calculations required in a real structural calculation, when considering the build-up of loading using his suggested procedure?

PROFESSOR J.D. GEDDES

At the present time it is not possible to give a direct

answer to the question posed. In the illustrative example used in my paper, the structure was divided into 30 elements and 30 positions of the displacement curve were adopted. It will also be noticed that half the elements are narrower than the others. This device was introduced to allow cracking to be localised for a study of post-cracking behaviour, which is not reported in the paper.

Because of the rapid decrease in load when slip reversal takes place, shown by the steepness of curves such as those of Figures 4 and 6 of the paper, it would appear to be necessary to take a relatively large number of displacement positions into account. Further work is being done on a range of structures and displacement patterns and it is hoped that this will lead to practical guidelines on how best to perform economic calculations.

MR.D. BURTON

Prior to attending the Conference, I used a computer to draw a vertical section above a 200 metre panel, at a depth of 600 metres, the computer drawing hexagons at 10 metre intervals, the sizes of which were proportional to the space absorbed by a 1 metre block as calculated by the machine. The calculation had to satisfy the known vertical and the known lateral displacements at various horizons between the mine working and the surface. Would Dr. Cox please indicate if the resulting pattern has relevance to the results of his model?

DR.D.W. COX - Department of Civil Engineering
Polytechnic of Central London

The graph presented at the end of my paper shows the distribution of bulking around the mine opening at various levels within the stratum.

PROFESSOR J.D. GEDDES

In my view, results of tests of the kind described by Dr. Cox when modelling brick structures, must be treated with caution even in the context of an illustrative rather than a quantitative tool. The lack of mortar between the bricks militates against a correct representation of what might happen in an actual situation. In recent work at UWIST, we have found, as others have found before us, that not only the presence of mortar but also its strength and creep properties have profound effects on crack patterns.

MR.B. LONGFOOT - New South Wales Institute of Technology,
Australia

I endorse the comment made by Professor Geddes. Actual brickwork, unlike that modelled on the friction table, is bonded together by mortar, and today it is standard practice to use cement-rich mortars. When these are matched with a suitable brick, excellent bond can be expected and tensile failure will generally be through alternate perpendicular joints and bricks (see Plate V of the paper) and not by bond failure on bed joints which would produce the stepped or toothed crack lines of the friction table models.

I would be hesitant, therefore, about applying the results of the model table tests to actual brick structures, particularly those of brickwork laid with significant quantities of cement in the mortar.

MR.J.E. GUMBEL - Mott, Hay and Anderson, Consulting Engineers

I have a fundamental question to put to Dr. Cox about the use of the friction table, with particular reference to his modelling of tunnel crown collapse and arching phenomena in general. In view of the demonstrated importance of lateral stresses in such cases, would Dr. Cox care to comment on the influence of the frictional resistance to horizontal movement of the blocks? Indeed, has he checked that the friction table is truly representative of gravitational loading, for example by comparison with centrifugal loading of his models?

DR.D.W. COX

In reply to Professor Geddes and Mr. Longfoot, I agree that mortar strength is an important factor in crack propagation in brick walls. There are other equally important factors which need to be considered. Firstly, clay bricks tend to expand after construction so that they become laterally prestressed in compression. On the other hand, silica bricks tend to shrink after construction so that they become prestressed in tension.

In response to Mr. Gumbel, the friction table is specifically built very long so that end effects are eliminated. I doubt whether centrifugal models can adequately reproduce the "stick-slip" type of movements of the blocks as on the friction table.

MR.B.L. DAVIES

I find the paper by Professor Klezhev and his colleagues particularly interesting in its description of the monitoring of reinforced concrete piled foundations subjected to mining movements. In the past when faced with the problem of designing bridge structures with piled foundations in a mining area, we have used driven steel piles, reliance being placed on their flexible and ductile behaviour to accommodate the ground movements. No attempt was made to monitor their behaviour. More recently, as an alternative, the vibroflotation technique has been adopted to improve the ground bearing capacity. This system makes use of a 0.5 m diameter vibrating poker to inject and compact vertical columns of 25 mm aggregate into the superficial strata. This technique can be far cheaper than piling, even when a mining slip layer is used under the foundation slab.

SESSION IV

Chairman : Professor R. Ciesielski, DrScEng
Technical University of Krakow, Poland

PROFESSOR P.B. ATTEWELL - Department of Geological Sciences
University of Durham

I wish to raise one or two questions concerning the paper by Professor Sagaseta and his colleagues. First, the values of Poisson's ratio that were used seem to be lower than one would measure in practice. In Figure 9 there is a better curve fit to actual settlement data points when the ratio is 0.2 than when it is 0.3. If higher ratios were to be used, then the data points would be well off line. How were the values of Poisson's ratio chosen? Second, how was the elastic deformation modulus chosen? It is difficult to make a reliable determination of the modulus in the laboratory. Third, do the authors consider that Equation (6) would have more general application in tunnelling, and if so what range of ψ-values would be taken? Finally, it should be noted that in Figure 10 the category marked 'ROCKS' should really be 'ROCKS, HARD CLAYS, AND <u>SANDS ABOVE THE WATER TABLE</u>'. One would have expected the authors' points to fall in that zone rather than in the shaded area on the graph.

In relation to the paper by Mr. Howland, what is the significance of the line Gl in his Figure 3?

PROFESSOR C. SAGASETA - University of Santander, Spain

Professor Attewell's first point concerns the values of Poisson's ratio used in our analysis. A parametric analysis using the finite element method was performed with ν = 0.1, 0.2, 0.3 and 0.4. These values cover practically all the possible situations in real soils and rocks, insofar as long-term settlements are concerned (Figures 1, 2 and 4). In the second part of our paper, the theoretical results are compared with actual measurements made in some tunnels of the Madrid subway. It was felt to be beyond the scope of the paper to give a complete description of the soils involved, which has been given elsewhere (Oteo and Moya, 1979; Escario et al., 1981). Thus only a brief outline is given, classifying them as 'hard marly clay' and 'sandy clay' layers. Referring to the specific point of the value of Poisson's ratio, no direct determinations have been made, but a value between 0.2 and 0.3 seems reasonable as an average value, in the context of effective stresses (see Table DI).

On the second question, we agree with Professor Attewell that the choice of a relevant value for this parameter is a difficult task, not only in relation to laboratory results but also from in-situ tests. Even if the soil mass surrounding a tunnel is homogeneous, the deformation modulus in practice will vary from point to point, due to the differing stress and strain paths (extension over the crown and vertical compression at the sides). In the references cited above, some information is given about the method of selection adopted for the Madrid subway, and a more thorough analysis has been presented by Salas (1979) in his panel contribution at the Montreux Congress on Rock Mechanics. For this particular application, the secant modulus at an axial strain of one percent in consolidated-drained triaxial tests has been found to be suitable in allowing for the complex pattern of zones in which different deformation moduli operate.

Professor Attewell also refers to the validity of our Equation (6), which allows for the overall effects of the construction procedure by means of a single coefficient ψ. Strictly speaking, this equation relates to the maximum settlement at the ground surface (at the tunnel centre-line). Nevertheless, in some test sections of the Madrid and Caracas subways, where comprehensive instrumentation was installed, with borehole extensometers and inclinometers, the equation has proved to be of more general use. Thus if an elastic finite element analysis is performed, on the assumption of an unlined tunnel, and the resulting displacement field is then scaled by the coefficient ψ (the value of which is determined by fitting the maximum settlement, δ_{max}, to the measured value), the resulting pattern of ground movements at the surface and within the soil mass agrees fairly well with the measured values. An exception is provided by points located very close to the tunnel, where local effects of the construction procedure are not adequately reproduced. In this sense, Equation (6), or at least the meaning of the coefficient ψ can be said to be more general than stated in the paper. Regarding application of this equation to other cases, Equation (6) is obviously completely and absolutely general, provided that a suitable value of ψ is taken for each case. Concerning this, we now have more data available than at the time the paper was prepared. The suggested value of 0.4 to 0.5 has proved to be good for machine tunnelling, with concrete segment lining, in hard soils. If the type of soil or the excavation and support methods are different, the appropriate value of ψ will vary. From our experience in the above-cited tunnels, with different construction procedures, the following values are suggested (for hard to medium-stiff soils) :

- With shield tunnelling, if the advance rate is extremely low, or if the excavation is stopped for several days, ψ can rise to 0.8-1.0 .
- In hand excavated tunnels, supported with steel ribs with intermediate lagging, ψ can also be about 0.8-1.0 , unless special measures are taken to reduce the ground loss.

Professor Attewell's final question refers to the position of the point of inflection on the transverse surface settlement trough. In Figure 10 of our paper, the word ROCKS followed by ... was intended to refer to the zone defined by Peck (1969) as correctly pointed out by Professor Attewell. In Figure 10, the Peck boundaries have only been shown to demonstrate that the more recently published data (i.e. since 1969) falls outside his proposed boundaries. If the new points are added to Peck's original figure and his curves are removed from the plot, it will be seen that the points fall in a relatively narrow band, with one exception - the point numbered 11 by Peck relating to the San Paolo tunnel, reported by Terzaghi (1950) (see, for instance Figure 36 of Attewell, (1977)). This narrow band agrees reasonably well with the shaded area given in our Figure 10, which follows from the theoretical calculations by finite elements.

TABLE DI
TYPICAL VALUES OF POISSON'S RATIO ν (after BOWLES, 1977)

Type of soil	ν
Clay, saturated	0.4-0.5
Clay, unsaturated	0.1-0.3
Sandy clay	0.2-0.3
Silt	0.3-0.35
Sand, dense	0.2-0.4
Coarse (void ratio 0.4-0.7)	0.15
Fine (void ratio 0.4-0.7)	0.25
Rock	0.1-0.4 (dependent on rock type)
Loess	0.1-0.3
Ice	0.36
Concrete	0.15

REFERENCES

P.B. Attewell - Ground movements caused by tunnelling in soil, Proceedings of the Conference on Large Ground Movements and Structures, UWIST, Cardiff, Pentech Press (1977) 812-948.

J.E. Bowles - Foundation Analysis and Design, McGraw Hill (1977).
V. Escario, J.M. Garcia Gonzalez, J.F. Moya, C. Oteo and C. Sagaseta - Geotechnical problems during the construction of the Madrid subway extension, Case History Volume, Japanese Society of Soil Mechanics and Foundation Engineering (1981).
C.S. Oteo and J.F. Moya - Estimation of the soil parameters of Madrid in relation to the tunnel construction, Proceedings of the VIIth European Conference on Soil Mechanics and Foundation Engineering, Brighton 3 (1979) 239-247.
R.B. Peck - Deep excavations and tunnelling in soft ground, State of the Art Report, Proceedings of the VIIth International Conference on Soil Mechanics and Foundation Engineering, Mexico (1969) 225-290.
J.A.J. Salas - Deux cas de subsidence en Espagne, Proceedings of the IVth Congress of the International Society for Rock Mechanics, Montreux (1979).

MR.A.F. HOWLAND - Harrison and Company, Norwich

In response to Professor Attewell, Figure 3 is intended to show diagrammatically the development of the 'balance point' at the crown of the tunnel and the way in which, as the tunnel pressure increases, the balance point moves down the sides of the tunnel to its invert.

Lines G1 and G2 represent the generalised distribution of the groundwater pressure adjacent to the tunnel. To simplify the case, and show the general pattern, they have been drawn as straight lines. The differing slopes represents the effect of underdrainage to the tunnels as the tunnel pressure varies and the rate of flow is changed. As the pressure in the tunnel is increased, the downward flow is reduced and G1 moves towards G2, so that essentially if the tunnel pressure were to be increased with the balance point maintained at the invert so that a no-flow condition was developed, line G2 would represent a hydrostatic groundwater distribution.

The diagram illustrates why the balance point develops in the crown at tunnel pressures less than the hydrostatic head of water above the tunnel.

PROFESSOR G.E. BAUER - Department of Civil Engineering
Carleton University, Ottawa, Canada

At first glance Mr. Howland's paper appears to contain a simple method to predict the magnitude and the distribution of settlements above tunnels in soft ground. He finds good agreement between his calculated values and those observed at two different tunnel sites, at Willington Quay and at Stockton on Tees. His method assumes that the tunnel acts as a free drain and surface settlement occurs due to the increase in effective stress caused by the change in the flow pattern. There are, however, several simplified assumptions involved in his paper which are not stated. In addition, there are a few statements which cannot remain unchallenged.

The author assumes that the level of the groundwater table remains unchanged during the steady seepage flow into the tunnel. This is rarely the case in clays or silts and is only true if a free water table is maintained above the soil surface as shown in Figure D4, or if in the case of the water table being below the ground surface, an unlimited supply is available, which sometimes will be the case in coarse sands or gravels. In the case of silty or clayey soils, as dealt with by the author, the water table will be drawn down towards the tunnel as indicated in Figure D5. In several locations of the Frankfurt subway extension in a heavily overconsolidated clay the water table was drawn down 10 m and on several occasions even below the invert of the tunnel (Bauer, 1980). Also Mr. Howland assumes a downward gradient caused by gravity flow as shown by his Figures 4a and 7a. This is rarely the case for clays overlying weathered rock or limestone. In the two case histories cited by the author, Willington Quay and Stockton on Tees, nothing is mentioned of the directions of the hydraulic gradients before construction of the tunnels was started. It is of importance to know if the porewater pressures were monitored before and during construction.

The loss of ground above tunnels in clays with a high water table can generally be attributed to three main factors which are briefly discussed now, in decreasing order of importance. The main factor producing ground subsidence is caused by the driving of the tunnel proper, that is, the squeezing of soil into the tunnel face, by the stress relief which occurs before the tunnel liner is installed and by any subsequent deformation the tunnel might undergo. The second largest component of settlement occurs due to the lowering or depression of the groundwater table as mentioned earlier. The third part of the settlement, and probably the smallest, arises from the increased effective stress caused by the seepage pressure of the water flowing toward the tunnel.

Mr. Howland has assumed that the total settlement is caused by the latter phenomenon. I am not convinced that the flow net assumed in the author's analysis is correct. Also the other two components of settlement, which are considered to be the major contributors, are neglected in the method outlined in the paper and therefore his good agreement between the calculated settlements and the total settlements observed at the two sites seems to be purely accidental.

REFERENCE

G.E. Bauer - Three techniques of subway construction with special reference to the Frankfurt Subway extension, Proceedings of the Canadian Society of Civil Engineers, (1980).

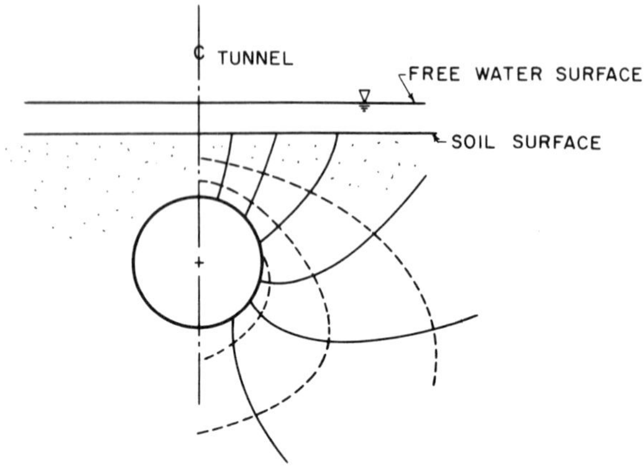

FLOW NET FOR FREE WATER TABLE ABOVE GROUND SURFACE

Figure D4

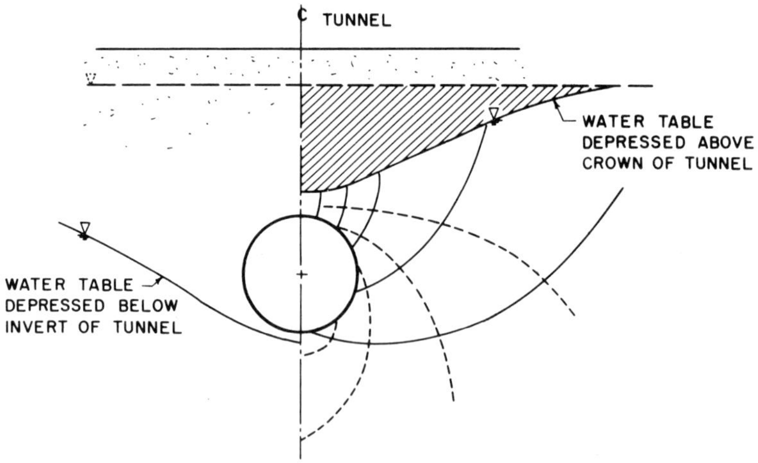

FLOW NET FOR DEPRESSED WATER TABLE

Figure D5

MR.A.F. HOWLAND

Professor Bauer has made a number of points that require comment.

He has remarked, as can be seen from the flow net, that the water table is assumed to be unchanged during the analysis and that this will only occur where continual recharge exists to the upper boundary, so that at other times a depression will develop in the water table. However, for any stable flow net to develop, some form of recharge is required and the depth of any depression will depend on the balance between the recharge and the loss to the tunnel.

It is implicit in the paper that the correct flow net must be used for the analysis. Obviously the greater the understanding of the local hydrology, the more exact can be the construction of the flow net. It should not be forgotten that the purpose of the flow net is to determine the distribution of the pore pressure change. The analysis would therefore still be valid when a groundwater depression exists, since the flow net will adjust accordingly to describe the correct changes in effective stress. In fine-grained soils a reduction in pore pressure will not necessarily mean a lowering of the local groundwater table.

With regard to the pore water pressures before and during construction, these were not monitored at Stockton on Tees, but piezometers were installed just above the crown and at axis level at Willington Quay (Attewell et al., 1978). These suggest a slight increase in pore pressure over the original, to a value approximately equal to the tunnel pressure. As the tunnel pressure was released a drop in pore pressure was noted. This subsequently rose in association with the high pressure grouting and caulking and became fairly stable.

This monitored variation is in accordance with the suggested model. Figure 4b of the paper shows that limited pore pressure reduction would be expected in the crown and axis levels. When the tunnel pressure is released, the groundwater will respond as modelled in Figure 4a, with a marked decrease in pore pressure adjacent to the tunnel. The final increase in pore pressure is due to the production of less-permeable lining by the grouting and caulking. As discussed in the paper, this allows a build-up of pressure in the natural ground due to the reduced seepage into the tunnel.

Neither of the two case histories presented shows an underlying bed of limestone or weathered rock, and so I am unclear as to the relevance of Professor Bauer's comments on this aspect.

Professor Bauer asserts that ground consolidation forms the smallest part of the settlement experienced above a tunnel.

A number of case histories specifically show the importance of the role of consolidation. Henry (1974) shows an increase in settlement from some 25 mm to 90 mm when the compressed air was removed, whilst a doubling of the settlement associated with the removal of compressed air is reported by Glossop and Farmer (1979) for a tunnel driven in soft organic silty clay. As is noted in my paper, the relevance of consolidation has been previously noted by Attewell et al. (1978) for the Willington Quay tunnel, and the importance of the process referred to in principle by Farmer (1977).

Whereas the three factors referred to by Professor Bauer are all no doubt active, his assertion of their relative importance may be too rigid. Around tunnels in soft ground, to which the paper specifically refers, it seems that the role of ground consolidation by general dewatering may be underestimated. The paper does not suggest that this relationship will necessarily be true for tunnels in stiff overconsolidated clay of the type quoted by Professor Bauer, since care should be exercised in arbitrarily extrapolating the behaviour of one material type to that of another.

REFERENCES

P.B. Attewell, I.W. Farmer and N.H. Glossop - Ground deformation caused by tunnelling in a silty alluvial clay, Ground Engineering 11 (1978) 32-41.

I.W. Farmer - Case histories of settlement above tunnels in clay, Proceedings of the Conference on Large Ground Movements and Structures, UWIST, Cardiff, Pentech Press (1977) 357-371.

N.H. Glossop and I.W. Farmer - Settlement associated with removal of compressed air during tunnelling in alluvial clay, Geotechnique 29 (1979) 67-72.

K. Henry - Grangemouth Sewer Tunnel, Tunnels and Tunnelling 6 (1974) 25-29.

MR.I.F. SYMONS - Transport and Road Research Laboratory

In the paper on the prediction of settlement above soft ground tunnels, Mr. Howland appears to have used values of the coefficient of consolidation c_v derived from settlement records. For prediction in advance of construction, use of c_v values from standard laboratory tests may lead to serious underestimates of the rate of settlement in soft alluvial deposits (Lewis et al., 1975). This is probably because the laboratory tests do not adequately reflect the natural drainage structure in the ground, and in such circumstances better predictions are likely to be obtained from the use of consolidation parameters based on in-situ measurements of permeability.

REFERENCE

W.A. Lewis, R.T. Murray and I.F. Symons - Settlement and

stability of embankments constructed on soft alluvial soils, Proceedings of the Institution of Civil Engineers 59 Part 2 (1975).

MR.A.F. HOWLAND

Mr. Symons' point concerning the discrepancies between in-situ and laboratory measurement is very valid for many material parameters, and his comments on the greater reliability of predictions based on in-situ measurements is noted.

PROFESSOR P.B. ATTEWELL

I have three questions to put to Mr. O'Reilly and his colleagues :

(1) Was the presence, or possible presence, of large boulders revealed during the site investigation? If not, could their presence have been inferred? If it had been inferred, do the authors think that the bentonite machine would still have been given its second trial at Acton Grange?

(2) Are the kinks on the Figure 3 traces related to pressure surges inside the bentonite bulkhead?

(3) What relative building structural effects did the settlement produce at the different measurement points?

MR.M.S.M. LEONARD - D'Appolonia and Partners (formerly with Hong Kong Mass Transport Railway Corpn.)

In their paper Mr. O'Reilly and his co-authors refer to the fact that ground treatment had caused settlements. On the Hong Kong Mass Transit Railway construction, extensive chemical grouting was used successfully to limit settlements. At first it was found that significant movements were being recorded as grouting work took place, and it was believed to be due to the injection process. However, careful observation showed that the settlement was produced by the drilling of the grout holes and when the method of drilling was altered, these "installation" settlements no longer occurred. Chemical grouting, using tube-a-manchette techniques allowed building settlements to be effectively controlled during the tunnelling work. Have the authors any comments to add?

MR.M.P. O'REILLY - Transport and Road Research Laboratory

In reply to Mr. Leonard we agree that it is the drilling of the holes for injecting the chemicals which is the source of problems. Soils, depending on their state of compaction, may densify when energy is applied to them and this process can be particularly marked in loose, cohesionless soils such as those encountered at Warrington.

The 100 mm diameter holes for the injection of the grouts there were driven in groups of five to a depth of 9 m at 1.2 m centres along the tunnel line, using a 75 hp air-flushed drill rig which advanced a steel casing with the drill bit (New). Settlements during drilling were 10-20 mm, much of which was recovered - sometimes over-recovered - during the injection process.

One can well imagine, as Mr. Leonard indicates, that changes in the method of drilling could improve matters, but there is also the question of how loose the in-situ soils are. A situation can be visualised in really loose ground where virtually any construction process inputting energy densifies the soil and some settlement is inevitable.

In reply to Professor Attewell's first question, we would agree that with the benefit of hindsight it does seem surprising that the presence of boulders was not detected during the site investigation. However, this is just another instance which serves to emphasise the need for thorough site investigation and indicates the kind of information and hence benefit which might be obtained from some trial pits or large-diameter man-entry boreholes to supplement the data from the small-diameter bore-holes generally used in site investigation.

Had the presence of boulders been known or suspected at the time of selection of the tunnelling method we are not really in a position to know what the outcome would have been, a factor militating against the choice of the bentonite shield process.

No records were kept of the instantaneous pressures in the bentonite slurry in the face of the machine but we would suspect, in answer to Professor Attewell's second question, that the peaks were associated with resumption of tunnelling after the machine had been at a standstill for lining erection.

No measurements of the structural effects caused by the settlements were made by TRRL during their studies at Warrington.

PROFESSOR P.B. ATTEWELL

Would Professor Wilson please explain Figures 3 and 4 of his paper a little more fully, and say in particular why the earth pressures were reduced to zero at the corners of the shaft?

PROFESSOR N.E. WILSON - Department of Civil Engineering and
Engineering Mechanics
McMaster University, Canada

In response to Professor Attewell's request for clarif-ication of the pressure diagram at the corners of the shaft, in the paper it has been suggested that the earth pressures acting on the buried structure (i.e. the pressures causing bending moments) have been reduced to zero at the corners. This is a modification, in the horizontal direction, of Peck's (1969) design envelopes based on semi-empirical rules.

The earth pressures under consideration are those which act on the buried structure, and take into account the flexibility of the structural members and the method of installation. In the case of a long trench excavation, Peck suggests (p. 279, Figure 33) that the earth pressure on the structure at the base of the excavation is essentially zero; this is due to the arching effect within the soil as the base of the excavation is considered to be undisturbed. It is not suggested that the earth pressure within the soil is zero, but rather that the earth pressures acting on the buried structure are zero. I have used this principle of arching in the horizontal direction to diminish the earth pressures at the corners to zero, on the proven assumption that the stability of the corner piles is maintained. From field experience, the earth pressures on the wood lagging adjacent to the corner piles were very low, while the stability of the corner piles was adequately maintained by heavy walers.

PROFESSOR G.E. BAUER

Professor Wilson gives an excellent account of the construction of a 20 ft x 20 ft x 75 ft deep shaft in dense till, with a layer of dense sand near the base of the shaft. The purpose of the shaft was to accommodate a tunnel boring machine. He gives an accurate description of the structural elements of the shaft and the many factors which have to be considered in the construction procedures. From this viewpoint the paper is a valuable addition to the literature; as also the author's section on the "philosophy of design for temporary structures" which gives a fresh and enlightening look at this problem.

However, I have considerable problems in understanding the rationale behind the "lintel-type" loading or earth pressure diagram which has been employed in the design of the structural elements.

The author has derived his loading diagram from that suggested by Peck (1969) for long braced cuts in stiff-fissured clays. Braced cuts generally experience sufficient deformation to bring about active conditions of earth pressure. Professor Wilson states in his paper that "it was found that the horizontal movements of the structure were barely measurable and consequently, the surface ground movements were insignificant". This probably means then, that the lateral pressures acting on the shaft were somewhere between the active and the at-rest conditions. Therefore, it is quite inconceivable how one can assume a further reduction of lateral pressures to 50 percent at the centre of the wales and a reduction to zero pressure at the corner piles. Figure D6(a) shows the lateral pressure distribution in a horizontal plane through a circular shaft of uniform stiffness. Figure D6(b) shows a similar stress distribution for a square shaft which, for the sake of argument, has been assumed to deform uniformly inward.

If, however, the corner piles are fairly rigid and unyielding, as stated by the author and this is reasonable since these piles are "braced" in two perpendicular directions, then a redistribution of lateral pressures will occur. This redistribution will occur, according to Terzaghi (1936), from the yielding centre parts to the less yielding corners of the excavation. Therefore, these corner piles will carry additional loads and should under no circumstances be considered to be pressure free as shown by Professor Wilson's Figure 3. A more realistic lateral earth pressure distribution would be in the form shown in Figure D7.

The trapezoidal earth pressure distribution assuming zero pressures at the base of the shaft, might also not be valid since the till was underlain by dense sand at this elevation.

REFERENCES

R.B. Peck - Deep excavations and tunnelling in soft ground, State-of-the-Art Report, Proceedings of the 7th International Conference on Soil Mechanics and Foundation Engineering, Mexico (1969) 225-288.

K. Terzaghi - A fundamental fallacy in earth pressure computations, Journal of the Boston Society of Civil Engineers, April (1936) 78.

PROFESSOR N.E. WILSON

I thank Dr. Bauer for his discussion on my paper. A critical discussion adds significantly to the value of a paper and Dr. Bauer's experience in earth pressure research is extensive. The design of buried structures is a controversial topic, as it is more empirical than theoretical.

The basic intent of the paper was to describe the field behaviour of the structural steel walers during the short lifetime of a temporary access shaft in dense till. The shaft had been designed and built on an empirical basis by a well-experienced tunnel contractor; after the shaft had been completed and without any evidence of ground movements or major deflections the Consultants to the Owner requested some substantiation for the empirical design. At this time, the shaft had been completed, but the "eyes" for the tunnel boring machine had not been cut.

As this shaft, which was similar to many other shafts in dense till constructed by the same contractor, had established its own satisfactory equilibrium, the main concern was for the

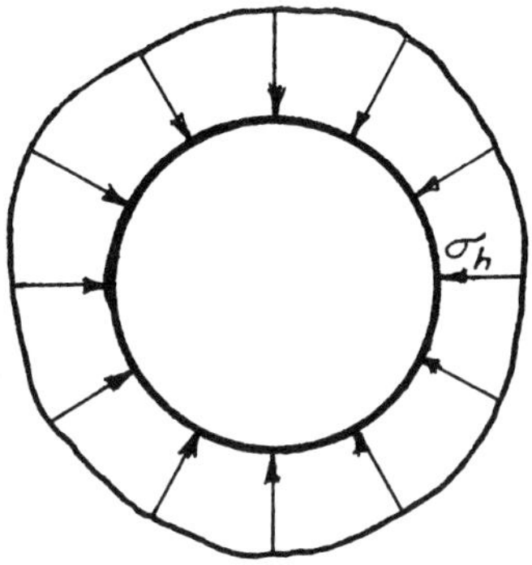

Lateral earth pressures on a vertical shaft (circular)

Figure D6a

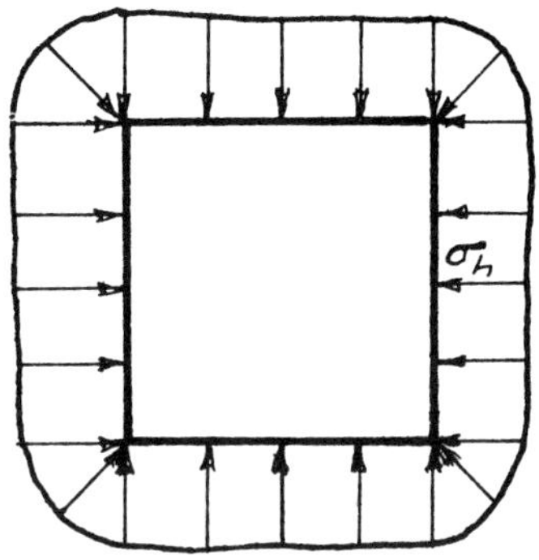

Probable lateral earth pressures on square shaft

Figure D6b

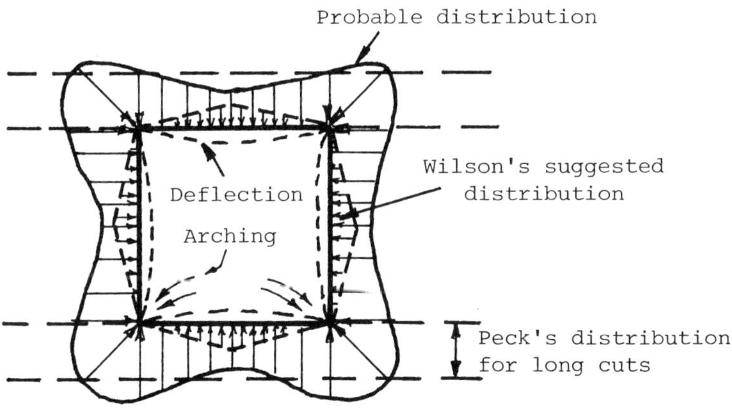

Lateral earth pressure distribution on square shaft with unyielding corner piles

Figure D7

stability of the corner soldier piles as the "eyes" were being cut, and as the soil was being removed from the proximity of the soldier piles. Additional considerations of bottom heave were relevant under these conditions.

An inspection of the shaft clearly indicated that the bending moments on the walers were small as the deflections were extremely small. The direct inference is that the soil pressures, even at great depths, were small and that redistribution of stress had taken place within the soil; the key-issue was that the stability of the corner piles had to be maintained to provide the "arching" action of the soil adjacent to the walers.

The reduction of earth pressure to zero at the base of an excavation (for particular soil types) has been recommended by Peck (1969); this reduction is related to the presence of undisturbed soil at the base of the excavation. Peck's recommendations, which related to a long trench excavation, involved redistribution in the vertical direction. In the case of a narrow shaft, it is suggested that these redistributions take place in the horizontal plane, as well as in the vertical plane.

The major concern is the maintenance of the stability of the corner soldier piles, and this concern has been under-scored in the paper. As further excavation was necessary at the base of this deepshaft for the installation of the tunnel-boring machines, there was a need to provide substantial corner braces at the "eyes" to increase the stiffness of the soldier piles. Thus, there is the direct implication that the soil pressures at the corners are the "at-rest" coefficient, K_o; the minor deflections of the walers and the necessary over-break during the installation of the wood lagging allows redistribution, without large associated bending moments.

Dr. Bauer's comments regarding the soil pressures on the corner piles is a vital hypothesis; the soil pressures causing structural deformations (i.e. bending moments) of the walers were far less than those suggested by Dr. Bauer, for this particular excavation. This case history and this discussion add to our knowledge of temporary structures, which is based on empirical designs.

MR.M.P. O'REILLY

In their paper Mr. Hanafy and Professor Emery consider the example of a 2.67 m o.d. tunnel at a depth of 93 m in a siltstone formation and apply their finite element analysis method to this. The lining analysed is a precast lining 170 mm thick, backed with 19 mm of Ethafoam. Using this support system the maximum computed stress in the lining is 327 kPa. Since a reasonable working stress for concrete is about 20-40 times this amount, it would have been of interest to know what stresses would have been imposed on the lining had the Ethafoam layer been omitted and replaced by conventional back-grouting.

And if, as I suspect, the stresses would still be within acceptable limits then I would have thought it preferable to stick with a simple conventional precast tunnel lining which is back-grouted after erection, rather than suggest the introduction of embellishments such as plastic foam layers at this early stage of introducing precast tunnel linings into North America.

SESSION V

Chairman : Professor F.W. Williams, MA,PhD,CEng,MIStructE,MICE
Department of Civil Engineering and Building Technology, UWIST

MR.J.C. FINNIGAN - Welsh Office, Cardiff

I have a question for Dr. Lovenbury concerning rock anchors. What is the life-span of his anchors in the limestone rock and of those in the shale strata? In the event of the failure of even a single anchor after a given period of time additional stress, or rather tension, would automatically be transferred to adjacent rock anchors. The additional load to be carried could then conceivably result in premature failure of these adjacent anchors and in that event a progressive and rapid failure of the entire group of anchors in any particular section could presumably follow, with catastrophic consequences. What I am of course referring to is failure by metal corrosion - even of a single anchor. I would also welcome some indication of the cost of these works.

MR.M.S.M. LEONARD - D'Appolonia SA, Brussels, Belgium

In the situation described by Dr. Lovenbury, I would be unhappy to use water flush drilling because it might cause an increase in pore pressures and hence a decrease in stability. Can further information be given on this point?

DR.H.T. LOVENBURY - Rofe, Kennard and Lapworth
Consulting Engineers

In reply to Mr. Finnigan, the corrosion aspect is particularly pertinent, since the 3-year design life of the temporary anchors has already been reached. It had been hoped that the slip would now be permanently secured by the formation of a berm across the valley at the downstream toe of the new dam. However, difficulties encountered during the driving of the tunnel have caused further delay. As a result, the berm is only partially complete and reliance is still being placed on the majority of the 100 tonne anchors to maintain stability.

With regard to corrosion, the main concern is at the anchor head where no protection was provided. Although surveys on site suggested that the anchors are generally sound, Ground Anchors Ltd. were asked to make their own assessment. Visual inspection showed that surprisingly little corrosion of the exposed tendons has occurred. Tests on an untensioned length of tendon indicated a reduction in section of up to 5 percent with a similar loss in tensile strength. The extent of corrosion on the stressed tendon between the anchor plate and the rock face is expected to be more severe. This, and the integrity of the overall anchor installation, can only be determined by check lifting. To assess the reserve of safety now available, testing of selected anchors to above working load would be required. However, because of the possible risk of anchor failure occurring and the uncertain effect of such a failure on the installation as a whole, no testing has been performed.

The direct cost of the temporary remedial works at 1967/77 prices totalled approximately £220,000. Of this, the installation of the rock anchors amounted to approximately £75,000, though this did not include the provision or installation of the anchor heads.

In answer to Mr. Leonard, the original attempts to drill the anchor holes were by air flush, but difficulties encountered in advancing the inclined holes through the sandstone/shale interface led to a change to water flush and casing. Full flush return was achieved using the modified technique. Unfortunately, there were no piezometers close enough to detect whether there was any local increase in pore water pressure on the slip surface during drilling. However, since the excavation was backfilled prior to commencement of drilling and the anchors were stressed as the installation proceeded, the use of water circulation probably had little influence on the stability of the slip mass.

MR.J.C. FINNIGAN

In the paper by Dr, Buist and Mr. Naylor concerning ground stability problems and how these affected the choice of route for the proposed Chapel-en-le-Frith by-pass, it would seem that the combination of landslipped material and limited mining subsidence along the northern route was preferred to the more extensive mining subsidence hazard of the southern route. Not a very enviable choice to make. I wonder though if the northern route would still have been preferred if the Mam Tor landslip, or any part of it, had been located on this route.

Although apprehensive about mining subsidence and the very serious damage and expense which can be caused as a result of it, I would be very cautious about building on landslipped ground. Admittedly it would seem that in the Chapel-en-le-Frith case no large-scale landslip, such as the one at Mam Tor, occurs along the route which has also been modified to avoid some smaller-scale landslips. Even so, this route has apparently not avoided all of the known landslipped ground.

In the 'valleys' of the South Wales coalfield we have a classic, if not unique, combination of landslippage and mining subsidence. There is not much choice of route here since most of the valleys, which are fairly steep-sided and themselves present a severe constraint on the siting of roads and other developments, have a history both of intensive coal mining and of landslippage. The choice of route, or of building etcetera, in many of the cases here is more often than not one of - 'landslippage and mining subsidence' or 'mining subsidence and landslippage'. Little or no choice in fact.

Over 500 landslips have been identified and many of the valleys have several of the large-scale Mam Tor type landslip, where sandstone beds at the crest are foundering on shale strata beneath, which in turn are slipping down the valley sides, by both 'rotational' and 'translational' movement. To complicate the situation, there are numerous 'flow' slides in the overlying debris and superficial deposits. Many of the landslides are currently active, but many are classed as 'dormant', simply because there has been no apparent evidence of ground movement over a number of years. However, in many of these cases, the stability factor is very often close to unity so that these 'dormant' landslips could well be potentially the more dangerous, particularly where building developments have encroached over the years on to seemingly stable ground. And, following periods of substantial rainfall - not uncommon in Wales - the resulting increase in groundwater and its consequences, can often be sufficient to reduce the stability factor in some of the cases dangerously close to, or even less than unity.

In addition, there is also the mining subsidence factor to contend with. We are all aware that mining subsidence by itself is a sufficient hazard of some concern, but in South Wales it presents a further problem in that this subsidence can disturb the equilibrium of an otherwise dormant landslide. In Session II we dealt at some length with the problem of mining subsidence and particularly with the consequences of 'pillar' collapse. In the South Wales coalfield, as indeed in other coalfields both here and abroad, most of the older mine workings were by the 'pillar and stall' method. Pillar collapse must obviously result in a reduction in the stability of any landslide in the vicinity of such workings. An added complication in the Valleys is that the Coal Measures succession extends well up the valley sides and mined seams occur above, as well as below, some of the landslipped ground. Collapse of these upper workings, as well as disturbing the ground equilibrium - mainly by causing the overlying sandstone strata to fracture and founder, thus overloading the upper reaches of the landslip with rock scree - permits easy access of groundwater into the upper sector of the landslipped ground, thus reducing the stability even further.

The authors have also referred to the importance of notifying the National Coal Board of any findings, and particularly of old records, of past mining which are discovered in the course of any investigations of this type. The NCB possess a fairly comprehensive data bank of records of mine workings and shafts but many of the older workings and shafts are not recorded . In these cases it is particularly important, I would say, to provide the NCB with any evidence of old workings which are discovered by exploratory site investigations and/or excavations.

DR.S.P. BENTLEY - Department of Civil Engineering and Building Technology, UWIST

I wish to add some comments to those made by Mr. Finnigan and to stress the importance of a full appreciation of all the factors which have led to landsliding in a given area in the past and which may lead to reactivation of movements in apparently stable slide material. I can illustrate this by reference to landsliding in the South Wales valleys, an area in which a departmental research group is currently working.

As Mr. Finnigan has indicated, landsliding is a common phenomenon on valley slopes in South Wales. Sequences of sedimentary rocks form a plunging syncline and the mass movements are concentrated in the Upper Coal Measures of the northern crop. In this area the upland plateau has been dissected by glacial erosion and the rocks outcropping in the steep valley sides are mainly sandstones and mudstones which dip at about 5° SW. Interbedded coal seams and seat-earths act as aquicludes and control groundwater flow.

Glacially derived colluvium and degraded argillaceous strata (Rhondda Beds) are involved in superficial sloughing and slump failures. Such movements are usually associated with the position of aquicludes on the valley sides. Rotational failures, involving 100,000 m^3 of rock material, also occur in the Rhondda Beds and incorporate the overlying Brithdir and Hughes sandstones. There are some striking examples of this type of failure in the upper reaches of the Rhymney and Ebbw Fach valleys. Secondary wedge failures may occur in the rotated rock elements (SO 137047) and at many localities mobile debris aprons have been generated (SO 127046), (SO 207068).

Many inter-related factors are responsible for creating these unstable slope conditions. An appreciation of these causative factors is clearly advantageous for those planners and engineers charged with advising on urban development. Oversteepening by glacial erosion and the subsequent stress relief must have modified the structure of the rock masses in the valley sides. In addition, the periglacial processes of mechanical weathering and solifluction have caused severe degradation of the mudstone strata. Such processes appear to have been more intense on the warmer western-facing slopes. It is in these areas that the larger debris aprons have developed. The incidence of mass movements was probably quite high in the immediate post-glacial period and other high incidences may relate to long-term fluctuations in climatic conditions. However, in recent times the incidence of slope failures has been remarkably high. It may be argued that this is simply a function of improved archival records, but in fact newspaper records show that most of the large rotational failures occurred during the last 100 years. If this is the case it may be useful to consider how the urbanisation of the South Wales valleys and other activities of man have influenced the stability of slopes.

The evolution of industry and townships during the nineteenth century resulted in significant man-made alterations to the physical environment. Natural drainage patterns were altered, additional loadings were imposed and in some instances natural deposits whose presence was vital for stability were removed. The advent of coal mining, both in hillside 'levels' and in large-scale subsurface operations is an additional factor which has caused modifications to stress patterns and significant ground movements in overlying strata. Excavations in hillside seams can lead to alteration in both horizontal and vertical flow paths, collapse and strata disaggregation can also significantly affect the secondary permeability and storage capacity of overlying strata. It is generally accepted that subsidence will cause a lowering and tilting of the ground surface, possible reactivation of faults, and extension/compression movements which may produce graben features in hillsides. In the South Wales situation such ground movements could have a detrimental effect if: (a) aquicludes are fractured and groundwater flow paths become altered, (b) well-defined tension zones promote the backsapping of rock slopes, and (c) hydrostatic pressures which develop locally beneath superficial deposits are influenced by tension zones.

In situations commonly occurring in the valleys, the ground movements can be far greater than those with which the engineer is customarily concerned. Since these ground movements cannot usually be predicted, successful engineering design depends on an appreciation of the nature of the movements involved and their interaction with structures. A long-term ground movement problem is associated with debris aprons which, generated by rotational failures, continue to creep or flow downslope. In some cases the debris apron is essentially a solifluction lobe which begins to move, perhaps along ancient surfaces, in response to strains from discrete rock failures higher up the slope and marked increases in water content. Although clearly mobile, there is no existing way of relating their rate of movement with a measure of stability. It is in such cases that conventional site investigation should be augmented by measurement of surface and subsurface movements. It is only when the seasonal movement patterns are fully understood that a decision can be made either to delay remedial works or to implement measures specifically designed for the problem.

PROFESSOR J.D. GEDDES - Department of Civil Engineering and Building Technology, UWIST

I have two questions to direct to Dr. Buist and his colleague. In their paper they refer to the range of factors, technical, sociological and environmental, which have to be considered when dealing with route location. Mr. Finnigan and Dr. Bentley have further amplified the technical problems which may be involved. However technical problems can often be overcome and the relative costs of alternative solutions can be estimated. The sociological and environmental aspects are much more difficult to quantify, particularly on a money basis. Can the authors give an indication of the weighting of the mining and geological factors in determining the choice in favour of the northern route? What are the estimated costs of the two routes referred to in the paper?

DR.D.S. BUIST - Department of Transport, Midland Road Construction Unit, Derbyshire

The alignment of the published northern route of the by-pass was determined to a great extent by the existing topographical features of the Blackbrook Valley, such as the Peak Forest Canal, the railway, the Buxworth Basin and the Blackbrook. Consequently, the socio-economic factors were of paramount importance in the choice of the alignment. These factors were also significant in the choice between the northern and southern routes. The greater environmental and economic benefits of the published route were the decisive factors in the choice between the alternatives.

Although both routes were affected by geological and mining problems, these had little influence on the overall choice of route, other than that their existence was reflected in the cost estimates. The estimated cost of the northern route was £10.79M and the Department's alternative southern route was £11.81M (May 1978 prices).

MR.J. WHITBREAD - British Rail, Derby

Dr. Duncan's paper is interesting in that it presents, amongst other things, results of vibration measurements on houses in different structural states. Unfortunately, the only common site of measurements reported in the paper is a ground floor window sill. The results show that the maximum effects measured are at about the same peak velocity in both cases, but at different frequencies (and hence amplitudes).

I would like Dr. Duncan to elaborate on these common sets of results and ask if he, or other delegates, know of other cases in which similar input stimuli have been examined as to their effect on buildings in different states of repair?

DR.E.G. PRATER - Swiss Federal Institute of Technology,Zurich

In relation to Dr. Duncan's valuable paper, the following comments on the disturbance to house occupants due to freight trains along a reconstructed section of railway line in the region of Zurich may be of interest to delegates.

In connection with the new rail link to Zurich Airport, the existing above-ground single track local line to Kloten has been replaced by a covered-in four track section with a new 300 m long station at Opfikon. Slurry-trench tied-back walls stabilize the cut and a 1 m thick stabilized gravel layer to protect against oil penetrating the ground underlies the track. The ground is morainic but in places the new section cuts through the marly bedrock.

Local residents were accustomed to the noise caused by passing trains, but with the opening of the new rail link in June 1980, the number of trains per day will increase from about 70 to 180, and thus the new construction was intended partly to reduce disturbance. In fact, however, it has had the opposite effect. The track running nearest to the houses i.e. at a minimum distance of about 6 m, was brought into operation at the end of 1978. Noise levels are indeed supposed to be lower, but now the residents of about ten houses are affected by vibrations, less due to their intensity than to the disquieting rumbling sounds, which the residents are not used to. The vibrational effects seem to be greater for the heavy goods trains. Early morning trains (5 a.m.) give rise to the greatest disturbance, many residents being awakened from their sleep.

After complaints and the threat of court action the Swiss Federal Railways (SBB) conducted a programme of measurements both of noise and vibration. Four measuring stations were chosen and at each of these the vibrations from about 35 trains were recorded on each of four successive days. Measured maximum velocities were generally well below the minimum to cause damage to any structure, which according to the new Swiss Standard (Schweizer Norm SN 640 312, 1979) would be 8 mm/s at frequencies around 60 Hz for the type of building involved.

The Swiss standard only caters for damage, there being no guidelines laid down for human reaction and discomfort. The German Standard DIN 4150/2 (cf. Table I of Dr. Duncan's paper) sets limits for the disturbance to people, which are partly exceeded in this case. These limits are quite stringent, e.g. for a mixed zone at night-time the KB value is 0.2 (~ 0.30 mm/s at f = 60 Hz), being 0.3 (~ 0.42 mm/s) for frequently occurring and 8 (~ 11 mm/s) for occasional vibrations (whereby the factor 0.18 in Equation (1) of Dr. Duncan's paper becomes 0.13 for peak values and non-periodic vibrations), but as Duncan concludes, they do indeed correspond to the discomfort experienced by persons, even if there is no damage to their health.

The vibrations appear to be most troublesome where the foundation material is rock (the stabilized gravel, diaphragm walls and anchorages also helping by acting as good transmission media). The SBB report, however, that the main cause of high vibration is due to the unevenness developed in the wheels of the new Gotthard locomotives type RE 6/6.

Up till now the demands of the residents have not been met, including their wish to have a rubber mat installed under the gravel base supporting the tracks. Reluctance to allow such a precedent case is no doubt dictated by the fear of a snowball effect.

DR.W.J. LARNACH - Department of Civil Engineering
University of Bristol

Dr. Duncan has raised some interesting points in his paper. I have been involved on a number of occasions in litigations in which it has been alleged that vibrations, particularly those induced by piling operations, have caused damage to adjacent buildings. However, at the time the vibrations were occurring, no measurements of their intensity had been made. Thus, at the time of my involvement, I was in the position of having to make estimates of what the vibration intensity might have been: given for example, general ground conditions; pile driver input energy; distance between piling and building. The work of Wiss (1967) and Attewell and Farmer (1973) has been of use in this respect because they have attempted to relate many of the parameters just mentioned with an envelope of peak particle velocity to be expected. But clearly, in my view, what is needed is a much more comprehensive collection of good case records so that general guidance can be refined. I fear that the present situation of non-collection of critical vibration details when vibrations are actually occurring may well continue, with subsequent difficulties in settling arguments in court or elsewhere.

REFERENCES

P.B. Attewell and I.W. Farmer - Attenuation of ground vibrations from pile driving, Ground Engineering 6 (1973).

J.F. Wiss - Damage Effects of Pile Driving Vibrations, Highway Research Record No. 155 (1967).

PROFESSOR J.D. GEDDES

I refer to the papers by Dr. Duncan and Professor

Ciesielski. In neither paper is information given on soil types and their in-situ conditions. Will the authors comment and perhaps give data on this aspect? It seems to me that since the vibrations dealt with are usually transmitted by the foundation soil, this is a fundamental omission. Some information on the foundation types and their depths below ground surface level would also be welcomed.

DR.W. DUNCAN - Department of Civil Engineering
University of Strathclyde

In reply to Mr. Whitbread, there were many other measurements made around and inside the buildings. Those given in the paper are typical examples.

Taking Dr. Larnach's point, there is always the problem, when faced with damaged property, of knowing whether the vibrations caused the damage, or if the vibration characteristics have been altered by the damage. In the particular case cited in the paper, the damaged house was semi-detached with only one half damaged. Readings in the undamaged half were very similar to those in the damaged part and this helped to point to the conclusion that the vibrations had not been a major cause of the damage.

Professor Geddes has raised the question of ground conditions. The interaction of the soil and the structure is important and there is always an uncertainty, in cases of prediction of vibrations, as to whether the transmission will be altered after a structure is built. We recently undertook some vibration monitoring near a railway where it was proposed to build private housing and we are hoping to go back after the houses are built to check if there is any change in the vibration characteristics.

The soil conditions at a particular site obviously have a marked effect on the vibration patterns. For example, the ground between house A and the railway line was the site of an old railway goods yard. There were several platforms with rubble infill to bring the track bed up to a general level. Since the surface cracks are very important these ground conditions will have a major effect on the vibration transmission. This makes it very difficult to extrapolate from one site to another.

PROFESSOR R. CIESIELSKI - Technical University of Krakow, Poland

There are certain rules which should be followed in the measurement of ground vibrations if value is to be obtained from them. Usually, measurements should be made on the ground if comparisons of behaviour are to be attempted. We have only measured the horizontal movements. The energy of vibration is certainly important and this should be recorded. The effect of soil conditions is illustrated in References 6 and 7 at the end of my paper. Soil conditions can change the vibrations by 100 percent, but this is only for surface waves of the Love type. The question of how the vibrations are transferred to the structure is important. Sometimes a coefficient of reduction of friction between the soil and structure is used. In the Japanese Code, this factor varies between 0.4 and unity, which is a wide variation.

I have some questions to put to Dr. Prater, who I congratulate on two good papers. Does he have any information about the vibration of the embankments in the longitudinal direction? This is sometimes important.

Regarding the tank problem, was consideration given to the stresses developed in the tank when it is half full and when waves are generated inside? These conditions may lead to worse results. Finally, for what category of seismicity have the calculations been made?

MR.M.P. O'REILLY - Transport and Road Research Laboratory

The problem of simulating soil behaviour, referred to by Dr. Prater towards the end of his paper on slope stability under earthquake loading, may be overcome by using the centrifuge modelling technique. The Department of Engineering at Cambridge University and, I understand, some U.S. organisations are proposing to use the latter techniques to study earthquake effects on dams.

DR.D. COX - Department of Civil Engineering
Polytechnic of Central London

With reference to the paper by Dr. Prater on slope stability, positive pore pressures could be expected to develop as a result of cyclic loading, over a wide zone within an earth dam. Negative pore pressures due to dilation during shear would only extend over the shear plane, would rapidly dissipate, and shear would then continue. It therefore seems to be difficult to compare the behaviour of a triaxial test sample with actual field behaviour without an assessment of the permeability as proposed.

DR.E.G. PRATER

Mr. O'Reilly has raised the question of the use of centrifuge equipment. Several investigations have been carried out on model dams placed on a shaking table. The drawback of such tests is the inability to model the intergranular pressures, which for gravel materials may be important because of grain crushing effects and the consequent influence on the Mohr strength envelope. The centrifuge would appear to overcome this problem. However, the influence of pore water pressure and its dissipation during the earthquake would be difficult, if not impossible, to model.

The problem of pore pressure development in dams under seismic loading has been discussed by Seed (1979), Prater and Studer (1979) and Bossoney et al. (1981). Pore pressures can be generated even in dense granular soils. The mechanism seems to depend on grain breakage and crushing at the asperities with some small re-arrangement of the structure. Cohesive soils, by way of contrast, develop pore suctions under cyclic loading if they are heavily compacted (or overconsolidated) and "dry" in the sense of critical state terms. If the slope were to fail by sliding along a thin shear surface, the pore pressures generated by the earthquake could be significant as pointed out by Dr. Cox. A complicated generation/dissipation analysis is then generally necessary to evaluate the danger to stability. On the other hand, if no well-defined failure surface is developed, the inherent dilatancy of the material would cause pore pressure reduction as strains are developed over a wide zone, thus inhibiting catastrophic failure if the slope were designed on the basis of the ultimate rather than the peak drained strength.

In reply to Professor Ciesielski, the longitudinal vibration of dams may be important in narrow valleys in which the three-dimensional nature of the problem deserves recognition. With regard to field observations, Stroppini reported, for the magnitude (M_L) 5.7 Oroville earthquake of 1975, a recorded maximum acceleration along the axis of the dam higher than that in the section. The result is, of course, also affected by the direction of approach of the ground waves. Other data is given by Okamoto (1973) (see reference in the speaker's paper). The natural periods parallel and perpendicular to the dam axis may be approximately the same. For several Japanese earth dams, the value is around 0.4 second. Analysis of the natural periods of vibration of a dam, treating it as a three-dimensional body using shear beam analysis or finite elements in more complicated cases can be carried out.

Concerning the sloshing effects and a more critical response when the tank is part full, in the present investigation only the full tank condition was finally considered, preliminary studies having shown that sloshing effects could be neglected. As pointed out by Professor Ciesielski, in some cases a more critical condition may be obtained when the tank is not full. This depends also on the characteristics of the selected seismogram, sloshing frequencies generally being rather small.

The categories of seismicity used have been based on a sine-sweep function. The interesting features of such a function are (i) its short duration i.e. 4 sec, (ii) its smooth response spectrum. The function used needs to be changed somewhat if it is to match the U.S. NRC spectrum more closely (see also Johnson and Epstein, 1976).

REFERENCES

C. Bossoney, E.G. Prater, M. Balissat, J. Studer and N. Zingg - Earthquake resistance of a rockfill dam, Recent Advances in Geotechnical Earthquake Engineering and Soil Dynamics, St. Louis (1981).

G.R. Johnson and H.I. Epstein - Short duration analytic earthquake, Journal of the Structural Division, ASCE, May (1976).

E.G. Prater - Cyclic shear resistance of non-cohesive soils, Journal of the Geotechnical Engineering Division, ASCE, 106 (1980) 111-116.

E.G. Prater and J. Studer - Some considerations on the seismic behaviour of rockfill dams, Proceedings of the XIIIth Congress on Large Dams, New Delhi (1979).

H.B. Seed - Considerations in the earthquake resistant design of earth and rockfill dams, Geotechnique 29 (1979) 215-263.

E.W. Stroppini - The Oroville earthquake and Oroville Dam, ASCE Publication "The Evaluation of Dam Safety" (1976) 301-328.

DR.W.J. LARNACH

On the issue of longitudinal vibrations, Professor Severn of Bristol University and staff of the Building Research Establishment have been collaborating in doing excitation vibration tests on some dams. Results of this work are to be published (Severn, Jeary and Ellis, 1980).

REFERENCE

R.T. Severn, A.P. Jeary and B.R. Ellis - Forced vibration tests and theoretical studies on dams, Proceedings of the Institution of Civil Engineers 69 Part 2 (1980) 605-634.

DR.J. WILSON - Mott, Hay and Anderson, Consulting Engineers

Mr. Joshi and Professor Emery's paper is to be welcomed, particularly because so little has been published on the subject. They do not, however, mention the recent work of Dowding and Rozen who have made a significant contribution in this field.

REFERENCES

C.H. Dowding - Lifeline viability of rock tunnels, empirical correlation and future research needs, Proceedings, The Current State of Knowledge of Lifeline Earthquake Engineering, ASCE (1977) 320-336.

C.H. Dowding and A. Rozen - Damage to rock tunnels from earthquake shaking, Jnl Geotech. Engg. Div. ASCE 104 (1978) 175-191.

A. Rozen - Response of Rock Tunnels to Earthquake Shaking, MSc Thesis, Massachusetts Institute of Technology (1976).

SESSION VI

Chairman : Mr.E.M. O'Leary, BE,CEng,FICE,MIHE,MConsE
Veryard and Partners, Consulting Engineers

MR.J. PRYKE - Pynford Limited

I noted with interest, the reference in Professor Ciesielski's paper to the use of jacked-in Mega piles in protective works. Will Professor Ciesielski give details of their size, safe load-bearing capacity, maximum driving force, the time required to insert them and the number of men employed in the driving team?

PROFESSOR R. CIESIELSKI - Technical University of Krakow, Poland

In response to Mr. Pryke's question, the following details will further illustrate the use of jacked-in Mega piles for protective works in the ancient part of the City of Krakow.

Mega piles were used for some buildings of which the largest was the Czartoryski Museum. Relevant details of the installation adopted are as follows:-

For protection against settlement, 190 Mega piles were used, of 25 and 35 cm diameter. Their lengths, which depended on the precise geotechnical conditions at various positions, ranged from 150 to 900 cm. The prefabricated, reinforced concrete elements forming the piles were 60 cm long. The pile spacing was 75 to 110 cm (Figs.D8 and D9). The in-situ bearing capacity, from field tests, was 30 t for the 25 cm piles and 45 t for those of 35 cm.

A jacking force of 40 to 50 t (400 to 500 kN) was used. The piles were placed under the brick foundations or near the foundations supported on steel grid beams (subsequently concreted). According to length, the installation time for a pile varied from 12 to 30 hours. The crew comprised three persons, one handling the jacking pump and the other two preparing the pile elements and generally assisting.

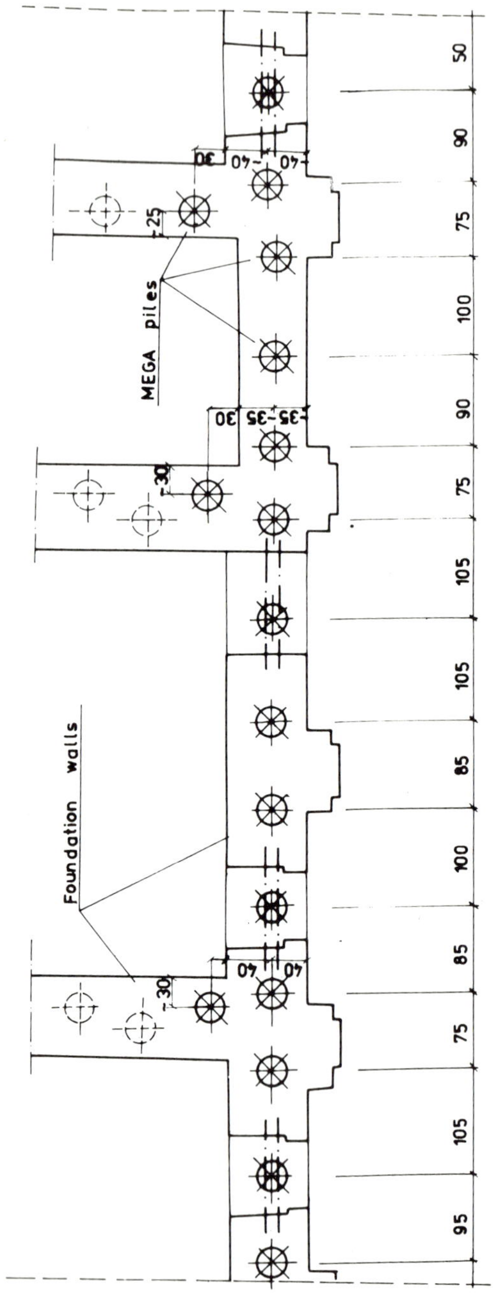

SITUATION OF MEGA-PILES

Figure D8

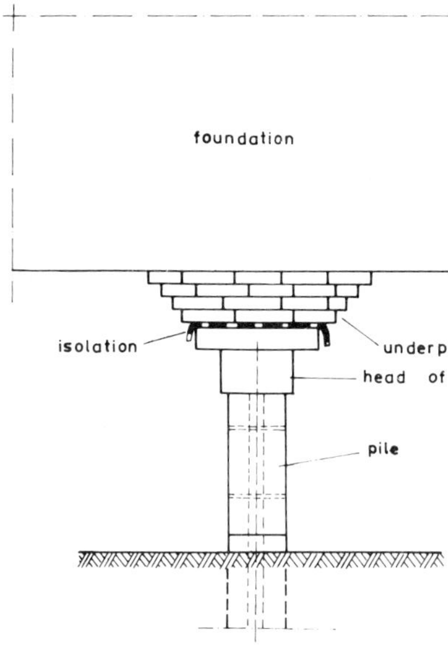

Figure D9a

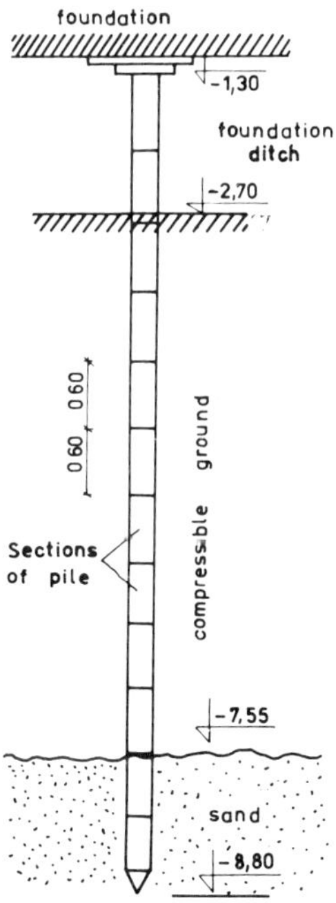

Figure D9b

Costs were a little higher than for other protective methods, but the method described was simple to carry out and the results achieved proved to be very successful.

MR.D.J. GOLDSMITH - Billiton Exploration U.K. Ltd., Cornwall

In view of the fact that the loading imposed on the floor of the warehouse described by Drs. Wood and Larnach is likely to decrease from time to time, are measurements being continued to study possible uplift phenomena?

DR.L.A. WOOD - Department of Civil Engineering
Queen Mary College, London

It is anticipated that the warehouse will remain more-or-less permanently loaded. This fact, together with the nature of the foundation conditions, leads us to believe that there will be little or no uplift. However, measurements are to be continued.

I will take this opportunity to add some further comments on our paper.

The method of analysis used has been developed over a number of years and is incorporated in a computer program RAFTS (Wood, 1978), which provides the engineer with a tool for the economic solution of soil/structure interaction problems associated with foundations. The particular example of the warehouse produced encouraging results as to the predictive capability of the soil model used (Wood, 1977). However, the ground bearing slab of the warehouse is a very flexible structure and its presence does not greatly influence the settlements obtained. Some of the results obtained from a similar analysis carried out on a 22-storey shear wall structure founded on gravel overlying London clay are shown in Fig.D10. In this case, the inclusion of the rigidity of the structure has a dramatic effect in reducing the predicted central settlement from 140 mm to 80 mm (differential settlements from 65 mm to 3 mm). Details of the structure and the analysis are given in Wood (1980). The agreement obtained between the measured and computed settlements is most encouraging and suggests that such a relatively economic and simple analysis would be adequate for design purposes.

REFERENCES

L.A. Wood - The economic analysis of raft foundations, International Journal of Numerical and Analytical Methods in Geomechanics 1 (1977) 397-405.

L.A. Wood - RAFTS -a program for the analysis of soil-structure interaction, Advances in Engineering Software, CML Publns I (1978) 11-17.

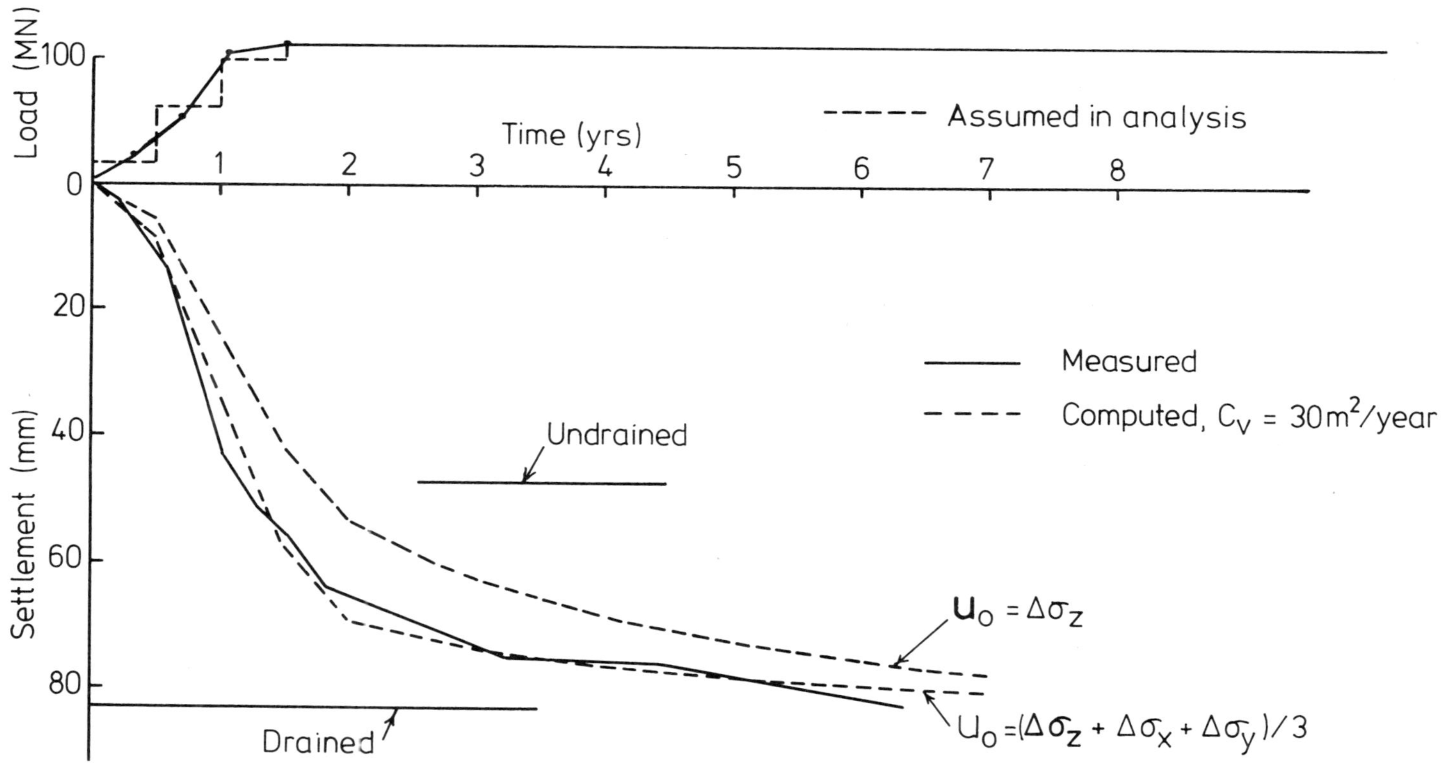

u_o = initial excess pore-water pressure, $\Delta\sigma_z$etc = increase in normal stresses

Figure D10 Settlement at centre of 22 storey residential block (after Wood, 1980)

L.A. Wood - Time dependent settlement of structures, New Developments in Boundary Element Methods, Ed. C.A. Brebbia, CML Publns (1980).

PROFESSOR J.D. GEDDES - Department of Civil Engineering and Building Technology, UWIST

From the description given by Drs. Wood and Larnach, it appears that there is likely to be a considerable element of transient loading provided on the warehouse floor by wheeled vehicles. This is bound to have a cumulative effect on the settlements experienced. Although measurements of such loadings, frequency and positioning would obviously present difficulties and involve a great deal of time, have such measurements been considered or implemented? It would clearly be useful when comparing predicted and actual settlements of the kind generated in the study described to have some idea of the allowance that might, or should, be made for transient loadings.

DR.W.J. LARNACH - Department of Civil Engineering
University of Bristol

To-date, no such measurements have been made. Although the suggestion is a valid one, we have not had the time to devote to such an exercise. However, in the paper, we do recognise the importance of transient loadings, and have attempted to quantify the effects of the delivery lorries.

MR.D. BECKETT - Sir Frederick Snow and Partners
Consulting Engineers

I wish to congratulate Mr. Withers on being courageous enough to deal with matters relating to structural design, taking into account soil/structure interaction. From his presentation, I am not clear how he has dealt with the problem of crack width control, if indeed this is necessary. I have considered this in relation to ground beams spanning between pad or piled supports.

For simplicity, let us consider a two-span continuous ground beam ABC with AB = BC = L, subjected to a nominal uniformly distributed loading q per unit length. An analysis for flexure at the ultimate limit state, assuming plastic behaviour and equal resistance moments in the support and span, will result in a design bending moment:

$$M_B^U = 0.086\ \gamma_f\ q\ L^2$$

where γ_f represents the global partial safety factor for loading. At service load the bending moment at support B will be given by:

$$M_B^S = 0.125\ q\ L^2$$

The crack width estimation is related to M_B^S. The latter can be close to M_B^U and thus may lead to excessive crack widths. For a settlement of δ at support B, the bending moment at B reduces to:

$$M_B = M_B^S - \frac{3\ E\ I\ \delta}{L^2} \qquad (1)$$

where E I represents the flexural rigidity of the reinforced concrete section. If the settlement of δ occurs at an outer support, then the bending moment at B increases to:

$$M_B = M_B^S + \frac{3\ E\ I\ \delta}{2\ L^2} \qquad (2)$$

The value of E I related to the cracked section can be obtained from the equation of bending and may be expressed in the form:

$$E\ I = \frac{P}{100}\ E_s\ (1 - n)(1 - \frac{n}{3})\ b\ d^3 \qquad (3)$$

where $P = 100\ A_s\ /\ b\ d$, A_s is the reinforcement area, E_s is the modulus of elasticity of the reinforcement, n is the neutral axis factor, b is the width and d is the depth of the section.

The neutral axis factor is given by:

$$P = \frac{50}{\alpha_e} \cdot \frac{n^2}{1 - n} \qquad (4)$$

where α_e is the ratio of the elastic moduli for steel and concrete, expressed as $E_s\ /\ E_c$.

Although Equation (1) shows a reduction in moment at B, there will be a corresponding increase in span moment which may result in a critical section for cracking related to a sagging moment.

In Appendix 6 of his paper, Mr. Withers gives allowable values of δ/L ranging from 1/1000 to 1/100 and thus Equations

(1) and (2) can be used to estimate maximum moments and thus crack widths. The influence of creep may be taken into account by adjusting the values of α_e , noting that Equation (3) is not particularly sensitive to large changes in α_e .

DR.W.J. LARNACH

In his paper Mr. Withers makes use of a three-point loading system. I suggest that such a distribution of loading is unrealistic. It would be of value to calibrate the method proposed against a full interactive analysis.

MR.N.J. WITHERS - Coffey and Partners, Queensland, Australia

Mr. Beckett asked a question to clarify how the crack limitation part of the slab design was made. He distinguished between serviceability limit state and ultimate limit state approaches and outlined a method for an elastic analysis of the slab allowing for a reduced modular ratio in the long term.

The cracking of the slab was catered for using the method advocated in CP 110, which uses the concept of a partially cracked section. The details are described in Appendix 5 of the paper.

As the serviceability limit state deflections for the brick or unreinforced blockwork superstructure placed on the slabs are more onerous than the serviceability limit state deflections of the slab, the former control the serviceability aspects of the design.

There are differing views regarding the necessity to restrict cracking of the slab. I have adopted a conservative approach which attempts to prevent any cracking from structural actions of the slab. Perusal of Figure 11 in the paper will show that using a mean load factor of 1.5 for the slab loads generally results in moments under working loads which are below the moment needed to crack the slab.

I was interested in the alternative approach outlined by Mr. Beckett and if it is a more efficient and economical means of slab design it would be welcome indeed. However, the analysis of the slab is rudimentary and therefore any suggestions for improvements which still keep within the overall budgetary control of £50 to £150 (for the site investigation, laboratory testing and design of residential foundations in Australia) are highly desirable.

Dr. Larnach made the point that the method proposed in the paper is no substitute for complete interaction analysis. I agree, but go further by pointing out that there is a gap between complete interaction analyses using computer methods, with the concomitant sophistication of soil testing to obtain parameters, and the semi-empirical approaches used by the majority of designers to-date. The method described in the paper is intended to fill that gap by providing an inexpensive but rational method of design. I would welcome any case histories with measurements of ground conditions and slab movements as well as the complete interaction analyses Dr. Larnach refers to, so that refinements can be made to the proposed method.

I will also take this opportunity to describe a procedure for modifying the laboratory results so that the field behaviour of the soil is more truly represented. Without this modification, crude as it is, field deformations under design conditions may be seriously underestimated.

In the laboratory, one sample is subjected to confined swell under saturated conditions. A second sample is air-dried to 10,000 kPa suction without confinement and thus shrinks three-dimensionally. In the field, such three-dimensional shrinkage can only occur once the soil has cracked into discrete units. This generally happens for suctions above 300 to 3000 kPa. Therefore the laboratory results need to be modified so that the difference between extreme dry and extreme wet deformations in the laboratory, (a) → (b) in Fig.D11a, are transformed to realistic "design" field deformations, (A) → (B) in Fig.D11b.

A quasi-elastic approach to quantify the relationship between one- and three-dimensional deformations can be justified on the grounds that the deformation of expansive soils appears reversible between constant limits (after several cycles) as they experience alternate wet and dry conditions. Other approaches are possible and following experimentation may prove to be preferable. For the moment, however, the following expression can be used:

$$\left(\frac{\Delta H}{H}\right)_{1D} = k.\left(\frac{\Delta H}{H}\right)_{3D}$$

where $k = (1 + 2K)$, K = coefficient of lateral stress. After many cycles of wetting and drying, one might expect K to be sensibly constant. Until sufficient testing is done to correlate laboratory tests and field behaviour it would seem prudent to use $K = K_o = 0.5$, say, resulting in $k = 2$.

Amending the laboratory results in this manner produces a kinked suction-deformation curve. To maintain the simplicity of expression in the design process it is considered that only a small loss of accuracy results from linking the end points of this curve (see Fig.D11b) to obtain the log linear expression derived in Appendix 2 of the paper.

Since the paper was written, I have produced programs for

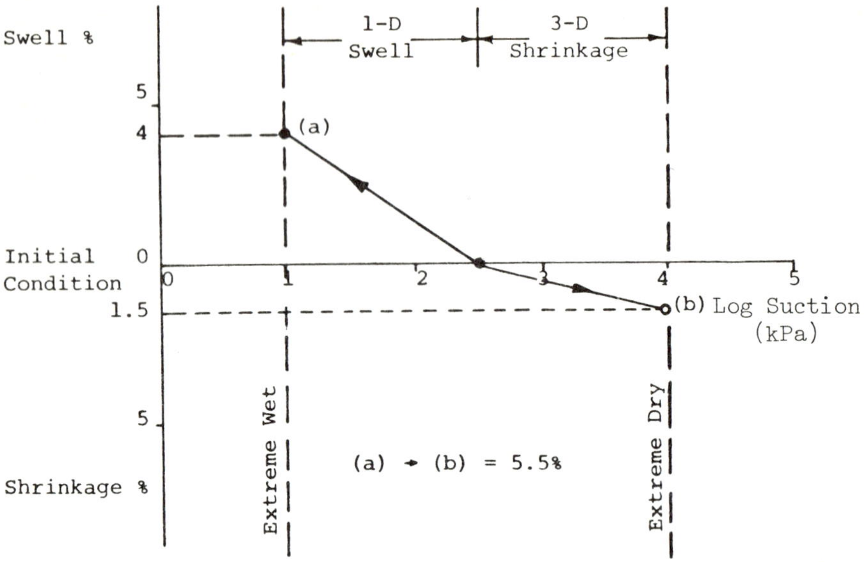

LABORATORY RESULTS - INITIAL EQUILIBRIUM CONDITION

Figure D11a

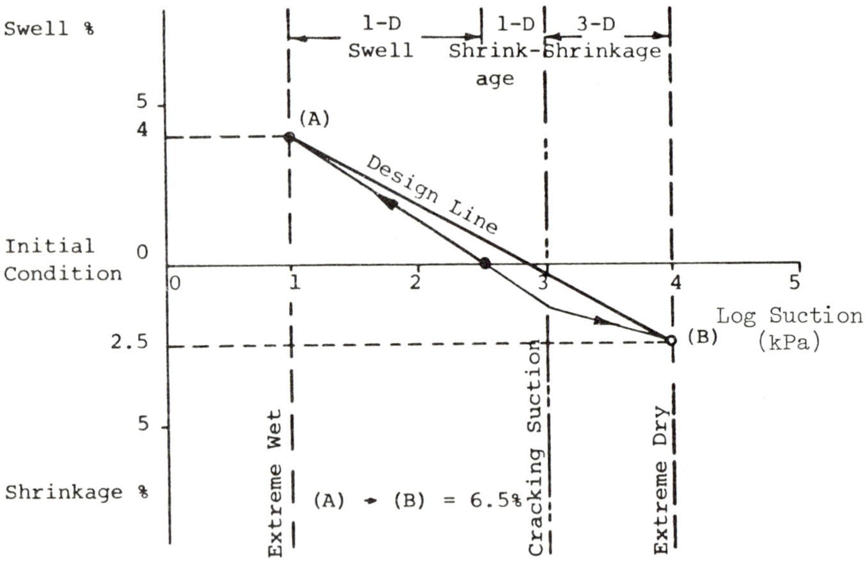

FIELD BEHAVIOUR - INITIAL EQUILIBRIUM CONDITION

Figure D11b

the HP 97 calculator which obviates the need for much hand calculation and thus the use of Figs. 8, 9 and 10 in the paper. The pattern of soil pressures beneath the slab has been made more sophisticated and the procedures of Appendix 4 of the paper made redundant. These programs have been incorporated into a draft manual which provides guidelines for the site investigation, laboratory testing procedures and instructions for the design process.

DR.M.J. CREED - Department of Civil Engineering
University College, Cork

Figure 9 of Dr. Wood's paper shows the results of a time-settlement calculation for particular points in the warehouse floor. There is little information in the paper on the method of computation utilised. Perhaps Dr. Wood would like to elaborate on the approach he used in making these time-settlement predictions.

DR.L.A. WOOD

I would refer Dr. Creed to my paper (Wood, 1980) for a detailed explanation of the analysis used. In our paper we have assumed vertical drainage of the silt layers, with the gravel and weathered slate layers considered as free-draining; in addition, during unloading/reloading, the soil stiffness has been increased.

PROFESSOR G.E. BAUER - Department of Civil Engineering
Carleton University, Ottawa, Canada

To overcome negative skin friction in Leda clay, piles are often coated down to the neutral point. Determination of this position is therefore of practical importance. Have any experiments been carried out by Dr. Symons and his colleagues on floating piles i.e. those without toe bearing of significance? What would be the neutral point position for piles in a bridge abutment? Also, would the authors care to comment on the possible scale effect in extrapolating neutral point determinations on model piles to the full-scale pile?

DR.M.V. SYMONS - Department of Civil Engineering and Building
Technology, UWIST

We have carried out experiments on floating piles but have measured the neutral point position only for the case of negative friction caused by self-weight consolidation of the clay stratum i.e. with no surcharge loading applied. It was found that the neutral point stabilised at an average level below the clay surface of 0.57 times the initial embedded length of the pile. A neutral point higher than that of the end-bearing case is bound to occur because floating piles undergo greater settlement than end-bearing piles, with the result that the soil settlement and pile settlement curves intersect at higher levels (compare with Fig.1(c) of our paper).

On the question of the neutral point position for piles to the bridge abutment described by Professor Bauer, it is difficult to give an opinion. The case is different to that described in the paper, with differential rates and magnitude of consolidation due to the presence of an embankment. It is a complicated practical case and simplifying assumptions would have to be made in order that design could proceed.

With respect to the extrapolation of model test results to full-scale piling, it has to be recognised that difficulties exist and care has to be exercised. This investigation simulated a practical case, however, and the approximate correspondence of our testing to those of the full-scale testing of Bozozuk and Labrecque (1969) suggests that the results obtained are realistic enough to be used in practice.

REFERENCE

M. Bozozuk and A. Labrecque - Downdrag measurements on a 270 ft composite pile, Proceedings of Symposium on Performance of Deep Foundations ASTM STP444 (1969) 15-40.

PROFESSOR J.D. GEDDES

Piles are infrequently used as single elements in a piled foundation. Has Dr. Symons any information to offer on the relationship between the neutral point position on a single pile and on similar piles arranged in a group?

DR.M.V. SYMONS

Research needs to be carried out to determine the relationship between the position of the neutral point for piles in a group and the single identical pile. The single pile neutral point position may well apply in the case of groups with high spacings but, at close spacings, pile interaction is bound to affect the neutral point. The presence of a pile cap could also influence the neutral point position within a group, but there appears to be no information extant on this aspect of negative friction.

DR.W.J. LARNACH

In the experiments described by Dr. Symons, the model piles were not driven or bored into position. Will not the method of placement be of practical significance in determining the subsequent behaviour and hence the position of the neutral point?

DR.M.V. SYMONS

The method of installation of the piles in this research programme (piles put in position with the clay and sand subsequently laid around them) is certainly different to the driving and boring procedures adopted in practice. This is not believed to affect the mechanics of consolidation of the clay stratum which is instrumental in producing the negative friction and it is considered that the subsequent build-up of negative friction and the position of the neutral point were not particularly dis-similar to what would have been obtained had the piles been installed differently. What was probably affected was the actual magnitude of the negative friction loading achieved, because the coefficients of earth pressure on the pile shaft would have been different had the piles been driven or bored into the clay and sand strata.

MR.B.I. TOLLIN - Howard Humphreys and Partners
Consulting Engineers

I wish to comment on Professor Bauer's question to Dr. Symons regarding the neutral point of piles for a bridge abutment in a Leda clay deposit. There is a significant difference between this problem and that investigated by Dr. Symons. In the latter case, the whole depth of the clay is subject to surcharge loading and settlement. In Professor Bauer's case, the thickness of clay subject to loading and settlement is a function of the plan dimensions of the embankment. The neutral points will therefore not be in similar locations.

In Professor Bauer's case, I would suggest that the use of piles is not particularly beneficial when compared with normal spread footings. Piles will only significantly reduce the structural settlements if they are supported on a more incompressible layer beneath the soil that is giving rise to the negative skin friction. This, of course, can itself result in the problem of the structure hanging up above the settling embankment fill.

SESSION VII

Chairman : Mr.C.J. Evans, MA,CEng,FIStructE,FICE,FIHE,FASCE
Vice-President, Institution of Structural Engineers
Chairman Wales Branch

DR.W.J. LARNACH - Department of Civil Engineering
University of Bristol

I wish to raise a number of questions relating to the paper by Dr. Creed and his colleagues. These are as follows.

With regard to modelling the behaviour of a diaphragm wall in London clay, did the model include the effects of interface slip between the wall and the soil?

Insofar as the treatment of the anchorages is concerned, were they, in the mathematical model, fixed in space, attached to finite element nodes, or given slip properties?

Finally, the value of K_o, the coefficient of earth pressure at rest, is important in such an analysis. How sensitive are the results to changes in the value of K_o?

DR.M.J. CREED - Department of Civil Engineering
University College, Cork

The wall was analysed for both no slip and no friction interface conditions. The results, as was to be expected, were different and it was found that the best correlation with the field data was obtained using the no slip condition.

In the model an anchorage consisted of a number of colinear finite element nodes, located in the region occupied by the anchorage in the field. As the anchors in the monitored wall were under-reamed, no slip was expected to occur between the anchors and the soil, and none was allowed for in the analytical model.

I agree that the value of K_o is significant, though not as critical as it might appear. For a saturated soil below the water table a reduction in K_o from say 3 to 2 gives rise to a reduction in lateral load of about 25 percent. Such a reduction in lateral load would reduce soil movements due to lateral load by a corresponding amount.

MR.B.I. TOLLIN - Howard Humphreys and Partners

I have questions to put to Dr. Wong and to Mr. Leonard and his colleagues on their respective papers.

I have noted in Dr. Wong's paper that the contractor was unable, using recharge wells, to maintain the groundwater at the specified level. I have encountered the same problem with the use of recharge wells; fortunately no damage resulted from the groundwater lowering. When a protective measure for structures, such as the use of recharge wells, is specified and does not work, it creates a difficult situation of confused responsibility between consultant and contractor, especially if damage occurs. In a case where any lowering of groundwater levels would seriously affect a structure, I would be hesitant in the future to rely on recharge wells to maintain groundwater levels. I ask Dr. Wong to comment.

Could Dr. Wong also clarify the penetration of the excavation sheeting? In Figure 1 of his paper, the sheeting does not extend below the base in the finite element idealisation.

I noted that recharge wells were also used in the construction of the Hong Kong transit system, discussed in Mr. Leonard's paper. I would ask him if these were successful in maintaining groundwater levels and if they were, what made them successful?

DR.M.J. CREED

In Mr. Leonard's paper it appears that a large proportion of lateral wall movement had occurred at an early stage during mass excavation. I would like to know if most of this movement occurred during the diaphragm wall construction?

I also have a question for Professor Stamatello, on the ground conditions for which his Equation (1) applies. What typically is the extent of the Quaternary surface deposits in Warsaw? Is there much fissuring present in the upper regions of the Pliocene clays?

DR.I.H. WONG - Ebasco Services Incorporated, New York, U.S.A.

I agree with Mr. Tollin's comment that a state of confused responsibility would arise between consultants and contractors if damage occurred to structures despite the use of recharge wells. Like many other tools available to the foundation engineer, it cannot be assured that the use of recharge wells will always eliminate settlements of nearby buildings. This lack of prior assurance of success should not preclude its use. What is required then is a monitoring programme involving piezometers and settlement measuring devices to determine the effectiveness of the system installed. If measurements show that the desired results are not being achieved alternative corrective measures can be taken.

On Mr. Tollin's question on sheeting penetration, it should be noted that the strata conditions shown in the paper were very variable. At the particular section of the excavation represented by Figure 1, rock was encountered above the bottom of the excavation thereby preventing full penetration of the sheeting. As described in the paper approximately 15.3 m (50 ft) north of the test section the contractor was able to install the sheeting to the design depth. In the finite element idealisation shown in Figure 4, the sheeting was assumed to extend below the bottom of the excavation.

MR.M.S.M. LEONARD - D'Appolonia SA, Brussels, Belgium

Of two experiences with the use of recharge wells on the Hong Kong Mass Transit Railway, one carried out by Ove Arup and Partners was highly successful. The other failed to meet the specification, possibly due to lack of experience on the part of the contractor concerned.

In reply to Dr. Creed, the wall movements were proportionately greatest in the early part of the excavation. This observation confirms the importance of providing adequate strutting at the top of the diaphragm walls. In order to limit the magnitude of such movements this requirement and the need to provide adequate control of the effects of dewatering are related aspects which are often overlooked. The need for proper design checking and work supervision is obvious.

PROFESSOR H. STAMATELLO - Technical University of Warsaw, Poland

Dr. Creed has queried the nature of the clays in Warsaw. The stratigraphy consists of Quaternary sediments, overlying Pliocene clays. Cracks were formed in the Pliocene clays due to movement of the over-riding glacier. No fissuring is present in the Quaternary deposits.

MR.I. COOPER - Water Research Centre, Henley

Professor Stamatello's paper on ground movements caused by deep excavations is particularly interesting because it refers to a large number of measurements taken over the years of his study. I wish to ask two questions.

(1) He mentions that surface measurements were taken. Were any "in ground" measurements made?

(2) On page four of his paper, Professor Stamatello states "little success has attended attempts to present these movements (movements caused by tunnelling) in the form of a mathematical formula, even though empirical in character". I would have thought that in the light of information of the kind referred to in Session IV and the current state of settlement prediction for soft ground tunnelling using a stochastic model, bearing in mind the large amount of data now available to put into the model, we should be able to predict with a high degree of accuracy, ground strains and movements. Would Professor Stamatello care to comment?

It is also interesting to compare Dr. Howe's approach to the assessment of the risk zone with Professor Stamatello's method. Would Dr. Howe comment on Professor Stamatello's prediction of 0.3 H for the extent of this zone compared with his own prediction of H + W/2, both assuming good construction techniques?

PROFESSOR H. STAMATELLO

In reply to Mr. Cooper, measurements were taken only at the ground surface and not within the soil mass.

The formulae put forward in my paper were derived purely on the basis of measurements made by me. Bearing in mind the highly variable conditions encountered, these are believed to be perfectly adequate for practical purposes.

MR.I.F. SYMONS - Tunnel and Underground Pipes Division
Transport and Road Research Laboratory

I have a contribution to make which relates to the papers by Messrs Gumbel and Wilson, Mr. Crofts and his colleagues and Dr. Howe and his co-authors.

As a part of a programme of research on the preservation of buried services, studies are being carried out by the Transport and Road Research Laboratory to assess the effect of deep trenching on shallow buried pipes. As a first step, observations are being made of the ground movements produced by trenching in a range of ground conditions. This involves full-scale field trenching trials together with measurements on a number of working schemes. The paper by Gumbel and Wilson describes a preliminary trenching trial in London Clay carried out for the Laboratory by Mott, Hay and Anderson. This formed a prelude to the first full-scale trial in similar ground. To ensure that the work on the main trial was as realistic as possible it was supervised by a Local Authority and executed by a contractor experienced in sewer trench construction. The trench was 25 m long, 5 m deep and 1 m wide. The work was carried out in five discrete sections, with delays between sections to allow for measurements to be made (Table DII).

TABLE DII

EXCAVATION AND BACKFILLING SEQUENCE

Bay number	Day(s) on which excavation was carried out	Day(s) on which backfilling was carried out
1	0 & 1	23
2	3	21 & 22
3	4	18
4	7	16
5	9	14 & 15

Within each 5 m long section the trench was excavated to full depth over half the length before steel sheeting was placed and timber walings installed and strutted at three levels. As excavation proceeded, pipes were installed in the trench and the excavated material stockpiled on one side. On completion of the excavation the trench was backfilled using the excavated material placed in layers and compacted. A plan showing the layout of the instrumentation is shown in Figure D12. This comprised surface stations for vertical and horizontal movements, inclinometers and magnetic rings for observations of movement at depth and piezometers to monitor pore pressures in the ground.

To illustrate the magnitude and distribution of movements developed at this site, examples of the preliminary results are given in Figures D13 to D17. Figure D13 gives settlement profiles along a section normal to the centre of the trench for three dates corresponding to half excavation, complete excavation and to three weeks after backfilling of the trench. These show that on this section most of the movement had occurred by completion of excavation and that the difference between the surcharged and unsurcharged sides amounted to only about 10 to 20 percent. Figure D14 shows the horizontal movements recorded by four inclinometers on the same section. The results for the two inclinometers located closest to the trench on opposite sides show that the lateral movement extended about 1.5 m beyond the bottom of the trench and that the movement on the surcharged side was about 30 to 40 percent greater than on the unsurcharged side.

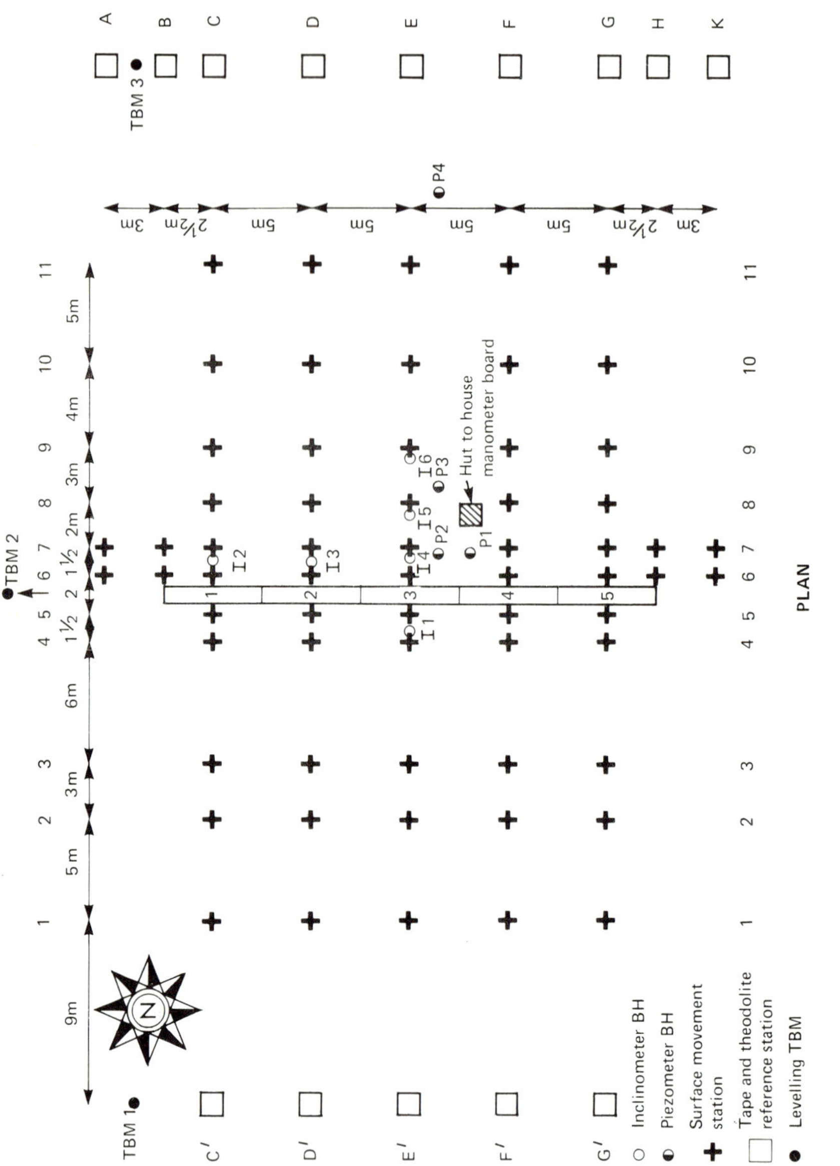

LAYOUT OF INSTRUMENTATION

Figure D12

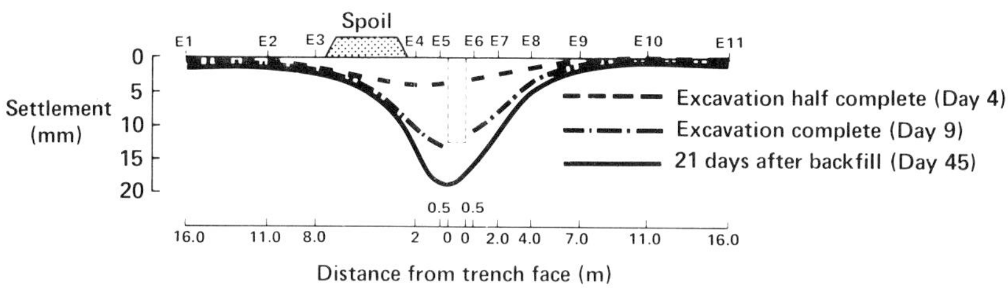

SURFACE SETTLEMENT AT CENTRELINE OF EXCAVATION (LINE E)

Figure D13

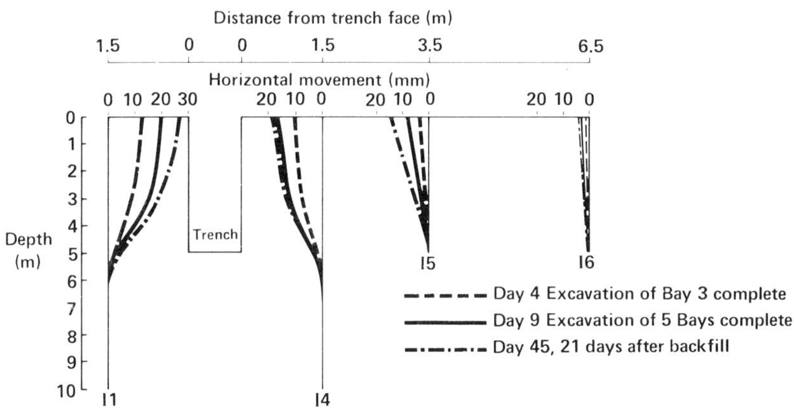

MOVEMENTS SHOWN BY INCLINOMETERS AT LINE E (CENTRE BAY 3)

Figure D14

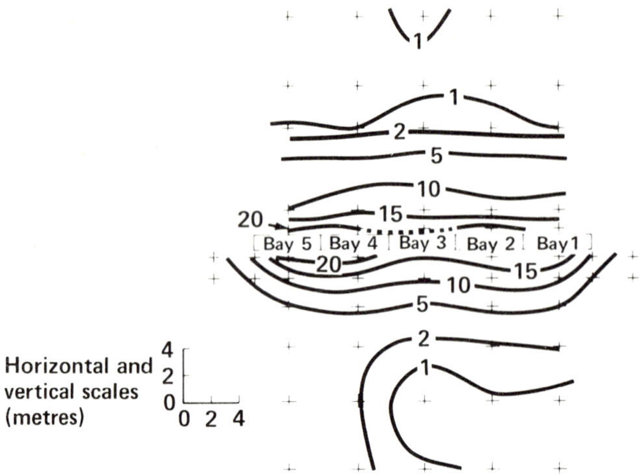

CONTOURS OF GROUND SETTLEMENT (DAY 45)

Figure D15

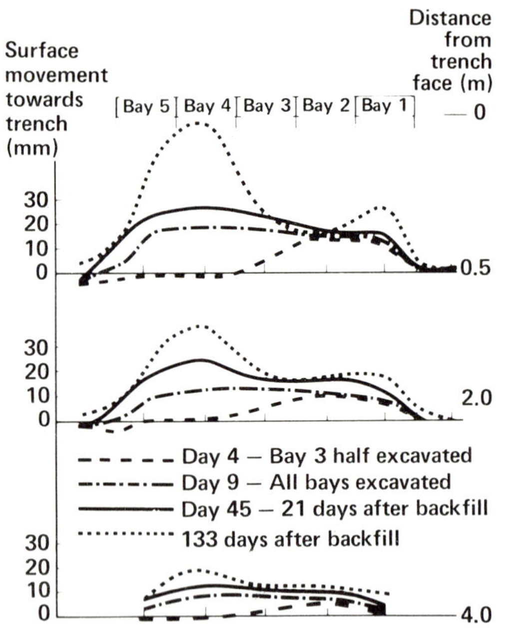

HORIZONTAL MOVEMENT OF THE GROUND SURFACE TOWARDS THE TRENCH

Figure D16

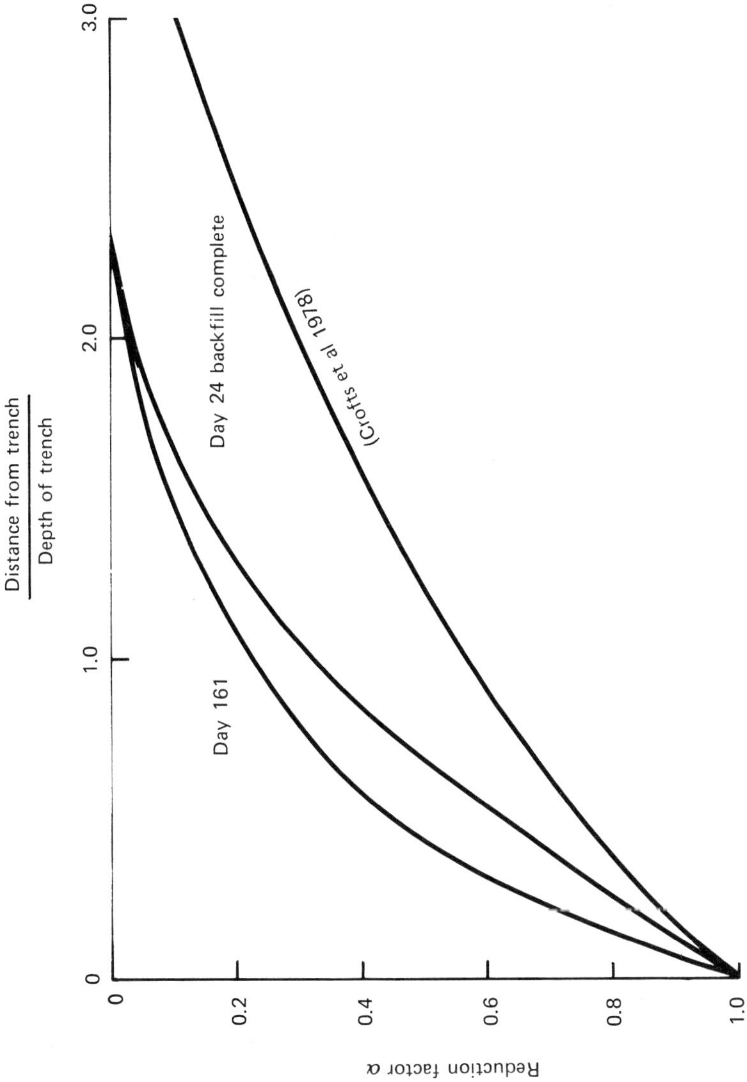

Figure D17 AVERAGE DISTRIBUTION OF HORIZONTAL MOVEMENT AT GROUND SURFACE FOR LINE F (CENTRE BAY 4)

Figure D15 shows contours of settlement recorded three weeks after backfilling. The greatest movement occurred in the vicinity of Bays 4 and 5 and probably reflects the lower densities achieved in backfilling these sections. This is also apparent in the pattern of horizontal movement recorded at ground surface along three lines parallel to the trench on the unsurcharged side (Figure D16). On completion of excavation the lateral movement on each line was fairly uniform within a zone between the centres of Bays 1 and 5. During the following month the increases in movement were greater in the vicinity of Bays 4 and 5. This trend continued and the measurements four months after backfilling indicate that the lateral movements of the ground surface were much smaller near Bays 2 and 3 where a higher standard of backfilling was achieved.

The distribution of movement with distance from the trench is likely to be of importance in considering the effect on adjacent buried services and structures. Examples of the average distribution of horizontal movement at the ground surface on both sides of the trench, on a line at right angles to the centre of Bay 4 are shown in Figure D17 as non-dimensional plots. Also shown is the distribution curve suggested by Crofts et al. which indicates a more extensive zone of movement than recorded at this site.

Finally it would be useful if Dr. Howe and his colleagues could give the derivation of their factors of safety for the proposed risk zones referred to in Section 5 and Figure 15 of their paper. They assumed a clay soil under undrained conditions for which ϕ_u is zero. A simple wedge analysis shows that the factor of safety along an arbitrary failure plane is then a function of the undrained cohesion c_u, the unit weight of the soil γ, the effective depth H and the angle of inclination of the failure plane to the horizontal θ. Such an analysis indicates smaller increases in factor of safety than suggested in the paper as the value of θ decreases from 45° to 34° and then to 27°.

MR.C. McCAUL - Tunnel and Underground Pipes Division
Transport and Road Research Laboratory

One of the working schemes where the Laboratory has carried out observations of ground movements associated with trenching, involved the renewal of storm and foul water sewers in an urban area. The sewers were laid in a combined trench 1.9 m wide by 4.5 m deep, which followed the line of a minor road. The ground profile at the site consisted of about 1.5 m of made ground and 4.5 m of loose to medium density fine sand with gravel, overlying stiff clay. A few days before excavation began, dewatering probes were installed and the groundwater level in the sand and gravel was lowered from a depth of about 2 m to a depth of about 5 m close to the trench. The trench was excavated in bays 2.5 to 3 m long and supported by linked steel sheeting driven to a depth of 5 m and braced by timber walings and struts at three levels.

On a line at right angles to the trench, survey stations were installed in the ground and the settlements and horizontal movements were measured as the trench was excavated and subsequently backfilled. In addition, subsurface ground movement was measured using borehole inclinometer and magnetic ring extensometer systems. A cross-section along the line of instrumentation showing the location of the survey stations close to the trench is given in Figure D18. Two other more remote stations, S11 and S12, were positioned at 9.9 m and 11.9 m respectively from the edge of the trench. Stations S1 to S5 were installed in the surface of the minor road, while the remaining stations were located in an adjoining field.

Examples of the short-term movements recorded at this site are shown in Figures D19 to D23. Figure D19 illustrates the development of horizontal movement at the ground surface. The trench sheeting protruded above ground level and this prevented the measurement of horizontal movement on stations S1 to S3 during construction. No detectable movement resulted from the dewatering and no significant movement was noted when the end of the trench was 2 m away from the survey line. During the next two days, when the bay immediately adjacent to the survey stations was excavated and supported, a maximum movement of 24 mm was measured on station S4. Little further increase in movements then occurred until the bay was backfilled and the trench supports withdrawn. This process then produced a somewhat smaller increase in the horizontal surface movement of stations S4 to S8 than did the excavation of the trench. After backfilling, the horizontal movements measured on stations S4 to S6 were much smaller than those for the equivalent stations S1 to S3 located in the road on the opposite side of the trench. These latter stations were subjected to loading from heavy construction traffic and the movements may have been influenced by the presence of an old storm sewer near to the trench (Figure D18).

Figure D20 shows the profile of horizontal movement at the ground surface measured three weeks after backfilling on the side of the trench not affected by construction traffic. Also shown is the profile of movement suggested by Crofts et al. The maximum movement at the trench wall estimated from the measurements is about 0.65 percent of the trench depth, which agrees well with the suggested value. However, the measured horizontal movements at ground surface decrease more rapidly with distance from the trench face. This is illustrated on Figure D21 which gives the same data replotted as a distribution curve. No movement is apparent beyond a distance of 1.75 times the trench depth from the edge of the excavation and at a distance of one trench depth the measured movement is only about 10 percent of the maximum.

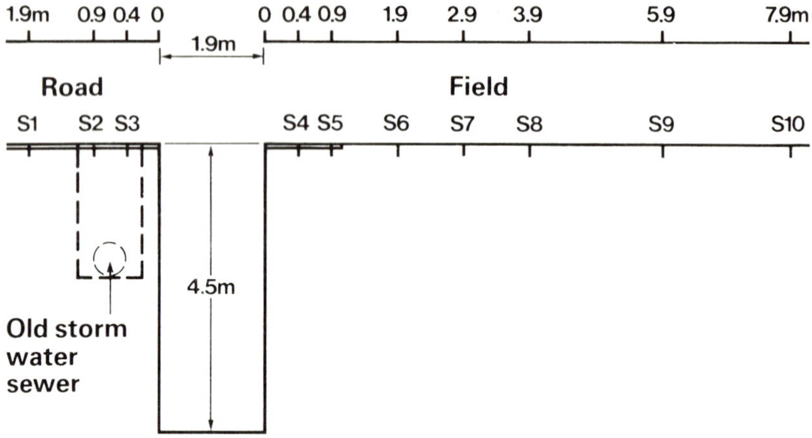

LOCATION OF SURVEY STATIONS ALONG THE LINE OF INSTRUMENTATION

Figure D18

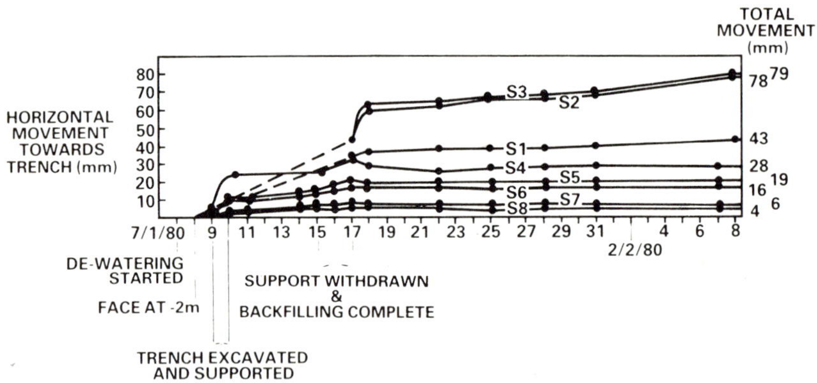

DEVELOPMENT OF HORIZONTAL SURFACE MOVEMENT WITH TIME

Figure D19

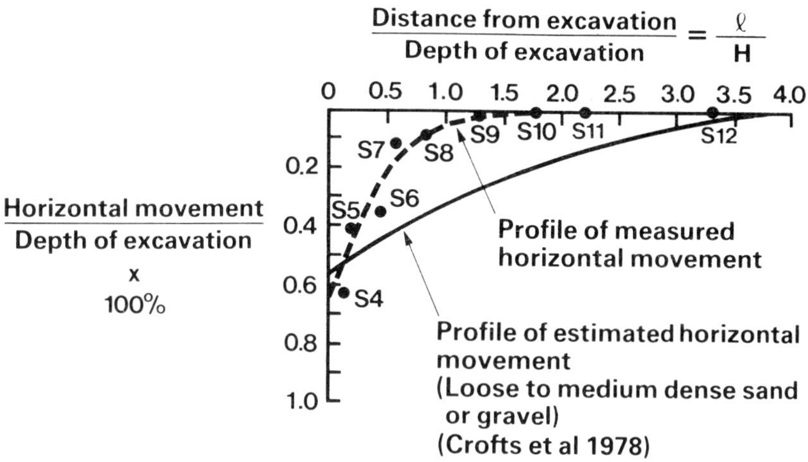

PROFILE OF HORIZONTAL SURFACE MOVEMENT THREE WEEKS AFTER BACKFILLING

Figure D20

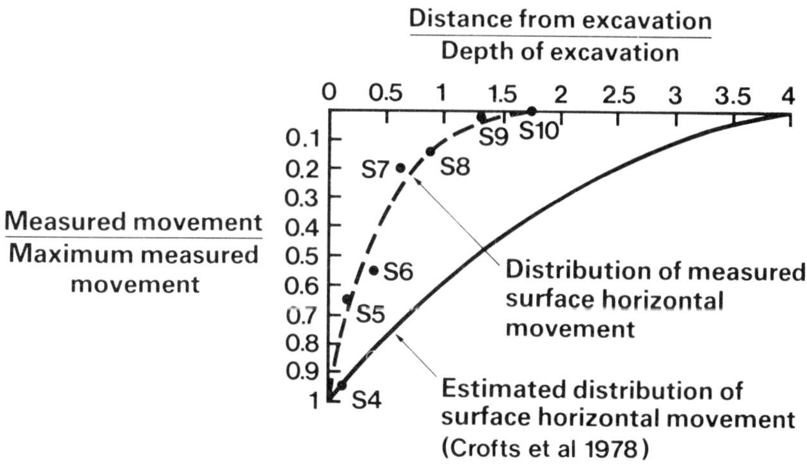

DISTRIBUTION OF HORIZONTAL SURFACE MOVEMENT THREE WEEKS AFTER BACKFILLING

Figure D21

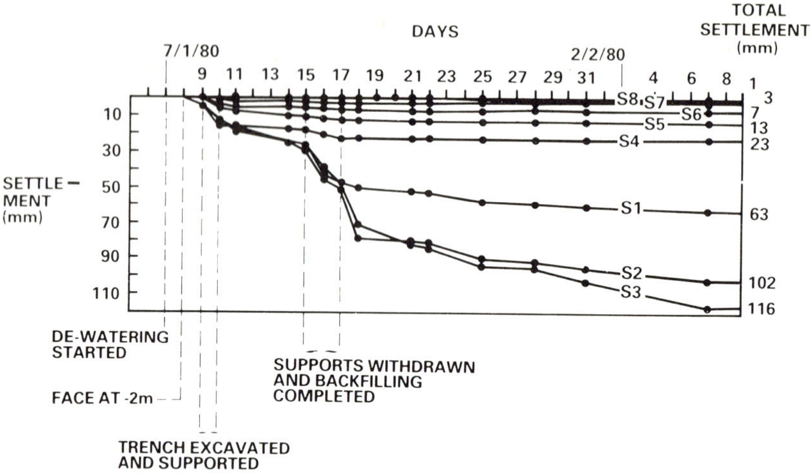

DEVELOPMENT OF SURFACE SETTLEMENT WITH TIME

Figure D22

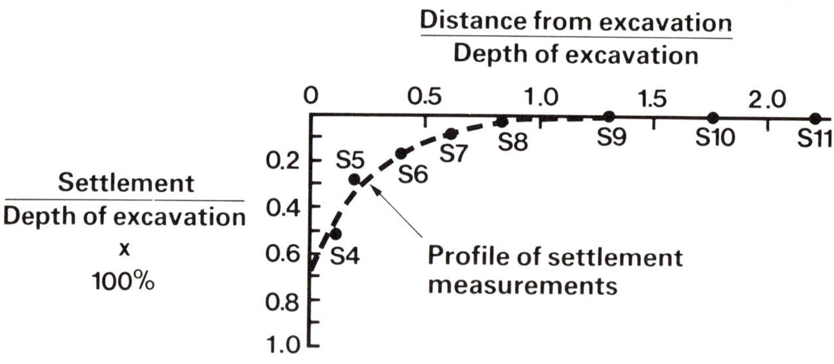

PROFILE OF SURFACE SETTLEMENT THREE WEEKS AFTER BACKFILLING

Figure D23

Figure D22 shows the development of surface settlement. The trends are similar to those of the horizontal movements, with settlements on the road side appreciably greater than on the opposite side of the trench. The profile of surface settlement three weeks after backfilling on the non-trafficked side is shown in Figure D23. The zone of settlement is less extensive than that for the horizontal movements with insignificant vertical movement beyond a distance of one trench depth from the face of the excavation.

MR.J.E. GUMBEL - Mott, Hay and Anderson, Consulting Engineers

I wish to suggest that conventional trench shoring systems installed after excavation cannot alone be relied on to prevent disturbance to adjacent buried services. The case records presented by Dr. Howe and his colleagues (Sites 1 and 2) show that large, potentially damaging ground movements can occur even where high standards of workmanship are achieved. The highest standards of support must of course be encouraged for the safety of those working in the trench; but other simple measures which could reduce the risk of pipe failures should not be overlooked.

One such measure is to remove spoil and divert traffic away from pipes buried near the trench. The data presented by Mr. Symons and Mr. McCaul confirm our own findings that surcharging of the trench shoulder increases ground movements significantly. Time is another important factor, so the trench should not be kept open longer than is strictly necessary. Also, as demonstrated by Mr. Symons' results, backfill should be well compacted to minimise long-term movement. All these measures have the advantage that they are relatively easy to control, whereas standards of trench support will always vary.

DR.M. HOWE - Engineering Research Station, British Gas

In reply to Mr. Cooper, who invited comment on Professor Stamatello's prediction of 0.3 H as the extent of a high-risk zone, we would consider the latter to be inadequate. Our work and that of the TRRL presented at this conference indicates that lateral movements only become insignificant at a distance from the trench greater than one trench depth (i.e. H).

Mr. Symons suggests that this distance can be as high as 2 H and and the work of Mr. McCaul shows that at a distance of H the horizontal movement is about 10 percent of the maximum. In our field work we have recorded displacements in excess of 120 mm at a distance of 0.27 H which, if the TRRL reduction factor is applied, would mean a movement of about 12 mm at a distance of H. This amount of movement can be sufficient to cause failure in cast-iron mains running both parallel to and crossing an excavation. Therefore a zone of high-risk ground must extend beyond distance H from the trench and we consider our estimate of H + W/2 to be realistic. The risk zones also have to allow for slight deviations in route and depth of excavation and for the possibility of wash out or other groundwater problems.

Mr. Cooper's reference to good construction techniques raises several points. We are very much in favour of improved standards of wall support for deep excavations and will encourage the introduction of detailed specifications and codes of practice defining the levels of support needed to prevent ground movement in deep excavations. The revised British Standard CP 2003, currently in draft form, is an example.

Any reconsideration of the zone of ground at risk could follow the elimination of ground movement by effective wall support practices at all times whilst the excavation is open and subsequently by backfilling to suitably high compaction standards.

Many current trench support techniques principally aim at protecting personnel and not specifically at preventing ground movement. Indeed one frequently observes inherently strong support systems employed in deep excavations but they are not in full contact with the sides of the trench and this allows movement to take place.

As risk has to be predicted often many months in advance of work beginning, the ability of principals to hold their deep excavation contractors to specific small limits of ground movement would also be a relevant step towards reducing the size of risk zones.

Finally in answer to Mr. Symons, Figure 15 of our paper represents a supported deep trench excavation in an urban area. Figure 15 applies to any soil and details areas within the ground where movements could be sufficient to damage cast-iron pipes. It was not intended that the boundaries of these risk zones be used to calculate the overall stability of the works. However, the paper does set out criteria to indicate that for most soils, possible slip surfaces would be contained within the 45° wedge. The factors that are applied are to guard against pipe damage and are not factors of safety against shear failure for unsupported soil masses.

MR.J.E. CROFTS - Nachshen, Crofts and Leggatt
Consulting Engineers

Using the approach described in our paper, Tables have been prepared from which the risk of breakage of cast iron mains can be predicted as a result of a proposed nearby excavation.

We are most grateful to Mr.J.A. Keith, Senior Civil Engineer of the British Gas Corporation for the following information. The method has been successfully used in Scotland. In general, category 3 cases are recommended for replacement and very few failures in C.I. mains have been reported subsequent to contractors starting works. In fact, only two cases have resulted in failure.

Examples :

(1) The proposed scheme was to sheet pile using trench sheeting driven ahead of the excavation through coarse gravel (medium to dense); 1 m clearance required to maintain a category 2 risk for the 6 in. C.I. main.

Once work started, reports came in of breakages and upon further investigation it was evident that the contractor was excavating a wider track than anticipated and the 6 in. C.I. L.P. gas main was only 300 to 600 mm clear of the sheeting due to the initial dig to prove no services prior to driving.

The method of excavation was changed (at the contractor's request) to a box shore, in an attempt to expedite work, and this initially seemed successful. However, a further break in the gas main was reported and this turned out to be due to over-excavating the trench below the toe of the box shore, and a trench wall collapse occurred carrying about 4.5 m of gas main with it. This happened once more during the contract. In both instances, the main was about 600 to 800 mm away from the excavation. Extensive diversions and abandonments were carried out over the 2 km route.

(2) At another site, the proposed scheme was to sheet pile, using trench sheeting driven ahead of the excavation through well-graded sand (loose to medium-dense; tidal); 2 m clearance was required for category 2 risk.

When work started, the method of excavation was described as "opening up to a depth of 900 mm to expose services, insert trench sheeting and first level of shoring, and knock down sheeting ahead of dig using a J.C.B. backacter." The proposed 1 m clearance turned out to be about 600 mm but the gas main should have been category 2 (it was bedded in the loose sand extending down to about 1.5 m). However, after five fractures occurred it was discovered that excavation was being carried out to considerably deeper depths before the sheeting and shoring were being placed, and upon recalculation on this new basis category 3 risk was achieved. This did not take into account the tidal effects noted in excavation, but the rapidity of the failures may indicate that the tides did have an effect. All five fractures occurred within one week of starting a new section and some were noticed the day after opening up a new track. The track was opened in 1.5 m lengths, by 2 m wide. Since the reason for the failures was brought to the attention of the Local Authority no further breakages have occurred.

(3) Another job had been started prior to checking the

risk and a failure of a 2 in. C.I. main had occurred. Upon checking on site, it was found that the main was about 600 mm from the track and about 400 mm over. The sewer track was dug through about 2 m of loose, dry sand overlying slightly damp and denser sand to a depth of about 4.3 m. The quality of shoring was very poor and plywood sheeting was used to retain the sand, with large voids behind, since the track was excavated and then sheeted and shored. The failure in the main was due to a collapse of the side of the track and not simply to support movement.

A check against the NCL document showed that much of the remainder of the route was category 3 and the old 2 in. C.I. main was replaced in P.E.

The authors believe that these examples show the proposed method to be workable and useful, even at an early stage of its development.

The distribution of ground movements adjacent to a trench is of crucial importance in determining the extent to which a nearby pipe is affected. This movement decays with distance from the trench face. This may be assessed by the curve set out in the paper, which was based on the data then available relating to deep, wide excavations, which were both propped and tied. In the absence of a data base relating to shallow, narrow, propped excavations, the general form of the decay of ground movement from a trench face was of necessity patterned on that portrayed by Peck (1969).

It may be expected, therefore, that the recent emergence of additional case histories at this conference in the contributions of Mr. Symons and Mr. McCaul relating to measurements of ground movements adjacent to trenches will enable a re-assessment to be made.

A recent parametric study, using finite elements, has shown that the decay of ground surface movement away from the face of a trench is sensitive to the degree of non-homogeneity i.e. to the variation with depth of Young's modulus of elasticity (Kyrou, 1980). From Figure D24 it may be seen that the x-curve proposed by Crofts et al (1978), is similar to the analytical prediction for a homogeneous soil, while the prediction for a non-homogeneous soil gives a more rapid change near the trench face similar to that observed by Symons and McCaul and reported to this conference. Accordingly, we feel that information relating to the degree of non-homogeneity at trench sites would give greater relevance to the ground movement records obtained from these sites.

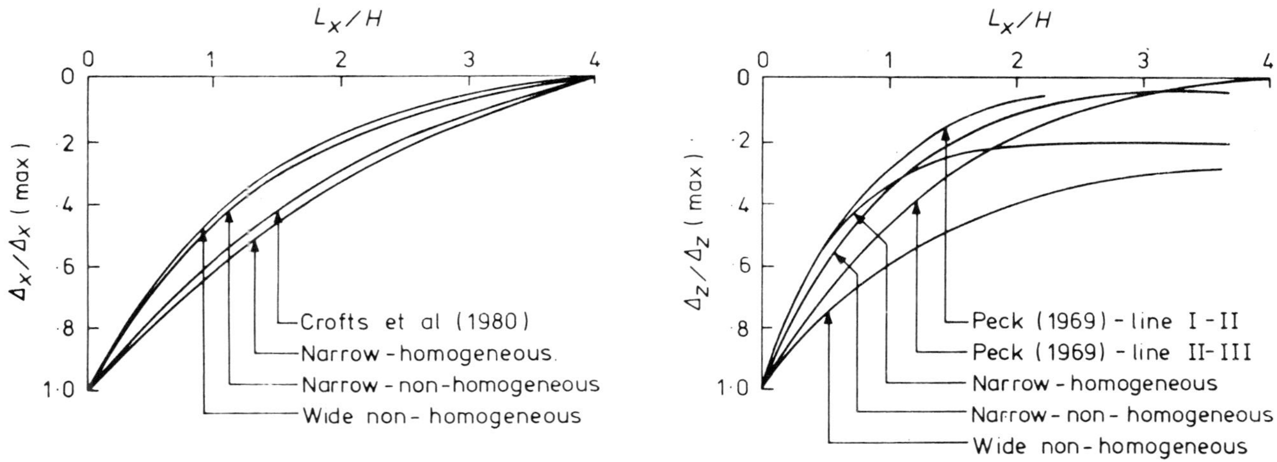

Horizontal ground surface movement	Δ_x
Vertical ground surface movement	Δ_z
Distance from excavation face	L_x
Depth of excavation	H

GROUND SURFACE MOVEMENTS CLOSE TO EXCAVATION FACES

Figure D24

REFERENCES

R.B. Peck - Deep excavations and tunnelling in soft ground, Proceedings of the 7th International Conference on Soil Mechanics and Foundation Engineering, Mexico (1969) 259-281.

K. Kyrou - The Effect of Trench Excavation Induced Ground Movements on Adjacent Buried Pipelines, PhD Thesis (1980) University of Surrey.